Mathematical Analysis, Differential Equations and Applications

Series on Computers and Operations Research

ISSN: 1793-7973

Series Editor: Panos M. Pardalos *(University of Florida, USA)*

Published

Vol. 11 *Mathematical Analysis, Differential Equations and Applications*
        edited by T. M. Rassias and P. M. Pardalos

Vol. 10 *Analysis, Cryptography and Information Science*
        edited by N. J. Daras, P. M. Pardalos and M. Th. Rassias

Vol. 9 *Analysis, Geometry, Nonlinear Optimization and Applications*
        edited by P. M. Pardalos and T. M. Rassias

Vol. 8 *Network Design and Optimization for Smart Cities*
        edited by K. Gakis and P. M. Pardalos

Vol. 7 *Computer Aided Methods in Optimal Design and Operations*
        edited by I. D. L. Bogle and J. Zilinskas

Vol. 6 *Recent Advances in Data Mining of Enterprise Data:
        Algorithms and Applications*
        edited by T. W. Liao and E. Triantaphyllou

Vol. 5 *Application of Quantitative Techniques for the Prediction
        of Bank Acquisition Targets*
        by F. Pasiouras, S. K. Tanna and C. Zopounidis

Vol. 4 *Theory and Algorithms for Cooperative Systems*
        edited by D. Grundel, R. Murphey and P. M. Pardalos

Vol. 3 *Marketing Trends for Organic Food in the 21st Century*
        edited by G. Baourakis

Vol. 2 *Supply Chain and Finance*
        edited by P. M. Pardalos, A. Migdalas and G. Baourakis

Vol. 1 *Optimization and Optimal Control*
        edited by P. M. Pardalos, I. Tseveendorj and R. Enkhbat

Series on Computers and Operations Research   Vol. 11

Mathematical Analysis, Differential Equations and Applications

Editors

Themistocles M Rassias
National Technical University of Athens, Greece

Panos M Pardalos
University of Florida, USA

 World Scientific

NEW JERSEY • LONDON • SINGAPORE • BEIJING • SHANGHAI • HONG KONG • TAIPEI • CHENNAI

Published by

World Scientific Publishing Co. Pte. Ltd.
5 Toh Tuck Link, Singapore 596224
USA office: 27 Warren Street, Suite 401-402, Hackensack, NJ 07601
UK office: 57 Shelton Street, Covent Garden, London WC2H 9HE

Library of Congress Cataloging-in-Publication Data
Names: Rassias, Themistocles M., 1951– editor.
Title: Mathematical analysis, differential equations and applications / editors,
    Themistocles M. Rassias, National Technical University of Athens, Greece,
    Panos M. Pardalos, University of Florida, USA.
Description: New Jersey : World Scientific, 2024. | Series: Series on computers and
    operations research, 1793-7973 ; Vol. 11 | Includes bibliographical references and index.
Identifiers: LCCN 2024000389 | ISBN 9789811267031 (hardcover) |
    ISBN 9789811267048 (ebook) | ISBN 9789811267055 (ebook other)
Subjects: LCSH: Mathematical analysis. | Differential equations.
Classification: LCC QA300 .M227 2024 | DDC 515/.35--dc23/eng/20240412
LC record available at https://lccn.loc.gov/2024000389

British Library Cataloguing-in-Publication Data
A catalogue record for this book is available from the British Library.

Copyright © 2024 by World Scientific Publishing Co. Pte. Ltd.

All rights reserved. This book, or parts thereof, may not be reproduced in any form or by any means, electronic or mechanical, including photocopying, recording or any information storage and retrieval system now known or to be invented, without written permission from the publisher.

For photocopying of material in this volume, please pay a copying fee through the Copyright Clearance Center, Inc., 222 Rosewood Drive, Danvers, MA 01923, USA. In this case permission to photocopy is not required from the publisher.

For any available supplementary material, please visit
https://www.worldscientific.com/worldscibooks/10.1142/13162#t=suppl

Desk Editors: Soundararajan Raghuraman/Steven Patt

Typeset by Stallion Press
Email: enquiries@stallionpress.com

© 2024 World Scientific Publishing Company
https://doi.org/10.1142/9789811267048_fmatter

Preface

This book presents essential mathematical results in a broad variety of topics in Mathematical Analysis, Differential Equations and their various applications. More specifically, it focuses on topics which include differential operators, Wardowski maps, low-oscillation functions, Galois and Pataki connections, power sets, Catalan-type numbers, functional equations, Ulam stability, Chebyshev polynomials, Bernstein type polynomials, Hermite–Hadamard type integral inequalities, the Cauchy–Schwarz–Bunyakovsky inequality, moment generating functions, variational inequalities, univalent contact problems, the equivariant minimax principle, biconvex functions, nonlinear fractional differential equations, inner product spaces, Banach Lie algebras, probabilistic normed spaces, finite element method, etc. Effort has been made for the book to have an interdisciplinary flavor and to feature various applications, such as applications in the restricted three-body problem, Network Games, machine learning, tomography, elastic scattering, fluid mechanics, etc.

The book will be particularly useful to graduate students and advanced research scientists in various branches of Mathematical Analysis as well as it will serve as an important reference for researchers in Mathematics, Physics and Engineering.

We would like to thank all the contributors of papers for participating in this collective effort. Last but not least, we wish to express our appreciation to the staff of World Scientific Publishing Co. for their very valuable assistance throughout the preparation of this volume.

Panos M. Pardalos
Florida, USA

Themistocles M. Rassias
Athens, Greece

About the Editors

Panos M. Pardalos serves as Professor Emeritus of industrial and systems engineering at the University of Florida. Additionally, he is the Paul and Heidi Brown Preeminent Professor of industrial and systems engineering. He is also an affiliated faculty member of the Computer and Information Science Department, the Hellenic Studies Center, and the biomedical engineering program, as well as the Director of the Center for Applied Optimization. Pardalos is a world-leading expert in global and combinatorial optimization. His recent research interests include network design problems, optimization in telecommunications, e-commerce, data mining, biomedical applications and massive computing.

Themistocles M. Rassias is a Professor at the National Technical University of Athens, Greece. He has published more than 300 papers, 10 research books and 45 edited volumes in research Mathematics as well as four textbooks in Mathematics (in Greek) for university students. He serves as a member of the Editorial Board of several international mathematical journals. His work extends over several fields of mathematical analysis. It includes nonlinear functional analysis, functional equations, approximation theory, analysis on manifolds, calculus of variations, inequalities, metric geometry and their applications. He has contributed a number of results in the stability of minimal submanifolds, in the solution of Ulam's Problem for approximate homomorphisms in Banach spaces, in the theory of isometric mappings in metric spaces and in complex analysis (Poincaré's inequality and harmonic mappings).

Contents

Preface		v
About the Editors		vii
1.	Galois and Pataki Connections on Power Sets	1
	Santanu Acharjee, Michael Th. Rassias, and Árpád Száz	
2.	A Functional Equation Related to Inner Product Spaces in Šerstnev Probabilistic Normed Spaces	77
	Ahmad Alinejad, Hamid Khodaei, Michael Th. Rassias, and Hamid Vosoughian	
3.	Hyers–Ulam–Rassias Stability of Set-Valued Functional Equations: A Fixed Point Approach	97
	H. Azadi Kenary	
4.	Two Heuristic Methods for Solving Generalized Nash Equilibrium Problems Using a Novel Penalty Function	111
	Benjamin Benteke, Monica Gabriela Cojocaru, Roie Fields, Mihai Nica, and Kira Tarasuk	

5. The Finite Element Method with
   Applications to Fluid Mechanics   139
   *Kyriaki N. Biraki, Konstantina C. Kyriakoudi,
   Anastasios C. Felias, and Michail A. Xenos*

6. Comparison between Two Descent
   SQP-ADMs for Structured Variational
   Inequalities   175
   Abdellah Bnouhachem and Themistocles M. Rassias

7. Solvability for a Class of Unilateral Contact
   Problems with Friction and Damage   201
   Oanh Chau, Arnaud Heibig, and Adrien Petrov

8. Vector Inequalities for Analytic Functions
   of Operators in Hilbert Spaces and
   Applications for Numerical Radius and
   p-Schatten Norm   225
   Silvestru Sever Dragomir

9. General Equivariant Minimax Principle
   and Fountain Theorem in the Presence
   of Nonadmissible Representations   275
   Lucas Fresse and Viorica V. Motreanu

10. On a Prey–Mesopredator–Predator System   297
    D. Goeleven and R. Oujja

11. Differential Operator Associated
    with the (q, k)-Symbol Raina's Function   321
    Rabha W. Ibrahim

12. Analysis and Solvability of Complex
    K-Symbol Liu Fractional Dynamical
    Systems   343
    Rabha W. Ibrahim

13.	Elastic Scattering by an Inhomogeneous Medium with Unknown Buried Obstacles	371
	Angeliki K. Kaiafa, George Kanakoudis, and Vassilios Sevroglou	
14.	Some New Bounds of Gauss–Jacobi and Hermite–Hadamard-Type Integral Inequalities	391
	Artion Kashuri	
15.	Payoff-Independent Action Update for Continuous Action Social Dilemmas: A Preliminary Investigation	427
	Ath. Kehagias	
16.	Applying Logarithm Sobolev Inequalities to Probability and Statistics	465
	Christos P. Kitsos	
17.	Some Certain Families of Catalan-Type Numbers and Polynomials: Analysis of Their Generating Function with Their Functional Equations	485
	Irem Kucukoglu and Yilmaz Simsek	
18.	On the Generalizations of the Cauchy–Schwarz–Bunyakovsky Inequality with Applications to Elasticity	541
	D. Labropoulou, T. Labropoulos, P. Vafeas, and D. M. Manias	
19.	Lie Bracket Derivations in Banach Lie Algebras	571
	Jung Rye Lee, Choonkil Park, and Michael Th. Rassias	

20. Aboodh Transform and Ulam Stability of Second-Order Linear Differential Equations 587

   Ramdoss Murali, Arumugam Ponmana Selvan, Sanmugam Baskaran, Choonkil Park, and Michael Th. Rassias

21. Some Classes of Extended General Variational Inequalities 617

   Muhammad Aslam Noor and Khalida Inayat Noor

22. Biconvex Functions and Bivariational Inequalities 657

   Muhammad Aslam Noor, Khalida Inayat Noor, Michael Th. Rassias, and Waseem Asghar Khan

23. Some Properties of a Class of Network Games with Strategic Complements or Substitutes 689

   Mauro Passacantando and Fabio Raciti

24. Chebyshev Polynomials of the First Kind and Applications in Tomography 711

   Nicholas E. Protonotarios, Vangelis Marinakis, Nikolaos Dikaios, and George A. Kastis

25. Bernstein-Type Polynomials Associated with Characteristic Function, Moment Generating Functions of Beta-Type Distribution and Their Approximation Applications 731

   Yilmaz Simsek and Fusun Yalcin

26. On Removing Diverse Data for Training Machine Learning Models 761

   Kim Thuyen Ton, Daniel Aloise, and Claudio Contardo

27.	Wardowski Maps Modeled by Low-Oscillation Functions	785
	Mihai Turinici	
28.	Contractive Maps on Relational MC-Quasimetric Spaces	821
	Mihai Turinici	
29.	Hyers–Ulam–Rassias Stability of the Nonlinear Fractional Differential Equations with ρ-Fractional Derivative	877
	Chun Wang	
30.	Out-of-Plane Equilibrium Points in the Restricted Three-Body Problem with Radiation Pressure, Poynting–Robertson Drag and Angular Velocity Variation	903
	Aguda Ekele Vincent, Angela E. Perdiou, and Jagadish Singh	

Index 935

© 2024 World Scientific Publishing Company
https://doi.org/10.1142/9789811267048_0001

Chapter 1

Galois and Pataki Connections on Power Sets

Santanu Acharjee[*,†,‡,**], Michael Th. Rassias[§,††], and Árpád Száz[‖,‡‡]

[*]*Department of Mathematics, Gauhati University, Guwahati, Assam, India*
[†]*Department of Mathematics and Engineering Sciences, Hellenic Military Academy, Vari Attikis, Greece*
[‡]*Institute for Advanced Study, Program in Interdisciplinary Studies, New Jersey, USA*
[§]*Institute of Mathematics, University of Zürich, Zürich, Switzerland*
[‖]*Institute of Mathematics, University of Debrecen, Debrecen, Hungary*

[**]*sacharjee326@gmail.com*
[††]*michail.rassias@math.uzh.ch*
[‡‡]*szaz@science.unideb.hu*

If X is a set and R is a relation on X, then having in mind the abbreviation poset by Birkhoff for a partially ordered set, the ordered pair $X(R) = (X, R)$ will be called a goset instead of a generalized ordered set.

In the sequel, we shall suppose that $X(R)$ and $Y(S)$ are gosets, φ is a function of X to itself, f is a function of X to Y and g is a function of Y to X. Thus, in particular, we may have $\varphi = g \circ f$.

Now, having in mind Schmidt's ingenious reformulation of the definition of Ore's Galois connections, we shall say that f is increasingly left (right) g-seminormal if

$$x \, R \, g(y) \implies f(x) \, S \, y \quad (f(x) \, S \, y \implies x \, R \, g(y))$$

for all $x \in X$ and $y \in Y$.

Moreover, having in mind the definition of increasing functions, Dedekind's reformulation of the definition of closure operations and the definition of Pataki's increasingness, we shall say that f is increasingly left (right) φ-semiregular if

$$u \, R \, \varphi(v) \implies f(u) \, S \, f(v) \quad (f(u) \, S \, f(v) \implies u \, R \, \varphi(v))$$

for all $u, v \in X$.

Thus, if f is increasingly left (right) g-seminormal and $\varphi = g \circ f$, then we can easily see that f is increasingly left (right) φ-semiregular. Therefore, semiregular functions have to be studied before the seminormal ones.

Moreover, if f is increasingly φ-regular in the sense that it is both left and right φ-semiregular, and R is transitive and S is reflexive, then we can see that f is increasing.

Furthermore, if R and S are preorders, then we can see that φ is a closure operation if and only if φ is increasingly φ-regular, or equivalently there is a function h of X to Y which is increasingly φ-regular.

In a former paper, the third author has also intensively investigated several further properties of semiregular and seminormal functions by gradually assuming some basic properties of the relations R and S.

In this paper, we are mainly interested in the particular case when the goset $X(R)$ is replaced by a power set. Though, it is usually enough to consider the case when both $X(R)$ and $Y(S)$ are replaced by power sets.

Namely, a function F of $\mathcal{P}(X)$ to $\mathcal{P}(Y)$ will be shown to be increasingly normal if and only if F is union-preserving in the sense that $F\left(\bigcup \mathcal{A}\right) = \bigcup F[\mathcal{A}]$ for all $\mathcal{A} \subseteq \mathcal{P}(X)$, where $F[\mathcal{A}] = \{F(A) : A \in \mathcal{A}\}$.

Moreover, it will be shown that the function F is union-preserving if and only if $F(A) = \bigcup F[A]$ for all $A \subseteq X$, where $F[A] = \{F(x) : x \in A\}$ with $F(x) = F(\{x\})$ for all $x \in X$.

If $\mathcal{R}$ is relator (family of relations) of X to Y, then by taking

$$\mathrm{Int}_{\mathcal{R}}(B) = \{A \subseteq X : \exists \, R \in \mathcal{R} : R[A] \subseteq B\}$$

for all $B \subseteq Y$, we may naturally define a relation $\mathrm{Int}_{\mathcal{R}}$ on $\mathcal{P}(Y)$ to $\mathcal{P}(X)$.

Thus, we evidently have $\mathrm{Int}_{\mathcal{R}} = \bigcup_{R \in \mathcal{R}} \mathrm{Int}_R$ with $\mathrm{Int}_R = \mathrm{Int}_{\{R\}}$. Therefore, by defining $F(\mathcal{R}) = \mathrm{Int}_{\mathcal{R}}$ for any relator $\mathcal{R}$ on X to Y, we can obtain an increasingly normal function F of $\mathcal{P}(X \times Y)$ to $\mathcal{P}(\mathcal{P}(Y) \times \mathcal{P}(X))$.

By finding the unique Galois adjoint G of this function F, we can already obtain a basic operation $\Phi = G \circ F$ for relators such that, under the notation $\mathcal{R}^\# = \Phi(\mathcal{R})$, for any two relators $\mathcal{R}$ and $\mathcal{S}$ on X to Y we have

$$\operatorname{Int}_\mathcal{R} \subseteq \operatorname{Int}_\mathcal{S} \iff \mathcal{R} \subseteq \mathcal{S}^\#.$$

Hence, by using the theory of regular functions, we can infer that $\#$ is a closure operation such that, for any relator $\mathcal{R}$ on X to Y, just $\mathcal{S} = \mathcal{R}^\#$ is the largest relator on X to Y such that $\operatorname{Int}_\mathcal{S} \subseteq \operatorname{Int}_\mathcal{R}$ ($\operatorname{Int}_\mathcal{S} = \operatorname{Int}_\mathcal{R}$).

Moreover, by using the theory normal functions, we can determine for which relations Int on $\mathcal{P}(Y)$ to $\mathcal{P}(X)$ there exists a relator $\mathcal{R}$ on X to Y such that the equality $\operatorname{Int} = \operatorname{Int}_\mathcal{R}$ could hold.

This result will show that proximity spaces and their obvious generalizations should not be studied without generalized uniformities. We shall see that the same statement can also be applied to other basic topological structures.

1. Introduction

An important particular case of "Galois connections" was already considered by Birkhoff in the first edition of his famous book "Lattice Theory", under the name "polarities" [4, p. 122].

More concretely, Birkhoff offered the following construction: "Let ρ be any binary relation between the members of two classes I and J. For any subsets $X \subset I$ and $Y \subset J$, define $X^* \subset J$ (the "polar" of X) as the set of all $y \in J$ such that $x\,\rho\,y$ for all $x \in X$, and we define $Y^\dagger \subset I$ (the "polar" of Y) as the set of all $x \in I$ such that $x\,\rho\,y$ for all $y \in Y$".

Thus, he established the following basic properties of the operations $*$ and $\dagger$:

(a) $X \subset X_1$ implies $X^* \supset X_1^*$; (c) $X \subset (X^*)^\dagger$ and $Y \subset (Y^\dagger)^*$.

(b) $Y \subset Y_1$ implies $Y^\dagger \supset Y_1^\dagger$;

Moreover, he derived some consequences of properties (a)–(c). In particular, he studied the particular case when $I = J$ and ρ is symmetric. And, he listed several remarkable illustrating examples for polarities. These already well indicated that polarities can be applied in a great variety of mathematical theories.

The above observations of Birkhoff were extended to posets (partially ordered sets) by Ore [38], who, having in mind the classical

Galois theory of algebraic equations, introduced the terms "Galois correspondences" and "Galois connexions".

More concretely, Ore offered the following definition: "Let P and Q denote two partially ordered sets. We shall assume that there exists a correspondence from P to Q, $p \to \mathfrak{D}(p)$, and also a correspondence from Q to P, $q \to \mathfrak{P}(q)$.

These two correspondences $\mathfrak{D}$ and $\mathfrak{P}$ together shall be called a Galois correspondence between P and Q provided that the two following conditions are fulfilled:

(α) When $p_1 \supset p_2$ are two elements in P or $q_1 \supset q_2$ are two elements in Q then

$$\mathfrak{D}(p_1) \subseteq \mathfrak{D}(p_2), \quad \mathfrak{P}(q_1) \subseteq \mathfrak{P}(q_2).$$

(β) For any element p in P or q in Q

$$\mathfrak{P}\mathfrak{D}(p) \supseteq p, \quad \mathfrak{D}\mathfrak{P}(q) \supseteq q.$$

We shall also say that there exists a Galois connexion between P and Q when a pair of Galois correspondences $\mathfrak{D}$ and $\mathfrak{P}$ has been defined".

The next important step in the theory of Galois connections was made by Schmidt [53]. Despite him being aware of the papers of Everett [21], Riquet [49] and Pickert [46] too, he was mainly interested in the original setting of Birkhoff.

By considering a relation R between two sets E_1 and E_2, he defined and investigated three important set functions defined such that, for any $M_1 \subseteq E_1$,

$$R(M_1) = \{x_2 : \exists\ x_1 \in M_1 : x_1 R x_2\},$$
$$R[M_1] = \{x_2 : \forall\ x_1 \in M_1 : x_1 R x_2\},$$
$$R)M_1(= \{x_2 : \forall\ x_1 R x_2 : x_1 \in M_1\}.$$

Moreover, having in mind these set functions generated by the relation R, he assumed that ω_1 is an arbitrary function of $\mathfrak{P}_1 = \mathcal{P}(E_1)$ to $\mathfrak{P}_2 = \mathcal{P}(E_2)$ and defined an associated set function ω_2 such that

$$\omega_2 M_2 = \max\{M_1 : M_2 \subseteq \omega_1 M_1\}$$

for all $M_2 \in \mathfrak{P}_2$ provided that the above maximum exists.

The investigations of the relationships between these two functions led him to the ingenious observation that properties (a)–(c) established by Birkhoff can be replaced by the single requirement that

$$M_2 \subseteq \omega_1 M_1 \iff M_1 \subseteq \omega_2 M_2$$

for all $M_1 \in \mathfrak{P}_1$ and $M_2 \in \mathfrak{P}_2$.

In addition, he also considered modifications and specializations of the above equivalence. For instance, he also investigated the more natural requirement that

$$\omega_1 M_1 \subseteq M_2 \iff M_1 \subseteq \omega_2 M_2$$

for all $M_1 \in \mathfrak{P}_1$ and $M_2 \in \mathfrak{P}_2$.

This shows that if f is a function of a goset (generalized ordered set) X to another Y and g is a function of Y to X such that

$$f(x) \leq y \iff x \leq g(y)$$

for all $x \in X$ and $y \in Y$, then we may naturally say that the pair (f, g) is a Galois connection between X and Y [15, p. 155].

Curiously enough, in Refs. [25, p. 18] and [31], the increasingness of the corresponding functions was also postulated even in the case when X and Y are posets. Namely, this is usually a consequence of the above equivalence.

Now, if f and g are as above, then by defining $\varphi = g \circ f$ and having in mind Pataki's ideas [41], we can see that

$$f(u) \leq f(v) \iff u \leq g(f(v)) \iff u \leq (g \circ f)(v) \iff u \leq \varphi(v)$$

for all $u, v \in X$.

This shows that before Galois connections, it is more convenient to investigate first another, more simple connection which usually lies between closure operations and Galois connections.

Thus, if φ is a function of the goset X to itself such that

$$f(u) \leq f(v) \iff u \leq \varphi(v)$$

for all $u, v \in X$, then the pair (f, φ) was called a Pataki connection between X and Y by the third author [67].

Namely, if $\mathfrak{F}$ is a structure (set-valued function) and $\square$ is an operation for relators (families of relations) on X, then Pataki [41] called the function $\mathfrak{F}$ to be $\square$-increasing if, for any two relators $\mathcal{R}$ and $\mathcal{S}$ on X, we have

$$\mathfrak{F}_\mathcal{R} \subseteq \mathfrak{F}_\mathcal{S} \iff \mathcal{R} \subseteq \mathcal{S}^\square.$$

Several particular cases of the above Galois and Pataki connections were formerly also considered by the third author [57]. Moreover, he also determined the Galois adjoint of some particular structures for relators [66].

In this chapter, we use a somewhat more convenient terminology. Moreover, we are mainly be interested in the particular case, when the goset X is replaced by a power set equipped with the ordinary set inclusion.

2. A Few Basic Facts on Relations

A subset F of a product set $X \times Y$ is called a *relation* on X to Y. In particular, a relation F on X to itself is simply called a relation on X. And, $\Delta_X = \{(x,x) : x \in X\}$ is called the *identity relation* on X.

If F is a relation on X to Y, then for any $x \in X$ and $A \subseteq X$ the sets $F(x) = \{y \in Y : (x,y) \in F\}$ and $F[A] = \bigcup_{a \in A} F(a)$ are called the *images* or *neighborhoods* of x and A under F, respectively.

If $(x,y) \in F$, then instead of $y \in F(x)$, we may also write $x \, F \, y$. However, instead of $F[A]$, we cannot write $F(A)$. Namely, it may occur that, in addition to $A \subseteq X$, we also have $A \in X$.

The sets $D_F = \{x \in X : F(x) \neq \emptyset\}$ and $R_F = F[X]$ are called the *domain* and *range* of F, respectively. And, if $D_F = X$, then we say that R is a *relation of X to Y* or that F is a *nonpartial relation on X to Y*.

If F is a relation on X to Y and $U \subseteq D_F$, then the relation $F\,|\,U = F \cap (U \times Y)$ is called the *restriction* of F to U. Moreover, if F and G are relations on X to Y such that $D_F \subseteq D_G$ and $F = G\,|\,D_F$, then G is called an *extension* of F.

In particular, a relation f on X to Y is called a *function* if for each $x \in D_f$ there exists $y \in Y$ such that $f(x) = \{y\}$. In this case, by identifying singletons with their elements, we may simply write $f(x) = y$ instead of $f(x) = \{y\}$.

Moreover, a function $\star$ of X to itself is called a *unary operation* on X. While, a function $*$ of X^2 to X is called a *binary operation* on X. And, for any $x, y \in X$, we usually write $x^\star$ and $x * y$ instead of $\star(x)$ and $*((x, y))$.

If F is a relation on X to Y, then a function f of D_F to Y is called a *selection function* of F if $f(x) \in F(x)$ for all $x \in D_F$. Thus, by the Axiom of Choice [18], we can see that every relation is the union of its selection functions.

For a relation F on X to Y, we may naturally define two *set-valued functions* φ_F of X to $\mathcal{P}(Y)$ and Φ_F of $\mathcal{P}(X)$ to $\mathcal{P}(Y)$ such that $\varphi_F(x) = F(x)$ for all $x \in X$ and $\Phi_F(A) = F[A]$ for all $A \subseteq X$.

Functions of X to $\mathcal{P}(Y)$ can be naturally identified with relations on X to Y. While, functions of $\mathcal{P}(X)$ to $\mathcal{P}(Y)$ are more powerful objects than relations on X to Y. In Ref. [78], they were briefly called *corelations on X to Y*.

However, if U is a relation on $\mathcal{P}(X)$ to Y and V is a relation on $\mathcal{P}(X)$ to $\mathcal{P}(Y)$, then it is better to say that U is a *super relation* and V is a *hyperrelation on X to Y* [83]. Thus, closures (proximities) [85] are super (hyper)relations.

Note that a super relation on X to Y is an arbitrary subset of $\mathcal{P}(X) \times Y$. While, a corelation on X to Y is a particular subset of $\mathcal{P}(X) \times \mathcal{P}(Y)$. Thus, set inclusion is a natural partial order for super relations but not for corelations.

For a relation F on X to Y, the relation $F^c = (X \times Y) \setminus F$ is called the *complement* of F. Thus, it can be shown that $F^c(x) = F(x)^c = Y \setminus F(x)$ for all $x \in X$, and $F^c[A]^c = \bigcap_{a \in A} F(a)$ for all $A \subseteq X$.

Moreover, the relation $F^{-1} = \{(y, x) : (x, y) \in F\}$ is called the *inverse* of F. Thus, it can be shown that $F^{-1}[B] = \{x \in X : F(x) \cap B \neq \emptyset\}$ for all $B \subseteq Y$, and in particular, $D_F = F^{-1}[Y]$.

If F is a relation on X to Y, then we have $F = \bigcup_{x \in X} \{x\} \times F(x)$. Therefore, the values $F(x)$, where $x \in X$, uniquely determine F. Thus, a relation F on X to Y can also be naturally defined by specifying $F(x)$ for all $x \in X$.

For instance, if G is a relation on Y to Z, then the *composition relation* $G \circ F$ can be naturally defined such that $(G \circ F)(x) = G[F(x)]$ for all $x \in X$. Thus, it can be shown that $(G \circ F)[A] = G[F[A]]$ for all $A \subseteq X$.

While, if G is a relation on Z to W, then the *box product* $F \boxtimes G$ can be defined such that $(F \boxtimes G)(x, z) = F(x) \times G(z)$ for all $x \in X$ and $z \in Z$. Thus, it can be shown that $(F \boxtimes G)[A] = G \circ A \circ F^{-1}$ for all $A \subseteq X \times Z$ [71].

Hence, by taking $A = \{(x, z)\}$, and $A = \Delta_Y$ if $Y = Z$, one can at once see that the box and composition products are actually equivalent tools. However, the box product can be immediately defined for any family of relations.

The above unary operation c is inversion compatible in the sense that $(F^c)^{-1} = (F^{-1})^c$. Moreover, concerning the above binary operations $\circ$ and $\boxtimes$, we can prove that $(G \circ F)^{-1} = F^{-1} \circ G^{-1}$ and $(F \boxtimes G)^{-1} = F^{-1} \boxtimes G^{-1}$.

3. Some Important Relational Properties

Now, a relation R on X may be briefly defined to be *reflexive* if $\Delta_X \subseteq R$ and *transitive* if $R \circ R \subseteq R$. Moreover, R may be briefly defined to be *symmetric* if $R \subseteq R^{-1}$, *antisymmetric* if $R \cap R^{-1} \subseteq \Delta_X$ and *total* if $X^2 \subseteq R \cup R^{-1}$.

In addition to the above well-known, basic properties, several further remarkable relational properties were also studied in Ref. [60] with the help of the self-closure and interior relations $R^- = R^{-1} \circ R$ and $R^\circ = R^{c-c} = (R^{-1} \circ R^c)^c$.

In the sequel, as it is usual, a reflexive and transitive (symmetric) relation will be called a *preorder (tolerance) relation*. And, a symmetric (antisymmetric) preorder relation will be called an *equivalence (partial order) relation*.

For a relation R on X, we may now also naturally define $R^0 = \Delta_X$, and $R^n = R \circ R^{n-1}$ if $n \in \mathbb{N}$. Moreover, we may also define $R^\infty = \bigcup_{n=0}^{\infty} R^n$. Thus, R^∞ is the smallest preorder relation on X containing R [26].

Now, in contrast to $(R^c)^c = R$ and $(R^{-1})^{-1} = R$, we have $(R^\infty)^\infty = R^\infty$. Moreover, analogously to $(R^c)^{-1} = (R^{-1})^c$, we also have $(R^\infty)^{-1} = (R^{-1})^\infty$. Thus, in particular R^{-1} is also a preorder on X if R is a preorder on X.

For $A \subseteq X$, the *Pervin relation* $R_A = A^2 \cup (A^c \times X)$ is an important preorder on X [45]. While, for a *pseudometric* d on X, the *Weil surrounding* $B_r^d = \{(x, y) \in X^2 : d(x, y) < r\}$, with $r > 0$, is an important tolerance on X [88].

Note that $S_A = R_A \cap R_A^{-1} = R_A \cap R_{A^c} = A^2 \cap (A^c)^2$ is already an equivalence relation on X. And, more generally if $\mathcal{A}$ is a *cover* (*partition*) of X, then $S_\mathcal{A} = \bigcup_{A \in \mathcal{A}} A^2$ is a tolerance (equivalence) relation on X.

Now, as a straightforward generalization of the Pervin relation R_A, for any $A \subseteq X$ and $B \subseteq Y$, we may also naturally consider the Hunsaker–Lindgren relation $R_{(A,B)} = (A \times B) \cup (A^c \times Y)$ [27].

However, it is now more important to note that if $\mathcal{A} = (A_n)_{n=1}^\infty$ is an increasing sequence in $\mathcal{P}(X)$, then the *Cantor relation* $R_\mathcal{A} = \Delta_X \cup \bigcup_{n=1}^\infty (A_n \times A_n^c)$ is also an important preorder on X [28, 39].

Note that if R is only a reflexive relation on X and $x \in X$, then $\mathcal{A}_R(x) = \left(R^n(x)\right)_{n=1}^\infty$ is already an increasing sequence in $\mathcal{P}(X)$. Thus, the preorder relation $R_{\mathcal{A}_R(x)}$ may also be naturally investigated.

Moreover, for a real function φ of X and a quasi-pseudometric d on X [22], the *Brøndsted relation* $R_{(\varphi,d)} = \{(x,y) \in X^2 : d(x,y) \leq \varphi(y) - \varphi(x)\}$ is also an important preorder on X [10].

From this relation, by letting φ and d to be the zero functions, we can obtain the *specialization* and *preference relations* $R_d = \{(x,y) \in X^2 : d(x,y) = 0\}$ and $R_\varphi = \{(x,y) \in X^2 : \varphi(x) \leq \varphi(y)\}$, respectively. (See Refs. [14, 87].)

If R is a relation on X to Y, then the ordered pair $(X,Y)(R) = ((X,Y), R)$ is usually called a *formal context* or *context space* [24]. However, it is better to call it a *relational space* or a *properly simple relator space* [40].

If in particular R is a relation on X, then having in mind a widely used terminology of Birkhoff [4] the ordered pair $X(R) = (X, R)$ may be called a *goset* (generalized ordered set) [74], instead of a *relational system* [13, 50].

If P is a relational property, then the goset $X(R)$ will be said to have property P if the relation R has this property. For instance, the goset $X(R)$ will be called *reflexive* if R is a reflexive relation on X.

In particular, the goset $X(R)$ will be called a *proset* (preordered set) if R is a preorder on X. Moreover, $X(R)$ will be called a *poset* (partially ordered set) if R is a partial order on X.

The terms "goset" and "proset" were perhaps first introduced by the third author. However, by Rudeanu [51], the abbreviations "toset" and "woset" for totally and well-ordered sets, respectively, were also used.

Thus, every set X is a poset with the identity relation Δ_X. Moreover, X is a proset with the universal relation X^2. And, the power set $\mathcal{P}(X) = \{A : A \subseteq X\}$ of X is a poset with the ordinary set inclusion $\subseteq$ and also with its inverse $\supseteq$.

Several definitions on posets can as well be applied to gosets. For instance, if $X(R)$ is a goset, then for any $Y \subseteq X$ the goset $Y(R \cap Y^2)$ is called a *subgoset* of $X(R)$. While, the goset $X'(R') = X(R^{-1})$ is called the *dual* of $X(R)$.

4. Lower and Upper Bounds in Gosets

Notation 1. In this and the following section, we shall assume that R is a relation on X.

Remark 1. Note that the subsequent definition can be immediately extended to the more general case when R is a relation on X to Y.

Moreover, an even more general case, when $\mathcal{R}$ is a relator on X to Y, was already treated in Ref. [64]. However, an immediate extension to super relators seems impossible.

Definition 1. For any $A, B \subseteq X$ and $x, y \in X$, we define

(1) $A \in \mathrm{Lb}_R(B)$ and $B \in \mathrm{Ub}_R(A)$ if $A \times B \subseteq R$;
(2) $x \in \mathrm{lb}_R(B)$ if $\{x\} \in \mathrm{Lb}_R(B)$; (4) $B \in \mathfrak{L}_R$ if $\mathrm{lb}_R(B) \neq \emptyset$;
(3) $y \in \mathrm{ub}_R(A)$ if $\{y\} \in \mathrm{Ub}_R(A)$; (5) $A \in \mathfrak{U}_R$ if $\mathrm{ub}_R(A) \neq \emptyset$.

Thus, for instance, we can easily prove the following two theorems:

Theorem 1. *We have*

(1) $\mathrm{Ub}_R = \mathrm{Lb}_{R^{-1}} = \mathrm{Lb}_R^{-1}$; (2) $\mathrm{ub}_\mathcal{R} = \mathrm{lb}_{R^{-1}}$; (3) $\mathfrak{U}_R = \mathfrak{L}_{R^{-1}}$.

Theorem 2. *For any $A, B \subseteq X$, we have*

(1) $A \in \mathrm{Lb}_R(B) \iff A \subseteq \mathrm{lb}_R(B)$;
(2) $B \in \mathrm{Ub}_R(A) \iff B \subseteq \mathrm{ub}_R(A)$.

Proof. By Definition 1, we have

$$A \in \mathrm{Lb}_R(B) \iff A \times B \subseteq R \iff \forall\, x \in A: \{x\} \times B \subseteq R$$
$$\iff \forall\, x \in A: \{x\} \in \mathrm{Lb}_R(B)$$
$$\iff \forall\, x \in A: x \in \mathrm{lb}_R(B)$$
$$\iff A \subseteq \mathrm{lb}_R(B).$$

Thus, assertion (1) is true.

From assertion (1), by using Theorem 1, we can easily see that assertion (2) is also true. □

Remark 2. The above two theorems show that the above lower and upper bound relations are actually equivalent tools in the goset $X(R)$.

Now, as an immediate consequence of Theorems 1 and 2, we can also state the following:

Corollary 1. *For any $A, B \subseteq X$, we have*

$$A \subseteq \mathrm{lb}_R(B) \iff B \subseteq \mathrm{ub}_R(A).$$

Proof. By Theorems 1 and 2, it is clear that

$$A \subseteq \mathrm{lb}_R(B) \iff A \in \mathrm{Lb}_R(B) \iff B \in \mathrm{Lb}_R^{-1}(A)$$
$$\iff B \in \mathrm{Ub}_R(A) \iff B \subseteq \mathrm{ub}_R(A). \quad \square$$

Hence, by identifying singletons with their elements, we can immediately derive the following:

Corollary 2. *For any $A, B \subseteq X$, we have*

(1) $\mathrm{lb}_R(B) = \{x \in X : B \subseteq \mathrm{ub}_R(x)\}$;
(2) $\mathrm{ub}_R(A) = \{y \in X : A \subseteq \mathrm{lb}_R(y)\}$.

Proof. For instance, by Corollary 1, for any $x \in X$ we have

$$x \in \mathrm{lb}_R(B) \iff \{x\} \subseteq \mathrm{lb}_R(B) \iff B \subseteq \mathrm{ub}_R(\{x\})$$
$$\iff B \subseteq \mathrm{ub}_R(x). \quad \square$$

Remark 3. However, it is now more important to note that by defining

$$F(A) = \mathrm{ub}_{\mathcal{R}}(A) \quad \text{and} \quad G(B) = \mathrm{lb}_{\mathcal{R}}(B),$$

for all $A \subseteq X$ and $B \subseteq Y$, we can at once see that

$$F(A) \subseteq' B \iff B \subseteq F(A) \iff B \subseteq \mathrm{ub}_{\mathcal{R}}(A)$$
$$\iff A \subseteq \mathrm{lb}_{\mathcal{R}}(B) \iff A \subseteq G(B)$$

for all $A \subseteq X$ and $B \subseteq Y$.

Thus, the functions F and G establish a *Galois connection* [15, p. 155] between the poset $\mathcal{P}(X)$ and the dual of the poset $\mathcal{P}(Y)$. Therefore, several properties of the super relations ub_R and lb_R can be derived from the extensive theory of Galois connections [7, 15, 16, 24, 25].

Thus, for instance, from Corollary 1 we can already derive the following theorem. However, it is usually more convenient to apply some direct proofs.

Theorem 3. *If $A \subseteq X$, then*

(1) $\mathrm{lb}_R(A) \subseteq \mathrm{lb}_R(B)$ *for all* $B \subseteq A$;

(2) $A \subseteq \mathrm{ub}_R\bigl(\mathrm{lb}_R(A)\bigr)$; (3) $\mathrm{lb}_R(A) = \mathrm{lb}_R\bigl(\mathrm{ub}_R\bigl(\mathrm{lb}_R(A)\bigr)\bigr)$.

In addition to Corollary 2, it is also worth proving the following:

Theorem 4. *For any $A, B \subseteq X$, we have*

(1) $\mathrm{ub}_R(A) = \bigcap_{x \in A} \mathrm{ub}_R(x)$; (2) $\mathrm{lb}_R(B) = \bigcap_{y \in B} \mathrm{lb}_R(y)$.

Remark 4. Assertion (1) can be generalized by showing that the relation $F = \mathrm{ub}_R$ is *union-reversing* in the sense that, for any $\mathcal{A} \subseteq \mathcal{P}(X)$, we have $F\bigl(\bigcup \mathcal{A}\bigr) = \bigcap_{A \in \mathcal{A}} F(A)$.

Now, by Theorem 4 and Corollary 2, we can also state the following:

Corollary 3. *For any $A, B \subseteq X$, we have*

(1) $\mathrm{ub}_R(A) = \bigcap_{x \in A} R(x)$;
(2) $\mathrm{lb}_R(B) = \{x \in X : B \subseteq R(x)\}$.

Proof. For any $x, y \in X$, we have
$$y \in \mathrm{ub}_R(x) \iff xRy \iff (x,y) \in R \iff y \in R(x).$$
Therefore, $\mathrm{ub}_R(x) = R(x)$. □

Remark 5. Assertion (1) can be briefly reformulated by stating that $\mathrm{ub}_R(A) = R^c[A]^c$ for all $A \subseteq X$.

Remark 6. If R is a relation on X to Y, then for any $A \subseteq X$ and $B \subseteq Y$, we may also naturally define the following:

(1) $A \in \mathrm{Int}_R(B)$ if $R[A] \subseteq B$;
(2) $A \in \mathrm{Cl}_\mathcal{R}(B)$ if $R[A] \cap B \neq \emptyset$.

However, these topological tools, and their particular cases, are not independent of the former algebraic ones. Namely, by Ref. [64], we have $\mathrm{Int}_R = \mathrm{Lb}_{R^c} \circ \mathcal{C}_X$.

5. Some Further Important Basic Tools in Gosets

Now, by using Definition 1, we may also naturally introduce the following:

Definition 2. For any $A \subseteq X$, we define

(1) $\min_R(A) = A \cap \mathrm{lb}_R(A)$;
(2) $\max_R(A) = A \cap \mathrm{ub}_R(A)$;
(3) $\mathrm{Min}_R(A) = \mathcal{P}(A) \cap \mathrm{Lb}_R(A)$;
(4) $\mathrm{Max}_R(A) = \mathcal{P}(A) \cap \mathrm{Ub}_R(A)$;
(5) $\inf_R(A) = \max_R(\mathrm{lb}_R(A))$;
(6) $\sup_R(A) = \min_R(\mathrm{ub}_R(A))$;
(7) $\mathrm{Inf}_R(A) = \mathrm{Max}_R[\mathrm{Lb}_R(A)]$;
(8) $\mathrm{Sup}_R(A) = \mathrm{Min}_R[\mathrm{Ub}_R(A)]$;
(9) $A \in \ell_R$ if $A \in \mathrm{Lb}_R(A)$;
(10) $A \in \mathcal{L}_R$ if $A \subseteq \mathrm{lb}_R(A)$.

By using this definition, for instance, we can prove the following theorems:

Theorem 5. *We have*

(1) $\mathrm{Max}_R = \mathrm{Min}_{R^{-1}}$;
(2) $\mathrm{Sup}_R = \mathrm{Inf}_{R^{-1}}$;
(3) $\ell_R = \ell_{R^{-1}}$;
(4) $\max_R = \min_{R^{-1}}$;
(5) $\sup_R = \inf_{R^{-1}}$;
(6) $\ell_R = \mathcal{L}_R$.

Theorem 6. *For any $A \subseteq X$, we have*

(1) $\max_R(A) = \bigcap_{x \in A} A \cap \mathrm{ub}_R(x)$;
(2) $\max_R(A) = \{x \in A : A \subseteq \mathrm{lb}_R(x)\}$.

Theorem 7. *For any $A \subseteq X$, we have*

(1) $\sup_R(A) = \mathrm{ub}_R(A) \cap \mathrm{lb}_R(\mathrm{ub}_R(A))$;
(2) $\max_R(A) = A \cap \sup_R(A)$;
(3) $\sup_R(A) = \inf_R(\mathrm{ub}_R(A))$.

Proof. To prove assertion (3), note that by assertion (1) and Theorem 3, and their duals, we have

$$\sup_R(A) = \mathrm{lb}_R(\mathrm{ub}_R(A)) \cap \mathrm{ub}_R(A)$$
$$= \mathrm{lb}_R(\mathrm{ub}_R(A)) \cap \mathrm{ub}_R(\mathrm{lb}_R(\mathrm{ub}_R(A))) = \inf_R(\mathrm{ub}_R(A)).$$
□

Theorem 8. *For any $A \subseteq X$, we have*

$$\sup_R(A) = \{x \in X : \mathrm{ub}_R(x) = \mathrm{ub}_R(A)\}$$
$$= \{x \in \mathrm{ub}_R(A) : \mathrm{ub}_R(A) \subseteq \mathrm{ub}_R(x)\}.$$

Theorem 9. *For any $A \subseteq X$, the following assertions are equivalent:*

(1) $A \in \mathcal{L}_R$;   (3) $A \in \mathrm{Min}_R(A)$;
(2) $A \in \mathrm{Ub}_R(A)$; (4) $A \in \mathrm{Max}_R(A)$.

Corollary 4. *For any $A \subseteq X$, the following assertions are equivalent:*

(1) $\mathrm{ub}_R(A) \in \mathcal{L}_R$;
(2) $\mathrm{ub}_R(A) = \sup_R(A)$;
(3) $\mathrm{ub}_R(A) \subseteq \mathrm{lb}_R(\mathrm{ub}_R(A))$.

Theorem 10. *We have*

$$\mathcal{L}_R = \{\min_R(A) : \quad A \subseteq X\} = \{\max_R(A) : \quad A \subseteq X\}.$$

Theorem 11. *If R is reflexive on X, then the following assertions are equivalent:*

(1) *R is antisymmetric;*
(2) *$\operatorname{card}(A) \leq 1$ for all $A \in \mathcal{L}_R$;*
(3) *$\max_R$ is a function;* (4) *$\sup_R$ is a function.*

Remark 7. The implications $(1) \Longrightarrow (3) \Longleftrightarrow (4)$ do not require the relation R to be reflexive.

Definition 3. *The relation R on X, or the goset $X(R)$, will be called:*

(1) *inf-complete if $\inf_R(A) \neq \emptyset$ for all $A \subseteq X$;*
(2) *min-complete if $\min_R(A) \neq \emptyset$ for all $\emptyset \neq A \subseteq X$.*

Remark 8. Thus, for instance, the set $\mathbb{Z}$ of all integers is min-complete but not inf-complete.
While, the set $\overline{\mathbb{R}} = \mathbb{R} \cup \{-\infty, +\infty\}$ of all extended real numbers is inf-complete but not min-complete.

Now, by letting A to be a singleton, and then a doubleton, we can obtain the following:

Theorem 12. *If R is min-complete, then R is reflexive and total.*

Moreover, by using Theorem 7, we can also easily prove the following:

Theorem 13. *The following assertions are equivalent:*

(1) *R is inf-complete;* (2) *R is sup-complete.*

Proof. By Theorem 7, for any $A \subseteq X$, we have $\sup_R(A) = \inf_R\bigl(\operatorname{ub}_R(A)\bigr)$. Hence, the implication $(1) \Longrightarrow (2)$ immediately follows. □

Remark 9. For several other reasonable order-theoretic completeness properties, and their relationships, see Refs. [8, 9].

Remark 10. Now, analogously to Definition 2, for any $A \subseteq X$ we may also naturally define the followings:

(1) $A \in \mathcal{T}_R$ if $A \in \operatorname{Int}_R(A)$; (3) $A \in \mathcal{E}_R$ if $\operatorname{int}_R(A) \neq \emptyset$;
(2) $A \in \mathcal{T}_R$ if $A \subseteq \operatorname{int}_R(A)$; (4) $A \in \mathcal{N}_R$ if $\operatorname{cl}_R(A) \notin \mathcal{E}_R$.

In a relator space $X(\mathcal{R})$, the family $\mathcal{E}_\mathcal{R}$ of all *fat sets* is frequently a more important tool than the family $\mathcal{T}_\mathcal{R}$ of all *topologically open sets*.

Note that $\mathcal{T}_R$ and $\mathcal{E}_R$ are just the families of all *ascending* and *residual subsets* of the goset $X(R)$, respectively.

Moreover, if, for instance, $X = \mathbb{R}$ and $R(x) = \{x-1\} \cup [x, +\infty[$ for all $x \in X$, then $\mathcal{T}_R\{\emptyset, X\}$, but $\mathcal{E}_R$ is quite a large family.

However, the importance of fat and dense sets lies mainly in the fact that they can be used to define convergence and adherence of nets to nets and points [47].

6. A Few Basic Facts on Increasing Functions

Notation 2. In this and the following section, we shall assume that f is a function of one goset $X(R)$ to another $Y(S)$.

Definition 4. The function f will be called *increasing* if, for all $u, v \in X$,

$$u\,R\,v \implies f(u)\,S\,f(v).$$

Remark 11. Now, the function f may be briefly defined to be *decreasing* if it is increasing as a function $X(R)$ to the dual $Y(S^{-1})$ of $Y(S)$.

Moreover, the function may, for instance, be briefly defined to be *strictly increasing* if it is increasing as a function of $X(R \setminus \Delta_X)$ to $Y(S \setminus \Delta_Y)$.

However, to define a strict form of the relation R, instead of $R \setminus \Delta_X$, the relation $R \setminus R^{-1}$ can also be well used (See, for instance, Patrone [44].)

The following theorem shows that the strictly increasing functions are closely related to the injective, increasing ones.

Theorem 14. *If R is total on X and S is reflexive on Y, then the following assertions are equivalent:*

(1) *f is strictly increasing;* (2) *f is injective and increasing.*

Remark 12. To prove the implication (2) $\implies$ (1), we do not need any extra conditions on the relations R and S.

While, if assertion (1) holds and f is onto Y, then to prove that f^{-1} is also strictly increasing, we have to assume that R is total and S is antisymmetric.

Concerning increasing functions, we can also prove the following theorems:

Theorem 15. *The following assertions are equivalent:*

(1) f is increasing;
(2) $f[\mathrm{ub}_R(x)] \subseteq \mathrm{ub}_S(f(x))$ for all $x \in X$;
(3) $f[\mathrm{ub}_R(A)] \subseteq \mathrm{ub}_S(f[A])$ for all $A \subseteq X$.

Theorem 16. *If R is reflexive on X, then the following assertions are equivalent:*

(1) f is increasing;
(2) $f[\max_R(A)] \subseteq \mathrm{ub}_S(f[A])$ for all $A \subseteq X$;
(3) $f[\max_R(A)] \subseteq \max_S(f[A])$ for all $A \subseteq X$.

From Theorem 15, by using Theorem 3, we can immediately derive the following:

Theorem 17. *If f is increasing, then for any $A \subseteq X$, we have*

$$\mathrm{lb}_S\left(\mathrm{ub}_S(f[A])\right) \subseteq \mathrm{lb}_S\left(f[\mathrm{ub}_R(A)]\right).$$

Moreover, by using Theorems 11 and 15, we can also prove the following:

Theorem 18. *If f is increasing and R and S are antisymmetric and sup-complete, then for any $A \subseteq X$ we have*

$$\sup\nolimits_S(f[A]) \, S \, f\left(\sup\nolimits_R(A)\right).$$

Finally, we note that, by the results of Ref. [80], the following theorems are also true. Therefore, instead of "increasing", we may also naturally say "continuous".

Theorem 19. *The following assertions are equivalent:*

(1) f is increasing;
(2) $(u,v) \in R$ implies $\big(f(u),\ f(v)\big) \in S$;
(3) $v \in R(u)$ implies $f(v) \in S\big(f(u)\big)$ for all $u \in X$.

Theorem 20. *The following assertions are equivalent:*

(1) f is increasing; (4) $f \circ R \circ f^{-1} \subseteq S$;
(2) $f \circ R \subseteq S \circ f$; (5) $R \circ f^{-1} \subseteq f^{-1} \circ S$.
(3) $R \subseteq f^{-1} \circ S \circ f$;

Theorem 21. *The following assertions are equivalent:*

(1) f is increasing;
(2) $(f \boxtimes f)[R] \subseteq S$;
(3) $(f \boxtimes R)[\Delta_X] \subseteq (S \boxtimes f)^{-1}[\Delta_Y]$;
(4) $R \subseteq (f \boxtimes f)^{-1}[S]$;
(5) $(R^{-1} \boxtimes f)[\Delta_X] \subseteq (f^{-1} \boxtimes S)[\Delta_Y]$.

Remark 13. Now, a relation F on the goset $X(R)$ to a set Y may be naturally called *increasing* if the associated set-valued function φ_F is increasing. That is, $u\,R\,v$ implies $F(u) \subseteq F(v)$ for all $u,v \in X$.

However, if F is a relation on $X(R)$ to $Y(S)$, then in addition to the above inclusion-increasingness of F, we may also define an *order-increasingness* of F by requiring the implication $u \in \mathrm{lb}_R(v) \implies F(u) \in \mathrm{Lb}_S\big(F(v)\big)$ for all $u,v \in X$.

Thus, we can show that F is inclusion-increasing if and only if $R \circ F^{-1} \subseteq F^{-1}$, or equivalently F^{-1} is ascending-valued. And, F is order-increasing if and only if $F \circ R \circ F^{-1} \subseteq S$, or equivalently $F[R(u)] \subseteq \mathrm{ub}_S\big(F(u)\big)$ for all $u \in X$ [80].

7. The Induced Order and Interior Relations

The following definition and some of the forthcoming theorems do not need the relation R postulated in Notation 2:

Definition 5. For each $u \in X$ and $y \in Y$, we define
$$\mathrm{Ord}_f(u) = \big\{v \in X : f(u)\,S\,f(v)\big\}$$
and
$$\mathrm{Int}_f(y) = \big\{x \in X : f(x)\,S\,y\big\}.$$

The relations Ord_f and Int_f will be called the *natural order* and the *proximal interior* induced by f, respectively.

Remark 14. If F is a relation on one set X to another Y, then by using the associated set-valued function φ_F, we may also naturally define $\mathrm{Ord}_F = \mathrm{Ord}_{\varphi_F}$ and $\mathrm{Int}_F = \mathrm{Int}_{\varphi_F}$.

If in particular U is super relation on X to Y, then, for instance, for any $B \subseteq Y$, we may also naturally define $\mathrm{int}_U(B) = \{x \in X : \{x\} \in \mathrm{Int}_U(B)\}$.

Concerning the relations Ord_f and Int_f, we can easily prove the following four theorems:

Theorem 22. Ord_f *is the largest relation on X making the function f to be increasing.*

Proof. If f is increasing with respect to the relations R and S, then

$$v \in R(u) \implies u\,R\,v \implies f(u)\,S\,f(v) \implies v \in \mathrm{Ord}_f(u),$$

and thus $R(u) \subseteq \mathrm{Ord}_f(u)$ for all $u \in X$. Therefore, $R \subseteq \mathrm{Ord}_f$ also holds. $\square$

Theorem 23. *The following assertions hold:*

(1) Ord_f *is a preorder on X if S is a preorder on Y;*
(2) Ord_f *is a partial order on X if f is injective and S is a partial order on Y.*

Theorem 24. *If S is a preorder, then the following assertions are equivalent:*

(1) f *is increasing;*
(2) Ord_f *is decreasing;*
(3) Ord_f *is ascending valued.*

Proof. If $u\,R\,v$ and (1) hold, then $f(u)\,S\,f(v)$. Moreover, if $w \in \mathrm{Ord}_f(v)$, then $f(v)\,S\,f(w)$. Hence, by the transitivity of S, we can infer that $f(u)\,S\,f(w)$, and thus $w \in \mathrm{Ord}_f(u)$. Therefore, $\mathrm{Ord}_f(v) \subseteq \mathrm{Ord}_f(u)$, and thus (2) also holds.

Conversely, if $u\,R\,v$ and (2) hold, then $\mathrm{Ord}_f(v) \subseteq \mathrm{Ord}_f(u)$. Moreover, by the reflexivity of S, we also have $f(v)\,S\,f(v)$, and thus

$v \in \mathrm{Ord}_f(v)$. Therefore, $v \in \mathrm{Ord}_f(u)$, and thus $f(u)\, S\, f(v)$ is also true. Consequently, (1) also holds. □

Theorem 25. *If S is transitive, then we have the following:*

(1) Int_f *is increasing;*
(2) Int_f *is descending valued if f is increasing.*

Proof. To prove (2), note that if $y \in Y$ and $x \in \mathrm{Int}_f(y)$, then $f(x)\, S\, y$. Moreover, if $u \in X$ such that $u\, R\, x$ and f increasing, then $f(u)\, S\, f(x)$. Thus, by the transitivity of S, we also have $f(u)\, S\, y$, and thus $u \in \mathrm{Int}_f(y)$. Therefore, $\mathrm{Int}_f(y)$ is a descending subset of X. □

The following two theorems show that the relations Ord_f and Int_f are not independent of each other, and they are also closely related to the relations lb_S and ub_S.

Theorem 26. *We have the following:*

(1) $\mathrm{Int}_f \circ f = \mathrm{Ord}_f^{-1}$; (2) $\mathrm{Ord}_f = f^{-1} \circ \mathrm{Int}_f^{-1}$.

Proof. By the corresponding definitions, for all $u, v \in X$, we have

$$v \in (\mathrm{Int}_f \circ f)(u) \iff v \in \mathrm{Int}_f(f(u))$$
$$\iff f(v) \leq f(u) \iff u \in \mathrm{Ord}_f(v)$$
$$\iff v \in \mathrm{Ord}_f^{-1}(u).$$

Therefore, assertion (1) is true, and thus assertion (2) is also true. □

Theorem 27. *We have the following:*

(1) $\mathrm{Int}_f = f^{-1} \circ \mathrm{lb}_S \circ \Delta_Y$; (2) $\mathrm{Int}_f^{-1} = \mathrm{ub}_S \circ f$.

Proof. By the corresponding definitions, for all $x \in X$ and $y \in Y$, we have

$$x \in \mathrm{Int}_f(y) \iff f(x)\, S\, y \iff f(x) \in \mathrm{lb}_S(y) \iff x \in f^{-1}[\mathrm{lb}_S(y)]$$
$$\iff x \in f^{-1}\left[\mathrm{lb}_S(\Delta_Y(y))\right]$$
$$\iff x \in \left(f^{-1} \circ \mathrm{Int}_f \circ \mathrm{lb}_S \circ \Delta_Y\right)(y).$$

Therefore, assertion (1) is true.

Moreover, quite similarly, we can also see that
$$y \in \operatorname{Int}_f^{-1}(x) \iff x \in \operatorname{Int}_f(y) \iff f(x)\, S\, y$$
$$\iff y \in \operatorname{ub}_S(f(x)) \iff y \in (\operatorname{ub}_S \circ f)(x).$$
Therefore, assertion (2) is also true. □

Remark 15. In this respect, it is also worth noting that
$$y \in \operatorname{ub}_S(f[\operatorname{Int}_f(y)])$$
for all $y \in Y$, namely, for every $x \in \operatorname{Int}_f(y)$, we have $f(x)\, S\, y$.

Now, we can also easily prove the following:

Theorem 28. *If*
$$f[\sup_R(A)] \subseteq \operatorname{lb}_S(\operatorname{ub}_S(f[A]))$$
for all $A \subseteq X$, then
$$\max_R(\operatorname{Int}_f(y)) = \sup_R(\operatorname{Int}_f(y))$$
for all $y \in Y$.

Proof. If $y \in Y$, then by Theorem 7 we have
$$\max_R(\operatorname{Int}_f(y)) \subseteq \sup_R(\operatorname{Int}_f(y)).$$
Therefore, we need actually prove only the converse inclusion.

For this, note that if $x \in \sup_R(\operatorname{Int}_f(y))$, then by the assumed property of f we have
$$f(x) \in f[\sup_R(\operatorname{Int}_f(y))] \subseteq \operatorname{lb}_S(\operatorname{ub}_S(f[\operatorname{Int}_f(y)])).$$
Moreover, by Remark 15, we also have $y \in \operatorname{ub}_S(f[\operatorname{Int}_f(y)])$. Therefore, we necessarily have $f(x) \leq y$, and thus $x \in \operatorname{Int}_f(y)$. Hence, by Theorem 7, we can see that
$$x \in \operatorname{Int}_f(y) \cap \sup_R(\operatorname{Int}_f(y)) = \max_R(\operatorname{Int}_f(y)).$$
Therefore, $\sup_R(\operatorname{Int}_f(y)) \subseteq \max_R(\operatorname{Int}_f(y))$, and thus the required equality is also true. □

Remark 16. Note that, by Theorem 7, for a subset A of the goset $X(R)$ we have $\max_R(A) = \sup_R(A)$ if and only if $\sup_R(A) \subseteq A$.

8. Extensive, Involutive and Idempotent Functions

Notation 3. In this and the following section, we shall assume that φ is a function of a goset $X(R)$ to itself.

Definition 6. The function φ will be called

(1) extensive if $\Delta_X \, R \, \varphi$;
(2) intensive if $\varphi \, R \, \Delta_X$;
(3) right semi-involutive if $\Delta_X \, R \, \varphi^2$;
(4) left semi-involutive if $\varphi^2 \, R \, \Delta_X$;
(5) right semi-idempotent if $\varphi \, R \, \varphi^2$;
(6) left semi-idempotent if $\varphi^2 \, R \, \varphi$.

Remark 17. Property (3), in detailed form, means only that $\Delta_X(x) \, R \, \varphi^2(x)$, i.e., $x \, R \, \varphi(\varphi(x))$ for all $x \in X$.

By using Definition 6, we can easily establish the following:

Theorem 29. *The following assertions hold:*

(1) *φ is right semi-idempotent if φ is extensive;*
(2) *φ is right semi-involutive if and only if φ^2 is extensive;*
(3) *φ is right semi-idempotent if and only if $\varphi \, | \, \varphi[X]$ is extensive.*

Proof. If φ is extensive, then $x \, R \, \varphi(x)$ for all $x \in X$. Hence, taking $u \in X$ and writing $\varphi(u)$ in place of x, we can infer that $\varphi(u) \, R \, \varphi^2(u)$. Thus, φ is right semi-idempotent.

Moreover, if $y \in \varphi[X]$, then there exists $x \in X$ such that $y = \varphi(x)$, and thus $\varphi(y) = \varphi^2(x)$. Moreover, if φ is right semi-idempotent, then $\varphi(x) \, R \, \varphi^2(x)$, and thus $y \, R \, \varphi(y)$. Therefore, the restriction $\varphi \, | \, \varphi[X]$ is extensive. □

Remark 18. In addition to the above observations, it is also worth noting that φ is extensive with respect to R if and only if $\varphi(x) \in R(x)$ for all $x \in X$. That is, φ is a selection function of R.

Thus, analogously to a relational reformulation of the Axiom of Choice, the following generalization of a theorem of Bourbaki [5, p. 4] may also be considered as a selection theorem.

Theorem 30. *If φ is strictly increasing and R is antisymmetric and min-complete, then φ is extensive.*

Proof. Assume on the contrary that φ is not extensive. Then, by Remark 18, φ is not a selection function of R. Thus,

$$A = \{x \in X : \varphi(x) \notin R(x)\} \neq \emptyset.$$

Therefore, by the min-completeness of R, there exists $a \in X$ such that $a \in \min_R(A)$. Hence, by the definition of $\min_R$, we can infer that

$$a \in A \quad \text{and} \quad a \in \operatorname{lb}_R(A),$$

and thus $a\,R\,x$ for all $x \in A$.

Now, since $a \in A$, we can also note that $a\,R\,a$, and thus $a \in R(a)$. Moreover, by the definition of A, we can also note that $\varphi(a) \notin R(a)$. Therefore, $\varphi(a) \neq a$. Moreover, from Theorem 12, we know that R is total. Thus, since $a\,R\,\varphi(a)$ does not hold, we necessarily have $\varphi(a)\,R\,a$.

Hence, by using that $\varphi(a) \neq a$ and φ is strictly increasing, we can infer that $\varphi(\varphi(a))\,R\,\varphi(a)$ and $\varphi(\varphi(a)) \neq \varphi(a)$. Thus, by the antisymmetry of R, $\varphi(a)\,R\,\varphi(\varphi(a))$ cannot hold. This shows that $\varphi(\varphi(a)) \notin R(\varphi(a))$, and thus $\varphi(a) \in A$. Hence, by using that $a\,R\,x$ for all $x \in A$, we can infer that $a\,R\,\varphi(a)$, and thus $\varphi(a) \in R(a)$. This contradiction shows that φ is extensive. □

Remark 19. Note that if φ is extensive, R is antisymmetric and x is a maximal element of $X(R)$ in the sense that $x\,R\,y$ implies $y\,R\,x$ for all $y \in X$, then x is already a fixed point of φ in the sense that $\varphi(x) = x$.

This simple but important fact was first explicitly stated by Brøndsted [11]. And, fixed point theorems for extensive maps (which were sometimes also called expansive, progressive, increasing or inflationary) were proved by several authors.

9. Involution, Projection and Closure Operations

Definition 7. The function φ will be called

(1) *involution operation* if it is increasing and both left and right semi-involutive;
(2) *projection operation* if it is increasing and both left and right semi-idempotent;

(3) *closure (interior) operation* if it is an extensive (intensive) projection operation.

Remark 20. Moreover, φ may, for instance, be called a

(1) *preclosure operation* if it is increasing and extensive;
(2) *semi-closure operation* if it is extensive and left semi-idempotent;
(3) *left semi-modification operation* if it is increasing and left semi-idempotent.

Note that, by Theorem 29, an extensive operation is right semi-idempotent. Moreover, the corresponding interior operations can be briefly defined by using the dual of $X(R)$.

In connection with Definition 6, it is also worth mentioning that if, for instance, φ is both left and right semi-idempotent and R is antisymmetric, then φ is idempotent in the sense that $\varphi^2 = \varphi$. However, if φ is idempotent and R is not reflexive, then φ need not be either left or right semi-idempotent.

Concerning closure operations, for instance, we can prove the following:

Theorem 31. *If φ is a closure operation and R is antisymmetric and inf-complete, then for any $A \subseteq X$ we have*

$$\inf{}_R(\varphi[A]) = \varphi\Big(\inf{}_R(\varphi[A])\Big).$$

Proof. By the dual of Theorem 18, we have

$$\inf{}_R(\varphi[A]) \in R\Big(\varphi\Big(\inf{}_R(A)\Big)\Big).$$

Hence, by writing $\varphi[A]$ in place of A, we can see that

$$\inf{}_R(\varphi[\varphi[A]]) \in R\Big(\varphi\Big(\inf{}_R(\varphi[A])\Big)\Big).$$

Moreover, because of the antisymmetry of R, we can note that φ is now idempotent. Therefore, $\varphi[\varphi[A]] = (\varphi \circ \varphi)[A] = \varphi^2[A] = \varphi[A]$.

Thus, we actually have

$$\inf_R (\varphi[A]) \in R\Big(\varphi\Big(\inf_R (\varphi[A])\Big)\Big).$$

Moreover, by extensivity of φ, the converse inclusion is also true. Hence, by using the antisymmetry of R, we can see that the required equality is also true. □

Remark 21. It can be easily seen that an operation φ on a set X is idempotent if and only if $\varphi[X]$ is the family of all fixed points of φ.

Therefore, by using Theorem 31, we can also prove the following:

Corollary 5. *Under the conditions of Theorem 31, for any $A \subseteq \varphi[X]$, we have*

$$\inf_R (A) = \varphi\Big(\inf_R(A)\Big).$$

Proof. Now, because of the antisymmetry of R, the operation φ is idempotent. Thus, by Remark 21, we have $\varphi(y) = y$ for all $y \in \varphi[X]$. Hence, by using the assumption $A \subseteq \varphi[X]$, we can see that $\varphi[A] = A$. Thus, Theorem 31 gives the required equality. □

Remark 22. Note that if φ is an extensive and left semi-idempotent, and R reflexive and antisymmetric, then $\varphi[X]$ is also the family of all elements x of X which are φ-closed in the sense that $\varphi(x) R x$.

Therefore, if in addition to the conditions of Theorem 31, R is reflexive, then the assertion of Corollary 5 can also be expressed by stating that the infimum of any family of φ-closed elements of $X(R)$ is also φ-closed.

Now, instead of a counterpart of Theorem 31, we can only prove the following:

Theorem 32. *If φ is a closure operation, and R is transitive, antisymmetric and sup-complete, then for any $A \subseteq X$ we have*

$$\varphi(\sup_R(A)) = \varphi(\sup_R (\varphi[A])).$$

Proof. Define $\alpha = \sup_R(A)$ and $\beta = \sup_R(\varphi[A])$. Then, by Theorem 18, we have $\beta\, R\, \varphi(\alpha)$. Hence, since φ is increasing, we can infer that $\varphi(\beta)\, R\, \varphi(\varphi(\alpha))$. Moreover, since φ is now idempotent, we also have $\varphi(\varphi(\alpha)) = \varphi(\alpha)$. Therefore, $\varphi(\beta)\, R\, \varphi(\alpha)$.

On the other hand, since φ is extensive, for any $x \in A$ we have $x\, R\, \varphi(x)$. Moreover, since $\beta \in \mathrm{ub}_R(\varphi[A])$, we also have $\varphi(x)\, R\, \beta$. Hence, by using the transitivity of R, we can infer that $x\, R\, \beta$. Therefore, $\beta \in \mathrm{ub}_R(A)$. Now, by using that $\alpha \in \mathrm{lb}_{RX}(\mathrm{ub}_X(A))$, we can see that $\alpha\, R\, \beta$. Hence, by using the increasingness of φ, we can infer that $\varphi(\alpha)\, R\, \varphi(\beta)$. Therefore, by the antisymmetry of R, we actually have $\varphi(\alpha) = \varphi(\beta)$, and thus the required equality is also true. □

By using this theorem, we only prove the following addition to Theorem 31:

Corollary 6. *Under the conditions of Theorem 32, for any $A \subseteq X$, the following assertions are equivalent:*

(1) $\sup_R(\varphi[A]) = \varphi(\sup_R(A))$; (2) $\sup_R(\varphi[A]) = \varphi\bigl(\sup_R(\varphi[A])\bigr)$.

10. Galois-Type Connections between Gosets

Notation 4. In this and the following eight sections, we shall assume the followings:

(1) $X(R)$ and $Y(S)$ are gosets;
(2) φ is a function of X to itself;
(3) f is a function of X to Y and g is a function of Y to X.

In Ref. [81], slightly extending the ideas of Ore [38], Schmidt [53, p. 209], Blyth and Janowitz [7, p. 11] and Száz [69] on Galois connections, residuated mappings and increasingly normal functions, the third author has introduced the following:

Definition 8. We say that the function f is

(1) *increasingly right g-seminormal* if

$$f(x)\, S\, y \implies x\, R\, g(y);$$

(2) *increasingly left g-seminormal* if

$$x\, R\, g(y) \implies f(x)\, S\, y,$$

for all $x \in X$ and $y \in Y$.

Remark 23. Now, the function f may be naturally called *increasingly g-normal* if it is both increasingly left and right g-seminormal.

Moreover, the function f may, for instance, be naturally called *increasingly normal* if it is increasingly g-normal for some function g.

Later, we shall see that the increasingly normal functions are usually increasing. Therefore, the function f may, for instance, be naturally called *decreasingly normal* if it is increasing normal as a function of $X(R)$ to $Y(S^{-1})$.

In this respect, it is also worth mentioning that the following simple dualization principle can be proved:

Theorem 33. *If f is an increasingly left (right) g-seminormal function of $X(R)$ to $Y(S)$, then g is an increasingly right (left) f-seminormal function of $Y(S^{-1})$ to $X(R^{-1})$.*

Proof. If f is increasingly right g-seminormal, then by the corresponding definitions it is clear that

$$y\, S^{-1} f(x) \implies f(x)\, S\, y \implies x\, R\, g(y) \implies g(y)\, R^{-1}\, x$$

for all $y \in Y$ and $x \in X$. Therefore, g is increasingly left f-seminormal as a function of $Y(S^{-1})$ to $X(R^{-1})$. □

Corollary 7. *If f is an increasingly g-normal function of $X(R)$ to $Y(S)$, then g is an increasingly f-normal function of $Y(S^{-1})$ to $X(R^{-1})$.*

Remark 24. By Theorem 33, the properties of the functions g and $f \circ g$ can be immediately derived from those of f and $g \circ f$. However, it is sometimes more convenient to apply some direct proofs.

In Ref. [81], having in mind the properties of the function $\varphi = g \circ f$, and slightly extending the ideas of Pataki [41] and Száz [69], the third author has also introduced the following:

Definition 9. We say that the function f is

(1) *increasingly right φ-semiregular* if
$$f(u)\,S\,f(v) \implies u\,R\,\varphi(v);$$

(2) *increasingly left φ-semiregular* if
$$u\,R\,\varphi(v) \implies f(u)\,S\,f(v),$$

for all $u, v \in X$.

Remark 25. Now, the function f may be naturally called *increasingly φ-regular* if it is both increasingly left and right φ-semiregular.

Moreover, the function f may, for instance, be naturally called *increasingly regular* if it is increasingly φ-regular for some function φ.

Analogously to Remark 23, the function f may, for instance, be naturally called *decreasingly regular* if it is increasingly regular as a function of $X(R)$ to $Y(S^{-1})$.

Unfortunately, now we do not have a counterpart of Theorem 33. However, to clarify the relationship between normal and regular functions, we can easily prove the following two theorems:

Theorem 34. *If f is increasingly left (right) g-seminormal and $\varphi = g \circ f$, then f is increasingly left (right) φ-semiregular.*

Corollary 8. *If f is increasingly g-normal and $\varphi = g \circ f$, then f is increasingly φ-regular.*

Theorem 35. *If f is increasingly left (right) φ-semiregular, f is onto Y, and $\varphi = g \circ f$, then f is increasingly left (right) g-seminormal.*

Proof. Suppose that $x \in X$ and $y \in Y$. Then, since $Y = f[X]$, there exists $v \in X$ such that $y = f(v)$.

Now, if f is increasingly right φ-semiregular, then we can easily see that
$$f(x)\,S\,y \implies f(x)\,S\,f(v) \implies x\,R\,\varphi(v)$$
$$\implies x\,R(g \circ f)(v) \implies x\,R \leq g\big(f(v)\big) \implies x\,R\,g(y).$$

Therefore, f is increasingly right g-seminormal too. □

Corollary 9. *If f is increasingly φ-regular, f is onto Y and $\varphi = g \circ f$, then f is increasingly g-normal.*

Remark 26. By Theorem 34, it is clear that several properties of the increasingly normal functions can be immediately derived from those of the increasingly regular ones. Therefore, the latter ones have to be studied before the former ones.

Moreover, from Theorem 35, we can see that the increasing regular functions are still less general objects than the increasingly normal ones. Later, we shall see that they are strictly between closure operations and increasingly normal functions.

11. Some Basic Properties of Increasingly Semiregular Functions

Now, as some immediate consequences of the corresponding definitions, we can prove the following theorems and their corollaries:

Theorem 36. *If f is increasingly right φ-semiregular and S is reflexive, then φ is extensive.*

Proof. Due to the reflexivity of S, for any $x \in X$, we have $f(x) S f(x)$. Hence, by using the assumed semiregularity of f, we can infer that $x R \varphi(x)$. Therefore, φ is extensive: □

Hence, by using Theorem 29, we can immediately infer the following:

Corollary 10. *Under the assumptions of Theorem 36, the function φ is right semi-idempotent.*

Theorem 37. *If f is increasingly φ-regular, R is transitive and S is reflexive, then f is increasing.*

Proof. By Theorem 36, we have $x R \varphi(x)$ for all $x \in X$. Therefore, if $u, v \in X$ such that $u R v$, then by the inequality $v R \varphi(v)$ and the transitivity of R, we also have $u R \varphi(v)$. Hence, by using the assumed left semiregularity of f, we can infer that $f(u) S f(v)$. Therefore, f is increasing. □

Now, as an immediate consequence of Theorems 36 and 37, we can also state the following:

Corollary 11. *Under the assumptions of Theorem 37, we have $f\,S\,(f \circ \varphi)$.*

Theorem 38. *If f is increasingly left φ-semiregular and R is reflexive, then $(f \circ \varphi)\,S\,f$.*

Proof. Due to the reflexivity of R, for any $x \in X$ we have $\varphi(x)\,R\,\varphi(x)$. Hence, by using the assumed semiregularity of f, we can infer $f(\varphi(x))\,S\,f(x)$. Therefore, $(f \circ \varphi)(x)\,S\,f(x)$, and thus $(f \circ \varphi)\,S\,f$. □

Now, combining Corollary 11 and Theorem 38, we can also state the following:

Corollary 12. *If f is increasingly φ-regular, R is a preorder and S is reflexive, then*

(1) $(f \circ \varphi)\,S\,f$;   (2) $f\,S\,(f \circ \varphi)$.

Remark 27. Thus, if in addition S is antisymmetric, then we can also state that $f = f \circ \varphi$.

Theorem 39. *If f is increasingly φ-regular, R is reflexive and S is transitive, then φ is left semi-idempotent.*

Proof. By Theorem 38, we have $(f \circ \varphi)\,S\,f$. Hence, by using the corresponding definitions, we can infer that $(f \circ \varphi^2)\,S\,(f \circ \varphi)$. Now, by the transitivity of Y, it is clear that $(f \circ \varphi^2)\,S\,f$ also holds. Therefore, for any $x \in X$, we have $f(\varphi^2(x))\,S\,f(x)$. Hence, by using the assumed right semiregularity of f, we can infer that $\varphi^2(x)\,R\,\varphi(x)$. Therefore, φ is left semi-idempotent. □

Now, as an immediate consequence of Theorems 36 and 39, we can also state the following:

Corollary 13. *If f is increasingly φ-regular, R is reflexive and S is a preorder, then φ is a semiclosure operation.*

Remark 28. Thus, φ is both left and right semi-idempotent. Therefore, if in addition S is antisymmetric, then φ is idempotent.

Theorem 40. *If f is increasingly φ-regular and R and S are preorders, then φ is a closure operation.*

Proof. By Corollary 13, we need only show that φ is also increasing. For this, note that if $u, v \in X$ such that $u\,R\,v$, then by Theorem 37 we have $f(u)\,S\,f(v)$. Moreover, by Theorem 38, we have $f\bigl(\varphi(u)\bigr)\,S\,f(u)$. Thus, by the transitivity of S, we also have $f\bigl(\varphi(u)\bigr)\,S\,f(v)$. Hence, by using the assumed right semiregularity of f, we can infer that $\varphi(u)\,R\,\varphi(v)$. Therefore, φ is increasing. □

Theorem 41. *If φ is extensive and R is transitive, then φ is increasingly right φ-semiregular.*

Proof. If $u, v \in X$ such that $\varphi(u)\,R\,\varphi(v)$, then because of $u\,R\,\varphi(v)$ and the transitivity of R we also have $u\,R\,\varphi(v)$. Therefore, φ is increasingly right φ-semiregular. □

Thus, by Theorem 36, we can also state the following:

Corollary 14. *If R is a preorder, then φ is extensive if and only if it is increasingly right φ-semiregular.*

Theorem 42. *If φ is a left semi-modification operation and R is transitive, then φ is increasingly left φ-semiregular.*

Proof. If $u, v \in X$ such that $u\,R\,\varphi(v)$, then by the increasingness of φ we also have $\varphi(u)\,R\,\varphi\bigl(\varphi(v)\bigr)$. Moreover, since φ is left semi-idempotent, we also have $\varphi\bigl(\varphi(v)\bigr)\,R\,\varphi(v)$. Hence, by using the transitivity of R, we can infer that $\varphi(u)\,R\,\varphi(v)$. Therefore, φ is increasingly left φ-semiregular. □

Thus, by Theorem 41, we can also state the following:

Corollary 15. *If φ is a closure operation and R is transitive, then φ is increasingly φ-regular.*

Now, combining this corollary and Theorem 40, we can also state the following:

Theorem 43. *If R and S are preorders, then the following assertions are equivalent:*

(1) *φ is a closure operation;*
(2) *φ is increasingly φ-regular;*

(3) there exists an increasingly φ-regular function h of $X(R)$ to a proset $Z(T)$.

Remark 29. Thus, increasingly regular functions are natural generalizations of closure operations. Moreover, every closure operation can be derived from increasingly regular functions.

From Theorem 43, by using the corresponding definitions, we can easily derive the following:

Corollary 16. *If R and S are preorders, then the following assertions are equivalent:*

(1) *f is increasingly φ-regular;*
(2) *φ is a closure operation and $\mathrm{Ord}_\varphi = \mathrm{Ord}_f$.*

Remark 30. Note that, by Definition 5, the equality $\mathrm{Ord}_\varphi = \mathrm{Ord}_f$ means only that, for any $u, v \in X$, we have

$$\varphi(u)\,R\,\varphi(v) \iff f(u)\,S\,f(v).$$

12. Some Basic Properties of Increasingly Seminormal Functions

From the results of Section 11, by using Theorem 34, for instance, we can immediately derive the following theorems. While, to prove their corollaries Theorem 33 can be applied.

Theorem 44. *If f is increasingly right g-seminormal and S is reflexive, then $g \circ f$ is extensive.*

Corollary 17. *If f is increasingly left g-seminormal and R is reflexive, then $f \circ g$ is intensive.*

Theorem 45. *If f is increasingly g-normal, R is transitive and S is reflexive, then f is increasing.*

Corollary 18. *If f is increasingly g-normal, R is reflexive and S is transitive, then g is increasing.*

Theorem 46. *If f is increasingly g-normal, R is a preorder and S is reflexive, then*

(1) $(f \circ g \circ f)\,S\,f;$ (2) $f\,S\,(f \circ g \circ f).$

Corollary 19. *If f is increasingly g–normal, R is reflexive and S is a preorder, then*

(1) $(g \circ f \circ g) \, R \, g;$ (2) $g \, R \, (g \circ f \circ g).$

Remark 31. If in addition S and R are antisymmetric, then the equalities

$$f = f \circ g \circ f \quad \text{and} \quad g = g \circ f \circ g$$

can also be stated in Theorem 46 and Corollary 19, respectively.

Theorem 47. *If f is increasingly g-normal and R and S are preorders, then*

(1) $g \circ f$ *is a closure operation;*   (2) $f \circ g$ *is an interior operation.*

Now, as a useful characterization of normal functions, we can also prove the following:

Theorem 48. *If R and S are preorders, then the following assertions are equivalent:*

(1) f *is g-normal;*
(2) f *and g are increasing, $g \circ f$ is extensive and $f \circ g$ is intensive.*

Proof. If assertion (1) holds, then from Theorems 44 and 45 and their corollaries, we can at once see that assertion (2) also holds. Therefore, we need actually prove the converse implication.

For this, assume that assertion (2) holds, $x \in X$ and $y \in Y$. Now, if $f(x) S y$, then by using the increasingness of g we can see that $g(f(x)) R g(y)$. Hence, by using that $x \, R (g \circ f)(x) = g(f(x))$, we can already infer that $x \, R g(y)$. Therefore, f is right g-seminormal.

Conversely, if $x \, R g(y)$, then by using the increasingness of f we can see that $f(x) S f(g(y))$. Hence, by using that $f(g(y)) = (f \circ g)(y) S y$, we can already infer that $f(x) S y$. Therefore, f is also left g-seminormal. Thus, assertion (1) also holds. $\square$

Remark 32. This theorem shows that the recent definition of Galois connections [15, p. 155], suggested by Schmidt [53, p. 209], is equivalent to the old one given by Ore [38].

Moreover, as a counterpart of Theorems 43, we can also prove the following:

Theorem 49. *If R is preorder, then the following assertions are equivalent:*

(1) φ *is an involution operation;* (2) φ *is increasingly φ-normal.*

Proof. If assertion (2) holds, then by Theorem 48, φ is increasing and φ^2 is both extensive and intensive. Thus, by Theorem 29 and its dual, φ is both right and left semi-involutive. Therefore, by Definition 7, assertion (1) also holds.

Conversely, if assertion (1) holds, then by Definition 7, φ is increasing and φ^2 is both extensive and intensive. Thus,

$$u\, R\, \varphi(\varphi(u)) \quad \text{and} \quad \varphi(\varphi(v))\, R\, v$$

for all $u, v \in X$. Hence, by using the transitivity of R, we can see that

$$\varphi(u)\, R\, v \implies \varphi(\varphi(u))\, R\, \varphi(v) \implies u\, R\, \varphi(v)$$

and

$$u\, R\, \varphi(v) \implies \varphi(u)\, R\, \varphi(\varphi(v)) \implies \varphi(u)\, R\, v.$$

Thus, by Definition 8 and Remark 23, assertion (2) also holds. □

13. Some Very Particular Properties of Increasingly Regular and Normal Functions

Theorem 50. *If f is increasingly φ-regular and R and S are partial orders, then the following assertions are equivalent:*

(1) $\varphi = \Delta_X$; (2) f *is injective.*

Proof. By Corollary 12 and the antisymmetry of S, for any $x \in X$, we have $f(\varphi(x)) = f(x)$. Hence, if assertion (2) holds, we can infer that $\varphi(x) = x = \Delta_X(x)$. Thus, assertion (1) also holds.

To prove the converse implication, suppose now that $u, v \in X$ such that $f(u) = f(v)$. Then, by the reflexivity of S, we also have $f(u)\, S\, f(v)$ and $f(v)\, S\, f(u)$. Hence, by using the right

φ-semiregularity of f, we can infer that $u\,R\,\varphi(v)$ and $v\,R\,\varphi(u)$. Hence, if assertion (2) holds, then we can infer that $u\,R\,v$ and $v\,R\,u$. Thus, by the antisymmetry of R, we also have $u = v$. Therefore, assertion (1) also holds. □

Remark 33. By the corresponding definitions, f is increasing if and only if f is increasingly left Δ_X-regular.

Theorem 51. *If f is increasingly g-normal, and R and S are partial orders, then the following assertions are equivalent:*

(1) f *is injective;* (2) $g \circ f = \Delta_X$; (3) g *is onto X.*

Proof. By Corollary 8, the function f is $g \circ f$-normal. Hence, by Theorem 50, we can see that assertions (1) and (2) are equivalent.

Moreover, by Corollary 19 and the antisymmetry of R, we have

$$g\left(f\left(g(y)\right)\right)(x) = g(y)$$

for all $y \in Y$. Hence, if assertion (3) holds, i.e., $g[Y] = X$, then we can infer that

$$g(f(x)) = x$$

for all $x \in X$. Therefore, assertion (1) also holds.

Conversely, if assertion (2) holds, then we can at once see that

$$X = \Delta_X[X] = g[f[X]] \subseteq g[Y].$$

Therefore, $X = g[Y]$, and thus assertion (3) also holds. □

From this theorem, by using Theorem 33, we can immediately derive the following:

Corollary 20. *If f is g-normal, and R and S are partial orders, then the following assertions are equivalent:*

(1) f *is onto Y;* (2) $f \circ g = \Delta_Y$; (3) g *is injective.*

Now, as an immediate consequence of Theorem 51 and Corollary 20, we can also state the following:

Corollary 21. *If f is g-normal, injective and onto Y, and R and S are partial orders, then $g = f^{-1}$.*

Remark 34. Thus, if f is g-normal, then g may be considered as a certain generalized inverse function of f.

Finally, we note that, by using Theorem 43 and Corollary 9, we can prove the following:

Theorem 52. *If φ is a closure operation on X, R and S are preorders, $Z = \varphi[X]$, $f = \varphi$ and $g = \Delta_Z$, then f is an increasingly g-normal function of $X(R)$ onto $Z(R \cap Z^2)$ such that $\varphi = g \circ f$.*

Proof. By Theorem 43, φ is an increasingly φ-regular function of $X(R)$ into itself. Hence, since $f = \varphi$ and $Z = \varphi[X]$, we can see that f is an increasingly φ-regular function of $X(R)$ onto $Z(R \cap Z^2)$. Moreover, we can also note that $g \circ f = \Delta_Z \circ \varphi = \varphi$. Therefore, by Corollary 9, we can state that f is an increasingly g-normal function of $X(R)$ onto $Z(R \cap Z^2)$. □

Remark 35. Thus, every closure operation can be derived from increasingly normal functions.

14. Characterizations of Increasingly Seminormal Functions

Simple reformulations of properties (1) and (2) in Definition 8 yield the next two theorems:

Theorem 53. *The following assertions are equivalent:*

(1) $\mathrm{lb}\big(g(y)\big) \subseteq \mathrm{Int}_f(y)\big)$ *for all $y \in Y$;*
(2) *f is an increasingly left g-seminormal.*

Proof. If assertion (2) holds, then by the corresponding definitions, for all $x \in X$ and $y \in Y$, we have

$$x \in \mathrm{lb}\big(g(y)\big) \implies x \, R \, g(y) \implies f(x) \, S \, y \implies x \in \mathrm{Int}_f(y).$$

Therefore, assertion (1) also holds. The converse implication can be proved quite similarly. □

Theorem 54. *The following assertions are equivalent:*

(1) $\operatorname{Int}_f(y) \subseteq \operatorname{lb}(g(y))$ *for all* $y \in Y$;
(2) $g(y) \in \operatorname{ub}(\operatorname{Int}_f(y))$ *for all* $y \in Y$;
(3) f *is an increasingly right g-seminormal.*

Proof. To prove the equivalence of assertions (1) and (2), note that by Corollary 1, for any $y \in X$, we have

$$\operatorname{Int}_f(y) \subseteq \operatorname{lb}(g(y)) \iff g(y) \in \operatorname{ub}(\operatorname{Int}_f(y)). \qquad \square$$

Now, as an immediate consequence of the above two theorems, we can also state the following:

Corollary 22. *The following assertions are equivalent:*

(1) f *is an increasingly g-normal;*
(2) $\operatorname{Int}_f(y) = \operatorname{lb}(g(y))$ *for all* $y \in Y$.

Hence, by using Theorem 49, we can immediately derive the following:

Corollary 23. *If R is a preorder, then the following assertions are equivalent:*

(1) φ *is an involution;*   (2) $\operatorname{Int}_\varphi(v) = \operatorname{lb}(\varphi(v))$ *for all* $v \in X$.

From Theorem 53, we can also immediately derive the following:

Theorem 55. *If f is an increasingly left g-seminormal and R is reflexive, then g is a selection of Int_f.*

Proof. By the corresponding definitions and Theorem 53, for all $y \in Y$, we have

$$g(y) \in \operatorname{lb}(g(y)) \subseteq \operatorname{Int}_f(y). \qquad \square$$

Thus, by Theorem 54 and Definition 2, we can also state the following:

Corollary 24. *If f is increasingly g-normal and R is reflexive, then for any $y \in Y$ we have*

$$g(y) \in \max(\operatorname{Int}_f(y)).$$

Remark 36. If in addition R is antisymmetric, then by Theorem 11 we may write $g(y) = \max(\text{Int}_f(y))$ in the above corollary.

Now, as a partial converse to Theorem 55, we can also prove the following:

Theorem 56. *If f is increasing, S is transitive and g is a selection of Int_f, then f is increasingly left g-seminormal.*

Proof. For any $y \in Y$, we have $g(y) \in \text{Int}_f(y)$, and thus $f(g(y)) S y$. Hence, by using the increasingness of f and the transitivity of S, we can see that

$$x\, R\, g(y) \implies f(x)\, S\, f(g(y)) \implies f(x)\, S\, y$$

for all $x \in X$. Thus, the required assertion is true. □

Thus, by Theorem 55, we can also state the following:

Corollary 25. *If f is increasing, R is reflexive and S is transitive, then the following assertions are equivalent:*

(1) *f is increasingly left g-seminormal;* (2) *g is a selection of Int_f.*

Now, by using our former results, we can also easily prove the following:

Theorem 57. *If R and S are preorders, then the following assertions are equivalent:*

(1) *f is increasingly g-normal;*
(2) *f is increasing and $g(y) \in \max(\text{Int}_f(y))$ for all $y \in Y$.*

Proof. If assertion (1) holds, then by Theorem 45 and Corollary 24 we can see that assertion (2) also holds.

While if assertion (2) holds, then by Definition 2 we have

$$g(y) \in \text{Int}_f(y) \quad \text{and} \quad g(y) \in \text{ub}(\text{Int}_f(y))$$

for all $y \in Y$. Hence, by using Theorem 54 and 56, we can see that f is increasingly left and right g-seminormal. Thus, assertion (1) also holds. □

From this theorem, by using Theorem 49, we can immediately derive the following:

Corollary 26. *If R is a preorder, then the following assertions are equivalent:*

(1) *φ is an involution operation;*
(2) *φ is an increasing and $\varphi(y) \in \max\bigl(\mathrm{Int}_\varphi(y)\bigr)$ for all $y \in Y$.*

Moreover, as an immediate consequence of Theorem 57, we can also state the following:

Theorem 58. *If R and S are preorders, then the following assertions are equivalent:*

(1) *f is increasingly normal;*
(2) *f is increasing and $\max\bigl(\mathrm{Int}_f(y)\bigr) \neq \emptyset$ for all $y \in Y$.*

Proof. If the second part of assertion (2) holds, then by the Axiom of Choice there exists a function h of Y to X such that $h(y) \in \max\bigl(\mathrm{Int}_f(y)\bigr)$ for all $y \in Y$. Thus, if in addition f is increasing, then by Theorem 57 we can see that f is h-normal. Thus, in particular assertion (1) also holds. $\square$

Corollary 27. *If R and S are preorders, and $X(R)$ is max-complete, then the following assertions are equivalent:*

(1) *f is increasingly normal;*
(2) *f is increasing and $f[X]$ is cofinal in $Y(S^{-1})$.*

Proof. Since $X(R)$ is max-complete, for any $y \in Y$, we have

$$\max\bigl(\mathrm{Int}_f(y)\bigr) \neq \emptyset \iff \mathrm{Int}_f(y) \neq \emptyset \iff \exists\, x \in X : \ x \in \mathrm{Int}_f(y)$$
$$\iff \exists\, x \in X : \ f(x)\, S\, y \iff \exists\, x \in X : \ y\, S^{-1} f(x).$$

Therefore,

$$\forall\, y \in Y : \ \max\bigl(\mathrm{Int}_f(y)\bigr) \neq \emptyset \iff \forall\, y \in Y : \ \exists\, x \in X : \ y\, S^{-1} f(x).$$

That is, $F[X]$ is cofinal in the dual proset $Y(S^{-1})$. Thus, Theorem 58 can be applied to obtain the required equivalence. $\square$

Now, in particular, we can also state the following:

Corollary 28. *If R and S are preorders, $X(R)$ is max-complete and f is onto Y, then the following assertions are equivalent:*

(1) f *is increasing;* (2) f *is increasingly normal.*

Remark 37. Due to the above results, the increasing normality of f may be naturally considered as a strong increasingness of f.

Remark 38. Moreover, by Theorem 57, we may naturally define a relation G_f on Y to X such that, for all $y \in Y$,

$$G_f(y) = \max\bigl(\operatorname{Int}_f(y)\bigr).$$

Thus, the second part of assertion (2) of Theorem 57 can be reformulated in form that g is a selection of G_f. Moreover, G_f may be studied separately.

15. Some Further Characterizations of Increasingly Normal Functions

The following theorem allows us to easily prove a useful supremum property of increasingly normal functions which fails to hold for increasing functions:

Theorem 59. *If f is increasingly normal, then for any $A \subseteq X$ we have*

$$f\bigl[\operatorname{lb}(\operatorname{ub}(A))\bigr] \subseteq \operatorname{lb}\bigl(\operatorname{ub}(f[A])\bigr).$$

Proof. If $y \in f\bigl[\operatorname{lb}(\operatorname{ub}(A))\bigr]$, then there exists $x \in \operatorname{lb}(\operatorname{ub}(A))$ such that $y = f(x)$. Moreover, if $b \in \operatorname{ub}(f[A])$, then for any $a \in A$ we have $f(a)\,S\,b$. Hence, by using that f is h-normal, for some function h of Y to X, we can infer that $a\,R\,h(b)$. Therefore, $h(b) \in \operatorname{ub}(A)$, and thus because of $x \in \operatorname{lb}(\operatorname{ub}(A))$ we have $x\,R\,h(b)$. Hence, by using that f is h-normal, we can infer that $f(x)\,S\,b$, and thus $y\,S\,b$. Therefore, $y \in \operatorname{lb}(\operatorname{ub}(f[A]))$ also holds. □

From this theorem, by using Theorem 28, we can derive the following:

Corollary 29. *If f is increasingly normal, then for any $y \in Y$ we have*
$$\max\bigl(\mathrm{Int}_f(y)\bigr) = \sup\bigl(\mathrm{Int}_f(y)\bigr).$$

Proof. By Theorems 7 and 59, we have
$$f\bigl[\sup(A)\bigr] = f\bigl[\mathrm{ub}(A) \cap \mathrm{lb}\bigl(\mathrm{ub}(A)\bigr)\bigr] \subseteq f\bigl[\mathrm{lb}\bigl(\mathrm{ub}(A)\bigr)\bigr]$$
$$\subseteq \mathrm{lb}\bigl(\mathrm{ub}\bigl(f[A]\bigr)\bigr)$$

for all $A \subseteq X$. Hence, by using Theorem 28, we can already see that the required equality is also true. □

Moreover, from Theorem 59, by using Theorems 15 and 45, we can also derive the following:

Theorem 60. *If f is increasingly normal, R is transitive and S is reflexive, then for any $A \subseteq X$ we have*
$$f[\sup(A)] \subseteq \sup\bigl(f[A]\bigr).$$

Proof. From Theorem 45, we know that f is increasing. Moreover, by using Theorems 7, 15 and 59, we can see that
$$f\bigl[\sup(A)\bigr] = f\bigl[\mathrm{ub}(A) \cap \mathrm{lb}\bigl(\mathrm{ub}(A)\bigr)\bigr] \subseteq f\bigl[\mathrm{ub}(A)\bigr] \cap f\bigl[\mathrm{lb}\bigl(\mathrm{ub}(A)\bigr)\bigr]$$
$$\subseteq \mathrm{ub}\bigl(f[A]\bigr) \cap \mathrm{lb}\bigl(\mathrm{ub}\bigl(f[A]\bigr)\bigr) = \sup\bigl(f[A]\bigr). \quad \square$$

Now, to obtain some partial converses of the above theorems, we can also prove the following:

Theorem 61. *If $X(R)$ is a sup-complete proset and $Y(S)$ is an arbitrary proset, then the following assertions are equivalent:*

(1) *f is increasingly normal;*
(2) *$f[\sup(A)] \subseteq \sup\bigl(f[A]\bigr)$ for all $A \subseteq X$;*

(3) f is increasing and $\sup(\operatorname{Int}_f(y)) \subseteq \operatorname{Int}_f(y))$ for all $y \in Y$;
(4) f is increasing and $\max(\operatorname{Int}_f(y)) = \sup(\operatorname{Int}_f(y))$ for all $y \in Y$;
(5) f is increasing and $f[\sup(A)] \subseteq \operatorname{lb}(\operatorname{ub}(f[A]))$ for all $A \subseteq X$;
(6) f is increasing and $f[\operatorname{lb}(\operatorname{ub}(A))] \subseteq \operatorname{lb}(\operatorname{ub}(f[A]))$ for all $A \subseteq X$.

Proof. From Theorem 60, we can see that (1) implies (2). Moreover, from Theorems 45 and 59, we can see that (1) also implies (6).

On the other hand, from Theorem 7, we can see that $\sup(A) \subseteq \operatorname{lb}(\operatorname{ub}(A)$, and thus

$$f[\sup(A)] \subseteq f[\operatorname{lb}(\operatorname{ub}(A))].$$

Therefore, (6) implies (5).

Moreover, if (5) holds, then by Theorem 28 we can see that (4) also holds. While, if (4) holds, then by sup-completeness of $X(R)$, we have

$$\max(\operatorname{Int}_f(y)) = \sup(\operatorname{Int}_f(y)) \neq \emptyset$$

for all $y \in Y$. Thus, from Theorem 58, we can see that (1) also holds.

On the other hand, if (2) holds, then by using Theorem 7 we can see that

$$f[\max(A)] \subseteq f[\sup(A)] \subseteq \sup(f[A]) \subseteq \operatorname{ub}(f[A])$$

for all $A \subseteq X$. Thus, by Theorem 16, f is increasing. Moreover, we can also note that

$$f[\sup(A)] \subseteq \sup(f[A]) \subseteq \operatorname{lb}(\operatorname{ub}(f[A]))$$

for all $A \subseteq X$. Therefore, (5), and thus (1), also holds.

Now, to complete the proof, it remains to note only that, because of Theorem 7, (3) and (4) are also equivalent. □

Remark 39. If in addition R is antisymmetric, then instead of (2) we may write that $f(\sup(A)) \in \sup(f[A])$ for all $A \subseteq X$.

While, if in addition both R and S are antisymmetric, then instead of (2) we may write that $f(\sup(A)) = \sup(f[A])$ for all $A \subseteq X$.

16. Characterizations of Semiregular Functions

Simple reformulations of properties (1) and (2) in Definition 9 yield the following two theorems.

Theorem 62. *The following assertions are equivalent:*

(1) f is increasingly left φ-semiregular;
(2) $\mathrm{lb}(\varphi(x)) \subseteq \mathrm{Int}_f(f(x))$ for all $x \in X$.

Theorem 63. *The following assertions are equivalent:*

(1) f is increasingly right φ-semiregular;
(2) $\mathrm{Int}_f(f(x)) \subseteq \mathrm{lb}(\varphi(x))$ for all $x \in X$;
(3) $\varphi(x) \in \mathrm{ub}(\mathrm{Int}_f(f(x)))$ for all $x \in X$.

Now, as an immediate consequence of the above two theorems, we can also state the following:

Corollary 30. *The following assertions are equivalent:*

(1) f is increasingly φ-semiregular;
(2) $\mathrm{Int}_f(f(x)) = \mathrm{lb}(\varphi(x))$ for all $x \in X$.

Thus, by Theorem 43, we can also state the following:

Corollary 31. *If R is a preorder, then the following assertions are equivalent:*

(1) φ is a closure operation;
(2) $\mathrm{Int}_\varphi(\varphi(x)) = \mathrm{lb}(\varphi(x))$ for all $x \in X$.

From Theorem 62, we can immediately derive the following:

Theorem 64. *If f is increasingly left φ-regular and R is reflexive, then for any $x \in X$ we have*

$$\varphi(x) \in \mathrm{Int}_f(f(x)).$$

Thus, by Theorem 63 and Definition 2, we can also state the following:

Corollary 32. *If f is increasingly φ-regular and R is reflexive, then for any $x \in X$ we have*

$$\varphi(x) \in \max(\mathrm{Int}_f(f(x))).$$

Remark 40. If in addition R is antisymmetric, then by Theorem 11 we may write $\varphi(x) = \max(\operatorname{Int}_f(f(x)))$ in the above corollary.

Now, as a partial converse to Theorem 64, we can also easily prove the following:

Theorem 65. *If f is increasing, S is transitive and*
$$\varphi(x) \in \operatorname{Int}_f(f(x))$$
for all $x \in X$, then f is increasingly left φ-semiregular.

Thus, by Theorem 64, we can also state the following:

Corollary 33. *If f is increasing, R is reflexive and S is transitive,*

(1) $\varphi(x) \in \operatorname{Int}_f(f(x))$ for all $x \in X$;
(2) f is increasingly left φ-semiregular.

Now, analogously to Theorem 57, we can also prove the following:

Theorem 66. *If R and S are preorders, then the following assertions are equivalent:*

(1) f is increasingly φ-regular;
(2) f is increasing and $\varphi(x) \in \max(\operatorname{Int}_f(f(x)))$ for all $x \in X$.

From this theorem, by using Theorem 43, we can immediately derive the following:

Corollary 34. *If R is a preorder, then the following assertions are equivalent:*

(1) φ is a closure operation;
(2) f is increasing and $\varphi(x) \in \max(\operatorname{Int}_\varphi(\varphi(x)))$ for all $x \in X$.

Moreover, as an immediate consequence of Theorem 66, we can also state the following:

Theorem 67. *If R and S are preorders, then the following assertions are equivalent:*

(1) f is increasingly regular;
(2) f is increasing and $\max(\operatorname{Int}_f(f(x))) \neq \emptyset$ for all $x \in X$.

Now, by using this theorem and Theorem 58, we can also prove the following:

Corollary 35. *If R and S are preorders and f is onto Y, then the following assertions are equivalent:*

(1) *f is increasingly regular;*   (2) *f is increasingly normal.*

Proof. If assertion (1) holds, then from Theorem 67 we can see that f is increasing and $\max(\operatorname{Int}_f(f(x))) \neq \emptyset$ for all $x \in X$. Hence, since $Y = f[X]$, we can infer that $\max(\operatorname{Int}_f(y)) \neq \emptyset$ for all $y \in Y$. Therefore, by Theorem 58, assertion (2) also holds. Moreover, by Corollary 8, the converse implication is always true. $\square$

Now, analogously to Theorem 61, we can also prove the following:

Theorem 68. *If $X(R)$ is a sup-complete proset, $Y(S)$ is an arbitrary proset and f is onto Y, then the following assertions are equivalent:*

(1) *f is increasingly regular;*
(2) *$f[\sup(A)] \subseteq \sup(f[A])$ for all $A \subseteq X$;*
(3) *f is increasing and $\sup(\operatorname{Int}_f(f(x))) \subseteq \operatorname{Int}_f(f(x))$ for all $x \in X$;*
(4) *f is increasing and $\max(\operatorname{Int}_f(f(x))) = \sup(\operatorname{Int}_f(f(x)))$ for all $x \in X$.*

From this theorem, by using Theorem 43, we can immediately derive the following:

Corollary 36. *If $X(R)$ is a sup-complete proset and φ is onto X, then the following assertions are equivalent:*

(1) *φ is a closure operation;*
(2) *$\varphi[\sup(A)] \subseteq \sup(\varphi[A])$ for all $A \subseteq X$;*
(3) *φ is increasing and $\sup(\operatorname{Int}_f(f(x))) \subseteq \operatorname{Int}_f(f(x))$ for all $x \in X$;*
(4) *φ is increasing and $\max(\operatorname{Int}_f(f(x))) = \sup(\operatorname{Int}_f(f(x)))$ for all $x \in X$.*

Remark 41. In Theorem 68 and Corollary 36, we may also write $\operatorname{Ord}_f^{-1}(x)$ and $\operatorname{Ord}_\varphi^{-1}(x)$ in place of $\operatorname{Int}_f(f(x))$ and $\operatorname{Int}_\varphi(\varphi(x))$, respectively.

Remark 42. Moreover, by Theorem 66, we may naturally define a relation Φ_f on X such that, for all $x \in X$,

$$\Phi_f(x) = \max(\operatorname{Int}_f(f(x))).$$

Thus, we have $\Phi_f = G_f \circ f$. Moreover, the second part of assertion (2) of Theorem 66 can be reformulated in the form that φ is a selection of Φ_f.

17. Relational Characterizations of Increasingly Seminormal Functions

Analogously to Theorem 20, we can also prove the following two theorems:

Theorem 69. *The following assertions are equivalent:*

(1) *f is increasingly right g-seminormal;*
(2) $S \circ f \subseteq g^{-1} \circ R$; (3) $g \circ S \circ f \subseteq R$.

Proof. For any $x \in X$ and $y \in Y$, the following assertions are equivalent:

$$f(x) S y \implies x R g(y),$$
$$y \in S(f(x)) \implies g(y) \in R(x),$$
$$y \in S(f(x)) \implies y \in g^{-1}[R(x)],$$
$$S(f(x)) \subseteq g^{-1}[R(x)]$$
$$(S \circ f)(x) \subseteq (g^{-1} \circ R)(x).$$

Therefore, by Definition 8, assertions (1) and (2) are equivalent.

Moreover, by using some basic properties of composition, we can see that

(2) $\implies g \circ S \circ f \subseteq g \circ g^{-1} \circ R \implies g \circ S \circ f \subseteq \Delta_X \circ R \implies$ (3)

and

(3) $\implies g^{-1} \circ g \circ S \circ f \subseteq g^{-1} \circ R \implies \Delta_Y \circ S \circ f \subseteq g^{-1} \circ R \implies$ (2).

Therefore, assertions (2) and (3) are also equivalent. □

Theorem 70. *The following assertions are equivalent:*

(1) f *is increasingly left g-seminormal;*

(2) $g^{-1} \circ R \subseteq S \circ f$; (3) $g^{-1} \circ R \circ f^{-1} \subseteq S$.

Proof. From the first part of the proof of the above theorem, it is clear that assertions (1) and (2) are equivalent.

Moreover, analogously to the second part of the proof of the above theorem, we can see that

$$(2) \implies g^{-1} \circ R \circ f^{-1} \subseteq S \circ f \circ f^{-1}$$
$$\implies g^{-1} \circ R \circ f^{-1} \subseteq S \circ \Delta_Y \implies (3)$$

and

$$(3) \implies g^{-1} \circ R \circ f^{-1} \circ f \subseteq S \circ f \implies g^{-1} \circ R \circ \Delta_X \subseteq S \circ f \implies (2).$$

Therefore, assertions (2) and (3) are also equivalent. □

Thus, as an immediate consequence of the above two theorems, we can state the following:

Corollary 37. *The following assertions are equivalent:*

(1) f *is increasingly g-normal;* (2) $S \circ f = g^{-1} \circ R$.

Hence, by using Theorem 49, we can immediately derive the following:

Corollary 38. *If R is a preorder, then following assertions are equivalent:*

(1) φ *is an involution;* (2) $R \circ \varphi = \varphi^{-1} \circ R$.

By using the box product of relations [71], the above results can be reformulated in the following forms:

Theorem 71. *The following assertions are equivalent:*

(1) f *is increasingly right g-seminormal;*

(2) $\left(f^{-1} \boxtimes S\right)[\Delta_Y] \subseteq \left(R^{-1} \boxtimes g^{-1}\right)[\Delta_X]$; (3) $\left(f^{-1} \boxtimes g\right)[S] \subseteq R$.

Theorem 72. *The following assertions are equivalent:*

(1) f *is increasingly left g-seminormal;*

(2) $\left(R^{-1} \boxtimes g^{-1}\right)[\Delta_X] \subseteq \left(f^{-1} \boxtimes S\right)[\Delta_Y]$; (3) $\left(f \boxtimes g^{-1}\right)[R] \subseteq S$.

Corollary 39. *The following assertions are equivalent:*

(1) *f is increasingly g-normal;*
(2) $(f^{-1} \boxtimes S)[\Delta_Y] = (R^{-1} \boxtimes g^{-1})[\Delta_X].$

Corollary 40. *If R is a preorder, then following assertions are equivalent:*

(1) *φ is an involution;* (2) $(\varphi^{-1} \boxtimes R)[\Delta_X] = (R^{-1} \boxtimes \varphi^{-1})[\Delta_X].$

18. Relational Characterizations of Increasingly Semiregular Functions

Theorem 73. *The following assertions are equivalent:*

(1) *f is increasingly right φ-semiregular;*
(2) $f^{-1} \circ S \circ f \subseteq \varphi^{-1} \circ R;$ (3) $\varphi \circ f^{-1} \circ S \circ f \subseteq R.$

Proof. For any $u, v \in X$, the following assertions are equivalent:

$$f(u) S f(v) \implies u R \varphi(v),$$
$$f(v) \in S(f(u)) \implies \varphi(v) \in R(u),$$
$$v \in f^{-1}[S(f(u))] \implies v \in \varphi^{-1}[R(u)],$$
$$f^{-1}[S(f(u))] \subseteq \varphi^{-1}[R(u)],$$
$$(f^{-1} \circ S \circ f)(u) \subseteq (\varphi^{-1} \circ R)(u).$$

Therefore, by Definition 9, assertions (1) and (2) are equivalent.

Moreover, by using some basic properties of composition, we can see that

$$(2) \implies \varphi \circ f^{-1} \circ S \circ f \subseteq \varphi \circ \varphi^{-1} \circ R$$
$$\implies \varphi \circ f^{-1} \circ S \circ f \subseteq \Delta_X \circ R \implies (3)$$

and

$$(3) \implies \varphi^{-1} \circ \varphi \circ f^{-1} \circ S \circ f \subseteq \varphi^{-1} \circ R$$
$$\implies \Delta_X \circ f^{-1} \circ S \circ f \subseteq \varphi^{-1} \circ R \implies (2).$$

Therefore, assertions (2) and (3) are also equivalent. □

Theorem 74. *The following assertions are equivalent:*

(1) f is increasingly left φ-semiregular;
(2) $\varphi^{-1} \circ R \subseteq f^{-1} \circ S \circ f$;
(3) $f \circ \varphi^{-1} \circ R \subseteq S \circ f$;   (4) $\varphi^{-1} \circ R \circ f^{-1} \subseteq f^{-1} \circ S$.

Proof. From the first part of the proof of the above theorem, it is clear that assertions (1) and (2) are equivalent.

Moreover, analogously to the second part of the proof of the above theorem, we can see that

$$(2) \implies f \circ \varphi^{-1} \circ R \subseteq f \circ f^{-1} \circ S \circ f$$
$$\implies f \circ \varphi^{-1} \circ R \subseteq \Delta_Y \circ S \circ f \implies (3)$$

and

$$(3) \implies f^{-1} \circ f \circ \varphi^{-1} \circ R \subseteq F^{-1} \circ S \circ f$$
$$\implies \Delta_X \circ \varphi^{-1} \circ R \subseteq F^{-1} \circ S \circ f \implies (2).$$

Therefore, assertions (2) and (3) are also equivalent.

Quite similarly, now we can also see that

$$(2) \implies \varphi^{-1} \circ R \circ f^{-1} \subseteq f^{-1} \circ S \circ f \circ f^{-1}$$
$$\implies \varphi^{-1} \circ R \circ f^{-1} \subseteq f^{-1} \circ S \circ \Delta_Y \implies (4)$$

and

$$(4) \implies \varphi^{-1} \circ R \circ f^{-1} \circ f \subseteq f^{-1} \circ S \circ f^{-1}$$
$$\implies \varphi^{-1} \circ R \circ \Delta_X \subseteq f^{-1} \circ S \circ f^{-1} \implies (2).$$

Therefore, assertions (2) and (4) are also equivalent. □

Thus, as an immediate consequence of the above two theorems, we can state the following:

Corollary 41. *The following assertions are equivalent:*

(1) f is increasingly φ-regular;   (2) $\varphi^{-1} \circ R = f^{-1} \circ S \circ f$.

Hence, by using Theorem 43, we can immediately derive the following:

Corollary 42. *If R is a preorder, then the following assertions are equivalent:*

(1) φ *is a closure operation;*   (2) $\varphi^{-1} \circ R = \varphi^{-1} \circ S \circ \varphi$.

By using the box product of relations [71], the above results can be reformulated in the following forms:

Theorem 75. *The following assertions are equivalent:*

(1) f *is increasingly right φ-semiregular;*
(2) $\left(f^{-1} \boxtimes f^{-1}\right)[S] \subseteq \left(R^{-1} \boxtimes \varphi^{-1}\right)[\Delta_X]$;
(3) $\varphi \circ \left(f^{-1} \boxtimes f^{-1}\right)[S] \subseteq R$.

Theorem 76. *The following assertions are equivalent:*

(1) f *is increasingly left φ-semiregular;*
(2) $\left(R^{-1} \boxtimes \varphi^{-1}\right)[\Delta_X] \subseteq \left(f^{-1} \boxtimes f^{-1}\right)[S]$;
(3) $\left(R^{-1} \boxtimes f\right)[\varphi^{-1}] \subseteq \left(f^{-1} \boxtimes S\right)[\Delta_Y]$;
(4) $\left(f \boxtimes \varphi\right)[R] \subseteq \left(S^{-1} \boxtimes f^{-1}\right)[\Delta_Y]$.

Corollary 43. *The following assertions are equivalent:*

(1) f *is increasingly φ-regular;*
(2) $\left(f^{-1} \boxtimes f^{-1}\right)[S] = \left(R^{-1} \boxtimes \varphi^{-1}\right)[\Delta_X]$.

Corollary 44. *If R is a preorder, then following assertions are equivalent:*

(1) φ *is a closure operation;*
(2) $\left(\varphi^{-1} \boxtimes \varphi^{-1}\right)[S] = \left(R^{-1} \boxtimes \varphi^{-1}\right)[\Delta_X]$.

19. Increasingly Seminormal Functions of Power Sets to Gosets

Notation 5. In this and the following four sections, we shall assume the following:

(1) F is a function of the poset $\mathcal{P}(X)$ to a goset Y;
(2) G_F is a function of Y to $\mathcal{P}(X)$ such that, for all $y \in Y$,

$$G_F(y) = \{x \in X : \quad F(\{x\}) \leq y\}.$$

Remark 43. Here, by using our former notation for super relations, we may naturally write $F^\triangleleft(x)$ instead of $F(\{x\})$.

However, by the identifying singletons with their elements, it seems now more convenient to write $F(x)$ instead of $F(\{x\})$.

Therefore, if $A \subseteq X$, then in addition to $F(A)$ we may also naturally use the notation $F[A] = \{F(x) : x \in A\}$.

Remark 44. By Definition 5, for all $y \in Y$ we have

$$\mathrm{Int}_F(y) = \{A \subseteq X : \quad F(A) \leq y\}.$$

Therefore, for any $x \in X$, we have

$$x \in G_F(y) \iff F(\{x\}) \leq y \iff \{x\} \in \mathrm{Int}_F(y) \iff x \in \mathrm{int}_F(y).$$

Thus, we actually have $G_F = \mathrm{int}_F$.

However, the appropriateness of our present notation G_F instead of int_F is already apparent from the following:

Theorem 77. *If G is a function of Y to $\mathcal{P}(X)$, then*

(1) $G \leq G_F$ *if F is increasingly left G-seminormal;*
(2) $G_F \leq G$ *if F is increasingly right G-seminormal.*

Proof. If F is increasingly right G-seminormal, then by the corresponding definitions, for any $y \in Y$,

$$x \in G_F(y) \implies F(\{x\}) \leq y \implies \{x\} \subseteq G(y) \implies x \in G(y).$$

Therefore, $G_F(y) \subseteq G(y)$ for all $y \in Y$, and thus $G_F \leq G$. This shows that assertion (2) is true.

Assertion (1) can be proved quite similarly by reversing the above argument. □

Now, as an immediate consequence of this theorem, we can also state the following:

Corollary 45. *If G is a function of Y to $\mathcal{P}(X)$ such that F is increasingly G-normal, then $G = G_F$.*

Thus, in particular, we can also state the following three corollaries:

Corollary 46. *There exists at most one function G of Y to $\mathcal{P}(X)$ such that F is increasingly G-normal.*

Corollary 47. *If F is an increasingly normal, then $G = G_F$ is the unique function of Y to $\mathcal{P}(X)$ such that F is increasingly G-normal.*

Corollary 48. *The following assertions are equivalent:*

(1) F *is increasingly normal;* (2) F *is increasingly G_F-normal.*

Hence, by using Corollary 18, we can immediately derive the following theorem which can also be easily proved directly.

Theorem 78. *If Y is transitive, then G_F is increasing.*

Now, as a certain converse to this theorem, we can also prove the following:

Theorem 79. *If G_F is increasing, F is onto Y and Y is reflexive, then Y is transitive.*

Proof. Suppose that $y_1, y_2, y_3 \in Y$ such that $y_1 \leq y_2$ and $y_2 \leq y_3$. Then, by the increasingness of G_F, we also have $G_F(y_1) \subseteq G_F(y_2)$ and $G_F(y_2) \subseteq G_F(y_3)$, and thus $G_F(y_1) \subseteq G_F(y_3)$.

Moreover, since $F[X] = Y$, there exists $x \in X$ such that $F(x) = y_1$. Hence, by using the reflexivity of Y, we can infer that $F(x) \leq y_1$, and thus $x \in G_F(y_1)$. Now, because of $G_F(y_1) \subseteq G_F(y_3)$, we can also state that $x \in G_F(y_3)$. Therefore, $F(x) \leq y_3$, and thus $y_1 \leq y_3$ also holds. This proves the required assertion. □

Thus, as an immediate consequence of the above two theorems, we can also state the following:

Corollary 49. *If F is onto Y and Y is reflexive, then the following assertions are equivalent:*

(1) G_F *is increasing;* (2) Y *is transitive.*

Remark 45. *If Y is a transitive, then we can easily see that the inverse of the relation associated with G_F is ascending valued.*

However, it is now more interesting that, by using of Remark 43, we can also prove the following:

Theorem 80. *If X is also a goset, Y is transitive and the restriction of F to X is increasing, then G_F is descending valued.*

Proof. Suppose that $y \in Y$, $x \in G_F(y)$ and $u \in X$ such that $u \leq x$. Then, by the assumed increasingness of F, we have $F(u) \leq F(x)$. Moreover, since $x \in G_F(y)$ we also have $F(x) \leq y$. Hence, by the transitivity of Y, it follows that $F(u) \leq y$, and thus $u \in G_F(y)$. Therefore, $G_F(y)$ is a descending subset of X, and thus the required assertion is also true. □

Moreover, as a certain converse to this theorem, we can also prove the following:

Theorem 81. *If X is a goset, Y is reflexive and G_F is descending valued, then the restriction of F to X is increasing.*

Proof. Suppose that $x_1, x_2 \in X$ such that $x_1 \leq x_2$. Then, by the reflexivity of Y, we have $F(x_2) \leq F(x_2)$, and thus $x_2 \in G_F(F(x_2))$. Hence, since $x_1 \leq x_2$ and $G_F(F(x_2))$ is descending, we can already infer that $x_1 \in G_F(F(x_2))$, and thus $F(x_1) \leq F(x_2)$. Therefore, the required assertion is also true. □

Thus, as the immediate consequence of the above two theorems, we can state the following:

Corollary 50. *If X is a goset, Y is a proset, then the following assertions are equivalent:*

(1) G_F *is descending valued;* (2) $F \,|\, X$ *is increasing.*

20. Characterizations of Increasing G_F-Seminormalities

From Theorem 53, we can immediately derive the following:

Theorem 82. *The following assertions assertions are equivalent:*

(1) $\mathcal{P}(G_F(y)) \subseteq \mathrm{Int}_F(y)$ *for all $y \in Y$;*
(2) F *is increasingly left G_F-seminormal.*

Proof. By Theorem 53, assertion (2) is equivalent to the following statement:

(a) $\mathrm{lb}\bigl(G_F(y)\bigr) \subseteq \mathrm{Int}_F(y)$ for all $y \in Y$.

Moreover, by the corresponding definitions, for any $A \subseteq X$ and $y \in Y$, we have

$$A \in \mathrm{lb}\bigl(G_F(y)\bigr) \iff A \subseteq G_F(y) \iff A \in \mathcal{P}\bigl(G_F(y)\bigr),$$

and thus $\mathrm{lb}\bigl(G_F(y)\bigr) = \mathcal{P}\bigl(G_F(y)\bigr)$.

Therefore, statement (a) is equivalent to assertion (1), and thus assertions (1) and (2) are also equivalent. $\square$

Remark 46. If Int_F is descending valued, then for any $y \in Y$ we have

$$\mathcal{P}\bigl(G_F(y)\bigr) \subseteq \mathrm{Int}_F(y) \iff G_F(y) \subseteq \mathrm{Int}_F(y).$$

Therefore, as an immediate consequence of Theorems 25 and 82, we can state the following:

Corollary 51. *If F is increasing and Y is transitive, then the following assertions are equivalent:*

(1) $G_F(y) \in \mathrm{Int}_F(y)$ for all $y \in Y$;
(2) F is increasingly left G_F-seminormal.

Quite similarly, from Theorem 54 we can easily derive the following:

Theorem 83. *The following assertions are equivalent:*

(1) $\bigcup \mathrm{Int}_F(y) \subseteq G_F(y)$ for all $y \in Y$;
(2) $\mathrm{Int}_F(y) \subseteq \mathcal{P}\bigl(G_F(y)\bigr)$ for all $y \in Y$;
(3) F is increasingly right G_F-seminormal.

Proof. By Theorem 54, assertion (3) is equivalent to the following statements:

(a) $\mathrm{Int}_F(y) \subseteq \mathrm{lb}\bigl(G_F(y)\bigr)$ for all $y \in Y$;
(b) $G_F(y) \in \mathrm{ub}\bigl(\mathrm{Int}_F(y)\bigr)$ for all $y \in Y$.

Moreover, from the proof of Theorem 82, we know that $\mathrm{lb}(G_F(y)) = \mathcal{P}(G_F(y))$. Therefore, statement (a) is equivalent to assertion (2), and thus assertions (2) and (3) are equivalent.

On the other hand, by the corresponding definitions, for any $y \in Y$, we have

$$G_F(y) \in \mathrm{ub}(\mathrm{Int}_F(y)) \iff \forall\, A \in \mathrm{Int}_F(y): \ A \subseteq G_F(y)$$
$$\iff \bigcup \mathrm{Int}_F(y) \subseteq G_F(y).$$

Therefore, statement (b) is equivalent to assertion (1), and thus (1) and (3) are also equivalent. □

Remark 47. If $y \in Y$, then by Remark 44, for any $x \in X$,

$$x \in G_F(y) \implies \{x\} \in \mathrm{Int}_F(y).$$

Therefore, the inclusion $G_F(y) \subseteq \bigcup \mathrm{Int}_F(y)$ is always true.

Thus, as an immediate consequence of Theorem 83, we can also state the following:

Corollary 52. *The following assertions are equivalent:*

(1) $G_F(y) = \bigcup \mathrm{Int}_F(y)$ *for all* $y \in Y$;
(2) F *is increasingly right* G_F-*seminormal.*

Remark 48. By Remark 44, assertion (1) can be written in the more instructive form that $\mathrm{int}_F(y) = \bigcup \mathrm{Int}_F(y)$ for all $y \in Y$.

Now, as an immediate consequence of Theorems 82 and 83, we can also state the following:

Theorem 84. *The following assertions are equivalent:*

(1) F *is increasingly* G_F-*normal;*
(2) $\mathrm{Int}_F(y) = \mathcal{P}(G_F(y))$ *for all* $y \in Y$.

Remark 49. Again, by Remark 44, assertion (2) can be written in the more instructive form that $\mathrm{Int}_F(y) = \mathcal{P}(\mathrm{int}_F(y))$ for all $y \in Y$.

Moreover, from Theorems 57, 58 and 61, by using Corollary 48, we can derive the following two theorems:

Theorem 85. *If Y is a proset, then the following assertions are equivalent:*

(1) *F is increasingly G_F-normal;*
(2) *F is increasing and $\max(\operatorname{Int}_F(y)) \neq \emptyset$ for all $y \in Y$;*
(3) *F is increasing and $G_F(y) = \max(\operatorname{Int}_F(y))$ for all $y \in Y$,*
(4) *F is increasing and $G_F(y) \in \operatorname{Int}_F(y) \subseteq \mathcal{P}(G_F(y))$ for all $y \in Y$;*
(5) *F is increasing and $G_F(y) = \bigcup \operatorname{Int}_F(y) \in \operatorname{Int}_F(y)$ for all $y \in Y$.*

Proof. To check the equivalence of assertions (3) and (4), note that, for any $y \in Y$, the following statements are equivalent:

(a) $G_F(y) = \max(\operatorname{Int}_F(y))$;
(b) $G_F(y) \in \operatorname{Int}_F(y)$ and $A \subseteq G_F(y)$ for all $A \in \operatorname{Int}_F(y)$. □

Theorem 86. *If Y is a proset, then the following assertions are equivalent:*

(1) *F is increasingly G_F-normal;*
(2) *$F\left(\bigcup \mathcal{A}\right) \in \sup(F[\mathcal{A}])$ for all $\mathcal{A} \subseteq \mathcal{P}(X)$;*
(3) *F is increasing and $\bigcup \operatorname{Int}_F(y) \in \operatorname{Int}_F(y)$ for all $y \in Y$;*
(4) *F is increasing and $F\left(\bigcup \mathcal{A}\right) \in \operatorname{lb}\left(\operatorname{ub}(F[\mathcal{A}])\right)$ for all $\mathcal{A} \subseteq \mathcal{P}(X)$;*
(5) *F is increasing and $F\left[\mathcal{P}(\bigcup \mathcal{A})\right] \in \operatorname{lb}\left(\operatorname{ub}(F[\mathcal{A}])\right)$ for all $\mathcal{A} \subseteq \mathcal{P}(X)$.*

Proof. To derive this theorem from Theorem 61, note that for any $\mathcal{A} \subseteq \mathcal{P}(X)$ we have $\sup(\mathcal{A}) = \bigcup \mathcal{A}$.

Moreover, we have $\operatorname{lb}(\mathcal{A}) = \mathcal{P}(\bigcap \mathcal{A})$ and $\operatorname{ub}(\mathcal{A}) = \mathcal{P}^{-1}(\bigcup \mathcal{A})$, and thus

$$\operatorname{lb}\left(\operatorname{ub}(\mathcal{A})\right) = \mathcal{P}\left(\bigcap \mathcal{P}^{-1}\left(\bigcup \mathcal{A}\right)\right) = \mathcal{P}\left(\bigcup \mathcal{A}\right).$$ □

Remark 50. Note that, by Corollary 48, in assertion (1) we may simply write "F is increasingly normal".

Moreover, if Y is a poset, then by Theorem 11, in assertion (2), we may simply write $F\left(\bigcup \mathcal{A}\right) = \sup(F[\mathcal{A}])$.

21. Some Further Characterizations of Increasing G_F-Seminormalities

Now, in addition to Theorem 82, we can also prove the following:

Theorem 87. *The following assertions are equivalent:*

(1) *F is increasingly left G_F-seminormal;*
(2) $\mathrm{ub}(F[A]) \subseteq \mathrm{ub}(F(A))$ *for all $A \subseteq X$;*
(3) $F(x) \leq y$ *for all $x \in A$ implies that $F(A) \leq y$.*

Proof. If $A \subseteq X$ and $y \in \mathrm{ub}(F[A])$, then because of $F[A] = \{F(x) : x \in A\}$ we have $F(x) \leq y$, and thus $x \in G_F(y)$ for all $x \in A$. Therefore, $A \subseteq G_F(y)$. Hence, if (1) holds, we can infer that $F(A) \leq y$, and thus $y \in \mathrm{ub}(F(A))$. Therefore, (2) also holds.

The converse implication can be proved quite similarly by reversing the above argument. Moreover, we can note that assertion (3) is only a detailed reformulation of assertion (2). □

From this theorem, by using Corollary 2, we can immediately derive the following:

Corollary 53. *The following assertions are equivalent:*

(1) *F is increasingly left G_F-seminormal,*
(2) $F(A) \in \mathrm{lb}\big(\mathrm{ub}(F[A])\big)$ *for all $A \subseteq X$.*

Analogously to Theorem 87, we can also prove the following:

Theorem 88. *The following assertions are equivalent:*

(1) *F is increasingly right G_F-seminormal;*
(2) $\mathrm{ub}(F(A)) \subseteq \mathrm{ub}(F[A])$ *for all $A \subseteq X$;*
(3) $F(A) \leq y$ *implies that $F(x) \leq y$ for all $x \in A$.*

Proof. If $A \subseteq X$ and $y \in Y$ such that $F(A) \leq y$, then we have $y \in \mathrm{ub}(F(A))$. Thus, if (2) holds, then we also have $y \in \mathrm{ub}(F[A])$. Hence, by using that $F[A] = \{F(x) : x \in A\}$, we can already infer that $F(x) \leq y$, and thus $x \in G_F(y)$ for all $x \in A$. Therefore, $A \subseteq G_F(y)$, and thus (1) also holds.

The converse implication can be proved quite similarly by reversing the above argument. Moreover, we can note that assertion (3) is only a detailed reformulation of assertion (2). □

From this theorem, by using Corollary 1, we can immediately derive the following:

Corollary 54. *The following assertions are equivalent:*

(1) F is increasingly right G_F-seminormal;
(2) $F[A] \subseteq \mathrm{lb}\,(\mathrm{ub}\,(F(A)))$ for all $A \subseteq X$.

Now, as an immediate consequence of Theorems 87 and 88, we can also state the following:

Theorem 89. *The following assertions are equivalent:*

(1) F is increasingly G_F-normal;
(2) $\mathrm{ub}\,(F(A)) = \mathrm{ub}\,(F[A])$ for all $A \subseteq X$;
(3) $F(A) \leq y$ if and only if $F(x) \leq y$ for all $x \in A$.

From this theorem, by using Corollary 1, we can immediately derive the following:

Corollary 55. *The following assertions are equivalent:*

(1) F is increasingly G_F-normal;
(2) $F(A) \in \mathrm{lb}\,(\mathrm{ub}\,(F[A]))$ and $F[A] \subseteq \mathrm{lb}\,(\mathrm{ub}\,(F(A)))$ for all $A \subseteq X$.

Remark 51. From the first part of assertion (2), by using Theorem 7 we can see that

$$\{F(A)\} \cap \mathrm{ub}\,(F[A]) \subseteq \mathrm{lb}\,(\mathrm{ub}\,(F[A])) \cap \mathrm{ub}\,(F[A]) = \sup(F[A])$$

for all $A \subseteq X$.

Hence, if $F(A) \in \mathrm{ub}\,(F[A])$, then we can infer that $F(A) \in \sup(F[A])$.

22. Increasingly Semiregular Functions of Power Sets to Gosets

Notation 6. In this and the following two sections, in addition to Notation 5, we shall assume that Φ_F is a function of $\mathcal{P}(X)$ to itself

such that
$$\Phi_F(A) = \{x \in X : F(x) \leq F(A)\}$$
for all $A \subseteq X$, where $F(x) = F(\{x\})$ for all $x \in X$.

Thus, by using the definition of G_F, we can easily establish the following:

Theorem 90. *We have*
$$\Phi_F = G_F \circ F.$$

Proof. For any $x \in X$ and $A \subset X$, namely, we have
$$x \in \Phi_F(A) \iff F(x) \leq F(A) \iff x \in G_F(F(A))$$
$$\iff x \in (G_F \circ F)(A).$$

Therefore, the required equality is also true. $\square$

Remark 52. Hence, by using Remark 44, we can immediately infer that
$$\Phi_F = \mathrm{int}_F \circ F.$$

Now, analogously to Theorem 77, we can also prove the following:

Theorem 91. *If Φ is a function of $\P(X)$ to itself, then*

(1) $\Phi \leq \Phi_F$ *if F is increasingly left Φ-semiregular;*
(2) $\Phi_F \leq \Phi$ *if F is increasingly right Φ-semiregular.*

Proof. If F is increasingly right Φ-semiregular, then by the corresponding definitions, for any $A \subseteq X$,
$$x \in \Phi_F(A) \implies F(\{x\}) \leq F(A) \implies \{x\} \subseteq \Phi(A) \implies x \in \Phi(A).$$

Therefore, $\Phi_F(A) \subseteq \Phi(A))$ for all $A \subseteq X$, and thus $\Phi_F \leq \Phi$ also holds. This proves assertion (1).

Assertion (2) can be proved quite similarly by reversing the above argument. $\square$

Now, as an immediate consequence of this theorem, we can also state the following:

Corollary 56. *If Φ is a function of $\mathcal{P}(X)$ to itself such that F is increasingly Φ-regular, then $\Phi = \Phi_F$.*

Hence, it is clear that in particular we also have the following three corollaries:

Corollary 57. *There exists at most one function Φ of $\mathcal{P}(X)$ to itself such that F is increasingly Φ-regular.*

Corollary 58. *If F is increasingly regular, then $\Phi = \Phi_F$ is the unique function of $\mathcal{P}(X)$ to itself such that F is increasingly Φ-regular.*

Corollary 59. *The following assertions are equivalent:*

(1) F is increasingly regular;   (2) F is increasingly Φ_F-regular.

From Theorem 90, by using Theorem 78, we can immediately derive the following:

Theorem 92. *If F is increasing and Y is transitive, then Φ_F is increasing.*

Remark 53. If Φ is a function of $\mathcal{P}(X)$ to itself such that F is increasingly Φ-regular, and Y is a proset, then by Theorem 36 the function Φ is increasing.

Now, as an immediate consequence of Theorem 92, we can also state the following:

Corollary 60. *If F is increasing and Y is transitive, then the inverse of the relation associated with Φ_F is ascending valued.*

Moreover, from Theorem 90, by using Theorem 80, we can immediately derive the following:

Theorem 93. *If X is also a goset, Y is transitive and the restriction of F to X is increasing, then Φ_F is descending valued.*

23. Characterizations of Φ_F-Semiregularities

From Theorem 62, we can immediately derive the following:

Theorem 94. *The following assertions are equivalent:*

(1) F *is increasingly left* Φ_F-*semiregular;*
(2) $\mathcal{P}(\Phi_F(A)) \subseteq \mathrm{Int}_F(F(A))$ *for all* $A \subseteq X$.

Proof. By Theorem 62, assertion (1) is equivalent to the following statement:

(a) $\mathrm{lb}(\Phi_F(A)) \subseteq \mathrm{Int}_F(F(A))$ for all $A \subseteq X$.

Moreover, by the corresponding definitions, for any $A, B \subseteq X$, we have

$$B \in \mathrm{lb}(\Phi_F(A)) \iff B \subseteq \Phi_F(A) \iff B \in \mathcal{P}(\Phi_F(A)),$$

and thus $\mathrm{lb}(\Phi_F(A)) = \mathcal{P}(\Phi_F(A))$.

Therefore, statement (a) is equivalent to assertion (2), and thus assertions (1) and (2) are also equivalent. □

Remark 54. If $\mathrm{Int}_F \circ F$ is descending valued, then for any $A \subseteq X$ we have

$$\mathcal{P}(\Phi_F(A)) \subseteq \mathrm{Int}_F(F(A)) \iff \Phi_F(A) \in \mathrm{Int}_F(F(A)).$$

Therefore, as an immediate consequence of Theorems 25 and 94, we can state the following:

Corollary 61. *If F is increasing and Y is transitive, then the following assertions are equivalent:*

(1) F *is increasingly left* Φ_F-*semiregular;*
(2) $\Phi_F(A) \in \mathrm{Int}_F(F(A))$ *for all* $A \subseteq X$.

Quite similarly, from Theorem 63, we can immediately derive the following:

Theorem 95. *The following assertions are equivalent:*

(1) F *is increasingly right* Φ_F-*semiregular;*
(2) $\bigcup \mathrm{Int}_F(F(A)) \subseteq \Phi_F(A)$ *for all* $A \subseteq X$;
(3) $\mathrm{Int}_F(F(A)) \subseteq \mathcal{P}(\Phi_F(A))$ *for all* $A \subseteq X$.

Proof. By Theorem 63, assertion (1) is equivalent to the following statements:

(a) $\operatorname{Int}_F(F(A)) \subseteq \operatorname{lb}(\Phi_F(A))$ for all $A \subseteq X$;
(b) $\Phi_F(A) \in \operatorname{ub}(\operatorname{Int}_F(F(A)))$ for all $A \subseteq X$.

Moreover, from the proof of Theorem 94, we know that $\operatorname{lb}(\Phi_F(A)) = \mathcal{P}(\Phi_F(A))$. Therefore, statement (a) is equivalent to assertion (3), and thus assertions (1) and (3) are equivalent.

On the other hand, by the corresponding definitions, for any $A \subseteq X$, we have

$$\Phi_F(A) \in \operatorname{ub}(\operatorname{Int}_F(F(A))) \iff \forall\, B \in \operatorname{Int}_F(F(A)) : \quad B \subseteq \Phi_F(A)$$
$$\iff \bigcup \operatorname{Int}_F(F(A)) \subseteq \Phi_F(A).$$

Therefore, statement (b) is equivalent to assertion (2), and thus assertions (2) and (1) are also equivalent. □

Remark 55. If $A \subseteq X$, then the corresponding definitions

$$x \in \Phi_F(A) \implies F(\{x\}) \leq F(A) \implies \{x\} \in \operatorname{Int}_F(F(A)).$$

Therefore, the inclusion $\Phi_F(A) \subseteq \bigcup \operatorname{Int}_F(F(A))$ is always true.

Thus, as an immediate consequence of Theorem 95, we can also state the following:

Corollary 62. *The following assertions are equivalent:*

(1) *F is increasingly right Φ_F-semiregular;*
(2) *$\Phi_F(A) = \bigcup \operatorname{Int}_F(F(A))$ for all $A \subseteq X$.*

Moreover, as an immediate consequence of Theorems 94 and 95, we can also state the following:

Theorem 96. *The following assertions are equivalent:*

(1) *F is increasingly Φ_F-regular;*
(2) *$\operatorname{Int}_F(F(A)) = \mathcal{P}(\Phi_F(A))$ for all $A \subseteq X$.*

Remark 56. By Remark 52, assertion (2) can be written in the more instructive form that $\operatorname{Int}_F(F(A)) = \mathcal{P}(\operatorname{int}_F(F(A)))$ for all $A \subseteq X$.

Finally, we note that from Corollary 35, and Theorems 66 and 68, by using Corollaries 48 and 59, we can immediately derive the following theorems:

Theorem 97. *The following assertions are equivalent:*

(1) F is increasingly Φ_F-regular;
(2) F is increasing and $\max(\text{Int}_F(F(A))) \neq \emptyset$ for all $A \subseteq X$;
(3) F is increasing and $\Phi_F(A) = \max(\text{Int}_F(F(A)))$ for all $A \subseteq X$;
(4) F is increasing and $\Phi_F(A) \in \text{Int}_F(F(A)) \subseteq \mathcal{P}(\Phi_F(A))$ for all $A \subseteq X$;
(5) F is increasing and $\Phi_F(A) = \bigcup \text{Int}_F(F(A)) \in \text{Int}_F(F(A))$ for all $A \subseteq X$.

Remark 57. By the definition of Int_F, for any $A, B \subseteq X$, we have

$$B \in \text{Int}_F(F(A)) \iff F(B) \leq F(A).$$

Therefore, if the second part of assertion (3) holds, then for each $A \subseteq X$, just $B = \Phi_F(A)$ is the largest subset of X such that $F(B) \leq F(A)$.

If assertion (1) holds, and in particular Y is poset, then by Corollary 12 we also have $F(\Phi_F(A)) = F(A)$. Therefore, in this case $B = \Phi_F(A)$ is also the largest subset of X such that $F(B) = F(A)$.

Theorem 98. *If F is onto Y and Y is a proset, then the following assertions are equivalent:*

(1) F is increasingly Φ_F-regular;   (2) F is increasingly G_F-normal.

Theorem 99. *If Y is a proset and F is onto Y, then the following assertions are equivalent:*

(1) F is increasingly Φ_F-regular;
(2) $F(\bigcup \mathcal{A}) \in \sup(F[\mathcal{A}])$ for all $\mathcal{A} \subseteq \mathcal{P}(X)$;
(3) F is increasing and $\bigcup \text{Int}_F(F(A)) \in \text{Int}_F(F(A))$ for all $A \subseteq X$;
(4) F is increasing and $F(\bigcup \mathcal{A}) \in \text{lb}(\text{ub}(F[\mathcal{A}]))$ for all $\mathcal{A} \subseteq \mathcal{P}(X)$;
(5) F is increasing and $F[\mathcal{P}(\bigcup \mathcal{A})] \in \text{lb}(\text{ub}(F[\mathcal{A}]))$ for all $\mathcal{A} \subseteq \mathcal{P}(X)$.

Remark 58. Note that, by Corollary 59, in assertion (1) we may simply write "F is increasingly regular".

Moreover, if Y is a poset, then by Theorem 11, in assertion (2), we may simply write $F(\bigcup \mathcal{A}) = \sup(F[\mathcal{A}])$.

24. Increasing Semiregularity and Seminormality Properties of Quasi-Increasing Functions

A particular case of the following definition has been first introduced in Ref. [71].

Definition 10. The function F will be called *quasi-increasing* if

$$F(x) \leq F(A)$$

for all $x \in X$ and $A \subseteq X$ with $x \in A$.

Remark 59. If F is increasing, then F is in particular quasi-increasing.

If $x \in A \subseteq X$, namely, then $\{x\} \subseteq A$. Hence, by using the increasingness of F, we can infer that $F(\{x\}) \leq F(A)$, and thus $F(x) \leq F(A)$.

Moreover, as a simple reformulation of the above definition we can also state the following.

Theorem 100. *The following assertions are equivalent:*

(1) F *is quasi-increasing;* (2) $F(A) \in \mathrm{ub}(F[A])$ *for all* $A \subseteq X$.

Proof. If $A \subseteq X$ and $y \in F[A]$, then because of $F[A] = \{F(x) : x \in A\}$ there exists $x \in A$ such that $y = F(x)$. Moreover, if assertion (1) holds, then we have $F(x) \leq F(A)$, and thus $y \leq F(A)$. This shows that $F(A) \in \mathrm{ub}(F[A])$, and thus assertion (2) also holds.

The converse implication (2) $\implies$ (1) can be proved quite similarly. □

From this theorem, by using Corollary 2, we can immediately derive the following:

Corollary 63. *The following assertions are equivalent:*

(1) F *is quasi-increasing;* (2) $F[A] \subseteq \mathrm{lb}(F(A))$ *for all* $A \subseteq X$.

However, it is now more important to note that, by using the corresponding definitions, we can also easily prove the following two theorems.

Theorem 101. *If F is increasingly right Φ_F-semiregular and Y is reflexive, then F is quasi-increasing.*

Proof. If $A \subseteq X$, then by the reflexivity of Y we have $F(A) \le F(A)$. Hence, by using the assumed right semiregularity of F, we can infer that $A \subseteq \Phi_F(A)$. Therefore, for any $x \in A$, we have $x \in \Phi_F(A)$, and hence $F(x) \le F(A)$. Thus, F is quasi-increasing. $\square$

Theorem 102. *If F is quasi-increasing and Y is transitive, then F is increasingly right Φ_F-semiregular.*

Proof. Suppose that $U, V \subseteq X$ such that $F(U) \le F(V)$. Then, by the quasi-increasingness of F, for any $x \in U$ we have $F(x) \le F(U)$. Hence, by using the transitivity of Y, we can infer that $F(x) \le F(V)$, and thus $x \in \Phi_F(V)$ for all $x \in U$. Therefore, $U \subseteq \Phi_F(V)$, and thus F is increasingly right Φ_F-semiregular. $\square$

Now, as an immediate consequence of this theorem, we can also state the following:

Corollary 64. *If F is quasi-increasing and Y is transitive, then the following assertions are equivalent:*

(1) *F is increasingly Φ_F-regular;*
(2) *F is increasingly left Φ_F-semiregular.*

From Theorems 101 and 102, by using Theorem 36 and a partial converse of it, we can also immediately derive the following:

Theorem 103. *If Y is a proset, then the following assertions are equivalent:*

(1) *Φ_F is extensive;*
(2) *F is quasi-increasing;*
(3) *F is increasingly right Φ_F-semiregular.*

Proof. By Theorems 101 and 102, it is clear that assertions (2) and (3) are equivalent.

Moreover, from Theorem 36, we can see that assertion (3) implies assertion (1) even if Y is only reflexive. Therefore, we need only show that if assertion (1) holds, then assertion (2) also holds.

For this, note that if $x \in A \subseteq X$, then by assertion (1) we have $A \subseteq \Phi_F(A)$, and thus also $x \in \Phi_F(A)$. Hence, by the definition of Φ_F, we can infer that $F(x) \leq F(A)$. Therefore, assertion (2) holds even if Y is an arbitrary goset. $\square$

Moreover, from Theorems 101 and 102, by using Theorems 34 and 90, we can also easily derive the following two theorems which can also be easily proved directly:

Theorem 104. *If F is increasingly right G_F-seminormal and Y is reflexive, then F is quasi-increasing.*

Theorem 105. *If F is quasi-increasing and Y is transitive, then F is increasingly right G_F-seminormal.*

Now, as an immediate consequence of this theorem, we can also state the following:

Corollary 65. *If F is quasi-increasing and Y is transitive, then the following assertions are equivalent:*

(1) *F is increasingly G_F-normal;*
(2) *F is increasingly left G_F-seminormal.*

Moreover, by Theorems 103 and 104, we can also state the following.

Theorem 106. *If Y is a proset, then the following assertions are equivalent:*

(1) *F is quasi-increasing;*
(2) *F is increasingly right G_F-seminormal.*

Now, as an immediate consequence of Theorems 7 and 100 and Corollary 53, we can also state the following.

Theorem 107. *The following assertions are equivalent:*

(1) *$F(A) \in \sup(F[A])$ for all $A \subseteq X$;*
(2) *F is quasi-increasing and increasingly left G_F-seminormal.*

Remark 60. If in particular Y is antisymmetric, then by Theorem 11 we may simply write $F(A) = \sup(F[A])$ in assertion (1).

Moreover, by Theorems 106 and 107, we can also state the following.

Theorem 108. *If Y is a proset, then the following assertions are equivalent:*

(1) F is increasingly G_F-normal,
(2) $F(A) \in \sup(F[A])$ for all $A \subseteq X$.

Remark 61. If in particular Y is a poset, then by Theorem 11 we may simply write $F(A) = \sup(F[A])$ in assertion (2).

25. A Few Basic Fats on Union-Preserving Functions

Notation 7. In this section, by specializing our former notations, we shall assume that F is a function of $\mathcal{P}(X)$ to $\mathcal{P}(Y)$.

Now, in addition to Definition 10, we may also naturally use the following.

Definition 11. The function F will be called *union-preserving* if

$$F\left(\bigcup \mathcal{A}\right) = \bigcup_{A \in \mathcal{A}} F(A)$$

for all $\mathcal{A} \subseteq \mathcal{P}(X)$.

Remark 62. Note that the following assertions are equivalent:

(1) F is increasing;
(2) $\bigcup_{A \in \mathcal{A}} F(A) \subseteq F(\bigcup \mathcal{A})$ for all $\mathcal{A} \subseteq \mathcal{P}(X)$;
(3) $F(A_1) \cup F(A_2) \subseteq F(A_1 \cup A_2)$ for all $A_1, A_2 \subseteq X$.

Thus, to clarify the relationship between increasingness and union-preservingness, we can at once state the following:

Theorem 109. *The following assertions are equivalent:*

(1) F is union-preserving;
(2) F is increasing and $F(\bigcup \mathcal{A}) \subseteq \bigcup_{A \in \mathcal{A}} F(A)$ for all $\mathcal{A} \subseteq \mathcal{P}(X)$.

However, it is now more important to note that we can also state the following theorem which has also been proved, in a different way, by Pataki [42].

Theorem 110. *The following assertions are equivalent:*

(1) *F is union-preserving;* (2) *$F(A) = \bigcup_{x \in A} F(x)$ for all $A \subseteq X$.*

Proof. Since $A = \bigcup_{x \in A} \{x\}$ for all $A \subseteq X$, it is clear that assertion (1) implies assertion (2).

While, if assertion (2) holds, then we can note that F is increasing. Therefore, to obtain assertion (1), by Theorem 109, we need only show that
$$F\left(\bigcup \mathcal{A}\right) \subseteq \bigcup_{A \in \mathcal{A}} F(A)$$
for all $\mathcal{A} \subseteq \mathcal{P}(X)$.

For this, note that if $\mathcal{A} \subseteq \mathcal{P}(X)$, then by assertion (2) we have
$$F\left(\bigcup \mathcal{A}\right) = \bigcup_{x \in \bigcup \mathcal{A}} F(x).$$

Therefore, if $y \in F(\bigcup \mathcal{A})$, then there exists $x \in \bigcup \mathcal{A}$ such that $y \in F(x)$. Thus, in particular, there exists $A_0 \in \mathcal{A}$ such that $x \in A_0$. Hence, by using the quasi-increasingness of F, we can already see that
$$y \in F(x) \subseteq F(A_0) \subseteq \bigcup_{A \in \mathcal{A}} F(A).$$

Therefore, $F(\bigcup \mathcal{A}) \subseteq \bigcup_{A \in \mathcal{A}} F(A)$ also holds. □

Remark 63. Note that the following assertions are equivalent:

(1) *F is quasi-increasing;* (2) *$\bigcup_{x \in A} F(x) \subseteq F(A)$ for all $A \subseteq X$.*

Thus, as an immediate consequence of Theorem 110, we can also state the following:

Corollary 66. *The following assertions are equivalent:*

(1) *F is union-preserving;*
(2) *F is quasi-increasing and $F(A) \subseteq \bigcup_{x \in A} F(x)$ for all $A \subseteq X$.*

By using Theorem 110, we can also easily establish the following:

Example 1. If R is a relation on X to Y and
$$R^\triangleright(A) = R[A]$$
for all $A \subseteq X$, then $U = R^\triangleright$ is the unique union-preserving function of $\mathcal{P}(X)$ to $\mathcal{P}(Y)$ such that $U(\{x\}) = R(x)$ for all $x \in X$.

Remark 64. Conversely, it can be shown that if U is a function of $\mathcal{P}(X)$ to $\mathcal{P}(y)$ and $U^\triangleleft$ is a relation on X to Y such that
$$U^\triangleleft(x) = U(\{x\})$$
for all $x \in X$, then under the notation $U^\circ = U^{\triangleleft\triangleright}$ the following assertions are equivalent:

(1) $U^\circ = U$;

(2) U is union-preserving;

(3) $U = R^\triangleright$ for some relation R on X to Y.

26. Increasing Normality and Regularity of Union-Preserving Functions

Notation 8. In this section, by specializing our former notations, we shall assume the following:

(1) F is a function of $\mathcal{P}(X)$ to $\mathcal{P}(Y)$;

(2) Φ_F is a function of $\mathcal{P}(Y)$ to itself such that, for all $A \subseteq X$,
$$\Phi_F(A) = \{x \in X : \quad F(x) \subseteq F(A)\};$$

(3) G_F is a function of $\mathcal{P}(Y)$ to $\mathcal{P}(X)$ such that, for all $B \subseteq Y$,
$$G_F(B) = \{x \in X : \quad F(x) \subseteq B\},$$
where $F(x) = F(\{x\})$ for all $x \in X$.

Remark 65. Thus, by Theorem 90, we can at once state that
$$\Phi_F = G_F \circ F.$$

Moreover, by Theorem 78, we can at once state the following:

Theorem 111. *The function G_F is increasing.*

Remark 66. Thus, if F is increasing, then Φ_F is also increasing.

Now, by Theorems 37 and 40 and Corollary 12, we can also state the following:

Theorem 112. *If F is increasingly Φ_F-regular, then*

(1) *F is increasing;* (2) *Φ_F is a closure operation;* (3) *$F = F \circ \Phi_F$.*

Remark 67. If F is increasingly G_F-normal, then by Corollary 8 and Remark 59 we can see that F is increasingly Φ_F-regular. Thus, the above assertions (1)–(3) also hold.

Moreover, by Corollary 48 and Theorems 58, 85 and 86, we can state the following:

Theorem 113. *The following assertions are equivalent:*

(1) *F is union-preserving;*
(2) *F is increasingly normal;*
(3) *F is increasingly G_F-normal;*
(4) *F is increasing and $\max(\mathrm{Int}_F(B)) \neq \emptyset$ for all $B \subseteq Y$;*
(5) *F is increasing and $G_F(B) = \max(\mathrm{Int}_F(B))$ for all $B \subseteq Y$.*

Hint. To prove the equivalence of assertions (1) and (3), note that, by Theorem 86, we have

$$(3) \iff \forall \; \mathcal{A} \subseteq \mathcal{P}(X): \quad F\left(\bigcup \mathcal{A}\right) = \sup(F[\mathcal{A}])$$
$$\iff \forall \; \mathcal{A} \subseteq \mathcal{P}(X): \quad F\left(\bigcup \mathcal{A}\right) = \bigcup F[\mathcal{A}]$$
$$\iff \forall \; \mathcal{A} \subseteq \mathcal{P}(X): \quad F\left(\bigcup \mathcal{A}\right) = \bigcup_{A \in \mathcal{A}} F(A) \iff (1).$$

□

From the above theorem, by using Theorem 110 and Corollary 66, we can immediately derive the following:

Corollary 67. *The following assertions are equivalent:*

(1) *F is increasingly normal;*
(2) *$F(A) = \bigcup_{x \in A} F(x)$ for all $A \subseteq X$;*
(3) *F is quasi-increasing and $F(A) \subseteq \bigcup_{x \in A} F(x)$ for all $A \subseteq X$.*

Remark 68. The implication (2) $\implies$ (1) can also be easily proved directly by noting that if assertion (2) holds, then for any $A \subseteq X$ and $B \subseteq Y$ we have

$$F(A) \subseteq B \iff \bigcup_{x \in A} F(x) \subseteq B \iff \forall\, x \in A:\ F(x) \subseteq B$$

$$\iff \forall\, x \in A:\ x \in G_F(B) \iff A \subseteq G_F(B).$$

Therefore, F is increasingly G_F-normal, and thus assertion (1) also holds.

Now, by Corollary 59 and Theorem 97, we can also state the following:

Theorem 114. *The following assertions are equivalent:*

(1) *F is increasingly regular;*
(2) *F is increasingly Φ_F-regular;*
(3) *F is increasing and $\max\bigl(\mathrm{Int}_F\bigl(F(A)\bigr)\bigr) \neq \emptyset$ for all $A \subseteq X$;*
(4) *F is increasing and $\Phi_F(A) = \max\bigl(\mathrm{Int}_F\bigl(F(A)\bigr)\bigr)$ for all $A \subseteq X$.*

Moreover, by Corollary 35, we can also state the following:

Theorem 115. *If F is onto $\mathcal{P}(Y)$, then the following assertions equivalent:*

(1) *F is increasingly regular;* (2) *F is increasingly normal.*

Remark 69. Thus, if F is onto $\mathcal{P}(Y)$, then an analog of Corollary 67 for the increasing regularity of F can also be stated.

References

[1] S. Acharjee, M. T. Rassias and Á. Száz, Galois and Pataki connections for ordinary functions and super relations. *Electron. J. Math.* **4**, 46–99 (2022).

[2] S. Acharjee, M. T. Rassias and Á. Száz, Applications of Galois and Pataki connections to relator spaces. In: Th. M. Rassias and P. M. Pardalos (eds.) *Global Optimization, Approximation and Applications*, World Scientific, to appear.

[3] H. Arianpoor, Preorder relators and generalized topologies, *J. Lin. Top. Algebra* **5**, 271–277 (2016).

[4] G. Birkhoff, *Lattice Theory.* Colloquium Publications, Vol. 25 (American Mathematical Society, Providence, 1967).
[5] F. Blanqui, A point on fixpoints in posets, HAL Archives, hal-01097809 (2014).
[6] T. S. Blyth, *Lattices and Ordered Algebraic Structures.* Universitext (Springer-Verlag, London, 2005).
[7] T. S. Blyth and M. F. Janowitz, *Residuation Theory* (Pergamon Press, Oxford, 1972).
[8] Z. Boros and Á. Száz, Finite and conditional completeness properties of generalized ordered sets, *Rostock. Math. Kolloq.* **59**, 75–86 (2005).
[9] Z. Boros and Á. Száz, Infimum and supremum completeness properties of ordered sets without axioms, *An. St. Univ. Ovidius Constanta Ser. Math.* **16**, 31–37 (2008).
[10] A. Brøndsted, On a lemma of Bishop and Phelps, *Pac. J. Math.* **55**, 335–341 (1974).
[11] A. Brøndsted, Fixed points and partial orders, *Proc. Am. Math. Soc.* **60**, 365–366 (1976).
[12] S. Buglyó and Á. Száz, A more important Galois connection between distance functions and inequality relations, *Sci. Ser. A Math. Sci. (N.S.)* **18**, 17–38 (2009).
[13] I. Chajda and H. Länger, Groupoids corresponding to relational systems, *Miskolc Math. Notes* **17**, 111–118 (2016).
[14] S. Cobzas, Ekeland, Takahashi and Caristi principles in quas–metric spaces, *Topol. Appl.* **265**, 1–22 (2019).
[15] B. A. Davey and H. A. Priestley, *Introduction to Lattices and Order* (Cambridge University Press, Cambridge, 2002).
[16] K. Denecke, M. Erné and S. L. Wismath (eds.) *Galois Connections and Applications* (Kluwer Academic Publisher, Dordrecht, 2004).
[17] J. C. Derderian, Galois connections and pair algebras, *Can. J. Math.* **21**, 498–501 (1969).
[18] T. Deskins, *The Axiom of Choice and Some Equivalences* (Department of Mathematics, Kenon College, 2012).
[19] M. Erné, Closure, *Contemp. Math.* **486**, 163–238 (2009).
[20] M. Erné, J. Koslowski, A. Melton and G. A. Strecker, A primer on Galois Connections, *Ann. New York Acad. Sci.* **704**, 103–125 (1993).
[21] C. J. Everett, Closure operators and Galois theory in lattices, *Trans. Am. Math. Soc.* **55**, 514–525 (1944).
[22] P. Fletcher and W. F. Lindgren, *Quasi-Uniform Spaces* (Marcel Dekker, New York, 1982).
[23] B. Ganter, Relational Galois connections. In S. O. Kuznetsov and S. Schmidt (eds.) *Formal Concept Analysis.* Lecture Notes in Artificial Intelligence, Vol. 4390 (Springer-Verlag, Berlin, 2007), pp. 1–17.

[24] B. Ganter and R. Wille, *Formal Concept Analysis* (Springer-Verlag, Berlin, 1999).
[25] G. Gierz, K. H. Hofmann, K. Keimel, J. D. Lawson, M. Mislove and D. S. Scott, *A Compendium of Continuous Lattices* (Springer-Verlag, Berlin, 1980).
[26] T. Glavosits, Generated preorders and equivalences, *Acta Acad. Paed. Agrienses Sect. Math.* **29**, 95–103 (2002).
[27] W. Hunsaker and W. Lindgren, Construction of quasi-uniformities, *Math. Ann.* **188**, 39–42 (1970).
[28] G. J. Jinag and Y. J. Cho, Cantor order and completeness, *Int. J. Pure Appl. Math.* **2**, 391–396 (2002).
[29] J. Kurdics and Á. Száz, Well-chained relator spaces, *Kyungpook Math. J.* **32**, 263–271 (1992).
[30] J. Kurdics and Á. Száz, Well-chainedness characterizations of connected relators, *Math. Pannon.* **4**, 37–45 (1993).
[31] J. Lambek, Some Galois connections in elementary number theory, *J. Number Theory* **47**, 371–377 (1994).
[32] N. Levine, On uniformities generated by equivalence relations, *Rend. Circ. Mat. Palermo* **18**, 62–70 (1969).
[33] N. Levine, On Pervin's quasi uniformity, *Math. J. Okayama Univ.* **14**, 97–102 (1970).
[34] J. Mala and Á. Száz, Properly topologically conjugated relators, *Pure Math. Appl. Ser. B*, **3**, 119–136 (1992).
[35] J. Mala and Á. Száz, Modifications of relators, *Acta Math. Hungar.* **77**, 69–81 (1997).
[36] K. Meyer and M. Nieger, Hüllenoperationen, Galoisverbindungen und Polaritäten, *Boll. Un. Mat. Ital.* **14**, 343–350 (1977).
[37] H. Nakano and K. Nakano, Connector theory, *Pac. J. Math.* **56**, 195–213 (1975).
[38] O. Ore, Galois connexions, *Trans. Am. Math. Soc.* **55**, 493–513 (1944).
[39] S. Park, Partial orders and metric completeness, *Proc. Coll. Natur. Sci. SNU* **12**, 11–17 (1987).
[40] G. Pataki, Supplementary notes to the theory of simple relators, *Radovi Mat.* **9**, 101–118 (1999).
[41] G. Pataki, On the extensions, refinements and modifications of relators, *Math. Balk.* **15**, 155–186 (2001).
[42] G. Pataki, On a generalized infimal convoltion of set functions, *Ann. Math. Sil.* **27**, 99–106 (2013).
[43] G. Pataki and Á. Száz, A unified treatment of well-chainedness and connectedness properties, *Acta Math. Acad. Paedagog. Nyházi. (N.S.)* **19**, 101–165 (2003).

[44] F. Petrone, Well-posed minimum problems for preorders, *Rend. Sem. Mat. Univ. Padova* **84**, 109–121 (1990).
[45] W. J. Pervin, Quasi-uniformization of topological spaces, *Math. Ann.* **147**, 316–317 (1962).
[46] G. Pickert, Bemerkungen über Galois-Verbindungen, *Arch. Math.* **3**, 285–289 (1952).
[47] M. T. Rassias and Á. Száz, Basic tools and continuity-like properties in relator spaces, *Contribut. Math.* **3**, 77–106 (2021).
[48] T. M. Rassias and Á. Száz, Ordinary, super and hyper relators can be used to treat the various generalized open sets in a unified way. In N. J. Daras and T. M. Rassias (eds.) *Approximation and Computation in Science and Engineering.* Springer Optimization and Its Applications, Vol. 180 (Springer Nature Switzerland AG, Cham, 2022), pp. 709–782.
[49] M. J. Riquet, Relations binaires, fermetures et correspondances de Galois, *Bull. Soc. Math. Fr.* **76**, 114–155 (1948).
[50] D. A. Romano, A characterization of some homomorphisms between relational systems. Preprint.
[51] S. Rudeanu, *Sets and Ordered Sets* (Bentham Science Publisher, 2012).
[52] M. Salih and Á. Száz, Generalizations of some ordinary and extreme connectedness properties of topological spaces to relator spaces, *Electron. Res. Arch.* **28**, 471–548 (2020).
[53] J. Schmidt, Bieträge zur Filtertheorie II, *Math. Nachr.* **10**, 197–232 (1953).
[54] Á. Száz, Basic tools and mild continuities in relator spaces, *Acta Math. Hungar.* **50**, 177–201 (1987).
[55] Á. Száz, Lebesgue relators, *Monathsh. Math.* **110**, 315–319 (1990).
[56] Á. Száz, Structures derivable from relators, *Singularité* **3**, 14–30 (1992).
[57] Á. Száz, Refinements of relators, *Tech. Rep. Inst. Math. Univ. Debrecen* **76** (1993). 19 p.
[58] Á. Száz, Cauchy nets and completeness in relator spaces, *Colloq. Math. Soc. János Bolyai* **55**, 479–489 (1993).
[59] Á. Száz, Relations refining and dividing each other, *Pure Math. Appl.* **6**, 385–394 (1995).
[60] Á. Száz, Topological characterizations of relational properties, *Grazer Math. Ber.* **327**, 37–52 (1996).
[61] Á. Száz, Uniformly, proximally and topologically compact relators, *Math. Pannon.* **8**, 103–116 (1997).
[62] Á. Száz, A Galois connection between distance functions and inequality relations, *Math. Bohem.* **127**, 437–448 (2002).
[63] Á. Száz, Somewhat continuity in a unified framework for continuities of relations, *Tatra Mt. Math. Publ.* **24**, 41–56 (2002).

[64] Á. Száz, Upper and lower bounds in relator spaces, *Serdica Math. J.* **29**, 239–270 (2003).
[65] Á. Száz, Rare and meager sets in relator spaces, *Tatra Mt. Math. Publ.* **28**, 75–95 (2004).
[66] Á. Száz, Galois-type connections on power sets and their applications to relators, *Tech. Rep. Inst. Math. Univ. Debrecen* **2005/2** (2005). 38 p.
[67] Á. Száz, Supremum properties of Galois–type connections, *Comment. Math. Univ. Carolin.* **47**, 569–583 (2006).
[68] Á. Száz, Minimal structures, generalized topologies, and ascending systems should not be studied without generalized uniformities, *Filomat* **21**, 87–97 (2007).
[69] Á. Száz, Galois type connections and closure operations on preordered sets, *Acta Math. Univ. Comenian. (N.S.)* **78**, 1–21 (2009).
[70] Á. Száz, Galois–type connections and continuities of pairs of relations, *J. Int. Math. Virt. Inst.* **2**, 39–66 (2012).
[71] Á. Száz, Inclusions for compositions and box products of relations, *J. Int. Math. Virt. Inst.* **3**, 97–125 (2013).
[72] Á. Száz, A particular Galois connection between relations and set functions, *Acta Univ. Sapientiae Math.* **6**, 73–91 (2014).
[73] Á. Száz, Generalizations of Galois and Pataki connections to relator spaces. *J. Int. Math. Virtual Inst.* **4**, 43–75 (2014).
[74] Á. Száz, Basic tools, increasing functions, and closure operations in generalized ordered sets. In P. M. Pardalos and T. M. Rassias (eds.) *Contributions in Mathematics and Engineering, In Honor of Constantion Caratheodory* (Springer, Cham, Switzerland, 2016), pp. 551–616.
[75] Á. Száz, A natural Galois connection between generalized norms and metrics, *Acta Univ. Sapientiae Math.* **9**, 360–373 (2017).
[76] Á. Száz, Four general continuity properties, for pairs of functions, relations and relators, whose particular cases could be investigated by hundreds of mathematicians, *Int. J. Math. Stat. Oper. Res.* **3**, 135–154 (2023).
[77] Á. Száz, The closure-interior Galois connection and its applications to relational equations and inclusions, *J. Int. Math. Virt. Inst.* **8**, 181–224 (2018).
[78] Á. Száz, Corelations are more powerful tools than relations. In T. M. Rassias (ed.) *Applications of Nonlinear Analysis.* Springer Optimization and Its Applications, Vol. 134 (Springer, Cham, Switzerland, 2018), pp. 711–779.
[79] Á. Száz, Relationships between inclusions for relations and inequalities for corelations, *Math. Pannon.* **26**, 15–31 (2018).

[80] Á. Száz, A unifying framework for studying continuity, increasingness, and Galois connections, *MathLab J.* **1**, 154–173 (2018).

[81] Á. Száz, Galois and Pataki connections on generalized ordered sets, *Earthline J. Math. Sci.* **2**, 283–323 (2019).

[82] Á. Száz, Birelator spaces are natural generalizations of not only bitopological spaces, but also ideal topological spaces, In T. M. Rassias and P. M. Pardalos (eds.) *Mathematical Analysis and Applications.* Springer Optimization and Its Applications, Vol. 154 (Springer Nature Switzerland AG, Cham, 2019), pp. 543–586.

[83] Á. Száz, Super and hyper products of super relations, *Tatra Mt. Math. Publ.* **78**, 85–118 (2021).

[84] Á. Száz, Non-conventional stacked three relator spaces. In: B. Hazarika, S. Acharjee and D. S. Djordjevć (eds.), *Advances in Topology and Dynamical Systems: An Interdisciplinary Applications*, Springer, 2023, to appear.

[85] W. J. Thron, *Topological Structures* (Holt, Rinehart and Winston, New York, 1966).

[86] J. Túri and Á. Száz, Comparisons and compositions of Galois–type connections, *Miskolc Math. Notes* **7**, 189–203 (2006).

[87] A. D. Wallace, Relations on topological spaces. In *General Topology and its Relations to Modern Analysis and Algebra, Proceedings of the Symposium Held in Prague in September 1961* (Academia Publishing House of the Czechoslovak Academy of Sciences, Prague, 1962), pp. 356–360.

[88] A. Weil, *Sur les espaces à structure uniforme et sur la topologie générale.* Actualités Scientifiques et Industrielles, Vol. 551 (Herman and Cie, Paris, 1937).

[89] W. Xia, Galois connections and formal concept analysis, *Demonst. Math.* **27**, 751–767 (1994).

© 2024 World Scientific Publishing Company
https://doi.org/10.1142/9789811267048_0002

Chapter 2

A Functional Equation Related to Inner Product Spaces in Šerstnev Probabilistic Normed Spaces

Ahmad Alinejad[*,∥], Hamid Khodaei[†,**,††], Michael Th. Rassias[‡,§,‡‡], and Hamid Vosoughian[¶,§§]

[*]*College of Farabi, University of Tehran, Tehran, Iran*
[†]*Department of Mathematics, Malayer University, Malayer, Iran*
[‡]*Department of Mathematics and Engineering Sciences,
Hellenic Military Academy, Greece*
[§]*Institute for Advanced Study, Program in Interdisciplinary Studies,
Princeton, NJ, USA*
[¶]*Department of Mathematics, Semnan University, Semnan, Iran*

[∥]*alinejad.ahmad@ut.ac.ir*
[**]*hkhodaei@malayeru.ac.ir*
[††]*hkhodaei.math@gmail.com*
[‡‡]*michail.rassias@math.uzh.ch*
[§§]*hvosoughian@yahoo.com*

We consider a functional equation related to inner product spaces in probabilistic normed spaces in the sense of Šerstnev. The work presents a relationship between various disciplines: the theory of metric linear spaces, the theory of triangular norms, the theory of probabilistic normed spaces, the theory of fixed points and the theory of functional equations.

1. Introduction

Triangular norms first appeared in the framework of probabilistic metric spaces in the work of Menger [27]. It also turns out that this is an essential operation in several fields. Triangular norms are an indispensable tool for the interpretation of conjunctions in fuzzy logic [14] and, subsequently, for the intersection of fuzzy sets [45] (for some further information, see Refs. [21–24]).

Šerstnev introduced the first definition of a probabilistic (random) normed space in a series of papers; [40–43] he was motivated by problems of best approximations in statistics. His definition follows the same path used to probabilize the notion of metric space and to introduce probabilistic normed spaces. A new, wider definition of a probabilistic normed space was introduced by Alsina et al. [3, 4]. Their definition quickly became the standard one, and to the best of the authors' knowledge, it has been adopted by all the researchers who, after them, have investigated the properties, uses, or applications of probabilistic normed spaces (see Refs. [12, 13, 25, 39]).

The question of how much a function satisfies an equation approximately (for example, a difference, differential, functional or integral equation) may differ from a solution to the equation arises naturally in applications of mathematics. The theory of Ulam stability provides some efficient tools to evaluate such errors; see Refs. [7, 15] for further details and references. The fixed point method for studying the stability of functional equations was used for the first time by Baker in 1991 (see Ref. [5]). Next, in 2003, Radu [36] delivered a lecture at the Seminar on Fixed Point Theory Cluj-Napoca, where he proved the stability of a functional equation using the fixed point method. Alsina [2] considered the stability of a functional equation in probabilistic normed spaces, and in 2008, Miheţ and Radu considered the stability of a Cauchy additive functional equation in random normed spaces using the fixed point method [30] (for some results on this topic, see Refs. [6, 8–10, 18–20, 28, 29, 31, 32, 34, 35, 44]).

In the literature, there are many characterizations of inner product spaces. In 1935, Jordan and von Neumann [16] showed that a normed space X is an inner product space if and only if the parallelogram identity (or the Jordan–von Neumann identity or Appolonius law),

$$\|x+y\|^2 + \|x-y\|^2 = 2\left(\|x\|^2 + \|y\|^2\right),$$

holds for all $x, y \in X$. This translates into a prominent functional equation known as the quadratic functional equation:

$$f(x+y) + f(x-y) = 2f(x) + 2f(y). \tag{1}$$

Let us mention that by a quadratic mapping, we mean each solution of (1). It is well known [1] that a mapping $f : X \to Y$ between two real vector spaces X and Y is quadratic if and only if there exists a unique symmetric biadditive mapping $B : X \times X \to Y$ such that $f(x) = B(x, x)$, for all $x \in X$. The biadditive function B is given by $B(x, y) = \frac{1}{4}(f(x+y) - f(x-y))$.

It was shown by Rassias [37] that a normed space $(X, \|\cdot\|)$ is an inner product space if and only if for any finite set of vectors $x_1, \ldots, x_n \in X$ and a fixed integer $n \geq 2$,

$$\sum_{i=1}^{n} \left\| x_i - \frac{1}{n} \sum_{j=1}^{n} x_j \right\|^2 = \sum_{i=1}^{n} \|x_i\|^2 - n \left\| \frac{1}{n} \sum_{i=1}^{n} x_i \right\|^2.$$

Employing the above identity, Najati and Rassias [33] obtained the functional equation

$$\sum_{i=1}^{n} f\left(x_i - \frac{1}{n} \sum_{j=1}^{n} x_j \right) = \sum_{i=1}^{n} f(x_i) - nf\left(\frac{1}{n} \sum_{i=1}^{n} x_i \right). \tag{2}$$

In Ref. [33, Theorem 2.3], it is shown that if a mapping $f : X \to Y$ between two real vector spaces X and Y satisfies functional equation (2), then the mapping f is the sum of an additive mapping and a quadratic mapping. The aim of this work is to extend the applications of the fixed point alternative method to provide stability for functional equation (2) related to inner product spaces in Šerstnev probabilistic normed spaces.

2. Triangular Norms and Probabilistic Normed Spaces

We start this section by presenting some notions and results about triangular norms.

Definition 1. A triangular norm (t-norm) T is a binary operation on the unit interval $T : [0, 1] \times [0, 1] \longrightarrow [0, 1]$ that is associative, commutative, increasing in each place, and which has 1 as identity, namely, for all x, y and z in $[0, 1]$, one has the following:

(T1) $T(T(x,y), z) = T(x, T(y, z))$;
(T2) $T(x, y) = T(y, x)$;
(T3) $T \mapsto T(t, x)$ and $t \mapsto T(x, t)$ are increasing;
(T4) $T(1, x) = T(x, 1) = x$.

Examples of *t*-norms are T_M, T_P, T_L, T_D, and T_Δ, defined as follows:

$$T_M(x, y) := \min\{x, y\};$$
$$T_P(x, y) := xy;$$
$$T_L(x, y) := \max\{x + y - 1, 0\};$$
$$T_D(x, y) := \begin{cases} \min\{x, y\}, & \text{if } \max\{x, y\} = 1, \\ 0, & \text{otherwise}; \end{cases}$$
$$T_\Delta(x, y) := \begin{cases} \frac{xy}{2}, & \text{if } \max\{x, y\} < 1, \\ xy, & \text{otherwise}. \end{cases}$$

If, for any two *t*-norms T_1 and T_2, the inequality $T_1(x, y) \leq T_2(x, y)$ holds for all $(x, y) \in [0, 1]^2$, then we say that T_1 is weaker than T_2, or, equivalently, T_2 is stronger than T_1.

For any *t*-norm T, we have $T_D \leq T \leq T_M$. Also, since $T_L < T_P$, we obtain $T_D < T_L < T_P < T_M$.

Proposition 1 ([12]). *A t-norm T is continuous if and only if it is continuous in its first component, i.e., for all $y \in [0, 1]$, if the one-place function*

$$T(\cdot, y) : [0, 1] \longrightarrow [0, 1], \quad x \mapsto T(x, y)$$

is continuous.

For example, the *t*-norms T_L and T_M are continuous, but the *t*-norm T_Δ is not continuous.

A *t*-norm T can be extended in a unique way to an *n*-array operation, taking, for any $(x_1, \ldots, x_n) \in [0, 1]^n$, the value $T(x_1, \ldots, x_n)$

defined by
$$T^0_{i=1} x_i = 1, \quad T^n_{i=1} x_i = T\left(T^{n-1}_{i=1} x_i, x_n\right) = T(x_1, \ldots, x_n).$$

The t-norm T can also be extended to a countable operation, taking, for any sequence $\{x_n\}$ in $[0,1]$, the value
$$T^\infty_{i=1} x_i = \lim_{n\to\infty} T^n_{i=1} x_i.$$

The limit on the right side of the last equality exists since the sequence $\{T^n_{i=1} x_i\}$ is nonincreasing and bounded from below.

It is known [13] that for $T \geq T_L$, the following implication holds:
$$\lim_{n\to\infty} T^\infty_{i=1} x_{n+i} = 1 \iff \sum_{n=1}^\infty (1 - x_n) < \infty.$$

Let Δ^+ denote the space of all distribution functions, that is, the space of all mappings $F : \mathbb{R} \cup \{-\infty, \infty\} \to [0,1]$ such that F is left-continuous, nondecreasing on $\mathbb{R}$, $F(0) = 0$ and $F(+\infty) = 1$. D^+ is a subset of Δ^+ consisting of all functions $F \in \Delta^+$ for which $l^- F(+\infty) = 1$, where $l^- f(x)$ denotes the left limit of the function f at the point x. The space Δ^+ is partially ordered by the usual pointwise ordering of functions, i.e., $F \leq G$ if and only if $F(t) \leq G(t)$, for all $t \in \mathbb{R}$. The maximal element for Δ^+ in this order is the distribution function ε_0 given by
$$\varepsilon_0(t) = \begin{cases} 0, & \text{if } t \leq 0, \\ 1, & \text{if } t > 0. \end{cases}$$

The function $F(t)$ defined by
$$F(t) = \begin{cases} 0, & \text{if } t \leq 0, \\ t, & \text{if } 0 < t < 1, \\ 1, & \text{if } t \geq 1 \end{cases}$$

is a distribution function. Since $\lim_{t\to\infty} F(t) = 1$, $F \in D^+$. Note that $F(t+s) \geq T_M(F(t), F(s))$, for all $t, s > 0$.

Definition 2 ([41]). A Šerstnev probabilistic normed space is a triple (X, Λ, T), where X is a vector space, Λ is a mapping from X into D^+ and T is a continuous t-norm, such that the following conditions hold:

(PN1) $\Lambda_x(t) = \varepsilon_0(t)$, for all $t > 0$, if and only if $x = 0$ (0 is the null vector in X, and Λ_x denotes the value of Λ at a point $x \in X$);
(PN2) $\Lambda_{\alpha x}(t) = \Lambda_x(\frac{t}{|\alpha|})$, for all $x \in X$, $\alpha \in \mathbb{R} \setminus \{0\}$ and $t \geq 0$;
(PN3) $\Lambda_{x+y}(t+s) \geq T(\Lambda_x(t), \Lambda_y(s))$, for all $x, y \in X$ and $t, s \geq 0$.

Every normed space $(X, \|\cdot\|)$ defines a random normed space (X, Λ, T_M), where $\Lambda_u(t) = \frac{t}{t+\|u\|}$, for all $t > 0$. This space is called an induced random normed space.

Definition 3. Let (X, Λ, T) be a Šerstnev probabilistic normed space.

(1) A sequence $\{x_n\}$ in X is said to be convergent to a point $x \in X$ if, for any $\epsilon > 0$ and $\lambda > 0$, there exists a positive integer N such that
$$\Lambda_{x_n - x}(\epsilon) > 1 - \lambda,$$
whenever $n \geq N$.
(2) A sequence $\{x_n\}$ in X is called a Cauchy sequence if, for any $\epsilon > 0$ and $\lambda > 0$, there exists a positive integer N such that
$$\Lambda_{x_n - x_m}(\epsilon) > 1 - \lambda,$$
whenever $n \geq m \geq N$.
(3) A Šerstnev probabilistic normed space (X, Λ, T) is said to be complete if and only if every Cauchy sequence in X is convergent to a point in X.

Theorem 1 ([38]). *If (X, Λ, T) is a Šerstnev probabilistic normed space and $\{x_n\}$ is a sequence such that $x_n \to x$, then $\lim_{n \to \infty} \Lambda_{x_n}(t) = \Lambda_x(t)$ almost everywhere.*

3. Generalized Metrics and Fixed Point

In Ref. [26], Luxemburg introduced the notion of a generalized metric space by allowing the value ∞ for distance mapping.

Definition 4. The pair (X, d) is called a generalized complete metric space if X is a nonempty set and $d : X^2 \longrightarrow [0, 1]$ satisfies the following conditions:

(a) $d(x, y) = 0$ if and only if $x = y$;
(b) $d(x, y) = d(y, x)$, for all $x, y \in X$;
(c) $d(x, z) \leq d(x, y) + d(y, z)$, for all $x, y, z \in X$;
(d) every d-Cauchy sequence in X is d-convergent to a point in X.

Definition 5. Let (X, d) be a generalized metric space. A mapping $J : X \to X$ satisfies a Lipschitz condition with Lipschitz constant $L \geq 0$ if

$$d(Jx, Jy) \leq L d(x, y),$$

for all $x, y \in X$. If the Lipschitz constant L is less than 1, then the mapping J is called a strictly contractive mapping.

The following theorem is proved in Ref. [17] (also see Ref. [11]).

Theorem 2 (Luxemburg–Jung Theorem). *Let (X, d) be a complete generalized metric space and $J : X \to X$ be a strictly contractive mapping with Lipschitz constant L such that $d(x_0, J(x_0)) < +\infty$, for some $x_0 \in X$. Then, J has a unique fixed point in the set $Y := \{y \in X : d(x_0, y) < \infty\}$, and the sequence $\{J^n(x)\}$ converges to the fixed point x^*, for all $x \in Y$. Moreover, $d(x_0, J(x_0)) \leq \delta$ implies $d(x^*, x_0) \leq \frac{\delta}{1-L}$.*

Let X be a linear space, (X, Λ, T_M) be a complete Šerstnev probabilistic normed space and G be a mapping from $X \times \mathbb{R}$ into $[0, 1]$ such that $G(x, \cdot) \in D^+$, for all $x \in X$. Consider the set

$$E := \{g : X \to Y : g(0) = 0\}$$

and the mapping d_G defined on $E \times E$ by

$$d_G(g, h) = \inf\{u \in \mathbb{R}^+ : \Lambda_{g(x)-h(x)}(ut) \geq G(x, t), \ \forall \ x \in X, \ t > 0\},$$

where, as usual, $\inf \emptyset = +\infty$.

Lemma 1 ([29, 30]). *d_G is a complete generalized metric on E.*

4. Stability of Functional Equation (2): An Even Case

In this section, we prove the probabilistic stability of functional equation (2) using the fixed point method for an even case.

Lemma 2 ([33, Lemma 2.2]). *Let V and W be real linear spaces. If an even mapping $f : V \to W$ satisfies (2), then the mapping f is quadratic, that is, f is a solution of $f(x + y) + f(x - y) = 2f(x) + 2f(y)$, for all $x, y \in V$.*

Hereinafter, let X be a linear space and (Y, Λ, T_M) be a complete Šerstnev probabilistic normed space. For convenience, we use the following abbreviation for a given function $f : X \to Y$:

$$\Delta f(x_1, \ldots, x_n) = \sum_{i=1}^{n} f\left(x_i - \frac{1}{n}\sum_{j=1}^{n} x_j\right) - \sum_{i=1}^{n} f(x_i) + nf\left(\frac{1}{n}\sum_{i=1}^{n} x_i\right),$$

for all $x_1, \ldots, x_n \in X$, where $n \geq 2$ is a fixed integer.

Theorem 3. *Let $\Phi : \underbrace{X \times X \times \cdots \times X}_{n\text{-terms}} \to D^+$ be a function ($\Phi(x_1, \ldots, x_n)$ is denoted by $\Phi_{x_1, \ldots, x_n}$) such that, for some $0 < \alpha < 4$,*

$$\Phi_{2x_1, \ldots, 2x_n}(\alpha t) \geq \Phi_{x_1, \ldots, x_n}(t), \tag{3}$$

for all $x_1, \ldots, x_n \in X$ and $t > 0$. Suppose that an even function $f : X \to Y$ with $f(0) = 0$ satisfies the inequality

$$\Lambda_{\Delta f(x_1, \ldots, x_n)}(t) \geq \Phi_{x_1, \ldots, x_n}(t), \tag{4}$$

for all $x_1, \ldots, x_n \in X$ and $t > 0$. Then, there exists a unique quadratic function $Q : X \to Y$ such that

$$\Lambda_{f(x)-Q(x)}(t) \geq \Psi_x^e\left((4-\alpha)t\right), \tag{5}$$

for all $x \in X$ and $t > 0$, where

$$\Psi_x^e(t) := T\bigg(\Phi_{nx,nx,\underbrace{0,\ldots,0}_{n-2}}((n-1)t), T\bigg(T\bigg(\Phi_{nx,\underbrace{0,\ldots,0}_{n-1}}\Big(\frac{(n-1)}{8}t\Big),$$

$$T\bigg(\Phi_{0,\underbrace{nx,\ldots,nx}_{n-1}}\Big(\frac{(n-1)}{16}t\Big), T\bigg(\Phi_{nx,\underbrace{0,\ldots,0}_{n-1}}\Big(\frac{(n-1)}{16}t\Big),$$

$$\Phi_{nx,\underbrace{0,\ldots,0}_{n-1}}\Big(\frac{(n-1)}{8n}t\Big)\bigg)\bigg)\bigg), T\bigg(\Phi_{x,(n-1)x,\underbrace{0,\ldots,0}_{n-2}}\Big(\frac{(n-1)}{8}t\Big),$$

$$T\bigg(\Phi_{0,\underbrace{nx,\ldots,nx}_{n-1}}\Big(\frac{(n-1)}{16}t\Big), T\bigg(\Phi_{nx,\underbrace{0,\ldots,0}_{n-1}}\Big(\frac{(n-1)}{16}t\Big),$$

$$\Phi_{nx,\underbrace{0,\ldots,0}_{n-1}}\Big(\frac{(n-1)}{8n}t\Big)\bigg)\bigg)\bigg)\bigg)\bigg).$$

Proof. Letting $x_1 = nx_1$ and $x_i = nx_2$ ($i = 2, \ldots, n$) in (4) and using the evenness of f, we get

$$\Lambda_{nf(x_1+(n-1)x_2)+f((n-1)(x_1-x_2))+(n-1)f(x_1-x_2)-f(nx_1)-(n-1)f(nx_2)}(t)$$
$$\geq \Phi_{nx_1,nx_2,\ldots,nx_2}(t), \qquad (6)$$

for all $x_1, x_2 \in X$ and $t > 0$. Interchanging x_1 and x_2 in (6) and using the evenness of f, we get

$$\Lambda_{nf((n-1)x_1+x_2)+f((n-1)(x_1-x_2))+(n-1)f(x_1-x_2)-(n-1)f(nx_1)-f(nx_2)}(t)$$
$$\geq \Phi_{nx_2,nx_1,\ldots,nx_1}(t), \qquad (7)$$

for all $x_1, x_2 \in X$ and $t > 0$. It follows from (6) and (7) that

$$\Lambda_{nf((n-1)x_1+x_2)+nf(x_1+(n-1)x_2)+2f((n-1)(x_1-x_2))+2(n-1)f(x_1-x_2)-nf(nx_1)-nf(nx_2)}(t)$$
$$\geq T\bigg(\Phi_{nx_1,nx_2,\ldots,nx_2}\Big(\frac{t}{2}\Big), \Phi_{nx_2,nx_1,\ldots,nx_1}\Big(\frac{t}{2}\Big)\bigg), \qquad (8)$$

for all $x_1, x_2 \in X$ and $t > 0$. Setting $x_1 = nx_1$, $x_2 = -nx_2$ and $x_i = 0$ ($i = 3, \ldots, n$) in (4) and using the evenness of f, we get

$$\Lambda_{f((n-1)x_1+x_2)+f(x_1+(n-1)x_2)+2(n-1)f(x_1-x_2)-f(nx_1)-f(nx_2)}(t) \geq \Phi_{nx_1,-nx_2,0,\ldots,0}(t), \tag{9}$$

for all $x_1, x_2 \in X$ and $t > 0$. We obtain from (8) and (9) that

$$\Lambda_{f((n-1)(x_1-x_2))-(n-1)^2 f(x_1-x_2)}(t)$$
$$\geq T\left(\Phi_{nx_1,-nx_2,0,\ldots,0}\left(\frac{t}{n}\right), T\left(\Phi_{nx_1,nx_2,\ldots,nx_2}\left(\frac{t}{2}\right),\right.\right.$$
$$\left.\left.\Phi_{nx_2,nx_1,\ldots,nx_1}\left(\frac{t}{2}\right)\right)\right), \tag{10}$$

for all $x_1, x_2 \in X$ and $t > 0$. So,

$$\Lambda_{f((n-1)x)-(n-1)^2 f(x)}(t) \geq T\left(\Phi_{0,nx,\ldots,nx}\left(\frac{t}{2}\right), T\left(\Phi_{nx,0,\ldots,0}\left(\frac{t}{2}\right),\right.\right.$$
$$\left.\left.\Phi_{nx,0,\ldots,0}\left(\frac{t}{n}\right)\right)\right), \tag{11}$$

for all $x \in X$ and $t > 0$. Putting $x_1 = nx$ and $x_i = 0$ ($i = 2, \ldots, n$) in (4), we obtain

$$\Lambda_{f(nx)-f((n-1)x)-(2n-1)f(x)}(t) \geq \Phi_{nx,0,\ldots,0}(t), \tag{12}$$

for all $x \in X$ and $t > 0$. It follows from (11) and (12) that

$$\Lambda_{f(nx)-n^2 f(x)}(t)$$
$$\geq T\left(\Phi_{nx,0,\ldots,0}\left(\frac{t}{2}\right), T\left(\Phi_{0,nx,\ldots,nx}\left(\frac{t}{4}\right),\right.\right.$$
$$\left.\left.T\left(\Phi_{nx,0,\ldots,0}\left(\frac{t}{4}\right), \Phi_{nx,0,\ldots,0}\left(\frac{t}{2n}\right)\right)\right)\right), \tag{13}$$

for all $x \in X$ and $t > 0$. Letting $x_2 = -(n-1)x_1$ in (9) and replacing x_1 by $\frac{x}{n}$ in the obtained inequality, we get

$$\Lambda_{f((n-1)x)-f((n-2)x)-(2n-3)f(x)}(t) \geq \Phi_{x,(n-1)x,0,\ldots,0}(t), \tag{14}$$

for all $x \in X$ and $t > 0$. It follows from (11) and (14) that

$$\Lambda_{f((n-2)x)-(n-2)^2 f(x)}(t)$$
$$\geq T\left(\Phi_{x,(n-1)x,0,\ldots,0}\left(\frac{t}{2}\right), T\left(\Phi_{0,nx,\ldots,nx}\left(\frac{t}{4}\right),\right.\right.$$
$$\left.\left.T\left(\Phi_{nx,0,\ldots,0}\left(\frac{t}{4}\right), \Phi_{nx,0,\ldots,0}\left(\frac{t}{2n}\right)\right)\right)\right), \quad (15)$$

for all $x \in X$ and $t > 0$. Applying (13) and (15), we get

$$\Lambda_{f(nx)-f((n-2)x)-4(n-1)f(x)}(t)$$
$$\geq T\left(T\left(\Phi_{nx,0,\ldots,0}\left(\frac{t}{4}\right), T\left(\Phi_{0,nx,\ldots,nx}\left(\frac{t}{8}\right),\right.\right.\right.$$
$$\left.\left.T\left(\Phi_{nx,0,\ldots,0}\left(\frac{t}{8}\right), \Phi_{nx,0,\ldots,0}\left(\frac{t}{4n}\right)\right)\right)\right),$$
$$T\left(\Phi_{x,(n-1)x,0,\ldots,0}\left(\frac{t}{4}\right), T\left(\Phi_{0,nx,\ldots,nx}\left(\frac{t}{8}\right),\right.\right.$$
$$\left.\left.\left.T\left(\Phi_{nx,0,\ldots,0}\left(\frac{t}{8}\right), \Phi_{nx,0,\ldots,0}\left(\frac{t}{4n}\right)\right)\right)\right)\right), \quad (16)$$

for all $x \in X$ and $t > 0$. Setting $x_1 = x_2 = nx$ and $x_i = 0$ ($i = 3, \ldots, n$) in (4), we obtain

$$\Lambda_{f((n-2)x)+(n-1)f(2x)-f(nx)}(t) \geq \Phi_{nx,nx,0,\ldots,0}(2t), \quad (17)$$

for all $x \in X$ and $t > 0$. It follows from (16) and (17) that

$$\Lambda_{f(2x)-4f(x)}(t)$$
$$\geq T\left(\Phi_{nx,nx,0,\ldots,0}((n-1)t), T\left(T\left(\Phi_{nx,0,\ldots,0}\left(\frac{(n-1)}{8}t\right),\right.\right.\right.$$
$$\left.\left.\left.T\left(\Phi_{0,nx,\ldots,nx}\left(\frac{(n-1)}{16}t\right), T\left(\Phi_{nx,0,\ldots,0}\left(\frac{(n-1)}{16}t\right),\right.\right.\right.\right.\right.$$

$$\Phi_{nx,0,\ldots,0}\left(\frac{(n-1)}{8n}t\right)\bigg)\bigg)\bigg)\bigg), T\left(\Phi_{x,(n-1)x,0,\ldots,0}\left(\frac{(n-1)}{8}t\right),\right.$$

$$T\left(\Phi_{0,nx,\ldots,nx}\left(\frac{(n-1)}{16}t\right), T\left(\Phi_{nx,0,\ldots,0}\left(\frac{(n-1)}{16}t\right),\right.\right.$$

$$\Phi_{nx,0,\ldots,0}\left(\frac{(n-1)}{8n}t\right)\bigg)\bigg)\bigg)\bigg)\bigg)\bigg),$$

for all $x \in X$ and $t > 0$. Therefore,

$$\Lambda_{\frac{f(2x)}{4}-f(x)}(t) \geq \Psi_x^e(4t), \tag{18}$$

for all $x \in X$ and all $t > 0$. Let S be the set of all even functions $h : X \to Y$ with $h(0) = 0$ and the mapping d be defined on $S \times S$ by

$$d(h,k) = \inf\left\{u \in \mathbb{R}^+ : \Lambda_{h(x)-k(x)}(ut) \geq \Psi_x^e(t),\ \forall x \in X,\ t > 0\right\}.$$

By Lemma 1, (S, d) is a complete generalized metric space.

Now, we consider the function $J : S \to S$ defined by

$$Jh(x) := \frac{h(2x)}{4},$$

for all $h \in S$ and $x \in X$. Let $f, g \in S$ such that $d(f, g) < \varepsilon$. Then,

$$\Lambda_{Jg(x)-Jf(x)}\left(\frac{\alpha u}{4}t\right) = \Lambda_{g(2x)-f(2x)}(\alpha u t) \geq \Psi_{2x}^e(\alpha t) \geq \Psi_x^e(t);$$

that is, if $d(f,g) < \varepsilon$, we have $d(Jf, Jg) < \frac{\alpha}{4}\varepsilon$. This means that

$$d(Jf, Jg) \leq \frac{\alpha}{4} d(f, g),$$

for all $f, g \in S$, that is, J is a strictly contractive self-function on S with the Lipschitz constant $\frac{\alpha}{4}$.

It follows from (18) that

$$\Lambda_{Jf(x)-f(x)}\left(\frac{t}{4}\right) \geq \Psi_x^e(t),$$

for all $x \in X$ and all $t > 0$, which implies that $d(Jf, f) \leq \frac{1}{4}$.

Due to Theorem 2, there exists a function $Q : X \to Y$ such that Q is a fixed point of J, i.e., $Q(2x) = 4Q(x)$, for all $x \in X$. Also, $d(J^m g, Q) \to 0$ as $m \to \infty$, which implies the equality

$$\lim_{m \to \infty} \frac{f(2^m x)}{4^m} = Q(x),$$

for all $x \in X$. If we replace $x_1, \ldots, x_n$ with $2^m x_1, \ldots, 2^m x_n$ in (4), respectively, and divide by 4^m, then it follows from (3) that

$$\Lambda_{\frac{\Delta f(2^m x_1, \ldots, 2^m x_n)}{4^m}}(t) \geq \Phi_{2^m x_1, \ldots, 2^m x_n}(4^m t)$$

$$= \Phi_{2^m x_1, \ldots, 2^m x_n}\left(\alpha^m \left(\frac{4}{\alpha}\right)^m t\right)$$

$$\geq \Phi_{x_1, \ldots, x_n}\left(\left(\frac{4}{\alpha}\right)^m t\right), \qquad (19)$$

for all $x_1, \ldots, x_n \in X$ and all $t > 0$. By letting $m \to \infty$ in (19), we find that

$$\Lambda_{\Delta Q(x_1, \ldots, x_n)}(t) = 1,$$

for all $x_1, \ldots, x_n \in X$ and all $t > 0$, which implies $\Delta Q(x_1, \ldots, x_n) = 0$. Thus, Q satisfies (2). By Lemma 2, $Q : X \to Y$ is quadratic.

According to the fixed point alternative, since Q is the unique fixed point of J in the set $\Omega = \{g \in S : d(f, g) < \infty\}$, Q is a unique function such that

$$\Lambda_{f(x) - Q(x)}(ut) \geq \Psi_x^e(t),$$

for all $x_1, \ldots, x_n \in X$ and all $t > 0$. Again, using the fixed point alternative gives

$$d(f, Q) \leq \frac{1}{1 - L} d(f, Jf) \leq \frac{1}{4(1 - L)} = \frac{1}{4(1 - \frac{\alpha}{4})}.$$

Thus,

$$\Lambda_{f(x) - Q(x)}\left(\frac{t}{4 - \alpha}\right) \geq \Psi_x^e(t),$$

for all $x \in X$ and $t > 0$. So,

$$\Lambda_{f(x) - Q(x)}(t) \geq \Psi_x^e((4 - \alpha)t),$$

for all $x \in X$ and $t > 0$. This completes the proof. □

5. Stability of Functional Equation (2): An Odd Case

In this section, we prove the probabilistic stability of functional equation (2) using the fixed point method for an odd case.

Lemma 3 ([33, Lemma 2.1]). *Let V and W be real linear spaces. If an odd mapping $f : V \to W$ satisfies (2), then the mapping f is additive, that is, f is a solution of $f(x+y) = f(x) + f(y)$, for all $x, y \in V$.*

Theorem 4. *Let $\Phi : \underbrace{X \times X \times \cdots \times X}_{n\text{-terms}} \to D^+$ be a function such that, for some $0 < \alpha < 2$, it satisfies (3), for all $x_1, \ldots, x_n \in X$ and $t > 0$. Suppose that an odd function $f : X \to Y$ satisfies (4), for all $x_1, \ldots, x_n \in X$ and $t > 0$. Then, there exists a unique additive function $A : X \to Y$ such that*

$$\Lambda_{f(x)-A(x)}(t) \geq \Psi_x^o((2-\alpha)t), \tag{20}$$

for all $x \in X$ and $t > 0$, where

$$\Psi_x^o(t) := T\left(\Phi_{2x,\underbrace{0,\ldots,0}_{n-1}}\left(\frac{t}{2}\right), T\left(\Phi_{x,x,\underbrace{0,\ldots,0}_{n-2}}\left(\frac{t}{2n}\right),\right.\right.$$

$$\left.\left. T\left(\Phi_{x,\underbrace{-x,\ldots,-x}_{n-1}}\left(\frac{t}{4}\right), \Phi_{-x,\underbrace{x,\ldots,x}_{n-1}}\left(\frac{t}{4}\right)\right)\right)\right).$$

Proof. Letting $x_1 = nx_1$ and $x_i = nx_1'$ ($i = 2, \ldots, n$) in (4) and using the oddness of f, we get

$$\Lambda_{nf(x_1+(n-1)x_1')+f((n-1)(x_1-x_1'))-(n-1)f(x_1-x_1')-f(nx_1)-(n-1)f(nx_1')}(t)$$
$$\geq \Phi_{nx_1,nx_1',\ldots,nx_1'}(t), \tag{21}$$

for all $x_1, x_1' \in X$ and $t > 0$. Interchanging x_1 and x_1' in (21) and using the oddness of f, we get

$$\Lambda_{nf((n-1)x_1+x_1')-f((n-1)(x_1-x_1'))+(n-1)f(x_1-x_1')-f(nx_1)-f(nx_1')}(t)$$
$$\geq \Phi_{nx_1',nx_1,\ldots,nx_1}(t), \tag{22}$$

for all $x_1, x_1' \in X$ and $t > 0$. Setting $x_1 = nx_1$, $x_2 = -nx_1'$ and $x_i = 0$ ($i = 3, \ldots, n$) in (4) and using the oddness of f, we get

$$\Lambda_{f((n-1)x_1+x_1')-f(x_1+(n-1)x_1')+2f(x_1-x_1')-f(nx_1)+f(nx_1')}(t)$$
$$\geq \Phi_{nx_1,-nx_1',0,\ldots,0}(t), \tag{23}$$

for all $x_1, x_1' \in X$ and $t > 0$. We obtain from (21)–(23) that

$$\Lambda_{f((n-1)(x_1-x_1'))+f(x_1-x_1')-f(nx_1)+f(nx_1')}(t)$$
$$\geq T\left(\Phi_{nx_1,-nx_1',0,\ldots,0}\left(\frac{t}{n}\right), T\left(\Phi_{nx_1,nx_1',\ldots,nx_1'}\left(\frac{t}{2}\right),\right.\right.$$
$$\left.\left.\Phi_{nx_1',nx_1,\ldots,nx_1}\left(\frac{t}{2}\right)\right)\right), \tag{24}$$

for all $x_1, x_1' \in X$ and $t > 0$. Putting $x_1 = n(x_1 - x_1')$ and $x_i = 0$ ($i = 2, \ldots, n$) in (4), we obtain

$$\Lambda_{f(n(x_1-x_1'))-f((n-1)(x_1-x_1'))-f((x_1-x_1'))}(t) \geq \Phi_{nx_1-nx_1',0,\ldots,0}(t), \tag{25}$$

for all $x_1, x_1' \in X$ and $t > 0$. It follows from (24) and (25) that

$$\Lambda_{f(n(x_1-x_1'))-f(nx_1)+f(nx_1')}(t)$$
$$\geq T\left(\Phi_{nx_1-nx_1',0,\ldots,0}\left(\frac{t}{2}\right), T\left(\Phi_{nx_1,-nx_1',0,\ldots,0}\left(\frac{t}{2n}\right),\right.\right.$$
$$\left.\left.T\left(\Phi_{nx_1,nx_1',\ldots,nx_1'}\left(\frac{t}{4}\right), \Phi_{nx_1',nx_1,\ldots,nx_1}\left(\frac{t}{4}\right)\right)\right)\right), \tag{26}$$

for all $x_1, x_1' \in X$ and $t > 0$. Replacing x_1 and x_1' by $\frac{x}{n}$ and $\frac{-x}{n}$ in (26), respectively, we obtain

$$\Lambda_{f(2x)-2f(x)}(t)T\left(\Phi_{2x,0,\ldots,0}\left(\frac{t}{2}\right), T\left(\Phi_{x,x,0,\ldots,0}\left(\frac{t}{2n}\right),\right.\right.$$
$$\left.\left.T\left(\Phi_{x,-x,\ldots,-x}\left(\frac{t}{4}\right), \Phi_{-x,x,\ldots,x}\left(\frac{t}{4}\right)\right)\right)\right),$$

for all $x \in X$ and $t > 0$. Therefore,

$$\Lambda_{\frac{f(2x)}{2}-f(x)}(t) \geq \Psi_x^o(2t), \tag{27}$$

for all $x \in X$ and all $t > 0$. Let S be the set of all odd functions $h : X \to Y$ and the mapping d be defined on $S \times S$ by

$$d(h, k) = \inf \left\{ u \in \mathbb{R}^+ : \Lambda_{h(x)-k(x)}(ut) \geq \Psi^o_x(t), \ \forall x \in X, \forall t > 0 \right\}.$$

By Lemma 1, (S, d) is a complete generalized metric space.

We consider the function $J : S \to S$ defined by $Jh(x) := \frac{h(2x)}{2}$, for all $h \in S$ and $x \in X$. Let $f, g \in S$ such that $d(f, g) < \varepsilon$. Then,

$$\Lambda_{Jg(x)-Jf(x)}\left(\frac{\alpha u}{2}t\right) = \Lambda_{g(2x)-f(2x)}(\alpha u t) \geq \Psi^o_{2x}(\alpha t) \geq \Psi^o_x(t);$$

that is, if $d(f, g) < \varepsilon$, we have $d(Jf, Jg) < \frac{\alpha}{2}\varepsilon$. This means that $d(Jf, Jg) \leq \frac{\alpha}{2} d(f, g)$, for all $f, g \in S$, that is, J is a strictly contractive self-function on S with the Lipschitz constant $\frac{\alpha}{2}$. It follows from (27) that

$$\Lambda_{Jf(x)-f(x)}\left(\frac{t}{2}\right) \geq \Psi^o_x(t),$$

for all $x \in X$ and all $t > 0$, which implies that $d(Jf, f) \leq \frac{1}{2}$.

Due to Theorem 2, there exists a function $A : X \to Y$ such that A is a fixed point of J, i.e., $A(2x) = 2A(x)$, for all $x \in X$.

Also, $d(J^m g, A) \to 0$ as $m \to \infty$, which implies $\lim_{m \to \infty} \frac{f(2^m x)}{2^m} = A(x)$, for all $x \in X$.

If we replace $x_1, \ldots, x_n$ with $2^m x_1, \ldots, 2^m x_n$ in (4), respectively, and divide by 2^m, then it follows from (3) that

$$\Lambda_{\frac{\Delta f(2^m x_1, \ldots, 2^m x_n)}{2^m}}(t) \geq \Phi_{2^m x_1, \ldots, 2^m x_n}(2^m t)$$

$$= \Phi_{2^m x_1, \ldots, 2^m x_n}\left(\alpha^m \left(\frac{2}{\alpha}\right)^m t\right)$$

$$\geq \Phi_{x_1, \ldots, x_n}\left(\left(\frac{2}{\alpha}\right)^m t\right), \tag{28}$$

for all $x_1, \ldots, x_n \in X$ and all $t > 0$. By letting $m \to \infty$ in (28), we find that

$$\Lambda_{\Delta A(x_1, \ldots, x_n)}(t) = 1,$$

for all $x_1, \ldots, x_n \in X$ and all $t > 0$, which implies $\Delta A(x_1, \ldots, x_n) = 0$. Thus, A satisfies (2). By Lemma 3, $A : X \to Y$ is additive. The rest of the proof is similar to the proof of Theorem 3. □

6. Stability of Functional Equation (2): A Mixed Case

Now, we prove the probabilistic stability of functional equation (2) using the fixed point method.

Lemma 4 ([33, Theorem 2.3]). *Let V and W be real linear spaces. A mapping $f : V \to W$ satisfies (2) if and only if f is the sum of an additive mapping and a quadratic mapping.*

Theorem 5. *Let $\Phi : \underbrace{X \times X \times \cdots \times X}_{n\text{-terms}} \to D^+$ be a function such that, for some $0 < \alpha < 2$, it satisfies (3) for all $x_1, \ldots, x_n \in X$ and $t > 0$. Suppose that a function $f : X \to Y$ with $f(0) = 0$ satisfies (4), for all $x_1, \ldots, x_n \in X$ and $t > 0$. Then, there exists a unique quadratic function $Q : X \to Y$ and a unique additive function $A : X \to Y$ such that*

$$\Lambda_{f(x)-Q(x)-A(x)}(t) \geq T\left(T\left(\Psi_x^e\left(\frac{(4-\alpha)}{2}t\right), \Psi_x^e\left(\frac{(4-\alpha)}{2}t\right)\right),\right.$$

$$\left. T\left(\Psi_x^o\left(\frac{(2-\alpha)}{2}t\right), \Psi_x^o\left(\frac{(2-\alpha)}{2}t\right)\right)\right), \quad (29)$$

for all $x \in X$ and $t > 0$, where $\Psi_x^e(t)$ and $\Psi_x^o(t)$ are defined as in Theorems 3 and 4.

Proof. Let $f_e(x) = \frac{1}{2}(f(x) + f(-x))$ for all $x \in X$. Then,

$$\Lambda_{\Delta f_e(x_1,\ldots,x_n)}(t) = \Lambda_{\frac{\Delta f(x_1,\ldots,x_n) + \Delta f(-x_1,\ldots,-x_n)}{2}}(t)$$

$$\geq T(\Lambda_{\Delta f(x_1,\ldots,x_n)}(t), \Lambda_{\Delta f(-x_1,\ldots,-x_n)}(t))$$

$$\geq T(\Phi_{x_1,\ldots,x_n}(t), \Phi_{-x_1,\ldots,-x_n}(t))$$

$$= T(\Phi_{x_1,\ldots,x_n}(t), \Phi_{x_1,\ldots,x_n}(t)),$$

for all $x_1, \ldots, x_n \in X$ and $t > 0$. Hence, in view of Theorem 3, there exists a unique quadratic function $Q : X \to Y$ such that

$$\Lambda_{f_e(x)-Q(x)}(t) \geq T(\Psi_x^e((4-\alpha)t), \Psi_x^e((4-\alpha)t)), \quad (30)$$

for all $x \in X$ and $t > 0$. On the other hand, let $f_o(x) = \frac{1}{2}(f(x) - f(-x))$, for all $x \in X$. Then, by using the above method from

Theorem 4, there exists a unique additive function $A: X \to Y$ such that

$$\Lambda_{f_o(x)-A(x)}(t) \geq T(\Psi_x^o((2-\alpha)t), \Psi_x^o((2-\alpha)t)), \tag{31}$$

for all $x \in X$ and $t > 0$. Hence, (29) follows from (30) and (31). □

References

[1] J. Aczél and J. Dhombres, *Functional Equations in Several Variables* (Cambridge University Press, 1989).

[2] C. Alsina, On the stability of a functional equation arising in probabilistic normed spaces. In: Walter, W. (eds.) *General Inequalities* 5. International Series of Numerical Mathematics/Internationale Schriftenreihe zur Numerischen Mathematik Série internationale d'Analyse numérique, Vol. 80. Birkhäuser Basel.

[3] C. Alsina, B. Schweizer and A. Sklar, On the definition of a probabilistic normed space, *Aequ. Math.* **46**, 91–98 (1993).

[4] C. Alsina, B. Schweizer and A. Sklar, Continuity properties of probabilistic norms, *J. Math. Anal. Appl.* **208**, 446–452 (1997).

[5] J. A. Baker, The stability of certain functional equations, *Proc. Am. Math. Soc.* **112**, 729–732 (1991).

[6] A. Bodaghi, T. M. Rassias and A. Zivari-Kazempour, A fixed point approach to the stability of additive-quadratic-quartic functional equations, *Int. J. Nonlinear Anal. Appl.* **11**, 17–28 (2020).

[7] J. Brzdęk, D. Popa, I. Raşa and B. Xu, *Ulam Stability of Operators* (Academic Press, Elsevier, Oxford, 2018).

[8] J. Brzdęk and M. Piszczek, Fixed points of some nonlinear operators in spaces of multifunctions and the Ulam stability, *J. Fixed Point Theory Appl.* **19**, 2441–2448 (2017).

[9] L. Cădariu and V. Radu, Remarks on the stability of monomial functional equations, *Fixed Point Theory.* **8**, 201–218 (2007).

[10] Y. J. Cho, T. M. Rassias and R. Saadati, *Stability of Functional Equations in Random Normed Spaces*. Springer Optimization and Its Applications, Vol. 86 (Springer, New York, 2013).

[11] J. B. Diaz and B. Margolis, A fixed point theorem of the alternative for contractions on the generalized complete metric space, *Bull. Am. Math. Soc.* **74**, 305–309 (1968).

[12] O. Hadžić and E. Pap, *Fixed Point Theory in PM Spaces.* (Kluwer Academic Publishers, Dordrecht, 2001).

[13] O. Hadžić, E. Pap and M. Budincević, Countable extension of triangular norms and their applications to the fixed point theory in probabilistic metric spaces, *Kybernetika.* **38**, 363–381 (2002).
[14] P. Hajek, *Metamathematics of Fuzzy Logic* (Kluwer Academic, Dordrecht, 1998).
[15] D. H. Hyers, G. Isac and T. M. Rassias, *Stability of Functional Equations in Several Variables* (Birkhäuser, Basel, 1998).
[16] P. Jordan and J. von Neumann, On inner products in linear, metric spaces, *Ann. Math.* **36**, 719–723 (1935).
[17] C. F. K. Jung, On generalized complete metric spaces, *Bull. Am. Math. Soc.* **75**, 113–116 (1969).
[18] H. Khodaei, Selections of generalized convex set-valued functions satisfying some inclusions, *J. Math. Anal. Appl.* **474**, 1104–1115 (2019).
[19] H. Khodaei, I. El-Fassi and B. Hayati, On selections of set-valued Euler-Lagrange inclusions with applications, *Acta Math. Sci.* **40**, 1–11 (2020).
[20] H. Khodaei and T. M. Rassias, Approximately generalized additive functions in several variables, *Int. J. Nonlinear Anal. Appl.* **1**, 22–41 (2010).
[21] E. P. Klement, R. Mesiar and E. Pap, *Triangular Norms* (Kluwer Academic, Dordrecht, 2000).
[22] E. P. Klement, R. Mesiar and E. Pap, Triangular norms, position paper I: Basic analytical and algebraic properties, *Fuzzy Sets Syst.* **143**, 5–26 (2004).
[23] E. P. Klement, R. Mesiar and E. Pap, Triangular norms, position paper II: General constructions and parameterized families, *Fuzzy Sets Syst.* **145**, 411–438 (2004).
[24] E. P. Klement, R. Mesiar and E. Pap, Triangular norms, position paper III: Continuous t-norms, *Fuzzy Sets Syst.* **145**, 439–454 (2004).
[25] B. Lafuerza Guillén and P. Harikrishnan, *Probabilistic Normed Spaces* (Imperial College Press, World Scientific, UK, London, 2014).
[26] W. A. J. Luxemburg, On the convergence of successive approximations in the theory of ordinary differential equations II, *Nederland Akad. Wetensch. Proc. Ser. A.* **61**; *Indag. Math.* **20**, 540–546 (1958).
[27] K. Menger, Statistical metrics, *Proc. Natl. Acad. Sci. USA.* **28**, 535–537 (1942).
[28] M. Miahi, F. Mirzaee and H. Khodaei, On convex-valued G-m-monomials with applications in stability theory, *RACSAM.* **115**, (2021). Article number 76.
[29] D. Miheţ, The fixed point method for fuzzy stability of the Jensen functional equation, *Fuzzy Sets Syst.* **160**, 1663–1667 (2009).

[30] D. Miheţ and V. Radu, On the stability of the additive Cauchy functional equation in random normed spaces, *J. Math. Anal. Appl.* **343**, 567–572 (2008).

[31] D. Miheţ, R. Saadati and S. M. Vaezpour, The stability of the quartic functional equation in random normed spaces, *Acta Appl. Math.* **110**, 797–803 (2010).

[32] D. Miheţ, R. Saadati and S. M. Vaezpour, The stability of an additive functional equation in Menger probabilistic φ-normed spaces, *Math. Slovaca.* **61**, 817–826 (2011).

[33] A. Najati and T. M. Rassias, Stability of a mixed functional equation in several variables on Banach modules, *Nonlinear Anal.* **72**, 1755–1767 (2010).

[34] C. Park, Fixed point method for set-valued functional equation, *J. Fixed Point Theory Appl.* **19**, 2297–2308 (2017).

[35] C. Park, V. Arasu and M. Angayarkanni, Stability of functional equations in Šerstnev probabilistic normed spaces, *J. Comput. Anal. Appl.* **26**, 42–49 (2019).

[36] V. Radu, The fixed point alternative and the stability of functional equations, *Fixed Point Theory.* **4**, 91–96 (2003).

[37] T. M. Rassias, New characterization of inner product spaces, *Bull. Sci. Math.* **108**, 95–99 (1984).

[38] B. Schweizer and A. Sklar, *Probabilistic Metric Spaces* (Elsevier, North Holand, New York, 1983).

[39] C. Sempi, A short and partial history of probabilistic normed spaces, *Mediterr. J. Math.* **3**, 283–300 (2006).

[40] A. N. Šerstnev, Random normed spaces: Problems of completeness, *Kazan Gos. Univ. Učen. Zap.* **122**, 3–20 (1962).

[41] A. N. Šerstnev, On the notion of a random normed space, *Dokl. Akad. Nauk SSSR.* **149**, 280–283 (1963).

[42] A. N. Šerstnev, Best approximation problem in random normed spaces, *Dokl. Akad. Nauk SSSR.* **149**, 539–542 (1963).

[43] A. N. Šerstnev, On a probabilistic generalization of a metric spaces, *Kazan Gos. Univ. Učen. Zap.* **124**, 3–11 (1964).

[44] Z. Wang, T. M. Rassias and M. Eshaghi Gordji, Stability of quadratic functional equations in Šerstnev probabilistic normed spaces, *UPB Sci. Bull. Ser. A.* **77**, 79–92 (2015).

[45] L. A. Zadeh, Fuzzy sets, *Inf. Control.* **8**, 338–353 (1965).

© 2024 World Scientific Publishing Company
https://doi.org/10.1142/9789811267048_0003

Chapter 3

Hyers–Ulam–Rassias Stability of Set-Valued Functional Equations: A Fixed Point Approach

H. Azadi Kenary

*Department of Mathematics, College of Science,
Yasouj University, Yasouj, Iran*
h.azadikenary@gmail.com

In this chapter, we prove the generalized Hyers–Ulam (or Hyers–Ulam–Rassias) stability of the following

$$f\left(\frac{\sum_{i=1}^{m-1} x_i}{m}\right) \oplus f\left(\frac{(m-1)x_1 - \sum_{j=2}^{m-2} x_j - mx_{m-1}}{m}\right)$$
$$\oplus f\left(\frac{mx_1 + (m-1)x_{m-1}}{m}\right) = 2f(x_1) \qquad (1)$$

set-valued functional equations by using the fixed point method.

The concept of Ulam–Hyers–Rassias stability originated from Rassias' stability theorem that appeared in his paper (Rassias in *Proc. Am. Math. Soc.* 72:297–300, 1978).

1. Introduction and Preliminaries

Set-valued functions in Banach spaces have been developed in the last decades. The pioneering works by Aumann [5] and Debreu [15] were inspired by problems arising in Control Theory and Mathematical Economics. We can refer to the works by Arrow and Debreu [3], McKenzie [28], the monographs by Hindenbrand [21], Aubin and Frankowska [4], Castaing and Valadier [8], Klein and Thompson [27] and the survey by Hess [20].

The stability problem of functional equations was originated from a question of Ulam [45] concerning the stability of group homomorphisms. Hyers [22] gave a first affirmative partial answer to the question of Ulam for Banach spaces. Hyers' theorem was generalized by Aoki [2] for additive mappings and by Rassias [40] for linear mappings by considering an unbounded Cauchy difference. The work of Rassias [40] has provided a lot of influence in the development of what we call *Hyers–Ulam stability* or as *Hyers–Ulam–Rassias stability* of functional equations. A generalization of the Rassias theorem was obtained by Găvruta [17] by replacing the unbounded Cauchy difference by a general control function in the spirit of Rassias' approach.

The functional equation

$$f(x+y) + f(x-y) = 2f(x) + 2f(y)$$

is called a *quadratic functional equation*. In particular, every solution of the quadratic functional equation is said to be a *quadratic mapping*. A generalized Hyers–Ulam stability problem for the quadratic functional equation was proved by Skof [44] for mappings $f : X \to Y$, where X is a normed space and Y is a Banach space. Cholewa [12] noticed that the theorem of Skof is still true if the relevant domain X is replaced by an Abelian group. Czerwik [14] proved the generalized Hyers–Ulam stability of the quadratic functional equation. The stability problems of several functional equations have been extensively investigated by a number of authors and there are many interesting results concerning this problem (see Refs. [1, 13, 17, 19, 23, 24, 26, 41–43]).

Let X be a set. A function $d : X \times X \to [0, \infty]$ is called a *generalized metric* on X if d satisfies

(1) $d(x, y) = 0$ if and only if $x = y$;
(2) $d(x, y) = d(y, x)$ for all $x, y \in X$;
(3) $d(x, z) \leq d(x, y) + d(y, z)$ for all $x, y, z \in X$.

Let (X, d) be a generalized metric space. An operator $T : X \to X$ satisfies a Lipschitz condition with Lipschitz constant L if there exists a constant $L \geq 0$ such that $d(Tx, Ty) \leq L d(x, y)$ for all $x, y \in X$. If the Lipschitz constant L is less than 1, then the operator T is called a strictly contractive operator. Note that the distinction between the generalized metric and the usual metric is that the range of the former is permitted to include the infinity. We recall the following theorem by Margolis and Diaz:

Theorem 1 ([9, 16]). *Let (X, d) be a complete generalized metric space and let $J : X \to X$ be a strictly contractive mapping with Lipschitz constant $L < 1$. Then for each given element $x \in X$, either $d(J^n x, J^{n+1} x) = \infty$ for all nonnegative integers n or there exists a positive integer n_0 such that*

(1) $d(J^n x, J^{n+1} x) < \infty, \quad \forall n \geq n_0$;
(2) *the sequence $\{J^n x\}$ converges to a fixed point y^* of J*;
(3) y^* *is the unique fixed point of J in the set $Y = \{y \in X \mid d(J^{n_0} x, y) < \infty\}$*;
(4) $d(y, y^*) \leq \frac{1}{1-L} d(y, Jy)$ *for all $y \in Y$*.

In 1996, Isac and Rassias [25] were the first to provide applications of stability theory of functional equations for the proof of new fixed point theorems with applications. By using fixed point methods, the stability problems of several functional equations have been extensively investigated by a number of authors (see Refs. [10, 11, 34, 35, 39]).

Let Y be a Banach space. We define the following:

2^Y: the set of all subsets of Y;
$C_b(Y)$: the set of all closed bounded subsets of Y;
$C_c(Y)$: the set of all closed convex subsets of Y;
$C_{cb}(Y)$: the set of all closed convex bounded subsets of Y.

On 2^Y we consider the addition and the scalar multiplication as follows:

$$C + C' = \{x + x' : x \in C, x' \in C'\}, \quad \lambda C = \{\lambda x : x \in C\},$$

where $C, C' \in 2^Y$ and $\lambda \in \mathbb{R}$. Further, if $C, C' \in C_c(Y)$, then we denote by $C \oplus C' = \overline{C + C'}$.

It is easy to check that

$$\lambda C + \lambda C' = \lambda(C + C'), \quad (\lambda + \mu)C \subseteq \lambda C + \mu C.$$

Furthermore, when C is convex, we obtain $(\lambda + \mu)C = \lambda C + \mu C$ for all $\lambda, \mu \in \mathbb{R}^+$.

For a given set $C \in 2^Y$, the distance function $d(\cdot, C)$ and the support function $s(\cdot, C)$ are respectively defined by

$$d(x, C) = \inf\{\|x - y\| : y \in C\}, \quad x \in Y,$$
$$s(x^*, C) = \sup\{\langle x^*, x \rangle : x \in C\}, \quad x^* \in Y^*.$$

For every pair $C, C' \in C_b(Y)$, we define the Hausdorff distance between C and C' by

$$h(C, C') = \inf\{\lambda > 0 : C \subseteq C' + \lambda B_Y, \ C' \subseteq C + \lambda B_Y\},$$

where B_Y is the closed unit ball in Y.

The following proposition reveals some properties of the Hausdorff distance:

Proposition 2. *For every $C, C', K, K' \in C_{cb}(Y)$ and $\lambda > 0$, the following properties hold:*

(a) $h(C \oplus C', K \oplus K') \leq h(C, K) + h(C', K')$;
(b) $h(\lambda C, \lambda K) = \lambda h(C, K)$.

Let $(C_{cb}(Y), \oplus, h)$ be endowed with the Hausdorff distance h. Since Y is a Banach space, $(C_{cb}(Y), \oplus, h)$ is a complete metric semigroup (see Ref. [8]). Debreu [15] proved that $(C_{cb}(Y), \oplus, h)$ is isometrically embedded in a Banach space as follows:

Lemma 3 ([15]). *Let $C(B_{Y^*})$ be the Banach space of continuous real-valued functions on B_{Y^*} endowed with the uniform norm $\|\cdot\|_u$. Then the mapping $j : (C_{cb}(Y), \oplus, h) \to C(B_{Y^*})$, given by $j(A) = s(\cdot, A)$, satisfies the following properties:*

(a) $j(A \oplus B) = j(A) + j(B)$;
(b) $j(\lambda A) = \lambda j(A)$;

(c) $h(A, B) = \|j(A) - j(B)\|_u$;
(d) $j(C_{cb}(Y))$ is closed in $C(B_{Y^*})$

for all $A, B \in C_{cb}(Y)$ and all $\lambda \geq 0$.

Let $f : \Omega \to (C_{cb}(Y), h)$ be a set-valued function from a complete finite measure space (Ω, Σ, ν) into $C_{cb}(Y)$. Then f is *Debreu integrable* if the composition $j \circ f$ is Bochner integrable (see Ref. [7]). In this case, the Debreu integral of f in Ω is the unique element $(D) \int_\Omega f d\nu \in C_{cb}(Y)$ such that $j((D) \int_\Omega f d\nu)$ is the Bochner integral of $j \circ f$. The set of Debreu integrable functions from Ω to $C_{cb}(Y)$ will be denoted by $D(\Omega, C_{cb}(Y))$. Furthermore, on $D(\Omega, C_{cb}(Y))$, we define $(f + g)(\omega) = f(\omega) \oplus g(\omega)$ for all $f, g \in D(\Omega, C_{cb}(Y))$. Then we obtain that $((\Omega, C_{cb}(Y)), +)$ is an abelian semigroup.

Set-valued functional equations have been extensively investigated by a number of authors and there are many interesting results concerning this problem (see Refs. [6, 30–33, 37, 38]).

In this chapter, using the fixed point method, we prove the Hyers–Ulam stability of an additive set-valued functional equation (1).

Throughout this chapter, let X be a real vector space and Y a Banach space.

2. Main Results

Let m be a fixed integer greater than 3, then we have the following lemma:

The main purpose of this chapter is to investigate the generalized Hyers–Ulam stability of the following set-valued m-dimensional additive functional equation:

$$f\left(\frac{\sum_{i=1}^{m-1} x_i}{m}\right) \oplus f\left(\frac{(m-1)x_1 - \sum_{j=2}^{m-2} x_j - mx_{m-1}}{m}\right)$$
$$\oplus f\left(\frac{mx_1 + (m-1)x_{m-1}}{m}\right) = 2f(x_1). \qquad (2)$$

Lemma 4. *Let V and W be linear spaces and let $f : V \to W$ be an odd mapping satisfying (1.1), for all $x_1, \ldots, x_{m-1} \in V$. Then f is additive.*

Proof. Put $x_1 = \cdots = x_{m-1} = 0$ in (1.1) to get $f(0) = 0$. By taking $x_1 = 0, w_1 = \frac{-x_2 - \cdots - x_{m-1}}{m}, w_2 = \frac{(m-1)x_{m-1}}{m}$ in (1.1), we lead to

$$f(-(w_1 + w_2)) = f(w_1) + f(w_2).$$

Hence f is additive. □

Now, using the fixed point method, we prove the Hyers–Ulam stability of the additive set-valued functional equation (1.1).

Definition 5. Let $f : X \to C_{cb}(Y)$. The generalized additive set-valued functional equation is defined by

$$f\left(\frac{\sum_{i=1}^{m-1} x_i}{m}\right) \oplus f\left(\frac{(m-1)x_1 - \sum_{j=2}^{m-2} x_j - mx_{m-1}}{m}\right)$$

$$\oplus f\left(\frac{mx_1 + (m-1)x_{m-1}}{m}\right) = 2f(x_1)$$

for all $x_1, \ldots, x_m \in X$. Every solution of the generalized additive set-valued functional equation is called an *generalized additive set-valued mapping*.

Theorem 6. *Let* $\varphi : X^n \to [0, \infty)$ *be a function such that there exists an* $L < 1$ *with*

$$\varphi\left(\frac{x}{n}, \frac{x_2}{n}, \ldots, \frac{x_n}{n}\right) \leq \frac{L\varphi(x_1, x_2, \ldots, x_n)}{n}$$

for all $x_1, x_2, \ldots, x_n \in X$. *Suppose that* $f : X \to (C_{cb}(Y), h)$ *is a mapping satisfying*

$$h\left(f\left(\frac{\sum_{i=1}^{n} x_i}{n}\right) \oplus \biguplus_{i=1}^{n} f\left(\frac{\sum_{i=1, i \neq j}^{n} x_i - (n-1)x_j}{n}\right), f(x_1)\right)$$

$$\leq \varphi(x_1, x_2, \ldots, x_n) \qquad (3)$$

for all $x_1, x_2, \ldots, x_n \in X$. *Then there exists a unique additive set-valued mapping* $A : X \to (C_{cb}(Y), h)$ *such that*

$$h(f(x), A(x)) \leq \frac{1}{1-L} \varphi(x, \underbrace{0, \ldots, 0}_{n-1}) \qquad (4)$$

for all $x \in X$.

Proof. Replace $(x_1, x_2, \ldots, x_n)$ by $(x, \underbrace{0, \ldots, 0}_{n-1})$ in (3). Since $f(x)$ is convex, we obtain

$$h\left(nf\left(\frac{x}{n}\right), f(x)\right) \leq \varphi\left(x, \underbrace{0, \ldots, 0}_{n-1}\right) \tag{5}$$

for all $x \in X$. Consider $S := \{g : g : X \to C_{cb}(Y), g(0) = \{0\}\}$ and introduce the generalized metric on X, $d(g, f) = inf\{\mu \in (0, \infty) : h(g(x), f(x)) \leq \mu\varphi(x, 0, \ldots, 0), x \in X\}$, where, as usual, $\inf \phi = +\infty$. It is easy to show that (S, d) is complete (see Refs. [18, Theorem 2.4] and [29, Lemma 2.1]).

Now we consider the linear mapping $J : S \to S$ such that

$$Jg(x) := ng\left(\frac{x}{n}\right)$$

for all $x \in X$. Let $g, f \in S$ be given such that $d(g, f) = \varepsilon$. Then

$$h(g(x), f(x)) \leq \varepsilon\varphi(x, \underbrace{0, \ldots, 0}_{n-1})$$

for all $x \in X$. Hence

$$h(Jg(x), Jf(x)) = h\left(ng\left(\frac{x}{n}\right), nf\left(\frac{x}{n}\right)\right)$$
$$= nh\left(g\left(\frac{x}{n}\right), f\left(\frac{x}{n}\right)\right) \leq L\varphi(x, \underbrace{0, \ldots, 0}_{n-1})$$

for all $x \in X$. So $d(g, f) = \varepsilon$ implies that $d(Jg, Jf) \leq L\varepsilon$. This means that $d(Jg, Jf) \leq Ld(g, f)$ for all $g, f \in S$. It follows from (5) that $d(f, Jf) \leq 1$.

By Theorem 1, there exists a mapping $A : X \to (C_{cb}(Y), h)$ satisfying the following:

(1) A is a fixed point of J, i.e.,

$$A\left(\frac{x}{n}\right) = \frac{1}{n}A(x) \tag{6}$$

for all $x \in X$. The mapping A is a unique fixed point of J in the set

$$M = \{g \in S : d(f, g) < \infty\}.$$

This implies that A is a unique mapping satisfying (6) such that there exists a $\mu \in (0, \infty)$ satisfying $h(f(x), A(x)) \leq \mu\varphi(x, 0, \ldots, 0)$ for all $x \in X$.

(2) $d(J^m f, A) \to 0$ as $n \to \infty$. This implies the equality

$$\lim_{m \to \infty} n^m f\left(\frac{x}{n^m}\right) = A(x)$$

for all $x \in X$.

(3) $d(f, A) \leq \frac{d(f, Jf)}{1-L}$, which implies the inequality $d(f, A) \leq \frac{1}{1-L}$. This implies that the inequality (4) holds.

By (3),

$$h\left(n^m f\left(\frac{\sum_{i=1}^n x_i}{n^{m+1}}\right) \oplus \biguplus_{i=1}^n n^m f\left(\frac{\sum_{i=1, i \neq j}^n x_i - (n-1)x_j}{n^{m+1}}\right),\right.$$

$$\left. n^m f\left(\frac{x_1}{n^m}\right)\right) \leq n^m \varphi\left(\frac{x_1}{n^m}, \frac{x_2}{n^m}, \ldots, \frac{x_n}{n^m}\right) \leq L^n \varphi(x_1, x_2, \ldots, x_n),$$

which tends to zero as $n \to \infty$ for all $x_1, x_2, \ldots, x_n \in X$. Thus,

$$A\left(\frac{\sum_{i=1}^n x_i}{n}\right) \oplus \biguplus_{i=1}^n A\left(\frac{\sum_{i=1, i \neq j}^n x_i - (n-1)x_j}{n}\right) = A(x_1).$$

This implies that $A : X \to C_{cb}(Y)$ is additive as desired. □

Corollary 7. *Let $p > 1$ and $\theta \geq 0$ be real numbers, and let X be a real normed space. Suppose that $f : X \to (C_{cb}(Y), h)$ is a mapping satisfying*

$$h\left(f\left(\frac{\sum_{i=1}^n x_i}{n}\right) \oplus \biguplus_{i=1}^n f\left(\frac{\sum_{i=1, i \neq j}^n x_i - (n-1)x_j}{n}\right), f(x_1)\right)$$

$$\leq \theta \left(\sum_{i=1}^n \|x_i\|^p\right) \tag{7}$$

for all $x_1, x_2, \ldots, x_n \in X$. Then there exists a unique additive set-valued mapping $A : X \to (C_{cb}(Y), h)$ satisfying

$$h(f(x), A(x)) \le \frac{n^p \theta}{n^p - n} ||x||^p$$

for all $x \in X$.

Proof. The proof follows from Theorem 2.2 by taking

$$\varphi(x_1, x_2, \ldots, x_n) := \left(\sum_{i=1}^n ||x_i||^p \right)$$

for all $x_1, x_2, \ldots, x_n \in X$. Then we can choose $L = n^{1-p}$ and we get the desired result. $\square$

Theorem 8. Let $\varphi : X^n \to [0, \infty)$ be a function such that there exists an $L < 1$ with

$$\varphi(x_1, x_2, \ldots, x_n) \le nL\varphi\left(\frac{x_1}{n}, \frac{x_2}{n}, \ldots, \frac{x_n}{n}\right)$$

for all $x_1, x_2, \ldots, x_n \in X$. Suppose that $f : X \to (C_{cb}(Y), h)$ is a mapping satisfying (3). Then there exists a unique additive set-valued mapping $A : X \to (C_{cb}(Y), h)$ such that

$$h(f(x), A(x)) \le \frac{L}{1-L} \varphi(x, \underbrace{0, \ldots, 0}_{n-1}) \tag{8}$$

for all $x \in X$.

Proof. It follows from (5) that

$$h\left(\frac{1}{n} f(nx), f(x)\right) \le \frac{1}{n} \varphi(nx, 0, \ldots, 0) \le L\varphi(x, 0, \ldots, 0) \tag{9}$$

for all $x \in X$. Let (S, d) be the generalized metric space introduced in Theorem 2.2.

Now we consider the linear mapping $J : S \to S$ such that

$$Jg(x) := \frac{1}{n} g(nx)$$

for all $x \in X$. Let $g, f \in S$ be given such that $d(g, f) = \varepsilon$. Then,

$$h(g(x), f(x)) \leq \varepsilon\varphi(x, \underbrace{0, \ldots, 0}_{n-1})$$

for all $x \in X$. Hence,

$$h(Jg(x), Jf(x)) = h\left(\frac{g(nx)}{n}, \frac{f(nx)}{n}\right) = \frac{h(g(nx), f(nx))}{n}$$
$$\leq L\varphi(x, \underbrace{0, \ldots, 0}_{n-1})$$

for all $x \in X$. So $d(g, f) = \varepsilon$ implies that $d(Jg, Jf) \leq L\varepsilon$. This means that $d(Jg, Jf) \leq Ld(g, f)$ for all $g, f \in S$. It follows from (9) that $d(f, Jf) \leq L$.

By Theorem 1, there exists a mapping $A : X \to (C_{cb}(Y), h)$ satisfying the following:

(1) A is a fixed point of J, i.e.,

$$\frac{A(nx)}{n} = A(x) \qquad (10)$$

for all $x \in X$. The mapping A is a unique fixed point of J in the set $M = \{g \in S : d(f, g) < \infty\}$. This implies that A is a unique mapping satisfying (10) such that there exists a $\mu \in (0, \infty)$ satisfying $h(f(x), A(x)) \leq \mu\varphi(x, 0, \ldots, 0)$ for all $x \in X$.

(2) $d(J^m f, A) \to 0$ as $n \to \infty$. This implies the equality

$$\lim_{m \to \infty} \frac{f(n^m x)}{n^m} = A(x)$$

for all $x \in X$.

(3) $d(f, A) \leq \frac{d(f, Jf)}{1-L}$, which implies the inequality $d(f, A) \leq \frac{L}{1-L}$. This implies that the inequality (8) holds.

The rest of the proof is similar to the proof of Theorem 2.2. □

Corollary 9. *Let $1 > p > 0$ and $\theta \geq 0$ be real numbers, and let X be a real normed space. Suppose that $f : X \to (C_{cb}(Y), h)$ is a mapping* satisfying (7). Then there exists a unique additive set-valued mapping $A : X \to Y$ satisfying

$$h(f(x), A(x)) \leq \frac{n^p \theta}{n - n^p} ||x||^p$$

for all $x \in X$.

Proof. The proof follows from Theorem 2.4 by taking

$$\varphi(x_1, x_2, \ldots, x_n) := \left(\sum_{i=1}^{n} ||x_i||^p \right)$$

for all $x_1, x_2, \ldots, x_n \in X$. Then we can choose $L = n^{p-1}$ and we get the desired result. □

References

[1] J. Aczel and J. Dhombres, *Functional Equations in Several Variables* (Cambridge University Press, Cambridge, 1989).
[2] T. Aoki, On the stability of the linear transformation in Banach spaces, *J. Math. Soc. Jpn.* **2**, 64–66 (1950).
[3] K. J. Arrow and G. Debreu, Existence of an equilibrium for a competitive economy, *Econometrica* **22**, 265–290 (1954).
[4] J. P. Aubin and H. Frankow, *Set-Valued Analysis* (Birkhäuser, Boston, 1990).
[5] R. J. Aumann, Integrals of set-valued functions, *J. Math. Anal. Appl.* **12**, 1–12 (1965).
[6] T. Cardinali, K. Nikodem and F. Papalini, Some results on stability and characterization of K-convexity of set-valued functions, *Ann. Polon. Math.* **58**, 185–192 (1993).
[7] T. Cascales and J. Rodriguez, Birkhoff integral for multi-valued functions, *J. Math. Anal. Appl.* **297**, 540–560 (2004).
[8] C. Castaing and M. Valadier, *Convex Analysis and Measurable Multifunctions*. Lecture Notes in Mathematics, Vol. 580 (Springer, Berlin, 1977).
[9] L. Cădariu and V. Radu, Fixed points and the stability of Jensen's functional equation, *J. Inequal. Pure Appl. Math.* **4**(1) (2003). Art. ID 4.
[10] L. Cădariu and V. Radu, On the stability of the Cauchy functional equation: A fixed point approach, *Grazer Math. Ber.* **346**, 43–52 (2004).

[11] L. Cădariu and V. Radu, Fixed point methods for the generalized stability of functional equations in a single variable, *Fixed Point Theory Appl.* **2008** (2008). Art. ID 749392.

[12] P. W. Cholewa, Remarks on the stability of functional equations, *Aequationes Math.* **27**, 76–86 (1984).

[13] C.-Y. Chou and J.-H. Tzeng, On approximate isomorphisms between Banach *-algebras or C^*-algebras, *Taiwan. J. Math.* **10**, 219–231 (2006).

[14] S. Czerwik, On the stability of the quadratic mapping in normed spaces, *Abh. Math. Sem. Univ. Hambg.* **62**, 59–64 (1992).

[15] G. Debreu, Integration of correspondences. In *Proceedings of Fifth Berkeley Symposium on Mathematical Statistics and Probability*, Vol. II, Part I (1966), pp. 351–372.

[16] J. Diaz and B. Margolis, A fixed point theorem of the alternative for contractions on a generalized complete metric space, *Bull. Am. Math. Soc.* **74**, 305–309 (1968).

[17] P. Găvruta, A generalization of the Hyers-Ulam-Rassias stability of approximately additive mappings, *J. Math. Anal. Appl.* **184**, 431–436 (1994).

[18] M. E. Gordji, C. Park and M. B. Savadkouhi, The stability of a quartic type functional equation with the fixed point alternative, *Fixed Point Theory* **11**, 265–272 (2010).

[19] M. E. Gordji and M. B. Savadkouhi, Stability of a mixed type cubic-quartic functional equation in non-Archimedean spaces, *Appl. Math. Lett.* **23**, 1198–1202 (2010).

[20] C. Hess, Set-valued integration and set-valued probability theory: An overview. In *Handbook of Measure Theory*, Vols. I, II (North-Holland, Amsterdam, 2002).

[21] W. Hindenbrand, *Core and Equilibria of a Large Economy* (Princeton University Press, Princeton, 1974).

[22] D. H. Hyers, On the stability of the linear functional equation, *Proc. Natl. Acad. Sci. USA* **27**, 222–224 (1941).

[23] D. H. Hyers, G. Isac and T. M. Rassias, *Stability of Functional Equations in Several Variables* (Birkhäuser, Basel, 1998).

[24] G. Isac and T. M. Rassias, On the Hyers-Ulam stability of ψ-additive mappings, *J. Approx. Theory.* **72**, 131–137 (1993).

[25] G. Isac and T. M. Rassias, Stability of ψ-additive mappings: Appications to nonlinear analysis, *Int. J. Math. Math. Sci.* **19**, 219–228 (1996).

[26] K. Jun and H. Kim, Approximate derivations mapping into the radicals of Banach algebras, *Taiwan. J. Math.* **11**, 277–288 (2007).

[27] E. Klein and A. Thompson, *Theory of Correspondence* (Wiley, New York, 1984).
[28] L. W. McKenzie, On the existence of general equilibrium for a competitive market, *Econometrica.* **27**, 54–71 (1959).
[29] D. Miheţ and V. Radu, On the stability of the additive Cauchy functional equation in random normed spaces, *J. Math. Anal. Appl.* **343**, 567–572 (2008).
[30] K. Nikodem, On quadratic set-valued functions, *Publ. Math. Debrecen* **30**, 297–301 (1984).
[31] K. Nikodem, On Jensen's functional equation for set-valued functions, *Radovi Mat.* **3**, 23–33 (1987).
[32] K. Nikodem, Set-valued solutions of the Pexider functional equation, *Funkcialaj Ekvacioj* **31**, 227–231 (1988).
[33] K. Nikodem, *K-Convex and K-Concave Set-Valued Functions*, Vol. 559 (Zeszyty Naukowe, Lodz, 1989).
[34] C. Park, Fixed points and Hyers-Ulam-Rassias stability of Cauchy-Jensen functional equations in Banach algebras, *Fixed Point Theory Appl.* **2007** (2007). Art. ID 50175.
[35] C. Park, Generalized Hyers-Ulam-Rassias stability of quadratic functional equations: A fixed point approach, *Fixed Point Theory and Appl.* **2008** (2008). Art. ID 493751.
[36] C. Park, Approximation of set-valued functional equations. Preprint.
[37] Y. J. Piao, The existence and uniqueness of additive selection for (α, β)-(β, α) type subadditive set-valued maps, *J. Northeast Normal Univ.* **41**, 38–40 (2009).
[38] D. Popa, Additive selections of (α, β)-subadditive set-valued maps, *Glas. Mat. Ser. III* **36**(56), 11–16 (2001).
[39] V. Radu, The fixed point alternative and the stability of functional equations, *Fixed Point Theory* **4**, 91–96 (2003).
[40] T. M. Rassias, On the stability of the linear mapping in Banach spaces, *Proc. Am. Math. Soc.* **72**, 297–300 (1978).
[41] T. M. Rassias (ed.) *Functional Equations and Inequalities* (Kluwer Academic, Dordrecht, 2000).
[42] T. M. Rassias, On the stability of functional equations in Banach spaces, *J. Math. Anal. Appl.* **251**, 264–284 (2000).
[43] T. M. Rassias, On the stability of functional equations and a problem of Ulam, *Acta Math. Appl.* **62**, 23–130 (2000).
[44] F. Skof, Proprietà locali e approssimazione di operatori, *Rend. Sem. Mat. Fis. Milano* **53**, 113–129 (1983).
[45] S. M. Ulam, *Problems in Modern Mathematics*, Chapter VI, Science edn. (Wiley, New York, 1940).

© 2024 World Scientific Publishing Company
https://doi.org/10.1142/9789811267048_0004

Chapter 4

Two Heuristic Methods for Solving Generalized Nash Equilibrium Problems Using a Novel Penalty Function

Benjamin Benteke[*], Monica Gabriela Cojocaru[†], Roie Fields[‡],
Mihai Nica[§], and Kira Tarasuk[¶]

University of Guelph, Guelph, Ontario, Canada

[*]*bbenteke@uoguelph.ca*
[†]*mcojocar@uoguelph.ca*
[‡]*rfields@uoguelph.ca*
[§]*nicam@uoguelph.ca*
[¶]*ktarasuk@uoguelph.ca*

In this chapter, we introduce two evolutionary algorithmic heuristics which utilize competitive selection, (linear and nonlinear) regression and stochastic gradient descent to evolve generations of new points toward identifying entire solution sets of generalized Nash games with jointly convex constraints. Both algorithms involve the so-called *shadow point function*, a novel penalty function for generalized Nash equilibrium problems (GNEPs), similar to the Nikaidô–Isoda penalty function in optimization. It is well known that GNEPs have in general entire sets of Nash equilibria, and the question of identification of the entire solution set for a given GNEP is still open, for large classes of GNEPs. Traditional optimization-based methods to solve GNEPs are based on

KKT systems, quasi-variational inequalities and projected differential equations/inclusions, which are, with a few exceptions, only designed to find one point in the solution set rather than the set itself. Our algorithms are evaluated on 2- and 3-player games in 2 and 3 dimensions, with both linear and nonlinear shared constraints. The success of these algorithms is discussed and compared, and future work is outlined.

1. Introduction

Game theory is a branch of applied mathematics that provides tools for analyzing situations in which parties, called players, make rational decisions that are interdependent. This interdependence causes each player to consider the other player's possible decisions, or strategies, in formulating their own strategy. A solution to a game describes the optimal decisions of the players, who may have similar, opposed or mixed interests, and the outcomes that may result from these decisions. Since these decisions affect one another, the optimal strategy for each player depends on the decision made by other players. As such, we seek to find strategy sets in which for each player, their objective cannot be further optimized assuming all other players do not change their already optimal decisions. In other words, every player's strategy is optimal given the optimal strategy sets of all other players. These strategy sets are referred to as generalized Nash equilibrium problems (GNEPs) and were introduced by Debreu in 1952 as a shared-constraint expansion of Nash equilibria [1]. It is worth reminding the reader that game theory was originally developed by John von Neumann (mathematician) and his colleague Oskar Morgenstern (economist) to solve problems in economics. It was further developed in the 1950s by Nash, who established the mathematical principles of game theory: a branch of mathematics that examines the rivalries between competitors with mixed interests. In addition to its use in economics [2–4], game theory has been expanded and utilized in other fields, such as animal behavior and cooperation [5–7], common pool resource use [8–10] and consumer markets, such as power and internet bandwidth allocation [11, 12].

GNEPs are traditionally solved using numerical methods, specifically optimization-based methods (see, for instance, Refs. [13–18]) and/or differential equation-based methods (see Refs. [19, 20]).

Cojocaru et al., following and inspired by the work of Nabetani et al. [21], have advanced the issue of computing the entire GNEP solution set to the forefront of the GNEP community circa 2017 onward, proposing both optimization-based methods [22–27] and heuristics for identifying their solution sets.

In mathematics, heuristic algorithms are methods which, on the basis of experience or judgment, seem likely to yield a good solution to a problem but which cannot be guaranteed to produce an optimum [28]. Many heuristic approaches have been taken to solving GNEPs. Nikaidô and Isoda developed a function to measure the fitness of any given point, known as the Nikaidô–Isoda function, which uses the difference between the objective function values at the given point versus at optimal points [29]. This function has been used by many other researchers as a basis with which to motivate algorithms that search for, most often, a single point which optimizes an objective function. Various gradient-type and relaxation methods have been used to compute equilibria in jointly convex GNEPs using the Nikaidô–Isoda function [30, 31]. Isolated solution points can be found by converting GNEPs into variational inequalities, however some of these methods do not allow for larger solution sets to be found and can only be used under certain conditions [15, 16]. In the last few years, complete methods developed based on the use of variational inequalities have been presented by Migot and Cojocaru, including for nonjointly convex GNEPs: Refs. [23, 32]. Other methods include transforming these generalized Nash equilibrium problems into Quasi-Variational Inequality Problems (QVIs) to allow for more solutions to be found [33], however QVIs are inefficient to solve, and the algorithms for solving QVI can be quite complicated and restricted to very well-behaved problems (Refs. [20, 34]). Wild developed an evolutionary technique to approximate solutions for GNEPs using the concept of Nash dominance to rank fitness of the agents [26], using advance evolutionary algorithmic techniques.

In this chapter, in Section 2, we remind the reader about the mathematical formalism of GNEP and the basic definitions. In Section 3, we introduce the *shadow point* function used to evaluate the fitness of points chosen in an expanded version of the feasible set for a jointly convex GNEP. In Sections 4 and 5, we use this function to introduce our two heuristics algorithms: *evolutionary-inspired algorithm* and *stochastic gradient descent*, and we show them in action on a list of examples known from literature. We close with remarks on the effectiveness of our algorithmic methods and future work.

2. Mathematical Formulation of Generalized Nash Equilibrium Problems

When Nash first introduced the Nash equilibrium problem, it involved an n-player game in which each player sought to minimize payoff (penalty) by choosing the optimal strategy from their strategy set [35, 36]. An extension to this game framework is to consider situations where strategy spaces of players are affected by the decisions of other players. This is needed when describing games where individuals have shared constraints, such as the use of common-pool resources or competitive markets. These considerations were first modeled by Arrow and Debreu, who introduced a version of Nash equilibrium problems, named generalized Nash equilibrium problems [1, 37].

Here, we formulate and define the GNEP similar to the many papers in the current standard literature. There are N players, with each player v controlling the variables $x^v \in \mathbb{R}^{n_v}$. The strategy vector formed by the decision variables of each player is denoted by x, where $x := (x^1, \ldots, x^N)^T$, with the decision vector for all the players *except for* player v being denoted by x^{-v}. Each of these players has an objective function $\theta_v : \mathbb{R}^N \to \mathbb{R}$ that depends on their own variables x^v, as well as other players' variables x^{-v}. The strategies of each player are constrained by the set $X_v(x^{-v}) \subseteq \mathbb{R}^{n_v}$, dependent on other player's strategies. The goal of a player v is to choose the strategy x^v such that it solves the minimization problem

$$\begin{aligned} &\text{minimize}_{x^v} \ \theta(x^v, x^{-v}) \\ &\text{subject to } x^v \in X_v(x^{-v}). \end{aligned} \quad (1)$$

For any x^{-v}, the solution set is denoted by $S_v(x^{-v})$. A solution of the GNEP is a vector $\hat{x} := (\hat{x}^1, \ldots, \hat{x}^N)^T$ such that $\hat{x}^v \in S_v(x^{-v}) \ \forall v \in (1, \ldots, N)$.

We call $\hat{x}$ an equilibrium if no player further optimizes their objective function by changing their strategy to any other point in the feasible set. Mathematically, if we let X be the set of all possible equilibria of a GNEP, then a strategy $\hat{x} \in X$ is a *generalized Nash equilibrium* iff

$$\theta_v(\hat{x}) \leq \theta_v(x^v, x^{-v}) \quad \forall v \in (1, \ldots, N) \quad \text{and} \quad \forall x^v \in X_v(x^{-v}). \quad (2)$$

By definition, when $X_v(x^{-v})$ are independent of other players' strategies (X_v), this reduces to a Nash equilibrium problem.

3. Shadow Point Penalty Function

In this chapter, we introduce two heuristic algorithms for solving GNEPs with jointly convex constraint sets, both different than the work in Ref. [26], where mutation and offspring generation are heavily featured in very technical undertakings. One of our goals is to propose the mechanism of evolutionary computation, but coupled with simple-to-solve optimization problems, that can achieve good results, and be *practitioner friendly*. Many researchers in economics, biology and social sciences are using Nash equilibrium concepts and models. One common barrier in using GNEP as the natural Nash equilibrium-type model (where players affect each other's payoffs and strategy sets) is that the numerical techniques involved in computing even one such point are heavily mathematically complex, to date. We seek to remove the barriers of mathematical complexity, if/when possible, with friendly, easy-to-use algorithms, which provide very reliable pictures of the sought solution set.

Since both methods require an evaluation of the fitness of points at any given stage, we introduce a shared penalty function across both algorithms. We call this function the *shadow point function*, and it is a variation of the Nikaidô–Isoda function. When designing a penalty function on our set of solutions, we wanted to ensure the following properties: (1) The function treats all solutions equally and (2) the output of the function should be easily interpretable under the definition of a generalized Nash equilibrium.

Let us assume a game with N players and a constraint set C, where by C we mean the jointly convex constraints of the game. Let x_v be a decision vector of variables controlled by player v. Then a point $P = (x_1, x_2, \ldots, x_N)$ is an N-tuple of decision vectors made by players.

Definition 1. We define $\hat{x}_v$, the **shadow coordinate** of x_v, as the optimum value of x_v under the constraint set $C(x^{-v})$, given the decision vectors of all other players x_{-v}. Then the **shadow point** $\hat{P} = (\hat{x}_1, \hat{x}_2, \ldots, \hat{x}_N)$ is the set of shadow coordinates associated with P.

Definition 2. We define $\text{penalty}(P) = ||P - \hat{P}||_2$, the Euclidean distance of P to its shadow point $\hat{P}$.

Let us look at an example. Let $N = 2$, $x_1 := x, x_2 =: y$ and

$$P_1 : \min_x (x-1)^2 \qquad P_2 : \min_y (y - \tfrac{1}{2})^2$$
$$x, y \geq 0 \quad \text{and} \quad x, y \geq 0$$
$$x + y \leq 1 \qquad x + y \leq 1.$$

Given a point $P = (x, y) = (0.4, 0.9)$, we solve for the shadow point $\hat{P} = (\hat{x}, \hat{y})$. For $y = 0.9$, the set $C(x^{-1})$ in this case becomes $\{x \geq 0, \ x + 0.9 \leq 1\}$. Thus,

$$\hat{x} = \min((x-1)^2) \quad \text{subject to } x + 0.9 \leq 1 \Rightarrow \hat{x} = 0.1. \quad (3)$$

Similarly, for $x = 0.4$, the set $C(x^{-2})$ in this case becomes $\{y \geq 0, \ 0.4 + y \leq 1\}$. Then,

$$\hat{y} = \min((y - 0.5)^2) \quad \text{subject to } y + 0.4 \leq 1 \Rightarrow \hat{y} = 0.5. \quad (4)$$

Thus, the shadow point is $\hat{P} = (\hat{x}, \hat{y}) = (0.1, 0.5)$, and

$$\text{penalty}(P) = ||P - \hat{P}||_2 = \sqrt{(0.4 - 0.1)^2 + (0.9 - 0.5)^2} = 0.5. \quad (5)$$

By the definition of a Nash equilibrium, a point P is a generalized Nash equilibrium, if and only if $x = \hat{x}$ and $y = \hat{y}$, thus $P = \hat{P}$ and $\text{penalty}(P) = 0$. Thus, any algorithm utilizing this penalty function should seek to minimize penalty in order to find solutions.

This penalty function is similar to the Nikaidô–Isoda function commonly used to solve GNEPs but varies in important ways. Let $u_i(P)$ represent the objective function of player i. The Nikaidô–Isoda function is given by

$$\text{penalty}(P) = \sum_{v=1}^{N} [\theta_v(\hat{x}_v) - \theta_v(x_v)]. \quad (6)$$

The advantage of the shadow point penalty over the Nikaidô–Isoda function is that it does not require convexity. The Nikaidô–Isoda function can give misleadingly good results when finding local optimums that are close to the global optimum in the u, even when the p used to obtain this local optimum is far from the global optimum of p. The shadow point function alleviates this problem, describing penalty as a Euclidean distance in the space created by P, as opposed to a scalar value judging relative quality of optimums according to the objective functions.

4. Evolutionary-Inspired Algorithm

The general process of an evolutionary-inspired algorithm (EIA) can be described by repeating 4 steps, with a 0th step to initialize the process in the first generation of points. We then present our method applied to six known examples of games from the literature, for which we know the solution set. Finally, we apply it to an example that does not have a known solution set.

Step 0: Initialization

Before we can begin our algorithm, we need a set of initial points. For each player, we randomly initialize points in the players' strategy set, respecting bounds for each variable but ignoring the shared constraints.

When solving complex problems, the feasible set can be difficult to calculate and describe. For this reason, initializing each dimension in a "best case scenario" global bounds allows for implementation of the algorithm without needing a full understanding and description of the feasible set.

Steps 1 and 2: Evaluation and Selection

To evaluate the fitness of each point, we use the penalty function introduced in Section 3. In the selection step, the existing points are ordered according to penalty, and a number $m > 0$ of lowest[a]

---

[a]Recall we aim for penalty values of 0, thus higher fitness means lower values of the penalties.

penalty scores is selected, while the unselected points are deleted. In keeping with the evolutionary idea, we now have to replenish our set of points and replace the deleted ones with a fresh selection. This selection is performed in the next step.

Steps 3 and 4: Regression and Replacement

We perform a regression of the selected points from the previous step, to achieve a curve of best fit for that generation of points. For problems with linear solution sets, a linear regression is sufficient. For problems with nonlinear constraints, we use a multivariate polynomial regression instead.

We now must replace the $N - m$ deleted points from step 2 using the regression curve found. To explain the explain the process better, let us look at the case $N = 2$, and x_1 and x_2 being 1-dimensional strategy vectors of players 1, respectively 2.

We first generate x_1 coordinates, and using the regression curve, as well as some manufactured noise, we find corresponding x_2 coordinates.

Since early iterations of the algorithm are unlikely to produce a large variety of x_1 values within the player 1's set, we must search past our current range of selected points. To do so, we take the current upper and lower x_1 limits of our selected points and extend them by a size of $\delta^{\text{generation}}$ of the way to the global bounds of x_1 as given by the player 1's set. In general, $0 < \delta < 1$, and *generation* is the current iteration of the algorithm.

Let $x_{\text{MinSelected}}$ and $x_{\text{MaxSelected}}$ represent the minimum and maximum x_1 values of our selected points, respectively, and let x_{GMin} and x_{GMax} represent the lower and upper bounds of x_1 based on the constraint set of P1, respectively.

Then, our new lower bound $x_{\text{MinSearch}}$ and new upper bound $x_{\text{MaxSearch}}$ are given by

$$x_{\text{MinSearch}} = x_{\text{MinSelected}} - (x_{\text{MinSelected}} - x_{\text{GMin}}) \cdot \delta^{\text{generation}}, \quad (7)$$

$$x_{\text{MaxSearch}} = x_{\text{MaxSelected}} + (x_{\text{GMax}} - x_{\text{MaxSelected}}) \cdot \delta^{\text{generation}}. \quad (8)$$

Once these new bounds are defined, we generate $N - m$ replacement x_1 values from a uniform distribution within this range $(x_{\text{MinSearch}}, x_{\text{MaxSearch}})$.

We select a δ very close to 1, such that in the early part of the algorithm, we generate new points from almost the entire global bounds of x, regardless of the coordinates and quality of the selected points. As the algorithm progresses and the generation count grows, the range of x coordinates in *selected* points should converge onto the true range of x in the solution set, allowing us to search a much more local area in later generations with a much lower risk of restricting our search space too quickly. As the confidence of the algorithm toward the range of solutions grows, $\delta^{\text{iterations}} \to 0$ and thus the extra space it searches beyond those bounds shrinks.

In the case of two or more dimensions for player 1's set, this process is repeated for all but one dimension.

After generating the x_1 coordinates, we are able to select corresponding x_2 coordinates (of player 2) using our regression curve. Since the points in early generations are not guaranteed to be of high quality, the regression curve resulting from an early generation is also not guaranteed to be representative of the true solution set. As such, it is important to add noise to the x_2 coordinates as well, beyond their location on the curve. This is to ensure that new locations are explored and that the regression does not converge too early and get stuck despite the selection of new points due to new x_1 coordinates.

This noise is accomplished in a similar manner as the expansion of the x_1 search space, by expanding the search space around the global bounds of x_2. Let $y_{\text{Regression}}$ represent the x_2 coordinate on the regression curve at the newly generated x_1 coordinates. Let y_{GMin} and y_{GMax} represent the global min and max bounds of x_2 based on the constraint set of player 2. Based on these definitions and this operation, our new lower bound $y_{\text{MinSearch}}$ and new upper bound $y_{\text{MaxSearch}}$ are given by

$$y_{\text{MinSearch}} = y_{\text{Regression}} - \left(y_{\text{Regression}} - y_{\text{GMin}}\right) \cdot \epsilon^{\text{generations}}, \qquad (9)$$

$$y_{\text{MaxSearch}} = y_{\text{Regression}} + \left(y_{\text{GMax}} - y_{\text{Regression}}\right) \cdot \epsilon^{\text{generations}}, \qquad (10)$$

with $0 < \epsilon < 1$. For each of the $N - m$ x_1-coordinates generated, a x_2-value is generated from a uniform distribution in $(y_{\text{MinSearch}}, y_{\text{MaxSearch}})$. Similar to the δ used previously, we select an $\epsilon < 1$ very close to 1. As the algorithm progresses and the regression becomes more representative of the true solution set, $\epsilon^{\text{generations}} \to 0$, and so we generate x_2 values much closer to the regression curve.

These four steps are repeated until threshold convergence is met, or until a maximum number of iterations are met. In order to get a more complete picture of the solution set, this algorithm is run from scratch many times, and the results of each model are attached to create a plot of the solution set.

4.1. *Implementation on known examples*

We now present our method applied to six known examples of games from the literature, for which we know the solution set, and to one example that does not have a known solution set. The algorithm is implemented in Python. Since the penalty function requires an optimization, we use the multivariable optimization function optimize.minimize_scalar built into the Python library SciPy, using the SLSQP method for bounded optimization.

Example 1 is the simplest of the problems. In this example, we have linear constraints, and the objective functions of each player are functions only of variables controlled by that player, albeit still with shared constraints. In Figure 1, we see the solutions found by the Evolutionary Inspired Algorithm (EIA) for this example represent the true solution set. When observing the heatmap in Figure 1, we see that solutions in the middle of the search space are more common than those toward the edges. This is a result of the point generation function described in Section 4. The algorithm generates a random number within a uniform distribution between the upper and lower bounds calculated by the bounds of each dimension, the proximity to those bounds and the current iteration count. When closer to the bounds, the search space is much larger on the opposite direction of the bound than the direction toward the bound. As such, the randomly generated number is biased toward the middle of the set, and thus more solutions are likely to be found in this region. This might be remedied by tweaking the generation function to choose a direction first (with equal chance in each direction) and then pick a value in the chosen direction based on the proximity to the bounds in that direction and iteration count as is currently performed. Out of the 5000 points initialized across 500 models, 4807 (96%) converged onto the solution set with penalty ≤0.1.

Example 2 is similar to Example 1 in that it has the same shared linear constraint, however now with objective functions that depend

Table 1. GNEP set we seek to solve, taken from Ref. [26].

Objective functions	Constraints	Optimal solutions
Example 1 $P_1 : \min_{x_1}(x_1 - 1)^2$ $P_2 : \min_{x_2}(x_2 - \frac{1}{2})^2$	$x_1, x_2 \geq 0$ $x_1 + x_2 \leq 1$	$\begin{pmatrix} t \\ 1-t \end{pmatrix}$, $\frac{1}{2} \leq t \leq 1$
Example 2 $P_1 : \min_x x^2 - xy - x$ $P_2 : \min_y y^2 - \frac{1}{2}xy - 2y$	$x, y \geq 0$ $x + y \leq 1$	$\begin{pmatrix} t \\ 1-t \end{pmatrix}$, $0 \leq t \leq \frac{2}{3}$
Example 3 $P_1 : \min_x x^2 - xy - x$ $P_2 : \min_y y^2 - \frac{1}{2}xy - 2y$	$x, y \geq 0$ $x^2 + y^2 \leq 1$	$\begin{pmatrix} t \\ \sqrt{1-t^2} \end{pmatrix}$, $0 \leq t \leq \frac{4}{5}$
Example 4. Harker's example $P_1 : \min_x x^2 + \frac{8}{3} - 34x$ $P_2 : \min_y y^2 - \frac{5}{4}xy - 24.25y$	$x, y \geq 0$ $x, y \leq 10$ $x + y \leq 15$	$\begin{pmatrix} 5 \\ 9 \end{pmatrix} \cup \begin{pmatrix} t \\ 15-t \end{pmatrix}$, $9 \leq t \leq 10$
Example 5 $P_1 : \min_{x,y} x^2 + xy + y^2 +$ $(x+y)z - 25x - 38y$ $P_2 : \min_z z^2 + (x+y)z - 25z$	$x, y, z \geq 0$ $x + 2y - z \leq 14$ $3x + 2y + z \leq 30$	$\begin{pmatrix} t \\ 11-t \\ 8-t \end{pmatrix}$, $0 \leq t \leq 2$
Example 6. River Basin example $P_1 : \min_{x_1}(\alpha_1 x_1 + \beta(x_1$ $+ x_2 + x_3) - \chi_1)x_1$ $P_2 : \min_{x_2}(\alpha_2 x_2 + \beta(x_1$ $+ x_2 + x_3) - \chi_2)x_2$ $P_3 : \min_{x_3}(\alpha_3 x_3 + \beta(x_1$ $+ x_2 + x_3) - \chi_3)x_3$	$x_1, x_2, x_3 \geq 0$ $3.25x_1 + 1.25x_2$ $+ 4.125x_3 \leq 100$ $\alpha_1 = 0.001, \alpha_2 = 0.05,$ $\alpha_3 = 0.01, \beta = 0.01,$ $\chi_1 = 2.9, \chi_2 = 2.88,$ $\chi_3 = 2.85$	Solution known numerically
Example 7 $P_1 : \min_x x^2 - xy - x$ $P_2 : \min_y y^2 - \frac{1}{2}xy - 2y$ $P_3 : \min_z (z - \frac{1}{2})^2$	$x, y, z \geq 0$ $x^2 + y^2 \leq 1$ $z + y \leq 1$	Solution not known

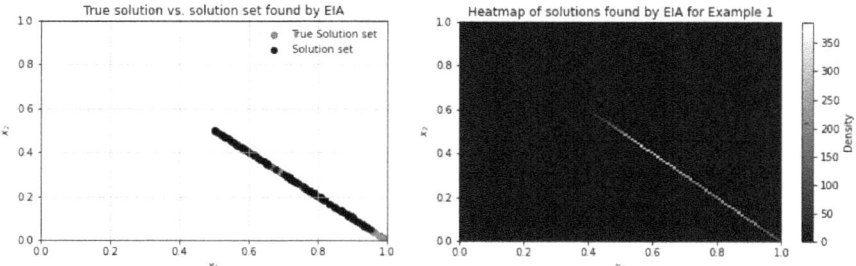

Figure 1. Solutions found by EIA for Example 1 and heatmap of solutions found by EIA for Example 1: The heatmap depicts areas of the density of points we discover in the solution set with this method.

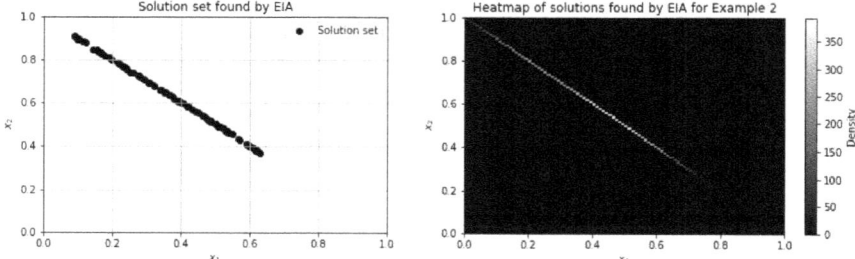

Figure 2. Solutions found by EIA for Example 2 and heatmap of the solutions.

on the strategies of both players. The EIA model was successful at finding solutions for this problem as well, as demonstrated in Figure 2. We observe that the set plotted represents the true solution set for this GNEP. Out of the 5000 points given by the 500 models, 4911 (98.22%) converged within a penalty ≤ 0.1. The points observed that are not touching the line are simply points so close to the solution set that they maintain a penalty within this 0.1 threshold. As is clear from the heatmap, the algorithm has some bias toward finding solutions closer to the center of the search space and does not find as many solutions close to the bounds. Similar to question 1, this is due to the next generation function having a much higher chance to generate points in the direction opposite the bounds when near the bounds, as there is less space there.

Example 3 introduces a key change in that the constraints are nonlinear, resulting in a nonlinear solution set. Since the EIA originally

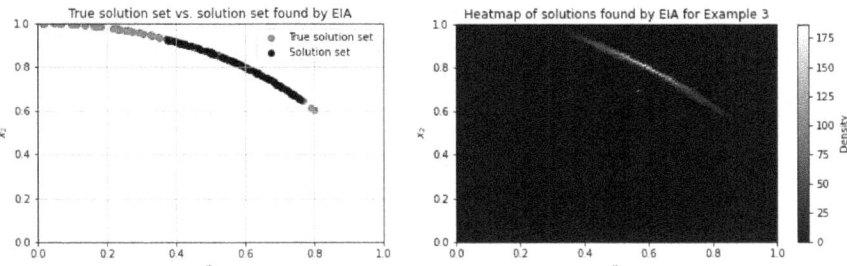

Figure 3. Solutions found by EIA for Example 3 and heatmap of the solution points density.

involved only linear regression, it was transformed to take nonlinear regression for the problem sets that have nonlinear constraints. Our method worked well, with the main limitation being that all of the points were in the solution set but missed parts of the solution set as seen in Figure 3. While our method is effective in finding points within the solution set, it misses values. We believe this is in part due to the nature of the EIA function, as the boundaries would shrink too much before the end of all the iterations, which would miss some other points in the optimal solution set.

Example 4 was formulated by Harker to demonstrate the ability of variational inequalities and quasi-variational inequalities to solve GNEPs [33], as illustrated in Figure 4. It is interesting in that in addition to a line segment, the solution set contains a "Rosen Point", a point that is isolated from the rest of the solution, as is seen in Figure 5.

This is problematic for the EIA method, as points near this Rosen point will have low penalty values. This causes the EIA to select points found near this solution in the selection step of the algorithm, where they are subsequently included in the linear regression. In this case, since the Rosen point is underneath the line created by the linear solution set, points found nearby this Rosen point "pull" our linear regression down, skewing its ability to find solutions on the linear segment.

When observing the heatmap, we can see that the solutions found near the Rosen point are so few that they cannot even be seen in the heatmap. As is the case for most of these examples, the points near the interior region of the search space are heavily favored over those near the bounds, for the same reason as previously.

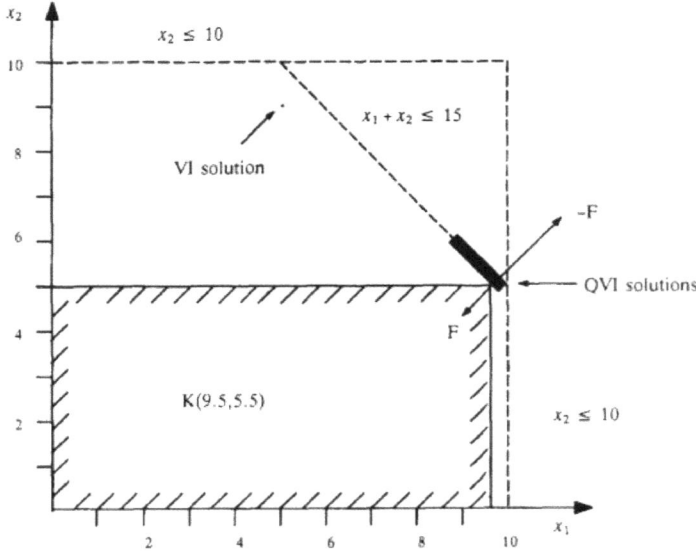

Figure 4. Solutions found by Harker for Example 4 [13].

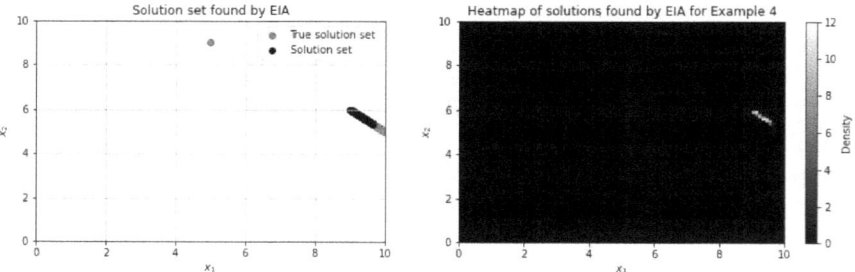

Figure 5. Solutions found by the EIA heuristics for the Harker example from Table 1.

Luckily, as demonstrated by Harker, this Rosen point can be computed relatively easily using variational inequalities [33]. After doing so, one can manually introduce an additional penalty for points found in the vicinity of the Rosen point, lowering their fitness and making it less likely for these points to be selected. This allows the algorithm to find solutions on the linear set, with the user having the knowledge that the additional solution point exists as well.

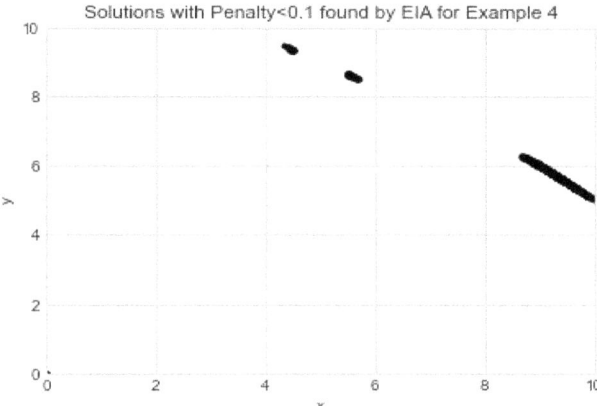

Figure 6. Solutions found by EIA for Example 4 with an additional penalty after the isolated Rosen point is accounted for.

In this example, a penalty of 1.0 was added to any point within 0.5 of the Rosen point. The results of running the EIA with this manufactured penalty are shown in Figure 6.

In total, 4996 of the 5000 points (99.99%) converged onto points with penalty ≤ 0.1. Some of these points still surround the Rosen point of (5, 9). These points are those which still have low penalty scores as they are near the Rosen point but are greater than 0.5 away and so manage to avoid the manufactured penalty. This can be remedied by increasing the region where we manufacture additional penalty. Fortunately, when observing the heatmap of solutions found in the EIA for Example 4, we see that almost all points found are within the line segment solution set. Example 5 starts to increase the complexity of these methods significantly, as it adds a third dimension, z. In this example, player 1 controls both variables x and y, and player 2 only has control over z. For this reason, we need to use a 2-dimensional optimizer when calculating the shadow point coordinates for player 1.

The EIA model does not run in parallel, and thus we can use built-in optimizers. Here, we use the SLSQP method in the optimize.minimize function in the SciPy Python library. When doing so, we find the same solution set as found by Ref. [26] as shown in Figure 7.

126   B. Benteke et al.

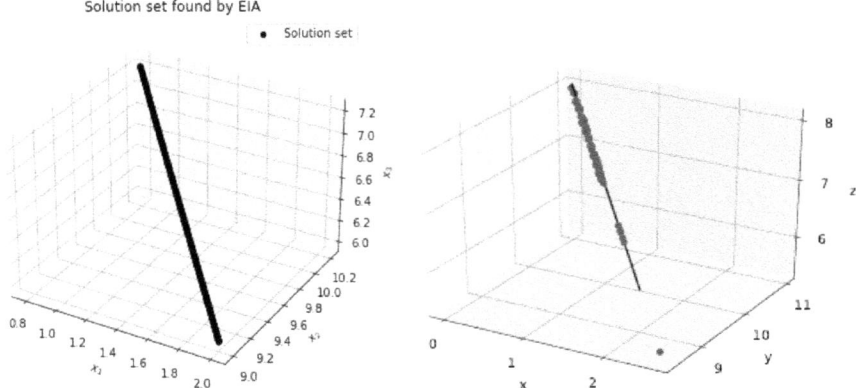

Figure 7. Solutions found for Example 5, as well as the solution set depiction from Ref. [26] — Right panel.

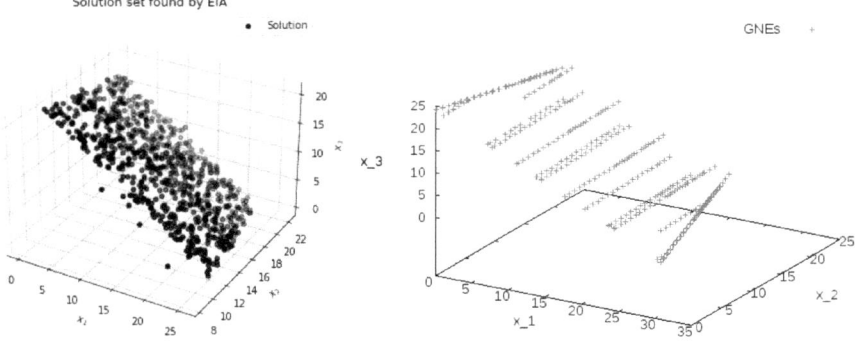

Figure 8. Solutions found by EIA for Example 6 versus the numerical solutions of this problem obtained by Migot and Cojocaru [32].

Example 6 actually has a real-world application as it is a river basin pollution game first studied by Krawczyk et al. [26, 31]. Mathematically, Example 6 is interesting in that it maintains the 3-dimensional challenge introduced by Example 5, but instead of having one player control 2 variables, it introduces a third player. So here we have a 3-player game with one variable controlled by each player.

Figure 8 shows the solutions found by the EIA method for Example 6. Each model made by the EIA is generally given by 10 points that lie on the same line. Since this solution set is given by a plane, there are many line segments that may be found each time the algorithm is run. Since we run 500 models and plot the results together, we observe here that the lines found by the EIA method are spread out enough that collectively they represent the majority of the true solution set. There are fewer points in the upper and lower regions of the solution set in the z direction, as well as in the lower end of the y direction. This is for the same reason as in the other examples; points are less likely to be initialized toward the bounds of the search space. We do not observe this phenomenon in the upper limits of the y direction since it is not near the bounds of the search space. In this example, 4999 of the 5000 points found by the EIA algorithm had a penalty ≤ 0.1. The high success rate of the algorithm for this example can likely be attributed to the larger size of the solution set making it easier to locate, as well as the the fact that the set is located in the relative center of the search space.

5. Stochastic Gradient Descent Algorithm

Unlike the EIA algorithm, the stochastic gradient descent algorithm operates on individual points of a generation. That is, the behavior of agents in this algorithm does not depend on the behavior of any other agents. The algorithm can be programmed such that it operates on a large number of points in parallel, allowing for a large number of solution to be found despite only running the algorithm once. The general process can be described by the repetition of 3 steps, with a 0th step to initialize our first generation of points.

Step 0: Initialization

This algorithm requires an initial "Parent" point to operate on. We initialize these parents in the same way as we do in Section 4. That is, we initialize each dimension within the best-case-scenario global bounds of the constraint set for that player.

Step 1: Exploration

In this step, each parent P generates a series of m "Children" in their local region. This is done by generating m steps, given by $\vec{\sigma} = (\sigma_1, \sigma_2, \ldots, \sigma_m)$ in $(-\alpha, \alpha)$ with $\sigma_i \in (-\alpha^{\text{generation}}, \alpha^{\text{generation}})$ and

$\alpha \in (0, 1)$ in each dimension with a Gaussian distribution and adding them to the parent P's coordinates to create a new set of points, the "children", given by $P + \sigma_i$. This allows the parent to search its local region without actually moving from its initial location. As the algorithm progresses, $\alpha^{\text{generation}}$ approaches 0, causing the children to search a more local area as confidence in the model grows.

Step 2: Evaluation

In the second step, the penalty values of each of the children are calculated using the shadow penalty function described in Section 3. In order to calculate this penalty, we used the SLSQP method that was also used in the EIA algorithm. This allows for far faster computation, allowing us to solve thousands of points at once instead of using an iterative approach.

Step 3: Transformation

In the third and final step of the algorithm, the fitness of the children, along with their respective step sizes away from their parent, is used to generate a step size with which to transform the parent point for the next generation. For each parent P, β, the step for the next generation, is a fitness-weighted average of the children's steps, given by

$$\beta = \frac{1}{m} \sum_{i=1}^{m} \sigma_i (\text{penalty}(P) - \text{penalty}(P + \sigma_i)) \qquad (11)$$

and thus the next generation of parents, P, is given by $P + \beta$ values. Similar to the evolutionary inspired algorithm, this process is repeated until threshold convergence, or until the maximum number of iterations are met.

5.1. Implementation on known examples: SGD

As mentioned previously, we test the SDG algorithm using the problem set described in Table 1. When applying the Stochastic Gradient Descent (SGD) model to Example 1, we find similar good solutions, as demonstrated in Figure 9. We only plot points with a penalty ≤ 0.1, allowing us to clearly see the solutions set. Out of the 10,000 points initialized in the set, 4787, or 47.87%, converged onto the solution set with a penalty ≤ 0.1.

The SGD model was less successful in terms of proportion of points that converged. This is likely due to the parent points operating in isolation. In the EIA model, many points are generated each iteration of the algorithm, and so poorest points are deleted, and even poor points that are selected are offset by fitter points in the averaging function of the regression. Additionally, points themselves do not move but rather the overall algorithm generates new points in the space. Lastly, these newly generated points can be generated in almost the entire search space in early iterations, allowing for the entire space to be searched earlier on, removing poor points early in the algorithm. In the SGD model, agents initialized with poor fitness levels must travel long distances in order to converge onto the solution set, and their travel is capped by the increasingly strong dampening effect applied by $\alpha^{\text{generation}}$. As such, many of these parent points do not make it all the way to the solution space before the $\alpha^{\text{generation}}$ becomes too low a value to move large enough distances. When observing the heatmap shown in Figure 9, we see that the solutions are biased toward the center of the space. This is due to the uniformly random initialization generating many points in the upper left, upper right and lower left regions of the space. Parents will evolve in the approximate direction that decreases their penalty the most, and so the nearest solution point for all parents in these spaces will be toward the center of the overall space. In order to converge near $(1, 0)$, a parent must be initialized nearby, otherwise

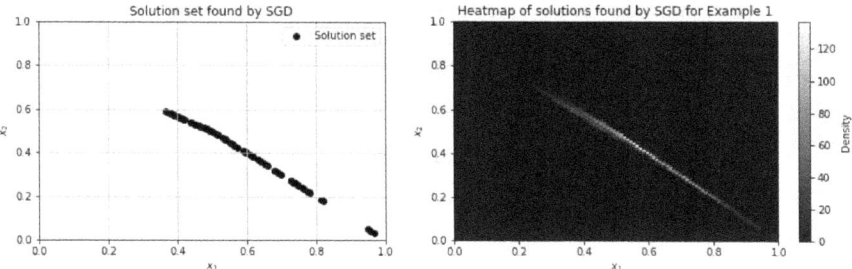

Figure 9. Solutions found by SGD for Example 1 and heatmap of solutions found by SGD for Example 1.

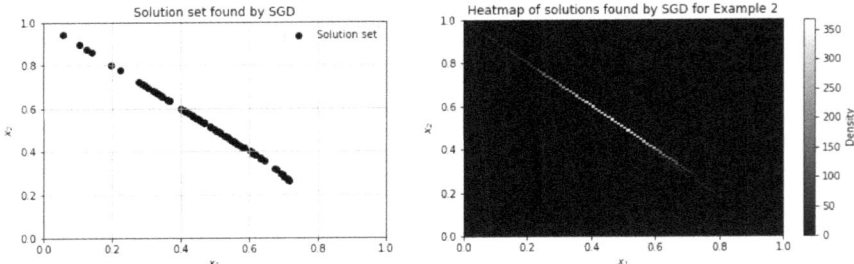

Figure 10. Solutions found by SGD for Example 2 and heatmap of solutions found by SGD for Example 2.

there is likely to be a closer point on the solution set that is closer to (0.5, 0.5).

When applying the SGD model to Example 2, we also achieve good results, as demonstrated in Figure 10. Out of the 10,000 points found by the SGD algorithm, 7418 (74.18%) had a penalty ≤0.1. The SGD model finds proportionally fewer solutions than the EIA model in Example 2 for similar reasons as in Example 1. The SGD algorithm is actually proportionally more successful in Example 2 than Example 1, due to the fact that the solution set spans a greater proportion of the search space than in Example 1. For this reason, we observe that the point (0.66, 0.33) is by far the most commonly found solution and that as you trend toward (0, 1), the number of solutions found becomes more sparse. This is again due to the same reason as in Example 1; the agents are designed to move toward the nearest solution, and so the all agents initialized to the bottom right of (0.66, 0.33) will converge onto that same point. Overall, the dependence of the objective function on both player's decisions does not appear to affect the ability of either of these algorithms to find the solution set. For Example 3, the SGD algorithm is still very successful when solving this problem, as demonstrated in Figure 11.

Out of 50,000 points, 25,539 (51.08%) converged onto the solution set within a penalty ≤0.1. This is because despite the fact that the solution set is much larger than in Example 1, the set is much closer to the border of the joint constraint set, and so there are still a large number of parents initialized who will be unable to reach the solution

Two Heuristic Methods for Solving Generalized Nash Equilibrium Problems 131

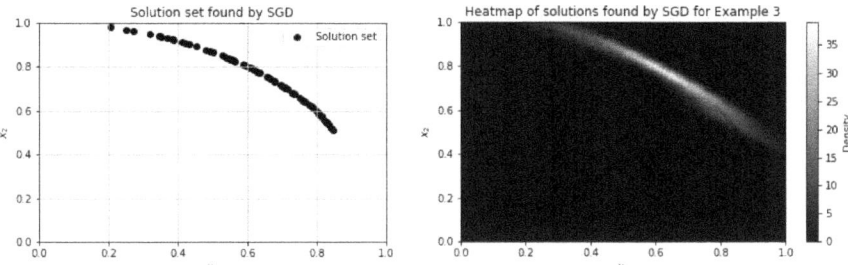

Figure 11. Solutions found by SGD for Example 3 and heatmap of solutions found by SGD for Example 3.

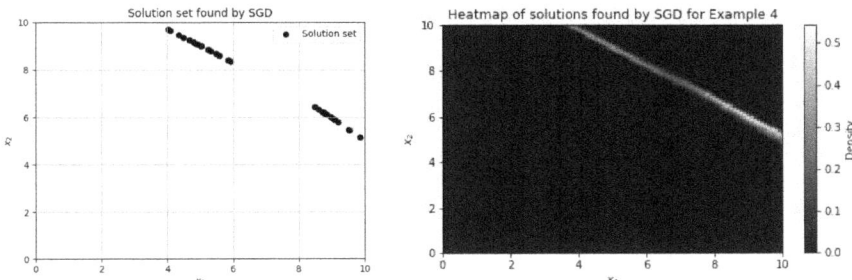

Figure 12. Solutions found by SGD for Example 4 and heatmap of solutions found by SGD for Example 4.

set before the $\delta^{\text{generations}}$ term becomes too powerful to take large enough steps reach the solution. This is especially true for points close to $(0, 0)$.

Harker's example (Example 4 here) poses some troubles for the SGD (as seen in Figure 12), specifically due to the isolated Rosen point, but is able to highlight the region of the Rosen equilibrium, as well as the line segment solutions. The SGD model was far less successful at solving Example 6, as seen in Figure 13. As one can see, far fewer solutions were found by this method than the EIA. In total, of the 10,000 agents placed in the space, only 188 converged onto a point with penalty ≤ 0.1. In addition, the algorithm was only able to find the lower y and bottom z boundaries of the solution set, without finding almost any interior points. The lack of interior points may be due to the behavior of the search algorithm. Parents search their vicinity for points that lower their penalty, however if

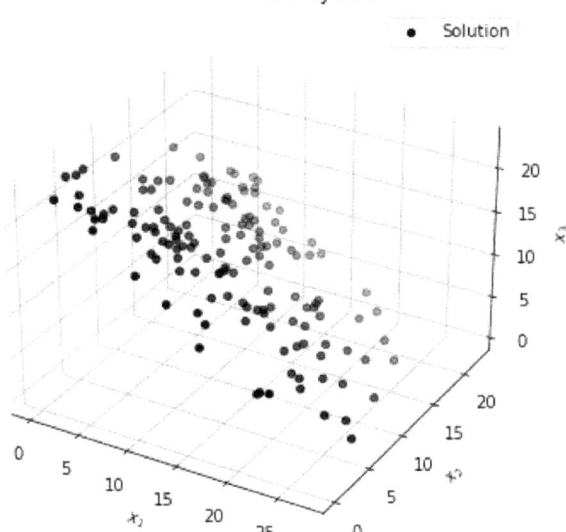

Figure 13. Solutions found by SGD for Example 6.

a point already lies on a solution, their is no benefit to moving. For this reason, points that find a boundary have no motivation to move and thus are likely to stay on the boundary, restricting them from finding the interior of the set.

Finally, Example 7 is a 3-player generalization of Example 3 and has no known solution set description. It features objective functions dependant on other player's strategies and objective functions dependant only on the player's own strategy. It also features a combination of both linear and nonlinear constraints, leading to both linear and nonlinear segments in the solution set. In Figure 14, we observe that the SGD algorithm was very successful at finding solutions to this GNEP but still lacked consistency. Of the 10,000 agents initialized in the search space, only 1959 (19.59%) converged onto points with penalty ≤ 0.1. While proportionally few of the agents found solutions, the entire solution set is well represented by the successful agents. This is due to only a small proportion of the solution set being near the bounds. The lower proportional success rate of this algorithm as compared to other GNEPs studied, Example 3 in particular, may be attributed to the fact that Example 7 is in 3-dimensions,

Two Heuristic Methods for Solving Generalized Nash Equilibrium Problems 133

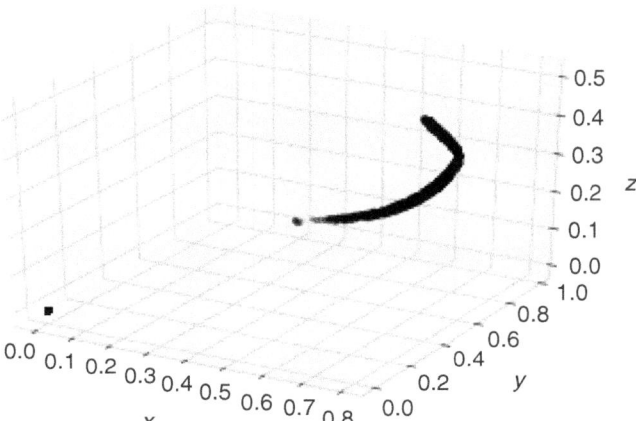

Figure 14. Solutions found by SGD for Example 7.

vastly increasing the size of the space needed to be searched to find solutions.

6. Limitations and Future Work

6.1. *EIA*

As seen throughout our presentation, neither of the algorithms act as a "jack of all trades". The EIA struggles to find solutions closer to the boundary of the solution set. One potential remedy for this would be to alter the new point generation function, to first make a binary choice in each dimension about whether to increase or decrease the value and only afterwards choose a value within the search space.

Further, two significant factors that need to be investigated are the number of points in each generation and the number of points selected in each generation, in such a way as to test improvement of performance as a function of the generation size. In this chapter, we use 100–500 points per generation. For this reason, ongoing work is looking to determine the point of "diminishing returns" for generation sizes, as to optimize the number of distinct Nash solutions for maximum computational efficiency. This value would likely be different depending on the complexity of the problem.

Raising the number of selected points decreases the "power" of any individual selected point in the linear regression. This would improve the algorithm's ability to work through outlier points and may improve the accuracy of the regression. This too would need to be optimized to the specific question the algorithm is used for.

Other variables that can be varied and their effects studied are the δ values we utilize to decrease the size of the search space over time, similar to a step size. δ values closer to 1 slow down the algorithm, requiring more iterations to converge onto the solution set, but make it much more likely the algorithm will find good solutions along the way. Lower δ values restrict the search space much faster, resulting in a much faster algorithm but one that risks converging too quickly and not finding the solution set before restricting the search space to a region that does not contain the solution.

6.2. *SGD*

The stochastic gradient descent method can also be further improved and optimized.

The algorithm struggled to find interior solution points to larger solution sets, such as planes. In these examples, the algorithm stops on the boundary of the set and has no motivation to move since the penalty is already minimized at 0. This would likely be replicated in solution sets such as these in higher dimensions as well, such as cubes; this is a current topic of investigation by Cojocaru *et al.*

The other major variable that should be studied is the step size α. A small α value results in smaller steps. The advantage of this is that the algorithm is very unlikely to "jump over" a solution and miss it. Unfortunately, since the α is dampened by $\alpha^{\text{generation}}$, an α value too small will impede some agents from being able to travel all the way to a solution point before the steps become too small to move there. Alternatively, a large value of α allows the agents to travel further distances before their movement is significantly impeded, however it also runs the risk of taking steps past the solution points and missing them. This value must be varied according to the problem set and is especially dependent on the size of the space needed to be searched.

Since we have so many parents working in parallel, it may be satisfying enough to the user if many or most of them never converge onto solutions, as long as enough parents do, so that the user gets a good understanding of the set.

Last but not least, one can vary the rate at which α is dampened by changing the dampening factor to another value. A value closer to 1 will slow down the rate at which α is dampened, allowing for more of the space to be searched but slowing down the time until convergence. A lower value will force α to zero much faster but may not give some parents a chance to find the solution set before they stop evolving. Once again, the optimal value will depend on the specific problem for which the algorithm is utilized. This is also the subject of ongoing work.

7. Conclusion

In this chapter, we introduce a novel penalty function, similar to the Nikaidô–Isoda function [29]. We utilize this function to build two heuristic models. The evolutionary-inspired algorithm selects the top points found in each iteration, performs a regression and uses that regression to generate replacement points within a space dependent on the current iteration count and the proximity to the global bounds of each dimension. The stochastic gradient descent algorithm initializes points (parents) randomly in the space and utilizes the penalty function and stochastic gradient descent to find solutions. The work presented here is based on Ref. [27], with two important advancements: the removal of the linear regression limitation and the parallelization of the EIA and SGD so that they can be run on generations of sizes 100–500 points.

The authors are working on the ongoing issues mentioned above, as well as working on a comparison between these algorithms and optimization-based methods of solving GNEPs, in order to (1) clarify the best use and power of the heuristics methods proposed and (2) expand their potential use to nonjointly convex GNE problems.

Acknowledgments

Funding from the National Sciences and Engineering Research Council (NSERC) of Canada is acknowledged through the Discovery Grant program (DG 400684) of Cojocaru (PI). The grant has funded the work of R. Fields, K. Tarasuk and B. Benteke. Nica acknowledges a similar NSERC Discovery grant that partially supported the work of K. Tarasuk.

References

[1] G. Debreu, A social equilibrium existence theorem, *Proc. Natl. Acad. Sci.* **38**(10), 886–893 (1952). doi: 10.1073/pnas.38.10.886. https://www.pnas.org/content/38/10/886.
[2] J. F. Nash, The bargaining problem, *Econometrica* (1950).
[3] A. R. Ofer and A. V. Thakor, A theory of stock price responses to alternative corporate cash disbursement methods: Stock repurchases and dividends, *J. Finance* **42**(2), 365–394 (1987).
[4] T. Karabiyik, O. Akal, and E. Aktas, Sustainable equilibrium in a stock market: Agent-based modeling with evolutionary game theory applied to traders, *Int. J. Bus. Account. Finance* **11**(1), 106–125 (2017).
[5] R. Bshary and R. F. Oliveira, Cooperation in animals: Toward a game theory within the framework of social competence, *Curr. Opin. Behav. Sci.* **3**, 31–37 (2015). https://doi.org/10.1016/j.cobeha.2015.01.008. https://www.sciencedirect.com/science/article/pii/S2352154615000182.
[6] C. Hadjichrysanthou and M. Broom, When should animals share food? Game theory applied to kleptoparasitic populations with food sharing, *Behav. Ecol.* **23**(5), 977–991 (2012). https://doi.org/10.1093/beheco/ars061.
[7] J. M. Smith, Game theory and the evolution of behaviour, *Behav. Brain Sci.* **7**(1), 95–101 (1984). doi: 10.1017/S0140525X00026327.
[8] G. Hardin, The tragedy of the commons, *J. Nat. Resour. Policy Res.* **1**(3), 243–253 (2009).
[9] E. Oftadeh, M. Shourian and B. Saghafian, An ultimate game theory based approach for basin scale water allocation conflict resolution, *Water Resour. Manag.* **31**(1), 4293–4308 (2017).
[10] Q. Han, G. Tan, X. Fu, Y. Mei and Z. Yang, Water resource optimal allocation based on multi agent game theory of Hanjiang river basin, *Water* **10**(9), 1184 (2018).
[11] L. Lin, C. Huang and L. Zhao, Application of cooperation game theory in reactive power market of hydraulic-power plant, *J. Coast. Res. Spec. Adv. Water Resour. Explor.* **93**, 578–581 (2019).

[12] K. Yamamoto, A comprehensive survey of potential game approaches to wireless networks, *IECE Trans. Commun.* **E98-B**(9), 1804–1823 (2015).
[13] P. T. Harker, Generalized nash games and quasi-variational inequalities, *Eur. J. Oper. Res.* **54**(1), 81–94 (1991).
[14] A. Von Heusinger and C. Kanzow, Optimization reformulations of the generalized nash equilibrium problem using nikaido-isoda-type functions, *Comput. Optim. Appl.* **43**, 353–377 (2009).
[15] D. Aussel and J. Dutta, Generalized nash equilibrium problem, variational inequality and quasiconvexity, *Oper. Res. Lett.* **36**(4), 461–464 (2008). https://doi.org/10.1016/j.orl.2008.01.002. https://www.sciencedirect.com/science/article/pii/S0167637708000072.
[16] F. Facchinei, A. Fischer and V. Piccialli, On generalized nash games and variational inequalities, *Oper. Res. Lett.* **35**(2), 159–164 (2007).
[17] A. Barbagallo and M.-G. Cojocaru, Dynamic vaccination games and variational inequalities on time-dependent sets, *J. Biol. Dyn.* **4**(6), 539–558 (2010).
[18] A. Dreves, Globally convergent algorithms for the solution of generalized Nash equilibrium problems, PhD Thesis, Universität Würzburg (2011).
[19] J. Lequyer and M.-G. Cojocaru, The replicator dynamics of generalized nash games, *Math. Appl. Sci. Eng.* **2**(2), 72–86 (2021).
[20] M.-G. Cojocaru, Dynamic equilibria of group vaccination strategies in a heterogeneous population, *J. Glob. Optim.* **40**, 51–63 (2008).
[21] K. Nabetani, Variational inequality approaches to generalized nash equilibrium problems, Technical Report, Department of Applied Mathematics, School of Informatics, Kyoto University (2008).
[22] M.-G. Cojocaru, E. Wild and A. Small, On describing the solution sets of generalized nash games with shared constraints, *Optim. Eng.* **19**, 845–870 (2018).
[23] T. Migot and M.-G. Cojocaru, A decomposition method for a class of convex generalized nash equilibrium problems, *Optim. Eng.* **22**(2021), 1–27 (2020).
[24] T. Migot and M.-G. Cojocaru, Nonsmooth dynamics of generalized nash games, *J. Nonlinear Var. Anal.* **1**(4), 27–44 (2020).
[25] T. Migot and M.-G. Cojocaru, On minty-variational inequalities and evolutionary stable states of generalized monotone games, *Oper. Res. Lett.* **49**(1), 96–100 (2021).
[26] E. Wild, A study of heuristic approaches for solving generalized Nash equilibrium problems and related games, PhD Thesis, University of Guelph (2017).

[27] R. Fields, Two heuristic methods for solving generalized Nash equilibrium problems using a novel penalty function, PhD Thesis, University of Guelph (2021).
[28] L. R. Foulds, The heuristic problem-solving approach, *J. Oper. Res. Soc.* **34**(10), 927–934 (1983).
[29] H. Nikaidô and K. Isoda, Note on non-cooperative convex games, *Pac. J. Math.* **5**(5), 807–815 (1955).
[30] J. B. Rosen, Existence and uniqueness of equilibrium points for concave n-person games, *Econometrica* **33**, 520–534 (1965).
[31] J. B. Krawczyk and S. Uryasev, Relaxation algorithms to find nash equilibria with economic applications, *Environ. Model. Assess.* **5**, 63–73 (2000).
[32] T. Migot and M.-G. Cojocaru, A parametrized variational inequality approach to track the solution set of a generalized nash equilibrium problem, *Eur. J. Oper. Res.* **283**(3), 1136–1147 (2020).
[33] P. T. Harker, Generalized nash games and quasi-variational inequalities, *Theory Methodol.* 81–84 (1988).
[34] B. K. Le, S. Adly and M.-G. Cojocaru, A decomposition method for a class of convex generalized nash equilibrium problems, Submitted: *JOTA* (2023).
[35] J. F. Nash, Equilibrium points in n-person games, *Proc. Natl. Acad. Sci.* **36**(1), 48–49 (1950). doi: 10.1073/pnas.36.1.48. https://www.pnas.org/content/36/1/48.
[36] J. F. Nash, Non-cooperative games, *Ann. Math.* **54**(2), 286–295 (1951).
[37] K. J. Arrow and G. Debreu, Existence of an equilibrium for a competitive economy, *Econometrica* **22**(3), 265–290 (1954).

© 2024 World Scientific Publishing Company
https://doi.org/10.1142/9789811267048_0005

Chapter 5

The Finite Element Method with Applications to Fluid Mechanics

Kyriaki N. Biraki[*], Konstantina C. Kyriakoudi[†], Anastasios C. Felias[‡], and Michail A. Xenos[§]

*Department of Mathematics,
University of Ioannina, Ioannina, Greece*

[*]*kiki.mpira@gmail.com*
[†]*k.kyriakoudi@uoi.gr*
[‡]*tsfelias1995@gmail.com*
[§]*mxenos@uoi.gr*

The finite element method (FEM) is a well-established approach for the numerical solution of ordinary differential equations (ODEs) and partial differential equations (PDEs). This method is a powerful tool in the study of various problems and has many applications, such as structural and fluid mechanics. In this review chapter, we mainly focus on applying the method to fluid mechanics problems. Initially, we present the FEM along with the basic theorems and examples. We analyze the error estimates for linear problems and the base functions that help distinguish the problem under consideration. We present the numerical solution of the Duffing equation, using the Galerkin FEM.

Additionally, we concentrate on the two-dimensional Stokes problem. We further introduce novel methods, such as the Discontinuous Galerkin (DG) FEM. The notion of adaptive mesh is also discussed. Lastly, we study the two-dimensional Navier–Stokes equations using the Galerkin

FEM. These advanced methods provide reliable numerical results in all studied cases. This is achieved with the application of FEM to "test problems", such as the backward-facing step. We obtain all the numerical results utilizing the software programs MATLAB and FEniCS.

1. Basic Principles of Finite Elements

1.1. *Introduction*

Nonlinear differential equations govern a plethora of biological, mechanical and physical phenomena. Most PDEs aren't analytically solvable, and to obtain a solution, numerical methods are utilized. The finite element method (FEM) is a well-established approach for the numerical solution of ODEs and PDEs. A large domain divides into smaller discrete cells, called finite elements, being simple polygonal shapes, forming the computational mesh of this domain. The method excels in its accurate representation of complex geometries, the finite elements of which are approximated by polynomials. Nowadays, FEM is arguably one of the most well-established and accurate computational techniques. There is a variety of applications in many fields, such as mechanical design, structural analysis, fluid flow, heat transfer and electromagnetism to computer programming aspects.

Origins of the FEM are found in the early approximation of π by considering a sequence of inscribed polygons, although the method was formally introduced in 1960 by Clough [27]. In terms of the present day notation, each side of the polygon represented an element, and as their number increases, the approximate values converge to the true one. Solving complex elasticity and structural analysis problems in civil and aeronautical engineering, for example, wings and fuselages are treated as assemblies of stringers, skins and sear panels, further developed the method.

In 1851, Schellback, in order to obtain a differential equation of a minimum surface area bounded by a specific closed curve, divided the surface into several triangles and used a finite difference expression to find the total discretized area [11]. The initial differential equation of a minimum surface area was then replaced by a system of algebraic equations. Until the 1900s, the behavior of structural frameworks, composed of several bars arranged in a regular pattern, has been approximated by one of an isotropic elastic body.

Ritz, in 1909, developed an efficient method finding approximate solution of deformable solid mechanics problems. His approach referred to an approximation of an energy functional by known functions multiplied with unknown coefficients. Minimizing the functional in relation to each unknown leads to a system determining those coefficients, satisfying the given boundary conditions.

In 1915, Boris Grigoryevich Galerkin derived an advanced method for the numerical solution of differential equations. His method with piecewise polynomial spaces is known as the finite element method. Technological advancements, further developed Galerkin's method. The approach traces back to variational principles introduced by several mathematicians, such as Leibniz, Euler, Lagrange, Dirichlet and many others. Hrenikoff, in 1941, introduced the framework method. In this approach, he replaced an elastic medium with a system of sticks and rods.

The FEM was introduced in the 1950s by structural engineers, especially in the aircraft industry, predicting stresses induced in aircraft wings, despite being independently proposed by Courant in 1943 [21]. He introduced special linear functions over triangular regions, obtained by dividing the cross-region and applying the method for the solution of torsional rigidity and hallow shaft. The latter introduced the Rayleigh–Ritz method. The functions introduced by Ritz did not need to satisfy the boundary conditions. Courant's theory could not be implemented due to the current absence of computers. Significant contributions to FEM were made by Turner, in 1956.

FEM obtained its main advancement in the 1960s and 1970s through developments of, among others, J. H. Argyris and collaborators. Clough, in 1960, introduced the term used until today, "finite element" in Ref. [3]. The first FEM book was published in 1967, by Zienkiewicz and Cheung [26]. The main motive behind the wide spread of FEM was the handling of big volume of numerical solution by computers [20].

In this section, the main focus is on the Galerkin FEM. As previously mentioned, any progress to FEM, regarding fluid mechanics applications, was significantly delayed due to nonlinear convection and solution instability originating from the element selection. In this section, we are analyzing basic principals of FEM, where more details can be found in the textbooks by Brenner and Scott [5].

In the following section, we discuss about the FEM for the Stokes and Navier–Stokes problems. Finally, we introduce the Discontinuous Galerkin (DG) FEM and the adaptive mesh refinement approach.

1.2. *Finite element theory*

1.2.1. *Basic concepts and definitions*

It is important to understand the basics of the finite element theory. Analyzing the following simple example helps the reader to understand the path followed to create the weak form of the problem. We consider the one-dimensional boundary value problem:

$$\begin{cases} -\dfrac{\mathrm{d}^2 u}{\mathrm{d}x^2} = f, & x \in (0,1) \\ u(0) = 0, & u(1) = 0. \end{cases} \quad (1)$$

Multiplying both parts of the equations with a test function v, with $v(0) = v(1) = 0$ and integrating by parts, we get

$$(f, v) := \int_0^1 f(x)v(x)\,\mathrm{d}x = \int_0^1 -u''(x)v(x)\,\mathrm{d}x = -u'(x)v(x)\big|_0^1$$

$$+ \int_0^1 u'(x)v'(x)\,\mathrm{d}x = \int_0^1 u'(x)v'(x)\,\mathrm{d}x := \alpha(u, v),$$

where $\alpha(u,v)$ is a bilinear form.

Definition 1 (Bilinear form). A bilinear form is a function $B: V \times V \to K$, where V is a vector space and K is a scalar field, that is linear in each argument separately:

(1) $B(u+v, w) = B(u,w) + B(v,w)$ and $B(\lambda u, v) = \lambda B(u,v)$;
(2) $B(u, v+w) = B(u,w) + B(u,w)$ and $B(u, \lambda v) = \lambda B(u,v)$.

Definition 2 (Square-integrable function). A square-integrable function is denoted as L^2 and is defined as, $f : [a,b] \to \mathbb{R}$, square-integrable on [a, b] $\iff \int_a^b |f(x)|^2\,\mathrm{d}x < \infty$.

Taking into consideration the above information, a function space can be defined as a test space:

$$V = \{v \in L^2(0,1) : \alpha(u,v) < \infty \text{ and } v(0)=v(1)=0\} \quad (2)$$

and
$$u \in V \quad \text{such that} \quad a(u,v) = (f,v), \quad \forall v \in V, \tag{3}$$
where $L^2(0,1)$ is the space of square integrable functions in $[0,1]$.

The function v, which multiplies the PDE, is referred as "test function". The unknown function u, that needs to be approximated, is called "trial function" [5]. The trial and test spaces, V and $\hat{V}$, are defined as
$$V = \{v \in H^1(\Omega) : v = u_0 \text{ on } \partial\Omega\},$$
$$\hat{V} = \{v \in H^1(\Omega) : v = 0 \text{ on } \partial\Omega\},$$
where $H^1(\Omega)$ is a Hilbert space.

Definition 3 (Hilbert space). Hilbert space is a vector space, whose topology is defined using an inner product. A Hilbert space example is, $L^2(0,1)$, with inner product $(\cdot,\cdot)$. Hilbert spaces are complete metric spaces.

Definition 4 (Sobolev space). Sobolev space is a vector space of functions equipped with a norm, that is, a combination of L^p-norms of the function and its derivatives, up to a given order. We define the Sobolev spaces as
$$W_p^k(\Omega) := \{f \in L^1_{loc}(\Omega) : \|f\|_{W_p^k(\Omega)}\} < \infty,$$
where $L^1_{loc}(\Omega)$ is the set of locally integrable functions. In the cases where, $p = 2$, Sobolev space is a Hilbert space.

The function v is an arbitrary function and has a natural interpretation in the setting of Hilbert spaces. For the linear case, it is known that, if the weak form is $a(u,v) = (f,v)$, where $a(\cdot,\cdot)$ is bilinear, and $u \in C^2[0,1]$ and $f \in C^0[0,1]$ satisfy the weak form, then u also satisfies the strong form, with appropriate initial conditions [5].

According to the Ritz–Galerkin approximation, we have that, if $S \subset V$ is any finite dimensional subspace and we consider that (3) with V is replaced by S, we get
$$u_S \in S \quad \text{such that} \quad a(u_S, v) = (f,v), \quad \forall v \in S. \tag{4}$$
With the above, we can define a discrete scheme for approximating (1) and it has been proven that given $f \in L^2(0,1)$, the equation has a unique solution.

Let's consider the two-dimensional Dirichlet problem,

$$u_{xx} + u_{yy} = f, \text{ in } \Omega, \ u = 0, \quad \text{on } \partial\Omega, \qquad (5)$$

where $u = u(x, y)$, $f = f(x, y)$, Ω is a connected open region and $\partial\Omega$ is the boundary of Ω. The weak form of this problem is

$$(f, v) := \int_\Omega fv \, ds = -\int_\Omega \nabla u \cdot \nabla v := -\alpha(u, v), \qquad (6)$$

where ∇ denotes the gradient and $\cdot$ denotes the dot product in the two-dimensional plane. Where $\alpha(u, v)$ can be turned into an inner product on a suitable space $H_0^1(\Omega)$ of once differentiable functions of Ω that are zero on $\partial\Omega$. Additionally, $v \in H_0^1(\Omega)$ which is a Hilbert space.

1.2.2. *Error estimates*

Definition 5 (Energy norm).

$$||v||_E = \sqrt{a(v, v)}, \quad \forall v \in V. \qquad (7)$$

A relationship between the energy norm and the inner product is

$$|a(u, v)| \leq ||u||_E ||v||_E. \qquad (8)$$

It's proven that

$$||u - u_S||_E = \min\{||u - v||_E : v \in S\}. \qquad (9)$$

The last equation is defined as the error estimator where u is the solution and u_s is the approximate solution.

Definition 6 ($L^2(0, 1)$ norm).

$$||v|| = (v, v)^{\frac{1}{2}} = \left(\int_0^1 v(x)^2 dx\right)^{\frac{1}{2}}. \qquad (10)$$

The size of the error $u - u_s$ in this norm has been proven to be

$$||u - u_S|| \leq \varepsilon \, ||u - u_S||_E \leq \varepsilon^2 \, ||u''|| = \varepsilon^2 ||f|| \qquad (11)$$

with ε being a small number. $||u - u_s||_E$ is of order ε, whereas $||u - u_s||$ is of order ε^2. We could conclude that the $L^2(0, 1)$ norm is weaker than the energy one.

1.2.3. Piecewise polynomial spaces

We introduce linear polynomials to construct the Galerkin FEM and for that purpose we should partition our domain. Let's consider a partition of $[0, 1]$, e.g., $0 = x_0 < x_1 < \cdots < x_n = 1$, and S be the linear space of functions v, such that,

$S = \{v : [0, 1] \to \mathbb{R} : v \text{ is continuous, } v|_{[x_n, x_{n+1}]} \text{ is linear polynomial for } i = 1, \ldots, n \text{ and } v(0) = v(1) = 0\}.$

For each $i = 1, \ldots, n$ we can define ϕ_i and $\phi_i(x_j) = \delta_{ij} = $ the Kronecker delta, i.e.,

$$\begin{cases} \phi_i(x) = \dfrac{x - x_{i-1}}{x_i - x_{i-1}}, & x \in [x_{i-1}, x_i] \\ \phi_i(x) = \dfrac{x_{i+1} - x}{x_{i+1} - x_i}, & x \in [x_i, x_{i+1}] \\ \phi_i(x) = 0, & \text{otherwise.} \end{cases}$$

- $\{\phi_i : 1 \leq i \leq n\}$ is a nodal basis from S and it's called the **nodal basis** of S. The set $\{\phi_i\}$ is linearly independent, and it is used to define the functions of a discrete space.
- $\{v(x_i)\}$ are the **nodal values** of function v.
- $\{x_i\}$ are the **nodes**.
- $v_I = \sum_{i=1}^n v(x_i)\phi_i$, for $v \in C^0([0,1])$ and $v_i \in S$, is the **interpolant** of v.

Remark 1. If $v \in S \implies v = v_I$ since $v - v_I$ is linear on each $[x_{i-1}, x_i]$, and zero at the endpoints.

Theorem 1. If $h = max_{1 \leq i \leq n}(x_i - x_{i-1})$, then $\|u - u_I\|_E \leq CH\|u''\|$, for all $u \in V$, where C is independent of h and u. The proof can be found in Ref. [5].

The interpolant of a continuous function, f, for the space of all piecewise linear functions is defined as

$$f_I := \sum_e \sum_{j=0}^1 f(x_{i(e,j)}) \phi_j^e \qquad (12)$$

where

- $i(e, j)$ denotes a numbering scheme called the **global-to-local index** and aims to convert the coordinates of each element into the interval of $[0, 1]$. For example, in the interval $[x_{e-1}, x_e]$ for $e = 1, \ldots, n$ and $j = 0, 1$ where 0 corresponds to left end of the interval and 1 to the right one,

$$i(e, j) = e + j - 1.$$

- $\{\phi_j^e\}$ with $j = 0, 1$ are the basis functions for the linear functions, on the interval $I_e = [x_{e-1}, x_e]$:

$$\phi_j^e(x) = \phi_j\left(\frac{x - x_{e-1}}{x_e - x_{e-1}}\right)$$

as far as it concerns the bilinear form, $a(v, w)$,

$$a(v, w) = \sum_e a_e(v, w),$$

where

$$\sum_e a_e(v, w) := \int_{I_e} v' w' \, \mathrm{d}x$$

$$= \frac{1}{(x_e - x_{e-1})} \int_0^1 \left(\sum_j v_{i(e,j)} \phi_j\right)' \left(\sum_j w_{i(e,j)} \phi_j\right)' \, \mathrm{d}x$$

$$= \frac{1}{(x_e - x_{e-1})} \begin{pmatrix} v_{i(e,0)} \\ v_{i(e,1)} \end{pmatrix}^t K \begin{pmatrix} w_{i(e,0)} \\ w_{i(e,1)} \end{pmatrix},$$

where K is the local stiffness matrix,

$$K_{i,j} := \int_0^1 \phi'_{i-1} \phi'_{j-1} \, \mathrm{d}x \quad \text{for } i, j = 1, 2.$$

1.2.4. *An application to the Duffing equation*

In this section, we present an example, the Duffing equation, using the Galerkin FEM. The Duffing equation serves as one of the simplest mathematical models for describing the nonlinear behavior.

The equation is a second-order nonlinear ODE, with constant coefficient, non homogeneous with a periodically forced function:

$$\begin{cases} \dfrac{d^2 u}{dx^2} + \delta \dfrac{du}{dx} + \alpha u - \beta u^3 = \gamma \cos(wt), \quad u = u(t), \quad t \in [0, L] \\ u(0) = 0 \quad \text{and} \quad u'(L) = 1. \end{cases} \quad (13)$$

The parameter α is the linear stiffness coefficient, and the parameter β represents the nonlinearity in the restoring force. If $\beta = 0$, the equation describes simple harmonic oscillation. Overall, the equation represents a nonlinear spring that does not obey Hooke's law. Initially, we demonstrate the steps of utilizing FEM in this equation and then we obtain numerical results [15, 19].

We discretize the domain of the equation into a small number of elements to highlight the numerical approach. We start this procedure by dividing the function into a small number of elements for presentation purposes (four elements, five nodes) in the domain $[0, 1]$. Let $\phi = \phi(x)$ be the basis or test function. For each element $e = 1, 2, 3, 4$, we have two nodes t_i and t_{i+1}:

$$\int_{t_i}^{t_{i+1}} \left(\ddot{u} + \delta \dot{u} + \alpha u - \beta u^3 \right) \phi \, dt - \int_{t_i}^{t_{i+1}} \gamma \phi \cos(wt) \, dt = 0, \quad (14)$$

where ϕ_i and ϕ_{i+1} were defined earlier. Let

$$\phi(x) = \sum_{j=1}^{2} \phi_j(t) \quad \text{and} \quad u(t) = \sum_{i=1}^{2} \phi_i(t).$$

Introducing the basis functions into equation (13), where

$$h = t_{i+1} - t_i, \quad (15)$$

the equation can be written as follows:

$$k_{ij}^e = g_i^e,$$

where k_{ij}^e is the stiffness matrix and it is given as

$$k_{ij}^e = \int_{s_i}^{s_{i+1}} \left(-\dot{\phi}_i, \dot{\phi}_j + \delta \phi_i \phi_j + \alpha \phi_i \phi_j - \beta \phi_i^3 \phi_j \right) dt$$

and

$$g_i^e = \gamma \int_{t_i}^{t_{i+1}} \phi_j \cos(\omega t) \mathrm{d}t - \phi_j \dot{u}|_{t_i}^{t_{i+1}}.$$

For a typical element, the stiffness matrix is

$$K^e = \begin{bmatrix} k_{11} & k_{12} \\ k_{21} & k_{22} \end{bmatrix}.$$

The global system of matrices equals

$$K = \begin{pmatrix} k_{11}^1 & k_{12}^1 & 0 & 0 & 0 \\ k_{21}^1 & k_{22}^1 + k_{11}^2 & k_{12}^2 & 0 & 0 \\ 0 & k_{21}^2 & k_{22}^2 + k_{11}^3 & k_{12}^3 & 0 \\ 0 & 0 & k_{21}^3 & k_{22}^3 + k_{11}^4 & k_{12}^4 \\ 0 & 0 & 0 & k_{21}^4 & k_{22}^4 \end{pmatrix}.$$

The force and boundary integral vectors are given as

$$F = \begin{pmatrix} F_1^1 \\ F_2^1 + F_1^2 \\ F_2^2 + F_1^3 \\ F_2^3 + F_1^4 \\ F_2^4 \end{pmatrix}, \quad G = \begin{pmatrix} G_1^1 \\ 0 \\ 0 \\ 0 \\ G_2^4 \end{pmatrix}.$$

Finally, the system we obtain is

$$\boldsymbol{KU = F + G}, \quad \text{where } \mathbf{U} = \begin{pmatrix} u_1 \\ u_2 \\ u_3 \\ u_4 \\ u_5 \end{pmatrix}.$$

Due to the essential boundary conditions, we already know that $u(0) = u_1 = 0$. We should not forget that the approximate solution is $u_h = \sum_{i=1}^n u(t_i)\phi_i$, where $u(t_i) = u_i$ and $\phi_i(t) = \phi_i$.

For $u(0) = \dot{u}(L) = 0$, $f(t) = 0.2\cos t$, $\delta = 0$, $\alpha = 0.06$, $\gamma = 0.2$, $w = 1$ and $\beta = 0.0001$, the numerical solution is shown for $N = 160$

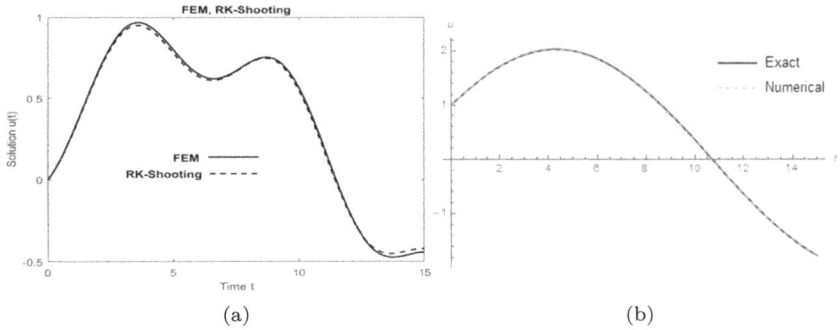

Figure 1. (a) Duffing equation with FEM and Runge–Kutta shooting method for $u(0) = u'(L) = 0$, $f(t) = 0.2\cos(\omega t)$, $\delta = 0$, $\alpha = 0.06$, $\gamma = 0.2$, $\omega = 1$ and $\beta = 0.0001$. (b) Comparing exact and numerical solution for the Duffing equation for $u(0) = 1, \dot{u}(0) = 0$, $f(t) = 0$, $\delta = 0$, $\alpha = 0.06$ and $\beta = 0.0001$.

nodes 119 elements and it is compared with the Runge–Kutta (RK) fourth-order shooting method for the same parameters, as depicted in Figure 1(a) [19, 24]. We can conclude that the FEM and RK numerical solutions, with these particular parameters, coincide.

For $u(0) = 1$, $\dot{u}(L) = -0.244$ and $f(t) = 0$ and all the other parameters as described above, we present the comparison between the exact solution and the corresponding numerical one, as shown in Figure 1(b):

$$\dot{u}(t) = \pm\sqrt{\alpha + \frac{\beta}{2} - \alpha u^2 - \frac{\beta}{2}u^4}. \tag{16}$$

All the numerical results of this section were obtained with the help of MATLAB and Mathematica software packages.

2. The Stokes and Navier–Stokes Problems

2.1. *The Stokes problem*

The linear Stokes equations are the limiting case for the Navier–Stokes equations, when Reynolds number, Re, tends to zero. Due to close relation with the nonlinear Navier–Stokes equations, the Stokes equations have attracted substantial attention from several researchers. We study the stationary Stokes problem for an incompressible fluid. Ω is a bounded open set of $\mathbb{R}^n$ (where $n = 2, 3$) with regular boundary and $\mathbf{f}$ is a square integrable function on Ω. We seek a solution $(\mathbf{u}, p) \in H_0^1(\Omega)^2 \times (L^2(\Omega)/\mathbb{R})$ of the problem [16, 20]

$$\begin{cases} -\nu \Delta \mathbf{u} + \nabla p = \mathbf{f} & \text{in } \Omega, \\ \text{div } \mathbf{u} = 0 & \text{in } \Omega, \\ \mathbf{u} = 0 & \text{on } \partial \Omega, \end{cases} \quad (17)$$

where $\mathbf{u} = (u, v)$ denotes the velocity vector, whereas $\mathbf{f} = (f_x, f_y)$ stands for the body force vector. Based on this problem, we introduce the error estimates (*a priori* and *a posteriori*) and we briefly discuss about the uniqueness of the solution for this problem. Our goal is to extend these arguments for the nonstationary case [4]. Following the finite element analysis, we obtain the following weak form:

$$\begin{cases} a(\mathbf{u}, \mathbf{v}) + b(p, \mathbf{v}) = (\mathbf{f}, \mathbf{v}), & \forall \mathbf{v} \in H_0^1(\Omega)^n, \ \mathbf{u} \in H_0^1(\Omega)^n, \\ b(\mathbf{u}, q) = a((u, p), (v, q)) & \forall q \in H^1(\Omega), \ p \in H^1(\Omega), \\ = \int_\Omega (\nabla \cdot u) q = 0, \ d\Omega = 0 \end{cases} \quad (18)$$

where $a(\mathbf{u}, \mathbf{v}) = \int_\Omega \nu \nabla \mathbf{u} \cdot \nabla \mathbf{v} \, d\Omega$ and $b(p, \mathbf{v}) = \int_\Omega p \, \nabla \mathbf{v} \, d\Omega$.

Given the finite dimensional subspaces, $V_h \subset H^1(\Omega)^n$ and $Q_h \subset H^1(\Omega)$, the discrete form is

$$\begin{cases} a(\mathbf{u}_h, \mathbf{v}_h) + b(p_h, \mathbf{v}_h) = (\mathbf{f}, \mathbf{v}_h), & \forall \mathbf{v}_h \in V_{0h}, \ \mathbf{u}_h \in V_{0h}, \\ b(\mathbf{u}_h, q_h) = 0, & \forall q_h \in Q_h, \ p_h \in Q_h, \end{cases} \quad (19)$$

where $V_{0h} = \{\mathbf{v}_h \in V_h : \mathbf{v}_h \vert_{\partial \Omega} = 0\}$.

We analyzed two different cases for triangular and quadrilateral elements, based on the number of nodes on each element [4]. We focus only on the Taylor–Hood method with six node triangular elements. The Taylor–Hood method utilizes second-order polynomials for the velocity and first-order polynomials for the pressure, at each element $(P_2 - P_1)$.

After finding a solution, for the problem under consideration, it is important to show that it is stable and how the input data can affect it. This can be done using the inf-sup condition, the Ladyzhenskaya–Babuska–Brezzi (LBB) condition. This is a condition for saddle point problems. Convergence is ensured for most discretization schemes for positive definite problems, but for saddle point problems, there are still discretizations that are unstable, due to spurious oscillations [22]. In such cases, a better approach is the local adaptation of the computational grid, briefly described in the following section [17]. To further discuss for the LBB condition, we introduce the following theorem.

Theorem 2. *If Ω is polygonal and $\Omega_h = \Omega$, $\Omega_h = \bigcup_i T_i$, where T_i are the triangles and h defines the length of greatest triangle side. If all triangles have at least one vertex which is not on $\partial\Omega$, and if V_h, Q_h are chosen as in the Taylor–Hood method, then there exists a constant C, independent of h, such that*

$$\sup_{\mathbf{v}_h \in V_{0h}} \frac{(\mathbf{v}_h, \nabla q_h)}{(\mathbf{v}_h, \mathbf{v}_h)^{\frac{1}{2}}} \geq C \left(\nabla q_h, \nabla q_h\right)^{\frac{1}{2}}, \quad \forall q_h \in Q_h. \tag{20}$$

This theorem follows the idea of the LBB condition, the proof depends on the choice of the elements and it can be found in Ref. [4]. An important question in solving such a problem is that of the existence and uniqueness of the solution of the problem. In this case, we focus on the discrete form of the problem under consideration, e.g., equation (19), where we can ensure the previous with the following Theorem [4].

Theorem 3. *Under the conditions of the previous theorem, Theorem 2, the discrete form, equation (19), has a unique solution $(\mathbf{u}_h, p_h)$ in $V_{0h} \times (Q_h/\mathbb{R})$.*

In the following section, we introduce and discuss error estimates of the Stokes problem.

2.1.1. A priori error estimates

The *a priori* error estimates express the error in terms of the regularity of the exact unknown solution. They provide important information about the order of convergence of an FEM. A *posteriori* error estimates express the error in terms of computable quantities, such as the residual error and the solution of an auxiliary dual problem. They contribute to the adaptation of the computational grid, as described in the following section [17]. A theorem that provides *a priori* error estimates for the discrete form of the stationary Stokes problem using Taylor–Hood elements $(P_2 - P_1)$ is as follows.

Theorem 4. *Let Ω be a polygon, and $\Omega_h = \Omega$, for all h. We assume that each element of $\mathcal{T}_h$ (set of triangles) has at least one vertex not on the boundary. Then the following inequalities are valid:*

$$\|\nabla(\mathbf{u} - \mathbf{u}_h)\| \leqslant h^2 K \left(\|\mathbf{u}\|_{H^3(\Omega)^N} + \|p\|_{H^2(\Omega)/R}\right),$$

$$\|\nabla(p - p_h)\| \leqslant h K \left(\|\mathbf{u}\|_{H^3(\Omega)^N} + \|p\|_{H^2(\Omega)/R}\right).$$

Similar inequalities can be found in the case where we have quadrilaterals [4].

2.1.2. The backward-facing step problem

The backward-facing step (BFS) is a "test problem" widely known for its application on internal flows. In this problem, flow separation is caused due to sudden changes in the geometry. This creates a recirculation zone close to the step wall, and downstream a reattachment point. In a two-dimensional BFS geometry, the fluid flow can be distinguished into three regions: the shear layer, the separation bubble and the reattachment zone [1, 6].

Due to the adverse pressure gradient that develops in the thin shear layer, the characteristics of a BFS flow begin with an upstream boundary layer that separates at the edge of the backward-facing step. The region where the shear layer develops is referred to as the shear layer region. This flow causes the formation of a recirculation zone, which is located between the shear layer and the adjacent wall. Eventually, the shear layer curves down toward the wall and reattaches at the so-called reattachment point. The horizontal distance between the step and the reattachment point is defined as the "reattachment length". Due to the oscillatory motion of the shear layer, the reattachment length is unsteady. Consequently, the reattachment point spreads within a zone, called reattachment zone [1, 6]. In this problem, the flow parameters of interest are

$$\begin{cases} u = \text{horizontal velocity component}, \quad v = \text{vertical velocity component}, \\ L = \text{length}, \quad h = \text{the step height}, \quad H = \text{the whole height}, \\ \mu = \text{viscosity}, \quad \rho = \text{density}, \end{cases}$$

with a typical set of boundary conditions, as shown in equation (21),

$$\begin{aligned} \mathbf{u} = \mathbf{u}_0 \quad &\text{on } \Gamma_D, \\ \nabla \mathbf{u} \cdot \mathbf{n} + p\mathbf{n} = \mathbf{g} \quad &\text{on } \Gamma_N, \end{aligned} \qquad (21)$$

where

- Γ_D are the **Dirichlet** conditions or the essential boundary conditions,
- Γ_N are the **Neumann** conditions or the natural boundary conditions.

For further analysis, it is easier to write equations (17) to the following form for finding $(u, p) \in W$ such that

$$a((\mathbf{u}, p), (\mathbf{v}, q)) = L(\mathbf{v}, q)$$

for all $(\mathbf{v}, q) \in W$, where

$$a((\mathbf{u}, p), (\mathbf{v}, q)) = \int_\Omega \nabla \mathbf{u} \cdot \nabla \mathbf{v} - (\nabla \cdot \mathbf{v})p + (\nabla \cdot \mathbf{u})q \, dx,$$

$$L(\mathbf{v}, q) = \int_\Omega \mathbf{f} \cdot \mathbf{v} \, dx + \int_{\partial \Omega_N} \mathbf{g} \cdot \mathbf{v} \, ds.$$

The space W is a mixed (product) function space, $W = V \times Q$, such that $\mathbf{u} \in V$ and $q \in Q$. We use the Galerkin FEM to analyze the velocity and pressure on this test problem. Figure 2(a) shows the domain Ω and the dimensions of the backward-facing step. Figure 2(b) depicts the computational mesh.

The numerical results are shown in Figure 3(a). The figure depicts the velocity magnitude for the backward-facing step problem. We also present the streamlines in the domain to visualize the recirculation zone close to the wall. It is observed that the maximum velocity is at the entrance of the channel and the velocity drops rapidly as the domain expands. In Figure 3(b), we visualize the pressure field with the classical Galerkin FEM. It can be observed that the pressure field is relatively smooth using the Taylor–Hood elements.

2.2. The Navier–Stokes problem

Most of every real situation in fluid flows is characterized by the Navier–Stokes equations that are the model of nonlinear PDEs, so the

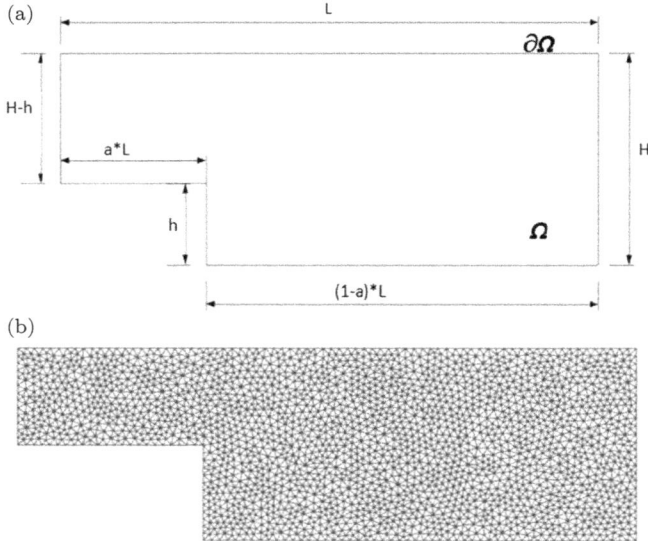

Figure 2. (a) The domain Ω and the dimensions of the backward step. (b) The computational mesh composed of approximately 6060 elements.

importance to solve this particular system of equations is recognizable. Due to the nonlinearity of the problems that are described by the Navier–Stokes equations, an exact solution is impossible to be obtained. However, it is very useful to describe and analyze the physics of fluid flow problems. The FEM constitutes an effective way to find a numerical solution to these equations.

The time-dependent and incompressible Navier–Stokes equations are given as

$$\begin{aligned} \frac{\partial \boldsymbol{u}}{\partial t} + (\boldsymbol{u} \cdot \nabla)\boldsymbol{u} - \nu \Delta \boldsymbol{u} + \nabla p &= \boldsymbol{f} \quad \text{in } \Omega \times (0, T), \\ \text{div } \boldsymbol{u} &= 0 \quad \text{in } \Omega \times (0, T), \\ \boldsymbol{u} &= \boldsymbol{0} \quad \text{on } \partial\Omega \times (0, T), \\ \boldsymbol{u}(\cdot, 0) &= \boldsymbol{u}_0 \quad \text{in } \Omega, \end{aligned} \qquad (22)$$

where $\boldsymbol{u}$ is the velocity field, p is the zero-mean pressure, $\boldsymbol{f}$ is an external force field and ν is the kinematic viscosity. These equations, equations (22), describe an incompressible fluid flow in Ω. Compared to the Stokes equations, we have here to deal with an additional nonlinearity and a time derivative. To obtain the weak formulation,

The Finite Element Method with Applications to Fluid Mechanics 155

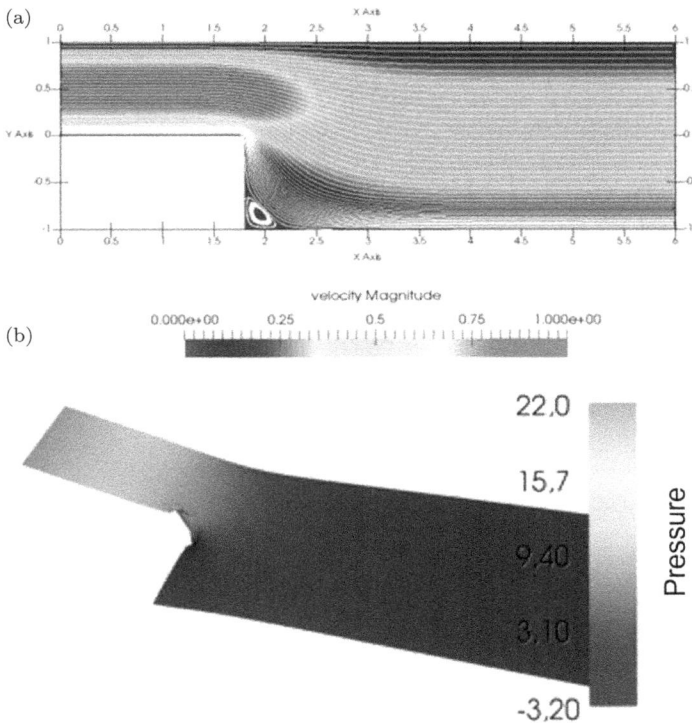

Figure 3. (a) The velocity magnitude of the Stokes equation on the backward step problem. (b) The pressure of the Stokes equation on the backward-facing step problem.

we multiply the momentum equations with a test function, v, defined in a suitable space V, and integrate both members with respect to the domain Ω:

$$\int_\Omega \frac{\partial u}{\partial t} \cdot v \mathrm{d}\Omega - \int_\Omega \nu \Delta u \cdot v \mathrm{d}\Omega + \int_\Omega (u \cdot \nabla) u \cdot v \mathrm{d}\Omega$$
$$+ \int_\Omega \nabla p \cdot v \mathrm{d}\Omega = \int_\Omega f \cdot v \mathrm{d}\Omega. \qquad (23)$$

Remark 2. Integrating by parts and using Gauss' divergence theorem:

$$- \int_\Omega \nu \Delta u \cdot v \mathrm{d}\Omega = \int_\Omega \nu \nabla u \cdot \nabla v \mathrm{d}\Omega - \int_{\partial\Omega} \nu \frac{\partial u}{\partial \hat{n}} \cdot v \mathrm{d}s$$

$$\int_\Omega \nabla p \cdot v \mathrm{d}\Omega = - \int_\Omega p \nabla \cdot v \mathrm{d}\Omega + \int_{\partial\Omega} p v \cdot \hat{n} \mathrm{d}s$$

Using these relations, (23) is rearranged to

$$\int_\Omega \frac{\partial u}{\partial t} \cdot v \mathrm{d}\Omega + \int_\Omega \nu \nabla u \cdot \nabla v \mathrm{d}\Omega + \int_\Omega (u \cdot \nabla) u \cdot v \mathrm{d}\Omega - \int_\Omega p \nabla \cdot v \mathrm{d}\Omega$$
$$= \int_\Omega f \cdot v \mathrm{d}\Omega + \int_{\partial\Omega} \left(\nu \frac{\partial u}{\partial \hat{n}} - p\hat{n} \right) \mathrm{d}s \cdot v \mathrm{d}\Omega \quad \forall v \in V, \qquad (24)$$

where $\hat{n}$ is the unit normal vector. In a similar manner, we multiply the continuity equation with a test function q, belonging to a space Q and integrated in the domain Ω:

$$\int_\Omega q \nabla \cdot u \mathrm{d}\Omega = 0. \quad \forall q \in Q.$$

The space functions are chosen as

$$V = \left[H_0^1(\Omega) \right]^d = \left\{ v \in \left[H^1(\Omega) \right]^d : v = 0 \text{ on } \Gamma_D \right\},$$
$$Q = L^2(\Omega).$$

Due to the set of boundary conditions,

$$\mathbf{u} = 0 \quad \text{on } \Gamma_D,$$
$$\nabla \mathbf{u} \cdot \mathbf{n} + p\mathbf{n} = \mathbf{g} \quad \text{on } \Gamma_N,$$

the integral on the boundary can be written as

$$\int_{\partial\Omega} \left(\nu \frac{\partial u}{\partial \hat{n}} - p\hat{n} \right) \cdot v \mathrm{d}s = \int_{\Gamma_D} \left(\nu \frac{\partial u}{\partial \hat{n}} - p\hat{n} \right) \cdot v \mathrm{d}s$$
$$+ \int_{\Gamma_N} \left(\nu \frac{\partial u}{\partial \hat{n}} - p\hat{n} \right) \cdot v \mathrm{d}s = \int_{\Gamma_N} g \cdot v \mathrm{d}s,$$

where

- $\int_{\Gamma_D} \left(\nu \frac{\partial u}{\partial \hat{n}} - p\hat{n} \right) \cdot v \mathrm{d}s = 0$,
- $\int_{\Gamma_N} \left(\nu \frac{\partial u}{\partial \hat{n}} - p\hat{n} \right) \cdot v \mathrm{d}s = -g \cdot v$.

Eventually, the weak form of the Navier–Stokes equations is expressed as

$$\begin{cases} \int_\Omega \dfrac{\partial \boldsymbol{u}}{\partial t} \cdot \boldsymbol{v} \mathrm{d}\Omega + \int_\Omega \nu \nabla \boldsymbol{u} \cdot \nabla \boldsymbol{v} \mathrm{d}\Omega + \int_\Omega (\boldsymbol{u} \cdot \nabla)\boldsymbol{u} \cdot \boldsymbol{v} \mathrm{d}\Omega - \int_\Omega p \nabla \cdot \boldsymbol{v} \mathrm{d}\Omega \\ = \int_\Omega \boldsymbol{f} \cdot \boldsymbol{v} \mathrm{d}\Omega + \int_{\Gamma_N} \boldsymbol{g} \cdot \boldsymbol{v} \mathrm{d}s \quad \forall \boldsymbol{v} \in V \\ \int_\Omega q \nabla \cdot \boldsymbol{u} \mathrm{d}\Omega = 0, \quad \forall q \in Q. \end{cases}$$

Existence and uniqueness of the solution of the problem is discussed in the following theorem:

Theorem 5. *If $f \in \left[L^2(0,T;V')\right]^d$ and $u_0 \in H$, there exists a weak solution to the Navier–Stokes equations, equation (22), that satisfies* [23]

$$\boldsymbol{u} \in L^2(0,T;\boldsymbol{V}) \cap L^\infty(0,T;\boldsymbol{H}),$$

where

$$\boldsymbol{V} = \left\{ v \in \left[H_0^1(\Omega)\right]^d : \mathrm{div}\, v = 0 \right\},$$

$$\boldsymbol{H} = \left\{ v \in \left[L_0^2(\Omega)\right]^d : \mathrm{div}\, v = 0 \right\}.$$

In the case of space dimension $d = 2$, this solution is unique and

$$\boldsymbol{u} \in C(0,T;\boldsymbol{H}),$$
$$\boldsymbol{u}' \in L^2(0,T;\boldsymbol{V}').$$

In three dimensions ($d = 3$), uniqueness is an open question.

2.2.1. An application to the Poiseuille flow

In this section, we study the incompressible Navier–Stokes flow between two long parallel plates, with no-slip condition on both walls which are spaced apart in height $2h$ (Poiseuille flow). We assume a constant flow, with density ρ and viscosity μ to be constant. Fluid is introduced with a parabolic velocity profile at the inlet of the domain, $u(y) = 4y(y-1)$. With the aforementioned assumptions, we note that

$$\begin{cases} y = -h \quad \text{or} \quad y = +h, \\ u(x,y) = 0, \quad v(x,y) = 0, \quad F_x = F_y = 0. \end{cases}$$

With the above boundary conditions, we obtain that

$$\frac{\partial u}{\partial x} = 0 \quad \text{i.e.,} \quad u = u(y), \tag{25}$$

$$\rho u_x \frac{\partial u_x}{\partial x} = -\frac{\partial p}{\partial x} + \mu \left(\frac{\partial^2 u_x}{\partial x^2} + \frac{\partial^2 u_x}{\partial y^2} \right), \tag{26}$$

$$-\frac{\partial p}{\partial y} = 0 \quad \text{i.e.,} \quad p = p(x). \tag{27}$$

Thus, we obtain

$$\frac{\partial p}{\partial x} = c = \mu \frac{d^2 u}{dy^2}. \tag{28}$$

Solving the differential equation (28), and applying the boundary conditions, yields to the analytical solution of the problem under consideration that is a parabolic profile as expected:

$$u(y) = -\frac{c}{2\mu}(h^2 - y^2). \tag{29}$$

The velocity has a parabolic profile, as shown in equation (29). If $c > 0$, the maximum velocity value is at the center of the domain, for $y = 0$ and $u_{\max} = -\frac{ch^2}{2\mu}$, that indicates a favorable pressure drop within this region, given by $\frac{\partial p}{\partial x} = c_2 = -\frac{ch^2}{2\mu}$ and depicted in the schematic of Figure 4(a).

We obtain the analytical solution for the Poiseuille problem, as shown in equation (29). Additionally, solving the same problem with the FEM and the software package FEniCS, we obtain the same parabolic profile, as shown in Figure 4(b). The FEM results were obtained for a computational mesh of approximately 2754 triangular elements. The maximum velocity is at the centerline of the domain,

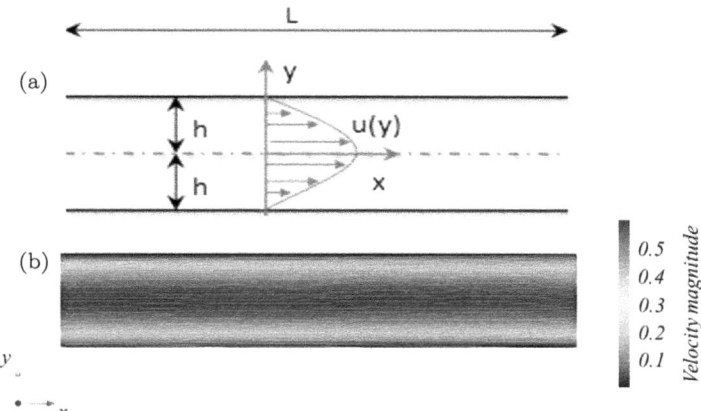

Figure 4. (a) Fluid flow schematics between two parallel plates providing a parabolic profile, known as Poiseuille flow. (b) The numerical solution of the Poiseuille flow, using FEM and the software package FEniCS.

and at the plates, we have zero velocity, no-slip condition, as discussed for the analytical solution.

3. Discontinuous and Adaptive FEM

3.1. *The discontinuous Galerkin FEM*

The discontinuous Galerkin (DG) FEMs are used in the numerical analysis of differential equations. They serve as an improvement to both the finite element and finite volume methods and they apply to a plethora of problems in fluid dynamics. The basis functions used are discontinuous. These methods, allowing discontinuities, apply with great flexibility and benefits, handling complex geometries, irregular meshes and polynomial approximations of different degrees in each element [7]. The discontinuous methods are distinguished from the continuous ones in integrating flux terms over interior faces.

The DG methods first arose in solving PDEs in the early 1970s, with continuous improvements on elliptic problems, throughout the decade [2, 13]. In the 1990s, extensions of the DG method, dealt with the compressible flow and nonlinear hyperbolic conservation laws.

The analysis and development of such methods are topics of active research [8].

The following example provides an approximate solution u_h of an ODE using the DG FEM. Consider the initial-value problem [7]:

$$\begin{cases} \dfrac{\mathrm{d}}{\mathrm{d}t}u(t) = f(t)u(t), & t \in (0,T), \\ u(0) = u_0. \end{cases} \qquad (30)$$

Initially, we divide the interval $I := (0,T)$ into subintervals $I_i := (t_i, t_{i+1})$, for $i = 0, 1, \ldots, N-1$. Next, we seek the approximate solution u_h, which on the interval I_i is a polynomial of degree at most k^i, requiring that

$$-\int_{I_i} u_h(s) \frac{\mathrm{d}}{\mathrm{d}s} v(s)\,\mathrm{d}s + \hat{u}_h v \big|_{t_i}^{t_{i+1}} = \int_I u(s)f(s)v(s)\,\mathrm{d}s, \qquad (31)$$

for polynomials v of degree at most k^i, where the quantity $\hat{u}_h$ is

$$\hat{u}_h := \begin{cases} u_0, & \text{if } t_i = 0 \\ \lim_{\varepsilon \to 0} u_h(t_i - \varepsilon), & \text{otherwise.} \end{cases}$$

The goal is to find a suitable definition of the *numerical trace*, $\hat{u}_h$, using discontinuous approximations, u_h, and applying the Galerkin weak formulation. The DG FEMs are *consistent* methods. So, when replacing the approximate solution u_h with the exact solution u, in the weak formulation of equation (31), the equation is satisfied. That can be applied if $\hat{u} = u$.

Multiplying the ODE with u and integrating on the domain $(0,T)$, we get

$$\frac{1}{2}u^2(T) - \frac{1}{2}u_0^2 = \int_0^T f(s)u^2(s)\,\mathrm{d}s.$$

Substituting $v = u_h$ in the weak formulation, equation (31), and integrating by parts, we obtain the following:

$$\sum_{i=0}^{N-1} \left(-\frac{1}{2}u_h^2 + \hat{u}_h u_h\right)\bigg|_{t_i}^{t_{i+1}} = \frac{1}{2}u_h^2(T^-) + \Theta_h(T')$$

$$-\frac{1}{2}u_0^2 = \int_0^T f(s)u_h^2(s)\,\mathrm{d}s,$$

where

$$\Theta_h(T) = -\frac{1}{2}u_h^2\left(T^-\right) + \sum_{i=0}^{N-1}\left(-\frac{1}{2}u_h^2 + \hat{u}_h u_h\right)\bigg|_{t_i}^{t_{i+1}} + \frac{1}{2}u_0^2.$$

The stability is gained when $\Theta_h(T) \geq 0$, so we set

$$u_h(t) = u_0, \quad t < 0.$$

We further introduce the following definitions:

Definition 7. We define as the average quantity of the discrete function u_h, $\{u_h\} = \frac{1}{2}\left(u_h^- + u_h^+\right)$. Additionally, we define the difference of the discrete function u_h on the faces of each element as $[u_h] = u_h^- - u_h^+$.

Definition 8. We define the limit of the discrete function u_h at the faces as $u_h^\pm(t) = \lim_{\varepsilon \to 0} u_h(t \pm \varepsilon)$.

Remark 3. The above definitions yield to the following equation:

$$\left[u_h^2\right] = 2\left\{u_h\right\}[u_h].$$

Taking into consideration the above definitions,

$$\Theta_h(T) = -\frac{1}{2}u_h^2\left(T^-\right) + \left(-\frac{1}{2}u_h^2\left(T^-\right) + \hat{u}_h(T)u_h\left(T^-\right)\right)$$

$$+ \sum_{i=1}^{N-1}\left(-\frac{1}{2}\left[u_h^2\right] + \hat{u}_h\left[u_h\right]\right)(t_i)$$

$$- \left(-\frac{1}{2}u_h^2\left(0^+\right) + \hat{u}_h(0)u_h\left(0^+\right)\right) + \frac{1}{2}u_0^2$$

$$= \left(\hat{u}_h(T) - u_h\left(T^-\right)\right)u_h\left(T^-\right) + \sum_{i=1}^{N-1}\left(\left(\hat{u}_h - \{u_h\}\right)[u_h]\right)(t_i)$$

$$- \left(\hat{u}_h(0) - u_0\right)u_h\left(0^+\right) + \frac{1}{2}[u_h]^2(0).$$

Based on the above, we have for $\widehat{u}_h$

$$\widehat{u}_h(t_i) = \begin{cases} u_0, & \text{if } t_i = 0 \\ (\{u_h\} + C^i [u_h])(t_i), & \text{if } t_i \in (0, T) \\ u_h(T-), & \text{if } t_i = T \end{cases}$$

$C^i \geq 0$ and $C^0 = 1/2$,

$$\Theta_h(T) = \sum_{i=0}^{N-1} C^i [u_h]^2 (t_i) \geq 0.$$

The accuracy of the DG method depends on the choice of C^i. The order of the DG method at the points t_i is $2k+1$, if we take $C^i = \dfrac{1}{2}$. For $C^i = 0$, it can be proven that the order of the DG method is $2k + 2$ [10, 13]. However, for $C^i \equiv 1/2$, DG methods are consistent and stable. As a consequence, we can handle different types of approximations in different elements.

For higher-order problems, the first step is the discretization of the domain of interest in triangles, denoted by $\mathcal{T}$, such triangulation. We further seek a discontinuous approximate solution u_h that in each element K, of the triangulation $\mathcal{T}$, belongs to space $V(K)$ [10, 13].

3.1.1. Application to the Poisson and Stokes equations

Next, we present the methodology of applying the Discontinuous Galerkin (DG) to the Poisson and Stokes equations [18]. For the Poisson equation, we consider a DG FEM. For this problem, we apply Dirichlet boundary conditions:

$$-\Delta u = f \quad \text{in } \Omega,$$
$$u = u_D \quad \text{on } \partial\Omega.$$

Next, we rewrite the Poisson equation to the weak form:

$$\int_\Omega -(\Delta u)v \, d\Omega = \int_\Omega fv \, d\Omega.$$

Assume that we have a mesh $\mathcal{T}$ of Ω with cells $\{K\}$ and split the left integral into sum over cell integrals:

$$\sum_{K \in \mathcal{T}} \int_K -(\Delta u) v \, \mathrm{d}K = \int_\Omega fv \, \mathrm{d}\Omega.$$

Now integrating by parts,

$$\sum_{K \in \mathcal{T}} \int_K \nabla u \cdot \nabla v \, \mathrm{d}K - \sum_{K \in \mathcal{T}} \int_{\partial K} (\nabla u \cdot n) v \, \mathrm{d}s = \int_\Omega fv \, \mathrm{d}\Omega,$$

where n is the outward unit normal vector. Before introducing the DG method, it is necessary to point out the definitions relevant to this method [9].

Definition 9 (Average and Jump operator). We define the average quantity of the function v as $\langle v \rangle = \frac{1}{2}(v^+ + v^-)$. We also define the difference, or jump, of the function v on the element faces as $[vn] = v^+ n - v^- n$ in Ω and $[vn] = vn$ on $\partial\Omega$.

Definition 10 (Jump identity). We define for the functions u and v, the jump identity, as $[uv] = [u]\langle v \rangle + \langle u \rangle [v]$ in Ω.

We consider a DG formulation to approximate the problem. For this formulation, the approximation space is made of piecewise discontinuous polynomials, such as

$$V = \left\{ v \in L^2(\Omega) : v|_K \in Q_p(K) \text{ for all } K \in \mathcal{T} \right\},$$

where $\mathcal{T}$ is the set of all cells K of the mesh and $Q_p(K)$ is a polynomial space of p-degree defined on a cell K. We need to introduce appropriate notation, in order to write the weak form of the problem under consideration. The sets of interior and boundary facets associated with the mesh $\mathcal{T}$ are denoted as $\mathcal{F}_i$ and $\mathcal{F}_e$, respectively. With v^+ and v^- being the restrictions of $v \in V$ to the cells K^+, K^- that share the same interior facet in $\mathcal{F}_i$ and n^+, n^-, the facet outward unit normals from the perspective of either K^+ or K^-, respectively [9].

With the above introduced notation, the weak form associated with the interior penalty formulation for the Poisson equation is presented as

$$a(v,u) = \sum_{K \in \mathcal{T}} \int_K \nabla v \cdot \nabla u \, \mathrm{d}s$$

$$- \sum_{K \in \mathcal{F}_e} \int_K v(\nabla u \cdot n) \mathrm{d}s - \sum_{K \in \mathcal{F}_e} \int_K (\nabla v \cdot n) u \, \mathrm{d}s$$

$$+ \sum_{K \in \mathcal{F}_e} \frac{\alpha}{h} \int_K v \cdot u \, \mathrm{d}s$$

$$- \sum_{K \in \mathcal{F}_i} \int_K [vn] \cdot \langle \nabla u \rangle \mathrm{d}s - \sum_{K \in \mathcal{F}_i} \int_K \langle \nabla v \rangle \cdot [un] \mathrm{d}s$$

$$+ \sum_{K \in \mathcal{F}_i} \frac{\alpha}{h} \int_K [vn] \cdot [un] \mathrm{d}s,$$

and the right-hand side is presented as

$$L(v) = \int_\Omega v f \, \mathrm{d}\Omega.$$

Remark 4. We provide the following equation for the element boundary:

$$\sum_{K \in \mathcal{T}} \int_{\partial K} (\nabla u \cdot n) v \, \mathrm{d}s = \sum_{K \in \mathcal{F}_i} \int_K \left(\nabla u^+ \cdot n^+ v^+ + \nabla u^- \cdot n^- v^- \right) \mathrm{d}s$$

$$+ \sum_{K \in \mathcal{F}_e} \int_K (\nabla u \cdot n) v \, \mathrm{d}s.$$

Remark 5. The terms, $\sum_{K \in \mathcal{F}_e} \frac{\alpha}{h} \int_K v \cdot u \, \mathrm{d}s$ and $\sum_{K \in \mathcal{F}_i} \frac{\alpha}{h} \int_K [vn] \cdot [un] \mathrm{d}s$, are artificially added in the formulation. The constant α is a stabilization parameter. This parameter should be chosen large enough such that the bilinear form $a(\cdot, \cdot)$ which is stable and continuous, where h is a measure for the average of the mesh size defined as $h = (h^+ + h^-)/2$, for the two neighboring cells K^+ and K^-, with the given interior facet.

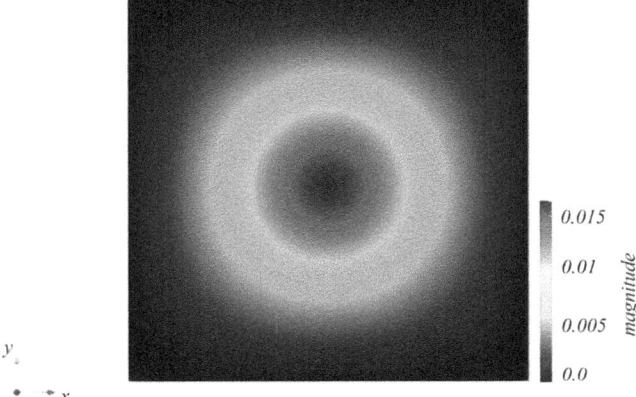

Figure 5. Numerical solution of the Poisson equation with DG FEM and homogeneous Dirichlet conditions.

When applying the above formulation, e.g., the DG FEM, using the software program FeniCS for the Poisson equation, we obtain the results shown in Figure 5. In the domain, we apply homogeneous Dirichlet conditions on the boundary. Internally, a Gaussian distribution is applied for the source term, defined by the function f, on the right-hand side of the Poisson equation, given by the following expression:

$$f(x,y) = c\exp\left(-\frac{(x-\alpha)^2 + (y-b)^2}{d}\right), \qquad (32)$$

where $c = 10$, $\alpha = b = 0.5$ and $d = 0.02$.

Following the same methodology as before, we use DG FEM for the Stokes problem:

$$\begin{cases} -\nu\Delta \mathbf{u} + \nabla p = \mathbf{f} & \text{in } \Omega, \\ \mathbf{u} = 0 & \text{on } \partial\Omega. \end{cases}$$

Consider the function spaces V, equipped with discontinuous functions, and Q, with continuous ones:

$$V = \left\{ \mathbf{v} \in \left(L^2(\Omega)\right)^d : v_i \in P_k(K) \quad \forall K \in \mathcal{T}, 1 \leq i \leq d \right\},$$
$$Q = \left\{ q \in H^1(\Omega) : q \in P_j(K) \quad \forall K \in \mathcal{T} \right\}.$$

We presented earlier the weak formulation:
$$a((\mathbf{u},p),(\mathbf{v},q)) = L(\mathbf{v},q)$$
for all $(\mathbf{v},q) \in W$, where
$$a((\mathbf{u},p),(\mathbf{v},q)) = \int_\Omega (\nu \nabla \mathbf{u} \cdot \nabla \mathbf{v} - \nabla p \cdot \mathbf{v} + \nabla q \cdot \mathbf{u}) d\Omega,$$
$$L((\mathbf{v},q)) = \int_\Omega \mathbf{f} \cdot \mathbf{v} \, d\Omega + \int_{\partial \Omega_N} \mathbf{g} \cdot \mathbf{v} \, ds.$$

The space W is considered as a mixed (product) function space, $W = V \times Q$, such that $u \in V$ and $q \in Q$.

In order to formulate the Stokes problem with the DG FEM, we consider both discontinuous functions as well as basis functions with varying polynomial orders. The particular bilinear and linear forms for the Stokes equation with DG method can be formulated as

$$a(\boldsymbol{v},q;\boldsymbol{u},p)$$
$$= \sum_{K \in \mathcal{T}} \int_K \nabla \boldsymbol{v} \cdot \nabla \boldsymbol{u} \, dK + \sum_{K \in \mathcal{T}} \int_K \boldsymbol{v} \cdot \nabla p \, dK - \sum_{K \in \mathcal{T}} \int_K \nabla q \cdot \boldsymbol{u} \, dK$$
$$+ \sum_{K \in \mathcal{F}_i} \int_K q[\boldsymbol{u} \cdot \boldsymbol{n}] \, ds - \sum_{K \in \mathcal{F}_i} \int_K \nu[\boldsymbol{v}] \cdot \langle \nabla \boldsymbol{u} \rangle \, ds$$
$$- \sum_{K \in \mathcal{F}_i} \int_K \nu \langle \nabla \boldsymbol{v} \rangle \cdot [\boldsymbol{u}] \, ds + \sum_{K \in \mathcal{F}_i} \int_K \nu \frac{\alpha}{h}[\boldsymbol{v}] \cdot [\boldsymbol{u}] \, ds$$
$$+ \sum_{K \in \mathcal{F}_e} \int_K q \boldsymbol{u} \cdot \boldsymbol{n} \, ds - \sum_{K \in \mathcal{F}_e} \int_K \nu \boldsymbol{v} \cdot \nabla \boldsymbol{u} \, ds$$
$$- \sum_{K \in \mathcal{F}_e} \int_K \nu \nabla \boldsymbol{v} \cdot \boldsymbol{u} \, ds + \sum_{K \in \mathcal{F}_e} \int_K \nu \frac{\alpha}{h} \boldsymbol{v} \cdot \boldsymbol{u} \, ds$$

and the right-hand side of the Stokes problem is given as
$$L(\boldsymbol{v},q) = \int_\Omega \boldsymbol{v} \cdot \boldsymbol{f} \, d\Omega.$$

The above formulation is constructed in a same manner as the Poisson equation with the terms $\sum_{K \in \mathcal{F}_i} \int_K \nu \frac{\alpha}{h}[\boldsymbol{v}] \cdot [\boldsymbol{u}] \, ds$ and

$\sum_{K \in \mathcal{F}_e} \int_K \nu \frac{\alpha}{h} \boldsymbol{v} \cdot \boldsymbol{u} \, \mathrm{d}s$ being artificially added for the stability of the equation.

3.2. Adaptive mesh refinement FEM

The adaptive mesh refinement (AMR) is an approach for increasing the accuracy of the numerical solution in certain sensitive regions of the discretized domain. Numerical solutions sometimes reveal accuracy problems to specific regions of the grid or mesh. However, some problems would be better suited if specific computational areas which needed precision could be refined only in the regions requiring the added precision rather than a uniform region. There are widely used methods that omit this problem, called Adaptive Finite Element Mesh Refinement Methods (AFEMs) with a range of applications to engineering problems. AFEMs can be classified into three categories [25]:

- In the h-refinement AFEM, we use the same type of finite element, but their sizes are continuously divided according to a geometric parameter, such as the element length or diameter. Among the three categories referred, this is the simplest and more common one to use.
- In the p-refinement AFEM, we increase the order of the polynomial basis functions, but the mesh element size is kept the same.
- In r-refinement AFEM, we keep the number of mesh nodes and elements the same, but the nodes are relocated to problematic areas needed to be optimized.

These methods can be combined, such as the hp-refinement method. We can use the AFEM to obtain a numerical solution of the desired accuracy, with less computing time, since lower degrees of freedom are needed. For the h-type AFEM, the main concept is to determine the regions to insert the nodes, so as to balance the numerical errors of the FEM solution through a local *a posteriori* error estimate procedure. The *a posteriori* error estimation obtains an estimated error for each element. This plays an important role in guiding the refinement procedures for AFEM.

The process consists of calculating the error indicators for each element of the mesh. Then, a selected number of elements in the domain, e.g., the elements with the largest error indicators, are finally refined. This process is repeated several times until the termination conditions are satisfied. Such conditions could be the maximum refinement number or maximum node number in the mesh. The elements can be refined with several methods, such as by bisection, trisection, regular refinement or any combinations of these methods [25].

More precisely, let's consider an *a posteriori* estimator for the Stokes problem. It can be shown that the discrete solution coincides with the continuous one. To define the Stokes reconstruction, we provide the definition of the Stokes reconstruction operator [12]. *A posteriori* error estimates express the error in terms of important quantities, such as the residual error equations and the solution of an auxiliary dual problem [14]. By using the classical Galerkin method, the FEM approximation $u_h \in V^h$ is the solution of

$$a(u_h, v) = L(v), \quad \forall v \in V^h.$$

The numerical error in the approximate solution u_h of u is defined as the function $e \in V$ such that

$$e = |u - u_h|.$$

The residual errors are defined as $r(v)$, where

$$\begin{aligned} r(v) &= L(v) - a(u_h, v) = a(u, v) - a(u_h, v) \\ &= a(u - u_h, v) \leqslant C \|u - u_h\|_V \|v\|_V, \quad v \in V. \end{aligned} \quad (33)$$

Furthermore,

$$\begin{aligned} \alpha \|u - u_h\|_V^2 &\leqslant a(u - u_h, u - u_h) \\ &= a(u, u - u_h) - a(u_h, u - u_h) \\ &= L(u - u_h) - a(u_h, u - u_h) = r(u - u_h). \end{aligned} \quad (34)$$

Remark 6. We note that the residual error $r(v)$ vanishes for all $v \in V^h$, i.e.,

$$r(v) = 0, \quad \forall v \in V^h.$$

This yields the following orthogonality property:

$$a(e, v) = 0, \quad \forall v \in V^h.$$

Combining (33), (34) yields

$$\alpha \|u - u_h\|_V \leqslant \|r\|_{V'} \leqslant C \|u - u_h\|_V, \tag{35}$$

where $\|r\|_{V'} = \sup_{v \in V, v \neq 0} r(v)/\|v\|_V$.

The *a posteriori* error estimates, equation (35), relate the numerical error with the residuals.

The main objective of goal-oriented error estimation is to evaluate the accuracy of FEM solutions in measures other than the energy norm [14]. In numerical applications, it is often necessary to control the error in a certain output functional $\mathcal{M} : V \to \mathbb{R}$ of the computed solution within a given tolerance $\varepsilon > 0$. Typical functionals are the quantities we are interested in. In such situations, one could ideally choose the finite element space $V_h \subset V$, such that the finite element solution u_h satisfies

$$\eta \equiv |\mathcal{M}(u) - \mathcal{M}(u_h)| \leqslant \varepsilon,$$

with minimal computational work. We assume that both the output functional and the variational problem are linear, but the analysis may be easily extended to the full nonlinear case. To estimate the error in the output functional, $\mathcal{M}$, we introduce an auxiliary dual problem, that is, find $z \in V^*$ such that

$$a^*(z, v) = \mathcal{M}(v), \quad \forall v \in \hat{V}^*.$$

We note here that the functional $\mathcal{M}$ is introduced as data in the dual problem. The dual (adjoint) bilinear form $a^* : V^* \times \hat{V}^* \to \mathbb{R}$ is defined by

$$a^*(v, w) = a(w, v), \quad \forall (v, w) \in V^* \times \hat{V}^*.$$

The dual trial and test spaces are given by

$$V^* = \hat{V},$$
$$\hat{V}^* = V_0 = \{v - w : v, w \in V\}.$$

The definition of the dual problem, leads to the following error representation:

$$\begin{aligned}\mathcal{M}(u) - \mathcal{M}(u_h) &= \mathcal{M}(u - u_h) \\ &= a^*(z, u - u_h) \\ &= a(u - u_h, z) \\ &= L(z) - a(u_h, z) \\ &= r(z) = r(z - z_h).\end{aligned}$$

So, the error is exactly represented by the residual of the dual solution:

$$\mathcal{M}(u) - \mathcal{M}(u_h) = r(z).$$

An adaptive algorithm seeks to determine a mesh size, $h = h(x)$, in a specific tolerance starting from an initial coarse mesh, and refine in those cells where the error indicator remains large.

For the Poisson equation, we take the goal functional to be defined as

$$\mathcal{M}(u) = \int_\Omega u \, dx. \tag{36}$$

In Figure 6, we present the solution of Poisson equation with zero Dirichlet conditions. At the middle of the domain, we apply a two-dimensional Gaussian distribution, given by the expression described in equation (32). We observe that locally refining the computational grid with the AFEM described above, and introducing the goal functional of equation (36), the results in the domain of interest are capturing in detail the numerical solution of the Poisson equation. The given tolerance for the simulations is 10^{-5} and the final number of cells is increased to 1052 from the initial number of cells which was 128. This provide accurate results for the Poisson problem.

Finally, we present similar results for the Stokes problem, where the mathematical formulation of the problem is discussed in a previous section. As depicted in Figure 7, we observe that locally refining the computational grid with the AFEM and introducing a similar goal functional of equation (36), the results in the backward-facing step test problem are substantially smoother, capturing in detail the

The Finite Element Method with Applications to Fluid Mechanics 171

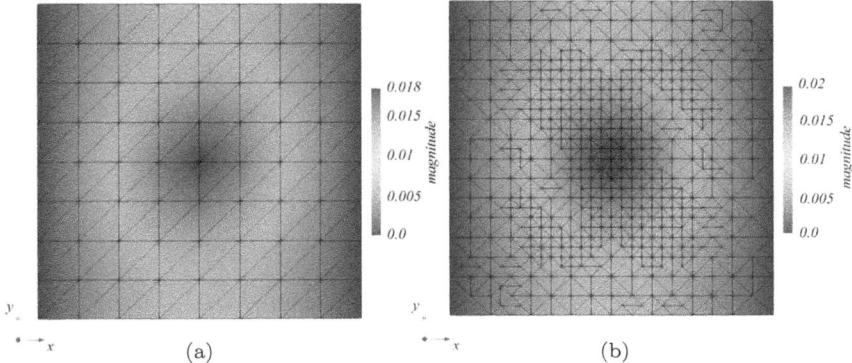

Figure 6. (a) Computational mesh and results for the Poisson equation. (b) Computational mesh and results for the Poisson equation after grid local refinement.

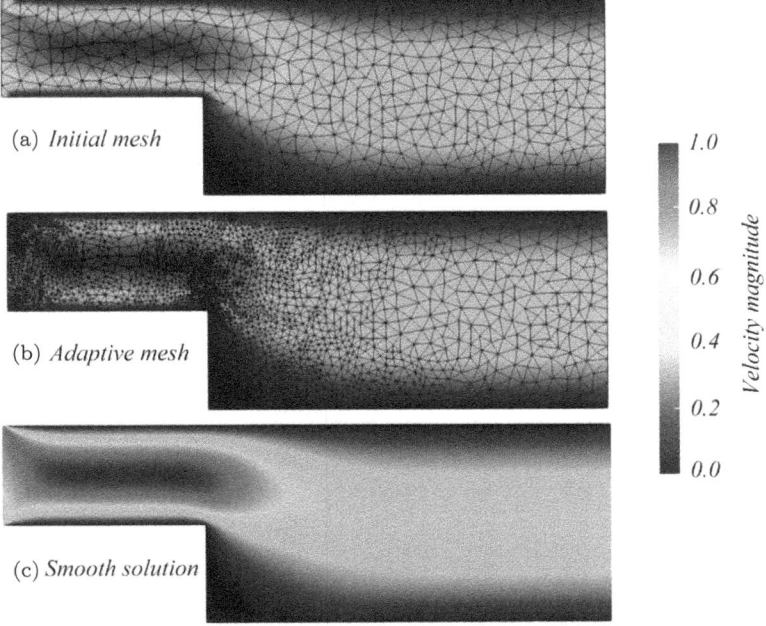

Figure 7. (a) The initial computational mesh for the backward-facing step test problem. (b) The final computational mesh for the test problem. (c) The obtained smoothed results obtained from the AFEM.

numerical solution of the Stokes equation for the unknown vector field (velocity), $\bar{q}$, and pressure field, p. The given tolerance for the adaptive simulation is 10^{-7}. The final number of cells is increased to 7464 from the initial number, which was 890 triangular cells, providing accurate and smooth results for the test problem.

Conclusions

In this review chapter, the FEM is utilized for the numerical solution of ODEs and PDEs. We mainly focus on applying the method to fluid mechanics problems. Initially, we present the method along with the basic theorems and examples. We analyze the error estimates for linear problems and the base functions to distinguish the problem under consideration. We present the numerical solution of the Duffing equation and compare this solution with the analytical one. We further concentrate our attention on the two-dimensional Stokes and Navier–Stokes problems. We finally focus on presenting novel FEM variants, such as the Discontinuous Galerkin (DG) method and adaptive methodologies (AFEM). These advanced methods provide reliable numerical results in all studied cases. This is achieved with the application of the FEM to "test problems", such as the backward-facing step. We obtain all the numerical results utilizing the software programs MATLAB and FEniCS.

Acknowledgment

This research was partially supported by project "Dioni: Computing Infrastructure for Big-Data Processing and Analysis" (MIS No. 5047222) co-funded by European Union (ERDF) and Greece through Operational Program "Competitiveness, Entrepreneurship and Innovation", NSRF 2014-2020.

References

[1] H. Antil, R. H. Hoppe and C. Linsenmann, Path-following primal-dual interior-point methods for shape optimization of stationary flow problems, *J. Numer. Math.* **15**, 81–100 (2007).

[2] G. A. Baker, Finite element methods for elliptic equations using nonconforming elements, *Math. Comput.* **31**(137), 45–59 (1977).

[3] E. Barkanov, *Introduction to the Finite Element Method* (Institute of Materials and Structures Faculty of Civil Engineering, Riga Technical University, 2001), pp. 1–70.

[4] M. Bercovier and O. Pironneau, Error estimates for finite element method solution of the stokes problem in the primitive variables, *Numer. Math.* **33**(2), 211–224 (1979).

[5] S. Brenner and R. Scott, *The Mathematical Theory of Finite Element Methods*, Vol. 15 (Springer Science & Business Media, New York, 2007).

[6] L. Chen, K. Asai, T. Nonomura, G. Xi and T. Liu, A review of backward-facing step (bfs) flow mechanisms, heat transfer and control, *Therm. Sci. Eng. Prog.* **6**, 194–216 (2018).

[7] B. Cockburn, Discontinuous galerkin methods. *ZAMM-J. Appl. Math. Mech./Z. Angew. Math. Mech. Appl. Math. Mech.* **83**(11), 731–754 (2003).

[8] B. Cockburn, G. E. Karniadakis and C. W. Shu, The development of discontinuous galerkin methods. In *Discontinuous Galerkin Methods* (Springer, Berlin, Heidelberg, 2000), pp. 3–50.

[9] B. Cockburn, G. E. Karniadakis and C. W. Shu (eds.) *Discontinuous Galerkin Methods* (Springer, Berlin, 2000). https://doi.org/10.1007/978-3-642-59721-3.

[10] M. Delfour, W. Hager and F. Trochu, Discontinuous galerkin methods for ordinary differential equations, *Math. Comput.* **36**(154), 455–473 (1981).

[11] M. J. Gander and G. Wanner, From Euler, Ritz, and Galerkin to modern computing, *SIAM Rev.* **54**(4), 627–666 (2012).

[12] F. Karakatsani and C. Makridakis, A posteriori estimates for approximations of time-dependent stokes equations, *IMA J. Numer. Anal.* **27**(4), 741–764 (2006).

[13] P. Lesaint and P. A. Raviart, On a finite element method for solving the neutron transport equation, *Publ. Math. Informa. Rennes* **S4**, 1–40 (1974).

[14] A. Logg, K. A. Mardal and G. Wells, *Automated Solution of Differential Equations by the Finite Element Method: The FEniCS Book*, Vol. 84 (Springer Science & Business Media, 2012).

[15] G. C. Mbah and K. K. Ibeh, Application of the finite element method to solving the duffing equation of ground motion, *Glob. J. Pure Appl. Sci.* **26**(1), 65–71 (2020).

[16] L. Mu and X. Ye, A simple finite element method for the stokes equations, *Adv. Comput. Math.* **43**(6), 1305–1324 (2017).

[17] R. H. Nochetto, K. G. Siebert and A. Veeser, Theory of adaptive finite element methods: An introduction. In *Multiscale, Nonlinear and Adaptive Approximation* (Springer, Berlin, Heidelberg, 2009), pp. 409–542.
[18] K. B. Ølgaard, A. Logg and G. N. Wells, Automated code generation for discontinuous Galerkin methods, *SIAM J. Sci. Comput.* **31**(2), 849–864 (2009).
[19] E. N. Petropoulou and M. A. Xenos, Qualitative, approximate and numerical approaches for the solution of nonlinear differential equations. In *Applications of Nonlinear Analysis* (Springer, Cham, Switzerland, 2018), pp. 611–664.
[20] A. Raptis, K. Kyriakoudi and M. A. Xenos, Finite element analysis in fluid mechanics. In *Mathematical Analysis and Applications* (Springer, Cham, Switzerland, 2019), pp. 481–510.
[21] J. Reddy, *An Introduction to the Finite Element Method*, Vol. 1221 (McGraw-Hill, New York, 2010).
[22] T. E. Tezduyar, S. Mittal, S. Ray and R. Shih, Incompressible flow computations with stabilized bilinear and linear equal-order-interpolation velocity-pressure elements, *Comput. Methods Appl. Mech. Eng.* **95**(2), 221–242 (1992).
[23] R. Thatcher, Locally mass-conserving taylor-hood elements for two- and three-dimensional flow, *Int. J. Numer. Methods Fluids* **11**(3), 341–353 (1990).
[24] M. A. Xenos and A. C. Felias, Nonlinear dynamics of the kdv-b equation and its biomedical applications. In *Nonlinear Analysis, Differential Equations, and Applications* (Springer, Cham, Switzerland, 2021), pp. 765–793.
[25] Y. Zhao, X. Zhang, S. L. Ho and W. Fu, An adaptive mesh method in transient finite element analysis of magnetic field using a novel error estimator, *IEEE Trans. Magn.* **48**(11), 4160–4163 (2012).
[26] O. C. Zienkiewicz and Y. K. Cheung, *Finite Element Method in Structural & Continuum Mechanics* (McGraw Hill, New York, 1967).
[27] O. C. Zienkiewicz, R. L. Taylor and J. Z. Zhu, *The Finite Element Method: Its Basis and Fundamentals* (Elsevier, Amsterdam, 2005).

© 2024 World Scientific Publishing Company
https://doi.org/10.1142/9789811267048_0006

Chapter 6

Comparison between Two Descent SQP-ADMs for Structured Variational Inequalities

Abdellah Bnouhachem[*,‡] and Themistocles M. Rassias[†,§]

[*]*Equipe MAISI, Ibn Zohr University, ENSA, Agadir, Morocco*
[†]*Department of Mathematics,
National Technical University of Athens,
Zografou Campus, Athens, Greece*
[‡]*babedallah@yahoo.com*
[§]*trassias@math.ntua.gr*

In this chapter, by combining the alternating direction method with the SQP method, we suggest and analyze two descent SQP alternating direction schemes for the separable constrained convex programming problem. Under certain conditions, the global convergence of both methods is proved. It is proved theoretically that the lower bound of the progress obtained by the second method is greater than that by the first one. Numerical results are given to verify the theoretical assertion.

1. Introduction

This chapter considers the constrained convex programming problem with the following separate structure:

$$\min \left\{ \theta_1(x) + \theta_2(y) | Ax + By = b, x \in \mathbb{R}^n_+, y \in \mathbb{R}^m_+ \right\}, \qquad (1)$$

where $\theta_1 : \mathbb{R}^n_+ \to \mathbb{R}$ and $\theta_2 : \mathbb{R}^m_+ \to \mathbb{R}$ are closed proper convex functions not necessarily smooth, $A \in \mathbb{R}^{l \times n}$, $B \in \mathbb{R}^{l \times m}$ are given matrices and $b \in \mathbb{R}^l$ is a given vector.

The alternating direction method (ADM) is an attractive tool for solving (1). It was first introduced in the mid-1970s by Glowinski and Marrocco [20] and Gabay and Mercier [18]. ADM has also regained a lot of attention [26, 34]. It updates the variables alternately by minimizing the augmented Lagrangian function with respect to the variables in a Gauss–Seidel manner. ADM is very effective at solving many practical optimization problems and has wide applications in areas, such as signal and image processing, machine learning, statistics, compressive sensing and operations research. We refer to Refs. [11, 27, 39, 40] for a few examples of applications. To make the ADM more efficient and practical, some strategies have been studied; for more details, one can refer to Refs. [7–9, 12, 15, 24, 28, 29, 35, 38].

Let $\partial(.)$ denote the subgradient operator of a convex function, and $f(x) \in \partial\theta_1(x)$ and $g(y) \in \partial\theta_2(y)$ are the subgradients of $\theta_1(x)$ and $\theta_2(y)$, respectively. By attaching a Lagrange multiplier vector $\lambda \in \mathbb{R}^l$ to the linear constraint $Ax + By = b$, problem (1) can be written in terms of finding $w \in \mathcal{W}$ such that

$$(w' - w)^\top Q(w) \geq 0, \quad \forall\, w' \in \mathcal{W}, \tag{2}$$

where

$$w = \begin{pmatrix} x \\ y \\ \lambda \end{pmatrix} \quad Q(w) = \begin{pmatrix} f(x) - A^\top \lambda \\ g(y) - B^\top \lambda \\ Ax + By - b \end{pmatrix}, \quad \mathcal{W} = \mathbb{R}^n_+ \times \mathbb{R}^m_+ \times \mathbb{R}^l. \tag{3}$$

Problems (2)–(3) are referred to as *structural variational inequality* (in short, SVI).

We now have a variety of techniques to suggest and analyze various ADMs with logarithmic-quadratic proximal (LQP) regularization [2–6, 10, 32, 35, 41] for solving SVI: for a given $w^k = (x^k, y^k, \lambda^k) \in \mathbb{R}^n_{++} \times \mathbb{R}^m_{++} \times \mathbb{R}^l$, and $\mu \in (0, 1)$, the predictor $\tilde{w}^k = (\tilde{x}^k, \tilde{y}^k, \tilde{\lambda}^k)$ is obtained via solving the following system:

$$f(x) - A^\top \left[\lambda^k - H(Ax + By^k - b)\right]$$
$$+ R\left[(x - x^k) + \mu(x^k - X_k^2 x^{-1})\right] = 0, \tag{4}$$

$$g(y) - B^\top \left[\lambda^k - H(Ax + By - b)\right]$$
$$+ S\left[(y - y^k) + \mu(y^k - Y_k^2 y^{-1})\right] = 0, \quad (5)$$
$$\tilde{\lambda}^k = \lambda^k - H(Ax + By - b), \quad (6)$$

where $H \in \mathbb{R}^{l \times l}$, $R \in \mathbb{R}^{n \times n}$ and $S \in \mathbb{R}^{m \times m}$ are symmetric positive definite, $X_k := \mathrm{diag}(x_1^k, x_2^k, \ldots, x_n^k)$, $Y_k := \mathrm{diag}(y_1^k, y_2^k, \ldots, y_n^k)$, x^{-1} be an n-vector whose jth element is $1/x_j$ and y^{-1} be an n-vector whose jth element is $1/y_j$.

Very recently, Bnouhachem and Rassias [8] suggested that the complementarity subproblems arising in ADM (4)–(5) could be regularized by the square quadratic proximal (SQP) regularization, from a given $w^k = (x^k, y^k, \lambda^k) \in \mathbb{R}_{++}^n \times \mathbb{R}_{++}^m \times \mathbb{R}^l$, $\tilde{w}^k = (\tilde{x}^k, \tilde{y}^k, \tilde{\lambda}^k) \in \mathbb{R}_{++}^n \times \mathbb{R}_{++}^m \times \mathbb{R}^l$ is obtained via solving the following system:

$$f(x) - A^\top \left[\lambda^k - H\left(\frac{1}{2}Ax^k + \frac{1}{2}Ax + By^k - b\right)\right]$$
$$+ R\left[\frac{1}{2}(x - x^k) + \mu(x^k - U_k(\sqrt{x})^{-1})\right] =: \xi_x^k \approx 0, \quad (7a)$$

$$g(y) - B^\top \left[\lambda^k - H\left(A\tilde{x}^k + \frac{1}{2}By + \frac{1}{2}By^k - b\right)\right]$$
$$+ S\left[\frac{1}{2}(y - v^k) + \mu(y^k - V_k(\sqrt{y})^{-1})\right] =: \xi_y^k \approx 0, \quad (7b)$$

$$\tilde{\lambda}^k = \lambda^k - H(A\tilde{x}^k + B\tilde{y}^k - b), \quad (7c)$$

where

$$|(w^k - \tilde{w}^k)^\top \xi^k| \leq \eta \|w^k - \tilde{w}^k\|_M^2, \quad \eta \in (0, 1), \quad (8)$$

$$M = \begin{pmatrix} \frac{1}{2}R & & \\ & \frac{1}{2}S & \\ & & \frac{1}{2}H^{-1} \end{pmatrix}, \quad (9)$$

$$\xi^k = \begin{pmatrix} \xi_x^k \\ \xi_y^k \\ 0 \end{pmatrix} = \begin{pmatrix} f(\tilde{x}^k) - f(x^k) - \frac{1}{2}A^\top HA(x^k - \tilde{x}^k) \\ g(\tilde{y}^k) - g(y^k) - \frac{1}{2}B^\top HB(y^k - \tilde{y}^k) \\ 0 \end{pmatrix}. \quad (10)$$

U_k and V_k are positive definite diagonal matrices defined by

$$U_k = \text{diag}\left(x_1^k\sqrt{x_1^k}, \ldots, x_n^k\sqrt{x_n^k}\right) := \begin{pmatrix} x_1^k\sqrt{x_1^k} & & \\ & \ddots & \\ & & x_n^k\sqrt{x_n^k} \end{pmatrix}$$

and

$$V_k = \text{diag}\left(y_1^k\sqrt{y_1^k}, \ldots, y_n^k\sqrt{y_n^k}\right),$$

$(\sqrt{x})^{-1} \in \mathbb{R}_{++}^n$ is a vector whose jth element is $1/\sqrt{x_j}$, $(\sqrt{y})^{-1} \in \mathbb{R}_{++}^m$ is a vector whose jth element is $1/\sqrt{y_j}$ and the new iterate $w^{k+1}(\alpha_k) = (x^{k+1}, y^{k+1}, \lambda^{k+1})$ is given by

$$w^{k+1}(\alpha_k) = (1-\sigma)w^k + \sigma P_{\mathcal{W}}[w^k - \alpha_k d(w^k, \tilde{w}^k)], \quad \sigma \in (0,1),$$

where

$$\alpha_k = \frac{\varphi_1(w^k, \tilde{w}^k)}{\|d(w^k, \tilde{w}^k)\|^2},$$

$$\varphi_1(w^k, \tilde{w}^k) = \|w^k - \tilde{w}^k\|_M^2 + (w^k - \tilde{w}^k)^\top G_1(w^k - \tilde{w}^k) + (w^k - \tilde{w}^k)^\top \xi^k,$$

$$d(w^k, \tilde{w}^k) = \begin{pmatrix} (1+\mu)R(x^k - \tilde{x}^k) - [f(x^k) - f(\tilde{x}^k)] \\ (1+\mu)S(y^k - \tilde{y}^k) - [g(y^k) - g(\tilde{y}^k)] + B^\top HA(x^k - \tilde{x}^k) \\ H^{-1}(\lambda^k - \tilde{\lambda}^k) \end{pmatrix}$$

and

$$G_1 = \begin{pmatrix} \frac{1}{2}A^\top HA & 0 & 0 \\ B^\top HA & \frac{1}{2}B^\top HB & 0 \\ A & B & \frac{1}{2}H^{-1} \end{pmatrix}.$$

We note that the convergence of this method was established under the assumption that the mappings $f(x)$ and $g(y)$ are Lipschitz continuous. However, in many applications, the mappings $f(x)$ and $g(y)$ may not be Lipschitz continuous. Inspired and motivated by the research going in this direction, we propose two descent SQP alternating direction methods for SVI. The global convergence of the proposed methods can be guaranteed by the continuity rather than the Lipschitz continuity of these mappings. It is proved theoretically that the lower bound of the progress obtained by the second method is greater than that by the first one. Preliminary numerical experiments show the efficiency of the proposed methods and support our theoretic analysis. Our results can be viewed as significant extensions of the previously known results.

2. SQP Alternating Direction Method

For any vector $u \in \mathbb{R}^n$, $\|u\|^2 = u^\top u$, $\|u\|_\infty = \max\{|u_1|\ldots,|u_n|\}$. Let $D \in \mathbb{R}^{n \times n}$ be a symmetric positive definite matrix and the operators $\lambda_l(D)$ and $\lambda_m(D)$ denote the largest eigenvalue and the smallest eigenvalue of D, respectively. We denote the D-norm of u by $\|u\|_D^2 = u^\top D u$.

Two important inequalities with respect to the projection operator are listed, and the proofs are available in many monographs, such as Ref. [16].

Lemma 1. *Let Ω be a nonempty closed convex subset of $\mathbb{R}^l$. Denote by $P_\Omega(.)$ the projection on Ω with respect to the Euclidean norm, that is,*

$$P_\Omega(v) = \operatorname{argmin}\{\|v - u\| : u \in \Omega\}.$$

Then, we have the following inequalities:

$$(z - P_\Omega[z])^\top (P_\Omega[z] - v) \geq 0, \quad \forall\, z \in \mathbb{R}^l, \quad v \in \Omega; \tag{11}$$

$$\|u - P_\Omega[z]\|^2 \leq \|z - u\|^2 - \|z - P_\Omega[z]\|^2, \quad \forall\, z \in \mathbb{R}^l, \quad u \in \Omega. \tag{12}$$

We introduce the following definitions which are useful in the sequel:

Definition 1. The mapping $T : \mathbb{R}^n \to \mathbb{R}^n$ is said to be

(a) monotone if

$$(Tx - Ty)^\top (x - y) \geq 0, \quad \forall x, y \in \mathbb{R}^n;$$

(b) k-Lipschitz continuous if there exists a constant $k > 0$ such that

$$\|Tx - Ty\| \leq k\|x - y\|, \quad \forall x, y \in \mathbb{R}^n.$$

We make the following standard assumptions:

Assumption A. f is monotone and continuous on $\mathbb{R}^n_+$ and g is monotone and continuous on $\mathbb{R}^m_+$.

Assumption B. The solution set of SVI, denoted by $\mathcal{W}^*$, is nonempty.

We propose the following SQP-ADMs for solving SVI:

Algorithm 1.

Step 0. *The initial step*:
Given $\varepsilon > 0$, $\mu \in (0,1)$, $\eta \in (0,1)$, $\rho \in (0, \frac{3}{2})$ and $w^0 = (x^0, y^0, \lambda^0) \in \mathbb{R}^n_{++} \times \mathbb{R}^m_{++} \times \mathbb{R}^l$.
$\beta_0 = \min\left\{ \frac{(1-\eta)\lambda_m(R)}{2(3-2\rho)\lambda_l(H)\|A\|^2}, \frac{(1-\eta)\lambda_m(S)}{2(3-2\rho)\lambda_l(H)\|B\|^2} \right\}$. Set $k = 0$.

Step 1. *Prediction step*:
Compute $\tilde{w}^k = (\tilde{x}^k, \tilde{y}^k, \tilde{\lambda}^k) \in \mathbb{R}^n_{++} \times \mathbb{R}^m_{++} \times \mathbb{R}^l$ by solving the following system:

$$\beta_k \left(f(x) - A^\top [\lambda^k - H(Ax^k + By^k - b)] \right)$$
$$+ R[\frac{1}{2}(x - x^k) + \mu(x^k - U_k(\sqrt{x})^{-1})] =: \xi^k_x \approx 0, \quad (13a)$$

$$\beta_k \left(g(y) - B^\top [\lambda^k - H(A\tilde{x}^k + By^k - b)] \right)$$
$$+ S[\frac{1}{2}(y - y^k) + \mu(y^k - V_k(\sqrt{y})^{-1})] =: \xi^k_y \approx 0, \quad (13b)$$

$$\tilde{\lambda}^k = \lambda^k - H(A\tilde{x}^k + B\tilde{y}^k - b), \quad (13c)$$

where β_k is a proper parameter which satisfies

$$\beta_k \|f(x^k) - f(\tilde{x}^k)\| \leq \frac{\eta r}{2} \|x^k - \tilde{x}^k\|,$$
$$\beta_k \|g(y^k) - g(\tilde{y}^k)\| \leq \frac{\eta s}{2} \|y^k - \tilde{y}^k\| \quad (14)$$

and

$$\xi^k = \begin{pmatrix} \xi^k_x \\ \xi^k_y \\ 0 \end{pmatrix} = \begin{pmatrix} \beta_k(f(\tilde{x}^k) - f(x^k)) + \rho \beta_k A^\top H A(x^k - \tilde{x}^k) \\ \beta_k(g(\tilde{y}^k) - g(y^k)) + \rho \beta_k B^\top H B(y^k - \tilde{y}^k) \\ 0 \end{pmatrix}. \quad (15)$$

Step 2. *Convergence verification:*
If $\max\{\|x^k - \tilde{x}^k\|_\infty, \|y^k - \tilde{y}^k\|_\infty, \|\lambda^k - \tilde{\lambda}^k\|_\infty\} < \epsilon$, then stop.

Step 3. *Correction step:*
The new iterate $w^{k+1}(\alpha_k) = (x^{k+1}, y^{k+1}, \lambda^{k+1})$ is given by
Correction I:

$$w_I^{k+1}(\alpha_k) = (1-\sigma)w^k + \sigma P_{\mathcal{W}}[w^k - \alpha_k d_1(w^k, \tilde{w}^k)],$$

$$\sigma \in (0,1) \tag{16}$$

or

Correction II:

$$w_{II}^{k+1}(\alpha_k) = (1-\sigma)w^k + \sigma P_{\mathcal{W}}[w^k - \alpha_k d_2(w^k, \tilde{w}^k)], \tag{17}$$

where

$$\alpha_k = \frac{\varphi(w^k, \tilde{w}^k)}{\|d_1(w^k, \tilde{w}^k)\|^2}, \tag{18}$$

$$d_2(w^k, \tilde{w}^k) = \begin{pmatrix} \beta_k(f(\tilde{x}^k) - A^\top \tilde{\lambda}^k) + \beta_k A^\top H(A(x^k - \tilde{x}^k) \\ + B(y^k - \tilde{y}^k)) \\ \beta_k(g(\tilde{y}^k) - B^\top \tilde{\lambda}^k) + \beta_k B^\top H(A(x^k - \tilde{x}^k) \\ + B(y^k - \tilde{y}^k)) \\ \beta_k(A\tilde{x}^k + B\tilde{y}^k - b) \end{pmatrix}, \tag{19}$$

$$\varphi(w^k, \tilde{w}^k) = (w^k - \tilde{w}^k)^\top d_1(w^k, \tilde{w}^k) - \frac{\mu}{2}\|x^k - \tilde{x}^k\|_R^2$$

$$-\frac{\mu}{2}\|y^k - \tilde{y}^k\|_S^2 + \beta_k(\lambda^k - \tilde{\lambda}^k)^\top (A(x^k - \tilde{x}^k)$$

$$+ B(y^k - \tilde{y}^k)), \tag{20}$$

$$d_1(w^k, \tilde{w}^k) = \begin{pmatrix} \frac{(1+\mu)}{2} R(x^k - \tilde{x}^k) - \beta_k[f(x^k) - f(\tilde{x}^k)] \\ +\rho\beta_k A^\top H A(x^k - \tilde{x}^k) \\ \frac{(1+\mu)}{2} S(y^k - \tilde{y}^k) - \beta_k[g(y^k) - g(\tilde{y}^k)] \\ +\beta_k B^\top H A(x^k - \tilde{x}^k) \\ +\rho\beta_k B^\top H B(y^k - \tilde{y}^k) \\ \beta_k H^{-1}(\lambda^k - \tilde{\lambda}^k) \end{pmatrix}. \tag{21}$$

Step 4. *Adjusting*:

Adaptive rule of choosing a suitable β_{k+1} as the start prediction step size for the next iteration

$$\beta_{k+1} := \begin{cases} \min\{\beta_0, \tau * \beta_k\} & \text{if } \max\{r_1, r_2\} \leq 0.5, \\ \beta_k & \text{otherwise,} \end{cases} \quad (22)$$

where

$$r_1 = \frac{2\beta_k \|f(x^k) - f(\tilde{x}^k)\|}{r\|x^k - \tilde{x}^k\|},$$

$$r_2 = \frac{2\beta_k \|g(y^k) - g(\tilde{y}^k)\|}{s\|y^k - \tilde{y}^k\|}$$

and $\tau > 1$.

Set $k := k+1$ and go to Step 1.

The following lemma was proved in Ref. [8]. We need this lemma to analyze the convergence for the ADM with SQP regularization.

Lemma 2. *Let $q(u) \in \mathbb{R}^n$ be a monotone mapping of u with respect to $\mathbb{R}_+^n$ and $R := \mathrm{diag}(r_1, \ldots, r_n) \in \mathbb{R}^{n \times n}$ be a positive definite diagonal matrix. For a given $u^k > 0$, $\mu > 0$, if $U_k := \mathrm{diag}(u_1^k \sqrt{u_1^k}, \ldots, u_n^k \sqrt{u_n^k})$, $\sqrt{u} = (\sqrt{u_1}, \ldots, \sqrt{u_n})$ and $(\sqrt{u})^{-1}$ be an n-vector whose jth element is $1/\sqrt{u_j}$, then the equation*

$$q(u) + R\left[\frac{1}{2}(u - u^k) + \mu(u^k - U_k(\sqrt{u})^{-1})\right] = 0 \quad (23)$$

has a unique positive solution u. Moreover, for any $v \geq 0$, we have

$$(v-u)^\top q(u) \geq \frac{1+\mu}{4}\left(\|u-v\|_R^2 - \|u^k - v\|_R^2\right) + \frac{1-\mu}{4}\|u^k - u\|_R^2. \quad (24)$$

In the following theorem, we show that the step size α_k of the new methods is bounded away from zero, which contributes much to the satisfactory efficiencies of the new methods.

Theorem 1. *For given $w^k \in \mathbb{R}_{++}^n \times \mathbb{R}_{++}^m \times \mathbb{R}^l$, let $\tilde{w}^k$ be generated by (13a)–(13c), then there exists two constants $\alpha_1 > 0$ and $\alpha_2 > 0$ such that*

$$\varphi(w^k, \tilde{w}^k) \geq \alpha_1 \|w^k - \tilde{w}^k\|^2 \quad (25)$$

and
$$\alpha_k \geq \frac{\alpha_1}{\alpha_2}. \tag{26}$$

Proof. Since

$$(\lambda^k - \tilde{\lambda}^k)^\top A(x^k - \tilde{x}^k) \geq -\frac{1}{4}\|\lambda^k - \tilde{\lambda}^k\|_{H^{-1}}^2 - \|A(x^k - \tilde{x}^k)\|_H^2,$$

$$(\lambda^k - \tilde{\lambda}^k)^\top B(y^k - \tilde{y}^k) \geq -\frac{1}{4}\|\lambda^k - \tilde{\lambda}^k\|_{H^{-1}}^2 - \|B(y^k - \tilde{y}^k)\|_H^2,$$

$$(A(x^k - \tilde{x}^k))^\top H B(y^k - \tilde{y}^k) \geq -\frac{1}{2}(\|A(x^k - \tilde{x}^k)\|_H^2 + \|B(y^k - \tilde{y}^k)\|_H^2).$$

It follows from the definition of $\varphi(w^k, \tilde{w}^k)$ that

$$\begin{aligned}\varphi(w^k, \tilde{w}^k) &= \frac{1}{2}\|x^k - \tilde{x}^k\|_R^2 + \frac{1}{2}\|y^k - \tilde{y}^k\|_S^2 + \beta_k\|\lambda^k - \tilde{\lambda}^k\|_{H^{-1}}^2 \\ &\quad + \beta_k(A(x^k - \tilde{x}^k))^\top H B(y^k - \tilde{y}^k) - \beta_k(x^k - \tilde{x}^k)^\top [f(x^k) \\ &\quad - f(\tilde{x}^k) - \rho A^\top H A(x^k - \tilde{x}^k)] \\ &\quad - \beta_k(y^k - \tilde{y}^k)^\top [g(y^k) - g(\tilde{y}^k) - \rho B^\top H B(y^k - \tilde{y}^k)] \\ &\quad + \beta_k(\lambda^k - \tilde{\lambda}^k)^\top (A(x^k - \tilde{x}^k) + B(y^k - \tilde{y}^k)) \\ &\geq \frac{1}{2}\|x^k - \tilde{x}^k\|_R^2 + \frac{1}{2}\|y^k - \tilde{y}^k\|_S^2 + \frac{1}{2}\beta_k\|\lambda^k - \tilde{\lambda}^k\|_{H^{-1}}^2 \\ &\quad - \frac{3}{2}\beta_k\|A(x^k - \tilde{x}^k)\|_H^2 - \frac{3}{2}\beta_k\|B(y^k - \tilde{y}^k)\|_H^2 \\ &\quad - \beta_k(x^k - \tilde{x}^k)^\top [f(x^k) - f(\tilde{x}^k) - \rho A^\top H A(x^k - \tilde{x}^k)] \\ &\quad - \beta_k(y^k - \tilde{y}^k)^\top [g(y^k) - g(\tilde{y}^k) - \rho B^\top H B(y^k - \tilde{y}^k)].\end{aligned}$$

By using Cauchy–Schwarz inequality in (14), we have

$$(w^k - \tilde{w}^k)^\top e_k \geq -\eta\|w^k - \tilde{w}^k\|_M^2,$$

where

$$e_k = \begin{pmatrix} \beta_k(f(\tilde{x}^k) - f(x^k)) \\ \beta_k(g(\tilde{y}^k) - g(y^k)) \\ 0 \end{pmatrix}$$

and
$$M = \begin{pmatrix} \frac{1}{2}R & & \\ & \frac{1}{2}S & \\ & & \frac{1}{2}\beta_k H^{-1} \end{pmatrix}. \tag{27}$$

Then, we have

$$\varphi(w^k, \tilde{w}^k) \geq \frac{1}{2}\|x^k - \tilde{x}^k\|_R^2 + \frac{1}{2}\|y^k - \tilde{y}^k\|_S^2 + \frac{1}{2}\beta_k\|\lambda^k - \tilde{\lambda}^k\|_{H^{-1}}^2$$
$$- \left(\frac{3}{2} - \rho\right)\beta_k\|A(x^k - \tilde{x}^k)\|_H^2$$
$$- \left(\frac{3}{2} - \rho\right)\beta_k\|B(y^k - \tilde{y}^k)\|_H^2 - \eta\|w^k - \tilde{w}^k\|_M^2$$
$$\geq \frac{(1-\eta)}{2}\|w^k - \tilde{w}^k\|_M^2 + \left(\frac{(1-\eta)\lambda_m(R)}{4}\right.$$
$$\left. - \left(\frac{3}{2} - \rho\right)\beta_k\lambda_l(H)\|A\|^2\right)\|x^k - \tilde{x}^k\|^2$$
$$+ \left(\frac{(1-\eta)\lambda_m(S)}{4} - \left(\frac{3}{2} - \rho\right)\beta_k\lambda_l(H)\|B\|^2\right)\|y^k - \tilde{y}^k\|^2$$
$$\geq \frac{(1-\eta)}{2}\|w^k - \tilde{w}^k\|_M^2 + \left(\frac{(1-\eta)\lambda_m(R)}{4}\right.$$
$$\left. - \left(\frac{3}{2} - \rho\right)\beta_0\lambda_l(H)\|A\|^2\right)\|x^k - \tilde{x}^k\|^2$$
$$+ \left(\frac{(1-\eta)\lambda_m(S)}{4} - \left(\frac{3}{2} - \rho\right)\beta_0\lambda_l(H)\|B\|^2\right)\|y^k - \tilde{y}^k\|^2$$
$$\geq \frac{(1-\eta)}{2}\|w^k - \tilde{w}^k\|_M^2$$
$$\geq \alpha_1\|w^k - \tilde{w}^k\|^2,$$

where $\alpha_1 > 0$ is a constant. The third inequality holds because $\beta_k \leq \beta_0$ for any k. The fourth inequality is obtained from the definition of β_0.

Recalling the definition in (21), we rewrite $d_1(w^k, \tilde{w}^k))$ as

$$d_1(w^k, \tilde{w}^k) = \xi_k + G(w^k - \tilde{w}^k),$$

where

$$G = \begin{pmatrix} \frac{(1+\mu)}{2}R & 0 & 0 \\ \beta_k B^\top H A & \frac{(1+\mu)}{2}S & 0 \\ 0 & 0 & \beta_k H^{-1} \end{pmatrix}.$$

Note that, for any $a, b \in \mathbb{R}^n$, we have

$$\|a + b\|^2 \leq 2\|a\|^2 + 2\|b\|^2.$$

It follows that

$$\begin{aligned}
\|d_1(w^k, \tilde{w}^k)\|^2 &\leq 2\|\xi_k\|^2 + 2\|G(w^k - \tilde{w}^k)\|^2 \\
&\leq 2\|\beta_k(f(\tilde{x}^k) - f(x^k)) + \rho\beta_k A^\top H A(x^k - \tilde{x}^k)\|^2 \\
&\quad + 2\|\beta_k(g(\tilde{y}^k) - g(y^k)) + \rho\beta_k B^\top H B(y^k - \tilde{y}^k)\|^2 \\
&\quad + 2\|G(w^k - \tilde{w}^k)\|^2 \\
&\leq (\eta^2 r^2 + 4\rho^2 \beta_0^2 \|A^\top H A\|)\|x^k - \tilde{x}^k\|^2 \\
&\quad + (\eta^2 s^2 + 4\rho^2 \beta_0^2 \|B^\top H B\|)\|y^k - \tilde{y}^k\|^2 \\
&\quad + 2\|G(w^k - \tilde{w}^k)\|^2 \\
&\leq \max\left(\eta^2 r^2 + 4\rho^2 \beta_0^2 \|A^\top H A\|, \eta^2 s^2 \right. \\
&\quad \left. + 4\rho^2 \beta_0^2 \|B^\top H B\|\right) \|w^k - \tilde{w}^k\|^2 \\
&\quad + 2\lambda_l(G^\top G)\|w^k - \tilde{w}^k\|^2 \\
&\leq \alpha_2 \|w^k - \tilde{w}^k\|^2,
\end{aligned}$$

where $\alpha_2 > 0$ is a constant. Therefore, it follows from (18) and (25) that

$$\alpha_k \geq \frac{\alpha_1}{\alpha_2}$$

and this completes the proof. $\square$

3. Basic Results

To prove the global convergence for the proposed methods, we first present some lemmas. The following results are due to applying Lemma 2 to the SQP systems in the prediction step of the proposed methods.

Lemma 3. *For given $w^k = (x^k, y^k, \lambda^k) \in \mathbb{R}_{++}^n \times \mathbb{R}_{++}^m \times \mathbb{R}^l$, let $\tilde{w}^k$ be generated by (13a)–(13c). Then for any $w = (x, y, \lambda) \in \mathcal{W}$, we have*

$$(w - \tilde{w}^k)^\top d_2(w^k, \tilde{w}^k) \geq (w - \tilde{w}^k)^\top d_1(w^k, \tilde{w}^k)$$
$$- \frac{\mu}{2}\|x^k - \tilde{x}^k\|_R^2 - \frac{\mu}{2}\|y^k - \tilde{y}^k\|_S^2. \quad (28)$$

Proof. Applying Lemma 2 to (13a) by setting $u^k = x^k$, $u = \tilde{x}^k$, $v = x$ in (24) and

$$q(u) = \beta_k \left(f(\tilde{x}^k) - A^\top[\lambda^k - H(Ax^k + By^k - b)] \right) - \xi_x^k,$$

we get

$$(x - \tilde{x}^k)^\top \left\{ \beta_k \left(f(\tilde{x}^k) - A^\top[\lambda^k - H(Ax^k + By^k - b)] \right) - \xi_x^k \right\}$$
$$\geq \frac{1+\mu}{4} \left(\|\tilde{x}^k - x\|_R^2 - \|x^k - x\|_R^2 \right) + \frac{1-\mu}{4}\|x^k - \tilde{x}^k\|_R^2. \quad (29)$$

Recall

$$\frac{1}{2}(x - \tilde{x}^k)^\top R(x^k - \tilde{x}^k) = \frac{1}{4}\left(\|\tilde{x}^k - x\|_R^2 - \|x^k - x\|_R^2 \right) + \frac{1}{4}\|x^k - \tilde{x}^k\|_R^2. \quad (30)$$

Adding (29) and (30), we obtain

$$(x - \tilde{x}^k)^\top \left\{ \frac{(1+\mu)}{2} R(x^k - \tilde{x}^k) - \beta_k f(\tilde{x}^k) + \beta_k A^\top \tilde{\lambda}^k + \xi_x^k \right.$$
$$\left. - \beta_k A^\top H \left(A(x^k - \tilde{x}^k) + B(y^k - \tilde{y}^k) \right) \right\} \leq \frac{\mu}{2}\|x^k - \tilde{x}^k\|_R^2. \quad (31)$$

Similarly, applying Lemma 2 to (13b), substituting $u^k = y^k$, $u = \tilde{y}^k$, $v = y$ and replacing R, n with S, m, respectively, in (24) and

$$q(u) = \beta_k(g(\tilde{y}^k) - B^\top[\lambda^k - H(A\tilde{x}^k + By^k - b)]) - \xi_y^k,$$

we get

$$(y - \tilde{y}^k)^\top \left\{ \beta_k(g(\tilde{y}^k) - B^\top[\lambda^k - H(A\tilde{x}^k + By^k - b)]) - \xi_y^k \right\}$$
$$\geq \frac{1+\mu}{4} \left(\|\tilde{y}^k - y\|_S^2 - \|y^k - y\|_S^2 \right) + \frac{1-\mu}{4} \|y^k - \tilde{y}^k\|_S^2. \quad (32)$$

Recall

$$\frac{1}{2}(y - \tilde{y}^k)^\top S(y^k - \tilde{y}^k) = \frac{1}{4} \left(\|\tilde{y}^k - y\|_S^2 - \|y^k - y\|_S^2 \right)$$
$$+ \frac{1}{4} \|y^k - \tilde{y}^k\|_S^2. \quad (33)$$

Adding (32) and (33), we have

$$(y - \tilde{y}^k)^\top \left\{ \frac{(1+\mu)}{2} S(y^k - \tilde{y}^k) - \beta_k g(\tilde{y}^k) + \beta_k B^\top \tilde{\lambda}^k + \xi_y^k \right.$$
$$\left. - \beta_k B^\top H B(y^k - \tilde{y}^k) \right\} \leq \frac{\mu}{2} \|y^k - \tilde{y}^k\|_S^2. \quad (34)$$

It follows from (31), (34), (13c) and (15) that

$$(w - \tilde{w}^k)^\top (d_1(w^k, \tilde{w}^k) - d_2(w^k, \tilde{w}^k)) - \frac{\mu}{2} \|x^k - \tilde{x}^k\|_R^2$$
$$- \frac{\mu}{2} \|y^k - \tilde{y}^k\|_S^2 \leq 0 \quad (35)$$

and the assertion of this lemma is proved. □

Lemma 4. *For given* $w^k = (x^k, y^k, \lambda^k) \in \mathbb{R}_{++}^n \times \mathbb{R}_{++}^m \times \mathbb{R}^l$, *let* $\tilde{w}^k$ *be generated by* (13a)–(13c). *Then for any* $w^* = (x, y, \lambda) \in \mathcal{W}^*$, *we have*

$$(\tilde{w}^k - w^*)^\top d_2(w^k, \tilde{w}^k) \geq \varphi(w^k, \tilde{w}^k) - (w^k - \tilde{w}^k)^\top d_1(w^k, \tilde{w}^k)$$
$$+ \frac{\mu}{2} \|x^k - \tilde{x}^k\|_R^2 + \frac{\mu}{2} \|y^k - \tilde{y}^k\|_S^2. \quad (36)$$

Proof. Recalling the definition in (20),

$$\varphi(w^k, \tilde{w}^k) = (w^k - \tilde{w}^k)^\top d_1(w^k, \tilde{w}^k) - \frac{\mu}{2} \|x^k - \tilde{x}^k\|_R^2 - \frac{\mu}{2} \|y^k - \tilde{y}^k\|_S^2$$
$$+ \beta_k(\lambda^k - \tilde{\lambda}^k)^\top \left(A(x^k - \tilde{x}^k) + B(y^k - \tilde{y}^k) \right). \quad (37)$$

Using the monotonicity of f and g, we obtain

$$\begin{pmatrix} \tilde{x}^k - x^* \\ \tilde{y}^k - y^* \\ \tilde{\lambda}^k - \lambda^* \end{pmatrix}^\top \begin{pmatrix} f(\tilde{x}^k) - A^\top \tilde{\lambda}^k \\ g(\tilde{y}^k) - B^\top \tilde{\lambda}^k \\ A\tilde{x}^k + B\tilde{y}^k - b \end{pmatrix} \geq \begin{pmatrix} \tilde{x}^k - x^* \\ \tilde{y}^k - y^* \\ \tilde{\lambda}^k - \lambda^* \end{pmatrix}^\top$$

$$\begin{pmatrix} f(x^*) - A^\top \lambda^* \\ g(y^*) - B^\top \lambda^* \\ Ax^* + By^* - b \end{pmatrix} \geq 0. \tag{38}$$

It follows from (38) that

$$(\tilde{w}^k - w^*)^\top d_2(w^k, \tilde{w}^k) \geq \beta_k(\tilde{w}^k - w^*)^\top$$

$$\times \begin{pmatrix} A^\top H(A(x^k - \tilde{x}^k) + B(y^k - \tilde{y}^k)) \\ B^\top H(A(x^k - \tilde{x}^k) + B(y^k - \tilde{y}^k)) \\ 0 \end{pmatrix}$$

$$= \beta(A\tilde{x}^k + B\tilde{x}^k - b)^\top$$
$$\times H\left(A(x^k - \tilde{x}^k) + B(y^k - \tilde{y}^k)\right)$$
$$= \beta_k(\lambda^k - \tilde{\lambda}^k)^\top \left(A(x^k - \tilde{x}^k) + B(y^k - \tilde{y}^k)\right). \tag{39}$$

Combining (37) and the above inequality, we can get the assertion of this lemma. □

The following theorem provides a unified framework for proving the convergence of the new algorithm:

Theorem 2. Let $w^* \in \mathcal{W}^*, w_I^{k+1}(\alpha_k)$ be defined by (16),

$$\Theta_I(\alpha_k) := \|w^k - w^*\|^2 - \|w_I^{k+1}(\alpha_k) - w^*\|^2 \tag{40}$$

and

$$w_*^k := P_\mathcal{W}[w^k - \alpha_k d_1(w^k, \tilde{w}^k)],$$

then

$$\Theta_I(\alpha_k) \geq \sigma(\|w^k - w_*^k - \alpha_k d_1(w^k, \tilde{w}^k)\|^2$$
$$+ 2\alpha_k \varphi(w^k, \tilde{w}^k) - \alpha_k^2 \|d_1(w^k, \tilde{w}^k)\|^2). \tag{41}$$

Proof. Since $w^* \in \mathcal{W}^*$, it follows from (12) that

$$\|w_*^k - w^*\|^2 \le \|w^k - \alpha_k d_1(w^k, \tilde{w}^k) - w^*\|^2$$
$$- \|w^k - \alpha_k d_1(w^k, \tilde{w}^k) - w_*^k\|^2. \tag{42}$$

From (16), we get

$$\|w_I^{k+1}(\alpha_k) - w^*\|^2 = \|(1-\sigma)(w^k - w^*) + \sigma(w_*^k - w^*)\|^2$$
$$= (1-\sigma)^2 \|w^k - w^*\|^2 + \sigma^2 \|w_*^k - w^*\|^2$$
$$+ 2\sigma(1-\sigma)(w^k - w^*)^\top (w_*^k - w^*).$$

Using the following identity

$$2(a+b)^\top b = \|a+b\|^2 - \|a\|^2 + \|b\|^2$$

for $a = w^k - w_*^k$, $b = w_*^k - w^*$ and (42), we obtain

$$\|w_I^{k+1}(\alpha_k) - w^*\|^2$$
$$= (1-\sigma)^2 \|w^k - w^*\|^2 + \sigma^2 \|w_*^k - w^*\|^2 + \sigma(1-\sigma)\{\|w^k - w^*\|^2$$
$$- \|w^k - w_*^k\|^2 + \|w_*^k - w^*\|^2\}$$
$$= (1-\sigma)\|w^k - w^*\|^2 + \sigma\|w_*^k - w^*\|^2 - \sigma(1-\sigma)\|w^k - w_*^k\|^2$$
$$\le (1-\sigma)\|w^k - w^*\|^2 + \sigma\|w^k - \alpha_k d_1(w^k, \tilde{w}^k) - w^*\|^2$$
$$- \sigma\|w^k - \alpha_k d_1(w^k, \tilde{w}^k) - w_*^k\|^2 - \sigma(1-\sigma)\|w^k - w_*^k\|^2$$
$$\le (1-\sigma)\|w^k - w^*\|^2 + \sigma\|w^k - \alpha_k d_1(w^k, \tilde{w}^k) - w^*\|^2$$
$$- \sigma\|w^k - \alpha_k d_1(w^k, \tilde{w}^k) - w_*^k\|^2. \tag{43}$$

Using the definition of $\Theta_I(\alpha_k)$ and (43), we get

$$\Theta_I(\alpha_k) \ge \sigma\|w^k - w_*^k\|^2 + 2\sigma\alpha_k(w_*^k - w^k)^\top d_1(w^k, \tilde{w}^k)$$
$$+ 2\sigma\alpha_k(w^k - w^*)^\top d_1(w^k, \tilde{w}^k). \tag{44}$$

It follows from (39) that

$$(w^k - w^*)^\top d_2(w^k, \tilde{w}^k) \ge (w^k - \tilde{w}^k)^\top d_2(w^k, \tilde{w}^k)$$
$$+ \beta_k(\lambda^k - \tilde{\lambda}^k)^\top \big(A(x^k - \tilde{x}^k) + B(y^k - \tilde{y}^k)\big). \tag{45}$$

Applying (28) (with $w = w^*$), we get
$$(w^* - \tilde{w}^k)^\top d_2(w^k, \tilde{w}^k) \geq (w^* - \tilde{w}^k)^\top d_1(w^k, \tilde{w}^k)$$
$$-\frac{\mu}{2}\|x^k - \tilde{x}^k\|_R^2 - \frac{\mu}{2}\|y^k - \tilde{y}^k\|_S^2. \quad (46)$$

Adding (45) and (46), we obtain
$$(w^k - w^*)^\top d_1(w^k, \tilde{w}^k) \geq (w^k - \tilde{w}^k)^\top d_1(w^k, \tilde{w}^k) + \beta_k(\lambda^k - \tilde{\lambda}^k)^\top$$
$$\times \left(A(x^k - \tilde{x}^k) + B(y^k - \tilde{y}^k)\right)$$
$$-\frac{\mu}{2}\|x^k - \tilde{x}^k\|_R^2 - \frac{\mu}{2}\|y^k - \tilde{y}^k\|_S^2. \quad (47)$$

Applying (47) to the last term on the right side of (44) and using the notation of $\varphi(w^k, \tilde{w}^k)$ in (37), we obtain
$$\Theta_I(\alpha_k) \geq \sigma\|w^k - w_*^k\|^2 + 2\sigma\alpha_k(w_*^k - w^k)^\top d_1(w^k, \tilde{w}^k)$$
$$+2\sigma\alpha_k\{(w^k - \tilde{w}^k)^\top d_1(w^k, \tilde{w}^k) + \beta_k(\lambda^k - \tilde{\lambda}^k)^\top$$
$$\times \left(A(x^k - \tilde{x}^k) + B(y^k - \tilde{y}^k)\right)$$
$$-\frac{\mu}{2}\|x^k - \tilde{x}^k\|_R^2 - \frac{\mu}{2}\|y^k - \tilde{y}^k\|_S^2\}$$
$$= \sigma\Big\{\|w^k - w_*^k\|^2 + 2\alpha_k(w_*^k - w^k)^\top d_1(w^k, \tilde{w}^k)$$
$$2\alpha_k(w^k - \tilde{w}^k)^\top d_1(w^k, \tilde{w}^k) + 2\alpha_k\beta_k(\lambda^k - \tilde{\lambda}^k)^\top$$
$$\times \left(A(x^k - \tilde{x}^k) + B(y^k - \tilde{y}^k)\right)$$
$$-\alpha_k\mu\|x^k - \tilde{x}^k\|_R^2 - \alpha_k\mu\|y^k - \tilde{y}^k\|_S^2\Big\}.$$
$$= \sigma\Big\{\|w^k - w_*^k - \alpha_k d_1(w^k, \tilde{w}^k)\|^2 - \alpha_k^2\|d_1(w^k, \tilde{w}^k)\|^2$$
$$+2\alpha_k\varphi(w^k, \tilde{w}^k)\Big\}$$

and the theorem is proved. □

Theorem 3. *Let* $w^* \in \mathcal{W}^*, w_{II}^{k+1}(\alpha_k)$ *be defined by* (17),
$$\Theta_{II}(\alpha_k) := \|w^k - w^*\|^2 - \|w_{II}^{k+1}(\alpha_k) - w^*\|^2 \quad (48)$$
and
$$w_p^k := P_{\mathcal{W}}[w^k - \alpha_k d_2(w^k, \tilde{w}^k)],$$

then

$$\Theta_{II}(\alpha_k) \geq \sigma(\|w^k - w_p^k - \alpha_k d_1(w^k, \tilde{w}^k)\|^2 \\ + 2\alpha_k \varphi(w^k, \tilde{w}^k) - \alpha_k^2 \|d_1(w^k, \tilde{w}^k)\|^2). \quad (49)$$

Proof. Similar as in (44), we can prove that

$$\Theta_{II}(\alpha_k) \geq \sigma \|w^k - w_p^k\|^2 + 2\sigma\alpha_k(w_p^k - w^*)^\top d_2(w^k, \tilde{w}^k). \quad (50)$$

Applying (28) (with $w = w_p^k$), we obtain

$$(w_p^k - \tilde{w}^k)^\top d_2(w^k, \tilde{w}^k) \geq (w_p^k - \tilde{w}^k)^\top d_1(w^k, \tilde{w}^k) \\ - \frac{\mu}{2}\|x^k - \tilde{x}^k\|_R^2 - \frac{\mu}{2}\|y^k - \tilde{y}^k\|_S^2. \quad (51)$$

Adding (36) and (51), we get

$$(w_p^k - w^*)^\top d_2(w^k, \tilde{w}^k) \geq (w_p^k - w^k)^\top d_1(w^k, w^k) + \varphi(w^k, \tilde{w}^k). \quad (52)$$

Applying (52) to the last term on the right side of (50), we obtain

$$\Theta_{II}(\alpha_k) \geq \sigma\|w^k - w_p^k\|^2 + 2\sigma\alpha_k(w_p^k - w^k)^\top d_1(w^k, \tilde{w}^k) \\ + 2\sigma\alpha_k\varphi(w^k, \tilde{w}^k) \\ = \sigma\{\|w^k - w_p^k - \alpha_k d_1(w^k, \tilde{w}^k)\|^2 - \alpha_k^2\|d_1(w^k, \tilde{w}^k)\|^2 \\ + 2\alpha_k\varphi(w^k, \tilde{w}^k)\}$$

and the theorem is proved. $\square$

4. Convergence of the Proposed Method

In this section, we prove the global convergence of the proposed method. The following result shows the contraction of the sequence generated by ADM with SQP regularization (13a)–(13c), based on which its global convergence can be established easily. From the computational point of view, a relaxation factor $\gamma \in (0, 2)$ is preferable in the correction.

Theorem 4. *Let $w^* \in \mathcal{W}^*$ be a solution of SVI and let $w^{k+1}(\gamma \alpha_k)$ be generated by* (16) *or* (17). *Then w^k and $\tilde{w}^k$ are bounded, and*

$$\|w^{k+1}(\gamma \alpha_k) - w^*\|^2 \leq \|w^k - w^*\|^2 - c\|w^k - \tilde{w}^k\|^2, \qquad (53)$$

where

$$c := \frac{\sigma \gamma (2 - \gamma) \alpha_1^2}{\alpha_2} > 0.$$

Proof. It follows from (41), (49), (25) and (26) that

$$\|w^{k+1}(\gamma \alpha_k) - w^*\|^2$$
$$\leq \|w^k - w^*\|^2 - \sigma(2\gamma \alpha_k \varphi(w^k, \tilde{w}^k) - \gamma^2 \alpha_k^2 \|d_1(w^k, \tilde{w}^k)\|^2)$$
$$= \|w^k - w^*\|^2 - \gamma(2 - \gamma) \alpha_k \sigma \varphi(w^k, \tilde{w}^k)$$
$$\leq \|w^k - w^*\|^2 - \frac{\sigma \gamma (2-\gamma) \alpha_1^2}{\alpha_2} \|w^k - \tilde{w}^k\|^2.$$

Since $\gamma \in (0, 2)$, we have

$$\|w^{k+1} - w^*\| \leq \|w^k - w^*\| \leq \cdots \leq \|w^0 - w^*\|,$$

and thus, $\{w^k\}$ is a bounded sequence.

It follows from (53) that

$$\sum_{k=0}^{\infty} c\|w^k - \tilde{w}^k\|^2 < +\infty,$$

which means that

$$\lim_{k \to \infty} \|w^k - \tilde{w}^k\| = 0. \qquad (54)$$

Since $\{w^k\}$ is a bounded sequence, we conclude that $\{\tilde{w}^k\}$ is also bounded. □

Now, we are ready to prove the convergence of the proposed methods.

Theorem 5. *The sequence $\{w^k\}$ generated by the proposed method converges to some w^∞ which is a solution of SVI.*

Proof. Since $\{w^k\}$ is bounded, it has at least one cluster point. Let w^∞ be a cluster point of $\{w^k\}$ and the subsequence $\{w^{k_j}\}$ converges to w^∞, since $\mathcal{W}$ is a closed set, we have $w^\infty \in \mathcal{W}$. By the construction of β_k, we have that $0 < \beta_k \le \beta_0, \forall k$. It follows from (54) that

$$\lim_{j\to\infty} d_1(w^{k_j}, \tilde{w}^{k_j}) = 0 \quad \text{and} \quad \lim_{j\to\infty} \frac{d_2(w^{k_j}, \tilde{w}^{k_j})}{\beta_{k_j}} = Q(w^\infty). \tag{55}$$

Moreover, (55) and (28) imply that

$$\lim_{j\to\infty} \frac{(w - w^{k_j})^\top d_2(w^{k_j}, \tilde{w}^{k_j})}{\beta_{k_j}} \ge 0, \quad \forall w \in \mathcal{W}, \tag{56}$$

and consequently

$$(w - w^\infty)^\top Q(w^\infty) \ge 0, \quad \forall w \in \mathcal{W},$$

which means that w^∞ is a solution of SVI.

Now we prove that the sequence $\{w^k\}$ converges to w^∞. Since

$$\lim_{k\to\infty} \|w^k - \tilde{w}^k\| = 0 \quad \text{and} \quad \{\tilde{w}^{k_j}\} \to w^\infty,$$

for any $\epsilon > 0$, there exists an $l > 0$ such that

$$\|\tilde{w}^{k_l} - w^\infty\| < \frac{\epsilon}{2} \quad \text{and} \quad \|w^{k_l} - \tilde{w}^{k_l}\| < \frac{\epsilon}{2}. \tag{57}$$

Therefore, for any $k \ge k_l$, it follows from (53) and (57) that

$$\|w^k - w^\infty\| \le \|w^{k_l} - w^\infty\| \le \|w^{k_l} - \tilde{w}^{k_l}\| + \|\tilde{w}^{k_l} - w^\infty\| < \epsilon.$$

This implies that the sequence $\{w^k\}$ converges to w^∞ which is a solution of SVI. $\square$

5. Comparison of Two Methods

Note that we use the identical step size α_k in both prediction–correction methods. Then we show the difference in the following theorem:

Theorem 6. Let $w^* \in \mathcal{W}^*$ be a solution of SVI. Let $\Theta_I(\alpha_k)$ and $\Theta_{II}(\alpha_k)$ be defined in (40) and (48), respectively. Then we have the following:

$$\Theta_I(\alpha_k) \geq \Upsilon_I(\alpha_k) := \sigma(\|w^k - w_*^k - \alpha_k d_1(w^k, \tilde{w}^k)\|^2$$
$$+ 2\alpha_k \varphi(w^k, \tilde{w}^k) - \alpha_k^2 \|d_1(w^k, \tilde{w}^k)\|^2), \quad (58)$$

$$\Theta_{II}(\alpha_k) \geq \Upsilon_{II}(\alpha_k) := \sigma(\|w^k - w_p^k - \alpha_k d_1(w^k, \tilde{w}^k)\|^2$$
$$+ 2\alpha_k \varphi(w^k, \tilde{w}^k) - \alpha_k^2 \|d_1(w^k, \tilde{w}^k)\|^2) \quad (59)$$

and

$$\Upsilon_I(\alpha_k) \leq \Upsilon_{II}(\alpha_k). \quad (60)$$

Proof. The assertions (58) and (59) follow from Theorems 2 and 3, respectively. Now, we prove the last assertion (60) of the theorem. Recall that $w_p^k \in \mathcal{W}$ and $w_*^k := P_{\mathcal{W}}[w^k - \alpha_k d_1(w^k, \tilde{w}^k)]$. We have

$$\|w_*^k - w_p^k\|^2 = \|P_{\mathcal{W}}[w^k - \alpha_k d_1(w^k, \tilde{w}^k)] - w_p^k\|^2$$
$$\leq \|w^k - \alpha_k d_1(w^k, \tilde{w}^k) - w_p^k\|^2$$
$$- \|\|w^k - \alpha_k d_1(w^k, \tilde{w}^k) - w_*^k\|^2, \quad (61)$$

which implies that the assertion (60) holds. □

Theorem 6 shows theoretically that the method with correction II is expected to make more progress than the method with correction I at each iteration.

6. Preliminary Computational Results

In order to verify the theoretical assertions, we consider the following optimization problem with matrix variables:

$$\min\left\{\frac{1}{2}\|X - C\|_F^2 : X \in S_+^n\right\}, \quad (62)$$

where $\|\cdot\|_F$ is the matrix Fröbenius norm, that is,

$$\|C\|_F = \left(\sum_{i=1}^{n}\sum_{j=1}^{n}|C_{ij}|^2\right)^{1/2},$$

$$S_+^n = \left\{H \in \mathbb{R}^{n\times n} : H^\top = H,\ H \succeq 0\right\}.$$

It has been shown [3] that problem (62) can be converted to the following variational inequality: find $u^* = (X^*, Y^*, Z^*) \in \mathcal{W} = S_+^n \times S_+^n \times \mathbb{R}^{n\times n}$ such that

$$\begin{cases} \langle X - X^*, (X^* - C) - Z^* \rangle \geq 0, \\ \langle Y - Y^*, (Y^* - C) + Z^* \rangle \geq 0, \quad \forall\, u = (X, Y, Z) \in \mathcal{W}, \\ X^* - Y^* = 0. \end{cases} \quad (63)$$

Problem (63) is a special case of (2)–(3) with matrix variables, where $A = I_{n\times n}$, $B = -I_{n\times n}$, $b = 0$, $f(X) = X - C$, $g(Y) = Y - C$ and $\mathcal{W} = S_+^n \times S_+^n \times \mathbb{R}^{n\times n}$.

For simplification, we take $R = rI_{n\times n}$, $S = sI_{n\times n}$ and $H = I_{n\times n}$, where $r > 0$ and $s > 0$ are scalars. In all tests we take $\gamma = 1.98$, $\mu = 0.1$, $\sigma = 0.95$, $C = \text{rand}(n)$ and $(X^0, Y^0, Z^0) = (I_{n\times n}, I_{n\times n}, 0_{n\times n})$ as the initial point in the test. The iteration is stopped as soon as

$$\max\left\{\|X^k - \tilde{X}^k\|, \|Y^k - \tilde{Y}^k\|, \|Z^k - \tilde{Z}^k\|\right\} \leq 10^{-5}.$$

All codes were written in MATLAB; we compare the proposed method with that in Ref. [32]. The iteration numbers, denoted by k, and the computational time for problem (62) with different dimensions are given in Tables 1–2.

Tables 1–2 show the efficiency of the proposed method and its superiority to the methods of Refs. [3, 6, 32] in terms of number of iteration and CPU time.

Table 1. Numerical results for the problem (62) with $r = s = 5$.

Dimension of the problem	The proposed method k	CPU (Sec.)	The method in Ref. [6] k	CPU (Sec.)	The method in Ref. [3] k	CPU (Sec.)	The method in Ref. [32] k	CPU (Sec.)
$n = 300$	11	3.9	78	6.47	98	7.43	100	8.79
$n = 500$	12	5.34	82	18.22	104	22.04	107	24.32
$n = 700$	12	8.81	84	48.82	106	50.61	110	60.91
$n = 800$	12	10.85	85	69.72	110	77.22	111	92.24

Table 2. Numerical results for the problem (62) with $r = s = 10$.

Dimension of the problem	The proposed method k	CPU (Sec.)	The method in Ref. [6] k	CPU (Sec.)	The method in Ref. [3] k	CPU (Sec.)	The method in Ref. [32] k	CPU (Sec.)
$n = 300$	10	3.24	139	9.27	161	10.32	173	12.61
$n = 500$	11	4.8	146	34.15	170	42.52	170	43.65
$n = 700$	12	7.51	151	70.46	178	79.75	183	81.85
$n = 800$	12	10.71	159	110.41	182	104.14	183	129.62

7. Conclusions

In this chapter, by combining the alternating direction method with the SQP method, we proposed two descent SQP alternating direction methods for solving variational inequality problems with separable structure. The main contributions of this chapter are as follows: first, the predictor is obtained via solving the SQP system approximately under significantly relaxed accuracy criterion, second, in order to ensure the efficiency of the algorithms, we adjust the penalty parameter by using a self-adaptive technique, it seems that the computational time and the iteration numbers are not sensitive to the problem size, and third, their numerical benefit is quite significant in comparisons with the previously known results.

References

[1] A. Auslender, M. Teboulle and S. Ben-Tiba, A logarithmic-quadratic proximal method for variational inequalities, *Comput. Optim. Appl.* **12**, 31–40 (1999).

[2] A. Bnouhachem, H. Benazza and M. Khalfaoui, An inexact alternating direction method for solving a class of structured variational inequalities, *Appl. Math. Comput.* **219**, 7837–7846 (2013).

[3] A. Bnouhachem, On LQP alternating direction method for solving variational, *J. Inequal. Appl.* **2014**(80), 1–15 (2014).

[4] A. Bnouhachem and M. H. Xu, An inexact LQP alternating direction method for solving a class of structured variational inequalities, *Comput. Math. Appl.* **67**, 671–680 (2014).

[5] A. Bnouhachem and Q. H. Ansari, A descent LQP alternating direction method for solving variational inequality problems with separable structure, *Appl. Math. Comput.* **246**, 519–532 (2014).

[6] A. Bnouhachem and A. Hamdi, Parallel LQP alternating direction method for solving variational inequality problems with separable structure, *J. Inequal. Appl.* **2014**(392), 1–14 (2014).

[7] A. Alhomaidan, A. Bnouhachem and A. Latif, An LQP-SQP alternating direction for solving variational inequality problems with separable structure, *J. Nonlinear Sci. Appl.* **10(12)**, 6246–6261 (2017).

[8] A. Bnouhachem and T. M. Rassias, An inexact alternating direction method with SQP regularization for the structured variational inequalities, *Int. J. Nonlinear Anal. Appl.* **8**(1), 269–289 (2017).

[9] A. Bnouhachem, Q. H. Ansari and S. Al Homidan, SQP alternating direction for structured variational inequality, *J. Nonlinear Convex Anal.* **19**(3), 461–476 (2018).

[10] A. Bnouhachem and T. M. Rassias, A new descent alternating direction method with LQP regularization for the structured variational inequalities, *Optim. Lett.* **13**(1), 175–192 (2019).

[11] S. Boyd, N. Parikh, E. Chu, B. Peleato and J. Eckstein, Distributed optimization and statistical learning via the alternating direction method of multipliers, *Found. Trends Mach. Learn.* **3**(1), 1–122 (2010).

[12] G. Chen and M. Teboulle, A proximal-based decomposition method for convex minimization problems, *Math. Program.* **64**, 81–101 (1994).

[13] J. Eckstein and D. P. Bertsekas, On the Douglas-Rachford splitting method and the proximal point algorithm for maximal monotone operators, *Math. Program.* **55**, 293–318 (1992).

[14] J. Eckstein, Some saddle-function splitting methods for convex programming, *Optim. Methods Softw.* **4**, 75–83 (1994).

[15] J. Eckstein and M. Fukushima, Reformulations and applications of the alternating direction method of multipliers, In W. W. Hager, D. W. Hearn and P. M. Pardalos (eds.) *Large Scale Optimization* (Springer US, 1994), pp. 115–134.
[16] F. Facchinei and J. S. Pang, *Finite-Dimensional Variational Inequalities and Complementarity Problems*. Springer Series in Operations Research, Vols. I and II (Springer, New York, 2003).
[17] M. Fortin and R. Glowinski (eds.) *Augmented Lagrangian Methods: Applications to the Solution of Boundary-Valued Problems* (North-Holland, Amsterdam, 1983).
[18] D. Gabay and B. Mercier, A dual algorithm for the solution of nonlinear variational problems via finite-element approximations, *Comput. Math. Appl.* **2**, 17–40 (1976).
[19] D. Gabay, Applications of the method of multipliers to variational inequalities. In M. Fortin and R. Glowinski (eds.) *Augmented Lagrange Methods: Applications to the Solution of Boundary-valued Problems* (North-Holland, Amsterdam, 1983), pp. 299–331.
[20] R. Glowinski and A. Marrocco, Sur l'approximation par éléments finis d'ordre un et la résolution par pénalisation-dualité d'une classe de problémes de Dirichlet nonlinéaires, *Rev. Fr. Autom. Inform. Rech. Opér. Anal. Numér.* **2**, 41–76 (1975).
[21] R. Glowinski, *Numerical Methods for Nonlinear Variational Problems* (Springer-Verlag, New York, 1984).
[22] R. Glowinski and P. Le Tallec, *Augmented Lagrangian and Operator-Splitting Methods in Nonlinear Mechanics*. SIAM Studies in Applied Mathematics (SIAM, Philadelphia, 1989).
[23] B. S. He, H. Yang and S. L. Wang, Alternating directions method with self-adaptive penalty parameters for monotone variational inequalities, *J. Optim. Theory Appl.* **106**, 349–368 (2000).
[24] B. S. He, L. Z. Liao, D. R. Han and H. Yang, A new inexact alternating directions method for monotone variational inequalities, *Math. Program.* **92**, 103–118 (2002).
[25] B. S. He, Parallel splitting augmented Lagrangian methods for monotone structured variational inequalities, *Comput. Optim. Appl.* **42**, 195–212 (2009).
[26] B. S. He, M. Tao and X. M. Yuan, Alternating direction method with Gaussian back substitution for separable convex programming, *SIAM J. Optim.* **22**, 313–340 (2012).
[27] H. Jiang, W. Deng and Z. Shen, Surveillance video processing using compressive sensing, *Inverse Probl. Imaging.* **6**(2), 201–214 (2012).
[28] Z. K. Jiang and A. Bnouhachem, A projection-based prediction-correction method for structured monotone variational inequalities, *Appl. Math. Comput.* **202**, 747–759 (2008).

[29] Z. K. Jiang and X. M. Yuan, New parallel descent-like method for solving a class of variational inequalities, *J. Optim. Theory Appl.* **145**, 311–323 (2010).

[30] S. Kontogiorgis and R. R. Meyer, A variable-penalty alternating directions method for convex optimization, *Math. Program.* **83**, 29–53 (1998).

[31] L. S. Hou, On the $O(1/t)$ convergence rate of the parallel descent-like method and parallel splitting augmented Lagrangian method for solving a class of variational inequalities, *Appl. Math. Comput.* **219**, 5862–5869 (2013).

[32] M. Li, A hybrid LQP-based method for structured variational inequalities, *Int. J. Comput. Math.* **89**(10), 1412–1425 (2012).

[33] A. Nagurney and P. Ramanujam, Transportation network policy modeling with goal targets and generalized penalty functions, *Transp. Sci.* **30**, 3–13 (1996).

[34] M. Tao and X. M. Yuan, Recovering low-rank and sparse components of matrices from incomplete and noisy observations, *SIAM J. Optim.* **21**(1), 57–81 (2011).

[35] M. Tao M. and X. M. Yuan, On the $O(1/t)$ convergence rate of alternating direction method with logarithmic-quadratic proximal regularization, *SIAM J. Optim.* **22**(4), 1431–1448 (2012).

[36] M. Teboulle, Convergence of proximal-like algorithms, *SIAM J. Optim.* **7**, 1069–1083 (1997).

[37] P. Tseng, Alternating projection-proximal methods for convex programming and variational inequalities, *SIAM J. Optim.* **7**, 951–965 (1997).

[38] K. Wang, L. L. Xu and D. R. Han, A new parallel splitting descent method for structured variational inequalities, *J. Ind. Manag. Optim.* **10**(2), 461–476 (2014).

[39] Y. Wang, J. Yang, W. Yin and Y. Zhang, A new alternating minimization algorithm for total variation image reconstruction, *SIAM J. Imaging Sci.* **1**(3), 248–272 (2008).

[40] J. Yang and Y. Zhang, Alternating direction algorithms for l1-problems in compressive sensing, *SIAM J. Sci. Comput.* **33**(1–2), 250–278 (2011).

[41] X. M. Yuan and M. Li, An LQP-based decomposition method for solving a class of variational inequalities, *SIAM J. Optim.* **21**(4), 1309–1318 (2011).

© 2024 World Scientific Publishing Company
https://doi.org/10.1142/9789811267048_0007

Chapter 7

Solvability for a Class of Unilateral Contact Problems with Friction and Damage

Oanh Chau[*,‡], Arnaud Heibig[†,§], and Adrien Petrov[†,¶]

[*]*University of La Réunion. Department of Mathematics, Saint-Denis, La Réunion, France*
[†]*Université de Lyon, CNRS, INSA de Lyon, Villeurbanne, France*
[‡]*oanh.chau@univ-reunion.fr*
[§]*arnaud.heibig@insa-lyon.fr*
[¶]*apetrov@math.univ-lyon1.fr*

This chapter deals with a class of frictional unilateral contact problems with damage, for viscoelastic bodies. This model is composed by the momentum equilibrium equation combined with a heat-transfer equation coupled to a flow rule describing the damage evolution. The contact and friction are modeled with contact conditions formulated in displacement velocities and Coulomb's dry friction law, respectively. The existence and uniqueness results are obtained by using some general results on the evolutionary variational inequalities, monotone operator theory and a fixed-point argument.

1. Introduction

Contact problems have applications in many fields of solid mechanics, namely, in manufacturing, machine dynamics and biomechanics. Such problems were intensively studied in the last decades and a considerable mathematical literature around the topics of modeling, mathematical and numerical analysis is devoted to dynamic and quasi-static frictional contact problems. Different types of frictional contact models with nonlinear viscoelastic or elasto-plastic materials in the framework of linearized infinitesimal deformations and using the abstract variational inequalities with monotonicity and convexity are considered in Refs. [5, 8, 10–12]. Some extensions to nonconvex contact conditions with nonmonotone and some possible multivalued constitutive laws led to the active domain of nonsmooth mechanic within the framework of the so-called hemivariational inequalities, see Ref. [13]. The reader is also referred to Ref. [9] as well as to the references therein.

The aim of this chapter is thus to prove an existence result for a thermoviscoelastodynamic with frictional unilateral contact boundary conditions and undergo the thermal expansion and damage. More precisely, we consider a viscoelastic body is subjected to unilateral contact, formulated in displacement velocities, with an obstacle on the other part of the boundary. The formulation of the contact conditions in velocities is rather realistic for a short time interval and under the assumption that the initial gap between the body and the obstacle vanishes; for further details, the reader is referred to Ref. [4]. The approach to prove the solvability is similar to the one developed in Ref. [3], the main novelty here lies in a general long memory material law, depending on the time and on the unknown temperature and damage fields. Note that the thermal expansion is described by a general nonlinear equation, involving the gradient of temperature, the deformation velocity and the associated boundary condition that are defined by a differential inclusion in a nonconvex framework. Another issue is that, when the clamped condition is removed, Korn's inequality cannot be used anymore. The problem considered in this chapter appears as a semicoercive and strongly nonlinear due to the friction. Note that semicoercive problems is originally studied in Ref. [5] for some Coulomb's friction models, where the inertial term of the dynamic process is used in order to compensate the coerciveness losses in *a priori* estimates. The variational formulation of the mechanical problem leads to a new nonstandard model of the system defined by a second-order quasi-variational inequality on the displacement field, coupled with one nonlinear inequality for the temperature and with a variational inequality on the damage field. Then by using classical results on evolution variational inequalities, with monotone operators and adopting fixed-point approach used in Ref. [2], the solvability to our problem follows.

This chapter is organized as follows. The equations are described in Section 2. The assumptions on the data and the weak formulation for the problem are introduced and the main result is presented (local existence result) in Section 3. Then, Section 4 is devoted to its proof. The proof of the solvability relies on some *a priori* estimates and a fixed-point argument.

2. The Mathematical Formulation

We consider a viscoelastic body occupying a domain Ω of $\mathbb{R}^d$ ($d = 2, 3$). The body is acted upon by time-dependent volume forces and surface tractions. The frictional contact may take place with an obstacle and thermal effects may occur. The boundary $\partial\Omega$ is supposed to be partitioned into two disjoint measurable parts Γ_F and Γ_C. Let $\boldsymbol{u}(t, x)$, $\theta(t, x)$ and $z(t, x)$ be the displacement, temperature and damage at time t of the material point of spatial coordinate $x \in \Omega$, respectively. We agree that if we write a function of time and space as a function of three variables, the first variable is the time, the second variable is the normal space variable x_1, the last variable is the tangential space variable x'. Let $\boldsymbol{f}_{\text{ext}}(t, x_1, x')$ and $\boldsymbol{f}_{\text{tra}}(t, x')$ denote the external forces density acting on Ω and the surface traction density applied on Γ_F, respectively. Then, given $T > 0$, the mathematical problem is formulated as follows:

$$\ddot{\boldsymbol{u}} - \operatorname{div} \boldsymbol{\sigma} = \boldsymbol{f}_{\text{ext}} \quad \text{in } (0, T) \times \Omega, \tag{1a}$$

$$\dot{\theta} - \operatorname{div}(\kappa(\cdot, \nabla \theta)) + \mathcal{N}(\cdot, \theta, \dot{\boldsymbol{u}}) = D_{\text{vis}}(\cdot, \boldsymbol{\varepsilon}(\dot{\boldsymbol{u}})) + q \quad \text{in } (0, T) \times \Omega, \tag{1b}$$

$$\dot{z} - \gamma \Delta z + \partial I_{[0,1]}(z) \ni \varphi(\boldsymbol{\sigma}, \boldsymbol{\varepsilon}(\boldsymbol{u}), z) \quad \text{in } (0, T) \times \Omega, \tag{1c}$$

where κ is the *heat conductivity*, $\mathcal{N}$ is a nonlinear term, q is the *heat source density*, D_{vis} is the *viscosity deformation heat* and $\boldsymbol{\varepsilon}(\boldsymbol{u}) \stackrel{\text{def}}{=} \frac{1}{2}(\nabla \boldsymbol{u} + \nabla \boldsymbol{u}^{\mathsf{T}})$ is the *infinitesimal strain tensor*.

As a typical example, we may consider $\kappa(\cdot, \nabla\theta) = (k_{ij}(\cdot))\nabla\theta$, $\mathcal{N}(\cdot, \theta, \dot{\boldsymbol{u}}) = k_1\theta + k_2\theta[\text{tr}(\varepsilon(\dot{\boldsymbol{u}}))]^{2/3}$ and $D_{\text{vis}}(\cdot, \boldsymbol{\varepsilon}(\dot{\boldsymbol{u}})) = -c_{ij}(\cdot)\varepsilon_{ij}(\dot{\boldsymbol{u}})$, where $k_{ij} \in L^{\infty}((0,T) \times \Omega, \mathbb{R}^+)$, $k_1, k_2 \in L^{\infty}(\Omega, \mathbb{R}^+)$ and $c_{ij} \in L^{\infty}((0,T) \times \Omega, \mathbb{R}^+)$ denote the conductivity tensor coefficients and the symmetric thermal expansion tensor coefficients.

To continue, $I_{[0,1]}$ is the *indicator function* of $[0,1]$, defined by $I_{[0,1]}(\xi) = +\infty$, $\forall \xi \in \mathbb{R} \setminus [0,1]$ and $I_{[0,1]}(\xi) = 0$, $\forall \xi \in [0,1]$, φ is a given constitutive function describing the sources of damage resulting from compression or tension and $\boldsymbol{\sigma}$ is the *stress tensor* defined as follows:

$$\boldsymbol{\sigma}(t,x) \stackrel{\text{def}}{=} \mathbf{E}(t,x,\boldsymbol{\varepsilon}(\boldsymbol{u}(t,x)), z(t,x)) + \mathbf{A}(t,x,\boldsymbol{\varepsilon}(\boldsymbol{u}(\dot{t},x)))$$

$$+ \int_0^t \mathbf{B}(t-s,x,\boldsymbol{\varepsilon}(\boldsymbol{u}(s,x)), z(s,x))\,\mathrm{d}s$$

$$+ C_{\exp}(t,x,\theta(t,x)) \quad \text{in} \in (0,T) \times \Omega, \tag{2}$$

where $\mathbf{A}$, $\mathbf{E}$, $\mathbf{B}$ and $C_{\exp}$ denote the *viscosity, elasticity, relaxation* and *thermal expansion tensors*, respectively. In the above, and throughout this chapter, notations $(\dot{\ })$ and ∂ denote the time derivative and the subdifferential in the sense of convex analysis (see Ref. [1]), respectively, and $(\cdot)^{\mathsf{T}}$ stands for the transpose of a tensor. Note that the evolution of the microscopic cracks responsible for the damage is described by the parabolic differential inclusion (1c) of the damage function z satisfying $0 \le z \le 1$, where γ is a positive constant and φ is a given constitutive function describing the damage source in the system. The value of the damage function represents the percentage of the safe part; $z = 1$ means that the body is undamaged while $z = 0$ means that the body is completely damaged. The reader is referred to Refs. [6, 7] for further details concerning the damage model considered in this chapter.

We now turn to define the boundary conditions. To this aim, let us introduce some notations; $u_\nu \stackrel{\text{def}}{=} \boldsymbol{u} \cdot \nu$ and $\sigma_\nu \stackrel{\text{def}}{=} \boldsymbol{u} \cdot \nu$ denote the *normal displacement* and *normal stress*, respectively, while $\boldsymbol{u}_\tau \stackrel{\text{def}}{=} \boldsymbol{u} - u_\nu \nu$, $\boldsymbol{\sigma}_\tau \stackrel{\text{def}}{=} \sigma\nu - \sigma_\nu \nu$ and $\mathscr{F}$ are the *tangential displacement, tangential stress* and *coefficient of friction*, respectively.

The boundary conditions are prescribed by

$$\boldsymbol{\sigma}\nu = \boldsymbol{f}_{\text{tra}} \quad \text{on } (0,T) \times \Gamma_{\text{F}}, \tag{3a}$$

$$0 \geq \dot{u}_\nu \perp \sigma_\nu \leq 0 \quad \text{on } (0,T) \times \Gamma_{\text{C}}, \tag{3b}$$

$$|\boldsymbol{\sigma}_\tau| \leq p_\tau(\cdot, u_\nu) \quad \text{on } (0,T) \times \Gamma_{\text{C}}, \tag{3c}$$

$$\begin{cases} \text{if } |\boldsymbol{\sigma}_\tau| < p_\tau(\cdot, u_\nu) \text{ then } \dot{\boldsymbol{u}}_\tau = \boldsymbol{0}, \\ \text{if } |\boldsymbol{\sigma}_\tau| = p_\tau(\cdot, u_\nu) \text{ then } \exists \lambda \geq 0 : \dot{\boldsymbol{u}}_\tau = -\lambda \boldsymbol{\sigma}_\tau, \end{cases}$$

$$\frac{\partial z}{\partial \nu} = 0 \quad \text{on } (0,T) \times \partial\Omega, \tag{3d}$$

$$-\kappa(\cdot, \nabla\theta)\nu \in \partial\mathscr{G}(\cdot, \theta) \quad \text{on } (0,T) \times \Gamma_{\text{C}}, \tag{3e}$$

$$\theta = 0 \quad \text{on } (0,T) \times \Gamma_{\text{F}}. \tag{3f}$$

The contact conditions (3b) are formulated in displacement velocities and they can be interpreted as a first-order approximation in time of the so-called Signorini's condition; the reader is referred to Ref. [14] for further details on this condition. The orthogonality has the natural meaning: if we have enough regularity, it means that the product $\dot{u}_\nu \sigma_\nu$ vanishes almost everywhere on $(0,T) \times \Gamma_{\text{C}}$. If we do not have enough regularity, the above inequality is integrated on an appropriate set of test functions, yielding a weak formulation for the unilateral condition (see Ref. [14]).

Remark 1. The contact condition (3b) could be replaced by the following condition:

$$u_\nu \leq 0, \quad \sigma_\nu \leq 0, \quad \sigma_\nu \dot{u}_\nu = 0 \quad \text{on } (0,T) \times \Gamma_{\text{C}}. \tag{4}$$

This condition can be interpreted as follows:

- If $\dot{u}_\nu(t,x) < 0$, then no contact takes place which implies that $\sigma_\nu(t,x) = 0$ and $(\dot{u}_\nu \sigma_\nu)(t,x) = 0$.
- If $u_\nu(t,x) = 0$, then the following situations should be considered:
  — If $u_\nu(t,x) > 0$, then $\frac{u_\nu(t+h,x) - u_\nu(t,x)}{h} \to \dot{u}_\nu(t,x)$, $h \to 0$, $h > 0$; $u_\nu(t+h,x) > 0$ for $h > 0$ small enough which contradicts the nonpenetration condition.
  — If $\dot{u}_\nu(t,x) = 0$, then $(\dot{u}_\nu \sigma_\nu)(x,t) = 0$.

— If $\dot{u}_\nu(t,x) < 0$, then $\frac{u_\nu(t+h,x)-u_\nu(t,x)}{h} \to \dot{u}_\nu(t,x)$, $h \to 0$, $h > 0$; $u_\nu(t+h,x) < 0$ for any $h > 0$ small enough, thus no contact takes place and $\sigma_\nu(t+h,x) = 0$ for any $h > 0$ small enough, and then the continuity regularity implies that $\sigma_\nu(t,x) = 0$ and $(\dot{u}_\nu \sigma_\nu)(t,x) = 0$.

Note that Theorem 1 remains valid when (3b) is replaced by (4).

The relations (3c) are the so-called Coulomb's dry friction law (see Ref. [5, p. 135]) and it means that the tangential shear should be smaller than the maximal frictional resistance $p_\tau(\cdot, u_\nu)$. If $|\boldsymbol{\sigma}_\tau| < p_\tau(\cdot, u_\nu)$, the body adheres to the obstacle (stick state), while if $|\boldsymbol{\sigma}_\tau| = p_\tau(\cdot, u_\nu)$, the body slips on the obstacle (slip state). A classical example for the normal frictional bound is $p_\tau(t,x,r) = \mathscr{F}(t,x) c_\nu(t,x) r^+$, where $\mathscr{F} \geq 0$ is the friction coefficient depending on $t \in (0,T)$ and $x \in \Gamma_C$. The identity (3d) represents the homogeneous Neumann boundary condition for the damage field; for further details, the reader is referred to Ref. [3, p. 241]. The boundary condition for the temperature is given by (3e), where $\mathscr{G}$ is some function defined on $(0,T) \times \Gamma_C \times \mathbb{R}$; $\mathscr{G}(\cdot, r)$ and $\partial\mathscr{G}(\cdot, r)$ denote its value, respectively, its subdifferential at $r \in \mathbb{R}$ with respect to the third variable.

Let $\mathscr{G}^0(\cdot, s; k)$ be the directional derivative at $r \in \mathbb{R}$ in the direction of $k \in \mathbb{R}$ defined by

$$\mathscr{G}^0(\cdot, r; k) \stackrel{\text{def}}{=} \limsup_{\ell \to 0, \ell > 0} \frac{\mathscr{G}(\cdot, r + k\ell) - \mathscr{G}(\cdot, r)}{\ell}.$$

For all $\rho \in \partial\mathscr{G}(\cdot, r)$, we have

$$\mathscr{G}^0(\cdot, r; k) \geq \rho k \quad \text{and} \quad |\mathscr{G}^0(\cdot, r; k)| \leq |\mathscr{G}^0(\cdot, s)| |k|,$$

where

$$\mathscr{G}^0(\cdot, r) \stackrel{\text{def}}{=} \limsup_{h \to 0, h \neq 0} \frac{\mathscr{G}(\cdot, r+h) - \mathscr{G}(\cdot, r)}{h}.$$

In particular, if $s \mapsto \mathscr{G}(\cdot, s)$ is convex on $\mathbb{R}$, we have

$$\mathscr{G}^0(\cdot, r; k) = \begin{cases} \mathscr{G}'_{\mathrm{L}}(\cdot, r)k & \text{if } k < 0, \\ \mathscr{G}'_{\mathrm{R}}(\cdot, r)k & \text{if } k > 0, \\ 0 & \text{if } k = 0, \end{cases}$$

$$\mathscr{G}^0(\cdot, r) = \max(\mathscr{G}'_{\mathrm{L}}(\cdot, s), \mathscr{G}'_{\mathrm{R}}(\cdot, s)) \quad \text{and} \quad \partial \mathscr{G}(\cdot, r) = [\mathscr{G}'_{\mathrm{L}}(\cdot, r), \mathscr{G}'_{\mathrm{R}}(\cdot, r)],$$

where $\mathscr{G}'_{\mathrm{L}}$ and $\mathscr{G}'_{\mathrm{R}}$ denote the left and right derivatives (with respect to the third variable), respectively.

Now let us define

$$\mathscr{J}(t, x, \theta(t, x)) \stackrel{\text{def}}{=} -\kappa(t, x, \nabla \theta(t, x))\nu(x), \quad (t, x) \in (0, T) \times \Gamma_{\mathrm{C}}.$$

Our problem is completed by the following initial conditions:

$$\boldsymbol{u}(0, x) = \boldsymbol{u}_0(x), \quad \dot{\boldsymbol{u}}(0, x) = \boldsymbol{v}_0(x), \quad z(0, x) = z_0(x), \quad \theta(0, x) = \theta_0(x) \tag{5}$$

for all $x \in \Omega$.

It is convenient to introduce some notations used throughout this chapter; let $\mathbb{S}^d$ stand for the space of second-order symmetric tensors on $\mathbb{R}^d$. The indices $i \in \{1, \ldots, d\}$ and $j \in \{1, \ldots, d\}$ and the summation over repeated indices are used in the sequel. Let the inner products on $\mathbb{R}^d$ and $\mathbb{S}^d$ be defined by

$$\boldsymbol{u} \cdot \boldsymbol{v} \stackrel{\text{def}}{=} u_i v_i \quad \text{and} \quad \boldsymbol{\sigma} \cdot \boldsymbol{\tau} \stackrel{\text{def}}{=} \sigma_{ij} \tau_{ij}$$

with the associated norm $|\boldsymbol{u}| \stackrel{\text{def}}{=} \sqrt{\boldsymbol{u} \cdot \boldsymbol{u}}$ and $|\boldsymbol{\sigma}| \stackrel{\text{def}}{=} \sqrt{\boldsymbol{\sigma} \cdot \boldsymbol{\sigma}}$ for all $(\boldsymbol{u}, \boldsymbol{v}) \in \mathbb{R}^d \times \mathbb{R}^d$ and $(\boldsymbol{\sigma}, \boldsymbol{\tau}) \in \mathbb{S}^d \times \mathbb{S}^d$. Furthermore, if X is a space of scalar functions, the bold-face notation $\mathbf{X}$ denotes the space X^d. Define the following real Hilbert spaces:

$$\mathrm{H} \stackrel{\text{def}}{=} \mathbf{L}^2(\Omega), \quad \mathscr{H} \stackrel{\text{def}}{=} \{\boldsymbol{\sigma} = (\sigma_{ij}) : \sigma_{ij} = \sigma_{ji} \in L^2(\Omega), \ 1 \leq i, j \leq d\},$$

$$\mathrm{V} \stackrel{\text{def}}{=} \{\boldsymbol{u} \in \mathbf{L}^2(\Omega) : \boldsymbol{\varepsilon}(\boldsymbol{u}) \in \mathscr{H}\},$$

which are endowed with the canonical inner products given by

$$\langle \boldsymbol{u}, \boldsymbol{v} \rangle_{\mathrm{H}} \stackrel{\text{def}}{=} \int_\Omega u_i v_j \, dx, \quad \langle \boldsymbol{\sigma}, \boldsymbol{\tau} \rangle_{\mathscr{H}} \stackrel{\text{def}}{=} \int_\Omega \sigma_{ij} \tau_{ij} \, dx,$$

$$\langle \boldsymbol{u}, \boldsymbol{v} \rangle_{\mathrm{V}} \stackrel{\text{def}}{=} \langle \boldsymbol{u}, \boldsymbol{v} \rangle_{\mathrm{H}} + \langle \boldsymbol{\varepsilon}(\boldsymbol{u}), \boldsymbol{\varepsilon}(\boldsymbol{v}) \rangle_{\mathscr{H}}$$

with the associated norms $\|\cdot\|_{\mathrm{H}}$, $\|\cdot\|_{\mathscr{H}}$ and $\|\cdot\|_{\mathrm{V}}$.

We recall that the norm $\|\cdot\|_{\mathrm{V}}$ is equivalent to the norm $\|\cdot\|_{(\mathrm{H}^1(\Omega))^d}$.

Identifying H with its own dual, we have an evolution triple (see, e.g., Ref. [15] II/A p. 416)

$$\mathrm{V} \subset \mathrm{H} \equiv \mathrm{H}' \subset \mathrm{V}',$$

where the inclusions are continuous and dense. Finally, we use the notation $\langle \cdot, \cdot \rangle_{\mathrm{V}' \times \mathrm{V}}$ to represent the duality pairing between V' and V, which means

$$\langle \boldsymbol{u}, \boldsymbol{v} \rangle_{\mathrm{V}' \times \mathrm{V}} = \langle \boldsymbol{u}, \boldsymbol{v} \rangle_{\mathrm{H}} \quad \text{for all } \boldsymbol{u} \in \mathrm{H} \text{ and } \boldsymbol{v} \in \mathrm{V}.$$

Moreover, by using Sobolev's trace theorem, there exists a constant $c_0 > 0$ depending only on Ω and Γ_{C} such that

$$\|\boldsymbol{v}\|_{\mathrm{L}^2(\Gamma_{\mathrm{C}})} \leq c_0 \|\boldsymbol{v}\|_{\mathrm{V}} \quad \text{for all } \boldsymbol{v} \in \mathrm{V}. \tag{6}$$

Let V_{Dir} and V_{Neu} be the admissible temperature and damage fields, respectively, defined as follows:

$$V_{\mathrm{Dir}} \stackrel{\text{def}}{=} \{\eta \in \mathrm{H}^1(\Omega) : \eta = 0 \text{ on } \Gamma_{\mathrm{F}}\},$$

$$V_{\mathrm{Neu}} \stackrel{\text{def}}{=} \{z \in \mathrm{H}^1(\Omega) \text{ on } \Gamma, \ 0 \leq z \leq 1 \text{ a.e. in } \Omega\}$$

endowed with the canonical inner products in $\mathrm{H}^1(\Omega)$. Last, the notation μ or $d\mu$ stands for the usual measure on $\partial\Omega$.

3. Statement of the Result

In this section, we introduce some assumptions on the constitutive functions of the problem and give a few simple consequences of these assumptions:

(A–1) The elasticity operator $\mathbf{E} : (0,T) \times \Omega \times \mathbb{S}^d \times \mathbb{R} \to \mathbb{S}^d$ satisfies the following:

(i) $\mathbf{E}(\cdot,\cdot,\boldsymbol{\tau},\lambda)$ is measurable on $(0,T)\times\Omega$ for all $\boldsymbol{\tau}\in\mathbb{S}^d$ and for all $\lambda\in\mathbb{R}$;

(ii) there exists $L_\mathbf{E}>0$ such that

$$|\mathbf{E}(t,x,\boldsymbol{\tau}_1,\lambda_1)-\mathbf{E}(t,x,\boldsymbol{\tau}_2,\lambda_2)|$$
$$\leq L_\mathbf{E}(|\boldsymbol{\tau}_1-\boldsymbol{\tau}_2|+|\lambda_1-\lambda_2|)$$

for all $\boldsymbol{\tau}_i\in\mathbb{S}^d$, for all $\lambda_i\in\mathbb{R}$, $i=1,2$, and for a.e. $(t,x)\in(0,T)\times\Omega$;

(iii) there exists $c_0^\mathbf{E}\in L^2((0,T)\times\Omega;\mathbb{R}^+)$ and $c_1^\mathbf{E},c_2^\mathbf{E}\geq 0$ such that

$$|\mathbf{E}(t,x,\boldsymbol{\tau},\lambda)|\leq c_0^\mathbf{E}(t,x)+c_1^\mathbf{E}|\boldsymbol{\tau}|+c_2^\mathbf{E}|\lambda|$$

for all $\boldsymbol{\tau}\in\mathbb{S}^d$, for all $\lambda\in\mathbb{R}$ and for a.e. $(t,x)\in(0,T)\times\Omega$.

(A–2) The viscosity operator $\mathbf{A}:(0,T)\times\Omega\times\mathbb{S}^d\to\mathbb{S}^d$ satisfies the following:

(i) $\mathbf{A}(\cdot,\cdot,\boldsymbol{\tau})$ is measurable on $(0,T)\times\Omega$ for all $\boldsymbol{\tau}\in\mathbb{S}^d$;

(ii) $\mathbf{A}(t,x,\cdot)$ is continuous on $\mathbb{S}^d$ for a.e. $(t,x)\in(0,T)\times\Omega$;

(iii) there exist $m_\mathbf{A}>0$ such that

$$(\mathbf{A}(t,x,\boldsymbol{\tau}_1)-\mathbf{A}(t,x,\boldsymbol{\tau}_2))\cdot(\boldsymbol{\tau}_1-\boldsymbol{\tau}_2)\geq m_\mathbf{A}|\boldsymbol{\tau}_1-\boldsymbol{\tau}_2|^2$$

for all $\boldsymbol{\tau}_1,\boldsymbol{\tau}_2\in\mathbb{S}^d$ and for a.e. $(t,x)\in(0,T)\times\Omega$;

(iv) there exists $c_0^\mathbf{A}\in L^2((0,T)\times\Omega;\mathbb{R}^+)$ and $c_1^\mathbf{A}>0$ such that

$$|\mathbf{A}(t,x,\boldsymbol{\tau})|\leq c_0^\mathbf{A}(t,x)+c_1^\mathbf{A}|\boldsymbol{\tau}|$$

for all $\boldsymbol{\tau}\in\mathbb{S}^d$ and for a.e. $(t,x)\in(0,T)\times\Omega$.

In the following, let us introduce for $t\in(0,T)$, $\mathbb{A}(t):V\to V'$ defined by for all $\boldsymbol{v},\boldsymbol{w}\in V$:

$$\langle\mathbb{A}(t)\boldsymbol{v},\boldsymbol{w}\rangle_{V'\times V}=(\mathbf{A}(t,\cdot,\varepsilon(\boldsymbol{v})),\varepsilon(\boldsymbol{w}))_{\mathcal{H}}.$$

(A–3) The relaxation tensor $\mathbf{B} : (0,T) \times \Omega \times \mathbb{S}^d \times \mathbb{R} \to \mathbb{S}^d$ satisfies the following:

(i) $\mathbf{B}(\cdot, \cdot, \boldsymbol{\tau}, \lambda) \in L^\infty((0,T) \times \Omega; \mathbb{S}^d)$ for all $\boldsymbol{\tau} \in \mathbb{S}^d$ and for all $\lambda \in \mathbb{R}$;

(ii) there exists $L_{\mathbf{B}} > 0$ such that
$$|\mathbf{B}(t, x, \boldsymbol{\tau}_1, \lambda_1) - \mathbf{B}(t, x, \boldsymbol{\tau}_2, \lambda_2)|$$
$$\leq L_{\mathbf{B}}(|\boldsymbol{\tau}_1 - \boldsymbol{\tau}_2| + |\lambda_1 - \lambda_2|)$$
for all $\boldsymbol{\tau}_i \in \mathbb{S}^d$, for all $\lambda_i \in \mathbb{R}$, $i = 1,2$ and for a.e. $(t,x) \in (0,T) \times \Omega$.

(A–4) The thermal expansion tensor $C_{\exp} : (0,T) \times \Omega \times \mathbb{R} \to \mathbb{S}^d$ verifies the following:

(i) $C_{\exp}(\cdot, \cdot, \vartheta)$ is measurable on $(0,T) \times \Omega$ for all $\vartheta \in \mathbb{R}$;

(ii) there exists $L_{C_{\exp}} > 0$ such that
$$|C_{\exp}(t, x, \vartheta_1) - C_{\exp}(t, x, \vartheta_2)| \leq L_{C_{\exp}}|\vartheta_1 - \vartheta_2|$$
for all $\vartheta_i \in \mathbb{R}$, $i = 1,2$, and for a.e. $(t, \boldsymbol{x}) \in (0,T) \times \Omega$;

(iii) there exist $c_0^{C_{\exp}} \in L^\infty((0,T) \times \Omega; \mathbb{R}^+)$ and $c_1^{C_{\exp}} > 0$ such that
$$|C_{\exp}(t, x, \vartheta)| \leq c_0^{C_{\exp}}(t, x) + c_1^{C_{\exp}}|\vartheta|$$
for all $\vartheta \in \mathbb{R}$ and for a.e. $(t, x) \in (0,T) \times \Omega$.

(A–5) The friction bound function $p_\tau : (0,T) \times \Gamma_C \times \mathbb{R} \to \mathbb{R}_+$ satisfies the following:

(i) there exists $L_\tau > 0$ such that
$$|p_\tau(t, x, r_1) - p_\tau(t, x, r_2)| \leq L_\tau |r_1 - r_2|$$
for all $r_i \in \mathbb{R}$, $i = 1,2$, and for a.e. $(t, x) \in (0,T) \times \Gamma_C$;

(ii) $p_\tau(\cdot, \cdot, r)$ is Lebesgue measurable on $(0,T) \times \Gamma_C$ for all $r \in \mathbb{R}$;

(iii) the mapping $p_\tau(\cdot, \cdot, r) = 0$ for all $r \leq 0$.

(A–6) The damage source $\varphi : \Omega \times \mathbb{S}^d \times \mathbb{S}^d \times [0,1] \to \mathbb{R}$ verifies the following:

(i) there exists $L_\varphi > 0$ such that

$$|\varphi(x, \sigma_1, \varepsilon_1, \xi_1) - \varphi(x, \sigma_2, \varepsilon_2, \xi_2)|$$
$$\leq L_\varphi (|\sigma_1 - \sigma_2| + |\varepsilon_1 - \varepsilon_2| + |\xi_1 - \xi_2|)$$

for all $(\sigma_i, \varepsilon_i) \in \mathbb{S}^d \times \mathbb{S}^d$, for all $\xi_i \in [0,1]$, $i = 1, 2$, and for a.e. $x \in \Omega$;

(ii) $\varphi(\cdot, \sigma, \varepsilon, \xi)$ is a Lebesgue measurable function on Ω for all $\sigma, \varepsilon \in \mathbb{S}^d$ and for all $\xi \in [0, 1]$;

(iii) $\varphi(\cdot, \mathbf{0}, \mathbf{0}, 0) \in \mathrm{L}^2(\Omega)$.

(A–7) The conductivity $\kappa : (0, T) \times \Omega \times \mathbb{R}^d \to \mathbb{R}^d$ satisfies the following:

(i) $\kappa(\cdot, \cdot, \xi)$ is measurable on $(0, T) \times \Omega$ for all $\xi \in \mathbb{R}^d$;
(ii) $\kappa(t, x, \cdot)$ is continuous on $\mathbb{R}^d$ for a.e. $(t, x) \in (0, T) \times \Omega$;
(iii) there exist $c_0^\kappa \in \mathrm{L}^2((0,T) \times \Omega; \mathbb{R}^+)$ and $c_1^\kappa \geq 0$ such that

$$|\kappa(t, x, \xi)| \leq c_0^\kappa(t, x) + c_1^\kappa |\xi|$$

for all $\xi \in \mathbb{R}^d$ and for a.e. $(t, x) \in (0, T) \times \Omega$;
(iv) there exists $m_1^\kappa > 0$ such that

$$(\kappa(t, x, \xi_1) - \kappa(t, x, \xi_2)) \cdot (\xi_1 - \xi_2) \geq m_1^\kappa |\xi_1 - \xi_2|^2$$

for all $\xi_i \in \mathbb{R}^d$, $i = 1, 2$ and for a.e. $(t, x) \in (0, T) \times \Omega$;
(v) there exists $m_2^\kappa > 0$ such that

$$\kappa(t, x, \xi) \cdot \xi \geq m_2^\kappa |\xi|^2$$

for all $\xi \in \mathbb{R}^d$ and a.e. $(t, x) \in (0, T) \times \Omega$.

(A–8) The nonlinear term $\mathscr{N} : (0, T) \times \Gamma_C \times \mathbb{R} \times \mathbb{R}^d \to \mathbb{R}$ satisfies the following:

(i) $\mathscr{N}(\cdot, \cdot, r, \xi)$ is measurable on $(0, T) \times \Gamma_C$, $\forall (r, \xi) \in \mathbb{R} \times \mathbb{R}^d$;

(ii) $\mathscr{N}(t, \boldsymbol{x}, \cdot, \cdot)$ is continuous on $\mathbb{R} \times \mathbb{R}^d$, a.e. $(t, \boldsymbol{x}) \in (0, T) \times \Gamma_C$;

(iii) $(\mathscr{N}(t, \boldsymbol{x}, r_1, \xi) - \mathscr{N}(t, \boldsymbol{x}, r_2, \xi))(r_1 - r_2) \geq 0$ for all $r_1, r_2 \in \mathbb{R}$, $\xi \in \mathbb{R}^d$, a.e. $(t, \boldsymbol{x}) \in (0, T) \times \Gamma_C$;

(iv) $\vartheta \in H^1(\Omega)$, $\boldsymbol{w} \in (H^1(\Omega))^d \implies \mathscr{N}(t, \cdot, \vartheta, \boldsymbol{w}) \in L^2(\Omega)$, a.e. $t \in (0, T)$.

(A–9) The viscosity–deformation heat function $D_{\text{vis}} : (0, T) \times \Omega \times \mathbb{S}^d \to \mathbb{R}$ satisfies the following:

(i) $D_{\text{vis}}(\cdot, \cdot, \boldsymbol{\tau})$ is measurable on $(0, T) \times \Omega$ for all $\boldsymbol{\tau} \in \mathbb{S}^d$;

(ii) $D_{\text{vis}}(t, x, \cdot)$ is continuous on $\mathbb{S}^d$ for a.e. $(t, x) \in (0, T) \times \Omega$;

(iii) there exist $C_i^{D_{\text{vis}}} > 0$, $i = 1, 2$ such that

$$|D_{\text{vis}}(t, x, \boldsymbol{\tau})| \leq C_0^{D_{\text{vis}}} + C_1^{D_{\text{vis}}} |\boldsymbol{\tau}|$$

for all $\boldsymbol{\tau} \in \mathbb{S}^d$ and a.e. $(t, x) \in (0, T) \times \Omega$.

(A–10) The nonlinear functions $\mathscr{G} : (0, T) \times \Gamma_C \times \mathbb{R} \to \mathbb{R}$ and $\mathscr{J} : (0, T) \times \Gamma_C \times \mathbb{R} \to \mathbb{R}$ satisfy the following:

(i) $\mathscr{G}(\cdot, \cdot, r)$ and $\mathscr{J}(\cdot, \cdot, r)$ are measurable on $(0, T) \times \Gamma_C$ for all $r \in \mathbb{R}$;

(ii) $\mathscr{G}(t, x, \cdot)$ is locally Lipschitz on $\mathbb{R}$ for a.e. $(t, x) \in (0, T) \times \Gamma_C$;

(iii) there exist $c_0^{\mathscr{G}} \in L^2((0, T) \times \Gamma_C; \mathbb{R}^+)$ and $c_1^{\mathscr{G}} \geq 0$ such that

$$|\mathscr{G}^0(t, x, r)| \leq c_0^{\mathscr{G}}(t, x) + c_1^{\mathscr{G}} |r|$$

for all $r \in \mathbb{R}$ and for a.e. $(t, x) \in (0, T) \times \Gamma_C$;

(iv) $(\mathscr{J}(t, x, r_1) - \mathscr{J}(t, x, r_2))(r_1 - r_2) \geq 0$ for all $r_i \in \mathbb{R}$, $i = 1, 2$, and for a.e. $(t, x) \in (0, T) \times \Gamma_C$.

(A–11) The density of the heat sources $q \in L^2(0, T; L^2(\Omega))$, the density of the external forces $\boldsymbol{f}_{\text{ext}} \in L^2(0, T; \mathbf{L}^2(\Omega))$ and the density of the surface traction forces $\boldsymbol{f}_{\text{tra}} \in L^2(0, T; \mathbf{L}^2(\Gamma_F))$.

(A–12) The initial data satisfy the following assumptions:

$$\boldsymbol{u}_0 \in V, \quad \boldsymbol{v}_0 \in V, \quad \theta_0 \in V_{\text{Dir}}, \quad z_0 \in V_{\text{Neu}}(\Omega).$$

Let us define $\boldsymbol{f} : [0, T] \to V'$ by

$$\langle \boldsymbol{f}(t), \boldsymbol{w} \rangle_{V' \times V} \stackrel{\text{def}}{=} \int_{\Omega} \boldsymbol{f}_{\text{ext}}(t) \cdot \boldsymbol{w} \, dx + \int_{\Gamma_F} \boldsymbol{f}_{\text{tra}}(t) \cdot \boldsymbol{w} \, d\mu \qquad (7)$$

for all $\boldsymbol{w} \in V$ and $t \in [0, T]$. It is clear that assumption (A–11) implies that $\boldsymbol{f} \in L^2(0, T; V')$. Let us define $j : V \times V : \mathbb{R}$ by

$$j(\boldsymbol{u}, \boldsymbol{w}) \stackrel{\text{def}}{=} \int_{\Gamma_C} \left(p_\tau(u_\nu) |\boldsymbol{w}_\tau|\right) d\mu.$$

According to assumptions (A–11), for any $\boldsymbol{w} \in V$, $p_\tau(u_\nu) \in V$. It comes that $j(\cdot, \cdot)$ is defined on $V \times V$. Let

$$\psi(t, \eta, \vartheta) \stackrel{\text{def}}{=} \int_{\Gamma_C} \mathscr{G}^0(t, x, \eta(x); \vartheta(x)) \, d\mu.$$

Note that assumptions (A–10) imply that $\psi(t, \eta; \vartheta)$ is well defined for all $\eta \in V_{\text{Dir}}$ and $\vartheta \in V_{\text{Dir}}$ for a.e. $t \in (0, T)$.

Denote by K the convex set of admissible displacements

$$K \stackrel{\text{def}}{=} \{\boldsymbol{u} \in V : u_\nu \leq 0 \text{ on } \Gamma_C\}.$$

Then, the weak formulation associated to problems (1)–(5) is given by the following:

Problem (WF): Find a displacement field $\boldsymbol{u} : [0, T] \to K$, a damage field $z : [0, T] \to V_{\text{Neu}}$ and a temperature field $\theta : [0, T] \to V_{\text{Dir}}$ such that

$$\langle \ddot{\boldsymbol{u}}(t), \boldsymbol{w} - \dot{\boldsymbol{u}}(t) \rangle_{V' \times V} + \langle \boldsymbol{\sigma}(t), \boldsymbol{\varepsilon}(\boldsymbol{w}) - \boldsymbol{\varepsilon}(\dot{\boldsymbol{u}}(t)) \rangle_{\mathscr{H}}$$
$$+ j(\boldsymbol{u}(t), \boldsymbol{w}) - j(\boldsymbol{u}(t), \dot{\boldsymbol{u}}(t))$$
$$\geq \langle \boldsymbol{f}(t), \boldsymbol{w} - \dot{\boldsymbol{u}}(t) \rangle_{V' \times V} \quad \text{for all } \boldsymbol{w} \in K \text{ and for a.e. } t \in (0, T), \qquad (8a)$$

$$(\dot{\theta}(t), \vartheta)_{L^2(\Omega)} + (\kappa(t, \nabla\theta(t)), \nabla\vartheta)_{L^2(\Omega)} + (\mathscr{N}(\theta(t), \dot{\boldsymbol{u}}(t)), \vartheta)_{L^2(\Omega)} \qquad (8b)$$

$$+ \psi(t, \theta(t), \vartheta) \geq (D_{\text{vis}}(t, \boldsymbol{\varepsilon}(\dot{\boldsymbol{u}}(t))), \vartheta)_{L^2(\Omega)} + (q(t), \vartheta)_{L^2(\Omega)}$$
$$\text{for all } \vartheta \in V_{\text{Dir}} \text{ and for a.e. } t \in (0, T), \qquad (8c)$$

$$\langle \dot{z}(t), \xi - z(t) \rangle_{L^2(\Omega)} + \gamma \langle \nabla z(t), \nabla \xi - \nabla z(t) \rangle_H$$

$$\geq \langle \varphi(\boldsymbol{\sigma}(t), \boldsymbol{\varepsilon}(\boldsymbol{u}(t)), z(t)), \xi - z(t) \rangle_{L^2(\Omega)}$$

for all $\xi \in V_{\text{Neu}}$ and for a.e. $t \in (0,T)$, (8d)

$$\boldsymbol{u}(0) = \boldsymbol{u}_0, \quad \dot{\boldsymbol{u}}(0) = \boldsymbol{v}_0, \quad z(0) = z_0, \quad \theta(0) = \theta_0 \quad \text{in } \Omega. \tag{8e}$$

If $s \mapsto \mathscr{G}(\cdot, s)$ is differentiable, then we may conclude that for all η, $\vartheta \in V_{\text{Dir}}$:

$$\psi(t, \eta, \vartheta) = \int_{\Gamma_C} \mathscr{G}'(t, \eta) \vartheta \, d\mu.$$

The following existence theorem can be proved by using some classical results for elliptic and parabolic variational inequalities and a fixed-point argument. The proof is given in the following section:

Theorem 1. *Assume that* (A–1)–(A–12) *hold. Assume that* $L_\tau < \frac{m_{\mathbf{A}}}{\sqrt{2T}c_0^2}$ *and let* $T > 0$. *Then the problem* (**WF**) *admits exactly one solution* $\{\boldsymbol{u}, \theta, z\}$ *such that*

$$\boldsymbol{u} \in C^1(0,T;H) \cap H^1(0,T;V) \cap W^{2,2}(0,T;V'), \tag{9a}$$

$$\theta \in C^0(0,T;L^2(\Omega)) \cap L^2(0,T;V_{\text{Dir}}) \cap H^1(0,T;V'_{\text{Dir}}), \tag{9b}$$

$$z \in H^1(0,T;L^2(\Omega)) \cap L^2(0,T;V_{\text{Neu}}). \tag{9c}$$

4. Proof of Theorem 1

This section is dedicated to the proof of existence and uniqueness results to problem (**WF**). To this aim, the first step consists in reformulating the problem (**WF**) in terms of the velocities instead of displacements. Then the monotonicity and fixed-point argument lead to the well-posedness of problem (**WF**). Let $\boldsymbol{v} \stackrel{\text{def}}{=} \dot{\boldsymbol{u}}$ be the velocity field. Since $\boldsymbol{u}(0) = \boldsymbol{u}_0$, we have

$$\boldsymbol{u}(t) = \boldsymbol{u}_0 + \int_0^t \boldsymbol{v}(s) \, ds \quad \text{for all } t \in [0,T]. \tag{10}$$

Consequently, relation (8a) can be rewritten as follows:

$$\langle \dot{v}(t), w - v(t)\rangle_{V'\times V} + \langle \sigma(t), \varepsilon(w) - \varepsilon(v(t))\rangle_{\mathscr{H}}$$
$$+ j(u(t), w) - j(u(t), v(t))$$
$$\geq \langle f(t), w - v(t)\rangle_{V'\times V} \quad \text{for all } w \in K \text{ and for a.e. } t \in (0, T).$$

We establish in the following the existence and uniqueness results for some auxiliary problems and also *a priori* estimates for the solutions of these auxiliary problems are exhibited. These *a priori* estimates play a crucial role in the proof of the unique solvability for the problem **(WF)**.

Lemma 1. *Assume that* (A–2), (A–5), (A–6), (A–11) *and* (A–12) *hold true. Then for all* $\eta \in L^2(0,T;V')$, *there exists an unique solution* $v_\eta \in C^0(0,T;H) \cap L^2(0,T;V) \cap H^1(0,T;V')$ *to problem:*

$$\langle \dot{v}_\eta(t), w - v_\eta(t)\rangle_{V'\times V} + \langle \mathbb{A}(t)v_\eta(t), w - v_\eta(t)\rangle_{V'\times V}$$
$$+ \langle \eta(t), w - v_\eta(t)\rangle_{V'\times V} + j(u_\eta(t), w) - j(u_\eta(t), v_\eta(t))$$
$$\geq \langle f(t), w - v_\eta(t)\rangle_{V'\times V} \quad \text{for all } w \in K \text{ and for a.e. } t \in (0, T), \tag{11a}$$

$$v_\eta(0) = v_0, \tag{11b}$$

where

$$u_\eta(t) \stackrel{\text{def}}{=} u_0 + \int_0^t v_\eta(s)\,ds \quad \text{for all } t \in [0, T].$$

Moreover, if $L_T < \frac{m_\mathbb{A}}{\sqrt{2T c_0^2}}$, *then there exists a constant* $C > 0$ *such that for all* $\eta_i \in L^2(0,T;V')$, $i = 1, 2$, *and for all* $t \in [0, T]$, *we have*

$$\|v_{\eta_2}(t) - v_{\eta_1}(t)\|_H^2 + \int_0^t \|v_{\eta_2} - v_{\eta_1}\|_V^2 \leq C \int_0^t \|\eta_1 - \eta_2\|_{V'}^2. \tag{12}$$

Proof. Let $\eta \in L^2(0,T;V')$ and $h \in C^0(0,T;V)$ be given. According to classical results for parabolic variational inequalities, we may prove that there we obtain the existence of a unique $v_{\eta,h} \in C^0(0,T;H) \cap L^2(0,T;V) \cap W^{1,2}(0,T;V')$ such that

$$\langle \dot{\boldsymbol{v}}_\eta(t), \boldsymbol{w}-\boldsymbol{v}_\eta(t)\rangle_{V'\times V} + \langle \mathbb{A}(t)\boldsymbol{v}_\eta(t), \boldsymbol{w}-\boldsymbol{v}_\eta(t)\rangle_{V'\times V}$$
$$+ \langle \eta(t), \boldsymbol{w}-\boldsymbol{v}_\eta(t)\rangle_{V'\times V} + j(h(t),\boldsymbol{w}) - j(h(t),\boldsymbol{v}_\eta(t))$$
$$\geq \langle \boldsymbol{f}(t), \boldsymbol{w}-\boldsymbol{v}_\eta(t)\rangle_{V'\times V} \quad \text{for all } \boldsymbol{w}\in K \text{ and for a.e. } t\in (0,T), \tag{13a}$$

$$\boldsymbol{v}_{\eta,x}(0) = \boldsymbol{v}_0 \tag{13b}$$

hold. The proof uses the techniques detailed in Ref. [15, II/B, p. 893]. Since it is quite a routine to adapt this proof to our case, we let the verification to the reader. Let us fix $\eta \in L^2(0,T;V')$. Set the operator $\Lambda_\eta : C^0(0,T;V) \to C^0(0,T;V)$ defined by

$$\Lambda_\eta h(t) \stackrel{\text{def}}{=} \boldsymbol{u}_0 + \int_0^t \boldsymbol{v}_{\eta,h}(s)\,\mathrm{d}s \quad \text{for all } h \in C^0(0,T;V).$$

By using algebraic manipulations and assumption (A–6), for all $(\boldsymbol{u}_i, \boldsymbol{w}_i) \in K \times K$, $i=1,2$, we find

$$j(\boldsymbol{u}_1, \boldsymbol{w}_2) - j(\boldsymbol{u}_1, \boldsymbol{w}_1) + j(\boldsymbol{u}_2, \boldsymbol{w}_1) - j(\boldsymbol{u}_2, \boldsymbol{w}_2)$$
$$\leq c_1 \|\boldsymbol{u}_2 - \boldsymbol{u}_1\|_V \|\boldsymbol{w}_2 - \boldsymbol{w}_1\|_V, \tag{14}$$

where $c_1 \stackrel{\text{def}}{=} L_\tau c_0^2$ (see (6)). Let $h_i \in C^0(0,T;V)$, $i=1,2$, be given. On the one hand, taking $h=h_1$ and $\boldsymbol{w} = \boldsymbol{v}_{\eta,h_2}$ in (13a) and on the other hand, taking $h=h_2$ and $\boldsymbol{w} = \boldsymbol{v}_{\eta,h_1}$ in (13a), then adding these inequalities and integrating over $(0,t)$ the resulting relation, we obtain

$$\|\boldsymbol{v}_{\eta,h_2}(t) - \boldsymbol{v}_{\eta,h_1}(t)\|_H^2 + \int_0^t \|\boldsymbol{v}_{\eta,h_2}(s) - \boldsymbol{v}_{\eta,h_1}(s)\|_V^2 \,\mathrm{d}s$$
$$\leq C \left(\int_0^t \|h_2(s) - h_1(s)\|_V^2 \,\mathrm{d}s + \int_0^t \|\boldsymbol{v}_{\eta,h_2}(s) - \boldsymbol{v}_{\eta,h_1}(s)\|_H^2 \,\mathrm{d}s \right)$$

for all $t \in [0,T]$. Grönwall's lemma (see, e.g., Ref. [2]) leads to

$$\|\Lambda_\eta h_2(t) - \Lambda_\eta h_1(t)\|_V^2 \leq C \int_0^t \|h_2(s) - h_1(s)\|_V^2 \,\mathrm{d}s$$

for all $h_i \in C^0(0,T;V)$, $i = 1,2$, and for all $t \in [0,T]$. Then we may conclude by using Banach's fixed-point principle that Λ_η has a unique fixed point denoted by h_η and clearly $\boldsymbol{v}_\eta = \boldsymbol{v}_{\eta,h_\eta}$ is the unique solution to problem (11).

We show in the following that (12) holds. To this aim, let $\eta_i \in L^2(0,T;V')$, $i = 1,2$. On the one hand, we choose $\eta = \eta_1$ and $\boldsymbol{w} = \boldsymbol{v}_{\eta_2}$ in (11a) and on the other hand, we choose $\eta = \eta_2$ and $\boldsymbol{w} = \boldsymbol{v}_{\eta_1}$ in (11a). Then adding these inequalities, integrating the resulting inequality over $(0,t)$ and using (14) and assumption (A–2), we get

$$\frac{1}{2}\|\boldsymbol{v}_{\eta_2}(t) - \boldsymbol{v}_{\eta_1}(t)\|_H^2 + m_{\mathbf{A}} \int_0^t \|\boldsymbol{v}_{\eta_2}(s) - \boldsymbol{v}_{\eta_1}(s)\|_V^2 \, ds$$

$$\leq m_{\mathbf{A}} \int_0^t \|\boldsymbol{v}_{\eta_2}(s) - \boldsymbol{v}_{\eta_1}(s)\|_H^2 \, ds$$

$$+ c_1 \int_0^t \|\boldsymbol{u}_{\eta_2}(s) - \boldsymbol{u}_{\eta_1}(s)\|_V \|\boldsymbol{v}_{\eta_2}(s) - \boldsymbol{v}_{\eta_1}(s)\|_V \, ds$$

$$+ \int_0^t |\langle \eta_2(s) - \eta_1(s), \boldsymbol{v}_{\eta_2}(s) - \boldsymbol{v}_{\eta_1}(s)\rangle_V| \, ds.$$

Since the product $|ab|$ can be estimated by $\frac{a^2}{4\alpha} + \alpha b^2$ for all $(a,b) \in \mathbb{R}^2$ and $\alpha > 0$, we obtain for all $t \in [0,T]$ the following inequality:

$$\frac{1}{2}\|\boldsymbol{v}_{\eta_2}(t) - \boldsymbol{v}_{\eta_1}(t)\|_H^2 + m_{\mathbf{A}} \int_0^t \|\boldsymbol{v}_{\eta_2}(s) - \boldsymbol{v}_{\eta_1}(s)\|_V^2 \, ds$$

$$\leq m_{\mathbf{A}} \int_0^t \|\boldsymbol{v}_{\eta_2}(s) - \boldsymbol{v}_{\eta_1}(s)\|_H^2 \, ds + \frac{c_1^2}{4\alpha} \int_0^t \|\boldsymbol{u}_{\eta_2}(s) - \boldsymbol{u}_{\eta_1}(s)\|_V^2 \, ds$$

$$+ 2\alpha \int_0^t \|\boldsymbol{v}_{\eta_2}(s) - \boldsymbol{v}_{\eta_1}(s)\|_V^2 \, ds + \frac{1}{4\alpha} \int_0^t \|\eta_2(s) - \eta_1(s)\|_{V'}^2 \, ds,$$

for all $z > 0$ and for all $t \in [0,T]$. Now verifying that

$$\int_0^t \|\boldsymbol{u}_{\eta_2}(s) - \boldsymbol{u}_{\eta_1}(s)\|_V^2 \, ds \leq T^2 \int_0^t \|\boldsymbol{v}_{\eta_2}(s) - \boldsymbol{v}_{\eta_1}(s)\|_V^2 \, ds,$$

observe that

$$\frac{1}{2}\|\boldsymbol{v}_{\eta_2}(t) - \boldsymbol{v}_{\eta_1}(t)\|_\mathrm{H}^2 + (m_\mathbf{A} - 2\alpha)\int_0^t \|\boldsymbol{v}_{\eta_2}(s) - \boldsymbol{v}_{\eta_1}(s)\|_\mathrm{V}^2\,ds$$

$$\leq m_\mathbf{A} \int_0^t \|\boldsymbol{v}_{\eta_2}(s) - \boldsymbol{v}_{\eta_1}(s)\|_\mathrm{H}^2\,ds + \frac{c_1^2 T^2}{4\alpha}\int_0^t \|\boldsymbol{v}_{\eta_2}(s) - \boldsymbol{v}_{\eta_1}(s)\|_\mathrm{V}^2\,ds$$

$$+ \frac{1}{4\alpha}\int_0^t \|\eta_2(s) - \eta_1(s)\|_{\mathrm{V}'}^2\,ds. \qquad (15)$$

The inequality (12) follows from Grönwall's lemma if we choose $\alpha > 0$ such that $\frac{c_1^2 T^2}{4\alpha} < m_\mathbf{A} - 2\alpha$, which gives $L_\mathcal{T} < \frac{m_\mathbf{A}}{Tc_0^2}\sqrt{2\varsigma(1-\varsigma)}$, where $\varsigma = \frac{2\alpha}{m_\mathbf{A}} \in (0,1)$.

The last inequality is equivalent to $L_\mathcal{T} < \frac{m_\mathbf{A}}{\sqrt{2}Tc_0^2}$. □

Here and in the following, we denote by $C > 0$ a generic constant, whose value may change from lines to lines.

Lemma 2. *Assume that* (A–8)–(A–12) *hold true. Then for all* $\eta \in \mathrm{L}^2(0,T;\mathrm{V}')$, *there exists a unique solution* $\theta_\eta \in \mathrm{C}^0(0,T;\mathrm{L}^2(\Omega)) \cap \mathrm{L}^2(0,T;\mathrm{V}_\mathrm{Dir}) \cap \mathrm{H}^1(0,T;\mathrm{V}'_\mathrm{Dir})$ *to problem:*

$$(\dot{\theta}_\eta(t), \vartheta)_{\mathrm{L}^2(\Omega)} + (\kappa(t, \nabla\theta_\eta(t)), \nabla\vartheta)_{\mathrm{L}^2(\Omega)} + (\mathscr{N}(\theta_\eta(t), \boldsymbol{v}_\eta(t)), \vartheta)_{\mathrm{L}^2(\Omega)}$$

$$+ \int_{\Gamma_C} \mathscr{J}(t, \theta_\eta(t))\vartheta\,d\mu = (D_\mathrm{vis}(t, \boldsymbol{\varepsilon}(\boldsymbol{v}_\eta(t))), \vartheta)_{\mathrm{L}^2(\Omega)} + (q(t), \vartheta)_{\mathrm{L}^2(\Omega)}$$

for all $\vartheta \in \mathrm{V}_\mathrm{Dir}$ and for a.e. $t \in (0,T)$, \hfill (16a)

$\theta_\eta(0) = \theta_0$. \hfill (16b)

Moreover, if $L_\mathcal{T} < \frac{m_\mathbf{A}}{\sqrt{2}Tc_0^2}$, *then there exists a constant* $C > 0$ *such that for all* $\eta_i \in \mathrm{L}^2(0,T;\mathrm{V}')$, $i = 1,2$, *we have*

$$\|\theta_{\eta_1}(t) - \theta_{\eta_2}(t)\|_{\mathrm{L}^2(\Omega)}^2 \leq C \int_0^t \|\eta_1 - \eta_2\|_{\mathrm{V}'}^2 \quad \text{for all } t \in [0,T]. \qquad (17)$$

Proof. Let $\eta \in L^2(0,T;V')$ be fixed.

Since $q \in L^2(0,T;L^2(\Omega))$, $\boldsymbol{v}_\eta \in L^2(0,T;V)$ and assumption (A–9) holds, we may infer that the function $\mathscr{R}(t) : (0,T) \to V'_{\text{Dir}}$ defined for a.e. $t \in (0,T)$ by

$$\langle \mathscr{R}(t), \vartheta \rangle_{V'_{\text{Dir}} \times V_{\text{Dir}}} = (D_{vis}(t, \boldsymbol{\varepsilon}(\boldsymbol{v}_\eta(t))), \vartheta)_{L^2(\Omega)} + (q(t), \vartheta)_{L^2(\Omega)}$$

satisfies $\mathscr{R} \in L^2(0,T;V'_{\text{Dir}})$.

On the other hand, assumptions (A–8) and (A–10) imply that the operator $\mathscr{K}(t) : V_{\text{Dir}} \to V'_{\text{Dir}}$ defined for a.e. $t \in (0,T)$ by

$$\langle \mathscr{K}(t)\xi, \vartheta \rangle_{V_{\text{Dir}}} \stackrel{\text{def}}{=} (\kappa(t,\nabla\xi), \nabla\vartheta)_{L^2(\Omega)} + (\mathscr{N}(\xi, \boldsymbol{v}_\eta(t)), \vartheta)_{L^2(\Omega)}$$
$$+ \int_{\Gamma_C} \mathscr{I}(t,\xi)\vartheta \, d\mu$$

for all $(\xi, \vartheta) \in V_{\text{Dir}} \times V_{\text{Dir}}$ is strongly monotone. Consequently, the solvability to problem (16) comes from a classical result on first-order evolution equation; the reader is referred to Ref. [11] for further details.

It remains to prove the inequality (17). To this aim, we observe that $\eta_i \in L^2(0,T;V')$, $i = 1, 2$, implying that

$$(\dot\theta_{\eta_1}(t) - \dot\theta_{\eta_2}(t), \theta_{\eta_1}(t) - \theta_{\eta_2}(t))_{L^2(\Omega)} + (\kappa(t, \nabla\theta_{\eta_1}(t))$$
$$- \kappa(t, \nabla\theta_{\eta_2}(t)), \nabla\theta_{\eta_1}(t) - \nabla\theta_{\eta_2}(t))_{L^2(\Omega)}$$
$$\leq (D_{vis}(t, \boldsymbol{v}_{\eta_1}(t)) - D_{vis}(t, \boldsymbol{v}_{\eta_2}(t)), \theta_{\eta_1}(t) - \theta_{\eta_2}(t))_{L^2(\Omega)} \quad (18)$$

for a.e. $t \in (0,T)$. We integrate (18) over $(0,t)$ and by using the strong monotonicity of $\kappa(\cdot, \nabla\theta)$ and the Lipschitz continuity of $D_{vis} : V \to V'_{\text{Dir}}$ independently of $t \in (0,T)$, we obtain

$$\|\theta_{\eta_1}(t) - \theta_{\eta_2}(t)\|^2_{L^2(\Omega)} \leq C \int_0^t \|\boldsymbol{v}_{\eta_1} - \boldsymbol{v}_{\eta_2}\|^2_V \quad \text{for all } t \in [0,T].$$

Finally, the inequality (17) follows from Lemma 1. □

Lemma 3. *Assume that* (A–12) *holds. Then for all* $\omega \in \mathrm{L}^2(0,T; \mathrm{L}^2(\Omega))$, *there exists an unique solution* $z_\omega \in \mathrm{H}^1(0,T;\mathrm{L}^2(\Omega)) \cap \mathrm{L}^2(0,T;\mathrm{V}_{\mathrm{Neu}})$ *to problem:*

$$\langle \dot{z}_\omega(t), \xi - z_\omega(t) \rangle_{\mathrm{L}^2(\Omega)} + \gamma \langle \nabla z_\omega(t), \nabla \xi - \nabla z_\omega(t) \rangle_{\mathbf{L}^2(\Omega)}$$
$$\geq \langle \omega(t), \xi - z_\omega(t) \rangle_{\mathrm{L}^2(\Omega)} \quad \text{for all } \xi \in \mathrm{V}_{\mathrm{Neu}} \text{ for a.e. } t \in (0,T), \tag{19a}$$

$$z_\omega(0) = z_0. \tag{19b}$$

Moreover, there exists a constant $C > 0$ *such that for all* $\omega_i \in \mathrm{L}^2(0,T;\mathrm{L}^2(\Omega))$, $i = 1,2$, *we have*

$$\|z_{\omega_2}(t) - z_{\omega_1}(t)\|_{\mathrm{L}^2(\Omega)}^2 \leq C \int_0^t \|\omega_1 - \omega_2\|_{\mathrm{L}^2(\Omega)}^2 \quad \text{for all } t \in [0,T]. \tag{20}$$

Proof. The inequality (19) follows from classical result on evolution variational inequalities.

Let $\omega_i \in \mathrm{L}^2(0,T;\mathrm{V}')$, $i = 1,2$. On the one hand, taking $\omega = \omega_1$ and $\xi = z_{\omega_2}$ in (19a) and on the other hand, taking $\omega = \omega_2$ and $\xi = z_{\omega_1}$ in (19a) and then adding the resulting inequalities and integrating over $(0,t)$, we obtain

$$\frac{1}{2}\|z_{\omega_1}(t) - z_{\omega_2}(t)\|_{\mathrm{L}^2(\Omega)}^2 + \gamma \int_0^t \|\nabla z_{\omega_1}(s) - \nabla z_{\omega_2}(s)\|_{\mathbf{L}^2(\Omega)}^2$$
$$\leq \int_0^t \|\omega_1(s) - \omega_2(s)\|_{\mathrm{L}^2(\Omega)} \|z_{\omega_1}(s) - z_{\omega_2}(s)\|_{\mathrm{L}^2(\Omega)} \, ds$$

for all $t \in [0,T]$.

Thus the inequality (20) follows from Grönwall's lemma. $\square$

Let $\mathrm{X} \stackrel{\text{def}}{=} \mathrm{L}^2(0,T;\mathrm{V}' \times \mathrm{L}^2(\Omega))$ and $\Lambda : \mathrm{X} \to \mathrm{X}$ be defined by $\Lambda(\eta,\omega) \stackrel{\text{def}}{=} (\Lambda_1(\eta,\omega), \Lambda_2(\eta,\omega))$ for all $(\eta,\omega) \in \mathrm{X}$ such that for a.e. $t \in (0,T)$, for all $\boldsymbol{w} \in \mathrm{V}$:

$$\langle \Lambda_1(\eta,\omega)(t), \boldsymbol{w} \rangle_{\mathrm{V}' \times \mathrm{V}} \stackrel{\text{def}}{=} (\mathbf{E}(t, \boldsymbol{u}_\eta(t), z_\omega(t)) + \mathscr{B}(t, \boldsymbol{u}_\eta, z_\omega)$$
$$+ C_{\exp}(t, \theta_\eta(t)), \varepsilon(\boldsymbol{w}))_{\mathscr{H}}$$
$$\Lambda_2(\eta,\omega)(t) \stackrel{\text{def}}{=} \varphi(\boldsymbol{\sigma}_{\eta,\omega}(t), \varepsilon(\boldsymbol{u}_\eta(t)), z_\omega(t)),$$

where

$$\mathscr{B}(t, \boldsymbol{u}_\eta, z_\omega) \stackrel{\text{def}}{=} \int_0^t \mathbf{B}(t-s, \boldsymbol{\varepsilon}(\boldsymbol{u}_\eta(s)), z_\omega(s))\,\mathrm{d}s$$

and

$$\boldsymbol{\sigma}_{\eta,\omega}(t) \stackrel{\text{def}}{=} \mathbf{E}(t, \boldsymbol{\varepsilon}(\boldsymbol{u}_\eta(t)), z_\omega(t)) + \mathbf{A}(t, \boldsymbol{\varepsilon}(\boldsymbol{v}_\eta(t))) + \mathscr{B}(t, \boldsymbol{u}_\eta, z_\omega) + C_{\exp}(t, \theta_\eta(t)).$$

Note that the functional space X is endowed with the following norm:

$$\|(\eta, \omega)\|_X^2 \stackrel{\text{def}}{=} \int_0^T \|(\eta(t), \omega(t))\|_{V' \times L^2(\Omega)}^2\,\mathrm{d}t,$$

where

$$\|(\eta(t), \omega(t))\|_{V' \times L^2(\Omega)}^2 \stackrel{\text{def}}{=} \|\eta(t)\|_{V'}^2 + \|\omega(t)\|_{L^2(\Omega)}^2.$$

Lemma 4. *Assume that (A–1)–(A–12) hold and assume that $L_\tau < \frac{m_\mathbf{A}}{\sqrt{2T}c_0^2}$. Then Λ has a unique fixed point (η^*, ω^*).*

Proof. Clearly, the assumptions (A–4) enable us to deduce that there exists a constant $C > 0$ such that

$$\|C_{\exp}(t, \xi_1) - C_{\exp}(t, \xi_2)\|_{V'} \leq C\|\xi_1 - \xi_2\|_{L^2(\Omega)}$$

for a.e. $t \in (0, T)$ and for all $\xi_i \in V_{\text{Dir}}, i = 1, 2$. Now let (η_i, ω_i), $i = 1, 2$, be given in X. We verify that

$$\|\Lambda(\eta_1, \omega_1)(t) - \Lambda(\eta_2, \omega_2)(t)\|_{V' \times L^2(\Omega)}^2$$
$$\leq C\big(\|\mathbf{E}(t, \boldsymbol{u}_{\eta_1}(t), z_{\omega_1}(t)) - \mathbf{E}(t, \boldsymbol{u}_{\eta_2}(t), z_{\omega_2}(t))\|_{V'}^2$$
$$+ \|\mathscr{B}(t, \boldsymbol{u}_{\eta_1}(t), z_{\omega_1}(t)) - \mathscr{B}(t, \boldsymbol{u}_{\eta_2}(t), z_{\omega_2}(t))\|_{V'}^2$$
$$+ \|C_{\exp}(t, \theta_{\eta_1}(t)) - C_{\exp}(t, \theta_{\eta_2}(t))\|_{V'}^2\big)$$
$$+ \|\varphi(\boldsymbol{\sigma}_{\eta_1(t),\omega_1}(t), \boldsymbol{\varepsilon}(\boldsymbol{u}_{\eta_1}(t)), z_{\omega_1}(t))$$
$$- \varphi(\boldsymbol{\sigma}_{\eta_2,\omega_2}(t), \boldsymbol{\varepsilon}(\boldsymbol{u}_{\eta_2}(t)), z_{\omega_2}(t))\|_{L^2(\Omega)}^2$$

for a.e. $t \in (0, T)$. Thus, it follows that

$$\|\Lambda(\eta_1, \omega_1)(t) - \Lambda(\eta_2, \omega_2)(t)\|_{V' \times L^2(\Omega)}^2$$
$$\leq C\big(\|\boldsymbol{u}_{\eta_1}(t) - \boldsymbol{u}_{\eta_2}(t)\|_V^2 + \|z_{\omega_1}(t) - z_{\omega_2}(t)\|_{L^2(\Omega)}^2$$
$$+ \|\theta_{\eta_1}(t) - \theta_{\eta_2}(t)\|_{L^2(\Omega)}^2 + \|\boldsymbol{v}_{\eta_1}(t) - \boldsymbol{v}_{\eta_2}(t)\|_H^2\big).$$

We may infer from Lemmas 1, 2 and 3 that if $L_\tau < \frac{m_\mathbf{A}}{\sqrt{2T}c_0^2}$, then there exists a constant $C > 0$ satisfying

$$\|\Lambda(\eta_1, \omega_1)(t) - \Lambda(\eta_2, \omega_2)(t)\|_{V' \times L^2(\Omega)}^2$$
$$\leq C\left(\int_0^t \|\eta_2(s) - \eta_1(s)\|_{V'}^2 \, ds + \int_0^t \|\omega_1(s) - \omega_2(s)\|_{L^2(\Omega)}^2 \, ds\right)$$

for all $(\eta_i, \omega_i) \in X$, $i = 1, 2$ and for all $t \in [0, T]$. Then using once again Banach's fixed-point argument, we conclude that Λ has an unique fixed point. □

We have now all the ingredients to prove Theorem 1. More precisely, it suffices to verify that $\boldsymbol{u} \stackrel{\text{def}}{=} \boldsymbol{u}_{\eta^*}$, $z \stackrel{\text{def}}{=} z_{\omega^*}$ and $\theta \stackrel{\text{def}}{=} \theta_{\eta^*}$ are solutions to problem **(WF)** such that (9) holds true while the uniqueness result follows from Lemmas 1, 2 and 3. Since it is quite a routine to establish the results mentioned above, the verification is let to the reader.

References

[1] H. Brezis, *Opérateurs maximaux monotones et semi-groupes de contractions dans les espaces de Hilbert*. North-Holland Mathematics Studies, No. 5. Notas de Matemática (50) (North-Holland Publishing Co., Amsterdam, 1973).

[2] O. Chau, PhD Thesis, Analyse variationnelle et numérique en mécanique du contact, University of Perpignan (2000).

[3] O. Chau, Habilitation Thesis, Quelques problèmes d'évolution en mécanique de contact et en biochimie, Université de la Réunion (2010).

[4] M. Cocu, A dynamic viscoelastic problem with friction and rate-dependent contact interaction, *Evol. Equ. Control Theory* **9**(4), 981–993 (2020).

[5] G. Duvaut and J. L. Lions, *Les Inéquations en mécanique et en physique* (Dunod, Paris, 1972).

[6] M. Frémond and B. Nedjar, Damage in concrete: The unilateral phenomenon, *Nucl. Eng. Des.* **156**, 323–335 (1995).
[7] M. Frémond and B. Nedjar, Damage, gradient of damage and principle of virtual work, *Int. J. Solids Struct.* **33**, 1083–1103 (1996).
[8] N. Kikuchi and J. T. Oden, *Contact Problems in Elasticity* (SIAM, Philadelphia, 1988).
[9] A. Kulig, Hyperbolic hemivariational inequalities for dynamic viscoelastic contact problems, *J. Elast.* **110**, 1–31 (2013).
[10] Y. Li and Z. Liu, A quasistatic contact problem for viscoelastic materials with friction and damage, *Nonlinear Anal.* **73**, 2221–2229 (2010).
[11] J. L. LIONS, *Quelques méthodes de résolution des problèmes aux limites non linéaires* (Dunod et Gauthier-Villars, 1969).
[12] P. D. Panagiotopoulos, *Inequality Problems in Meechanics and Applications* (Birkhäuser, Basel, 1985).
[13] P. D. Panagiotopoulos, *Hemivariational Inequalities, Applications in Mechanics and Engineering* (Springer-Verlag, Heidelberg, 1993).
[14] A. Signorini, Questioni di elasticità non linearizzata e semilinearizzata, *Rend. Mat. Appl.* **18**(5), 95–139 (1959).
[15] E. Zeidler, *Nonlinear Functional Analysis and its Applications, II/A* (Springer Verlag, New York, 1997).
[16] $\mathcal{A}\mathcal{M}S$, $\mathcal{A}\mathcal{M}S$-$\mathbb{A}T_{E}X$ *Version 2 User's Guide* (American Mathematical Society, Providence, 2004). http://www.ams.org/tex/amslatex.html.
[17] D. W. Baker and N. L. Carter, Seismic velocity anisotropy calculated for ultramafic minerals and aggregates, *Geophys. Mono.* (American Geophysical Union) **16**, 157–166 (1972).
[18] F. Benhamou and A. Colmerauer (eds.) *Constraint Logic Programming, Selected Research* (MIT Press, Cambridge, Massachusetts, 1993).

© 2024 World Scientific Publishing Company
https://doi.org/10.1142/9789811267048_0008

Chapter 8

Vector Inequalities for Analytic Functions of Operators in Hilbert Spaces and Applications for Numerical Radius and p-Schatten Norm

Silvestru Sever Dragomir

*Applied Mathematics Research Group, ISILC,
Victoria University, Melbourne City, Australia
School of Computer Science & Applied Mathematics,
University of the Witwatersrand,
Johannesburg, South Africa
sever.dragomir@vu.edu.au
http://rgmia.org/dragomir*

Let H be a complex Hilbert space, f be an analytic function on the domain G with $0 \in G$, A be an operator with $\mathrm{Sp}(A) \subset G$ and γ be a closed path in $G \setminus \{0\}$ with $\mathrm{Sp}(A) \subset \mathrm{ins}(\gamma)$, then for $B, C \in B(H)$ we have

$$|\langle C^* [f(A) - f(0) I] Bx, y \rangle|$$

$$\leq M(f, A; \gamma) \left\langle ||A|^\alpha B|^2 x, x \right\rangle^{1/2} \left\langle \left||A^*|^{1-\alpha} C\right|^2 y, y \right\rangle^{1/2}$$

225

for $x, y \in H$, where

$$M(f, A; \gamma) := \frac{1}{2\pi} \int_\gamma \frac{|f(\xi)| |d\xi|}{|\xi|(|\xi| - \|A\|)}.$$

Some natural applications for *numerical radius* and *p-Schatten norm* are also provided.

1. Introduction

In 1988, Kittaneh obtained the following generalization of Schwarz inequality [17]:

Theorem 1. *Assume that h and g are nonnegative functions on $[0, \infty)$ which are continuous and satisfying the relation $h(t)g(t) = t$ for all $t \in [0, \infty)$. For any $T \in \mathcal{B}(H)$,*

$$|\langle Tx, y \rangle| \leq \|h(|T|)x\| \|g(|T^*|)y\| \tag{1}$$

for all $x, y \in H$.

If we take $h(t) = t^\alpha$, $g(t) = t^{1-\alpha}$ with $\alpha \in [0, 1]$, then we obtain Kato's inequality

$$|\langle Tx, y \rangle| \leq \||T|^\alpha x\| \||T^*|^{1-\alpha} y\| \quad \text{for all } x, y \in H. \tag{2}$$

The *numerical radius* $w(T)$ of an operator T on H is given by

$$w(T) = \sup\{|\langle Tx, x \rangle|, \|x\| = 1\}. \tag{3}$$

Obviously, by (3), for any $x \in H$ one has

$$|\langle Tx, x \rangle| \leq w(T) \|x\|^2. \tag{4}$$

It is well known that $w(\cdot)$ is a norm on the Banach algebra $B(H)$ of all bounded linear operators $T : H \to H$, i.e.,

(i) $\omega(T) \geq 0$ for any $T \in B(H)$ and $\omega(T) = 0$ if and only if $T = 0$;
(ii) $\omega(\lambda T) = |\lambda| \omega(T)$ for any $\lambda \in \mathbb{C}$ and $T \in B(H)$;
(iii) $\omega(T + V) \leq \omega(T) + \omega(V)$ for any $T, V \in B(H)$.

This norm is equivalent with the operator norm. In fact, the following more precise result holds:

$$w(T) \le \|T\| \le 2w(T) \tag{5}$$

for any $T \in B(H)$.

Kittaneh [18], showed that for any operator $T \in B(H)$ we have the following refinement of the first inequality in (5):

$$w(T) \le \frac{1}{2}\left(\|T\| + \|T^2\|^{1/2}\right). \tag{6}$$

Utilizing the Cartesian decomposition for operators, Kittaneh [19] improved the inequality (5) as follows:

$$\frac{1}{4}\|T^*T + TT^*\| \le w^2(T) \le \frac{1}{2}\|T^*T + TT^*\| \tag{7}$$

for any operator $T \in B(H)$.

For powers of the absolute value of operators, one can state the following results obtained by El-Haddad and Kittaneh [13]:

If for an operator $T \in B(H)$, we denote $|T| := (T^*T)^{1/2}$, then

$$w^r(T) \le \frac{1}{2}\left\||T|^{2\alpha r} + |T^*|^{2(1-\alpha)r}\right\| \tag{8}$$

and

$$w^{2r}(T) \le \left\|\alpha|T|^{2r} + (1-\alpha)|T^*|^{2r}\right\|, \tag{9}$$

where $\alpha \in (0,1)$ and $r \ge 1$.

If we take $\alpha = \frac{1}{2}$ and $r = 1$, we get from (8) that

$$w(T) \le \frac{1}{2}\||T| + |T^*|\| \tag{10}$$

and from (9) that

$$w^2(T) \le \frac{1}{2}\left\||T|^2 + |T^*|^2\right\|. \tag{11}$$

For more related results, see the recent books on inequalities for numerical radii [5, 10].

Let $(H; \langle \cdot, \cdot \rangle)$ be a complex Hilbert space and $\mathcal{B}(H)$ the Banach algebra of all bounded linear operators on H. If $\{e_i\}_{i \in I}$ is an orthonormal basis of H, we say that $A \in \mathcal{B}(H)$ is of *trace class* if

$$\|A\|_1 := \sum_{i \in I} \langle |A| e_i, e_i \rangle < \infty. \tag{12}$$

The definition of $\|A\|_1$ does not depend on the choice of the orthonormal basis $\{e_i\}_{i \in I}$. We denote by $\mathcal{B}_1(H)$ the set of trace class operators in $\mathcal{B}(H)$.

We define the *trace* of a trace class operator $A \in \mathcal{B}_1(H)$ to be

$$\operatorname{tr}(A) := \sum_{i \in I} \langle A e_i, e_i \rangle, \tag{13}$$

where $\{e_i\}_{i \in I}$ is an orthonormal basis of H. Note that this coincides with the usual definition of the trace if H is finite-dimensional. We observe that the series (13) converges absolutely and it is independent from the choice of basis.

The following result collects some properties of the trace:

Theorem 2. *We have the following:*

(i) *If $A \in \mathcal{B}_1(H)$, then $A^* \in \mathcal{B}_1(H)$ and*

$$\operatorname{tr}(A^*) = \overline{\operatorname{tr}(A)}. \tag{14}$$

(ii) *If $A \in \mathcal{B}_1(H)$ and $T \in \mathcal{B}(H)$, then $AT, TA \in \mathcal{B}_1(H)$ and*

$$\operatorname{tr}(AT) = \operatorname{tr}(TA) \quad \text{and} \quad |\operatorname{tr}(AT)| \le \|A\|_1 \|T\|. \tag{15}$$

(iii) $\operatorname{tr}(\cdot)$ *is a bounded linear functional on $\mathcal{B}_1(H)$ with $\|\operatorname{tr}\| = 1$.*
(iv) *If $A, B \in \mathcal{B}_2(H)$ then $AB, BA \in \mathcal{B}_1(H)$ and $\operatorname{tr}(AB) = \operatorname{tr}(BA)$.*
(v) $\mathcal{B}_{\text{fin}}(H)$, *the space of operators of finite rank, is a dense subspace of $\mathcal{B}_1(H)$.*

For a large number of results concerning trace inequalities, see the recent survey paper [11].

An operator $A \in \mathcal{B}(H)$ is said to belong to the *von Neumann–Schatten class* $\mathcal{B}_p(H)$, $1 \leq p < \infty$, if the *p-Schatten norm* is finite [25, pp. 60–64]:

$$\|A\|_p := [\operatorname{tr}(|A|^p)]^{1/p} = \left(\sum_{i \in I} \langle |A|^p e_i, e_i \rangle\right)^{1/p} < \infty.$$

For $1 < p < q < \infty$, we have that

$$\mathcal{B}_1(H) \subset \mathcal{B}_p(H) \subset \mathcal{B}_q(H) \subset \mathcal{B}(H) \tag{16}$$

and

$$\|A\|_1 \geq \|A\|_p \geq \|A\|_q \geq \|A\|. \tag{17}$$

For $p \geq 1$, the functional $\|\cdot\|_p$ is a *norm* on the $*$-ideal $\mathcal{B}_p(H)$ and $\left(\mathcal{B}_p(H), \|\cdot\|_p\right)$ is a Banach space.

Also, see, for instance, [25, pp. 60–64]

$$\|A\|_p = \|A^*\|_p, \quad A \in \mathcal{B}_p(H), \tag{18}$$

$$\|AB\|_p \leq \|A\|_p \|B\|_p, \quad A, B \in \mathcal{B}_p(H) \tag{19}$$

and

$$\|AB\|_p \leq \|A\|_p \|B\|, \quad \|BA\|_p \leq \|B\| \|A\|_p, \quad A \in \mathcal{B}_p(H), \ B \in \mathcal{B}(H). \tag{20}$$

This implies that

$$\|CAB\|_p \leq \|C\| \|A\|_p \|B\|, \quad A \in \mathcal{B}_p(H), \quad B, C \in \mathcal{B}(H). \tag{21}$$

In terms of *p-Schatten norm* we have the *Hölder inequality* for $p, q > 1$ with $\frac{1}{p} + \frac{1}{q} = 1$:

$$(|\operatorname{tr}(AB)| \leq) \|AB\|_1 \leq \|A\|_p \|B\|_q, \quad A \in \mathcal{B}_p(H), \quad B \in \mathcal{B}_q(H). \tag{22}$$

For the theory of trace functionals and their applications, the reader is referred to Refs. [24, 25].

For $\mathcal{E} := \{e_i\}_{i \in I}$ an orthonormal basis of H, we define for $A \in \mathcal{B}_p(H)$, $p \geq 1$,

$$\|A\|_{\mathcal{E},p} := \left(\sum_{i \in I} |\langle Ae_i, e_i\rangle|^p\right)^{1/p}.$$

We observe that $\|\cdot\|_{\mathcal{E},p}$ is a norm on $\mathcal{B}_p(H)$ and

$$\|A\|_{\mathcal{E},p} \leq \|A\|_p \quad \text{for } A \in \mathcal{B}_p(H).$$

Further, we can take the supremum over all orthonormal basis in H, and we can also define, for $A \in \mathcal{B}_p(H)$, that

$$\omega_p(A) := \sup_{\mathcal{E}} \|A\|_{\mathcal{E},p} \leq \|A\|_p,$$

which is a *norm* on $\mathcal{B}_p(H)$.

It is also known that, if $\mathcal{E} = \{e_i\}_{i \in I}$ and $\mathcal{F} = \{f_i\}_{i \in I}$ are orthonormal basis, then [22]

$$\sup_{\mathcal{E},\mathcal{F}} \sum_{i \in I} |\langle Te_i, f_i\rangle|^s = \|T\|_s^s \quad \text{for } s \geq 1. \tag{23}$$

Let $\mathcal{B}$ be a unital Banach algebra, $A \in \mathcal{B}$ and G be a domain of $\mathbb{C}$ with $\mathrm{Sp}(A) \subset G$. If $f : G \to \mathbb{C}$ is analytic on G, we define an element $f(A)$ in $\mathcal{B}$ by

$$f(A) := \frac{1}{2\pi i} \int_\delta f(\xi)(\xi - A)^{-1} d\xi, \tag{24}$$

where $\delta \subset G$ is taken to be a closed rectifiable curve in G such that $\mathrm{Sp}(A) \subset \mathrm{ins}(\delta)$, the inside of δ.

It is well known (see, for instance, Ref. [7, pp. 201–204]) that $f(A)$ does not depend on the choice of δ and the *Spectral Mapping Theorem* (SMT)

$$\mathrm{Sp}(f(A)) = f(\mathrm{Sp}(A)) \tag{25}$$

holds.

Concerning other basic definitions and facts in the theory of Banach algebras, the reader can consult the classical books [9, 23].

2. Vector Inequalities

In 1988, Kittaneh [17, Corollary 7] obtained the following Schwarz-type inequality for natural powers of operators:

Lemma 1. *Let $T \in \mathcal{B}(H)$ and $\alpha \in [0,1]$. Then for natural number $n \geq 1$ we have*

$$|\langle T^n x, y \rangle|^2 \leq \|T\|^{2n-2} \left\langle |T|^{2\alpha} x, x \right\rangle \left\langle |T^*|^{2(1-\alpha)} y, y \right\rangle \quad (26)$$

for all $x, y \in H$.

Our first main result is as follows:

Theorem 3. *Let f be an analytic function on the domain G with $0 \in G$, A an operator with $\mathrm{Sp}(A) \subset G$ and γ a closed path in $G \setminus \{0\}$ with $\mathrm{Sp}(A) \subset \mathrm{ins}(\gamma)$, then for $B, C \in \mathcal{B}(H)$ we have*

$$|\langle C^* [f(A) - f(0) I] Bx, y \rangle|$$

$$\leq M(f, A; \gamma) \left\langle \||A|^\alpha B|^2 x, x \right\rangle^{1/2} \left\langle \||A^*|^{1-\alpha} C|^2 y, y \right\rangle^{1/2} \quad (27)$$

for $x, y \in H$, where $\alpha \in [0,1]$ and

$$M(f, A; \gamma) := \frac{1}{2\pi} \int_\gamma \frac{|f(\xi)| |d\xi|}{|\xi| (|\xi| - \|A\|)}.$$

In particular, we have

$$|\langle C^* [f(A) - f(0) I] Bx, y \rangle|$$

$$\leq M(f, A; \gamma) \left\langle \||A|^{1/2} B|^2 x, x \right\rangle^{1/2} \left\langle \||A^*|^{1/2} C|^2 y, y \right\rangle^{1/2} \quad (28)$$

for $x, y \in H$.

Proof. From Kittaneh's result (26), we have that

$$|\langle T^n x, y \rangle| \leq \|T\|^{n-1} \left\langle |T|^{2\alpha} x, x \right\rangle^{1/2} \left\langle |T^*|^{2(1-\alpha)} y, y \right\rangle^{1/2} \quad (29)$$

for all $x, y \in H$.

If we sum from $n = 1$ to $n = m+1$, then we get

$$\left|\langle T(1+T+\cdots+T^m)x,y\rangle\right|$$
$$\leq (1+\|T\|+\cdots+\|T\|^m)\langle|T|^{2\alpha}x,x\rangle^{1/2}\langle|T^*|^{2(1-\alpha)}y,y\rangle^{1/2} \tag{30}$$

for all $x, y \in H$.

If we take $m \to \infty$ in (30) and take into account that $\sum_{m=0}^{\infty} T^m = (I-T)^{-1}$ and $\sum_{m=0}^{\infty}\|T\|^m = (I-\|T\|)^{-1}$ for $\|T\| < 1$, then we get

$$\left|\langle T(I-T)^{-1}x,y\rangle\right|$$
$$\leq (I-\|T\|)^{-1}\langle|T|^{2\alpha}x,x\rangle^{1/2}\langle|T^*|^{2(1-\alpha)}y,y\rangle^{1/2} \tag{31}$$

for all $x, y \in H$.

Moreover, if we replace x by Bx and y by Cy in (31), then we obtain

$$\left|\langle C^*T(I-T)^{-1}Bx,y\rangle\right|$$
$$\leq (I-\|T\|)^{-1}\langle\||T|^{\alpha}B|^2 x,x\rangle^{1/2}\langle\||T^*|^{1-\alpha}C\|^2 y,y\rangle^{1/2} \tag{32}$$

for all $x, y \in H$.

By the analytic functional calculus, we have

$$f(A) - f(0)I = \frac{1}{2\pi i}\int_\gamma f(\xi)(\xi I - A)^{-1} d\xi - \frac{1}{2\pi i}\int_\gamma f(\xi)\xi^{-1} I d\xi$$

$$= \frac{1}{2\pi i}\int_\gamma f(\xi)\left[(\xi I - A)^{-1} - \xi^{-1} I\right] d\xi$$

$$= \frac{1}{2\pi i}\int_\gamma \xi^{-1} f(\xi) A(\xi I - A)^{-1} d\xi.$$

Therefore,

$$\langle C^* [f(A) - f(0)I] Bx, y \rangle$$
$$= \frac{1}{2\pi i} \int_\gamma \xi^{-1} f(\xi) \langle C^* A (\xi I - A)^{-1} Bx, y \rangle d\xi \qquad (33)$$

for $x, y \in H$.

By taking the modulus and using the properties of the integral, we obtain by (33) that

$$|\langle C^* [f(A) - f(0)I] Bx, y \rangle|$$
$$\leq \frac{1}{2\pi} \int_\gamma |\xi|^{-1} |f(\xi)| \left| \langle C^* A (\xi I - A)^{-1} Bx, y \rangle \right| |d\xi|$$
$$= \frac{1}{2\pi} \int_\gamma |\xi|^{-1} |f(\xi)| \left| \left\langle C^* \frac{A}{\xi} \left(I - \frac{A}{\xi} \right)^{-1} Bx, y \right\rangle \right| |d\xi|. \qquad (34)$$

If $\xi \in \gamma$, then $\left\| \frac{A}{\xi} \right\| < 1$ and by (32) for $T = \frac{A}{\xi}$ we get

$$\left| \left\langle C^* \frac{A}{\xi} \left(I - \frac{A}{\xi} \right)^{-1} Bx, y \right\rangle \right|$$
$$\leq \left(I - \left\| \frac{A}{\xi} \right\| \right)^{-1} \left\langle \left| \frac{A}{\xi} \right|^\alpha B \right|^2 x, x \right\rangle^{1/2} \left\langle \left| \left(\frac{A}{\xi} \right)^* \right|^{1-\alpha} C \right|^2 y, y \right\rangle^{1/2}$$
$$= \left(\frac{|\xi| - \|A\|}{|\xi|} \right)^{-1} \frac{1}{|\xi|^\alpha |\xi|^{1-\alpha}} \left\langle ||A|^\alpha B|^2 x, x \right\rangle^{1/2}$$
$$\times \left\langle ||A^*|^{1-\alpha} C|^2 y, y \right\rangle^{1/2}$$
$$= (|\xi| - \|A\|)^{-1} \left\langle ||A|^\alpha B|^2 x, x \right\rangle^{1/2} \left\langle ||A^*|^{1-\alpha} C|^2 y, y \right\rangle^{1/2} \qquad (35)$$

for $x, y \in H$.

Therefore, by (35) we get

$$\frac{1}{2\pi}\int_{\gamma}|\xi|^{-1}|f(\xi)|\left|\left\langle C^*\frac{A}{\xi}\left(I-\frac{A}{\xi}\right)^{-1}Bx,y\right\rangle\right||d\xi|$$

$$\leq \frac{1}{2\pi}\int_{\gamma}|\xi|^{-1}|f(\xi)|(|\xi|-\|A\|)^{-1}|d\xi|$$

$$\times \left\langle \||A|^{\alpha}B|^2 x,x\right\rangle^{1/2}\left\langle \||A^*|^{1-\alpha}C|^2 y,y\right\rangle^{1/2}$$

$$= M(f,A;\gamma)\left\langle \||A|^{\alpha}B|^2 x,x\right\rangle^{1/2}\left\langle \||A^*|^{1-\alpha}C|^2 y,y\right\rangle^{1/2}$$

for $x, y \in H$, which by (34) gives the desired result (27). □

Remark 1. With the assumptions of Theorem 3 and if $f(0) = 0$, then we get

$$|\langle C^* f(A) Bx, y\rangle|$$

$$\leq M(f,A;\gamma)\left\langle \||A|^{\alpha}B|^2 x,x\right\rangle^{1/2}\left\langle \||A^*|^{1-\alpha}C|^2 y,y\right\rangle^{1/2} \quad (36)$$

and

$$|\langle C^* f(A) Bx, y\rangle|$$

$$\leq M(f,A;\gamma)\left\langle \||A|^{1/2}B|^2 x,x\right\rangle^{1/2}\left\langle \||A^*|^{1/2}C|^2 y,y\right\rangle^{1/2} \quad (37)$$

for $x, y \in H$.

For $B = C = I$ in (36) and (37), we get the one operator inequalities

$$|\langle f(A) x, y\rangle| \leq M(f,A;\gamma)\left\langle |A|^{2\alpha} x,x\right\rangle^{1/2}\left\langle |A^*|^{2(1-\alpha)} y,y\right\rangle^{1/2} \quad (38)$$

and

$$|\langle f(A) x, y\rangle| \leq M(f,A;\gamma)\langle |A| x,x\rangle^{1/2}\langle |A^*| y,y\rangle^{1/2} \quad (39)$$

for $x, y \in H$.

If $A > 0$ and we take $B = A^{-\beta}$, $C = A^{-1+\beta}$, $\beta \in [0,1]$, in (36) and (37), then we get

$$|\langle f(A) A^{-1}x, y \rangle| \leq M(f, A; \gamma) \left\langle A^{2(\alpha-\beta)}x, x \right\rangle^{1/2} \left\langle A^{2(\beta-\alpha)}y, y \right\rangle^{1/2} \tag{40}$$

and

$$|\langle f(A) A^{-1}x, y \rangle| \leq M(f, A; \gamma) \left\langle A^{2(1/2-\beta)}x, x \right\rangle^{1/2} \left\langle A^{2(\beta-1/2)}y, y \right\rangle^{1/2} \tag{41}$$

for $x, y \in H$.

Corollary 1. *With the assumptions of Theorem 3 and if*

$$\|f\|_{\gamma,\infty} := \sup_{\xi \in \gamma} |f(\xi)| < \infty,$$

then, by denoting

$$M_\infty(f, A; \gamma) := \frac{1}{2\pi} \|f\|_{\gamma,\infty} \int_\gamma \frac{|f(\xi)| \, |d\xi|}{|\xi| (|\xi| - \|A\|)},$$

we have

$$|\langle C^* [f(A) - f(0) I] Bx, y \rangle|$$
$$\leq M_\infty(f, A; \gamma) \left\langle \left||A|^\alpha B\right|^2 x, x \right\rangle^{1/2} \left\langle \left||A^*|^{1-\alpha} C\right|^2 y, y \right\rangle^{1/2} \tag{42}$$

for $x, y \in H$.
In particular, we have

$$|\langle C^* [f(A) - f(0) I] Bx, y \rangle|$$
$$\leq M_\infty(f, A; \gamma) \left\langle \left||A|^{1/2} B\right|^2 x, x \right\rangle^{1/2} \left\langle \left||A^*|^{1/2} C\right|^2 y, y \right\rangle^{1/2} \tag{43}$$

for $x, y \in H$.

Remark 2. If we assume that $f : G \subset \mathbb{C} \to \mathbb{C}$ is an analytic function on the domain G and $A \in \mathcal{B}(H)$ with $\mathrm{Sp}(A) \subset D(0,R) \subset D$ where $D(0,R)$ is an open disk centered in 0 and of radius R, then by taking γ parametrized by $\xi(t) = Re^{2\pi i t}$ where $t \in [0,1]$, then $\mathrm{d}\xi(t) = 2\pi i R e^{2\pi i t} \mathrm{d}t$, $|\mathrm{d}\xi(t)| = 2\pi R \mathrm{d}t$, $|\xi| = R$ and by (40) we get for $A, B \in \mathcal{B}(H)$ that

$$|\langle C^*[f(A) - f(0)I]Bx, y\rangle|$$
$$\leq M(f, A; R) \left\langle \left||A|^\alpha B\right|^2 x, x \right\rangle^{1/2} \left\langle \left||A^*|^{1-\alpha} C\right|^2 y, y \right\rangle^{1/2} \quad (44)$$

for $x, y \in H$, where $\alpha \in [0, 1]$ and

$$M(f, A; R) := \frac{1}{R - \|A\|} \int_0^1 \left|f\left(Re^{2\pi i t}\right)\right| \mathrm{d}t.$$

In particular,

$$|\langle C^*[f(A) - f(0)I]Bx, y\rangle|$$
$$\leq M(f, A; R) \left\langle \left||A|^{1/2} B\right|^2 x, x \right\rangle^{1/2} \left\langle \left||A^*|^{1/2} C\right|^2 y, y \right\rangle^{1/2} \quad (45)$$

for $x, y \in H$.

Moreover, if $\|f\|_{R,\infty} := \sup_{t \in [0,1]} \left|f\left(Re^{2\pi i t}\right)\right| < \infty$, then we have the simpler inequalities

$$|\langle C^*[f(A) - f(0)I]Bx, y\rangle|$$
$$\leq \frac{\|f\|_{R,\infty}}{R - \|A\|} \left\langle \left||A|^\alpha B\right|^2 x, x \right\rangle^{1/2} \left\langle \left||A^*|^{1-\alpha} C\right|^2 y, y \right\rangle^{1/2} \quad (46)$$

for $x, y \in H$, where $\alpha \in [0, 1]$ and

$$|\langle C^*[f(A) - f(0)I]Bx, y\rangle|$$
$$\leq \frac{\|f\|_{R,\infty}}{R - \|A\|} \left\langle \left||A|^{1/2} B\right|^2 x, x \right\rangle^{1/2} \left\langle \left||A^*|^{1/2} C\right|^2 y, y \right\rangle^{1/2}. \quad (47)$$

3. Norm and Numerical Radius Inequalities

The following vector inequality for positive operators $A \geq 0$, obtained by McCarthy [21], is well known:

$$\langle Ax, x\rangle^p \leq \langle A^p x, x\rangle, \quad p \geq 1$$

for $x \in H$, $\|x\| = 1$.

Buzano's inequality [6]

$$\frac{1}{2}[\|x\|\,\|y\| + |\langle x, y\rangle|] \geq |\langle x, e\rangle \langle e, y\rangle| \tag{48}$$

that holds for any $x, y, e \in H$ with $\|e\| = 1$ will also be used in the sequel.

We also have the following norm and numerical radius inequalities:

Theorem 4. *Let $f : G \subset \mathbb{C} \to \mathbb{C}$ be an analytic function on the domain G with $0 \in G$ and $A \in \mathcal{B}(H)$ with $\mathrm{Sp}(A) \subset G$ and γ a closed rectifiable path in $G \setminus \{0\}$ and such that $\mathrm{Sp}(A) \subset \mathrm{ins}(\gamma)$. If $B, C \in \mathcal{B}(H)$, then we have the norm inequality*

$$\|C^*[f(A) - f(0)I]B\| \leq M(f, A; \gamma) \||A|^\alpha B\| \left\||A^*|^{1-\alpha} C\right\|. \tag{49}$$

We also have the numerical radius inequalities

$$\omega(C^*[f(A) - f(0)I]B) \leq \frac{1}{2} M(f, A; \gamma) \left\| \||A|^\alpha B|^2 + \left||A^*|^{1-\alpha} C\right|^2 \right\| \tag{50}$$

and

$$\omega^2(C^*[f(A) - f(0)I]B) \leq \frac{1}{2} M^2(f, A; \gamma)$$
$$\times \left[\||A|^\alpha B\|^2 \left\||A^*|^{1-\alpha} C\right\|^2 + \omega\left(\left||A^*|^{1-\alpha} C\right|^2 \||A|^\alpha B|^2\right) \right]. \tag{51}$$

Proof. We have from (27), by taking the supremum over $\|x\| = \|y\| = 1$, that

$$\|C^* [f(A) - f(0) I] B\|^2$$

$$= \sup_{\|x\|=\|y\|=1} |\langle C^* [f(A) - f(0) I] Bx, y \rangle|^2$$

$$\leq M^2 (f, A; \gamma) \sup_{\|x\|=1} \left\langle \||A|^\alpha B|^2 x, x \right\rangle \sup_{\|y\|=1} \left\langle \||A^*|^{1-\alpha} C\|^2 y, y \right\rangle$$

$$= M^2 (f, A; \gamma) \left\| \||A|^\alpha B|^2 \right\| \left\| \||A^*|^{1-\alpha} C\|^2 \right\|$$

$$= M^2 (f, A; \gamma) \||A|^\alpha B\|^2 \left\| |A^*|^{1-\alpha} C \right\|^2,$$

which gives (49).

From (27) we get, by taking $y = x$, the square root and using the A-G-mean inequality, that

$$|\langle C^* [f(A) - f(0) I] Bx, x \rangle|$$

$$\leq M(f, A; \gamma) \left\langle \||A|^\alpha B|^2 x, x \right\rangle^{1/2} \left\langle \||A^*|^{1-\alpha} C\|^2 x, x \right\rangle^{1/2}$$

$$\leq \frac{1}{2} M(f, A; \gamma) \left(\left\langle \||A|^\alpha B|^2 x, x \right\rangle + \left\langle \||A^*|^{1-\alpha} C\|^2 x, x \right\rangle \right)$$

$$= \frac{1}{2} M(f, A; \gamma) \left\langle \left(\||A|^\alpha B|^2 + \left||A^*|^{1-\alpha} C\right|^2 \right) x, x \right\rangle \quad (52)$$

for all $x \in H$.

By taking the supremum over $\|x\| = 1$ in (52), we get that

$$\omega \left(C^* [f(A) - f(0) I] B \right)$$

$$= \sup_{\|x\|=1} |\langle C^* [f(A) - f(0) I] Bx, x \rangle|$$

$$\leq \frac{1}{2} M(f, A; \gamma) \sup_{\|x\|=1} \left\langle \left(\||A|^\alpha B|^2 + \left||A^*|^{1-\alpha} C\right|^2 \right) x, x \right\rangle$$

$$= \frac{1}{2} M(f, A; \gamma) \left\| \||A|^\alpha B|^2 + \left||A^*|^{1-\alpha} C\right|^2 \right\|,$$

which proves (50).

From (27) for $y = x$ and Buzano's inequality, we derive that

$$|\langle C^* [f(A) - f(0)I] Bx, x \rangle|^2$$

$$\leq M^2(f, A; \gamma) \langle \||A|^\alpha B|^2 x, x \rangle \langle x, \||A^*|^{1-\alpha} C|^2 x \rangle$$

$$\leq \frac{1}{2} M^2(f, A; \gamma)$$

$$\times \left[\left\| \||A|^\alpha B|^2 x \right\| \left\| \||A^*|^{1-\alpha} C|^2 x \right\| + \left| \left\langle \||A|^\alpha B|^2 x, \||A^*|^{1-\alpha} C|^2 x \right\rangle \right| \right]$$

$$= \frac{1}{2} M^2(f, A; \gamma)$$

$$\times \left[\left\| \||A|^\alpha B|^2 x \right\| \left\| \||A^*|^{1-\alpha} C|^2 x \right\| + \left| \left\langle \||A^*|^{1-\alpha} C|^2 \||A|^\alpha B|^2 x, x \right\rangle \right| \right] \tag{53}$$

for all $x \in H$.

By taking the supremum over $\|x\| = 1$ in (53), we get that

$$\omega^2 (C^* [f(A) - f(0)I] B)$$

$$= \sup_{\|x\|=1} |\langle C^* [f(A) - f(0)I] Bx, x \rangle|^2$$

$$\leq \frac{1}{2} M^2(f, A; \gamma)$$

$$\times \sup_{\|x\|=1} \left[\left\| \||A|^\alpha B|^2 x \right\| \left\| \||A^*|^{1-\alpha} C|^2 x \right\| + \left| \left\langle \||A^*|^{1-\alpha} C|^2 \||A|^\alpha B|^2 x, x \right\rangle \right| \right]$$

$$\leq \frac{1}{2} M^2(f, A; \gamma)$$

$$\times \left[\sup_{\|x\|=1} \left\{ \left\| \||A|^\alpha B|^2 x \right\| \left\| \||A^*|^{1-\alpha} C|^2 x \right\| \right\} \right.$$

$$+ \sup_{\|x\|=1} \left| \left\langle \||A^*|^{1-\alpha} C|^2 \||A|^\alpha B|^2 x, x \right\rangle \right| \right]$$

$$\leq \frac{1}{2} M^2(f, A; \gamma)$$

$$\times \left[\sup_{\|x\|=1} \left\| ||A|^\alpha B|^2 x \right\| \sup_{\|x\|=1} \left\| ||A^*|^{1-\alpha} C|^2 x \right\| \right.$$

$$\left. + \sup_{\|x\|=1} \left| \left\langle ||A^*|^{1-\alpha} C|^2 ||A|^\alpha B|^2 x, x \right\rangle \right| \right]$$

$$= \frac{1}{2} M^2(f, A; \gamma)$$

$$\times \left[\left\| ||A|^\alpha B|^2 \right\| \left\| ||A^*|^{1-\alpha} C|^2 \right\| + \omega \left(||A^*|^{1-\alpha} C|^2 ||A|^\alpha B|^2 \right) \right]$$

$$= \frac{1}{2} M^2(f, A; \gamma) \left[\left\| ||A|^\alpha B|^2 \right\| \left\| ||A^*|^{1-\alpha} C \right\|^2 + \omega \left(||A^*|^{1-\alpha} C|^2 ||A|^\alpha B|^2 \right) \right],$$

which proves (124). □

Remark 3. If we take $\alpha = 1/2$ in Theorem 4, then we get the norm inequality

$$\|C^* [f(A) - f(0) I] B\| \leq M(f, A; \gamma) \left\| |A|^{1/2} B \right\| \left\| |A^*|^{1/2} C \right\| \tag{54}$$

and the numerical radius inequalities

$$\omega \left(C^* [f(A) - f(0) I] B \right) \leq \frac{1}{2} M(f, A; \gamma) \left\| \left| |A|^{1/2} B \right|^2 + \left| |A^*|^{1/2} C \right|^2 \right\| \tag{55}$$

and

$$\omega^2 \left(C^* [f(A) - f(0) I] B \right)$$

$$\leq \frac{1}{2} M^2(f, A; \gamma)$$

$$\times \left[\left\| |A|^{1/2} B \right\|^2 \left\| |A^*|^{1/2} C \right\|^2 + \omega \left(\left| |A^*|^{1/2} C \right|^2 \left| |A|^{1/2} B \right|^2 \right) \right]. \tag{56}$$

The second main result is as follows:

Theorem 5. *Assume that the conditions of Theorem 4 are satisfied. If $\alpha \in [0,1]$, $r > 0$, $p, q > 1$ with $\frac{1}{p} + \frac{1}{q} = 1$ and $pr, qr \geq 1$, then*

$$\omega^{2r} \left(C^* \left[f(A) - f(0) I \right] B \right)$$
$$\leq M^{2r}(f, A; \gamma) \left\| \frac{1}{p} \||A|^\alpha B|^{2rp} + \frac{1}{q} \left||A^*|^{1-\alpha} C\right|^{2rq} \right\|. \tag{57}$$

If $r \geq 1$, then

$$\omega^{2r} \left(C^* \left[f(A) - f(0) I \right] B \right)$$
$$\leq \frac{1}{2} M^{2r}(f, A; \gamma) \left[\||A|^\alpha B\|^{2r} \left\||A^*|^{1-\alpha} C\right\|^{2r} \right.$$
$$\left. + \omega^r \left(\left||A^*|^{1-\alpha} C\right|^2 \||A|^\alpha B|^2 \right) \right]. \tag{58}$$

If $r \geq 1$, $p, q > 1$ with $\frac{1}{p} + \frac{1}{q} = 1$ and $pr, qr \geq 2$, then also

$$\omega^{2r} \left(C^* \left[f(A) - f(0) I \right] B \right)$$
$$\leq \frac{1}{2} M^{2r}(f, A; \gamma) \left(\left\| \frac{1}{p} \||A|^\alpha B|^{2pr} + \frac{1}{q} \left||A^*|^{1-\alpha} C\right|^{2qr} \right\| \right.$$
$$\left. + \omega^r \left(\left||A^*|^{1-\alpha} C\right|^2 \||A|^\alpha B|^2 \right) \right). \tag{59}$$

Proof. If we take the power $r > 0$ in (27) written for $y = x$, then we get, by Young and McCarthy inequalities, that

$$\left| \langle C^* \left[f(A) - f(0) I \right] B x, x \rangle \right|^{2r}$$
$$\leq M^{2r}(f, A; \gamma) \left\langle \||A|^\alpha B|^2 x, x \right\rangle^r \left\langle \left||A^*|^{1-\alpha} C\right|^2 x, x \right\rangle^r$$
$$\leq M^{2r}(f, A; \gamma) \left[\frac{1}{p} \left\langle \||A|^\alpha B|^2 x, x \right\rangle^{rp} + \frac{1}{q} \left\langle \left||A^*|^{1-\alpha} C\right|^2 x, x \right\rangle^{rq} \right]$$
$$\leq M^{2r}(f, A; \gamma) \left[\frac{1}{p} \left\langle \||A|^\alpha B|^{2rp} x, x \right\rangle + \frac{1}{q} \left\langle \left||A^*|^{1-\alpha} C\right|^{2rq} x, x \right\rangle \right]$$
$$= M^{2r}(f, A; \gamma) \left[\left\langle \frac{1}{p} \||A|^\alpha B|^{2rp} + \frac{1}{q} \left||A^*|^{1-\alpha} C\right|^{2rq} x, x \right\rangle \right]$$

for $x \in H$ with $\|x\| = 1$.

By taking the supremum over $\|x\| = 1$, we get that

$$\omega^{2r}\left(C^*\left[f(A) - f(0)I\right]B\right)$$
$$= \sup_{\|x\|=1} |\langle C^*\left[f(A) - f(0)I\right]Bx, x\rangle|^{2r}$$
$$\leq M^{2r}(f, A; \gamma) \sup_{\|x\|=1}\left[\left\langle \left(\frac{1}{p}\||A|^\alpha B|^{2rp} + \frac{1}{q}\||A^*|^{1-\alpha}C\right|^{2rq}\right)x, x\right\rangle\right]$$
$$= M^{2r}(f, A; \gamma) \left\|\frac{1}{p}\||A|^\alpha B|^{2rp} + \frac{1}{q}\||A^*|^{1-\alpha}C\right|^{2rq}\right\|,$$

which proves (57).

If we take the power $r \geq 1$ in (53) and by using the convexity of the power function, we get

$$|\langle C^*\left[f(A) - f(0)I\right]Bx, x\rangle|^{2r}$$
$$= M^{2r}(f, A; \gamma)$$
$$\times \left[\frac{\left\|\||A|^\alpha B|^2 x\right\|\left\|\||A^*|^{1-\alpha}C\right|^2 x\right\| + \left|\left\langle \||A^*|^{1-\alpha}C\right|^2 \||A|^\alpha B|^2 x, x\right\rangle\right|}{2}\right]^r$$
$$\leq M^{2r}(f, A; \gamma)$$
$$\times \frac{\left\|\||A|^\alpha B|^2 x\right\|^r\left\|\||A^*|^{1-\alpha}C\right|^2 x\right\|^r + \left|\left\langle \||A^*|^{1-\alpha}C\right|^2 \||A|^\alpha B|^2 x, x\right\rangle\right|^r}{2}$$
(60)

for $x \in H$ with $\|x\| = 1$.

By taking the supremum over $\|x\| = 1$, we get that

$$\omega^{2r}\left(C^*\left[f(A) - f(0)I\right]B\right)$$
$$\leq M^{2r}(f, A; \gamma)$$
$$\times \frac{\left\|\||A|^\alpha B|^2\right\|^r\left\|\||A^*|^{1-\alpha}C\right|^2\right\|^r + \omega^r\left(\||A^*|^{1-\alpha}C\right|^2 \||A|^\alpha B|^2\right)}{2}$$

$$= M^{2r}(f, A; \gamma)$$

$$\times \frac{\||A|^\alpha B\|^{2r} \||A^*|^{1-\alpha} C\|^{2r} + \omega^r \left(\left||A^*|^{1-\alpha} C\right|^2 \||A|^\alpha B|^2\right)}{2},$$

which proves (58).

Also, observe that

$$\left\|\left||A|^\alpha B\right|^2 x\right\|^r \left\|\left||A^*|^{1-\alpha} C\right|^2 x\right\|^r$$

$$\leq \frac{1}{p}\left\|\left||A|^\alpha B\right|^2 x\right\|^{pr} + \frac{1}{q}\left\|\left||A^*|^{1-\alpha} C\right|^2 x\right\|^{qr}$$

$$= \frac{1}{p}\left\|\left||A|^\alpha B\right|^2 x\right\|^{2\frac{pr}{2}} + \frac{1}{q}\left\|\left||A^*|^{1-\alpha} C\right|^2 x\right\|^{2\frac{qr}{2}}$$

$$= \frac{1}{p}\left\langle \left||A|^\alpha B\right|^4 x, x\right\rangle^{\frac{pr}{2}} + \frac{1}{q}\left\langle \left||A^*|^{1-\alpha} C\right|^4 x, x\right\rangle^{\frac{qr}{2}}$$

$$\leq \frac{1}{p}\left\langle \left||A|^\alpha B\right|^{2pr} x, x\right\rangle + \frac{1}{q}\left\langle \left||A^*|^{1-\alpha} C\right|^{2qr} x, x\right\rangle$$

$$= \left\langle \left(\frac{1}{p}\left||A|^\alpha B\right|^{2pr} + \frac{1}{q}\left||A^*|^{1-\alpha} C\right|^{2qr}\right) x, x\right\rangle,$$

for $x \in H$ with $\|x\| = 1$. Then

$$\frac{\left\|\left||A|^\alpha B\right|^2 x\right\|^r \left\|\left||A^*|^{1-\alpha} C\right|^2 x\right\|^r + \left|\left\langle \left||A^*|^{1-\alpha} C\right|^2 \left||A|^\alpha B\right|^2 x, x\right\rangle\right|^r}{2}$$

$$\leq \frac{1}{2}\left[\left\langle \left(\frac{1}{p}\left||A|^\alpha B\right|^{2pr} + \frac{1}{q}\left||A^*|^{1-\alpha} C\right|^{2qr}\right) x, x\right\rangle\right.$$

$$\left. + \left|\left\langle \left||A^*|^{1-\alpha} C\right|^2 \left||A|^\alpha B\right|^2 x, x\right\rangle\right|^r\right]$$

and by (60)

$$|\langle C^* [f(A) - f(0)I]Bx, x\rangle|^{2r}$$
$$\leq \frac{1}{2}B^{2r}(f, A; \gamma)\left[\left\langle \left(\frac{1}{p}||A|^\alpha B|^{2pr} + \frac{1}{q}\left||A^*|^{1-\alpha}C\right|^{2qr}\right)x, x\right\rangle\right.$$
$$\left. + \left|\left\langle \left||A^*|^{1-\alpha}C\right|^2 ||A|^\alpha B|^2 x, x\right\rangle\right|^r\right]$$

for $x \in H$ with $\|x\| = 1$.

By taking the supremum over $\|x\| = 1$, we derive (59). □

Remark 4. If we take $r = 1$ and $p, q > 1$ with $\frac{1}{p} + \frac{1}{q} = 1$ in (57), then we obtain

$$\omega^2 (C^* [f(A) - f(0)I]B)$$
$$\leq M^2(f, A; \gamma) \left\|\frac{1}{p}||A|^\alpha B|^{2p} + \frac{1}{q}\left||A^*|^{1-\alpha}C\right|^{2q}\right\|, \quad (61)$$

which for $p = q = 2$ gives

$$\omega^2 (C^* [f(A) - f(0)I]B)$$
$$\leq \frac{1}{2}M^2(f, A; \gamma) \left\|||A|^\alpha B|^4 + \left||A^*|^{1-\alpha}C\right|^4\right\|. \quad (62)$$

If we take $r = 1$ and $p = q = 2$ in (59), then we get

$$\omega^2 (C^* [f(A) - f(0)I]B)$$
$$\leq \frac{1}{2}M^2(f, A; \gamma) \left(\frac{1}{2}\left\|||A|^\alpha B|^4 + \left||A^*|^{1-\alpha}C\right|^4\right\|\right.$$
$$\left. + \omega\left(\left||A^*|^{1-\alpha}C\right|^2 ||A|^\alpha B|^2\right)\right). \quad (63)$$

If we take $r = 2$ and $p, q > 1$ with $\frac{1}{p} + \frac{1}{q} = 1$ in (59), then we get

$$\omega^4 (C^* [f(A) - f(0)I]B)$$
$$\leq \frac{1}{2}M^4(f, A; \gamma) \left(\left\|\frac{1}{p}||A|^\alpha B|^{4p} + \frac{1}{q}\left||A^*|^{1-\alpha}C\right|^{4q}\right\|\right.$$
$$\left. + \omega^2\left(\left||A^*|^{1-\alpha}C\right|^2 ||A|^\alpha B|^2\right)\right). \quad (64)$$

We also have the following:

Theorem 6. *With the assumptions of Theorem 4, we have for $r \geq 1$, $\lambda \in [0,1]$ that*
$$\omega^2\left(C^*\left[f(A) - f(0)I\right]B\right)$$
$$\leq M^2(f, A; \gamma) \left\| (1-\lambda) \||A|^\alpha B|^{2r} + \lambda \left||A^*|^{1-\alpha} C\right|^{2r} \right\|^{1/r}$$
$$\times \||A|^\alpha B\|^{2\lambda} \left\||A^*|^{1-\alpha} C\right\|^{2(1-\lambda)} \tag{65}$$

for all $\alpha \in [0,1]$.

Also, we have
$$\omega^2\left(C^*\left[f(A) - f(0)I\right]B\right)$$
$$\leq M^2(f, A; \gamma) \left\| (1-\lambda) \||A|^\alpha B|^{2r} + \lambda \left||A^*|^{1-\alpha} C\right|^{2r} \right\|^{1/r}$$
$$\times \left\| \lambda \||A|^\alpha B|^{2r} + (1-\lambda) \left||A^*|^{1-\alpha} C\right|^{2r} \right\|^{1/r} \tag{66}$$

for all $\alpha \in [0,1]$ and $r \geq 1$.

Proof. From the first part of (53), we have
$$|\langle C^*[f(A) - f(0)I]Bx, x\rangle|^2$$
$$\leq M^2(f, A; \gamma) \left\langle \||A|^\alpha B|^2 x, x\right\rangle \left\langle x, \left||A^*|^{1-\alpha} C\right|^2 x\right\rangle$$
$$= M^2(f, A; \gamma) \left\langle \||A|^\alpha B|^2 x, x\right\rangle^{1-\lambda} \left\langle x, \left||A^*|^{1-\alpha} C\right|^2 x\right\rangle^\lambda$$
$$\times \left\langle \||A|^\alpha B|^2 x, x\right\rangle^\alpha \left\langle x, \left||A^*|^{1-\alpha} C\right|^2 x\right\rangle^{1-\lambda}$$
$$\leq M^2(f, A; \gamma) \left[(1-\lambda)\left\langle \||A|^\alpha B|^2 x, x\right\rangle + \lambda\left\langle x, \left||A^*|^{1-\alpha} C\right|^2 x\right\rangle\right]$$
$$\times \left\langle \||A|^\alpha B|^2 x, x\right\rangle^\lambda \left\langle x, \left||A^*|^{1-\alpha} C\right|^2 x\right\rangle^{1-\lambda}$$

for all $x \in H$, $\|x\| = 1$.

If we take the power $r \geq 1$, then we get by the convexity of power r that

$$|\langle C^* [f(A) - f(0)I] Bx, x \rangle|^{2r}$$
$$\leq M^{2r}(f, A; \gamma) \left[(1-\lambda) \langle ||A|^\alpha B|^2 x, x \rangle + \lambda \langle x, ||A^*|^{1-\alpha} C|^2 x \rangle \right]^r$$
$$\times \langle ||A|^\alpha B|^2 x, x \rangle^{r\lambda} \langle x, ||A^*|^{1-\alpha} C|^2 x \rangle^{r(1-\lambda)}$$
$$\leq M^{2r}(f, A; \gamma) \left[(1-\lambda) \langle ||A|^\alpha B|^2 x, x \rangle^r + \lambda \langle x, ||A^*|^{1-\alpha} C|^2 x \rangle^r \right]$$
$$\times \langle ||A|^\alpha B|^2 x, x \rangle^{r\lambda} \langle x, ||A^*|^{1-\alpha} C|^2 x \rangle^{r(1-\lambda)} \qquad (67)$$

for all $x \in H$, $\|x\| = 1$.

If we use McCarthy inequality for power $r \geq 1$, then we get

$$(1-\lambda) \langle ||A|^\alpha B|^2 x, x \rangle^r + \lambda \langle x, ||A^*|^{1-\alpha} C|^2 x \rangle^r$$
$$\leq (1-\lambda) \langle ||A|^\alpha B|^{2r} x, x \rangle + \lambda \langle x, ||A^*|^{1-\alpha} C|^{2r} x \rangle$$
$$= \langle \left[(1-\lambda) ||A|^\alpha B|^{2r} + \lambda ||A^*|^{1-\alpha} C|^{2r} \right] x, x \rangle$$

and by (67)

$$|\langle C^* [f(A) - f(0)I] Bx, x \rangle|^{2r}$$
$$\leq M^{2r}(f, A; \gamma) \left[\langle \left[(1-\lambda) ||A|^\alpha B|^{2r} + \lambda ||A^*|^{1-\alpha} C|^{2r} \right] x, x \rangle \right]$$
$$\times \langle ||A|^\alpha B|^2 x, x \rangle^{r\lambda} \langle x, ||A^*|^{1-\alpha} C|^2 x \rangle^{r(1-\lambda)} \qquad (68)$$

for all $x \in H$, $\|x\| = 1$.

If we take the supremum over $\|x\| = 1$, then we get

$$\omega^{2r}\left(C^*\left[f(A) - f(0)I\right]B\right)$$
$$= \sup_{\|x\|=1} \left|\langle C^*\left[f(A) - f(0)I\right]Bx, x\rangle\right|^{2r}$$
$$\leq M^{2r}(f, A; \gamma) \sup_{\|x\|=1} \left[\left\langle\left[(1-\lambda)\left||A|^\alpha B\right|^{2r} + \lambda\left||A^*|^{1-\alpha}C\right|^{2r}\right]x, x\right\rangle\right]$$
$$\times \sup_{\|x\|=1} \left\langle\left||A|^\alpha B\right|^2 x, x\right\rangle^{r\lambda} \sup_{\|x\|=1} \left\langle x, \left||A^*|^{1-\alpha}C\right|^2 x\right\rangle^{r(1-\lambda)}$$
$$= M^{2r}(f, A; \gamma) \left\|(1-\lambda)\left||A|^\alpha B\right|^{2r} + \lambda\left||A^*|^{1-\alpha}C\right|^{2r}\right\|$$
$$\times \left\||A|^\alpha B\right\|^{2r\lambda} \left\||A^*|^{1-\alpha}C\right\|^{2r(1-\lambda)},$$

which gives (65).

We also have

$$\left|\langle C^*\left[f(A) - f(0)I\right]Bx, x\rangle\right|^{2r}$$
$$\leq M^{2r}(f, A; \gamma) \left[\left\langle\left[(1-\lambda)\left||A|^\alpha B\right|^{2r} + \lambda\left||A^*|^{1-\alpha}C\right|^{2r}\right]x, x\right\rangle\right]$$
$$\times \left[\left\langle\left[\lambda\left||A|^\alpha B\right|^{2r} + (1-\lambda)\left||A^*|^{1-\alpha}C\right|^{2r}\right]x, x\right\rangle\right]$$

for all $x \in H$, $\|x\| = 1$, which proves (66). $\square$

Remark 5. If we take $r = 1$ in Theorem 6, then we get

$$\omega^2\left(C^*\left[f(A) - f(0)I\right]B\right)$$
$$\leq M^2(f, A; \gamma) \left\|(1-\lambda)\left||A|^\alpha B\right|^2 + \lambda\left||A^*|^{1-\alpha}C\right|^{2r}\right\|$$
$$\times \left\||A|^\alpha B\right\|^{2\lambda} \left\||A^*|^{1-\alpha}C\right\|^{2(1-\lambda)} \tag{69}$$

and

$$\omega^2 \left(C^* \left[f(A) - f(0) I \right] B \right)$$
$$\leq M^2 (f, A; \gamma) \left\| (1 - \lambda) \| A |^\alpha B |^2 + \lambda \left| |A^*|^{1-\alpha} C \right|^2 \right\|$$
$$\times \left\| \lambda \| A |^\alpha B |^2 + (1 - \lambda) \left| |A^*|^{1-\alpha} C \right|^2 \right\| \qquad (70)$$

for all $\alpha, \lambda \in [0, 1]$.

If we take $\lambda = 1/2$ in (69), then we obtain

$$\omega^2 \left(C^* \left[f(A) - f(0) I \right] B \right)$$
$$\leq \frac{1}{2} M^2 (f, A; \gamma) \left\| \| A |^\alpha B |^2 + \left| |A^*|^{1-\alpha} C \right|^{2r} \right\| \| |A|^\alpha B \| \left\| |A^*|^{1-\alpha} C \right\|. \qquad (71)$$

If we take $r = 2$ in Theorem 6, then we get

$$\omega^2 \left(C^* \left[f(A) - f(0) I \right] B \right)$$
$$\leq M^2 (f, A; \gamma) \left\| (1 - \lambda) \| A |^\alpha B |^4 + \lambda \left| |A^*|^{1-\alpha} C \right|^4 \right\|^{1/2}$$
$$\times \| |A|^\alpha B \|^{2\lambda} \left\| |A^*|^{1-\alpha} C \right\|^{2(1-\lambda)} \qquad (72)$$

and

$$\omega^2 \left(C^* \left[f(A) - f(0) I \right] B \right)$$
$$\leq M^2 (f, A; \gamma) \left\| (1 - \lambda) \| A |^\alpha B |^4 + \lambda \left| |A^*|^{1-\alpha} C \right|^4 \right\|^{1/2}$$
$$\times \left\| \lambda \| A |^\alpha B |^4 + (1 - \lambda) \left| |A^*|^{1-\alpha} C \right|^4 \right\|^{1/2} \qquad (73)$$

for all $\alpha, \lambda \in [0, 1]$.

If we take $\lambda = 1/2$ in (72), then we obtain

$$\omega^2 \left(C^* \left[f(A) - f(0) I \right] B \right)$$
$$\leq \frac{\sqrt{2}}{2} M^2(f, A; \gamma) \left\| \left| |A|^\alpha B \right|^4 + \left| |A^*|^{1-\alpha} C \right|^4 \right\|^{1/2}$$
$$\times \left\| |A|^\alpha B \right\| \left\| |A^*|^{1-\alpha} C \right\|. \tag{74}$$

4. Inequalities for Trace of Operators

We have the following result for trace of operators:

Theorem 7. *Let $r \geq 1/2$, $p, q > 1$ with $\frac{1}{p} + \frac{1}{q} = 1$ and $pr, qr \geq 1$. Let $f : G \subset \mathbb{C} \to \mathbb{C}$ be an analytic function on the domain G with $0 \in G$ and $A \in \mathcal{B}(H)$ with $\mathrm{Sp}(A) \subset G$ and γ a closed rectifiable path in $G \setminus \{0\}$ such that $\mathrm{Sp}(A) \subset \mathrm{ins}(\gamma)$. If $B, C \in \mathcal{B}(H)$ with $|A|^\alpha B \in \mathcal{B}_{2pr}(H)$ and $|A^*|^{1-\alpha} C \in \mathcal{B}_{2qr}(H)$ for $\alpha \in [0, 1]$, then $C^* \left[f(A) - f(0) I \right] B \in \mathcal{B}_{2r}(H)$ and*

$$\left\| C^* \left[f(A) - f(0) I \right] B \right\|_{2r} \leq M(f, A; \gamma) \left\| |A|^\alpha B \right\|_{2pr} \left\| |A^*|^{1-\alpha} C \right\|_{2qr}. \tag{75}$$

In particular,

$$\left\| C^* \left[f(A) - f(0) I \right] B \right\|_{2r} \leq M(f, A; \gamma) \left\| |A|^{1/2} B \right\|_{2pr} \left\| |A^*|^{1/2} C \right\|_{2qr} \tag{76}$$

for $|A|^{1/2} B \in \mathcal{B}_{2pr}(H)$ and $|A^*|^{1/2} C \in \mathcal{B}_{2qr}(H)$.

Proof. If we take in (27) the power $r > 0$ and $x = e_i$, $y = f_i$, where $\mathcal{E} = \{e_i\}_{i \in I}$ and $\mathcal{F} = \{f_i\}_{i \in I}$ are orthonormal basis and sum, then we get

$$\sum_{i \in I} \left| \langle C^* \left[f(A) - f(0) I \right] B e_i, f_i \rangle \right|^{2r}$$
$$\leq M^{2r}(f, A; \gamma) \sum_{i \in I} \left\langle \left| |A|^\alpha B \right|^2 e_i, e_i \right\rangle^r \left\langle \left| |A^*|^{1-\alpha} C \right|^2 f_i, f_i \right\rangle^r. \tag{77}$$

If we use Hölder's inequality for $p, q > 1$ with $\frac{1}{p} + \frac{1}{q} = 1$, then we get

$$\sum_{i \in I} \left\langle \left| |A|^\alpha B \right|^2 e_i, e_i \right\rangle^r \left\langle \left| |A^*|^{1-\alpha} C \right|^2 f_i, f_i \right\rangle^r$$

$$\leq \left(\sum_{i \in I} \left\langle \left| |A|^\alpha B \right|^2 e_i, e_i \right\rangle^{pr} \right)^{1/p} \left(\sum_{i \in I} \left\langle \left| |A^*|^{1-\alpha} C \right|^2 f_i, f_i \right\rangle^{qr} \right)^{1/q}.$$
(78)

By the McCarthy inequality for $pr, qr \geq 1$, we have

$$\sum_{i \in I} \left\langle \left| |A|^\alpha B \right|^2 e_i, e_i \right\rangle^{pr} \leq \sum_{i \in I} \left\langle \left| |A|^\alpha B \right|^{2pr} e_i, e_i \right\rangle$$

and

$$\sum_{i \in I} \left\langle \left| |A^*|^{1-\alpha} C \right|^2 f_i, f_i \right\rangle^{qr} \leq \sum_{i \in I} \left\langle \left| |A^*|^{1-\alpha} C \right|^{2qr} f_i, f_i \right\rangle,$$

therefore

$$\left(\sum_{i \in I} \left\langle \left| |A|^\alpha B \right|^2 e_i, e_i \right\rangle^{pr} \right)^{1/p} \left(\sum_{i \in I} \left\langle \left| |A^*|^{1-\alpha} C \right|^2 f_i, f_i \right\rangle^{qr} \right)^{1/q}$$

$$\leq \left(\sum_{i \in I} \left\langle \left| |A|^\alpha B \right|^{2pr} e_i, e_i \right\rangle \right)^{1/p} \left(\sum_{i \in I} \left\langle \left| |A^*|^{1-\alpha} C \right|^{2qr} f_i, f_i \right\rangle \right)^{1/q}$$

$$= \left(\left\| |A|^\alpha B \right\|_{2pr}^{2pr} \right)^{1/p} \left(\left\| |A^*|^{1-\alpha} C \right\|_{2qr}^{2qr} \right)^{1/q}$$

$$= \left\| |A|^\alpha B \right\|_{2pr}^{2r} \left\| |A^*|^{1-\alpha} C \right\|_{2qr}^{2r}.$$

By (77) and (78), we derive

$$\sum_{i \in I} |\langle C^* [f(A) - f(0) I] B e_i, f_i \rangle|^{2r}$$

$$\leq M^{2r}(f, A; \gamma) \left\| |A|^\alpha B \right\|_{2pr}^{2r} \left\| |A^*|^{1-\alpha} C \right\|_{2qr}^{2r}.$$
(79)

Now, if we take the supremum over $\mathcal{E}$ and $\mathcal{F}$ in (34), then by (23) we get

$$\|C^*[f(A) - f(0)I]B\|_{2r}^{2r} \leq M^{2r}(f, A; \gamma) \||A|^\alpha B\|_{2pr}^{2r} \||A^*|^{1-\alpha}C\|_{2qr}^{2r}$$

and inequality (75) is obtained. □

Remark 6. If we take $r = 1/2$ and $p = q = 2$, then by (75) we get

$$\|C^*[f(A) - f(0)I]B\|_1 \leq M(f, A; \gamma) \||A|^\alpha B\|_2 \||A^*|^{1-\alpha}C\|_2 \qquad (80)$$

provided that $|A|^\alpha B \in \mathcal{B}_2(H)$ and $|A^*|^{1-\alpha}C \in \mathcal{B}_2(H)$ for $\alpha \in [0, 1]$.
Also, if $r = 1$ and $p, q > 1$ with $\frac{1}{p} + \frac{1}{q} = 1$, then by (75) we get

$$\|C^*[f(A) - f(0)I]B\|_2 \leq M(f, A; \gamma) \||A|^\alpha B\|_{2p} \||A^*|^{1-\alpha}C\|_{2q} \qquad (81)$$

provided that $|A|^\alpha B \in \mathcal{B}_{2p}(H)$ and $|A^*|^{1-\alpha}C \in \mathcal{B}_{2q}(H)$ for $\alpha \in [0, 1]$.

We also have the following:

Theorem 8. Let $r \geq 1/2$, $p, q \geq 1$ with $\frac{1}{p} + \frac{1}{q} = \frac{1}{r}$. Let $f : G \subset \mathbb{C} \to \mathbb{C}$ be an analytic function on the domain G with $0 \in G$ and $A \in B(H)$ with $\mathrm{Sp}(A) \subset G$ and γ a closed rectifiable path in $G \setminus \{0\}$ such that $\mathrm{Sp}(A) \subset \mathrm{ins}(\gamma)$. If $|A|^\alpha B \in \mathcal{B}_{2p}(H)$ and $|A^*|^{1-\alpha}C \in \mathcal{B}_{2q}(H)$ for $\alpha \in [0, 1]$, then $C^*[f(A) - f(0)I]B \in \mathcal{B}_{2r}(H)$ and

$$\|C^*[f(A) - f(0)I]B\|_{2r} \leq M(f, A; \gamma) \||A|^\alpha B\|_{2p} \||A^*|^{1-\alpha}C\|_{2q}. \qquad (82)$$

In particular,

$$\|C^*[f(A) - f(0)I]B\|_{2r} \leq M(f, A; \gamma) \||A|^{1/2}B\|_{2p} \||A^*|^{1/2}C\|_{2q} \qquad (83)$$

for $|A|^{1/2}B \in \mathcal{B}_{2p}(H)$ and $|A^*|^{1/2}C \in \mathcal{B}_{2q}(H)$.

Proof. Assume that $\mathcal{E} = \{e_i\}_{i \in I}$ and $\mathcal{F} = \{f_i\}_{i \in I}$ are orthonormal basis in H. Observe that we have $\frac{1}{\frac{p}{r}} + \frac{1}{\frac{q}{r}} = 1$ and by Hölder's inequality for $\frac{p}{r}$ and $\frac{q}{r}$ we have

$$\sum_{i \in I} \left\langle |A|^\alpha B|^2 e_i, e_i \right\rangle^r \left\langle \left||A^*|^{1-\alpha} C\right|^2 f_i, f_i \right\rangle^r$$

$$= \sum_{i \in I} \left[\left\langle |A|^\alpha B|^2 e_i, e_i \right\rangle^p \right]^{\frac{r}{p}} \left[\left\langle \left||A^*|^{1-\alpha} C\right|^2 f_i, f_i \right\rangle^q \right]^{\frac{r}{q}}$$

$$\leq \left(\sum_{i \in I} \left\langle |A|^\alpha B|^2 e_i, e_i \right\rangle^p \right)^{r/p} \left(\sum_{i \in I} \left\langle \left||A^*|^{1-\alpha} C\right|^2 f_i, f_i \right\rangle^q \right)^{r/q}.$$
(84)

By McCarthy inequality for $p, q > 1$, we get

$$\sum_{i \in I} \left\langle |A|^\alpha B|^2 e_i, e_i \right\rangle^p \leq \sum_{i \in I} \left\langle |A|^\alpha B|^{2p} e_i, e_i \right\rangle$$

and

$$\sum_{i \in I} \left\langle \left||A^*|^{1-\alpha} C\right|^2 f_i, f_i \right\rangle^q \leq \sum_{i \in I} \left\langle \left||A^*|^{1-\alpha} C\right|^{2q} f_i, f_i \right\rangle$$

and by (84)

$$\sum_{i \in I} \left\langle |A|^\alpha B|^2 e_i, e_i \right\rangle^r \left\langle \left||A^*|^{1-\alpha} C\right|^2 f_i, f_i \right\rangle^r$$

$$\leq \left(\sum_{i \in I} \left\langle |A|^\alpha B|^{2p} e_i, e_i \right\rangle \right)^{r/p} \left(\sum_{i \in I} \left\langle \left||A^*|^{1-\alpha} C\right|^{2q} f_i, f_i \right\rangle \right)^{r/q}$$

$$= \left\||A|^\alpha B\right\|_{2p}^{2r} \left\||A^*|^{1-\alpha} C\right\|_{2q}^{2r}.$$
(85)

By (77) and (85), we get

$$\sum_{i \in I} \left| \langle C^* \left[f(A) - f(0) I \right] B e_i, f_i \rangle \right|^{2r}$$

$$\leq M^{2r}(f, A; \gamma) \left\||A|^\alpha B\right\|_{2p}^{2r} \left\||A^*|^{1-\alpha} C\right\|_{2q}^{2r}.$$
(86)

Now, if we take the supremum over $\mathcal{E}$ and $\mathcal{F}$ in (86), we get

$$\|C^*\left[f(A) - f(0)I\right]B\|_{2r}^{2r} \leq M^{2r}(f, A; \gamma) \||A|^\alpha B\|_{2p}^{2r} \left\||A^*|^{1-\alpha}C\right\|_{2q}^{2r}$$

and inequality (82) is thus proved. □

Remark 7. If we take $p = q = 2r = s \geq 1$, then by (82) we get

$$\|C^*\left[f(A) - f(0)I\right]B\|_s \leq M(f, A; \gamma) \||A|^\alpha B\|_{2s} \left\||A^*|^{1-\alpha}C\right\|_{2s} \quad (87)$$

provided that $|A|^\alpha B \in \mathcal{B}_{2s}(H)$ and $|A^*|^{1-\alpha}C \in \mathcal{B}_{2s}(H)$ for $\alpha \in [0, 1]$.

For $\alpha = 1/2$, we have

$$\|C^*\left[f(A) - f(0)I\right]B\|_s \leq M(f, A; \gamma) \left\||A|^{1/2}B\right\|_{2s} \left\||A^*|^{1/2}C\right\|_{2s} \quad (88)$$

provided that $|A|^{1/2}B \in \mathcal{B}_{2s}(H)$ and $|A^*|^{1/2}C \in \mathcal{B}_{2s}(H)$.

If $r = 2$ and $p, q > 1$ with $\frac{1}{p} + \frac{1}{q} = \frac{1}{2}$, then

$$\|C^*\left[f(A) - f(0)I\right]B\|_4 \leq M(f, A; \gamma) \||A|^\alpha B\|_{2p} \left\||A^*|^{1-\alpha}C\right\|_{2q} \quad (89)$$

provided that $|A|^\alpha B \in \mathcal{B}_{2p}(H)$ and $|A^*|^{1-\alpha}C \in \mathcal{B}_{2q}(H)$ for $\alpha \in [0, 1]$.

In particular,

$$\|C^*\left[f(A) - f(0)I\right]B\|_4 \leq M(f, A; \gamma) \left\||A|^{1/2}B\right\|_{2p} \left\||A^*|^{1/2}C\right\|_{2q} \quad (90)$$

for $|A|^{1/2}B \in \mathcal{B}_{2p}(H)$ and $|A^*|^{1/2}C \in \mathcal{B}_{2q}(H)$.

Theorem 9. *Let $f : G \subset \mathbb{C} \to \mathbb{C}$ be an analytic function on the domain G with $0 \in G$ and $A \in \mathcal{B}(H)$ with $\mathrm{Sp}(A) \subset G$ and γ a closed rectifiable path in $G \setminus \{0\}$ such that $\mathrm{Sp}(A) \subset \mathrm{ins}(\gamma)$. If $r \geq 1/2$, p,*

$q > 1$ with $\frac{1}{p} + \frac{1}{q} = 1$, pr, $qr \geq 1$ and $\||A|^\alpha B|^{2pr}$, $\left||A^*|^{1-\alpha} C\right|^{2qr} \in \mathcal{B}_1(H)$, then $C^*[f(A) - f(0)I]B \in \mathcal{B}_{2r}(H)$ and

$$\omega_{2r}^{2r}(C^*[f(A) - f(0)I]B)$$
$$\leq M^{2r}(f, A; \gamma) \operatorname{tr}\left(\frac{1}{p}\||A|^\alpha B|^{2pr} + \frac{1}{q}\left||A^*|^{1-\alpha} C\right|^{2qr}\right). \quad (91)$$

If $r \geq 1$ and $|A|^\alpha B, |A^*|^{1-\alpha} C \in \mathcal{B}_{4r}(H)$, then $C^*[f(A) - f(0)I]B \in \mathcal{B}_{2r}(H)$ and

$$\omega_{2r}^{2r}(C^*[f(A) - f(0)I]B)$$
$$\leq \frac{1}{2} M^{2r}(f, A; \gamma)$$
$$\times \left(\||A|^\alpha B\|_{4r}^{2r} \left\||A^*|^{1-\alpha} C\right\|_{4r}^{2r} + \omega_r^r\left(\left||A^*|^{1-\alpha} C\right|^2 \||A|^\alpha B|^2\right)\right)$$
$$\leq \frac{1}{2} M^{2r}(f, A; \gamma)$$
$$\times \left(\||A|^\alpha B\|_{4r}^{2r} \left\||A^*|^{1-\alpha} C\right\|_{4r}^{2r} + \left\|\left||A^*|^{1-\alpha} C\right|^2 \||A|^\alpha B|^2\right\|_r^r\right). \quad (92)$$

If $r \geq 1$, $p, q > 1$ with $\frac{1}{p} + \frac{1}{q} = 1$, pr, $qr \geq 2$, then

$$\omega_{2r}^{2r}(C^*[f(A) - f(0)I]B)$$
$$\leq \frac{1}{2} M^{2r}(f, A; \gamma) \left[\operatorname{tr}\left(\frac{1}{p}\||A|^\alpha B|^{2pr} + \frac{1}{q}\left||A^*|^{1-\alpha} C\right|^{2qr}\right)\right.$$
$$\left. + \omega_r^r\left(\left||A^*|^{1-\alpha} C\right|^2 \||A|^\alpha B|^2\right)\right]$$
$$\leq \frac{1}{2} M^{2r}(f, A; \gamma) \left[\operatorname{tr}\left(\frac{1}{p}\||A|^\alpha B|^{2pr} + \frac{1}{q}\left||A^*|^{1-\alpha} C\right|^{2qr}\right)\right.$$
$$\left. + \left\|\left||A^*|^{1-\alpha} C\right|^2 \||A|^\alpha B|^2\right\|_r^r\right]. \quad (93)$$

Vector Inequalities for Analytic Functions of Operators in Hilbert Spaces 255

Proof. From (27) for $y = x$, we have that

$$|\langle C^* [f(A) - f(0) I] Bx, x \rangle|^2$$
$$\leq M^2 (f, A; \gamma) \langle ||A|^\alpha B|^2 x, x \rangle \langle \left||A^*|^{1-\alpha} C\right|^2 x, x \rangle \qquad (94)$$

for $x \in H$ with $||x|| = 1$.

If we take the power $r > 0$, we get, by Young and McCarthy inequalities, that

$$|\langle C^* [f(A) - f(0) I] Bx, x \rangle|^{2r}$$
$$\leq M^{2r} (f, A; \gamma) \langle ||A|^\alpha B|^2 x, x \rangle^r \langle \left||A^*|^{1-\alpha} C\right|^2 x, x \rangle^r$$
$$\leq M^{2r} (f, A; \gamma) \left[\frac{1}{p} \langle ||A|^\alpha B|^2 x, x \rangle^{pr} + \frac{1}{q} \langle \left||A^*|^{1-\alpha} C\right|^2 x, x \rangle^{qr} \right]$$
$$\leq M^{2r} (f, A; \gamma) \left[\frac{1}{p} \langle ||A|^\alpha B|^{2pr} x, x \rangle + \frac{1}{q} \langle \left||A^*|^{1-\alpha} C\right|^{2qr} x, x \rangle \right]$$
$$= M^{2r} (f, A; \gamma) \langle \left(\frac{1}{p} ||A|^\alpha B|^{2pr} + \frac{1}{q} \left||A^*|^{1-\alpha} C\right|^{2qr} \right) x, x \rangle$$

for $x \in H$ with $||x|| = 1$.

If $\mathcal{E} = \{e_i\}_{i \in I}$ is an orthonormal basis, then by taking $x = e_i$ and summing over $i \in I$ we get

$$||C^* [f(A) - f(0) I] B||_{\mathcal{E}, 2r}^{2r}$$
$$= \sum_{i \in I} |\langle C^* [f(A) - f(0) I] Be_i, e_i \rangle|^{2r}$$
$$\leq M^{2r} (f, A; \gamma) \sum_{i \in I} \langle \left(\frac{1}{p} ||A|^\alpha B|^{2pr} + \frac{1}{q} \left||A^*|^{1-\alpha} C\right|^{2qr} \right) e_i, e_i \rangle$$
$$= M^{2r} (f, A; \gamma) \operatorname{tr} \left(\frac{1}{p} ||A|^\alpha B|^{2pr} + \frac{1}{q} \left||A^*|^{1-\alpha} C\right|^{2qr} \right),$$

which, by taking the supremum over $\mathcal{E}$, proves (91).

By Buzano's inequality, we have

$$\left\langle \left||A|^\alpha B\right|^2 x, x\right\rangle \left\langle x, \left||A^*|^{1-\alpha} C\right|^2 x\right\rangle$$

$$\leq \frac{1}{2}\left[\left\|\left||A|^\alpha B\right|^2 x\right\|\left\|\left||A^*|^{1-\alpha} C\right|^2 x\right\| + \left|\left\langle \left||A|^\alpha B\right|^2 x, \left||A^*|^{1-\alpha} C\right|^2 x\right\rangle\right|\right]$$

$$= \frac{1}{2}\left[\left\|\left||A|^\alpha B\right|^2 x\right\|\left\|\left||A^*|^{1-\alpha} C\right|^2 x\right\| + \left|\left\langle \left||A^*|^{1-\alpha} C\right|^2 \left||A|^\alpha B\right|^2 x, x\right\rangle\right|\right]$$

for $x \in H$ with $\|x\| = 1$.

If we take the power $r \geq 1$ and use the convexity of power function, then we get

$$\left\langle \left||A|^\alpha B\right|^2 x, x\right\rangle^r \left\langle x, \left||A^*|^{1-\alpha} C\right|^2 x\right\rangle^r$$

$$\leq \left[\frac{\left\|\left||A|^\alpha B\right|^2 x\right\|\left\|\left||A^*|^{1-\alpha} C\right|^2 x\right\| + \left|\left\langle \left||A^*|^{1-\alpha} C\right|^2 \left||A|^\alpha B\right|^2 x, x\right\rangle\right|}{2}\right]^r$$

$$\leq \frac{\left\|\left||A|^\alpha B\right|^2 x\right\|^r\left\|\left||A^*|^{1-\alpha} C\right|^2 x\right\|^r + \left|\left\langle \left||A^*|^{1-\alpha} C\right|^2 \left||A|^\alpha B\right|^2 x, x\right\rangle\right|^r}{2}$$

$$= \frac{\left\|\left||A|^\alpha B\right|^2 x\right\|^{2\frac{r}{2}}\left\|\left||A^*|^{1-\alpha} C\right|^2 x\right\|^{2\frac{r}{2}} + \left|\left\langle \left||A^*|^{1-\alpha} C\right|^2 \left||A|^\alpha B\right|^2 x, x\right\rangle\right|^r}{2}$$

$$= \frac{\left\langle \left||A|^\alpha B\right|^4 x, x\right\rangle^{\frac{r}{2}}\left\langle \left||A^*|^{1-\alpha} C\right|^4 x, x\right\rangle^{\frac{r}{2}} + \left|\left\langle \left||A^*|^{1-\alpha} C\right|^2 \left||A|^\alpha B\right|^2 x, x\right\rangle\right|^r}{2}$$

for $x \in H$ with $\|x\| = 1$.

Therefore,

$$\|C^*[f(A) - f(0)I]B\|_{\mathcal{E},2r}^{2r}$$

$$= \sum_{i \in I} |\langle C^*[f(A) - f(0)I]Be_i, e_i\rangle|^{2r}$$

$$\leq M^{2r}(f, A; \gamma) \sum_{i \in I} \left\langle \left||A|^\alpha B\right|^2 e_i, e_i\right\rangle^r \left\langle e_i, \left||A^*|^{1-\alpha} C\right|^2 e_i\right\rangle^r$$

$$\leq \frac{1}{2} M^{2r}(f, A; \gamma) \left[\sum_{i \in I} \left\langle \left||A|^{\alpha} B \right|^{4} e_{i}, e_{i} \right\rangle^{\frac{r}{2}} \left\langle \left||A^{*}|^{1-\alpha} C \right|^{4} e_{i}, e_{i} \right\rangle^{\frac{r}{2}} \right.$$

$$\left. + \sum_{i \in I} \left| \left\langle \left||A^{*}|^{1-\alpha} C \right|^{2} \left||A|^{\alpha} B \right|^{2} e_{i}, e_{i} \right\rangle \right|^{r} \right]. \tag{95}$$

Using Cauchy–Schwarz inequality, we have

$$\sum_{i \in I} \left\langle \left||A|^{\alpha} B \right|^{4} e_{i}, e_{i} \right\rangle^{\frac{r}{2}} \left\langle \left||A^{*}|^{1-\alpha} C \right|^{4} e_{i}, e_{i} \right\rangle^{\frac{r}{2}}$$

$$\leq \left(\sum_{i \in I} \left\langle \left||A|^{\alpha} B \right|^{4} e_{i}, e_{i} \right\rangle^{r} \right)^{1/2} \left(\sum_{i \in I} \left\langle \left||A^{*}|^{1-\alpha} C \right|^{4} e_{i}, e_{i} \right\rangle^{r} \right)^{1/2}$$

$$\leq \left(\sum_{i \in I} \left\langle \left||A|^{\alpha} B \right|^{4r} e_{i}, e_{i} \right\rangle \right)^{1/2} \left(\sum_{i \in I} \left\langle \left||A^{*}|^{1-\alpha} C \right|^{4r} e_{i}, e_{i} \right\rangle \right)^{1/2}$$

$$= \left\| |A|^{\alpha} B \right\|_{4r}^{2r} \left\| |A^{*}|^{1-\alpha} C \right\|_{4r}^{2r},$$

where for the last inequality we used McCarthy's result for $r \geq 1$. This proves (92).

Further, if we use Young's inequality for $p, q > 1$ with $\frac{1}{p} + \frac{1}{q} = 1$,

$$ab \leq \frac{1}{p} a^{p} + \frac{1}{q} b^{q}, \quad a, b \geq 0,$$

then we get

$$\left\| \left||A|^{\alpha} B \right|^{2} x \right\|^{r} \left\| \left||A^{*}|^{1-\alpha} C \right|^{2} x \right\|^{r}$$

$$\leq \frac{1}{p} \left\| \left||A|^{\alpha} B \right|^{2} x \right\|^{pr} + \frac{1}{q} \left\| \left||A^{*}|^{1-\alpha} C \right|^{2} x \right\|^{qr}$$

$$= \frac{1}{p} \left\| \left||A|^{\alpha} B \right|^{2} x \right\|^{2 \frac{pr}{2}} + \frac{1}{q} \left\| \left||A^{*}|^{1-\alpha} C \right|^{2} x \right\|^{2 \frac{qr}{2}}$$

$$= \frac{1}{p} \left\langle \left||A|^{\alpha} B \right|^{4} x, x \right\rangle^{\frac{pr}{2}} + \frac{1}{q} \left\langle \left||A^{*}|^{1-\alpha} C \right|^{4} x, x \right\rangle^{\frac{qr}{2}}$$

$$\leq \frac{1}{p}\left\langle ||A|^\alpha B|^{2pr} x, x\right\rangle + \frac{1}{q}\left\langle \left||A^*|^{1-\alpha} C\right|^{2qr} x, x\right\rangle$$

$$= \left\langle \left(\frac{1}{p}||A|^\alpha B|^{2pr} + \frac{1}{q}\left||A^*|^{1-\alpha} C\right|^{2qr}\right) x, x\right\rangle$$

for $x \in H$ with $\|x\| = 1$.
Therefore,

$$\|C^*[f(A) - f(0)I]B\|_{\mathcal{E},2r}^{2r}$$

$$= \sum_{i \in I} |\langle C^*[f(A) - f(0)I]Be_i, e_i\rangle|^{2r}$$

$$\leq M^{2r}(f, A; \gamma) \sum_{i \in I} \left\langle ||A|^\alpha B|^2 e_i, e_i\right\rangle^r \left\langle e_i, \left||A^*|^{1-\alpha} C\right|^2 e_i\right\rangle^r$$

$$\leq \frac{1}{2} M^{2r}(f, A; \gamma) \left[\sum_{i \in I} \left\langle \left(\frac{1}{p}||A|^\alpha B|^{2pr} + \frac{1}{q}\left||A^*|^{1-\alpha} C\right|^{2qr}\right) e_i, e_i\right\rangle\right.$$

$$\left. + \sum_{i \in I} \left|\left\langle \left||A^*|^{1-\alpha} C\right|^2 ||A|^\alpha B|^2 e_i, e_i\right\rangle\right|^r\right]$$

$$= \frac{1}{2} M^{2r}(f, A; \gamma) \left[\operatorname{tr}\left(\frac{1}{p}||A|^\alpha B|^{2pr} + \frac{1}{q}\left||A^*|^{1-\alpha} C\right|^{2qr}\right)\right.$$

$$\left. + \left\|\left||A^*|^{1-\alpha} C\right|^2 ||A|^\alpha B|^2\right\|_{\mathcal{E},r}^r\right],$$

which proves, by taking the supremum over $\mathcal{E}$, the desired inequality (93). □

Remark 8. Let $\alpha \in [0, 1]$. If $r = 1/2$, $p, q = 2$ and $||A|^\alpha B|^2$, $\left||A^*|^{1-\alpha} C\right|^2 \in \mathcal{B}_1(H)$, then $C^*[f(A) - f(0)I]B \in \mathcal{B}_1(H)$ and by (91) we get

$$\omega_1\left(C^*[f(A) - f(0)I]B\right)$$

$$\leq \frac{1}{2} M(f, A; \gamma) \operatorname{tr}\left(||A|^\alpha B|^2 + \left||A^*|^{1-\alpha} C\right|^2\right). \qquad (96)$$

If $r = 1$ and $p, q > 1$ with $\frac{1}{p} + \frac{1}{q} = 1$, then by (91) we obtain

$$\omega_2^2 \left(C^* \left[f(A) - f(0) I \right] B \right)$$
$$\leq M^2 (f, A; \gamma) \operatorname{tr} \left(\frac{1}{p} \||A|^\alpha B|^{2p} + \frac{1}{q} \left| |A^*|^{1-\alpha} C \right|^{2q} \right), \quad (97)$$

provided that $\||A|^\alpha B|^{2p}, \left| |A^*|^{1-\alpha} C \right|^{2q} \in \mathcal{B}_1(H)$.

If we take $r = 1$ in (92), then we get

$$\omega_2^2 \left(C^* \left[f(A) - f(0) I \right] B \right)$$
$$\leq \frac{1}{2} M^2 (f, A; \gamma)$$
$$\times \left(\||A|^\alpha B\|_4^2 \left\| |A^*|^{1-\alpha} C \right\|_4^2 + \omega_1 \left(\left| |A^*|^{1-\alpha} C \right|^2 ||A|^\alpha B|^2 \right) \right)$$
$$\leq \frac{1}{2} M^2 (f, A; \gamma)$$
$$\times \left(\||A|^\alpha B\|_4^2 \left\| |A^*|^{1-\alpha} C \right\|_4^2 + \left\| \left| |A^*|^{1-\alpha} C \right|^2 ||A|^\alpha B|^2 \right\|_1 \right), \quad (98)$$

provided that $|A|^\alpha B, |A^*|^{1-\alpha} C \in \mathcal{B}_4(H)$.

If $r = 1$ and $p = q = 2$ in (93), then we get for $\||A|^\alpha B|^{2p}$, $\left| |A^*|^{1-\alpha} C \right|^{2q} \in \mathcal{B}_1(H)$ that

$$\omega_2^2 \left(C^* \left[f(A) - f(0) I \right] B \right)$$
$$\leq \frac{1}{4} M^2 (f, A; \gamma) \left[\operatorname{tr} \left(\||A|^\alpha B|^{2p} + \left| |A^*|^{1-\alpha} C \right|^{2q} \right) \right.$$
$$+ \frac{1}{2} M^2 (f, A; \gamma) \omega_1 \left(\left| |A^*|^{1-\alpha} C \right|^2 ||A|^\alpha B|^2 \right)$$
$$\leq \frac{1}{4} M^2 (f, A; \gamma) \operatorname{tr} \left(\||A|^\alpha B|^{2p} + \left| |A^*|^{1-\alpha} C \right|^{2q} \right)$$
$$+ \frac{1}{2} M^2 (f, A; \gamma) \left\| \left| |A^*|^{1-\alpha} C \right|^2 ||A|^\alpha B|^2 \right\|_1. \quad (99)$$

We also have the following:

Theorem 10. *With the assumptions of Theorem 9, we have for $r \geq 1$, $\lambda \in [0,1]$ that*

$$\omega_{2r}^{2r}\left(C^*\left[f(A) - f(0)I\right]B\right)$$
$$\leq M^{2r}(f, A; \gamma) \left\|(1-\lambda)\,|A|^\alpha B|^{2r} + \lambda\,\left||A^*|^{1-\alpha} C\right|^{2r}\right\|$$
$$\times \||A|^\alpha B\|_{2r}^{2r\lambda} \left\||A^*|^{1-\alpha} C\right\|_{2r}^{2r(1-\lambda)}, \tag{100}$$

provided that $|A|^\alpha B$, $|A^*|^{1-\alpha} C \in \mathcal{B}_{2r}(H)$.

In particular,

$$\omega_{2r}^{2r}\left(C^*\left[f(A) - f(0)I\right]B\right)$$
$$\leq \frac{1}{2} M^{2r}(f, A; \gamma) \left\|\,|A|^\alpha B|^{2r} + \left||A^*|^{1-\alpha} C\right|^{2r}\right\|$$
$$\times \||A|^\alpha B\|_{2r}^{r} \left\||A^*|^{1-\alpha} C\right\|_{2r}^{r}. \tag{101}$$

Proof. If $\mathcal{E} = \{e_i\}_{i \in I}$ is an orthonormal basis, then by taking $x = e_i$ in (68) and summing over $i \in I$ we get

$$\sum_{i \in I} |\langle C^*\left[f(A) - f(0)I\right] B e_i, e_i \rangle|^{2r}$$
$$\leq M^{2r}(f, A; \gamma) \sum_{i \in I} \left[\left\langle \left[(1-\lambda)\,|A|^\alpha B|^{2r} + \lambda\,\left||A^*|^{1-\alpha} C\right|^{2r}\right] e_i, e_i \right\rangle\right]$$
$$\times \left\langle |A|^\alpha B|^2 e_i, e_i \right\rangle^{r\lambda} \left\langle \left||A^*|^{1-\alpha} C\right|^2 e_i, e_i \right\rangle^{r(1-\lambda)}$$
$$\leq M^{2r}(f, A; \gamma) \left\|(1-\lambda)\,|A|^\alpha B|^{2r} + \lambda\,\left||A^*|^{1-\alpha} C\right|^{2r}\right\|$$
$$\times \sum_{i \in I} \left\langle |A|^\alpha B|^2 e_i, e_i \right\rangle^{r\lambda} \left\langle \left||A^*|^{1-\alpha} C\right|^2 e_i, e_i \right\rangle^{r(1-\lambda)}. \tag{102}$$

If we use Hölder's inequality for $p = \frac{1}{\lambda}$, $q = \frac{1}{1-\lambda}$, then we have

$$\sum_{i \in I} \left\langle |A|^\alpha B|^2 e_i, e_i \right\rangle^{r\lambda} \left\langle \left||A^*|^{1-\alpha} C\right|^2 e_i, e_i \right\rangle^{r(1-\lambda)}$$

$$\leq \left(\sum_{i \in I} \left\langle |A|^\alpha B|^2 e_i, e_i \right\rangle^r \right)^\lambda \left(\sum_{i \in I} \left\langle \left||A^*|^{1-\alpha} C\right|^2 e_i, e_i \right\rangle^r \right)^{1-\lambda}$$

$$\leq \left(\sum_{i \in I} \left\langle |A|^\alpha B|^{2r} e_i, e_i \right\rangle \right)^\lambda \left(\sum_{i \in I} \left\langle \left||A^*|^{1-\alpha} C\right|^{2r} e_i, e_i \right\rangle \right)^{1-\lambda}$$

$$= \left\||A|^\alpha B\right\|_{2r}^{2r\lambda} \left\||A^*|^{1-\alpha} C\right\|_{2r}^{2r(1-\lambda)},$$

which proves (100). □

Remark 9. If we take $r = 1$ in Theorem 10, then we get for $\alpha \in [0, 1]$ that

$$\omega_2^2 \left(C^* [f(A) - f(0) I] B \right)$$

$$\leq M^2(f, A; \gamma) \left\| (1-\lambda) |A|^\alpha B|^2 + \lambda \left||A^*|^{1-\alpha} C\right|^2 \right\|$$

$$\times \left\||A|^\alpha B\right\|_2^{2\lambda} \left\||A^*|^{1-\alpha} C\right\|_2^{2(1-\lambda)}, \tag{103}$$

provided that $|A|^\alpha B, |A^*|^{1-\alpha} C \in \mathcal{B}_2(H)$.

In particular,

$$\omega_2^2 \left(C^* [f(A) - f(0) I] B \right)$$

$$\leq \frac{1}{2} M^2(f, A; \gamma) \left\| |A|^\alpha B|^2 + \left||A^*|^{1-\alpha} C\right|^2 \right\| \left\||A|^\alpha B\right\|_2 \left\||A^*|^{1-\alpha} C\right\|_2. \tag{104}$$

5. Applications for Complex Perspectives

Let f be a continuous function defined on the interval I of real numbers, Q a selfadjoint operator on the Hilbert space H and P a positive invertible operator on H. Assume that the spectrum

$\operatorname{Sp}\left(P^{-1/2}QP^{-1/2}\right) \subset \mathring{I}$. Then by using the continuous functional calculus, we can define the *perspective* $\mathcal{P}_f(Q, P)$ by setting

$$\mathcal{P}_f(Q, P) := P^{1/2} f\left(P^{-1/2}QP^{-1/2}\right) P^{1/2}.$$

If P and B are commutative, then

$$\mathcal{P}_\Phi(Q, P) = P f\left(QP^{-1}\right)$$

provided $\operatorname{Sp}\left(QP^{-1}\right) \subset \mathring{I}$.

Let f be an analytic function on the domain G, $P > 0$ and $T \in B(H)$ with $\operatorname{Sp}\left(P^{-1/2}TP^{-1/2}\right) \subset G$ and γ a closed path in G with $\operatorname{Sp}\left(P^{-1/2}TP^{-1/2}\right) \subset \operatorname{ins}(\gamma)$. We can define the *complex operator valued perspective* by

$$\mathcal{P}_f(T, P) := P^{1/2} f\left(P^{-1/2}TP^{-1/2}\right) P^{1/2}.$$

By using the analytic functional calculus, we have the representation

$$\mathcal{P}_f(T, P) := \frac{1}{2\pi i} \int_\gamma f(\xi) P^{1/2} \left(\xi I - P^{-1/2}TP^{-1/2}\right)^{-1} P^{1/2} d\xi.$$

We have the following bounds:

Proposition 1. *Let f be an analytic function on the domain G, $0 \in G$ and $f(0) = 0$. If $P > 0$ and $T \in B(H)$ with $\operatorname{Sp}\left(P^{-1/2}TP^{-1/2}\right) \subset G$ and γ a closed path in $G \setminus \{0\}$ with $\operatorname{Sp}\left(P^{-1/2}TP^{-1/2}\right) \subset \operatorname{ins}(\gamma)$, then*

$$|\langle \mathcal{P}_f(T, P) x, y \rangle| \leq M\left(f, P^{-1/2}TP^{-1/2}; \gamma\right)$$

$$\times \left\langle \left|\left|P^{-1/2}TP^{-1/2}\right|^\alpha P^{1/2}\right|^2 x, x \right\rangle^{1/2}$$

$$\times \left\langle \left|\left|P^{-1/2}T^*P^{-1/2}\right|^{1-\alpha} P^{1/2}\right|^2 y, y \right\rangle^{1/2}$$

$$\leq M\left(f, P^{-1/2}TP^{-1/2}; \gamma\right)$$

$$\times \left\|P^{-1/2}TP^{-1/2}\right\| \langle Px, x \rangle^{1/2} \langle Py, y \rangle^{1/2} \quad (105)$$

for all $\alpha \in [0, 1]$ and $x, y \in H$. Here

$$M\left(f, P^{-1/2}TP^{-1/2}; \gamma\right) = \frac{1}{2\pi} \int_\gamma \frac{|f(\xi)| \, |d\xi|}{|\xi| \left(|\xi| - \|P^{-1/2}TP^{-1/2}\|\right)}.$$

Proof. Observe that for $x, y \in H$,

$$\left\langle \left|P^{-1/2}TP^{-1/2}\right|^\alpha P^{1/2}\right|^2 x, x\right\rangle = \left\|\left|P^{-1/2}TP^{-1/2}\right|^\alpha P^{1/2}x\right\|^2$$

$$\leq \left\|\left|P^{-1/2}TP^{-1/2}\right|^\alpha\right\|^2 \left\|P^{1/2}x\right\|^2$$

$$= \left\|P^{-1/2}TP^{-1/2}\right\|^{2\alpha} \langle Px, x\rangle$$

and

$$\left\langle \left|P^{-1/2}T^*P^{-1/2}\right|^{1-\alpha} P^{1/2}\right|^2 y, y\right\rangle = \left\|\left|P^{-1/2}T^*P^{-1/2}\right|^{1-\alpha} P^{1/2}y\right\|^2$$

$$\leq \left\|\left|P^{-1/2}T^*P^{-1/2}\right|^{1-\alpha}\right\|^2$$

$$\times \left\|P^{1/2}y\right\|^2$$

$$= \left\|P^{-1/2}TP^{-1/2}\right\|^{2(1-\alpha)} \langle Py, y\rangle.$$

From (36), we have for $B = C = P^{1/2}$ and $A = P^{-1/2}TP^{-1/2}$ that

$$\left|\left\langle P^{1/2} f\left(P^{-1/2}TP^{-1/2}\right) P^{1/2}x, y\right\rangle\right|$$

$$\leq M\left(f, P^{-1/2}TP^{-1/2}; \gamma\right) \left\langle \left|P^{-1/2}TP^{-1/2}\right|^\alpha P^{1/2}\right|^2 x, x\right\rangle^{1/2}$$

$$\times \left\langle \left|P^{-1/2}T^*P^{-1/2}\right|^{1-\alpha} P^{1/2}\right|^2 y, y\right\rangle^{1/2}$$

$$\leq M\left(f, P^{-1/2}TP^{-1/2}; \gamma\right) \left\|P^{-1/2}TP^{-1/2}\right\| \langle Px, x\rangle^{1/2} \langle Py, y\rangle^{1/2}$$

for $x, y \in H$. This proves the desired result (105). $\square$

Corollary 2. *Let f be an analytic function on the domain G, $0 \in G$ and $f(0) = 0$. If $P > 0$ and $T \in B(H)$ with $\mathrm{Sp}\left(P^{-1/2}TP^{-1/2}\right) \subset D(0,R) \subset G$, then*

$$|\langle \mathcal{P}_f(T,P)x, y \rangle| \leq M\left(f, P^{-1/2}TP^{-1/2}; R\right)$$

$$\times \left\langle \left|\left|P^{-1/2}TP^{-1/2}\right|^{\alpha} P^{1/2}\right|^2 x, x \right\rangle^{1/2}$$

$$\times \left\langle \left|\left|P^{-1/2}T^*P^{-1/2}\right|^{1-\alpha} P^{1/2}\right|^2 y, y \right\rangle^{1/2}$$

$$\leq M\left(f, P^{-1/2}TP^{-1/2}; R\right)$$

$$\times \left\|P^{-1/2}TP^{-1/2}\right\| \langle Px, x \rangle^{1/2} \langle Py, y \rangle^{1/2}$$
(106)

for all $\alpha \in [0,1]$ and $x, y \in H$.

Moreover, if $\|f\|_{R,\infty} := \sup_{t \in [0,1]} \left|f\left(Re^{2\pi it}\right)\right| < \infty$, then we have the simpler inequalities

$$|\langle \mathcal{P}_f(T,P)x, y \rangle| \leq \frac{\|f\|_{R,\infty}}{R - \|P^{-1/2}TP^{-1/2}\|}$$

$$\times \left\langle \left|\left|P^{-1/2}TP^{-1/2}\right|^{\alpha} P^{1/2}\right|^2 x, x \right\rangle^{1/2}$$

$$\times \left\langle \left|\left|P^{-1/2}T^*P^{-1/2}\right|^{1-\alpha} P^{1/2}\right|^2 y, y \right\rangle^{1/2}$$

$$\leq \frac{\|f\|_{R,\infty} \|P^{-1/2}TP^{-1/2}\|}{R - \|P^{-1/2}TP^{-1/2}\|} \langle Px, x \rangle^{1/2} \langle Py, y \rangle^{1/2}$$
(107)

for all $\alpha \in [0,1]$ and $x, y \in H$.

Observe that

$$\left|\left|P^{-1/2}TP^{-1/2}\right|^{\alpha} P^{1/2}\right|^2 = P\left|P^{-1/2}TP^{-1/2}\right|^{2\alpha} P$$

$$= P\left(P^{-1/2}T^*P^{-1}TP^{-1/2}\right)^{\alpha} P.$$

Since $0 < P^{-1} \leq \|P^{-1}\| I$, then $0 < T^*P^{-1}T \leq \|P^{-1}\| |T|^2$ and
$$P^{-1/2}T^*P^{-1}TP^{-1/2} \leq \|P^{-1}\| P^{-1/2} |T|^2 P^{-1/2} \leq \|P^{-1}\|^2 |T|^2.$$
If we use Heinz inequality for $\alpha \in [0,1]$, then we get
$$\left(P^{-1/2}T^*P^{-1}TP^{-1/2}\right)^\alpha \leq \|P^{-1}\|^{2\alpha} |T|^{2\alpha},$$
which implies that
$$\left\| \left|P^{-1/2}TP^{-1/2}\right|^\alpha P^{1/2} \right\|^2 = P\left(P^{-1/2}T^*P^{-1}TP^{-1/2}\right)^\alpha P$$
$$\leq \|P^{-1}\|^{2\alpha} P|T|^{2\alpha} P.$$
Similarly,
$$\left\| \left|P^{-1/2}T^*P^{-1/2}\right|^{1-\alpha} P^{1/2} \right\|^2 \leq \|P^{-1}\|^{2(1-\alpha)} P|T^*|^{2(1-\alpha)} P.$$
Therefore,
$$0 < \left\| \left|P^{-1/2}TP^{-1/2}\right|^\alpha P^{1/2} \right\|^2 + \left\| \left|P^{-1/2}T^*P^{-1/2}\right|^{1-\alpha} P^{1/2} \right\|^2$$
$$\leq \|P^{-1}\|^{2\alpha} P|T|^{2\alpha} P + \|P^{-1}\|^{2(1-\alpha)} P|T^*|^{2(1-\alpha)} P$$
$$= P\left(\|P^{-1}\|^{2\alpha} |T|^{2\alpha} + \|P^{-1}\|^{2(1-\alpha)} |T^*|^{2(1-\alpha)}\right) P,$$
which implies that
$$\left\| \left|P^{-1/2}TP^{-1/2}\right|^\alpha P^{1/2} \right\|^2 + \left\| \left|P^{-1/2}T^*P^{-1/2}\right|^{1-\alpha} P^{1/2} \right\|^2 \right\|$$
$$\leq \left\| P\left(\|P^{-1}\|^{2\alpha} |T|^{2\alpha} + \|P^{-1}\|^{2(1-\alpha)} |T^*|^{2(1-\alpha)}\right) P \right\|$$
$$\leq \left\| \|P^{-1}\|^{2\alpha} |T|^{2\alpha} + \|P^{-1}\|^{2(1-\alpha)} |T^*|^{2(1-\alpha)} \right\| \|P\|^2.$$
From (50), we get
$$\omega\left(\mathcal{P}_f(T,P)\right) \leq \frac{1}{2} M\left(f, P^{-1/2}TP^{-1/2}; \gamma\right)$$
$$\times \left\| \|P^{-1}\|^{2\alpha} |T|^{2\alpha} + \|P^{-1}\|^{2(1-\alpha)} |T^*|^{2(1-\alpha)} \right\| \|P\|^2 \quad (108)$$
for $\alpha \in [0,1]$.

For $\alpha = 1/2$, we derive

$$\omega\left(\mathcal{P}_f(T,P)\right) \leq \frac{1}{2} M\left(f, P^{-1/2}TP^{-1/2}; \gamma\right) \||T| + |T^*|\| \|P^{-1}\| \|P\|^2. \tag{109}$$

If $P > 0$ and $T \in B(H)$ with $\text{Sp}\left(P^{-1/2}TP^{-1/2}\right) \subset D(0,R)$ and $\|f\|_{R,\infty} := \sup_{t \in [0,1]} \left|f\left(Re^{2\pi it}\right)\right| < \infty$, then we have the simpler inequalities

$$\omega\left(\mathcal{P}_f(T,P)\right) \leq \frac{\|f\|_{R,\infty}}{R - \|P^{-1/2}TP^{-1/2}\|}$$

$$\times \left\| \|P^{-1}\|^{2\alpha} |T|^{2\alpha} + \|P^{-1}\|^{2(1-\alpha)} |T^*|^{2(1-\alpha)} \right\| \|P\|^2 \tag{110}$$

for $\alpha \in [0,1]$.

For $\alpha = 1/2$, we derive

$$\omega\left(\mathcal{P}_f(T,P)\right) \leq \frac{\|f\|_{R,\infty}}{R - \|P^{-1/2}TP^{-1/2}\|} \||T| + |T^*|\| \|P^{-1}\| \|P\|^2. \tag{111}$$

If we use now the inequality (75) for $C = B = P^{1/2}$ and $A = P^{-1/2}TP^{-1/2}$, we get that

$$\|\mathcal{P}_f(T,P)\|_{2r} \leq M\left(f, P^{-1/2}TP^{-1/2}; \gamma\right)$$

$$\times \left\| \left|P^{-1/2}TP^{-1/2}\right|^\alpha P^{1/2} \right\|_{2pr}$$

$$\times \left\| \left|P^{-1/2}T^*P^{-1/2}\right|^{1-\alpha} P^{1/2} \right\|_{2qr} \tag{112}$$

provided that $\alpha \in [0,1]$, $r \geq 1/2$, $p, q > 1$ with $\frac{1}{p} + \frac{1}{q} = 1$ and $pr, qr \geq 1$ while $\left|P^{-1/2}TP^{-1/2}\right|^\alpha P^{1/2} \in \mathcal{B}_{2pr}(H)$ and $\left|P^{-1/2}T^*P^{-1/2}\right|^{1-\alpha} P^{1/2} \in \mathcal{B}_{2qr}(H)$.

Now, we observe that

$$\left\| \left| P^{-1/2}TP^{-1/2} \right|^\alpha P^{1/2} \right\|_{2pr} \leq \left\| \left| P^{-1/2}TP^{-1/2} \right|^\alpha \right\| \left\| P^{1/2} \right\|_{2pr}$$

$$= \left\| P^{-1/2}TP^{-1/2} \right\|^\alpha \|P\|_{pr}$$

and

$$\left\| \left| P^{-1/2}T^*P^{-1/2} \right|^{1-\alpha} P^{1/2} \right\|_{2qr} \leq \left\| \left| P^{-1/2}T^*P^{-1/2} \right|^{1-\alpha} \right\| \left\| P^{1/2} \right\|_{2qr}$$

$$= \left\| P^{-1/2}T^*P^{-1/2} \right\|^{1-\alpha} \|P\|_{qr}.$$

These imply that

$$\left\| \left| P^{-1/2}TP^{-1/2} \right|^\alpha P^{1/2} \right\|_{2pr} \left\| \left| P^{-1/2}T^*P^{-1/2} \right|^{1-\alpha} P^{1/2} \right\|_{2qr}$$

$$\leq \left\| P^{-1/2}TP^{-1/2} \right\|^\alpha \|P\|_{pr} \left\| P^{-1/2}T^*P^{-1/2} \right\|^{1-\alpha} \|P\|_{qr}$$

$$= \left\| P^{-1/2}TP^{-1/2} \right\|^\alpha \left\| P^{-1/2}TP^{-1/2} \right\|^{1-\alpha} \|P\|_{pr} \|P\|_{qr}$$

$$= \left\| P^{-1/2}TP^{-1/2} \right\| \|P\|_{pr} \|P\|_{qr}$$

and by (112) we obtain the simpler inequality

$$\|\mathcal{P}_f(T,P)\|_{2r} \leq M\left(f, P^{-1/2}TP^{-1/2}; \gamma\right)$$

$$\times \left\| P^{-1/2}TP^{-1/2} \right\| \|P\|_{pr} \|P\|_{qr}. \tag{113}$$

If $P > 0$ and $T \in B(H)$ with $\mathrm{Sp}\left(P^{-1/2}TP^{-1/2}\right) \subset D(0,R)$ and $\|f\|_{R,\infty} := \sup_{t \in [0,1]} \left| f\left(Re^{2\pi it}\right) \right| < \infty$, then we have the simpler inequality

$$\|\mathcal{P}_f(T,P)\|_{2r} \leq \frac{\|f\|_{R,\infty}}{R - \left\| P^{-1/2}TP^{-1/2} \right\|} \left\| P^{-1/2}TP^{-1/2} \right\| \|P\|_{pr} \|P\|_{qr} \tag{114}$$

for $r \geq 1/2$, $p, q > 1$ with $\frac{1}{p} + \frac{1}{q} = 1$ and $pr, qr \geq 1$ and $P \in \mathcal{B}_{pr}(H) \cap \mathcal{B}_{qr}(H)$.

6. Two Examples

Consider the exponential function $f(A) = \exp A$, $A \in \mathcal{B}(H)$. Assume that $A \in \mathcal{B}(H)$ and $\|A\| < R$ for some $R > 0$. Observe that for $t \in [0, 1]$,

$$\left|\exp\left(Re^{2\pi it}\right)\right| = |\exp[R(\cos(2\pi t) + i\sin(2\pi t))]| = \exp[R\cos(2\pi t)]$$

and then by (44) we get for $B, C \in \mathcal{B}(H)$ that

$$|\langle C^*[\exp(A) - I]Bx, y\rangle|$$
$$\leq \frac{1}{R - \|A\|} \int_0^1 \exp[R\cos(2\pi t)]\,dt$$
$$\times \left\langle \|A\|^\alpha B|^2 x, x\right\rangle^{1/2} \left\langle \left||A^*|^{1-\alpha} C\right|^2 y, y\right\rangle^{1/2} \quad (115)$$

for $x, y \in H$.

The *modified Bessel function of the first kind* $I_\nu(z)$ for real number ν can be defined by the power series as [2, p. 376]

$$I_\nu(z) = \left(\frac{1}{2}z\right)^\nu \sum_{k=0}^\infty \frac{\left(\frac{1}{4}z^2\right)^k}{k!\,\Gamma(\nu + k + 1)},$$

where Γ is the *gamma function*. For $n = 0$, we have $I_0(z)$ given by

$$I_0(z) = \sum_{k=0}^\infty \frac{\left(\frac{1}{4}z^2\right)^k}{(k!)^2}.$$

An integral formula for real number ν is

$$I_\nu(z) = \frac{1}{\pi} \int_0^\pi e^{z\cos\theta} \cos(\nu\theta)\,d\theta - \frac{\sin(\nu\pi)}{\pi} \int_0^\infty e^{-z\cosh t - \nu t}\,dt,$$

which simplifies for ν an integer n to

$$I_n(z) = \frac{1}{\pi} \int_0^\pi e^{z\cos\theta} \cos(n\theta)\,d\theta.$$

For $n = 0$, we have

$$I_0(z) = \frac{1}{\pi} \int_0^\pi e^{z \cos \theta} d\theta.$$

If we change the variable $\theta = 2\pi t$, then $dt = \frac{1}{2\pi} d\theta$ and

$$\int_0^1 \exp[R \cos(2\pi t)] dt = \frac{1}{2\pi} \int_0^{2\pi} \exp[R \cos \theta] d\theta$$
$$= \frac{1}{2} \left(\frac{1}{\pi} \int_0^\pi \exp[R \cos \theta] d\theta \right.$$
$$\left. + \frac{1}{\pi} \int_\pi^{2\pi} \exp[R \cos \theta] d\theta \right)$$
$$= \frac{1}{2} (I_0(R) + I_0(-R)) = I_0(R).$$

From (115), we then get

$$|\langle C^* [\exp(A) - I] Bx, y \rangle|$$
$$\leq \frac{I_0(R)}{R - \|A\|} \left\langle \||A|^\alpha B|^2 x, x \right\rangle^{1/2} \left\langle \||A^*|^{1-\alpha} C\|^2 y, y \right\rangle^{1/2} \quad (116)$$

for $\alpha \in [0, 1]$, $x, y \in H$, $A, B, C \in \mathcal{B}(H)$ with $\|A\| < R$.

By taking $B = C = I$ in (116), we get for $\|A\| < R$ that

$$|\langle [\exp(A) - I] x, y \rangle| \leq \frac{I_0(R)}{R - \|A\|} \left\langle |A|^{2\alpha} x, x \right\rangle^{1/2} \left\langle |A^*|^{2(1-\alpha)} y, y \right\rangle^{1/2}$$
$$(117)$$

for $x, y \in H$. In particular,

$$|\langle [\exp(A) - I] x, y \rangle| \leq \frac{I_0(R)}{R - \|A\|} \langle |A| x, x \rangle^{1/2} \langle |A^*| y, y \rangle^{1/2} \quad (118)$$

for $x, y \in H$.

By Theorem 4, we get the norm inequality

$$\|C^* [\exp(A) - I] BB\| \leq \frac{I_0(R)}{R - \|A\|} \||A|^\alpha B\| \||A^*|^{1-\alpha} C\|. \quad (119)$$

We also have the numerical radius inequalities

$$\omega(C^* [\exp(A) - I] B) \leq \frac{1}{2} \frac{I_0(R)}{R - \|A\|} \left\| ||A|^\alpha B|^2 + ||A^*|^{1-\alpha} C|^2 \right\| \quad (120)$$

and

$$\omega^2(C^* [\exp(A) - I] B) \leq \frac{1}{2} \left(\frac{I_0(R)}{R - \|A\|} \right)^2$$

$$\times \left[\||A|^\alpha B\|^2 \||A^*|^{1-\alpha} C\|^2 + \omega \left(||A^*|^{1-\alpha} C|^2 ||A|^\alpha B|^2 \right) \right]. \quad (121)$$

Let $r \geq 1/2$, $p, q > 1$ with $\frac{1}{p} + \frac{1}{q} = 1$ and $pr, qr \geq 1$. If B, $C \in \mathcal{B}(H)$ with $|A|^\alpha B \in \mathcal{B}_{2pr}(H)$ and $|A^*|^{1-\alpha} C \in \mathcal{B}_{2qr}(H)$ for $\alpha \in [0,1]$, then $C^* [\exp(A) - I] B \in \mathcal{B}_{2r}(H)$ and by (75)

$$\|C^* [\exp(A) - I] B\|_{2r} \leq \frac{I_0(R)}{R - \|A\|} \||A|^\alpha B\|_{2pr} \||A^*|^{1-\alpha} C\|_{2qr}. \quad (122)$$

If $r \geq 1/2$, $p, q > 1$ with $\frac{1}{p} + \frac{1}{q} = 1$, $pr, qr \geq 1$ and $||A|^\alpha B|^{2pr}$, $||A^*|^{1-\alpha} C|^{2qr} \in \mathcal{B}_1(H)$, then $C^* [\exp(A) - I] B \in \mathcal{B}_{2r}(H)$ and by (91)

$$\omega_{2r}^{2r}(C^* [\exp(A) - I] B)$$

$$\leq \left(\frac{I_0(R)}{R - \|A\|} \right)^{2r} \operatorname{tr} \left(\frac{1}{p} ||A|^\alpha B|^{2pr} + \frac{1}{q} ||A^*|^{1-\alpha} C|^{2qr} \right). \quad (123)$$

Consider the power series

$$f(z) := \ln(1-z)^{-1} = \sum_{n=1}^{\infty} \frac{1}{n} z^n.$$

We observe that for $|z| < 1$

$$\left| \ln(1-z)^{-1} \right| \leq \sum_{n=1}^{\infty} \frac{1}{n} |z|^n = \ln(1-|z|)^{-1}.$$

Now if we assume that $A, B, C \in \mathcal{B}(H)$ and $\|A\| < R < 1$, then by (42) we get

$$\left| \left\langle C^* \ln(1-A)^{-1} Bx, y \right\rangle \right|$$
$$\leq \frac{\ln(1-R)^{-1}}{R - \|A\|} \left\langle |A|^\alpha B|^2 x, x \right\rangle^{1/2} \left\langle \left||A^*|^{1-\alpha} C\right|^2 y, y \right\rangle^{1/2} \quad (124)$$

for $\alpha \in [0,1]$, $x, y \in H$.

By taking $B = C = I$ in (124), we get for $\|A\| < R < 1$ that

$$\left| \left\langle \ln(1-A)^{-1} x, y \right\rangle \right|$$
$$\leq \frac{\ln(1-R)^{-1}}{R - \|A\|} \left\langle |A|^{2\alpha} x, x \right\rangle^{1/2} \left\langle |A^*|^{2(1-\alpha)} y, y \right\rangle^{1/2} \quad (125)$$

for $x, y \in H$. In particular,

$$\left| \left\langle \ln(1-A)^{-1} x, y \right\rangle \right| \leq \frac{\ln(1-R)^{-1}}{R - \|A\|} \left\langle |A| x, x \right\rangle^{1/2} \left\langle |A^*| y, y \right\rangle^{1/2}$$
$$(126)$$

for $x, y \in H$.

One can state some norm, numerical radius and p-Schatten norm inequalities for $\ln(1-A)^{-1}$, however the details are omitted.

References

[1] A. Aluthge, Some generalized theorems on p-hyponormal operators, *Integral Equ. Oper. Theory* **24**, 497–501 (1996).

[2] M. Abramowitz and I. A. Stegun, *Handbook of Mathematical Functions*. Applied Mathematics Series, Vol. 55 (National Bureau of Standards, USA, 1972).

[3] A. Abu-Omar and F. Kittaneh, A numerical radius inequality involving the generalized Aluthge transform, *Studia Math.* **216**(1), 69–75 (2013).

[4] P. Bhunia, S. Bag and K. Paul, Numerical radius inequalities and its applications in estimation of zeros of polynomials, *Linear Algebra Its Appl.* **573**, 166–177 (2019).

[5] P. Bhunia, S. S. Dragomir, M. S. Moslehian and K. Paul, *Lectures on Numerical Radius Inequalities* (Springer, Cham, 2022). https://doi.org/10.1007/978-3-031-13670-2.

[6] M. L. Buzano, Generalizzazione della diseguaglianza di Cauchy-Schwarz (Italian), *Rend. Sem. Mat. Univ. Politech. Torino* **31**(1971/73), 405–409 (1974).

[7] J. B. Conway, *A Course in Functional Analysis*, 2nd edn. (Springer-Verlag, New York, 1990).

[8] M. Cho and K. Tanahashi, Spectral relations for Aluthge transform, *Sci. Math. Jpn.* **55**(1), 77–83 (2002).

[9] R. Douglas, *Banach Algebra Techniques in Operator Theory* (Academic Press, USA, 1972).

[10] S. S. Dragomir, *Inequalities for the Numerical Radius of Linear Operators in Hilbert Spaces* (SpringerBriefs in Mathematics Springer, USA, 2013). https://doi.org/10.1007/978-3-319-01448-7.

[11] S. S. Dragomir, Trace inequalities for operators in Hilbert spaces: A survey of recent results, *Aust. J. Math. Anal. Appl.* **19**(1) (2022). Art. 1. 202 p.

[12] S. S. Dragomir, Generalizations of Furuta's inequality, **61**(5), 617–626 (2013).

[13] M. El-Haddad and F. Kittaneh, Numerical radius inequalities for Hilbert space operators. II, *Studia Math.* **182**(2), 133–140 (2007).

[14] T. Furuta, An extension of the Heinz-Kato theorem, *Proc. Am. Math. Soc.* **120**(3), 785–787 (1994).

[15] G. Helmberg, *Introduction to Spectral Theory in Hilbert Space* (John Wiley & Sons Inc., New York, 1969).

[16] T. Kato, Notes on some inequalities for linear operators, *Math. Ann.* **125**, 208–212 (1952).

[17] F. Kittaneh, Notes on some inequalities for Hilbert space operators, *Publ. Res. Inst. Math. Sci.* **24**(2), 283–293 (1988).

[18] F. Kittaneh, A numerical radius inequality and an estimate for the numerical radius of the Frobenius companion matrix, *Stud. Math.* **158**(1), 11–17 (2003).

[19] F. Kittaneh, Numerical radius inequalities for Hilbert space operators, *Stud. Math.* **168**(1), 73–80 (2005).

[20] T. Kato, Notes on some inequalities for linear operators, *Math. Ann.* **125**, 208–212 (1952).

[21] C. A. McCarthy, C_p, *Isr. J. Math.* **5**, 249–271 (1967).

[22] J. R. Ringrose, *Compact Non-self-adjoint Operators* (Van Nostrand Reinhold, New York, 1971).
[23] W. Rudin, *Functional Analysis* (McGraw Hill, United Kingdom, 1973).
[24] B. Simon, *Trace Ideals and Their Applications* (Cambridge University Press, Cambridge, 1979).
[25] V. A. Zagrebnov, *Gibbs Semigroups*. Operator Theory: Advances and Applications, Vol. 273 (Birkhäuser, Switzerland, 2019).

© 2024 World Scientific Publishing Company
https://doi.org/10.1142/9789811267048_0009

Chapter 9

General Equivariant Minimax Principle and Fountain Theorem in the Presence of Nonadmissible Representations

Lucas Fresse[*,‡] and Viorica V. Motreanu[†,§]

*Institut Élie Cartan, Université de Lorraine,
Vandoeuvre-lès-Nancy, France

†Lycée Varoquaux, Tomblaine, France
‡lucas.fresse@univ-lorraine.fr
§vmotreanu@gmail.com

Given an isometric representation of a compact topological group on a Banach space and an invariant locally Lipschitz map, we establish general results including a deformation theorem and a minimax principle based on a notion of equivariant linking. As a byproduct, we generalize Bartsch's fountain theorem. The main novelty is that we allow the Banach space to contain also nonadmissible representations. Moreover, our results are established under a Cerami-type condition more general than the usual one. Finally, we provide new properties and examples of admissible representations.

1. Introduction

We consider a Banach space X endowed with an isometric representation of a compact topological group G. Then let $f : X \to \mathbb{R}$ be a G-invariant locally Lipschitz map. This is a favorable situation for proving multiplicity results guaranteeing the existence of infinitely many critical points for f: see Refs. [1, 2, 8–10] and the references therein. Often one restricts to considering the group $G = \{-1, 1\}$, which permits the use of Krasnosel'skii genus, but an alternative for considering arbitrary groups is the notion of admissible representations introduced in Ref. [1]. In Section 6, we give new properties and examples of admissible representations, in particular a complete classification of the admissible representations of the symmetric group $\mathfrak{S}_5$.

In the setting presented above, we first establish an equivariant deformation theorem (Theorem 1). Then, we derive a general equivariant minimax principle (Theorem 2), which involves a notion of equivariant linking introduced in Definition 2 (weaker than usual linking).

Our main result (Theorem 3) is a generalization of the so-called fountain theorem of Ref. [1]. We outline the main novelties of our result:

- f is a locally Lipschitz map, whereas the map in Ref. [1] is of class C^1.
- Instead of the Palais–Smale condition, we assume a more general, localized Cerami-type condition.
- In Ref. [1], it is assumed that X involves only one admissible representation V repeated infinitely many times; we relax this condition by allowing X to contain other, possibly nonadmissible, representations (see also Example 2).

In Ref. [9, Corollary 2.7] and [2], nonsmooth versions of fountain theorem are also given, but for usual Palais–Smale condition and the same assumption on the representation as in Ref. [1]. Our proof is also somewhat different from the proof given in the aforementioned references, as we are able to apply our general minimax principle (Theorem 2) instead of an *ad hoc* deformation theorem.

Notation 1. For a Banach space $(X, \|\cdot\|)$, its topological dual will be denoted by X^*, and it is a Banach space for the norm $\|\phi\|_* := \sup_{\|x\|\leq 1} |\langle \phi, x\rangle|$. Hereafter, $\langle \cdot, \cdot \rangle$ stand for the duality brackets for the pair (X^*, X). For $C, D \subset X$, we set $\mathrm{dist}(C, D) = \inf_{(x,y)\in C\times D} \|x-y\|$ and, if $C = \{x\}$, we write for simplicity $\mathrm{dist}(x, D) := \mathrm{dist}(\{x\}, D)$. Moreover, for $\delta > 0$, we denote $D_\delta = \{x \in X : \mathrm{dist}(x, D) < \delta\}$, with the convention that $D_\delta = \emptyset$ whenever $D = \emptyset$. For $f : X \to \mathbb{R}$ and $a \in \mathbb{R}$, we denote $f^a = \{x \in X : f(x) \leq a\}$.

We recall basic notions related to group actions. Given a group G, we call G-space a set X endowed with an action of G. We say that a map $\phi : X \to Y$ is G-invariant if $\phi(gx) = \phi(x)$ for all $(g, x) \in G \times X$. When Y is another G-space, we say that ϕ is G-equivariant if $\phi(gx) = g\phi(x)$ for all $(g, x) \in G \times X$. A subset $X' \subset X$ is said to be G-stable if $gx' \in X'$ for all $(g, x') \in G \times X'$. Finally, if G is a topological group, by topological G-space we mean a topological space X endowed with a continuous action of G, i.e., the map $G \times X \to X$, $(g, x) \mapsto gx$ is continuous.

2. Isometric Representation on a Banach Space

If $(X, \|\cdot\|)$ is a Banach space, let $O(X)$ be the group of isometric linear automorphisms, i.e., linear automorphisms $\varphi : X \to X$ such that $\|\varphi(x)\| = \|x\|$ for all $x \in X$. If X is a Hilbert space, then the elements of $O(X)$ are equivalently the linear automorphisms which preserve the scalar product.

Let G be a compact topological group. Fix an isometric representation of G on X, that is, a continuous morphism $\rho : G \to O(X)$. For simplicity, we denote $gx := \rho(g)(x)$. We get an isometric (dual) representation of G on X^* by letting $g\phi : x \mapsto \langle \phi, g^{-1}x\rangle$ whenever $\phi \in X^*$. Thus,

$$\langle g\phi, gx\rangle = \langle \phi, x\rangle \quad \text{for all } (\phi, x) \in X^* \times X, \text{ all } g \in G.$$

We consider maps $f : X \to \mathbb{R}$ which are locally Lipschitz. By $\partial f(x)$, we denote the subdifferential of f at $x \in X$, defined by

$$\partial f(x) = \{x^* \in X^* : \forall h \in X, \langle x^*, h\rangle \leq f^0(x; h)\},$$

where

$$f^0(x;h) := \limsup_{\substack{x' \to x \\ t \to 0^+}} \frac{f(x'+th) - f(x')}{t} \qquad (1)$$

is the generalized directional derivative of f at x in the direction h. For all $x \in X$, the set $\partial f(x)$ is nonempty, convex and weakly*-compact, which implies that the infimum

$$\lambda_f(x) := \inf_{x^* \in \partial f(x)} \|x^*\|_*$$

is attained. Moreover, the mapping $\lambda_f : X \to [0, +\infty)$ is lower semicontinuous. In general, the subdifferential ∂f is multivalued, but if f is of class C^1, then $\partial f(x) = \{f'(x)\}$, hence we recover the usual differential.

We say that x is a critical point of f if $0 \in \partial f(x)$, which equivalently means that $\lambda_f(x) = 0$. By $K_f \subset X$, we denote the set of critical points of f and, for $c \in \mathbb{R}$, we set

$$K_f^c := \{x \in K_f : f(x) = c\}.$$

Note that K_f and K_f^c are closed subsets of X (since λ_f is lower semicontinuous).

Lemma 1. *If f is G-invariant, we have the following:*

(a) $f^0(gx; h) = f^0(x; g^{-1}h)$ *for all* $x, h \in X$, *all* $g \in G$;
(b) $\partial f(gx) = g\partial f(x) := \{gx^* : x^* \in \partial f(x)\}$ *and* $\lambda_f(gx) = \lambda_f(x)$ *for all* $x \in X$, *all* $g \in G$;
(c) *the sets K_f and K_f^c (for all $c \in \mathbb{R}$) are G-stable.*

Proof. Part (a) follows from the calculation

$$f^0(x; g^{-1}h) = \limsup_{\substack{x'' \to x \\ t \to 0^+}} \frac{f(gx'' + th) - f(gx'')}{t}$$

$$= \limsup_{\substack{x' \to gx \\ t \to 0^+}} \frac{f(x'+th) - f(x')}{t}$$

where we use (1), the G-invariance of f, and the fact that $(x_n)_{n \geq 1} \mapsto (gx_n)_{n \geq 1}$ realizes a bijection between sequences that converge to x and gx, respectively.

Letting $x^* \in \partial f(x)$, we have

$$\langle gx^*, h \rangle = \langle x^*, g^{-1}h \rangle \leq f^0(x; g^{-1}h) = f^0(gx; h) \quad \text{for all } h \in X,$$

which implies that $gx^* \in \partial f(gx)$. Whence $g\partial f(x) \subset \partial f(gx)$. The reversed inclusion holds as we have similarly $g^{-1}\partial f(gx) \subset \partial f(x)$. Finally, the equality $\lambda_f(gx) = \lambda_f(x)$ immediately follows from the fact that $\partial f(gx) = g\partial f(x)$, knowing that the action of G on X^* is isometric. This shows (b).

Since $K_f = \{x \in X : \lambda_f(x) = 0\}$ and $K_f^c = \{x \in X : \lambda_f(x) = 0$ and $f(x) = c\}$, part (c) easily follows from (b) and the G-invariance of f. □

3. An Equivariant Deformation Theorem

We use the following compactness-type condition relative to a locally Lipschitz map $f : X \to \mathbb{R}$:

Definition 1. Given $c \in \mathbb{R}$, a subset $Z \subset X$ and a Lipschitz map $\varphi : Z \to [1, +\infty)$, we say that f satisfies the $(\varphi - \mathrm{C})_{Z,c}$ condition if every sequence $(x_n)_{n \geq 1}$ in Z such that

$$f(x_n) \to c \quad \text{and} \quad \varphi(x_n)\lambda_f(x_n) \to 0$$

has a convergent subsequence.

Remark 1. The same condition is used in Ref. [4]. It combines conditions introduced in Refs. [6] and [8, Section 5.2], which themselves generalize the classical Palais–Smale and Cerami conditions. The former, resp., the latter, is recovered by taking $Z = X$ and $\varphi = \varphi_1 := 1$, resp., $\varphi = \varphi_2 := 1 + \|\cdot\|$.

When the map f is G-invariant for an isometric representation of G on X, it will be natural to expect the set Z to be G-stable and the map $\varphi : Z \to [1, +\infty)$ to be G-invariant, which is the case of φ_1 and φ_2 above.

Theorem 1. *Let X be a Banach space endowed with an isometric representation of a compact topological group G. We consider the following:*

- *a G-invariant locally Lipschitz map $f : X \to \mathbb{R}$;*
- *a G-stable subset $Z \subset X$ whose complement $S := X \setminus Z$ is bounded;*

- a G-invariant Lipschitz map $\varphi : Z \to [1, +\infty)$;
- $c \in \mathbb{R}$, and we assume that f satisfies the $(\varphi - \mathrm{C})_{Z,c}$ condition.

Then, for every open subset $U \subset X$ with

$$K_f^c \cup S \subset U, \quad \mathrm{dist}(S, X \setminus U) > 0, \tag{2}$$

and every $\varepsilon_0, \theta \in (0, +\infty)$, there exist $\varepsilon \in (0, \varepsilon_0)$ and a continuous map $h : [0,1] \times X \to X$ such that, for every $(t, x) \in [0, 1] \times X$, we have the following:

(a) $x \in Z \Leftrightarrow h(t, x) \in Z$;
(b) $\|h(t, x) - x\| \leq \theta \varphi(x) t$ if $x \in Z$; $h(t, x) = x$ if $x \in S$;
(c) $f(h(t, x)) \leq f(x)$;
(d) $h(t, x) \neq x \Rightarrow f(h(t, x)) < f(x)$;
(e) $|f(x) - c| \geq \varepsilon_0 \Rightarrow h(t, x) = x$;
(f) $h(1, f^{c+\varepsilon}) \subset f^{c-\varepsilon} \cup U$ and $h(1, f^{c+\varepsilon} \setminus U) \subset f^{c-\varepsilon}$;
(g) $h(t, gx) = gh(t, x)$ for all $g \in G$.

Part (g) of the theorem means that, for every $t \in [0, 1]$, the mapping $h(t, \cdot) : X \to X$ is G-equivariant. Theorem 1 is an equivariant version of the deformation theorem obtained in Ref. [4, Theorem 5], and the proof given below follows the same scheme of arguments therein. Note that the open set U involved in the statement is not assumed to be G-invariant.

It is also worth noting that in the conclusion of the theorem, (c) holds with the more precise assertion that, for all $x \in X$, the map $t \mapsto f(h(t, x))$ is nonincreasing on $[0, 1]$. This follows from formula (8) in the proof of Theorem 5 in Ref. [4].

Proof of Theorem 1. The key step of the proof of Ref. [4, Theorem 5] is the construction of a suitable pseudogradient (see Lemma 2 below). The key point of the present proof is to show that this pseudogradient can be chosen G-equivariant (Lemma 3).

First, we summarize the preliminary steps. Fix U, ε_0 and θ as in the statement. Relying on (2), we can choose $\delta > 0$ such that

$$(K_f^c \cup S)_{3\delta} \subset U.$$

By exploiting the fact that f satisfies the $(\varphi - \mathrm{C})_{Z,c}$ condition, we can show that there are $\varepsilon_1 \in (0, \varepsilon_0)$ and $m \in (0, +\infty)$ such that

$$\bigl(|f(x) - c| \leq \varepsilon_1 \text{ and } \mathrm{dist}(x, K_f^c \cup S) \geq \delta\bigr) \Longrightarrow \varphi(x)\lambda_f(x) > m.$$

We define the closed subsets

$$A := \{x \in X : |f(x) - c| \leq \varepsilon_1 \text{ and } x \notin (K_f^c \cup S)_\delta\},$$
$$A_0 := \left\{x \in X : |f(x) - c| \leq \frac{\varepsilon_1}{2} \text{ and } x \notin (K_f^c \cup S)_{2\delta}\right\},$$
$$B := \left\{x \in X : |f(x) - c| \geq \varepsilon_1 \text{ or } x \in \overline{(K_f^c \cup S)_\delta}\right\}$$

and the locally Lipschitz map

$$\gamma : X \to [0,1], \quad x \mapsto \frac{\text{dist}(x,B)}{\text{dist}(x,A_0) + \text{dist}(x,B)}.$$

Finally, let $\hat{\varphi} : X \to [0, +\infty)$ be the extension by zero of φ.

Since the maps f, φ are G-invariant and the subsets K_f^c, Z and thus $S = X \setminus Z$ are G-stable (by Lemma 1 and the assumption of the theorem), we obtain that

$$A, A_0, B \text{ are } G\text{-stable} \quad \text{and} \quad \gamma, \hat{\varphi} \text{ are } G\text{-invariant.} \tag{3}$$

Lemma 2 (see Ref. [4, Claim 9]). *There is a locally Lipschitz map $V : X \to X$, with $V|_B = 0$, such that*

$$\|V(x)\| \leq \gamma(x)\hat{\varphi}(x) \quad \text{and}$$
$$\langle x^*, V(x) \rangle \geq m\gamma(x) \quad \text{for all } x \in X, \; x^* \in \partial f(x).$$

Lemma 3. *The locally Lipschitz map V of Lemma 2 can be chosen G-equivariant, that is, such that $V(gx) = gV(x)$ for all $(g,x) \in G \times X$.*

Proof. Since G is a compact topological group, it is equipped with a Haar measure, i.e., a G-invariant measure μ such that $\int_G d\mu(g) = 1$. Then, for V as in Lemma 2, we define

$$V_0 : X \to X, \; x \mapsto \int_G g^{-1} V(gx) \, d\mu(g).$$

This map is well defined because V is continuous and the map $g \mapsto gx$ is continuous (since the representation $G \to O(X)$ is continuous).

By construction, V_0 is G-equivariant since

$$V_0(g'x) = \int_G g^{-1} V(gg'x) \, d\mu(g)$$

$$= g' \int_G (gg')^{-1} V(gg'x) \, d\mu(gg') = g' V_0(x)$$

for all $(g', x) \in G \times X$, where we have used that the Haar measure is G-invariant. For showing the lemma, it now suffices to show that V_0 still fulfills the conditions of Lemma 2.

For showing that V_0 is locally Lipschitz, fix $x \in X$. For every $g \in G$, using that V is locally Lipschitz at gx, we find a neighborhood W_g of gx and a constant $\ell_g > 0$ such that

$$z, z' \in W_g \Rightarrow \|V(z) - V(z')\| \leq \ell_g \|z - z'\|.$$

Moreover, using the continuity of the action, there exist neighborhoods $U_g \subset G$ of g and $U'_g \subset X$ of x such that

$$(h, y) \in U_g \times U'_g \Rightarrow hy \in W_g.$$

Using that G is compact, we find a finite subcovering $G = U_{g_1} \cup \cdots \cup U_{g_m}$ and set $\ell = \max\{\ell_{g_1}, \ldots, \ell_{g_m}\}$ and $U' = U'_{g_1} \cap \cdots \cap U'_{g_m}$, which is a neighborhood of x in X. Let $y, y' \in U'$. For every $g \in G$, say $g \in U_{g_i}$, we get $gy, gy' \in W_{g_i}$, hence

$$\|V(gy) - V(gy')\| \leq \ell_{g_i} \|gy - gy'\| \leq \ell \|y - y'\|,$$

where we also use that G acts isometrically on X. Therefore,

$$\|V_0(y) - V_0(y')\| \leq \int_G \|V(gy) - V(gy')\| \, d\mu(g)$$

$$\leq \ell \int_G \|y - y'\| \, d\mu(g) = \ell \|y - y'\|.$$

Thus, V_0 is Lipschitz near every $x \in X$.

Since B is G-stable (see (3)) and $V|_B \equiv 0$, given $x \in B$ we have $V(gx) = 0$ for all $g \in G$, whence $V_0(x) = 0$.

Again since γ and $\hat{\varphi}$ are G-invariant (see (3)), given $x \in X$, we have $\|g^{-1} V(gx)\| = \|V(gx)\| \leq \gamma(gx) \hat{\varphi}(gx) = \gamma(x) \hat{\varphi}(x)$ for all $g \in G$, which easily yields $\|V_0(x)\| \leq \gamma(x) \hat{\varphi}(x)$.

Finally, given $x^* \in \partial f(x)$, knowing that $gx^* \in g\partial f(x) = \partial f(gx)$ (see Lemma 1) and γ is G-invariant (see (3)), we find that

$$\langle x^*, g^{-1}V(gx)\rangle = \langle gx^*, V(gx)\rangle \geq m\gamma(gx) = m\gamma(x) \quad \text{for all } g \in G,$$

which implies that $\langle x^*, V_0(x)\rangle \geq m\gamma(x)$. □

Now that we have $V : X \to X$ G-equivariant fulfilling Lemma 2, choose $\eta > 0$ small enough so that $e^{\eta\kappa} - 1 \leq \kappa\theta$, where $\kappa > 0$ stands for the Lipschitz constant for φ and $\theta > 0$ is as fixed at the beginning of the proof. For every $x \in X$, we consider the Cauchy problem

$$\begin{cases} u'(t) = -\eta V(u(t)) & \text{in } [0,1], \\ u(0) = x. \end{cases} \quad (4)$$

The properties of V stated in Lemma 2 imply that (4) has a unique global solution $h(\cdot, x) := u_x \in C^1([0,1], X)$. Moreover, the map $h : [0,1] \times X \to X$ is continuous (as it is the flow of the equation). In Ref. [4], it is shown that this map h fulfills the conditions (a)–(f) of the theorem. Then, it remains to check condition (g). To do this, let $g \in G$ and $x \in X$. We have to show that $u_{gx} = gu_x$, where $gu_x : [0,1] \to X$, $t \mapsto gu_x(t)$. Note that gu_x is a C^1-map which satisfies $gu_x(0) = gx$ and for all $t \in [0,1]$,

$$(gu_x)'(t) = gu'_x(t) = -\eta gV(u_x(t)) = -\eta V(gu_x(t)),$$

where we have used that V is G-equivariant. By definition of u_{gx}, we conclude that $u_{gx} = gu_x$ as asserted, and the proof of Theorem 1 is complete. □

4. An Equivariant Minimax Theorem

Throughout this section, we assume that $(X, \|\cdot\|)$ is a Banach space equipped with an isometric representation of a compact topological group G. We rely on a notion of equivariant linking pair.

Definition 2. Given

- E a topological G-space equipped with a G-equivariant continuous map $\gamma_0 : E_0 \to X$ defined on a (nonempty) G-stable subset $E_0 \subset E$,
- D a subset of X,

we say that $((E,\gamma_0), D)$ is an *equivariant linking pair* if

(a) $\gamma_0(E_0) \cap D = \emptyset$,
(b) $\Gamma_G((E,\gamma_0), X) := \{\gamma : E \to X \text{ continuous, } G\text{-equivariant: } \gamma|_{E_0} = \gamma_0\} \neq \emptyset$ and $\gamma(E) \cap D \neq \emptyset$ for every $\gamma \in \Gamma_G((E,\gamma_0), X)$.

Remark 2.

(a) It is worth noting that in the definition the subset $D \subset X$ is not assumed to be G-stable.
(b) Letting $\Gamma((E,\gamma_0), X) := \{\gamma : E \to X \text{ continuous: } \gamma|_{E_0} = \gamma_0\}$, one can see that

$$\Gamma((E,\gamma_0), X) \neq \emptyset \Rightarrow \Gamma_G((E,\gamma_0), X) \neq \emptyset.$$

Indeed, taking $\gamma \in \Gamma((E,\gamma_0), X)$, we have that $\hat{\gamma} : x \mapsto \int_G g^{-1}\gamma(gx)\, d\mu(g)$ is an element of $\Gamma_G((E,\gamma_0), X)$, where μ stands for the Haar measure on G. This can be justified as follows: due to the continuity of γ and of the action of G on E and X, the map $\hat{\gamma} : E \to X$ is well defined and continuous (for the latter fact, we also use that G is compact). Next, $\hat{\gamma}$ is G-equivariant since

$$\hat{\gamma}(g'x) = \int_G g^{-1}\gamma(gg'x)\, d\mu(g)$$
$$= g'\int_G (gg')^{-1}\gamma(gg'x)\, d\mu(gg') = g'\hat{\gamma}(x)$$

for all $(g', x) \in G \times E$, where we use that the Haar measure is G-invariant. Finally, since E_0 is G-stable and γ_0 is G-equivariant, for every $x \in E_0$ we have $\gamma(gx) = \gamma_0(gx) = g\gamma_0(x)$ for all $g \in G$, which implies that $\hat{\gamma}(x) = \gamma_0(x)$.
(c) The sets $\Gamma((E,\gamma_0), X)$ and $\Gamma_G((E,\gamma_0), X)$ are nonempty, for instance, if $E_0 \subset E \subset X$ and $\gamma_0 = \mathrm{id}_{E_0}$.
(d) Another case where we can guarantee that $\Gamma((E,\gamma_0), X)$ and thus $\Gamma_G((E,\gamma_0), X)$ are nonempty is when E is a metric space and $E_0 \subset E$ is closed. Indeed, in this case, it follows from Ref. [3, Theorem 6.1 in Chapter IX] that $\gamma_0 : E_0 \to X$ has a continuous extension $\gamma : E \to X$ (whose image is contained in the convex hull of $\gamma_0(E_0)$).

Example 1.

(a) Let X be a Banach space endowed with the action of $G = \{-1, 1\}$ given by $\pm 1 \mapsto \pm \mathrm{id}_X$. Let $D = \{x \in X : \|x\| = r,\ \phi(x) \geq 0\}$ for some $r > 0$ and $\phi \in X^*$. Let $E = \{(t, 0) : -1 \leq t \leq 1\} \cup \{(t, 1 - |t|) : -1 \leq t \leq 1\} \subset \mathbb{R}^2$ be a triangle on which we let G act through the symmetry by the vertical axis. Finally, let $E_0 = \{(-1, 0), (1, 0), (0, 1)\}$ and let $\gamma_0 : E_0 \to X$ be given by $\gamma_0((0, 1)) = 0$ and $\gamma_0((\pm 1, 0)) = \pm x_0$, where $x_0 \in X$, $\|x_0\| > r$. Then, $((E, \gamma_0), D)$ is an equivariant linking pair. Note that it is not a linking pair in the sense of Ref. [4, Definition 10] because one can find $\gamma \in \Gamma((E, \gamma_0), X)$ with $\gamma(E) \cap D = \emptyset$.

(b) Let X be a Hilbert space, Y a finite-dimensional subspace and $V = Y^\perp$. We equip X with the linear action of $G = \{-1, 1\} \times \{-1, 1\}$, where $g = (\epsilon_1, \epsilon_2) \in G$ acts on $x = y + v$ ($y \in Y$, $v \in V$) by $gx = \epsilon_1 y + \epsilon_2 v$. Let $E_0 = \{y \in Y : \|y\| = r\} \subset E = \{y \in Y : \|y\| \leq r\}$, $\gamma_0 = \mathrm{id}_{E_0}$ and $D = V$, then $((E, \gamma_0), D)$ is an equivariant linking pair.

(c) As a more abstract example, let $D = X^G := \{x \in X : \forall g \in G, gx = x\}$ be the subspace of fixed points for the action of G. Assume that the topological G-space E has at least one fixed point x_0, the G-stable subset $E_0 \subset E$ contains no fixed point, the G-equivariant, continuous map $\gamma_0 : E_0 \to X$ is injective and $\Gamma_G((E, \gamma_0), X) \neq \emptyset$. Then, $((E, \gamma_0), D)$ is an equivariant linking pair. Indeed, the conditions imply $\gamma_0(E_0) \subset X \setminus X^G$ and $\gamma(x_0) \in X^G = D$ for all $\gamma \in \Gamma_G((E, \gamma_0), X)$.

Theorem 2. *Assume that*

- $f : X \to \mathbb{R}$ *is a G-invariant locally Lipschitz map*,
- $((E, \gamma_0), D)$ *is an equivariant linking pair in the sense of Definition 2,*
- $a := \sup_{\gamma_0(E_0)} f \leq b := \inf_D f$ *and* $c := \inf_{\gamma \in \Gamma_G((E, \gamma_0), X)} \sup_{x \in E} f(\gamma(x)) \in \mathbb{R}$.

Note that $c \geq b$, due to the linking assumption. Let $S \subset X$ be a G-stable bounded set with

- $\mathrm{dist}(S, \{x \in X : f(x) \geq d\}) > 0$ *for some $d \in (b, c)$ if $c > b$ or*
- $\gamma_0(E_0) \subset S$ *and* $\mathrm{dist}(S, D) > 0$ *if $c = b$,*

and assume that f satisfies the $(\varphi - C)_{X \setminus S, c}$ condition with respect to some G-invariant Lipschitz map $\varphi : X \setminus S \to [1, +\infty)$.

Then, $K_f^c \neq \emptyset$. Moreover, if $c = b$, then $K_f^c \cap \overline{D} \neq \emptyset$.

Remark 3.

(a) The assumption that $c \in \mathbb{R}$ is automatically satisfied, e.g., when E is compact. It is satisfied, also, in the case where $E_0 \subset E \subset X$ and $\gamma_0 = \mathrm{id}_{E_0}$, provided that $\sup_E f < +\infty$.
(b) If $c > b$, the bounded set S can be empty. If $c = b$, we must have $\gamma_0(E_0)$ bounded and $\mathrm{dist}(\gamma_0(E_0), D) > 0$ for applying the theorem, in which case we can take $S = \gamma_0(E_0)$.
(c) Theorem 2 is an equivariant version of Ref. [4, Theorem 12], which is recovered when G is the trivial group and $E \subset X$. Note that in this special case, for having that $c \in \mathbb{R}$, it suffices to have that E_0 is closed and f is majorated on the convex hull of $\gamma_0(E_0)$ (see Remark 2(d)).

Remark 4. Theorem 2 can ensure in general multiplicity of critical points. Indeed, as pointed out in Lemma 1, the subset K_f^c is G-stable, so if $x \in K_f^c$, then the whole orbit $Gx = \{gx : g \in G\}$ consists of critical points. We can guarantee that there are several points in this orbit and thus several critical points if, for instance, 0 is the only fixed point of G in X and $c \neq f(0)$.

Proof of Theorem 2. (First case: $c > b$): Arguing by contradiction, assume that $K_f^c = \emptyset$. Taking $\varepsilon_0 \in (0, c - d)$, $U = \{x \in X : f(x) < d\}$ and $\theta = 1$, there are $\varepsilon \in (0, \varepsilon_0)$ and a continuous map $h : [0, 1] \times X \to X$ satisfying conditions (a)–(g) of Theorem 1. Using the definition of c, we find $\gamma \in \Gamma_G((E, \gamma_0), X)$ with $\sup_E f \circ \gamma \leq c + \varepsilon$ and then, we define the continuous map

$$\gamma_1 := h(1, \gamma(\cdot)) : E \to X.$$

Invoking Theorem 1(g), we have that γ_1 is G-equivariant. Since $\sup_{E_0} f \circ \gamma \leq a < c - \varepsilon_0$, Theorem 1(e) yields $\gamma_1|_{E_0} = \gamma_0$, thus $\gamma_1 \in \Gamma_G((E, \gamma_0), X)$. Now, due to Theorem 1(f), we have $\sup_E f \circ \gamma_1 \leq c - \varepsilon$, which contradicts the definition of c.

(Second case: $c = b$): Arguing by contradiction, we assume that $K_f^c \cap \overline{D} = \emptyset$. This time we apply Theorem 1 with $\varepsilon_0 \in (0, c)$, $U := X \setminus \overline{D}$ and $\theta = 1$, which yields $\varepsilon \in (0, \varepsilon_0)$ and a continuous map $h : [0,1] \times X \to X$ satisfying conditions (a)–(g). Choose again $\gamma \in \Gamma_G((E, \gamma_0), X)$ such that $\sup_E f \circ \gamma \leq c+\varepsilon$ and define $\gamma_1 := h(1, \gamma(\cdot))$. Again γ_1 is continuous, G-equivariant, and we have $\gamma_1|_{E_0} = \gamma_0$ due to Theorem 1(b) and the fact that $\gamma(E_0) \subset S$. Finally, Theorem 1(f) shows that $\gamma_1(E) \subset f^{c-\varepsilon} \cup U$ which implies that $\gamma_1(E) \cap D = \emptyset$, since $c - \varepsilon < b = \inf_D f$ and $U = X \setminus \overline{D}$. We have reached a contradiction with the assumption that $((E, \gamma_0), D)$ is an equivariant linking pair. The proof of the theorem is complete. □

5. A Nonsmooth Version of Fountain Theorem

Throughout this section, G is a compact topological group with isometric representation on the Banach space X. We recall the notion of admissible representation in the sense of Ref. [1].

Definition 3. We say that a finite-dimensional (real) representation V of G is admissible if, for every $k \geq 2$, considering the diagonal action of G on V^k, we have that for every G-stable open, bounded neighborhood $U \subset V^k$ of 0 and every G-equivariant continuous map $\phi : \partial U \to V^{k-1}$, there is $x \in \partial U$ with $\phi(x) = 0$.

In Section 6, we give properties and examples of admissible representations. We now state our main result, which extends the fountain theorem of Ref. [1].

Theorem 3. *Assume that*

- $X = \overline{Y \oplus \bigoplus_{k \geq 1} X_k}$, *where* Y, X_k *(for* $k \geq 1$*) are G-stable subspaces of X,*
- *for every $k \geq 1$, X_k is isomorphic to V as a representation of G, where V is an admissible (finite-dimensional) representation of G,*
- *$f : X \to \mathbb{R}$ is a G-invariant locally Lipschitz map which satisfies the $(\varphi - \mathrm{C})_{X \setminus S, c}$ condition for all $c \in (c_0, +\infty)$, where $c_0 > 0$, with respect to a G-stable bounded subset $S \subset X$ with $\mathrm{dist}(S, \{x \in X : f(x) \geq c_0\}) > 0$ and a G-invariant Lipschitz map $\varphi : X \setminus S \to [1, +\infty)$,*

- for every $\ell \geq 1$, there exist $r_\ell, R_\ell \in (0, +\infty)$ with $r_\ell < R_\ell$ such that

$$\sup_{\ell \geq 1} \max\{f(x) : x \in \bigoplus_{k=1}^{\ell} X_k, \ \|x\| = R_\ell\} < +\infty, \qquad (5)$$

$$\sup_{\ell \geq 1} \inf\{f(x) : x \in \overline{Y \oplus \bigoplus_{k \geq \ell} X_k}, \ \|x\| = r_\ell\} = +\infty. \qquad (6)$$

Then, f has an unbounded sequence of critical values.

We emphasize that a significant difference with the result in Ref. [1, Theorem 2.5] is that we allow X to contain a subrepresentation Y which is possibly non admissible (see also Example 2).

Proof. For $\ell \geq 1$, we denote

$$E_\ell = \left\{ x \in \bigoplus_{k=1}^{\ell} X_k : \|x\| \leq R_\ell \right\} \supset E_{\ell,0} = \left\{ x \in \bigoplus_{k=1}^{\ell} X_k : \|x\| = R_\ell \right\},$$

$$D_\ell = \left\{ x \in \overline{Y \oplus \bigoplus_{k \geq \ell} X_k} : \|x\| = r_\ell \right\}.$$

Note that these are G-stable subsets of X, moreover E_ℓ and $E_{\ell,0}$ are compact.

Claim 1. *For every $\ell \geq 1$, the pair $((E_\ell, \mathrm{id}_{E_{\ell,0}}), D_\ell)$ is an equivariant linking pair.*

Proof of Claim 1. Clearly, $E_{\ell,0} \cap D_\ell = \emptyset$ and we have that $\Gamma_G((E_\ell, \mathrm{id}_{E_{\ell,0}}), X)$ is nonempty since it contains id_{E_ℓ}. Let $\gamma \in \Gamma_G((E_\ell, \mathrm{id}_{E_{\ell,0}}), X)$. We have to show that $\gamma(E_\ell) \cap D_\ell \neq \emptyset$.

Let $U = \{x \in E_\ell : \|\gamma(x)\| < r_\ell\}$. Since $\|\gamma(x)\| = \|x\| = R_\ell > r_\ell$ for all $x \in E_{\ell,0}$, we have $U \subset E_\ell \setminus E_{\ell,0}$, which implies that U is an open, bounded subset of $\bigoplus_{k=1}^{\ell} X_k \cong V^\ell$ whose boundary (within this subspace) satisfies $\partial U \subset \{x \in E_\ell : \|\gamma(x)\| = r_\ell\}$. Moreover, U is G-stable (because E_ℓ is G-stable and γ is G-equivariant) and it contains 0 because $0 \in E_\ell$ and $\gamma(0) = 0$. The latter fact can be justified as follows: since γ is G-equivariant and 0 is a fixed point for the action of G, we know that $\gamma(0)$ is also a fixed point, but since V is admissible, its only fixed point is 0 (as it is explained at the beginning of Section 6).

Let $\phi : \partial U \to \bigoplus_{k=1}^{\ell-1} X_k \cong V^{\ell-1}$, $x \mapsto \pi_\ell(\gamma(x))$, where $\pi_\ell : X \to \bigoplus_{k=1}^{\ell-1} X_k$ stands for the linear projection. Again ϕ is G-equivariant and continuous. Since V is admissible, there is $x \in \partial U$ such that $\phi(x) = 0$. So we have $x \in E_\ell$, $\|\gamma(x)\| = r_\ell$ and $\pi_\ell(\gamma(x)) = 0$, that is, $\gamma(x) \in \ker \pi_\ell = \overline{Y \oplus \bigoplus_{k \geq \ell} X_k}$. Altogether we conclude that $\gamma(x) \in \gamma(E_\ell) \cap D_\ell$. The proof of the claim is complete. □

For showing that f has an infinite sequence of critical values, we show the following:

$$\text{for every } \alpha > 0, \text{ there is } c \in (\alpha, +\infty) \text{ such that } K_f^c \neq \emptyset. \quad (7)$$

Given $\alpha > 0$, and by virtue of (5) and (6), we fix $\ell \geq 1$ such that

$$b := \inf_{D_\ell} f > \max\left\{ \max_{E_{\ell,0}} f, c_0, \alpha \right\}.$$

Then we let

$$c = \inf_{\gamma \in \Gamma_G((E_\ell, \mathrm{id}_{E_{\ell,0}}), X)} \sup_{x \in E_\ell} f(\gamma(x)).$$

Since $\Gamma_G((E_\ell, \mathrm{id}_{E_{\ell,0}}), X) \neq \emptyset$ and E_ℓ is compact, we have $c \in \mathbb{R}$, and in fact $c \geq b$ due to Claim 1. By assumption, if $c > b$, we have

$$\mathrm{dist}(S, \{x \in X : f(x) \geq d\}) \geq \mathrm{dist}(S, \{x \in X : f(x) \geq c_0\}) > 0$$

for all $d \in (b, c)$. If $c = b$, we note that $S \cup E_{\ell,0}$ is bounded, G-stable and

$$\mathrm{dist}(S \cup E_{\ell,0}, D_\ell)$$
$$\geq \min\{\mathrm{dist}(S, \{x \in X : f(x) \geq c_0\}), \mathrm{dist}(E_{\ell,0}, D_\ell)\} > 0$$

since $D_\ell \subset \{x \in X : f(x) \geq c_0\}$, $E_{\ell,0} \cap D_\ell = \emptyset$, with $E_{\ell,0}$ compact and D_ℓ closed. In both cases, we can apply Theorem 2, which implies that $K_f^c \neq \emptyset$. Since $c \geq b > \alpha$, we have shown (7), and the proof of the theorem is complete. □

Example 2. Let X be a real Hilbert space, Y a finite-dimensional subspace and $Z = Y^\perp$. We consider the symmetry $s_Y : X \to X$, $x = y + z \mapsto y - z$ ($y \in Y$, $z \in Z$), and the isometric representation $\rho : G = \{-1, 1\} \to O(X)$ corresponding to this symmetry, i.e., $\rho(1) = \mathrm{id}_X$ and $\rho(-1) = s_Y$.

We can choose a Hilbert basis $(e_n)_{n\geq 1}$ of X such that $Y = \mathrm{Span}\{e_1, \ldots, e_d\}$ and $Z = \overline{\mathrm{Span}\{e_n : n \geq d+1\}}$. Then Y is a subrepresentation of X on which G operates trivially, while $Z = \overline{\bigoplus_{k\geq 1} X_k}$, where $X_k := \mathbb{R}e_{d+k}$ is a subrepresentation of X isomorphic to the representation of G on $V = \mathbb{R}$ given by $\pm 1 \mapsto \pm\mathrm{id}_{\mathbb{R}}$. The latter representation being admissible (see Section 6), Theorem 3 can be applied to the present situation.

Note that Ref. [1, Theorem 2.5] cannot be applied to the present example because X is not the closure of the sum of infinitely many copies of the same admissible representation of G. We even have that X contains the subrepresentation Y, which is not admissible (see Section 6).

In the following section, we give examples of admissible representations, which suggest more examples where Theorem 3 could be applied.

6. Admissible Representations

In this section, we give properties and examples related to the notion of admissible representation introduced in Ref. [1] and considered in Definition 3 and Theorem 3. Throughout this section, by "representation", we always mean a real, finite-dimensional representation.

As a first basic example of admissible representation, one has the representation of the group $\{-1, 1\}$ on $V = \mathbb{R}$ given by $\rho : \pm 1 \mapsto \pm\mathrm{id}_{\mathbb{R}}$: admissibility is implied by the Borsuk–Ulam theorem (see, e.g., Ref. [8, Theorem 4.15]).

When V is a representation of G, let $V^G := \{x \in V : \forall g \in G, gx = x\}$ denote the subspace of fixed points. We say that the representation is fixed-point-free if $V^G = 0$. If V is not fixed-point-free, i.e., V^G contains $x_0 \neq 0$, then it cannot be admissible because $\phi : V^2 \to V$, $(x, y) \mapsto x_0$ is continuous, G-equivariant and never vanishes. For instance, the trivial representation of $\{-1, 1\}$ on $\mathbb{R}$ is not admissible.

More generally, for $G = \mathbb{Z}/p\mathbb{Z}$ with p a prime number, a representation of G is admissible if and only if it is fixed-point-free: see Ref. [7].

The following lemma presents facts deduced from the definition. If $H \subset G$ is a subgroup and V is a representation of G, we denote by $\text{Res}_H(V)$ the representation of H obtained by restriction. If H has finite index in G and W is a representation of H, then we denote by $\text{Ind}_H^G(W)$ the representation of G obtained by induction. We refer to Refs. [12, 13] for the precise definition and a general background on representation theory of finite and compact groups.

Lemma 4. *Let G be a compact topological group and V a representation of G. Let $H \subset G$ be a subgroup.*

(a) *If V is admissible, then V^k is admissible for all $k \geq 1$.*
(b) *If $\text{Res}_H(V)$ is admissible for H, then V is admissible for G.*
(c) *If H has finite index in G and $V = \text{Ind}_H^G(W)$ for a representation W of H which is not admissible, then V is not admissible.*

Proof.

(a) If $\ell \geq 2$, $U \subset V^{k\ell}$ is a G-stable, bounded, open neighborhood of 0, and $\phi : \partial U \to V^{k(\ell-1)}$ is continuous and G-equivariant, then composing with the (G-equivariant) embedding $V^{k(\ell-1)} \subset V^{k\ell-1}$, we deduce that ϕ must have a zero by virtue of the admissibility of V.

(b) If $U \subset V^k$ is a G-stable, bounded, open neighborhood of 0 and $\phi : \partial U \to V^{k-1}$ is G-equivariant and continuous, then *a fortiori* U is H-stable and ϕ is H-equivariant. If we know that $\text{Res}_H(V)$ is admissible, then we conclude that ϕ must have a zero.

(c) By assumption, there exist $k \geq 2$, an H-stable, bounded, open subset $U \subset W^k$ containing 0 and a continuous, H-equivariant map $\phi : \partial U \to W^{k-1}$ such that $\phi(x) \neq 0$ for all $x \in \partial U$.

By definition of an induced representation (see Ref. [12, Section 3.3]), there is a decomposition

$$V = g_1 W \oplus g_2 W \oplus \cdots \oplus g_n W,$$

where n is the index of H in G and $g_1 = 1, g_2, \ldots, g_n \in G$ are representatives of the cosets in G/H. It follows that $V^k = \bigoplus_{i=1}^n g_i W^k$. Writing any element $x \in V^k$ in the form $x = g_1 x_1 + \cdots + g_n x_n$ with $x_i \in W^k$ ($i = 1, \ldots, n$), we define the open, bounded subset $U' = \{x \in V^k : x_1, \ldots, x_n \in U\}$ and the continuous map

$$\phi' : \partial U' \to V^{k-1} = \bigoplus_{i=1}^{n} g_i W^{k-1}, \quad x \mapsto \sum_{i=1}^{n} g_i \phi(x_i).$$

It is straightforward to check that U' is G-stable and ϕ' is G-equivariant. Moreover, for $x \in \partial U'$, we have $x_i \in \partial U$, thus $\phi(x_i) \neq 0$, for all $i \in \{1, \ldots, n\}$. We conclude that ϕ' has no zero. Thus, V is not admissible. □

Remark 5. Lemma 4(c) is stated (without proof) in Ref. [1, Remark 4.4], in the case of finite groups.

Admissible representations have the following purely algebraic characterization, given in Ref. [1, Theorem 2.4 and Remark 4.6]:

Proposition 1. *Let G be a compact Lie group and let V be a representation of G. The following conditions are equivalent:*

(i) *V is admissible;*
(ii) *there exist a subgroup $H \subset G$ and a normal subgroup K of H of finite index such that H/K is solvable, $V^H = 0$ and $V^K \neq 0$.*

In particular, the proposition implies that every fixed-point-free representation of a finite solvable group is admissible. This includes all abelian groups, dihedral groups and the symmetric groups $\mathfrak{S}_n$ for $n \leq 4$, for instance.

Also, as a consequence of Proposition 1, we get the following property, stronger than Lemma 4(a), which expresses that being admissible is a multiplicity-free property:

Proposition 2. *Let $V_1, \ldots, V_m$ be representations of a compact Lie group G and let $(\alpha_1, \ldots, \alpha_m)$ be a sequence of positive integers. Then, $V_1 \oplus \cdots \oplus V_m$ is admissible if and only if $V_1^{\alpha_1} \oplus \cdots \oplus V_m^{\alpha_m}$ is admissible.*

Proof. For every subgroup $H \subset G$, we have $(V_1^{\alpha_1} \oplus \cdots \oplus V_m^{\alpha_m})^H = (V_1^H)^{\alpha_1} \oplus \cdots \oplus (V_m^H)^{\alpha_m}$. Thus, the equality $(V_1^{\alpha_1} \oplus \cdots \oplus V_m^{\alpha_m})^H = 0$ is equivalent to the condition that $V_i^H = 0$ for all $i \in \{1, \ldots, m\}$. The result follows from Proposition 1 and this observation. □

From now on, we suppose that G is a finite group. It admits a finite collection $\hat{G} = \{V_1, \ldots, V_m\}$ of pairwise nonisomorphic irreducible representations, such that every representation V decomposes as

$$V \cong V_1^{\alpha_1} \oplus \cdots \oplus V_m^{\alpha_m} \quad (G\text{-equivariant isomorphism})$$

for a unique sequence of nonnegative integers $(\alpha_1, \ldots, \alpha_m)$, called the multiplicities of the irreducible representations V_i in V. Proposition 2 shows that the admissibility of V only depends on the set $\{i : \alpha_i \neq 0\}$, i.e., on the irreducible representations which actually occur in V.

Example 3. As an illustration, we characterize the admissible representations of the symmetric group $\mathfrak{S}_5$. We refer to Ref. [5] for the necessary background on the structure and the representation theory of symmetric groups.

The group $\mathfrak{S}_5$ has seven irreducible representations parametrized by the partitions of 5, i.e., sequences of positive integers $\lambda = (\lambda_1 \geq \cdots \geq \lambda_k)$ with $\lambda_1 + \cdots + \lambda_k = 5$:

$$V_1 = V_{(5)}, \ V_2 = V_{(4,1)}, \ V_3 = V_{(3,2)}, \ V_4 = V_{(3,1^2)},$$
$$V_5 = V_{(2^2,1)}, \ V_6 = V_{(2,1^3)}, \ V_7 = V_{(1^5)}$$

(where a^ℓ means a repeated ℓ times). The representations $V_1 = V_{(5)}$ and $V_7 = V_{(1^5)}$ are of dimension one: $V_{(5)}$ is the trivial representation (every $\sigma \in \mathfrak{S}_5$ acts on $V_{(5)}$ as the identity), while $V_{(1^5)}$ is the signature representation: every σ acts on $V_{(1^5)}$ by multiplication by the signature $\varepsilon(\sigma) \in \{-1, 1\}$. Our argument now displays into several steps:

(a) A representation V is fixed-point-free if and only if $V_{(5)}$ does not occur in V, i.e., the multiplicity of $V_{(5)}$ in V is zero. Only such representations can be admissible.

(b) The group $\mathfrak{S}_5$ has only one nontrivial, proper, normal subgroup: the alternating group $\mathfrak{A}_5 = \ker \varepsilon$. By definition, $V_{(1^5)}^{\mathfrak{A}_5} = V_{(1^5)} \neq 0$. Moreover, $\mathfrak{S}_5/\mathfrak{A}_5$ is solvable, in fact isomorphic to $\mathbb{Z}/2\mathbb{Z}$. In view of Proposition 1, we deduce that every representation V of G where $V_{(1^5)}$ occurs and $V_{(5)}$ does not occur is admissible.

(c) Now we turn our attention to representations

$$V = V_2^{\alpha_2} \oplus \cdots \oplus V_6^{\alpha_6}$$

in which the trivial representation $V_1 = V_{(5)}$ and the signature representation $V_7 = V_{(1^5)}$ do not occur. Since $V_{(5)}$ and $V_{(1^5)}$ are the only irreducible representations with fixed points for $\mathfrak{A}_5$, we have $V^{\mathfrak{A}_5} = 0$. Note also that $\mathfrak{A}_5$ is simple (its only proper normal subgroup is trivial) and is not solvable. It then follows from Proposition 1 that V is admissible if and only if there is a nontrivial subgroup $H \subset \mathfrak{S}_5$ other than $\mathfrak{S}_5$ and $\mathfrak{A}_5$ such that $\mathrm{Res}_H(V)$ is admissible (as a representation of H).

Apart from $\mathfrak{S}_5$ and $\mathfrak{A}_5$, the group $\mathfrak{S}_5$ has 16 nontrivial subgroups up to conjugacy and each such subgroup H is solvable; see Ref. [11]. Thus, $\mathrm{Res}_H(V)$ is admissible if and only if $V^H = 0$.

One of the subgroups is $H_0 := \{g \in \mathfrak{S}_5 : g(5) = 5\} \cong \mathfrak{S}_4$. It follows from the branching rule in Ref. [5, Theorem 9.2] that $V^{H_0} = 0$ if and only if $\alpha_2 = 0$. This means that V is admissible whenever $\alpha_2 = 0$.

It remains to consider the case where $\alpha_2 \geq 1$ and subgroups H which are not conjugate to a subgroup of H_0. There are exactly six such subgroups H explicitly listed in Ref. [11, Table 3]. The condition $V^H = 0$ holds if and only if $V_i^H = 0$ for all $i \in \{2, \ldots, 6\}$ with $\alpha_i \geq 1$. Moreover, if χ_i denotes the character of the irreducible representation V_i of $\mathfrak{S}_5$, then the equality $V_i^H = 0$ holds if and only if $\sum_{h \in H} \chi_i(h) = 0$ (see Ref. [12, Section 2.3]). Relying on these observations and using the table of characters of $\mathfrak{S}_5$ in Ref. [5, Section 6.3], we obtain that $V = V_2^{\alpha_2} \oplus \cdots \oplus V_6^{\alpha_6}$ is admissible if and only if $\alpha_5 = 0$ (in the case $\alpha_2 \geq 1$).

Altogether, we finally conclude the following:

$$V_1^{\alpha_1} \oplus \cdots \oplus V_7^{\alpha_7} \text{ is admissible} \Leftrightarrow \alpha_1 = 0$$

and

$$(\alpha_7 \neq 0 \text{ or } \alpha_2 = 0 \text{ or } \alpha_5 = 0).$$

References

[1] T. Bartsch, Infinitely many solutions of a symmetric Dirichlet problem, *Nonlinear Anal.* **20**(10), 1205–1216 (1993).

[2] G. Dai, Nonsmooth version of Fountain theorem and its application to a Dirichlet-type differential inclusion problem, *Nonlinear Anal.* **72**(3–4), 1454–1461 (2010).

[3] J. Dugundji, *Topology* (Allyn and Bacon Inc., Boston, 1966).

[4] L. Fresse and V. V. Motreanu, Localized Cerami condition and a deformation theorem. In: *Analysis, Geometry, Nonlinear Optimization and Applications*, pp. 405–416 (World Scientific Publishing Co. Pte. Ltd., Hackensack, New Jersey, 2023).
[5] G. D. James, *The Representation Theory of the Symmetric Groups*. Lecture Notes in Mathematics, Vol. 682 (Springer, Berlin, 1978).
[6] A. Kristály, V. V. Motreanu and C. Varga, A minimax principle with a general Palais-Smale condition, *Commun. Appl. Anal.* **9**(2), 285–297 (2005).
[7] R. Michalek, A $\mathbb{Z}^p$ Borsuk–Ulam theorem and index theory with a multiplicity result in partial differential equations, *Nonlinear Anal.* **13**(8), 957–968 (1989).
[8] D. Motreanu, V. V. Motreanu and N. Papageorgiou, *Topological and Variational Methods with Applications to Nonlinear Boundary Value Problems* (Springer, New York, 2014).
[9] D. Motreanu and P. D. Panagiotopoulos, *Minimax Theorems and Qualitative Properties of the Solutions of Hemivariational Inequalities* (Kluwer Academic Publishers, Dordrecht, 1999).
[10] P. H. Rabinowitz, *Minimax Methods in Critical Point Theory with Applications to Differential Equations* (American Mathematical Society, Providence, 1986).
[11] D. Samaila, Counting the subgroups of the one-headed group S_5 up to automorphism, *IOSR J. Math.* **8**(3), 87–93 (2013).
[12] J-P. Serre, *Linear Representations of Finite Groups*. Graduate Texts in Mathematics, Vol. 42 (Springer-Verlag, New York, 1977).
[13] E. B. Vinberg, *Linear Representations of Groups*. Modern Birkhäuser Classics (Birkhäuser/Springer, New York, 2010).

© 2024 World Scientific Publishing Company
https://doi.org/10.1142/9789811267048_0010

Chapter 10

On a Prey–Mesopredator–Predator System

D. Goeleven[*] and R. Oujja[†]

*University of La Réunion,
Saint-Denis, La Réunion, France*
[*]*daniel.goeleven@univ-reunion.fr*
[†]*rachid.oujja@univ-reunion.fr*

The aim of this chapter is to model the interactions in some island environment between a seabird (the prey), a rodent (the mesopredator) and a raptor (the predator). The objective is also to take into account the massive arrival and departure of another seabird S which only frequents the island during some part of the year to reproduce and lay eggs and which thus temporarily affects the interactions between the resident species.

1. Introduction

Prey–predator systems are an essential aspect of ecology where two or more species interact with each other in a food chain or web. These systems are a dynamic and complex balance between the population of predators and their prey. Predators are organisms that feed on other organisms, while prey are the organisms that are consumed by predators. The relationship between predators and prey is vital for the survival of the entire ecosystem. If the population of predators increases, the population of prey decreases. Similarly, if the population of prey increases, the population of predators will increase as well.

The study of prey–predator systems is crucial for understanding the behavior of populations in an ecosystem. It helps in predicting the changes that might occur due to environmental factors or human intervention. The complexity of these systems makes it challenging to study them, but with the advancement in technology and mathematical modeling, researchers have been able to gain a better understanding of these systems.

One of the most well-known examples of a prey–predator system is the relationship between wolves and deer. Wolves are predators that feed on deer, and the population of wolves is dependent on the availability of deer. If the population of deer decreases, the population of wolves will decrease as well. Similarly, if the population of deer increases, the population of wolves will increase. This balance is crucial for the survival of both species and the ecosystem as a whole. Prey–predator systems play an essential role in maintaining the balance of an ecosystem. The study of these systems is crucial for understanding the behavior of populations and predicting changes that might occur. It is essential to conserve and protect these systems to ensure the survival of all species and maintain a healthy ecosystem.

Prey–mesopredator–predator systems are a more complex version of the traditional prey–predator system. In this system, there is a third component known as the mesopredator, which is a predator that feeds on smaller prey. The mesopredator is then preyed upon by a larger predator, which is at the top of the food chain.

The relationship between the three components of this system is complex and dynamic. The mesopredator is dependent on the smaller prey for its survival, and the larger predator is dependent on the mesopredator for its food source. The smaller prey is also indirectly affected by the larger predator, as the presence of the larger predator can affect the behavior and distribution of the smaller prey.

One well-known example of a prey–mesopredator–predator system is the relationship between hawks, snakes and lizards. In this system, hawks are the top predator, while snakes are the mesopredator, and lizards are the smaller prey. The hawks prey on the snakes, and the snakes prey on the lizards.

The dynamics of this system are influenced by factors, such as habitat, climate and human activities. For example, human activities such as hunting, deforestation and development can alter the habitat and distribution of the species, affecting their survival and the balance of the system.

Understanding the dynamics of prey–mesopredator–predator systems is crucial for conservation and management of ecosystems. It helps in predicting the impact of changes in the environment or human activities on the populations of the species and the ecosystem as a whole. It also highlights the importance of maintaining a balance between the populations of different species to ensure the survival of the ecosystem and all its components.

Prey–predator models and prey–mesopredator–predator models have been the subject of numerous publications (see, e.g., Refs. [3–8, 10, 11, 13]). They are also often presented in mathematics courses as applications of systems of differential equations (see, e.g., Refs. [1, 2]). The study of prey–predator models provides indeed natural connections across mathematics and biology. It is therefore a subject at the interface between two disciplines which allow us to develop effective teaching methods and strategies to improve students' learning outcomes.

The aim of this chapter is to model the interactions in some island environment between three different species: a seabird (like a stern) of the species $\mathcal{B}$, a rodent (like a rat) of the species $\mathcal{R}$ and a raptor (like an owl) of the species $\mathcal{C}$. The seabird is a prey, the rodent is a mesopredator and the raptor is a predator. The objective is also to take into account the massive arrival and departure of another seabird S which only frequents the island during some part of the year to reproduce and lay eggs and which thus temporarily affects the interactions between the resident species. We provide in this chapter a model that brings some original elements but above all develops an educational approach with a detailed presentation of the different stages of modeling. We use a compartment model to describe the dynamics of prey–mesopredator–predator systems. The compartments represent the populations of the different species involved in the system. Our main model divides the populations into different compartments based on their status and tracks the movement of individuals between these compartments over time. The model uses a set of differential equations to describe the flow of individuals between the compartments. The rate of movement between compartments is determined by factors, such as the birth rate, death rate and predation rate. Such a compartment model is useful for understanding the dynamics of prey–mesopredator–predator systems and predicting their behavior under different conditions. It can be used to evaluate the effectiveness of different management strategies like a rodent control program.

2. Preliminaries

The principle of the compartmental model consists of representing the populations of the different species. Let us here first represent the population of seabirds $\mathcal{B}$ by a compartment. We then draw up an input–output balance in $\mathcal{B}$ in this compartment at time t ($t \geq 0$). In this chapter, we take the year as the unit of time. Newborns are entrants. The new dead are leavers. For the compartment of Figure 1, and denoting by $B(t)$ the size of the population in $\mathcal{B}$ at time t, we can write

$$B'(t) = f_B B(t) - m_B B(t),$$

where f_B is the birth rate of seabird $\mathcal{B}$ and m_B is the natural death rate of seabird $\mathcal{B}$. Natural mortality refers to these natural causes

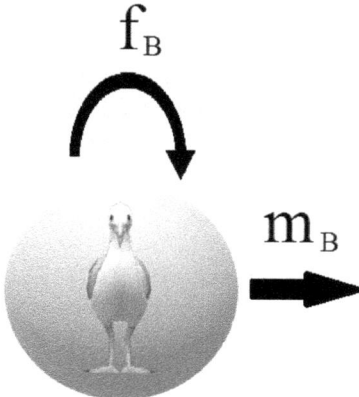

Figure 1. One seabird — One compartment.

other than predation, such as diseases, parasites and pathologies linked to aging. The terms $B'(t)$ and $f_B B(t) - m_B B(t)$ are two different mathematical expressions that measure the instantaneous rate of change in seabirds $\mathcal{B}$ at time t. The model is completed with an initial condition

$$B(0) = B_0,$$

where $B_0 \geq 0$ is the size of the initial population (at time $t = 0$).

We have thus

$$B'(t) = (f_B - m_B) B(t). \tag{1}$$

Note that the term $r_B = f_B - m_B$ is the reproduction rate of the seabird $\mathcal{B}$.

2.1. The model of Malthus

It is assumed here that the seabird $\mathcal{B}$ lives in paradisiacal conditions: it has unlimited resources to achieve its optimal rate of reproduction, and this, in the absence of predators. We assume constant birth and natural death rates, i.e., $f_B = f_B^* > 0$ and $m_B = m_B^* > 0$. We set $r_B = r_B^*$, where $r_B^* = f_B^* - m_B^*$ is the optimal reproduction rate of the seabird $\mathcal{B}$. If $B_0 \geq 0$ is the initial seabird population $\mathcal{B}$, then the solution of the differential equation (1) which satisfies the initial condition $B(0) = B_0$ is given by

$$B(t) = B_0 \, e^{r_B^* t}.$$

This simple model refers to the Malthusian model, also known as the exponential model. It was developed by Thomas Robert Malthus [9], an English economist and demographer. The original model describes the relationship between population growth and resource availability. This model is taught in undergraduate biology courses.

Remark 2.1.1. We have in particular

$$X(t+1) = e^{r_B^*} X(t).$$

It results that the size of the population increases by a factor $e^{r_B^*}$ from one year to the next one (See Figure 2).

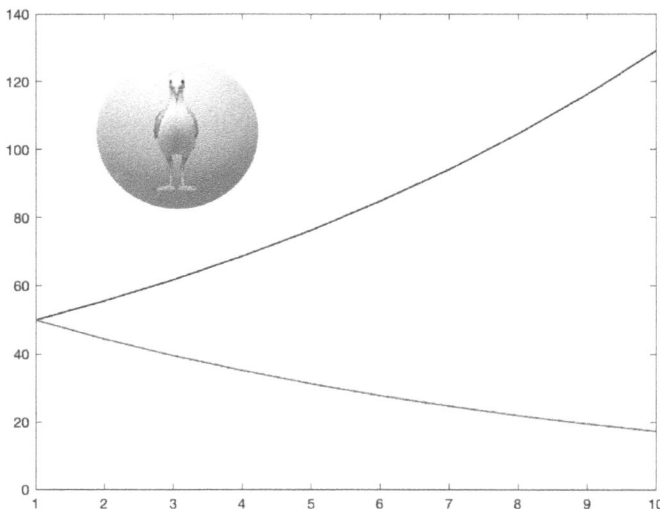

Figure 2. $B(0) = 50, f_B^* = 0.5$. Increasing curve: $m_B^* = 0.4 < f_B^*$/Decreasing curve: $m_B^* = 0.6 > f_B^e$.

2.2. Verhulst model with a constant rate predation

Under real conditions, resources are limited. It is here assumed that resource limitation affects the birth rate and natural mortality of the seabird $\mathcal{B}$. The reproduction rate of the seabird $\mathcal{B}$ is then represented by a Verhulst model:

$$r_B = r_B^* \left(1 - \frac{B}{K_B^*}\right),$$

where $r_B^* > 0$ is the optimal birth rate of seabird $\mathcal{B}$ and $K_B^* > 0$ is the carrying capacity of the island in seabirds $\mathcal{B}$. The parameter K_B^* is equal to the maximum number of seabirds $\mathcal{B}$ that the island can accommodate. The model expresses that the reproduction rate is all the lower as the size of the population approaches the carrying capacity of the island. In fact, the closer the population size approaches the carrying capacity, the less food there is available for each seabird and the fewer nesting sites there are for each pair. In addition, the spread of diseases and parasites is facilitated by overcrowding. The graph of r_B is a straight line passing through the points $(0, r_B^*)$ and $(K_B^*, 0)$. The optimal value of r_B on the interval $[0, K_B^*]$ is equal to r_B^* (See Figure 3).

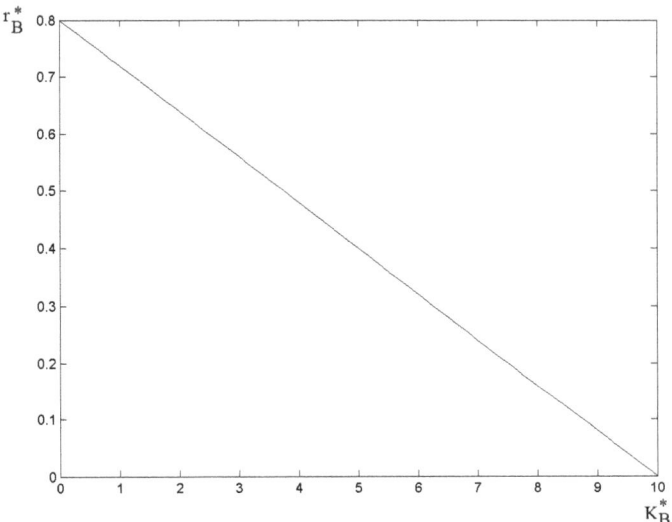

Figure 3. Verhulst model with $r_B^* = 0.8$ and $K_B^* = 10$: $r_B = 0.8(1 - \frac{B}{10})$.

We also assume that the seabird $\mathcal{B}$ is subject to predation by the rodent $\mathcal{R}$. The mortality rate μ_B of the seabird $\mathcal{B}$ by predation by rodents $\mathcal{R}$ is assumed here to be constant, i.e., $\mu_B = \mu_B^* > 0$ with

$$\mu_B^* = \lambda_{R,B}^* R,$$

where $\lambda_{R,B}^* > 0$ is the predation rate of the rodent $\mathcal{R}$ on the seabird population $\mathcal{B}$ (proportion of birds $\mathcal{B}$ killed per rodent $\mathcal{R}$) and R is the number (assumed here constant) of rodents $\mathcal{R}$ roaming in the island. The model can thus be written as

$$B'(t) = r_B^* \left(1 - \frac{B(t)}{K_B^*}\right) B(t) - \lambda_{R,B}^* R B(t). \tag{2}$$

The last model refers to the logistic growth model which was developed by the Belgian mathematician Pierre François Verhulst in the mid-1800s as a modification of the Malthusian model [12].

Remark 1. We have

$$B(t) = B_0 \, e^{\int_0^t \left(r_B^*\left(1 - \frac{B(s)}{K_B^*}\right) - \mu_B^*\right) ds}.$$

And thus
$$B_0 = 0 \implies (\forall t \geq 0) : B(t) = 0,$$
$$B_0 > 0 \implies (\forall t \geq 0) : B(t) > 0.$$

If $r_B^* \neq \mu_B^*$, the solution of the differential equation (2) satisfies the initial condition $B(0) = B_0 > 0$ which is given by

$$B(t) = \frac{1}{e^{-(r_B^* - \mu_B^*)t} \left(\frac{1}{B_0} - \frac{r_B^*}{(r_B^* - \mu_B^*)K_B^*} \right) + \frac{r_B^*}{(r_B^* - \mu_B^*)K_B^*}}. \tag{3}$$

If $r_B^* = \mu_B^*$, then the solution of the differential equation (2) which satisfies $B(0) = B_0 > 0$ is given by

$$B(t) = \frac{1}{\frac{1}{B_0} + \frac{r_B^*}{K_B^*} t}. \tag{4}$$

Remark 2. If $r_B^* > \mu_B^*$, then
$$\lim_{t \to +\infty} B(t) = \frac{r_B^* - \mu_B^*}{r_B^*} K_B^*.$$

If $r_B^* \leq \mu_B^*$, then
$$\lim_{t \to +\infty} B(t) = 0.$$

(See Figure 4)

2.3. Verhulst model with variable rate predation

Let us here consider the model

$$B'(t) = r_B^* \left(1 - \frac{B(t)}{K_B^*} \right) B(t) - \mu_B B(t).$$

The mortality rate of seabird $\mathcal{B}$ by predation of rodents $\mathcal{R}$ is given by

$$\mu_B = \lambda_{R,B} R,$$

where R is the number of rodents $\mathcal{R}$ and $\lambda_{R,B}$ is the annual rodent predation rate on seabirds $\mathcal{B}$, i.e., the proportion of seabirds $\mathcal{B}$ killed

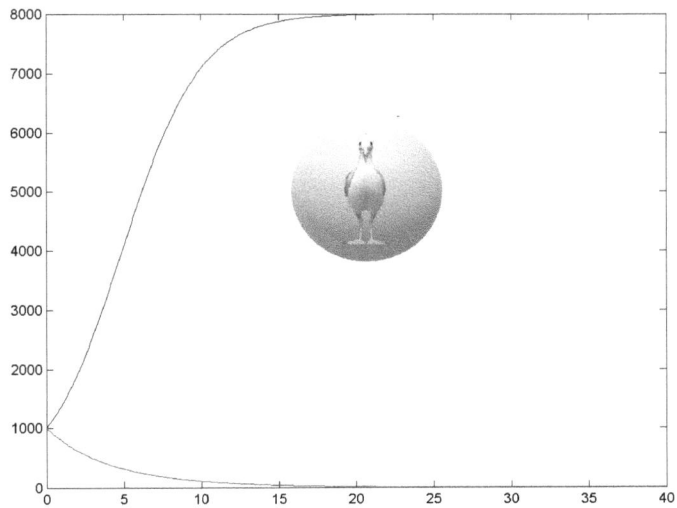

Figure 4. $B(0) = 1000/r_B^* = 0.5, K_B^* = 10000$/Increasing curve: $\mu_B^* = 0.1$/ Decreasing curve: $\mu_B^* = 0.7$.

per rodent per year. The model can thus be written as

$$B'(t) = r_B^* \left(1 - \frac{B(t)}{K_B^*}\right) B(t) - \lambda_{R,B} R B(t).$$

The goal is now to propose a model for $\lambda_{R,B}$. We note that

$$N(B) = \lambda_{R,B} B$$

is the number of seabirds killed per rodent in a population of size B.

Remark 3. The assumption of a constant predation rate $\lambda_{R,B} = \lambda_{R,B}^*$ as in the previous model has certain weaknesses. The predation rate model does not take into account the existence of another food source available to the rodent. And also, the number of seabirds killed per rodent in a population of size B is given by $N(B) = \lambda_{R,B}^* B$. The function $B \to N(B)$ is increasing and $N(B) \to +\infty$ when $B \to +\infty$. This indicates that a rodent kills in proportion to the quantity available. This does however not make biological sense if the quantity available for the year greatly exceeds the energy needs by the rodent.

Remark 4. Let us consider the model

$$\lambda_{R,B}(B) = \frac{\alpha}{A+B},$$

where $A > 0$ and $\alpha > 0$. The parameter A represents the number of units of another food source $\mathcal{A}$ available to the rodent. Note that the latter must be expressed in seabird equivalents! This model makes it possible to correct the two weaknesses of the constant rate predation model. It takes indeed into account the existence of another source of food available to the rodent. Besides,

$$N(B) = \lambda_{R,B} B = \frac{\alpha B}{A+B}$$

and therefore $N(B) \to \alpha$ when $B \to +\infty$. The number of seabirds killed per rodent therefore does not exceed the value α which therefore represents the maximum number of seabirds that a rodent kills in one year. This roughly corresponds to the number of seabirds needed to cover the annual energy needs of a rodent. However, the model has a weakness that the model whose weaknesses it corrects did not have. The function $B \to \lambda_{R,B}(B)$ is decreasing. The highest value of the predation rate corresponds to $B = 0$. It is therefore when seabirds are rare that the predation pressure of rodents on seabirds is the greatest. The predation rate should not be decreasing for $B < A$ because the rodent is a generalist predator. When birds are rare, they are scattered and difficult to spot. The generalist predator therefore concentrates on the most abundant resource. On the contrary, the model indicates that the rodent concentrates on the less abundant resource.

Considering the bad properties of the models discussed in the previous remarks, we consider another model:

$$\lambda_{R,B}(B) = \frac{\eta B}{A^2 + B^2},$$

where $\eta > 0$ and $A > 0$ are parameters. The coefficient A is the number of units of another food source $\mathcal{A}$ available for the rodent $\mathcal{R}$. The coefficient A is expressed in seabird equivalents. The model takes thus into account the existence of another source of food. The model expresses that the more the units of $\mathcal{A}$ available for the rodent $\mathcal{R}$, the lower the predation of the rodent on the seabird $\mathcal{B}$. And conversely, the fewer the units of $\mathcal{A}$ available for the rodent $\mathcal{R}$, the greater the predation of the rodent $\mathcal{R}$ on the seabird $\mathcal{B}$.

We have (See Figure 5)

$$N(B) = \lambda_{R,B}(B) B = \frac{\eta B^2}{A^2 + B^2}.$$

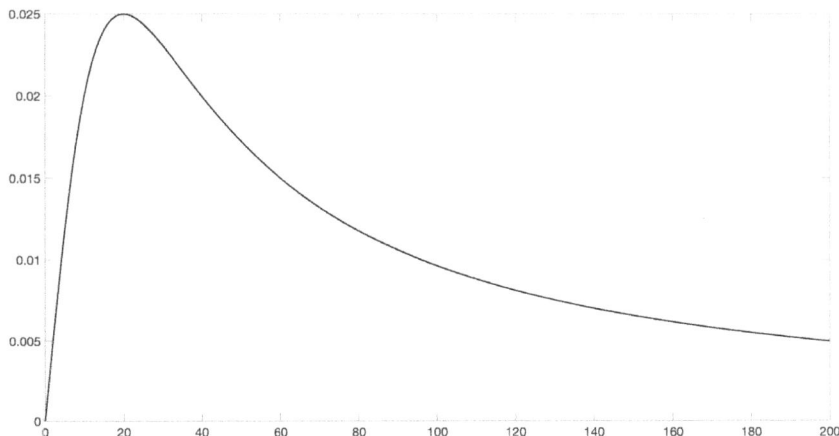

Figure 5. Model for rodent predation rate on seabirds with $A = 20$ and $\eta = 1$.

And thus $N(B) \to \eta$ when $B \to +\infty$. The number of seabirds killed per rodent does not exceed the value η which therefore represents the maximum number of seabirds that a rodent kills in one year to cover its energy needs.

We have $N(A) = \frac{\eta}{2}$. This indicates that when the number of seabirds is equal to the number of seabird equivalents of the other food source $\mathcal{A}$, the number of seabirds killed per rodent is equal to half the maximum number of seabirds a rodent kills in one year. This shows that when $B = A$, the rodent covers its energy needs with seabirds and seabird equivalents in equal parts. This supposes that the rodent has no preference for one or the other of the two resources when they are available in equal parts. Note that it is possible to parameterize the model with a preference coefficient for one of the two resources if necessary.

We have $\lambda_{R,B}(0) = 0$. Predation pressure on the seabird population is zero in the absence of seabirds. The function $\lambda_{R,B}$ is increasing on the interval $[0, A]$. Predation pressure on the seabird population indeed increases with the number of seabirds. In fact, the more they are, the easier it is for the rodent to find one. When $B = A$, there are as many seabirds as seabird equivalents. Seabirds are as easy to find as their seabird equivalents. The total prey population for the rodent is equal to $B + A = 2A$ and the rodent's predation pressure on the seabird population is equal to $\frac{\eta}{2A}$. This is indeed the value given by the model since $\lambda_{R,B}(A) = \frac{\eta}{2A}$. The function $\lambda_{R,B}$ is decreasing on

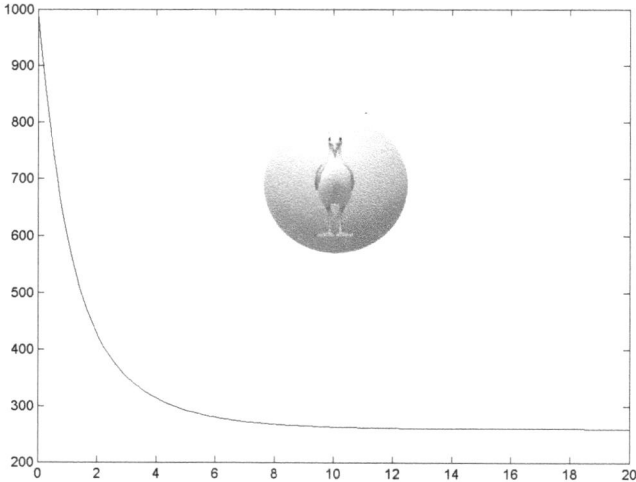

Figure 6. $B(0) = 1000/r_B^* = 0.4, K_B^* = 10000, R = 10, A = 1000, \eta = 200$.

the interval $[A, +\infty[$. When $B > A$, the rodent continues to take the same share of seabirds as when B was equal to A. The predation pressure on the seabird population decreases as the number of seabirds killed by the rodent varies relatively to the total number of seabirds available which increases more and more.

A first simulation of the model is given in Figure 6. We assume for the following simulation that the other food source for the rodent is composed of a resource available all year round and another that is seasonal, i.e., $A(t) = A + V(t)$, where $A > 0$ is the number of units of the resource available all year round (expressed in seabird equivalents) and $V(t) \geq 0$ is the number of units of the resource available in season only (expressed in seabird equivalents). The function V is a periodic function of period 1 (annual) which cancels out of season (see Figure 7).

3. Three-Compartment Model

We are now in position to propose a three-compartment model (Figures 8 and 9). We denote by $C(t)$ the size of the population in predators $\mathcal{C}$ at time t, $R(t)$ the size of the population of rodents $\mathcal{R}$ at time t and $B(t)$ the size of the population of seabirds $\mathcal{B}$ at

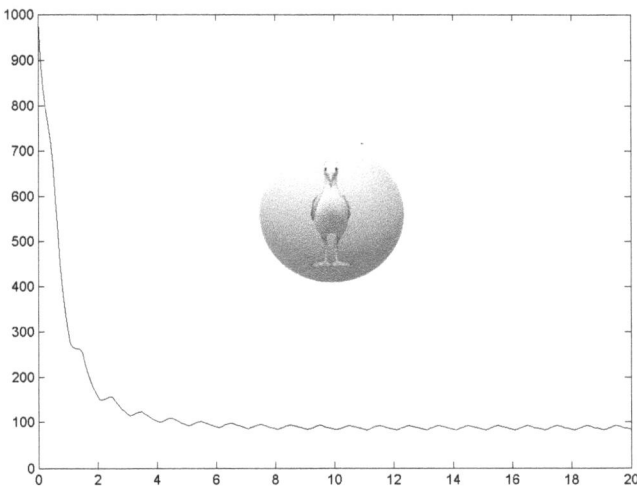

Figure 7. $B(0) = 1000/r_B^* = 0.4, K_B^* = 10000, R = 10, A(t) = 500 + 500 \max\{0, \sin(\pi t)\}, \eta = 200$.

time t. Let us here recall that we take the year as the unit of time. The initial populations of the three species are noted respectively by C_0, R_0 and B_0.

We also take into account the massive arrival and massive departure of a seabird $\mathcal{V}$ which only frequents the island during part of the year to reproduce and lays eggs. We denote by $V(t)$ the size of the population in $\mathcal{V}$ at time t. The function $t \mapsto V(t)$ is periodic with period 1 (annual). And we have $V(t) = 0$ outside the breeding season. Suppose that the bird $\mathcal{V}$ stays on the island during the months $N, N+1, \ldots, N+M$ of the year $(N \in \{1, 2, \ldots, 12\})$, $(M \in \mathbb{N}, M + N \leq 12)$. The function V is constructed by periodic extension of period 1 of the function:

$$V(t) = \begin{cases} V^*(t) & \text{si } t \in]\frac{N}{12}, \frac{N+M}{12}[\\ 0 & \text{si } t \in \left[0, \frac{N}{12}\right] \cup \left[\frac{N+M}{12}, 1\right], \end{cases}$$

where $V^*(t)$ is the number of seabirds $\mathcal{V}$ at time t in the breeding season. The function V^* is continuous and such that $V^*(\frac{N}{12}) = V^*(\frac{N+M}{12}) = 0$.

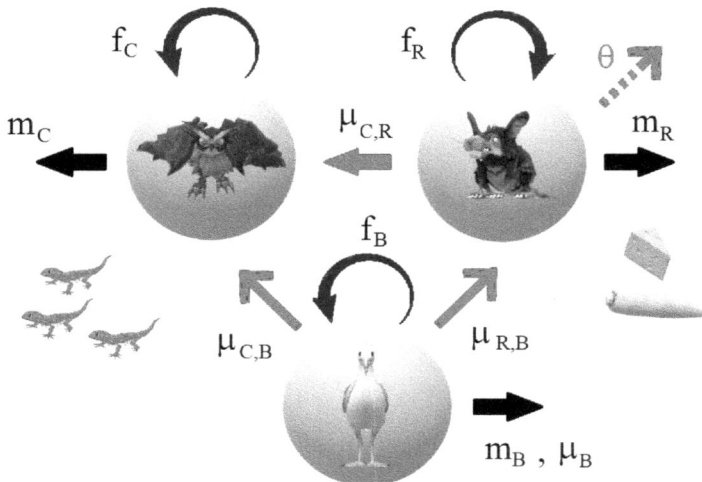

Figure 8. Three-compartment model.

In order to model the interactions between the predator, the mesopredator and the prey, we consider the model of Figure 8 with the following notations:

f_B	Birth rate of the seabird $\mathcal{B}$
m_B	Natural mortality rate of the seabird $\mathcal{B}$
r_B	Reproductive rate of the seabird $\mathcal{B}$ ($r_B = f_B - m_B$)
f_R	Birth rate of the rodent $\mathcal{R}$
m_R	Natural mortality rate of the rodent $\mathcal{R}$
r_R	Reproduction rate of the rodent $\mathcal{R}$: $r_R = f_R - m_R$
f_C	Birth rate of the raptor $\mathcal{C}$
m_C	Natural mortality rate of the raptor $\mathcal{C}$
r_C	Reproduction rate of the raptor $\mathcal{C}$: $r_C = f_C - m_C$
$\mu_{R,B}$	Mortality rate of seabird $\mathcal{B}$ by predation of rodents $\mathcal{R}$
$\mu_{C,B}$	Mortality rate of seabird $\mathcal{B}$ by predation of raptors $\mathcal{C}$
$\mu_{C,R}$	Mortality rate of rodent $\mathcal{R}$ by predation by raptors $\mathcal{C}$
μ_B	Mortality rate of seabird $\mathcal{B}$ by predation at sea
θ	Controlled harvest rate in rodents $\mathcal{R}$

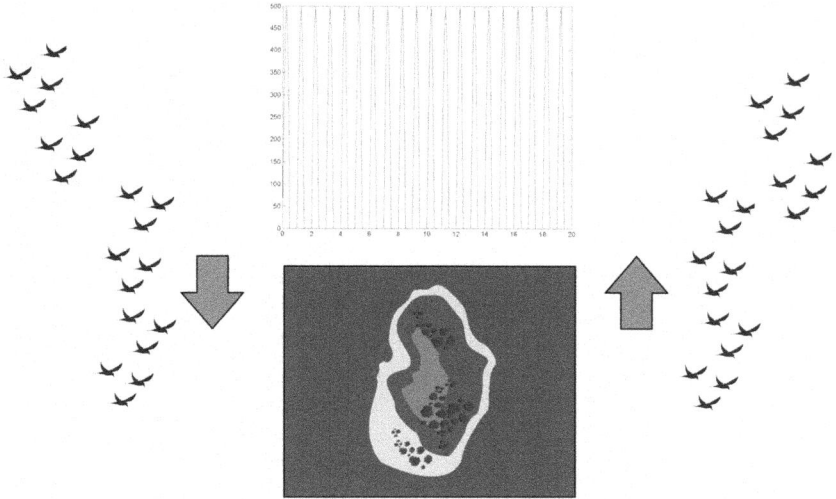

Figure 9. Seasonal arrivals and departures of a seabird of the species $\mathcal{V}$.

We obtain the following dynamical system:
$$\begin{cases} B'(t) = (r_B - \mu_{R,B} - \mu_{C,B} - \mu_B)B(t), \\ R'(t) = (r_R - \mu_{C,R} - \theta)R(t), \\ C'(t) = r_C C(t), \end{cases}$$
with the initial conditions: $B(0) = B_0$, $R(0) = R_0$, $C(0) = C_0$.

The reproduction rate of the seabird $\mathcal{B}$ is represented by the Verhulst model:
$$r_B = r_B^* \left(1 - \frac{B}{K_B^*}\right),$$
where $r_B^* > 0$ is the optimal reproduction rate of the seabird $\mathcal{B}$ and $K_B^* > 0$ is the carrying capacity of the island in seabirds $\mathcal{B}$ (number of seabirds $\mathcal{B}$ that the island can support each year).

The reproductive rate of the rodent is represented by the Verhulst model:
$$r_R = r_R^* \left(1 - \frac{R}{\frac{S_R^*}{\eta_{R,S}^*} + \frac{B}{\eta_{R,B}^*} + \frac{V}{\eta_{R,V}^*}}\right),$$

where $r_R^* > 0$ is the optimal reproduction rate of the rodent. The carrying capacity K_R^* of the island in rodents is here modeled by

$$K_R^* = \frac{S_R^*}{\eta_{R,S}^*} + \frac{B}{\eta_{R,B}^*} + \frac{V}{\eta_{R,V}^*}.$$

We note here by $S_R^* > 0$ the number of units of a food source $\mathcal{S}_R$ available each year for the rodent, excluding seabirds $\mathcal{B}$ and $\mathcal{V}$. The coefficient $\eta_{R,S}^* > 0$ is the number of units of the resource $\mathcal{S}_R$ necessary to cover the annual energy needs of one rodent. The coefficient $\eta_{R,B}^* > 0$ is the number of birds $\mathcal{B}$ necessary to satisfy the annual energy needs of one rodent. And the coefficient $\eta_{R,V}^* > 0$ is the number of birds $\mathcal{V}$ necessary to cover the annual energy needs of one rodent.

Remark 5.

(i) The resource $\mathcal{S}_R$ in quantity S_R^* alone allows feeding each year $\frac{S_R^*}{\eta_{R,S}^*}$ rodents $\mathcal{R}$. The seabird $\mathcal{B}$ in quantity B alone can feed $\frac{B}{\eta_{R,S}^*}$ rodents $\mathcal{R}$. The other seabird $\mathcal{V}$ in quantity V alone can feed $\frac{V}{\eta_{R,V}^*}$ rodents $\mathcal{R}$. The resource $\mathcal{S}_R$ and the two prey items $\mathcal{B}$ and $\mathcal{V}$ can thus cover the energy needs of K_R^* rodents $\mathcal{R}$ with

$$K_R^* = \frac{S_R^*}{\eta_{R,S}^*} + \frac{B}{\eta_{R,B}^*} + \frac{V}{\eta_{R,V}^*}.$$

(ii) Suppose, for example, that the food source $\mathcal{S}_R$ consists of $S_{R,1}^*$ apples and $S_{R,2}^*$ cherries. If η_{R,S_1}^* apples or η_{R,S_2}^* cherries meet the annual energy needs of a rodent, then

$$\frac{S_R^*}{\eta_{R,S}^*} = \frac{S_{R,1}^*}{\eta_{R,S_1}^*} + \frac{S_{R,2}^*}{\eta_{R,S_2}^*}.$$

We can then set

$$S_R^* = \eta_{R,S_2}^* S_{R,1}^* + \eta_{R,S_1}^* S_{R,2}^*$$

and

$$\eta_{R,S}^* = \eta_{R,S_1}^* \eta_{R,S_2}^*.$$

(iii) If a rodent needs Q calories per year and a unit of $\mathcal{S}_R$ provides α calories, then $\eta^*_{R,S} = \frac{Q}{\alpha}$.

(iv) The carrying capacity model can be adapted to constraints other than those related to energy needs. Some rodents like rats live in family groups who aggressively defend a territory. Let us suppose that average group size of related rats is D. If the island offers living space for rats of A square meters and if the territory defended by a family group is a square meters, then the carrying capacity of the island in terms of living space for rats is $K^*_A = \frac{A}{a} D$ rats. Taking into account both energy needs and territorial constraints, we may use the model

$$K^*_R = \min\left\{ K^*_A, \frac{S^*_R}{\eta^*_{R,S}} + \frac{B}{\eta^*_{R,B}} + \frac{V}{\eta^*_{R,V}} \right\}.$$

The raptor's reproductive rate is given by the Verhulst model:

$$r_C = r^*_C \left(1 - \frac{C}{\frac{S^*_C}{\eta^*_{C,S}} + \frac{R}{\eta^*_{C,R}} + \frac{B}{\eta^*_{C,B}} + \frac{V}{\eta^*_{C,V}}} \right),$$

where $r^*_C > 0$ is the raptor's optimal reproduction rate. The carrying capacity K^*_C of the island in raptors is here modeled by

$$K^*_C = \frac{S^*_C}{\eta^*_{C,S}} + \frac{R}{\eta^*_{C,R}} + \frac{B}{\eta^*_{C,B}} + \frac{V}{\eta^*_{C,V}}.$$

We denote by $S^*_C > 0$ the number of units of a food source $\mathcal{S}_C$ available for the raptors, excluding rodents $\mathcal{R}$ and seabirds $\mathcal{B}$ and $\mathcal{V}$. The coefficient $\eta^*_{C,S} > 0$ is the number of units of $\mathcal{S}_C$ necessary to cover the annual energy needs of a raptor $\mathcal{C}$. The coefficient $\eta^*_{C,B} > 0$ denotes the number of seabirds $\mathcal{B}$ necessary to cover the annual energy needs of a raptor $\mathcal{C}$. The coefficient $\eta^*_{C,R} > 0$ is the number of rats $\mathcal{R}$ necessary to cover the annual energy needs of a raptor $\mathcal{C}$. And the coefficient $\eta^*_{C,V} > 0$ denotes the number of seabirds $\mathcal{V}$ needed to cover the annual energy needs of a raptor $\mathcal{C}$.

Remark 6. The carrying capacity model can be adapted to constraints other than those related to energy needs. Raptors form territorial pairs. If the island offers a living area for raptors of Λ square meters and if the territory defended by a pair of raptors is l square meters, then the carrying capacity of the island in terms of living space for raptors is $K_\Lambda^* = 2\frac{\Lambda}{l}$ raptors. We can thus set

$$K_C^* = \min\left\{K_\Lambda^*, \frac{S_C^*}{\eta_{C,S}^*} + \frac{R}{\eta_{C,R}^*} + \frac{B}{\eta_{C,B}^*} + \frac{V}{\eta_{C,V}^*}\right\}.$$

The mortality rate of seabird $\mathcal{B}$ by predation of rodents $\mathcal{R}$ is modeled by the term

$$\mu_{R,B} = \lambda_{R,B}^* \frac{BR}{\left(\frac{\eta_{R,B}^*}{\eta_{R,S}^*}S_R^* + \frac{\eta_{R,B}^*}{\eta_{R,G}^*}G\right)^2 + B^2},$$

where $\lambda_{R,B}^* > 0$ is the maximum number of seabirds $\mathcal{B}$ killed per rodent $\mathcal{R}$ and per year.

Remark 7.

(i) The term $A = \frac{\eta_{R,B}^*}{\eta_{R,S}^*}S_R^* + \frac{\eta_{R,B}^*}{\eta_{R,G}^*}V$ is the total number of nonseabird prey $\mathcal{B}$ available to rodents expressed in seabird equivalents $\mathcal{B}$. It is important to express the different food sources for the rat in bird equivalents $\mathcal{B}$. To fail to do so consists to add apples to birds without considering the energy value of each food resource. If it takes 3 apples or 1 bird to feed a rodent, 1 apple is only worth a third of a bird.

(ii) One unit of the non-avian resource $\mathcal{S}_R$ covers the food needs of $\frac{1}{\eta_{R,S}^*}$ rodent per year. A seabird $\mathcal{B}$ covers the food needs of $\frac{1}{\eta_{R,B}^*}$ rodent per year. So one non-avian resource unit $\mathcal{S}_R$ is equivalent to $\frac{1}{\eta_{R,S}^*}/\frac{1}{\eta_{R,B}^*} = \frac{\eta_{R,B}^*}{\eta_{R,S}^*}$ seabirds $\mathcal{B}$.

(iii) The predation rate of the rodent $\mathcal{R}$ on the seabird population $\mathcal{B}$ is given by

$$\lambda_{R,B} = \lambda_{R,B}^* \frac{B}{\left(\frac{\eta_{R,B}^*}{\eta_{R,S}^*}S_R^* + \frac{\eta_{R,B}^*}{\eta_{R,V}^*}V\right)^2 + B^2}.$$

We have

$$\mu_{R,B} = \lambda_{R,B} R.$$

The mortality rate of seabird $\mathcal{B}$ by predation of raptors $\mathcal{C}$ is modeled by the term

$$\mu_{C,B} = \lambda^*_{C,B} \frac{BC}{\left(\frac{\eta^*_{C,B}}{\eta^*_{C,S}}S^*_C + \frac{\eta^*_{C,B}}{\eta^*_{C,R}}R + \frac{\eta^*_{C,B}}{\eta^*_{C,V}}V\right)^2 + B^2},$$

where $\lambda^*_{C,B} > 0$ is the maximum number of seabirds $\mathcal{B}$ killed per raptor $\mathcal{C}$ and per year.

Remark 8. The predation rate of the raptor $\mathcal{C}$ on the seabird population $\mathcal{B}$ is given by

$$\lambda_{C,B} = \lambda^*_{C,B} \frac{B}{\left(\frac{\eta^*_{C,B}}{\eta^*_{C,S}}S^*_C + \frac{\eta^*_{C,B}}{\eta^*_{C,R}}R + \frac{\eta^*_{C,B}}{\eta^*_{C,V}}V\right)^2 + B^2}.$$

The rate of mortality of the seabird $\mathcal{B}$ by predation at sea is assumed to be constant:

$$\mu_B = \mu^*_B > 0.$$

The mortality rate of the rodent $\mathcal{R}$ by predation of the raptors $\mathcal{C}$ is modeled by the term

$$\mu_{C,R} = \lambda^*_{C,R} \frac{RC}{\left(\frac{\eta^*_{C,R}}{\eta^*_{C,S}}S^*_C + \frac{\eta^*_{C,R}}{\eta^*_{C,B}}B + \frac{\eta^*_{C,R}}{eta^*_{C,V}}V\right)^2 + R^2},$$

where $\lambda^*_{C,R} > 0$ is the maximum number of rodents $\mathcal{R}$ killed per raptor $\mathcal{C}$ and per year.

Remark 9. The raptor–predation rate on rats is given by

$$\lambda_{C,R} = \lambda^*_{C,R} \frac{R}{\left(\frac{\eta^*_{C,R}}{\eta^*_{C,S}}S^*_C + \frac{\eta^*_{C,R}}{\eta^*_{C,B}}B + \frac{\eta^*_{C,R}}{\eta^*_{C,V}}V\right)^2 + R^2}.$$

The controlled harvest rate in rodents $\mathcal{B}$ is assumed to be constant from year a_0:

$$(\forall t \in [0, a_0[) : \theta(t) = 0$$

and

$$(\forall t \geq a_0) : \theta(t) = \theta^*.$$

The model that describes the interactions between the predator, the mesopredator and the prey is therefore given by

$$\begin{cases} B'(t) = r_B^* \left(1 - \dfrac{B(t)}{K_B}\right) B(t) - \dfrac{\lambda_{R,B}^* R(t) B^2(t)}{\left(\dfrac{\eta_{R,B}^*}{\eta_{R,S}^*} S_R^* + \dfrac{\eta_{R,B}^*}{\eta_{R,V}^*} V(t)\right)^2 + B^2(t)} \\ \qquad - \dfrac{\lambda_{C,B}^* C(t) B^2(t)}{\left(\dfrac{\eta_{C,B}^*}{\eta_{C,S}^*} S_C^* + \dfrac{\eta_{C,B}^*}{\eta_{C,R}^*} R(t) + \dfrac{\eta_{C,B}^*}{\eta_{C,V}^*} V(t)\right)^2 + B^2(t)} - \mu_B^* B(t), \\ R'(t) = r_R^* \left(1 - \dfrac{R(t)}{\dfrac{S_R^*}{\eta_{R,S}^*} + \dfrac{B(t)}{\eta_{R,B}^*} + \dfrac{V(t)}{\eta_{R,V}^*}}\right) R(t) \\ \qquad - \lambda_{C,R}^* \dfrac{C(t) R^2(t)}{\left(\dfrac{\eta_{C,R}^*}{\eta_{C,S}^*} S_C^* + \dfrac{\eta_{C,R}^*}{\eta_{C,B}^*} B(t) + \dfrac{\eta_{C,R}^*}{\eta_{C,V}^*} V(t)\right)^2 + R^2(t)} - \theta(t) R(t), \\ C'(t) = r_C^* \left(1 - \dfrac{C(t)}{\dfrac{S_C^*}{\eta_{C,S}^*} + \dfrac{R(t)}{\eta_{C,R}^*} + \dfrac{B(t)}{\eta_{C,B}^*} + \dfrac{V(t)}{\eta_{C,V}^*}}\right) C(t), \end{cases}$$

with the initial conditions $B(0) = B_0$, $R(0) = R_0$ and $C(0) = C_0$.

Example 1. Let us here test the model with the data:

$r_B^* = 0.5$	$K_B^* = 20000$	$r_R^* = 6$	$S_R^* = 100000$	$r_C^* = 0.4$	$S_C^* = 20000$
$\eta_{R,B}^* = 100$	$\eta_{R,V}^* = 100$	$\eta_{C,B}^* = 100$	$\eta_{C,R}^* = 200$	$\eta_{C,V}^* = 100$	$\eta_{R,S}^* = 100$
$\eta_{C,S}^* = 100$	$\mu_B^* = 0.01$	$\lambda_{R,B}^* = 100$	$\lambda_{C,B}^* = 100$	$\lambda_{C,R}^* = 200$	$a_0 = 0, \theta^* = 0$

We set

$$V(t) = 50000 \max\{0, \sin(2\pi t)\}.$$

No control effort on the rodent is made (Figure 10).

On a Prey–Mesopredator–Predator System 317

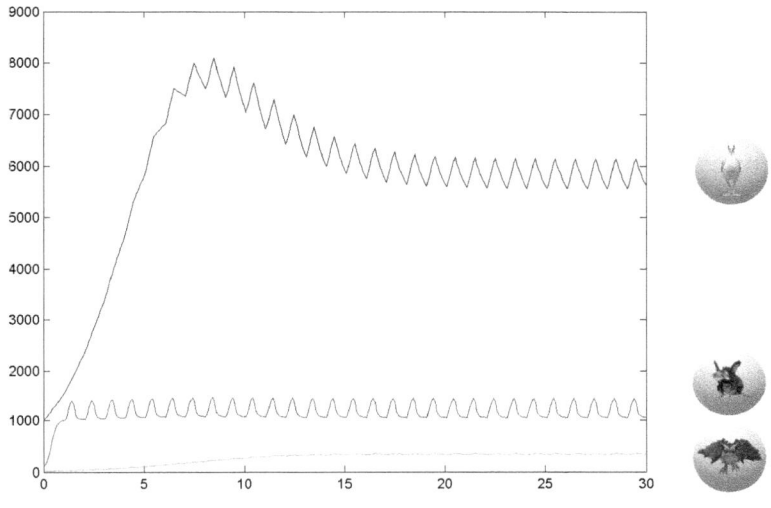

Figure 10. $B(0) = 1000$, $R(0) = 100$, $C(0) = 20$.

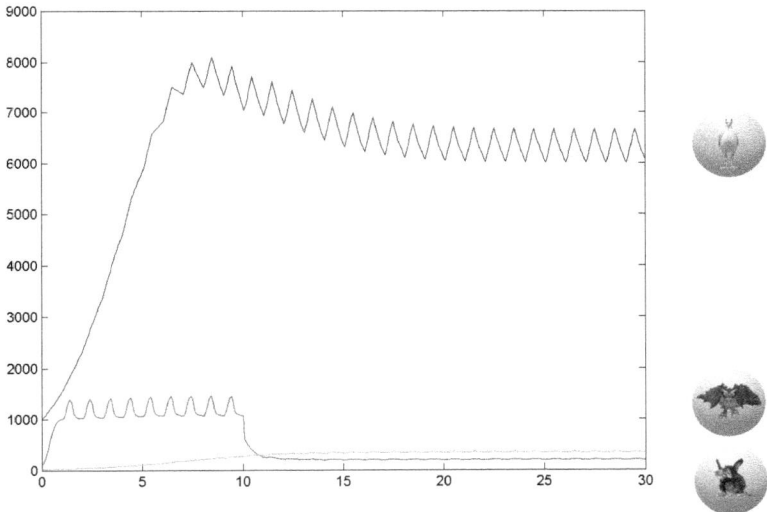

Figure 11. $B(0) = 1000$, $R(0) = 100$, $C(0) = 20$.

Example 2. In this example, a control effort on the rat is operated ($\theta^* = 5$) from the year $a_0 = 10$. The other data are the same as for the previous example (Figure 11).

$r_B^* = 0.5$	$K_B^* = 20000$	$r_R^* = 6$	$S_R^* = 100000$	$r_C^* = 0.4$	$S_C^* = 20000$
$\eta_{R,B}^* = 100$	$\eta_{R,V}^* = 100$	$\eta_{C,B}^* = 100$	$\eta_{C,R}^* = 200$	$\eta_{C,V}^* = 100$	$\eta_{R,S}^* = 100$
$\eta_{C,S}^* = 100$	$\mu_B^* = 0.01$	$\lambda_{R,B}^* = 100$	$\lambda_{C,B}^* = 100$	$\lambda_{C,R}^* = 200$	$a_0 = 10, \theta^* = 5$

We set

$$V(t) = 50000 \max\{0, \sin(2\pi t)\}.$$

Source of the 3D models: Microsoft Paint 3D

References

[1] K. Addi, D. Goeleven and R. Oujja, *Principes mathématiques pour Biologistes, Chimistes et Bioingénieurs, Applications et Exercices Corrigés* (Edition Ellipses, 2013).
[2] B. Anselme, *Biomathématiques, Outils, Méthodes et Exemples* (Dunod, Paris, 2015).
[3] F. Courchamp, M. Langlais and G. Sugihara, Cats protecting birds: Modelling the mesopredator release effect, *J. Anim. Ecol.* **68**, 282–292 (1999).
[4] F. Courchamp, M. Langlais and G. Sugihara, Rabbits killing birds: Modelling the hyperpredation process, *J. Anim. Ecol.* **69**, 154–164 (2000).
[5] K. R. Crooks and M. E. Soulé, Mesopredator release and avifaunal extinctions in a fragmented system, *Nature* **400**, 563–566 (1999).
[6] A. Edwin, Modeling and Analysis of a Two Prey — One Predator System with Harvesting, Holling-Type II and Ratio-Dependent Responses, Dissertation for the Award of the Degree of Master of Science in Mathematics of Makerere University, Ouganda (2010).
[7] A. J. Lotka, Undamped oscillations derived from the law of mass action, *J. Am. Chem. Soc.* **42**, 1595–1599 (1920).
[8] A. J. Lotka, *Element of Physical Biology* (Williams and Wilkins, Baltimore, 1925).
[9] T. R. Malthus, *An Essay on the Principle of Population* (Penguin Books, England, 1970). From the original edition of 1798.
[10] J. D. Murray, *Mathematical Biology I: An Introduction*. Interdisciplinary Applied Mathematics (Springer-Verlag, Berlin, 2002).

[11] M. Shatalov, J. C. Greef, S. V. Joubert and I. Fedotov, Parametric identification of the model with one predator and two prey species. In *Buffelspoort TIME2008 Peer-Reviewed Conference Proceedings*, 22–28 September (2008), pp. 101–109.

[12] P. F. Verhulst, Notice sur la Loi que la Population Poursuit dans son Accroissement, *Corresp. Math. Phys.* **10**, 113–121 (1838).

[13] V. Volterra, Variazionie Fluttuazioni del Numero d'Individui in Specie Animali Conviventi, *Mem. Acad. Lincei.* **2**, 31–113 (1926).

© 2024 World Scientific Publishing Company
https://doi.org/10.1142/9789811267048_0011

Chapter 11

Differential Operator Associated with the (q, k)-Symbol Raina's Function

Rabha W. Ibrahim

*Near East University, Mathematics Research Center,
Department of Mathematics, Near East Boulevard,
Nicosia/Mersin, Turkey
Department of Computer Science and Mathematics,
Lebanese American University,
Beirut, Lebanon
Information and Communication Technology Research Group,
Scientific Research Center, Al-Ayen University,
Thi-Qar, Iraq*

rabhaibrahim@yahoo.com

There are numerous unresolved issues and difficulties in the highly active field of quantum calculus (QC). The creation of q-analogs for specific functions, the investigation of q-difference equations, and the use of q-calculus in mathematical physics and quantum information theory are some of the most current study areas. This notion has recently gained traction in the field of geometric function theory. This chapter deals with the QC (Jackson's calculus) to define the q-differential operator related to the q-Raina function. With this newly created operator, we create a new subclass of analytical functions in conical domains. We examine one of the most well-liked characteristics of the suggested operator geometrically. Our strategy was motivated by the differential subordination hypothesis. We identified some well-known corollaries of our main conclusions as the outcome.

1. Introduction

A few instances of special functions include integrals and the results of numerous distinct kinds of differential equations. As an outcome, the most basic integrals or, at the very least, the integral representation of special functions are incorporated in the explanations of special functions in the majority of integral collections. The theory of special functions is closely related to numerous mathematical physics subjects [1] because differential operators are significant in both physics and mathematics. These special functions have been generalized by using different kinds of fractional calculus, Jackson calculus (quantum calculus), conformable calculus and fractal. Differentiation and integration are the fundamental procedures in classical calculus that can be employed to investigate continuous functions. These operations are expanded in quantum calculus to include operators on noncommutative spaces, where the sequence of operations is critical. As a result, quantum mechanics and other quantum systems can be studied. Q-derivatives, q-integrals and q-exponential, which are analogs of their classical counterparts, are some of the fundamental ideas in quantum calculus. These ideas have uses in physics, computer science and cryptography, among other fields.

The quantum calculus (QC) has recently been used to describe a variety of use cases in number theory and science, such as conformable quantum mechanics, nuclear and high-energy physics, inner energy and specific heat. It is an entirely new area of mathematical analysis and its applications, with uses in both science and mathematics. Jackson was the one who initially defined and improved the functions of q-differentiation and q-integration [2, 3]. Following that, Ismail *et al.* [4] considered the geometric function theory of q-calculus ($q \in (0, 1)$). QC is now being used by investigators to recommend and improve different types of Ma and Minda classes [5]. For example, researchers considered q-stralikeness criteria (see Refs. [6–8]). QC is furthermore utilized to specify a diversity of differential and integral operators, containing special functions (see Refs. [9–12]).

For example, based on the q-Mittag–Leffler function, Noor and Razzaque developed a q-differential operator [9]. In the open unit disk, Aldawish and Ibrahim presented q-inequalities involving a quantum differential operator [13]. q-higher order derivatives are given in different studies [14, 15]. In the subject of geometric function theory, numerous more investigations are familiarized, involving the Mittag–Leffler function and its generalities (see Refs. [16–18]).

Since it enables us to symbolize a whole class of fractional derivative operators with a single representation, the k-symbol notation is helpful. Different fractional derivative operators, such as the well-known Riemann–Liouville and Caputo derivatives, can be obtained by varying the value of k. The chain rule, product rule and integration by parts formula can all be expressed using the k-symbol format for fractional derivatives. Overall, the k-symbol is a helpful tool for fractional calculus in describing and working with fractional derivatives (see Refs. [19–21]).

We proceed the investigation in QC and special functions. Based on the generalization of Raina's function together with QC, we aim to expand a differential operator in the open unit disk in this note. The fractional differential operator is utilized to clarify a variety of advanced normalized analytic functions. A set of differential inequalities is studied by means of the concept of differential subordination and superordination. We correspondingly investigate the geometric possessions of a set of analytic functions. The advanced operator is utilized in a variability of applications.

2. Procedures

We request the following concepts.

2.1. *Geometric techniques*

Conformal mappings, which maintain angles and orientation, harmonic functions, which are solutions to the Laplace equation, and quasiconformal mappings, which are more general than conformal mappings and preserve specific geometric features of the domain, constitute some of the fundamental ideas in geometric function theory. Further significant subjects involve the theory of Teichmuller spaces, which offers a geometric interpretation of the moduli space of Riemann surfaces, and the Riemann mapping theorem, which asserts that every single merely linked subdomain in the complex plane may be conformally translated to the unit disk.

We begin with the fundamentals of geometric function theory, which are discussed in this book [22].

Definition 1. The complex domain $\mathbb{I} := \{\eta \in \mathbb{C} : |\eta| < 1\}$ is represented as the open unit disk. The analytic functions v_1, v_2 in $\mathbb{I}$ are subordinated $v_1 \prec v_2$ or $v_1(\eta) \prec v_2(\eta), \eta \in \mathbb{I}$ if for an analytic function $\psi, |\psi| \leq |\eta| < 1$ realizing $v_1(\eta) = v_2(\psi(\eta)), \eta \in \mathbb{I}$.

Diverse classes of analytic functions, which include univalent functions and convex functions, can be studied using differential subordination and superordination as useful tools. They additionally have applications in conformal mapping and the theory of flexibility among other branches of mathematics and physics.

On the opposite end of the spectrum, if a third function h occurs and violates the inequality, then a function f can be considered to be differentially superordinate to another function g:

$$|h(\eta)| \leq \frac{|f(\eta) - a|}{|g(\eta) - a|}, \quad |\eta| < 1$$

for any η in the range of f and g, for some complex number a. Due to this inequality, the modulus of h is "smaller" than the proportion of f and g distances from a. This implies that f is "closer" to a than g and that h determines the amount of proximity.

Definition 2. A normalization class of analytic function denoting by Π is given by the series

$$\phi(\eta) = \eta + \sum_{n=2}^{\infty} a_n \eta^n, \quad \eta \in \mathbb{I}, \quad \phi(0) = \phi'(0) - 1 = 0.$$

In addition, two functions $\sigma, \varsigma \in \Pi$ are convoluted $(\sigma * \varsigma)$ if they recognize the product [23]

$$(\sigma * \varsigma)(\eta) = \left(\eta + \sum_{n=2}^{\infty} \sigma_n \eta^n\right) * \left(\eta + \sum_{n=2}^{\infty} \varsigma_n \eta^n\right)$$

$$= \eta + \sum_{n=2}^{\infty} \sigma_n \varsigma_n \eta^n.$$

Definition 3. The class $\mathcal{S}^*$ of starlike functions and the class $\mathcal{C}$ of convex functions are both associated to Π. Lastly, the class $\mathcal{P} := \{\varrho : \varrho(\eta) = 1 + \varrho_1\eta + \varrho_2\eta^2 + \cdots, \eta \in \mathbb{I}\}$ is a special class of analytic functions in $\mathbb{I}$ with positive real part in $\mathbb{I}$ and $\varrho(0) = 1$.

2.2. Jackson calculus

Beginning in the 20th century, Jackson developed the q-calculus to investigate specific special functions that come up in the subject matter of physics, which include the q-analogs of the traditional orthogonal polynomials like the Hermite, Laguerre, and Jacobi polynomials. Similar to how standard calculus gives a framework for working common polynomials and functions, the Jackson calculus uses the aforementioned definitions to create a framework for manipulating q-polynomials and other q-functions. Numerous physics topics, such as quantum mechanics, statistical mechanics, and quantum field theory, can be studied using the Jackson calculus. Additionally, it has uses in algebraic geometry, number theory, and combinatorics.

Let's introduce the following definition

Definition 4. The variation operator can be used to display the Jackson derivative as demonstrated in the following:

$$(\Delta_q)\,\phi(\eta) = \frac{\phi(\eta) - \phi(q\,\eta)}{\eta(1-q)}, \quad q \in (0,1) \tag{1}$$

such that

$$\Delta_q\,(\eta^\kappa) = \left(\frac{1-q^k}{1-q}\right)\eta^{\kappa-1}, \quad k \in \mathbb{R}.$$

The sum of the numbers additionally appears in the series expression in the manner that follows:

$$(\Delta_q\,\phi)\,(\eta) = \sum_{n=0}^{\infty} a_n\,[n]_q\,\eta^{n-1}, \tag{2}$$

where
$$[n]_q := \frac{1-q^n}{1-q}.$$

Note that
$$\Delta_q C = 0, \quad \lim_{q \to 1^-} (\Delta_q \phi)(\eta) = \phi'(\eta),$$
where C is a constant.

2.3. Raina's function

Special functions are mathematical constructs with unique qualities that lend them valuable in a variety of scientific and technical domains. They are designed to address particular mathematical issues. These functions, like polynomials, exponential, and trigonometric functions, are frequently more complicated than simpler functions. Numerous disciplines, including engineering, physics, the field of finance, and a variety of others, use special functions extensively. They are crucial resources for resolving difficult mathematical puzzles and are widely employed in both scientific study and engineering planning.

Let's begin with the Mittag–Leffler function, which is a well-known special function.

Definition 5. The following series indicates the Raina's function [24]:
$$_\nu\mathcal{A}_{a,b}(\eta) = \sum_{n=0}^{\infty} \frac{\nu(n)}{\Gamma(an+b)} \eta^n, \quad \eta \in \mathbb{I},$$
where $a, b \in \mathbb{C}, \Re(a) > 0, \Re(b) > 0$ and $\nu := \{\nu(0), \nu(1), \ldots, \nu(n)\}$ is a bounded sequence of arbitrary real or complex numbers.

Remark 1.

- If for all $n \geq 0$, $\nu(n) = 1$, then we obtain the Mittag–Leffler function
$$\mathcal{A}_{a,b}(\eta) = \sum_{n=0}^{\infty} \frac{\eta^n}{\Gamma(an+b)}.$$

- If $a = 1, b = 1, \nu(n) = \dfrac{(a)_n(b)_n}{(c)_n}$, then we get the hypergeometric function

$$_2F_1(a,b;c;\eta) = \sum_{n=0}^{\infty} \dfrac{(a)_n(b)_n}{(c)_n} \dfrac{\eta^n}{\Gamma(n+1)}.$$

The enhanced Raina's function expression is then developed in the manner described in the following:

Definition 6. The k-symbol gamma function is frequently referred to as the motivated gamma function Γ_k which is prepared as follows [25]:

$$\Gamma_k(\chi) = \lim_{n \to \infty} \dfrac{n! k^n (nk)^{\frac{\chi}{k}-1}}{(\chi)_{n,k}},$$

$$(\chi)_{n,k} := \chi(\chi+k)(\chi+2k)\dots(\chi+(n-1)k),$$

where

$$(\chi)_{n,k} = \dfrac{\Gamma_k(\chi+nk)}{\Gamma_k(\chi)}.$$

Consequently, the k-symbol Raina's function is defined as follows:

$$_{\nu,k}\mathcal{A}_{a,b}(\eta) = \sum_{n=0}^{\infty} \dfrac{\nu(n)}{\Gamma_k(an+b)} \eta^n, \quad \eta \in \mathbb{I},$$

where $a, b \in \mathbb{C}, \Re(a) > 0, \Re(b) > 0$ and $\nu := \{\nu(0), \nu(1), \dots, \nu(n)\}$ is a bounded sequence of arbitrary real or complex numbers. The recent generalization of k-symbol Raina' s function is given by Ref. [26]:

$$_{\nu,k}\mathcal{A}_{a,b}(\eta) = \sum_{n=0}^{\infty} \dfrac{\nu(n)}{k\Gamma_k(ank+b)} \eta^n, \quad \eta \in \mathbb{I}.$$

Proceeding to define the normalization form of $_{\nu,k}\mathcal{A}_{a,b}(\eta)$, as follows:

$$_{\nu,k}\mathbb{A}_{a,b}(\eta) := \left(\frac{k\Gamma_k(b)}{\nu(0)}\right) \eta_{\nu,k}\mathcal{A}_{a,b}(\eta)$$

$$= \eta + \sum_{n=2}^{\infty} \left(\frac{\nu(n-1)k\Gamma_k(b)}{\nu(0)k\Gamma_k(ak(n-1)+b)}\right)\eta^n, \quad \eta \in \mathbb{I}$$

$$:= \eta + \sum_{n=2}^{\infty} \omega_n \eta^n, \tag{3}$$

where

$$\omega_n := \left(\frac{\nu(n-1)\Gamma_k(b)}{\nu(0)\Gamma_k(ak(n-1)+b)}\right), \quad \nu(0) \neq 0.$$

Note that the operator (3) includes the Sàlàgean differentiation and integration operators [27].

Definition 7. We provide the elements as follows: (q,k)-Raina differential operator, depending on the normalized function ϕ, in light of the quantum operator Δ_q:

$$_{\nu,k}\Delta_q^0(a,b)\phi(\eta) = \phi(\eta) * {}_{\nu,k}\mathbb{A}_{a,b}(\eta)$$
$$_{\nu,k}\Delta_q^1(a,b)\phi(\eta) = \eta\Delta_q\left(\phi(\eta) * {}_{\nu,k}\mathbb{A}_{a,b}(\eta)\right)$$
$$_{\nu,k}\Delta_q^2(a,b)\phi(\eta) = {}_{\nu,k}\Delta_q^1(a,b)\phi(\eta)\left({}_{\nu,k}\Delta_q^1(a,b)\phi(\eta)\right) \tag{4}$$
$$\vdots$$
$$_{\nu,k}\Delta_q^\kappa(a,b)\phi(\eta) = {}_{\nu,k}\Delta_q^1(a,b)\phi(\eta)\left({}_{\nu,k}\Delta_q^{\kappa-1}(a,b)\phi(\eta)\right),$$

where

$$_{\nu,k}\Delta_q^\kappa(a,b)\phi(\eta) = \eta + \sum_{n=2}^{\infty}[n]_q^\kappa \left(\frac{\nu(n-1)k\Gamma_{k,q}(b)}{\nu(0)k\Gamma_{k,q}(ak(n-1)+b)}\right) a_n \eta^n$$

$$:= \eta + \sum_{n=2}^{\infty}[n]_q^\kappa \, \Omega_n(a,b,\nu,q,k) a_n \eta^n, \quad \eta \in \mathbb{I},$$

$$\Omega_n(a,b,\nu,q,k) = \left(\frac{\nu(n-1)k\Gamma_{k,q}(b)}{\nu(0)k\Gamma_{k,q}(ak(n-1)+b)}\right),$$

where
$$\lim_{q \to 1\pm} \Gamma_{1,q}(\chi) = \Gamma(\chi).$$

We note that, for $k = 1, a = 0, \nu(n-1) = 1$ for all $n \geq 1$, we get the Sàlàgean q-differential operator [28]. And, when $k = 1, \nu(n-1) = 1$ for all $n \geq 1$, we obtain the q-differential operator in Ref. [9]. To study the suggested (k,q)-differential operator, we aim to put it in terms of a new class of analytic functions.

Definition 8. Define a convex analytic function $\mathfrak{P}_{\jmath,\wp}$ as follows:

$$\mathfrak{P}_{\jmath,\wp} = \begin{cases} \dfrac{1+\eta}{1-\eta}, & \text{if } \jmath = 0 \\ p_1(\jmath,\wp), & \text{if } \jmath = 1 \\ p_2(\jmath,\wp), & \text{if } 0 < \jmath < 1 \\ p_3(\jmath,\wp), & \text{if } \jmath > 1 \end{cases}$$

where, for analytic function $\varphi(\eta), \eta \in \mathbb{I}$ and $\wp \in \mathbb{C}\setminus\{0\}$, we have the following functions [29]:

$$p_1(\jmath,\wp) = 1 + \frac{2\wp}{\pi^2}\left(\log\left(\frac{1+\sqrt{\eta}}{1-\sqrt{\eta}}\right)\right)^2,$$

$$p_2(\jmath,\wp) = 1 + \frac{2\wp}{1-\jmath^2}\sinh^2\left(\frac{2}{\pi}\arccos(\jmath)\arctan h(\sqrt{\eta})\right),$$

$$p_3(\jmath,\wp) = 1 + \frac{\wp}{1-\jmath^2} + \frac{\wp}{\jmath^2-1}$$
$$\times \sin\left(\frac{\pi}{2R(t)}\int_0^{\varphi(\eta)/\sqrt{t}} \frac{d\alpha}{\sqrt{1-\alpha^2}\sqrt{1-(\alpha t)^2}}\right).$$

The function $\phi \in \Pi$ is a member in the class $\jmath - \mathcal{S}^{\kappa}_{q,k,\wp}(a,b)$ if and only if

$$\frac{\eta\Delta_q\left(\nu_{,k}\Delta^{\kappa}_q(a,b)\phi(\eta)\right)}{\nu_{,k}\Delta^{\kappa}_q(a,b)\phi(\eta)} \prec \mathfrak{P}_{\jmath,\wp}, \quad \eta \in \mathbb{I}, \tag{5}$$

where [29, 30]

$$\mathfrak{P}_{\jmath,\wp} = 1 + c_1\eta + c_2\eta^2 + \cdots$$

then

$$c_1 = \begin{cases} \dfrac{2\wp\Lambda^2}{1-\jmath^2}, & \text{if } 0 \leq \jmath < 1 \\ \dfrac{8\wp}{\pi^2}, & \text{if } \jmath = 1 \\ \dfrac{\pi^2\wp}{4(1+\tau)\sqrt{\tau}(\jmath^2-1)\Sigma^2(\tau)}, & \text{if } \jmath > 1, \end{cases}$$

$$c_2 = \begin{cases} \dfrac{\Lambda^2+2}{3}c_1, & \text{if } 0 \leq \jmath < 1 \\ \dfrac{2}{3}c_1, & \text{if } \jmath = 1 \\ \dfrac{4\Sigma^2(\tau)(\tau^2+6\tau+1)-\pi^2}{24(1+\tau)\sqrt{\tau}(\jmath^2-1)\Sigma^2(\tau)}c_1, & \text{if } \jmath > 1, \end{cases}$$

where $\Sigma(\tau)$ is Legendre's complete elliptic integral and $\Lambda := 2/\pi \arccos(t)$.

Remark 2.

- When $k = 1, \nu(n) = 1, \forall n$ the class $\jmath - \mathcal{S}^\kappa_{q,1,\wp}(a,b)$ is studied in Ref. [9].
- When $k = 1, \nu(n) = 1, \forall n$ the class $\jmath - \mathcal{S}^\kappa_{q,1,\wp}(0,1)$ is investigated in Ref. [31].

2.4. *Arguments*

In the sequel, we refer to the following preliminaries:

Lemma 1 ([32]). Assume that $\Psi(z) = \sum_{n=0}^\infty \psi_n \eta^n$ is univalent convex in $\mathbb{I}$ such that

$$\Xi(\eta) = \sum_{n=0}^\infty \chi_n \eta^n \prec \Psi(\eta)$$

then $|\chi_n| \leq |\psi_1|$ for all $n \geq 1$.

Lemma 2 ([5]). Consume that $F(\eta) = 1 + \sum_{n=1}^{\infty} f_n \eta^n$ is analytic in $\mathbb{I}$ under the condition

$$\Re(F(\eta)) > 0, \quad \eta \in \mathbb{I}.$$

Then

$$|f_2 - \Bbbk f_1^2| \leq 2\max\{1, |2\Bbbk - 1|\}, \quad \Bbbk \in \mathbb{C}.$$

3. Findings

We start our first result, as follows:

Theorem 1. If $\phi \in \jmath - \mathcal{S}_{q,k,\wp}^{\kappa}(a,b), q \to 1^-$ then

$$_{\nu,k}\Delta_q^{\kappa}(a,b)\phi(\eta) \prec \eta \exp\left(\int_0^{\eta} \frac{\mathfrak{P}_{\jmath,\wp}(\omega(\chi))}{\chi} d\chi\right),$$

where ω satisfies $\omega(0) = 0$ and $|\omega(\eta)| < 1$. In addition, when $|\eta| := u < 1$ we obtain

$$\exp\left(\int_0^1 \frac{\mathfrak{P}_{\jmath,\wp}(-u)}{u} du\right) \leq \left|\frac{_{\nu,k}\Delta_q^{\kappa}(a,b)\phi(\eta)}{\eta}\right| \leq \exp\left(\int_0^1 \frac{\mathfrak{P}_{\jmath,\wp}(u)}{u} du\right).$$

Proof. Since $\phi \in \jmath - \mathcal{S}_{q,k,\wp}^{\kappa}(a,b), q \to 1^-$ thus we have

$$\frac{\left(_{\nu,k}\Delta_q^{\kappa}(a,b)\phi(\eta)\right)'}{_{\nu,k}\Delta_q^{\kappa}(a,b)\phi(\eta)} - \frac{1}{\eta} = \frac{\mathfrak{P}_{\jmath,\wp}(\omega(\eta)) - 1}{\eta}.$$

Integration yields

$$_{\nu,k}\Delta_q^{\kappa}(a,b)\phi(\eta) \prec \eta \exp\left(\int_0^{\eta} \frac{\mathfrak{P}_{\jmath,\wp}(\omega(\chi))}{\chi} d\chi\right),$$

which gives

$$\frac{_{\nu,k}\Delta_q^{\kappa}(a,b)\phi(\eta)}{\eta} \prec \exp\left(\int_0^{\eta} \frac{\mathfrak{P}_{\jmath,\wp}(\omega(\chi))}{\chi} d\chi\right).$$

But, by the relation
$$\mathfrak{P}_{\jmath,\wp}(-u|\eta|) \leq \Re\left(\mathfrak{P}_{\jmath,\wp}(\omega(\eta u))\right) \leq \mathfrak{P}_{\jmath,\wp}(u|\eta|),$$
we then obtain
$$\int_0^1 \frac{\mathfrak{P}_{\jmath,\wp}(-u|\eta|)}{\varrho}du \leq \int_0^1 \frac{\Re\left(\mathfrak{P}_{\jmath,\wp}(\omega(zu))\right)}{u}du \leq \int_0^1 \frac{\mathfrak{P}_{\jmath,\wp}(u|\eta|)}{u}du.$$
Adding the aforementioned disparities, we get
$$\int_0^1 \frac{\mathfrak{P}_{\jmath,\wp}(-u|\eta|)}{\varrho}du \leq \log\left|\frac{\nu,k\Delta_q^\kappa(a,b)\phi(\eta)}{\eta}\right| \leq \int_0^1 \frac{\mathfrak{P}_{\jmath,\wp}(u|\eta|)}{u}du$$
which implies
$$\exp\left(\int_0^1 \frac{\mathfrak{P}_{\jmath,\wp}(-u)}{u}du\right) \leq \left|\frac{\nu,k\Delta_q^\kappa(a,b)\phi(\eta)}{\eta}\right| \leq \exp\left(\int_0^1 \frac{\mathfrak{P}_{\jmath,\wp}(u)}{u}du\right).$$
□

Corollary 1 ([9]). *If $k = \nu(n) = 1, \forall n \geq 1$ thus*
$$\nu,k\Delta_q^\kappa(a,b)\phi(\eta) \prec \eta \exp\left(\int_0^\eta \frac{\mathfrak{P}_{\jmath,\wp}(\omega(\chi))}{\chi}d\chi\right),$$
where ω admits $\omega(0) = 0$ and $|\omega(\eta)| < 1$. In addition, for $|\eta| := u < 1$ yields
$$\exp\left(\int_0^1 \frac{\mathfrak{P}_{\jmath,\wp}(-u)}{u}du\right) \leq \left|\frac{\nu,k\Delta_q^\kappa(a,b)\phi(\eta)}{\eta}\right| \leq \exp\left(\int_0^1 \frac{\mathfrak{P}_{\jmath,\wp}(u)}{u}du\right).$$

Corollary 2 ([31]). *If $k = \nu(n) = 1, \forall n \geq 1$ and $a = 0, b = 1$; thus,*
$$\nu,1\Delta_q^\kappa(a,b)\phi(\eta) \prec \eta \exp\left(\int_0^\eta \frac{\mathfrak{P}_{\jmath,\wp}(\omega(\chi))}{\chi}d\chi\right),$$
where ω achieves $\omega(0) = 0$ and $|\omega(\eta)| < 1$. Also, when $|\eta| := u < 1$, this implies that
$$\exp\left(\int_0^1 \frac{\mathfrak{P}_{\jmath,\wp}(-u)}{u}du\right) \leq \left|\frac{\nu,1\Delta_q^\kappa(0,1)\phi(\eta)}{\eta}\right| \leq \exp\left(\int_0^1 \frac{\mathfrak{P}_{\jmath,\wp}(u)}{u}du\right).$$

Theorem 2. If $\phi \in \jmath - \mathcal{S}_{q,k,\wp}^{\kappa}(a,b)$, then for $\mathfrak{P}(\eta) = 1 + c_1\eta + c_2\eta^2 + \cdots$, the inequalities

$$|a_2| \leq \frac{|c_1|}{[2]_q^k \Omega_2(a,b,\nu,q,k)([2]_q - 1)},$$

$$|a_n| \leq \frac{|c_1|}{[n]_q^\kappa \Omega_n(a,b,\nu,q,k)([n]_q - 1)} \prod_{\imath=1}^{n-2} \left(1 + \frac{|c_1|}{[\imath+1]_q - 1}\right), \quad n \geq 3$$

hold.

Proof. Putting

$$\frac{\eta \Delta_q \left(_{\nu,k}\Delta_q^\kappa(a,b)\phi(\eta)\right)}{_{\nu,k}\Delta_q^\kappa(a,b)\phi(\eta)} = T(\eta)$$

$$= 1 + \sum_{n=1}^{\infty} t_n \eta^n.$$

This leads to

$$\eta + \sum_{n=2}^{\infty} [n]_q^{\kappa+1} \Omega_n(a,b,\nu,q,k) a_n \eta^n$$

$$= \left(\eta + \sum_{n=2}^{\infty} [n]_q^\kappa \Omega_n(a,b,\nu,q,k) a_n \eta^n\right) \left(1 + \sum_{n=1}^{\infty} t_n \eta^n\right)$$

$$= \left(\sum_{n=0}^{\infty} t_n \eta^{n+1}\right) \left(\sum_{n=0}^{\infty} t_n \eta^n\right) \left(\sum_{n=2}^{\infty} [n]_q^\kappa \Omega_n(a,b,\nu,q,k) a_n \eta^n\right).$$

A comparison on the connects of η^n, we obtain

$$[n]_q^{\kappa+1} \Omega_n(a,b,\nu,q,k) a_n = [n]_q^\kappa \Omega_n(a,b,\nu,q,k) a_n$$

$$+ \sum_{\imath=1}^{n-1} [\imath]_q^\kappa \frac{\nu(\imath-1)\Gamma_{q,k}(b)}{\nu(0)\Gamma_{q,k}(a(\imath-1)+b)} a_\imath c_{n-\imath}.$$

Accordingly, we obtain

$$[n]_q^\kappa ([n]_q - 1) \Omega_n(a,b,\nu,q,k) a_n = \sum_{\imath=1}^{n-1} [\imath]_q^\kappa \frac{\nu(\imath-1)\Gamma_q(b)}{\nu(0)\Gamma_q(a(\imath-1)+b)} a_\imath p_{n-\imath}.$$

A calculation implies that

$$a_n = \frac{1}{[n]_q^\kappa ([n]_q - 1) \Omega_n(a, b, \nu, q, k)}$$

$$\times \sum_{\imath=1}^{n-1} [\imath]_q^\kappa \frac{\nu(\imath - 1)\Gamma_{k,q}(b)}{\nu(0)\Gamma_{k,q}(a(\imath - 1) + b)} a_\imath t_{n-\imath}.$$

But, in view of Lemma 1, we conclude that $|t_n| \leq |c_1|$. Therefore, this imposes that

$$|a_n| \leq \frac{|c_1|}{[n]_q^\kappa ([n]_q - 1) \Omega_n(a, b, \nu, q, k)} \sum_{\imath=1}^{n-1} [\imath]_q^\kappa \frac{\nu(\imath - 1)\Gamma_{k,q}(b)}{\nu(0)\Gamma_{k,q}(a(\imath - 1) + b)} |a_\imath|.$$

For $n = 2$, we have

$$|a_2| \leq \frac{|c_1|}{[2]_q^\kappa ([2]_q - 1) \Omega_2(a, b, \nu, q, k)} \sum_{\imath=1}^{1} [\imath]_q^\kappa \frac{\nu(\imath - 1)\Gamma_{k,q}(b)}{\nu(0)\Gamma_{k,q}(a(\imath - 1) + b)} |a_\imath|$$

$$= \frac{|c_1|}{[2]_q^\kappa \Omega_2(a, b, \nu, q, k)([2]_q - 1)}.$$

Now by mathematical induction, for $n = 3$

$$|a_3| \leq |\frac{|c_1|}{[3]_q^k \Omega_3(a, b, \nu, q, k)([3]_q - 1)} \left(1 + [2]_q^\kappa \Omega_2(a, b, \nu, q, k)|a_2|\right).$$

The result of aggregating the previous two disparities is

$$|a_3| \leq |\frac{|c_1|}{[3]_q^\kappa \Omega_3(a, b, \nu, q, k)([3]_q - 1)}$$

$$\times \left(1 + [2]_q^\kappa \Omega_2(a, b, \nu, q, k) \left(\frac{|c_1|}{[2]_q^\kappa \Omega_2(a, b, \nu, q, k)([2]_q - 1)}\right)\right)$$

$$= \frac{|c_1|}{[3]_q^\kappa \Omega_3(a, b, \nu, q, k)([3]_q - 1)} \left(1 + \frac{|c_1|}{[2]_q - 1}\right).$$

Suppose it is valid for j that is

$$|a_j| \leq \frac{|\rho_1|}{[j]_q^\kappa \Omega_j(a, b, \nu, q, k)([j]_q - 1)} \prod_{\imath=1}^{j-2} \left(1 + \frac{|c_1|}{[\imath + 1]_q - 1}\right), \quad n \geq 3.$$

Therefore, we receive

$$[|a_{j+1}| \leq \frac{|c_1|}{[j+1]_q^\kappa \Omega_{j+1}(a,b,\nu,q,k)([j+1]_q - 1)}$$

$$\times \left(1 + \frac{|c_1|}{[2]_q - 1} + \frac{|c_1|}{[3]_q - 1}\left(1 + \frac{|c_1|}{[2]_q - 1}\right)\right.$$

$$\left. + \cdots + \frac{|c_1|}{[j]_q^\kappa \Omega_j(a,b,\nu,q,k)([j]_q - 1)} \prod_{i=1}^{j-2}\left(1 + \frac{|c_1|}{[i+1]_q - 1}\right)\right)$$

$$\leq \frac{|c_1|}{[j]_q^\kappa \Omega_j(a,b,\nu,q,k)([j]_q - 1)} \prod_{i=1}^{j-2}\left(1 + \frac{|c_1|}{[i+1]_q - 1}\right),$$

which completes the proof. □

Corollary 3 ([9]). *If* $k = \nu(n) = 1, \forall n \geq 1$, *then*

$$|a_2| \leq \frac{|c_1|}{[2]_q^\kappa \Omega_2(a,b,1,q)([2]_q - 1)},$$

$$|a_n| \leq \frac{|c_1|}{[n]_q^\kappa \Omega_n(a,b,1,q,1)([n]_q - 1)} \prod_{i=1}^{n-2}\left(1 + \frac{|c_1|}{[i+1]_q - 1}\right), \quad n \geq 3.$$

Corollary 4 ([31]). *If* $k = \nu(n) = 1, \forall n \geq 1$ *and* $a = 0, b = 1$, *thus*

$$|a_2| \leq \frac{|c_1|}{[2]_q^\kappa \Omega_2(0,1,1,q,1)([2]_q - 1)},$$

$$|a_n| \leq \frac{|c_1|}{[n]_q^\kappa \Omega_n(0,1,1,q,1)([n]_q - 1)} \prod_{i=1}^{n-2}\left(1 + \frac{|c_1|}{[i+1]_q - 1}\right), \quad n \geq 3.$$

Theorem 3. *If* $\phi \in \jmath - \mathcal{S}_{q,k,\wp}^\kappa(a,b)$, *then for* $\mathfrak{P}(\eta) = 1 + c_1\eta + c_2\eta^2 + \cdots$, *the inequality*

$$|a_3 - \beta a_2^2| \leq \frac{|c_1|}{2[3]_q^\kappa \Omega_3(a,b,\nu,q,k)([3]_q - 1)} \max\{1, |2B - 1|\},$$

where $\beta \in \mathbb{C}$ and

$$B := \frac{1}{2}\left(1 - \frac{c_2}{c_1} - c_1\left(\frac{1}{[2]_q - 1} - \beta\frac{[3]_q^\kappa([3]_q - 1)}{2\Omega_2(a, b, \nu, q, k)\left([2]_q^\kappa([2]_q - 1)\right)^2}\right)\right), \tag{6}$$

holds.

Proof. By the fact that $\phi \in \jmath - \mathcal{S}_{q,k,\wp}^\kappa(a,b)$, we have

$$\frac{\eta \Delta_q\left(_{\nu,k}\Delta_q^\kappa(a,b)\phi(\eta)\right)}{_{\nu,k}\Delta_q^\kappa(a,b)\phi(\eta)} = \mathfrak{P}_{\jmath,\wp}(\omega(\eta)).$$

Suppose that the function $T(\eta) = 1 + t_1\eta + t_2\eta^2 + \cdots$ satisfies the series

$$\omega(\eta) = \frac{t_1}{2}\eta + \frac{1}{2}\left(t_2 - \frac{t_1^2}{2}\eta^2 + \cdots\right)$$

and

$$\mathfrak{P}_{\jmath,\wp}(\omega(\eta)) = 1 + \frac{c_1 t_1}{2}\eta + \left(\frac{c_2 t_1^2}{4} + \frac{1}{2}\left(t_2 - \frac{t_1^2}{2}\right)c_1\right)\eta^2 + \cdots.$$

Thus, we get

$$\frac{\eta \Delta_q\left(_{\nu,k}\Delta_q^\kappa(a,b)\phi(\eta)\right)}{_{\nu,k}\Delta_q^\kappa(a,b)\phi(\eta)} = 1 + [2]_q^\kappa \Omega_2(a,b,\nu,q,k)\left([2]_q - 1\right)a_2\eta$$
$$+ \left([3]_q^\kappa \Omega_3(a,b,\nu,q,k)\left([2]_q - 1\right)a_3\right.$$
$$\left. - \left([2]_q^\kappa \Omega_2(a,b,\nu,q,k)\right)^2\left([2]_q - 1\right)a_2^2\right)\eta^2.$$

A calculation brings

$$a_2 = \frac{c_1 t_1}{2[2]_q^\kappa \Omega_2(a,b,m,q,k)([2]_q - 1)},$$

$$a_3 = \frac{1}{[3]_q^\kappa \Omega_3(a,b,\nu,q,k)([3]_q - 1)}$$
$$\times \left(\frac{c_1 t_2}{2} + \frac{t_1^2}{4}\left(c_2 - c_1 + \frac{c_1^2}{[2]_q - 1}\right)\right),$$

$$a_3 - \beta a_2^2 = \frac{1}{[3]_q^\kappa \Omega_3(a,b,\nu,q,k)([3]_q - 1)}$$
$$\times \left(\frac{c_1 t_2}{2} + \frac{t_1^2}{4}\left(c_2 - c_1 + \frac{c_1^2}{[2]_q - 1}\right)\right)$$
$$- \beta\left(\frac{c_1 t_1}{2[2]_q^\kappa \Omega_2(a,b,\nu,q,k)([2]_q - 1)}\right)^2.$$

Hence, for any $b \in \mathbb{C}$, we have

$$a_3 - \beta a_2^2 = \frac{c_1}{2[3]_q^\kappa \Omega_3(a,b,\nu,q,k)([3]_q - 1)}(t_2 - Bt_1^2),$$

where B is formulated in (6). According to Lemma 2, we reach the desired conclusion. $\square$

Corollary 5 ([9]). If $\phi \in \jmath - \mathcal{S}_{q,k,\wp}^\kappa(a,b)$ with $k = \nu(n) = 1$, $\forall n$, then

$$|a_3 - ba_2^2| \leq \frac{|c_1|}{2[3]_q^\kappa \Omega_3(a,b,1,q,1)([3]_q - 1)} \max\{1, |2B - 1|\},$$

where $b \in \mathbb{C}$ and

$$B := \frac{1}{2}\left(1 - \frac{c_2}{c_1} - c_1\left(\frac{1}{[2]_q - 1} - \beta\frac{[3]_q^\kappa([3]_q - 1)}{2\Omega_2(a,b,1,q,1)\left([2]_q^\kappa([2]_q - 1)\right)^2}\right)\right). \tag{7}$$

Corollary 6 ([31]). If $\phi \in \jmath - \mathcal{S}_{q,k,\wp}^\kappa(0,1)$ with $k = \nu(n) = 1$, $\forall n$ and $a = 0$, $b = 1$, then

$$|a_3 - \beta a_2^2| \leq \frac{|\rho_1|}{2[3]_q^k \Omega_3(0,1,1,q,1)([3]_q - 1)} \max\{1, |2B - 1|\},$$

where $\psi \in \mathbb{C}$ and

$$B := \frac{1}{2}\left(1 - \frac{c_2}{c_1} - c_1\left(\frac{1}{[2]_q - 1} - \beta\frac{[3]_q^\kappa([3]_q - 1)}{2\Omega_2(0,1,1,q,1)\left([2]_q^\kappa([2]_q - 1)\right)^2}\right)\right). \tag{8}$$

Theorem 4. Let $\phi \in \Pi$. If
$$\sum_{n=2}^{\infty} (([n]_q - 1)(\jmath + 1) + |\wp|)|\Omega_n(a,b,\nu,q,k)|[n]_q^\kappa |a_n| \leq |\wp|,$$
then $\phi \in \jmath - \mathcal{S}_{q,k,\wp}^\kappa(a,b)$.

Proof. A computation yields

$$\left| \frac{\eta \Delta_q \left({}_{\nu,k}\Delta_q^\kappa(a,b)\phi(\eta) \right)}{{}_{\nu,k}\Delta_q^\kappa(a,b)\phi(\eta)} - 1 \right|$$

$$= \left| \frac{\eta \Delta_q \left({}_{\nu,k}\Delta_q^\kappa(a,b)\phi(\eta) \right) - {}_{\nu,k}\Delta_q^\kappa(a,b)\phi(\eta)}{{}_{\nu,k}\Delta_q^\kappa(a,b)\phi(\eta)} \right|$$

$$= \left| \frac{\sum_{n=2}^{\infty} ([n]_q - 1)[n]_q^\kappa \Omega_n(a,b,\nu,q,k) a_n \eta^n}{\eta + \sum_{n=2}^{\infty} [n]_q^\kappa \Omega_n(a,b,\nu,q,k) a_n \eta^n} \right|$$

$$\leq \frac{\sum_{n=2}^{\infty} |([n]_q - 1)[n]_q^\kappa \Omega_n(a,b,\nu,q,k)||a_n|}{1 - \sum_{n=2}^{\infty} |[n]_q^\kappa \Omega_n(a,b,\nu,q,k)||a_n|}.$$

Given the theorem's circumstances, we obtain

$$1 - \sum_{n=2}^{\infty} |[n]_q^\kappa \Omega_n(a,b,\nu,q,k)||a_n| > 0.$$

Since

$$\left| \frac{\jmath}{\wp} \left(\frac{\eta \Delta_q \left({}_{\nu,k}\Delta_q^\kappa(a,b)\phi(\eta) \right)}{{}_{\nu,k}\Delta_q^\kappa(a,b)\phi(\eta)} - 1 \right) \right|$$

$$- \Re \left(\frac{1}{\wp} \left(\frac{\eta \Delta_q \left({}_{\nu,k}\Delta_q^\kappa(a,b)\phi(\eta) \right)}{{}_{\nu,k}\Delta_q^\kappa(a,b)\phi(\eta)} - 1 \right) \right)$$

$$\leq \frac{\jmath}{|\wp|} \left| \left(\frac{\eta \Delta_q \left({}_{\nu,k}\Delta_q^\kappa(a,b)\phi(\eta) \right)}{{}_{\nu,k}\Delta_q^\kappa(a,b)\phi(\eta)} - 1 \right) \right|$$

$$+ \left| \frac{1}{\wp} \right| \left| \frac{\eta \Delta_q \left({}_{\nu,k}\Delta_q^\kappa(a,b)\phi(\eta) \right)}{{}_{\nu,k}\Delta_q^\kappa(a,b)\phi(\eta)} - 1 \right|$$

$$= \frac{\jmath+1}{|\wp|}\left|\left(\frac{\eta\Delta_q\left(\nu,_k\Delta_q^\kappa(a,b)\phi(\eta)\right)}{\nu,_k\Delta_q^\kappa(a,b)\phi(\eta)}-1\right)\right|$$

$$= \frac{\jmath+1}{|\wp|}\left|\frac{\eta\Delta_q\left(\nu,_k\Delta_q^\kappa(a,b)\phi(\eta)\right)-\Delta_q\left(\nu,_k\Delta_q^\kappa(a,b)\phi(\eta)\right)}{\Delta_q\left(\nu,_k\Delta_q^\kappa(a,b)\phi(\eta)\right)}\right|$$

$$\leq \frac{\jmath+1}{|\wp|}\left(\frac{\sum_{n=2}^{\infty}|([n]_q-1)[n]_q^\kappa\,\Omega_n(a,b,\nu,q,k)||a_n|}{1-\sum_{n=2}^{\infty}|[n]_q^\kappa\,\Omega_n(a,b,\nu,q,k)||a_n|}\right)$$

$$\leq 1,$$

this means that $\phi \in \jmath - \mathcal{S}_{q,k,\wp}^{\kappa}(a,b)$. □

Corollary 7 ([9]). *Let $\phi \in \Pi$. If*

$$\sum_{n=2}^{\infty}(([n]_q-1)(\jmath+1)+|\wp|)\,|\Omega_n(a,b,1,q,1)|[n]_q^\kappa|a_n| \leq |\wp|,$$

then

$$\frac{\eta\Delta_q\left(\nu,_1\Delta_q^\kappa(a,b)\phi(\eta)\right)}{\nu,_1\Delta_q^\kappa(a,b)\phi(\eta)} \prec \mathfrak{P}_{\jmath,\wp}, \quad \eta \in \mathbb{I}.$$

That is, $\phi \in \jmath - \mathcal{S}_{q,1,\wp}^{\kappa}(a,b)$, when $k=\nu(n)=1, \forall n$.

Corollary 8 ([31]). *Let $\phi \in \Pi$. If*

$$\sum_{n=2}^{\infty}(([n]_q-1)(\jmath+1)+|\wp|)\,|\Omega_n(0,1,1,q,1)|[n]_q^\kappa|a_n| \leq |\wp|,$$

thus we get

$$\frac{\eta\Delta_q\left(\nu,_1\Delta_q^\kappa(0,1)\phi(\eta)\right)}{\nu,_1\Delta_q^\kappa(0,1)\phi(\eta)} \prec \mathfrak{P}_{\jmath,\wp}, \quad \eta \in \mathbb{I}.$$

That is, $\phi \in \jmath - \mathcal{S}_{q,k,\wp}^{\kappa}(0,1)$, when $k=\nu(n)=1, \forall n$.

4. Conclusion

Based on the Jackson calculus and the k-symbol, Raina's function in $\mathbb{I}$ is formulated. The suggested (q,k)-symbol differential operator was acted on the normalized subclass. As a consequence of this process, we explored the most important behavior of the operator geometrically. Corollaries are presented covering the work in Refs. [9, 31]. Moreover, Theorem 4 yields the sufficient condition of the class. For future works, one can consider the symmetric formula of the suggested operator by using the symmetric differential operator in Ref. [33]:

$$D_\gamma \phi(\eta) = \left(\frac{\gamma}{\bar{\gamma}}\right) \eta\, \phi'(\eta) - \left(1 - \frac{\gamma}{\bar{\gamma}}\right) \eta\, \phi'(-\eta).$$

References

[1] A. Gil, J. Segura and N. M. Temme, *Numerical Methods for Special Functions* (Society for Industrial and Applied Mathematics, Philadelphia, USA, 2007).

[2] F. H. Jackson, XI.–On q-functions and a certain difference operator, *Earth Environ. Sci. Trans. R. Soc. Edinb.* **46**(2), 253–281 (1909).

[3] D. O. Jackson, T. Fukuda, O. Dunn and E. Majors, On q-definite integrals *Q. J. Pure Appl. Math.* **41**, 193–203 (1910).

[4] M. E. H. Ismail, E. Merkes and D. Styer, A generalization of starlike functions, *Complex Var. Theory Appl. Int. J.* **14**(1–4), 77–84 (1990).

[5] W. Ma and D. Minda, A unified treatment of some special classes of univalent functions. In Z. Li, F. Ren, L. Yang and S. Zhang (eds.) *Proceedings of the Conference on Complex Analysis*, Tianjin, China, 19–23 June 1992 (International Press, New York, 1994), pp. 157–169.

[6] T. M. Seoudy and M. K. Aouf, Coefficient estimates of new classes of q-starlike and q-convex functions of complex order, *J. Math. Inequal.* **10**(1), 135–145 (2016).

[7] S. Zainab, A. Shakeel, M. Imran, N. Muhammad, H. Naz, S. N. Malik and M. Arif, Sufficiency criteria for starlike functions associated with cardioid, *J. Funct. Spaces.* **2021**, 1–9 (2021).

[8] S. B. Hadid, R. W. Ibrahim and S. Momani, A new measure of quantum starlike functions connected with Julia functions, *J. Funct. Spaces.* **2022**, 1–9 (2022).

[9] S. Noor and A. Razzaque, New subclass of analytic function involving-Mittag–Leffler function in conic domains, *J. Funct. Spaces.* **2022**, 1–9 (2022).
[10] R. W. Ibrahim and D. Baleanu, On quantum hybrid fractional conformable differential and integral operators in a complex domain, *Rev. Real Acad. Cienc. Exactas Fis. Nat. Ser. A. Mat.* **115**(1), 1–13 (2021).
[11] R. W. Ibrahim and D. Baleanu, Convoluted fractional differentials of various forms utilizing the generalized Raina's function description with applications, *J. Taibah Univ. Sci.* **16**(1), 432–441 (2022).
[12] H. Tang, S. Khan, S. Hussain and N. Khan, Hankel and Toeplitz determinant for a subclass of multivalent q-starlike functions of order α, *AIMS Math.* **6**, 5421–5439 (2021).
[13] I. Aldawish and R. W. Ibrahim, Solvability of a new q-differential equation related to q-differential inequality of a special type of analytic functions, *Fractal Fract.* **5**(4), 228 (2021).
[14] K. R. Karthikeyan, S. Lakshmi, S. Varadharajan, D. Mohankumar and E. Umadevi, Starlike functions of complex order with respect to symmetric points defined using higher order derivatives, *Fractal Fract.* **6**(2), 116 (2022).
[15] S. Riaz, U. A. Nisar, Q. Xin, S. N. Malik and A. Raheem, On starlike functions of negative order defined by q-fractional derivative, *Fractal Fract.* **6**(1), 30 (2022).
[16] A. K. Shukla and J. C. Prajapati, On a generalization of Mittag-Leffler function and its properties. *J. Math. Anal. Appl.* **336**(2), 797–811 (2007).
[17] H. J. Haubold, A. M. Mathai and R. K. Saxena, Mittag–Leffler functions and their applications, *J. Appl. Math.* **2011**, 1–52 (2011). Article ID 298628. https://doi.org/10.1155/2011/298628.
[18] P. A. Ryapolov and E. B. Postnikov, Mittag-Leffler function as an approximant to the concentrated ferrofluid's magnetization curve, *Fractal Fract.* **5**(4), 147 (2021).
[19] M. H. AlSheikh, N. M. G. Al-Saidi and R. W. Ibrahim, Dental X-ray identification system based on association rules extracted by k-Symbol fractional haar functions, *Fractal Fract.* **6**(11), 669 (2022).
[20] S. B. Hadid and R. W. Ibrahim, Geometric study of 2D-wave equations in view of K-symbol airy functions, *Axioms* **11**(11), 590 (2022).
[21] R. W. Ibrahim, K-symbol fractional order discrete-time models of Lozi system, *J. Differ. Equ. Appl.* **12**, 1–20 (2022).
[22] S. S. Miller and P. T. Mocanu, *Differential Subordinations: Theory and Applications* (CRC Press, Canada, 2000).
[23] S. Ruscheweyh, *Convolutions in Geometric Function Theory* (Les Presses De L'Universite De Montreal, Montreal, 1982).

[24] R. K. Raina, On generalized Wright's hypergeometric functions and fractional calculus operators, *East Asian Math. J.* **21**, 191–203 (2005).
[25] R. Diaz and E. Pariguan, On hypergeometric functions and Pochhammer k-symbol (2004). *arXiv preprint* math/0405596.
[26] S.-B. Chen, S. Rashid, Z. Hammouch, M. A. Noor, R. Ashraf and Y.-M. Chu, Integral inequalities via RainaâĂŹs fractional integrals operator with respect to a monotone function, *Adv. Differ. Equ.* **2020**(1), 1–20 (2020).
[27] G. S. Sàlàgean, Subclasses of univalent functions. In *Complex Analysis-Fifth Romanian-Finnish Seminar*, Part 1, Bucharest, 1981. Lecture Notes in Mathematics, Vol. 1013 (Springer, Berlin, 1983), pp. 362–372.
[28] M. Govindaraj and S. Sivasubramanian, On a class of analytic functions related to conic domains involving q-calculus, *Anal. Math.* **43**(3), 475–487 (2017).
[29] S. Kanas and S. Altinkaya, Functions of bounded variation related to domains bounded by conic sections, *Math. Slovaca* **69**(4), 833–842 (2019).
[30] S. Kanas, Techniques of the differential subordination for domains bounded by conic sections, *Int. J. Math. Math. Sci.* **2003**(38), 2389–2400 (2003).
[31] S. Hussain, S. Khan, M. A. Zaighum and M. Darus, Certain subclass of analytic functions related with conic domains and associated with Salagean q-differential operator, *AIMS Math.* **2**(4), 622–634 (2017).
[32] W. Rogosinski, On the coefficients of subordinate functions, *Proc. Lond. Math. Soc.* **2**(1), 48–82 (1945).
[33] R. W. Ibrahim, Normalized symmetric differential operators in the open unit disk. In *Approximation and Computation in Science and Engineering* (Springer, Cham, 2022), pp. 417–434.

© 2024 World Scientific Publishing Company
https://doi.org/10.1142/9789811267048_0012

Chapter 12

Analysis and Solvability of Complex K-Symbol Liu Fractional Dynamical Systems

Rabha W. Ibrahim

*Near East University, Mathematics Research Center,
Department of Mathematics, Near East Boulevard,
Nicosia/Mersin, Turkey
Department of Computer Science and Mathematics,
Lebanese American University,
Beirut, Lebanon
Information and Communication Technology Research Group,
Scientific Research Center, Al-Ayen University,
Thi-Qar, Iraq*

rabhaibrahim@yahoo.com

A strong and adaptable mathematical paradigm is provided by dynamic systems with complex variables, which finds use in a number of scientific, engineering and mathematical fields. They promote clearer understanding of intricate phenomena, make mathematical expressions simpler and provide a concise manner to depict oscillatory processes. In this effort, we extend the complex Liu system into K-symbol fractional calculus.

Then we examine the stability and stabilization of complex variables' K-symbol fractional Liu system based on the extended well-known theory of stability. Calculations involving fractions are conceptualized in terms of real and complex K-symbol Caputo derivatives. The approach relies on fractional-order system stability concepts. It forces numerical solutions, as well as the solvability of the extended system (existence and uniqueness). In addition, certain requirements for the solution are developed, such as founding the upper solution of the suggested system. We shall indicate that the upper solution is formulated by the K-symbol Mittag–Leffler function. In numerous mathematical and scientific fields, investigation on the Mittag–Leffler function is underway. It is a flexible tool for characterizing intricate and unusual behavior in a variety of situations. Finally, examples for finding the solutions are illustrated covering the theory on this effort. The solutions are formulated in terms of the K-symbol Mittag–Leffler function.

1. Introduction

A complex dynamic system is described by the Complex Liu system [1], a mathematical representation. The well-known Liu system, a set of ordinary differential equations that displays chaotic behavior, is a derivative of this system. The complex Liu system increases complexity and dynamics by including a complex term to the original Liu system. The fractional model is suggested in the sense of Caputo differential operator by the researchers in Refs. [2, 3]. Chaotic attractions, limit cycles and bifurcations are just a few of the dynamical behaviors that the complex Liu system displays. The system's chaotic behavior results from its vulnerability to the starting circumstances, which implies that slight modifications in the initial state can have a significant impact on the trajectory of the system. Understanding the dynamics of the complicated Liu system can help scientists and engineers better understand nonlinear systems, chaos theory and complicated behavior. It is often utilized as a test to create control plans, chaotic synchronization approaches and other nonlinear analytic methods [4–6].

The act of fractional orders is included in the generalization of classical fractional calculus known as K-symbol fractional calculus [7]. The derivative and integral operators in conventional fractional calculus have been adapted to noninteger orders, enabling the investigation of nonlocal and memory-dependent phenomena [8–10]. The fractional orders themselves are regarded as variables in K-symbol fractional calculus and can have many values at once. As a mathematical tool to represent complicated systems with memory impacts that cannot be fully described by single fractional orders, the K-symbol fractional calculus framework was created. It offers a more adaptable and flexible method for modeling and assessing these systems. A group of fractional orders, each linked to a weight or coefficient, are represented by the K-symbol notation. K-symbol fractional calculus facilitates the modeling of systems with various memory characteristics or heterogeneous qualities by taking into account numerous fractional orders. It enables the integration of numerous phenomenological or physical factors into a coherent framework. This can be especially helpful in fields like engineering, physics and processing signals where memory-rich complicated systems are present [11, 12]. Comparable to conventional fractional calculus, K-symbol fractional calculus includes fractional differentiation and integration as mathematical operations. However, the fractional orders are now denoted by the K-symbol and are no longer permanent values. Algebraically, the K-symbol can be used to perform operations on the systems and functions that fractional differential equations reflect [13, 14].

The aim of the recent study is to investigate the stability and stabilization of a challenging fractional K-symbol Liu system. Real and complex K-symbol Caputo derivatives are used to comprehend calculations using fractions. The method is based on ideas of fractional-order system stability. It demands numerical answers. Also, specified are a few prerequisites for the proposed approach including a set of conditions for the existence and uniqueness and the upper bound solution. Examples and observations are illustrated in the sequel. The computations in this effort are coded by the software MATHEMATICA 13.3. The K-symbol gamma function is calculated by this software. This chapter is organized as follows: Section 2 deals with the methodology that is used. Section 3 includes our main results and Section 4 admits the final conclusion of this analysis.

2. K-Symbol Methods

The gamma function of the motivation, as well as commonly known as the K-symbol gamma function, can be described as follows [7] (see Figure 1):

$$\Gamma_k(v) = \lim_{m \to \infty} \frac{m! k^m (mk)^{\frac{v}{k}-1}}{(v)_{m,k}}, \quad k > 0,$$

where

$$(v)_{m,k} := v(v+k)(v+2k)\ldots(v+(m-1)k)$$

and

$$(v)_{m,k} = \frac{\Gamma_k(v+mk)}{\Gamma_k(v)}.$$

Moreover,

$$\Gamma(v) = \lim_{k \to 1} \Gamma_k(v); \quad \Gamma_k(v) = k^{v/k-1}\Gamma(v/k); \quad \Gamma_k(v+k) = v\Gamma_k(v).$$

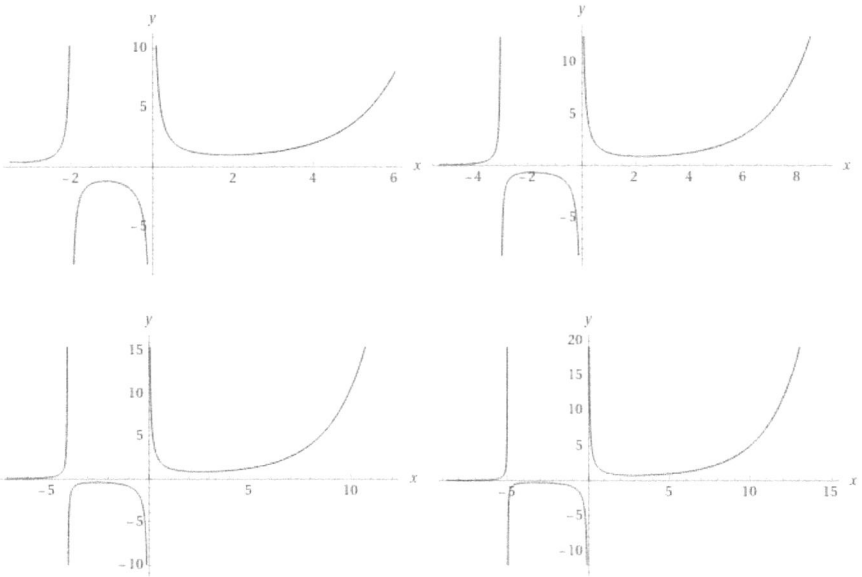

Figure 1. The plot of $\Gamma_k(x), k = 2, 3, 4$ accordingly.

Definition 1. The K-symbol Riemann–Liouville fractional integral has the following structure (see Figure 2):

$$\Upsilon^\nu_{a,k} h(\chi) = \frac{1}{k\Gamma_k(\nu)} \int_a^\chi (\chi - x)^{\nu/k-1} h(x) dx, \quad \nu > 0.$$

In addition, for $\nu \in (0,1)$, the formula for the K-Riemann–Liouville singular kernel is

$$\kappa^\nu_k(\chi) = \frac{(\chi)^{\nu/k-1}}{k\Gamma_k(\nu)}, \quad a = 0.$$

Example 1. For example, let $h(\chi) = 1$, then

$$\Upsilon^\nu_{a,k} h(\chi) = \frac{(\chi - a)^{\nu/k}}{\Gamma_k(\nu + k)}, \quad \nu \in (0,1].$$

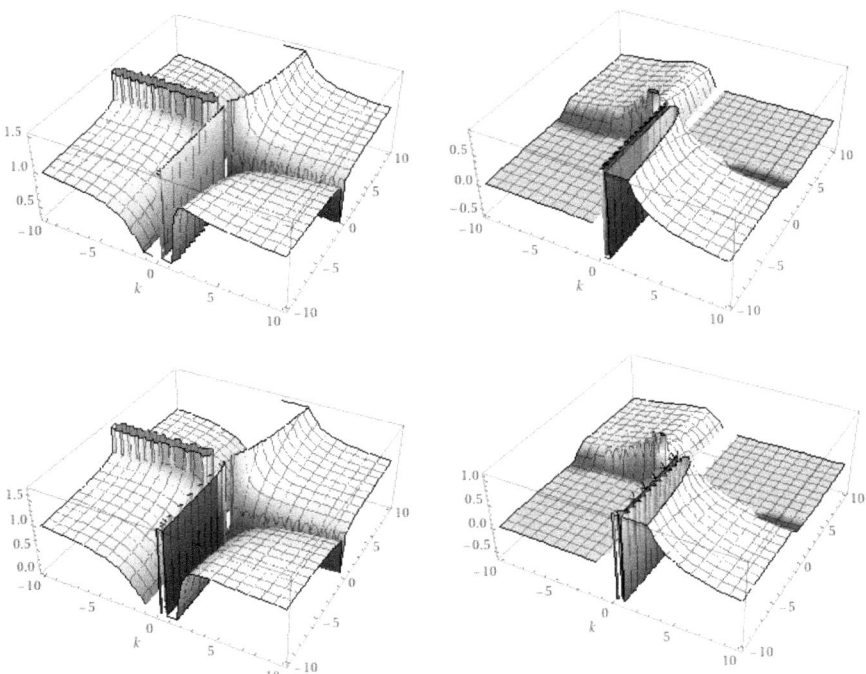

Figure 2. The plot of the K-symbol integral $\Upsilon^{1/2}_k$ and $\Upsilon^{3/4}_k$ accordingly.

For $a = 0$, we get

$$\Upsilon_k^\nu h(\chi) = \frac{1}{k\Gamma_k(\nu)} \int_0^\chi (\chi - x)^{\nu/k-1} h(x) \mathrm{d}x.$$

Hence, we have

$$\Upsilon_k^\nu h(\chi) = \frac{\chi^{\nu/k}}{\Gamma_k(\nu + k)}, \quad \nu \in (0, 1]$$

$$= \frac{\chi^{\frac{\nu}{k}}}{k^{\frac{\nu+k}{k}-1} \Gamma\left(\frac{\nu+k}{k}\right)}.$$

Example 2. Let $h(\chi) = \chi^{\frac{\mu}{k}}$

$$\Upsilon_k^\nu h(\chi) = \frac{\Gamma_k(\mu + k)}{\Gamma_k(\mu + \nu + k)} y^{\frac{\mu+\nu}{k}}, \quad \mu > 0$$

$$= \frac{k^{\frac{\mu+k}{k}-1} \Gamma\left(\frac{\mu+k}{k}\right)}{k^{\frac{\mu+\nu+k}{k}-1} \Gamma\left(\frac{\mu+\nu+k}{k}\right)} \chi^{\frac{\mu+\nu}{k}}$$

$$= k^{\frac{-\nu}{k}} \left(\frac{\Gamma\left(\frac{\mu+k}{k}\right)}{\Gamma\left(\frac{\mu+\nu+k}{k}\right)}\right) \chi^{\frac{\mu+\nu}{k}}.$$

It is obvious that we obtain the typical Riemann–Liouville fractional integral feature for $k = 1$:

$$\Upsilon_1^\nu h(\chi) = \left(\frac{\Gamma(\mu + 1)}{\Gamma(\mu + \nu + 1)}\right) \chi^{\mu+\nu}.$$

Definition 2. The Riemann–Liouville fractional derivative for the K-symbol calculus (see Figure 3) is indicated as follows, when $\nu \in (0, 1]$:

$$D_k^\nu h(\chi) = \frac{\mathrm{d}}{\mathrm{d}\chi} \left(\Upsilon_k^{1-\nu} g(y)\right).$$

Analysis and Solvability

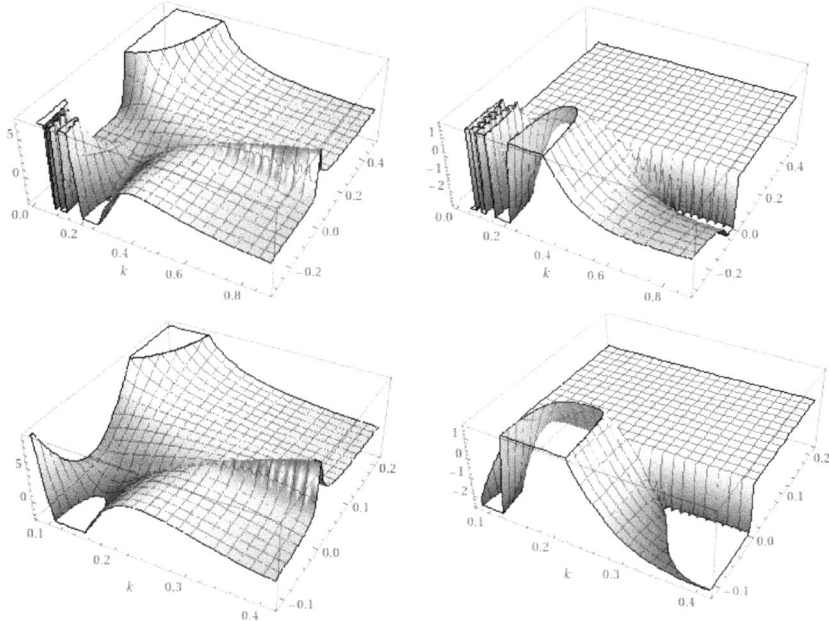

Figure 3. The plot of the K-symbol derivative $D_k^{1/2}$ and $D_k^{3/4}$ accordingly.

Or, in an extra generalized situation,
$$D_k^\nu h(\chi) = \frac{\mathrm{d}}{\mathrm{d}\chi}\left(k\,\Upsilon_k^{k-\nu} h(\chi)\right).$$

Example 3. Let $h(\chi) = 1$; then
$$D_k^\gamma h(\chi) = \frac{(1-\nu)y^{\frac{-(k+\nu-1)}{k}}}{k\Gamma_k(1-\nu+k)}$$
$$= \frac{(1-\nu)y^{\frac{-(k+\nu-1)}{k}}}{k^{\frac{1-\nu+k}{k}}\Gamma\left(\frac{1-\nu+k}{k}\right)}.$$

While, for $h(\chi) = \chi^{\mu/k}$ we have
$$D_k^\nu h(\chi) = (\mu+1-\nu)k^{\frac{\nu-1-k}{k}}\left(\frac{\Gamma\left(\frac{\mu+k}{k}\right)}{\Gamma\left(\frac{\mu-\nu+k+1}{k}\right)}\right)\chi^{\frac{\mu-\nu}{k}}.$$

Obviously, when $k = 1$, we get the usual fractional derivative

$$D_1^\nu h(\chi) = \left(\frac{\Gamma(\mu+1)}{\Gamma(\mu-\nu+1)}\right)\chi^{\mu-\nu}, \quad \nu \in (0,1].$$

Definition 3. The K-symbol Caputo calculus can be viewed by the following structure:

$$^C D_k^\nu h(\chi) = \begin{cases} k^\varpi \Upsilon_k^{\varpi k - \nu} h^{(\varpi)}(\chi) = \dfrac{k^{\varpi-1}}{\Gamma_k(\varpi k - \nu)} \int_0^\chi \dfrac{h^{(\varpi)}(x)}{(\chi-x)^{\nu/k-\varpi+1}}\,\mathrm{d}x \\ \dfrac{d^\varpi}{d\chi^\varpi} h(\chi), \quad \varpi k = \nu, \end{cases}$$

$$\big((\varpi-1)k < \nu < \varpi k,\ \chi > 0\big).$$

Remember that for $k = 1$, we have the classic Caputo derivative.

2.1. K-symbol Mittag–Leffler function

A particular mathematical function called the Mittag–Leffler function extends the idea of exponential functions to complex analysis and fractional calculus. In the late 19th and start of the 20th centuries, Swedish mathematicians Gosta Mittag–Leffler and Magnus Gustaf Mittag–Leffler developed it. The concept of special functions, fractional calculus, complex analysis and other areas of mathematics all heavily rely on this function.

The Mittag–Leffler function is a key component of fractional calculus's solutions to fractional differential equations and fractional integro-differential equations. It offers a means of bringing noninteger orders inside the scope of the differentiation and integration idea. It is utilized in complex analysis for both investigation of unique functions and the analysis of complicated functions. It can be used to solve specific kinds of integral equations and in a variety of other situations, including the theory of complete functions. It is discussed in relation to fractional Brownian motion in probability, which is utilized to explain specific categories of random processes with long-range dependency. Fractional differential equations are very useful for solving and studying fractional differential equations because they may precisely define a wide variety of physical processes.

Likewise depending on the K-symbol gamma function, the Mittag–Leffler form function can be explained as follows:

$$\Xi_{a,b}^k(\chi) = \sum_{\iota=0}^{\infty} \frac{\chi^\iota}{\Gamma_k(a\iota + b)}$$

$$= \sum_{\iota=0}^{\infty} \frac{\chi^\iota}{k^{\frac{a\iota+b}{k}-1} \Gamma\left(\frac{a\iota+b}{k}\right)}. \quad (1)$$

This function plays an important role in determining the upper solution of the suggested K-symbol dynamical system.

3. Results

For complex variables $\zeta_1 = z_1 + iz_2, \zeta_1^* = z_1 - iz_2, \zeta_2 = z_3 + iz_4$ and a real variable $\zeta_3 = z_5$, the complex Liu system is defined by

$$\dot{\zeta_1} = \sigma(-\zeta_1 + \zeta_2),$$
$$\dot{\zeta_2} = \rho\zeta_1 - \delta\zeta_1\zeta_3,$$
$$\dot{\zeta_3} = -\gamma\zeta_3 + \eta\zeta_1\zeta_1^*.$$

This system has equilibrium points as follows:

$$Q = \left\{ (0,0,0), \left(\pm\sqrt{\frac{\gamma\rho}{\eta\delta}}, \pm\sqrt{\frac{\gamma\rho}{\eta\delta}}, \frac{\rho}{\delta} \right) \right\},$$

where $\eta \neq 0$ and $\delta \neq 0$. Also, there are three nonzero eigenvalues:

$$\lambda_1 = -0.577216,$$
$$\lambda_2 = 0.5(-\sqrt{\sigma}\sqrt{4\rho + \sigma} - \sigma),$$
$$\lambda_3 = 0.5(\sqrt{\sigma}\sqrt{4\rho + \sigma} - \sigma),$$

where $4\rho + \sigma \geq 0$. Corresponding eigenvectors are

$$V_1 = (0, 0, 1),$$

$$V_2 = \left(-\frac{(\sigma + \sqrt{\sigma}\sqrt{4\rho + \sigma})}{(2\rho)}, 1, 0\right),$$

$$V_3 = \left(-\frac{(\sigma - \sqrt{\sigma}\sqrt{4\rho + \sigma})}{(2\rho)}, 1, 0\right), \quad \rho \neq 0.$$

In this chapter, using the K-symbol Caputo derivative, we proceed with the K-symbol fractional complex Liu system (see Figure 4):

$$\begin{aligned}
{}^C D_k^{\nu_1} \zeta_1 &= \sigma(-\zeta_1 + \zeta_2), \\
{}^C D_k^{\nu_2} \zeta_2 &= \rho \zeta_1 - \delta \zeta_1 \zeta_3, \\
{}^C D_k^{\nu_3} \zeta_3 &= -\gamma \zeta_3 + \eta \zeta_1 \zeta_1^*,
\end{aligned} \quad (2)$$

where $0 < \nu_j \leq 1$, $j = 1, 2, 3$. The system can be converted into the following 5D-real variables system:

$$\begin{aligned}
{}^C D_k^{\nu_1} z_1 &= \sigma(-z_1 + z_3), \\
{}^C D_k^{\nu_1} z_2 &= \sigma(-z_2 + z_4), \\
{}^C D_k^{\nu_2} z_3 &= \rho z_1 - \delta z_1 z_5, \\
{}^C D_k^{\nu_2} z_4 &= \rho z_2 - \delta z_2 z_5, \\
{}^C D_k^{\nu_3} z_5 &= -\gamma z_5 + \eta(z_1^2 + z_2^2).
\end{aligned} \quad (3)$$

System (3) is studied in the next investigation.

Definition 4. Consider the equation ${}^C D_k^{\nu} \varphi = \Phi(\chi, \varphi(\chi))$, $\nu \in (0, 1]$, has a zero solution such that for any initial values φ_0, there is a positive constant $\epsilon > 0$ which satisfies the inequality $\|\varphi(\chi)\| \leq \epsilon$, $\forall \chi > \chi_0$. Then the zero solution is called stable. If it is stable with the limit

$$\lim_{\chi \to \infty} \|\varphi(\chi)\| = 0,$$

then it is called asymptotically stable.

Analysis and Solvability 353

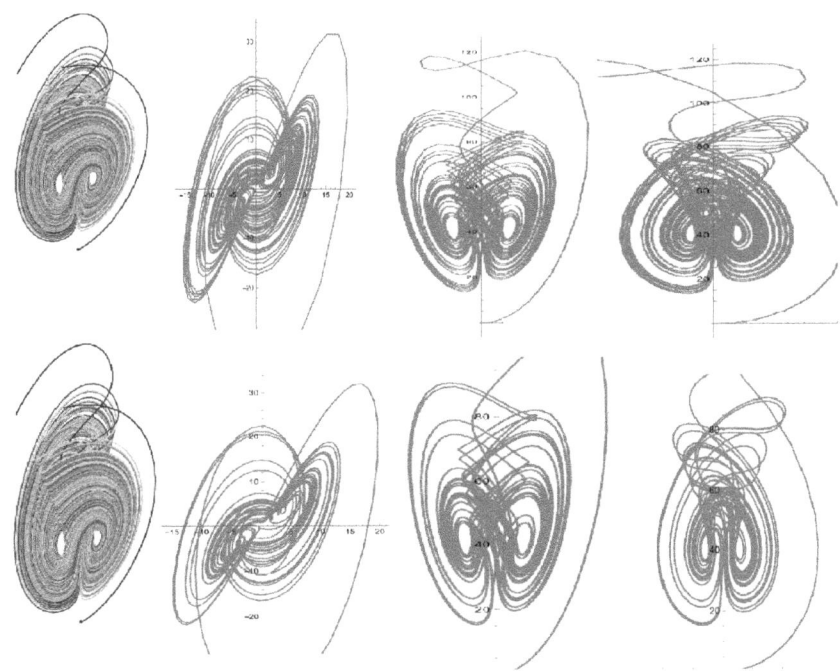

Figure 4. The plot of System (2), when $k = 1$, $\nu_1 = 14/17, \nu_2 = 15/17, \nu_3 = 16/17$ in the first row and $\nu_1 = \nu_2 = \nu_3 = 16/17$ in the second row. Also, the projections of (ζ_1, ζ_2), (ζ_1, ζ_3) and (ζ_2, ζ_3), respectively.

We have the following result, where the proof is quite similar when $k = 1$ (see Ref. [15]).

Lemma 1. *Consider the following system:*

$$^C D_k^\nu \varphi(\chi) = \Lambda \varphi(\chi) + \Omega(\varphi(\chi)), \quad \nu \in (0,1], \quad k \geq 1, \qquad (4)$$

where $\varphi(\chi) = (\varphi_1, \ldots, \varphi_n)^\top \in \mathbb{R}^n$, $\Lambda \in \mathbb{R}^{n \times n}$. If

- $|\arg(\mathrm{spec}(\Lambda))| > \frac{\nu \pi}{2k}$, *where $\mathrm{spec}(\Lambda)$ indicates the eigenvalues of Λ,*
- $\nu \|\Lambda\| > 1$,
-
$$\lim_{\varphi \to 0} \frac{\|\Omega(\varphi(\chi))\|}{\|\varphi(\chi)\|} = 0,$$

then (4) *admits asymptotically stable solution.*

Lemma 2. *Consider the following formula with a controller*
$$^C D_k^\nu \varphi(\chi) = (\Lambda + PQ)\varphi(\chi) + \Omega(\varphi(\chi)), \quad \nu \in (0,1], \quad k \geq 1, \quad (5)$$
where $Q \in \mathbb{R}^{1 \times n}$ *is a feedback,* $P \in \mathbb{R}^{n \times 1}$, $\varphi(\chi) = (\varphi_1, \ldots, \varphi_n)^\top \in \mathbb{R}^n$, $\Lambda \in \mathbb{R}^{n \times n}$. *If*

- $|\arg(\text{spec}(\Lambda))| > \frac{\pi\nu}{2k}$,
- $\nu \|\Lambda + PQ\| > 1$,
- $\lim_{\varphi \to 0} \frac{\|\Omega(\varphi(\chi))\|}{\|\varphi(\chi)\|} = 0$,

then (5) *admits asymptotically stable solution.*

Theorem 3. *For positive constants* $\gamma > 0$, $\sigma > 0$, *System* (3) *admits asymptotically stable behavior at the origin.*

Proof. It is possible to identify System (3) as having a structure (4), whereby

$$\Lambda = \begin{pmatrix} -\sigma & 0 & \sigma & 0 & 0 \\ 0 & -\sigma & 0 & \sigma & 0 \\ \rho & 0 & 0 & 0 & 0 \\ 0 & \rho & 0 & 0 & 0 \\ 0 & 0 & 0 & 0 & -\gamma \end{pmatrix} \quad \text{and} \quad \Omega(z) = \begin{pmatrix} 0 \\ 0 \\ -\delta z_1 z_5 \\ -\delta z_2 z_5 \\ \eta(z_1^2 + z_2^2) \end{pmatrix}.$$

Obviously, $h(u)$ satisfies

$$\lim_{z \to 0} \frac{\|\Omega(z(\chi))\|}{\|z(\chi)\|} = \lim_{z \to 0} \frac{\sqrt{(-\delta z_1 z_5)^2 + (-\delta z_2 z_5)^2 + \eta^2(z_1^2 + z_2^2)^2}}{\sqrt{(z_1)^2 + (z_2)^2 + (z_3)^2 + (z_4)^2 + (z_5)^2}}$$

$$\leq \lim_{z \to 0} \frac{\sqrt{\delta^2 z_5^2 (z_1^2 + z_2^2) + \eta^2 (z_1^2 + z_2^2)^2}}{\sqrt{z_1^2 + z_2^2}}$$

$$= \lim_{z \to 0} \sqrt{\delta^2 z_5^2 + \eta^2(z_1^2 + z_2^2)}$$

$$= 0.$$

Additionally, the system's characteristic equation is fulfilled:
$$(\gamma + \lambda)[\lambda^2 + \sigma\lambda - \rho\sigma]^2 = 0.$$

Since $\gamma > 0$, $\sigma > 0$, then

$$|\arg(\lambda_i)| > \frac{\pi}{2} > \frac{\pi\nu}{2k}, \quad \nu = \max(\nu_i, \, i = 1, 2, 3), \; k \geq 1. \quad (6)$$

In addition, for some $\gamma > 0$, $\sigma > 0$, we get $\|\Lambda\|\nu/k > 1$, $\nu/k := \min(\nu_i/k)$. It indicates that the state of equilibrium point (the origin) of System (3) is asymptotically stable as determined by Lemma 1. The demonstration is ended for. $\square$

Theorem 4. *For positive constants* $\gamma > 0$, $\sigma > 0$, *controlled System* (3) *admits asymptotically stable behavior at the origin.*

Proof. By putting the following input: $v(\chi) = PQz(\chi)$ on System (3) such that $P = (1,1,1,1,1)^\top$ and $Q = (1,1,1,1,1)$ and

$$|\arg(\operatorname{spec}(\Lambda + PQ))| > \frac{\nu\pi}{2k}; \quad \frac{\nu}{k}\|\Lambda + PQ\| > 1,$$

and

$$\lim_{z \to 0} \frac{\|\Omega(z(\chi))\|}{\|z(\chi)\|} = 0.$$

The origin of System (3), which is the equilibrium point, is inferred to be asymptotically stable by Lemma 2. The proof is finally complete. $\square$

3.1. *Stabilizing the origin*

According to the unique features of the system, one can employ a variety of control strategies to stabilize the origin of a fractional dynamic system. One can present an all-encompassing method here that makes use of fractional-order proportional-integral-derivative (FOPID) control.

Stability Analysis: Examine the closed-loop system's stability characteristics. In comparison to integer-order systems, stability criteria for fractional-order systems can be more complicated. Stability analysis can make use of tools like fractional Lyapunov stability theory.

Assessment and Simulation: Utilizing the created FOPID controller, simulate the closed-loop system. By looking at important parameters like settling time, overshoot and steady-state error, one may assess the system's performance and stability.

Improvement and Iteration: If essential, improve the controller structure based on the findings of the simulation. This can entail changing the controller gains or taking into account other control strategies.

Here, we establish a controller for (3) with K-symbol fractional order using the derivative of K-symbol fractional order. The proof is similar to the case $k = 1$.

Lemma 5. *The K-symbol fractional equation*

$$^{C}D_k^\nu \varphi(\chi) = \Omega(\varphi), \ \nu \in (0, 1], \tag{7}$$

admits asymptotically stable behavior if all eigenvalues of the Jacobian matrix computed at the fixed points achieve

$$|\arg[\lambda]| > \frac{\pi\nu}{2k}, \quad 0 < \nu < 1, \varphi \in R^n, \Omega : \mathbb{R}^n \to \mathbb{R}^n.$$

Theorem 6. *Suppose the Liu chaotic system with regulated K-symbol fractional order*

$$^{C}D_k^{\nu_1} z_1 = -\sigma z_1 + \sigma z_3,$$
$$^{C}D_k^{\nu_1} z_2 = -\sigma z_2 + \sigma z_4,$$
$$^{C}D_k^{\nu_2} z_3 = \rho z_1 - \delta z_1 z_5 + \Omega_1(z_1),$$
$$^{C}D_k^{\nu_2} z_4 = \rho z_2 - \delta z_2 z_5,$$
$$^{C}D_k^{\nu_3} z_5 = -\gamma z_5 + \eta(z_1^2 + z_2^2), \tag{8}$$

where

$$\Omega_1(z_1) = -\kappa_{11} \ ^{C}D_k^{\nu_1} z_1 - \kappa_{12} z_1$$

is the K-symbol fractional-order controller and κ_{1i}, $i = 1, 2$, is the feedback connection. If $\eta > 0, \sigma > 0$,

$$1 + \kappa_{11} > 0, \quad \kappa_{12} = \rho + \kappa_{11}\sigma,$$

then System (8) admits asymptotically behavior converge toward the origin.

Proof. The Jacobi matrix of (8) at the origin can be formulated in the next matrix

$$J = \begin{pmatrix} -\sigma & 0 & \sigma & 0 & 0 \\ 0 & -\sigma & 0 & \sigma & 0 \\ \rho + \kappa_{11}\sigma - \kappa_{12} & 0 & -\kappa_{11}\sigma & 0 & 0 \\ 0 & \rho & 0 & 0 & 0 \\ 0 & 0 & 0 & 0 & -\gamma \end{pmatrix}.$$

Since $\kappa_{12} = \rho + \kappa_{11}\sigma$, we have

$$J = \begin{pmatrix} -\sigma & 0 & \sigma & 0 & 0 \\ 0 & -\sigma & 0 & \sigma & 0 \\ 0 & 0 & -\kappa_{11}\sigma & 0 & 0 \\ 0 & \rho & 0 & 0 & 0 \\ 0 & 0 & 0 & 0 & -\gamma \end{pmatrix}.$$

Consequently, the characteristic equation has the following character:

$$(\gamma + \lambda)(\sigma + \lambda)[\lambda^3 + \sigma(\kappa_{11} + 1)\lambda^2 + \lambda\sigma(\kappa_{11}\sigma - \sigma) - \rho\sigma^2\kappa_{11}] = 0.$$

Since $\gamma > 0, \sigma > 0$ and $1 + \kappa_{11} > 0$, thus

$$|\arg[\lambda_i]| > \frac{\pi\nu}{2k}, \quad i = 1,\ldots,5,$$

where $\nu := \max \nu_j$, $j = 1, 2, 3$. The equilibrium point (the origin) of System (8) is asymptotically stable, which means that System (8) at the origin can be stabilized using a fractional-order derivative, as determined by Lemma 5. The evidence is conclusive. □

Similarly, we get the next outcomes:

Theorem 7. *Consider the controlled K-symbol chaotic system*

$$\begin{aligned} {}^C D_k^{\mu_1} z_1 &= -\sigma z_1 + \sigma z_3, \\ {}^C D_k^{\mu_1} z_2 &= -\sigma z_2 + \sigma z_4, \\ {}^C D_k^{\mu_2} z_3 &= \rho z_1 - \delta z_1 z_5, \\ {}^C D_k^{\mu_2} z_4 &= \rho z_2 - \delta z_2 z_5 + \Omega_2(z_2), \\ {}^C D_k^{\mu_3} z_5 &= -\gamma z_5 + \eta(z_1^2 + z_2^2), \end{aligned} \quad (9)$$

where
$$\Omega_2(z_2) = -\kappa_{21}{}^C D_k^{\nu_1} z_2 - \kappa_{22} z_2$$

is the controller of fractional power and κ_{2i}, $i = 1, 2$, is the feedback connection. If $\gamma > 0, \sigma > 0$,
$$1 + \kappa_{21} > 0, \quad \kappa_{22} = \rho + \kappa_{21}\sigma,$$

then unstable equilibrium point will be reached by System (9) as it asymptotically converges.

Theorem 8. *Consider the chaotic system with regulated K-symbols:*
$$\begin{aligned}
{}^C D_k^{\nu_1} z_1 &= -\sigma z_1 + \sigma z_3 + \Omega_3(z_3), \\
{}^C D_k^{\nu_1} z_2 &= -\sigma z_2 + \sigma z_4, \\
{}^C D_k^{\nu_2} z_3 &= \rho z_1 - \delta z_1 z_5, \\
{}^C D_k^{\nu_2} z_4 &= \rho z_2 - \delta z_2 z_5, \\
{}^C D_k^{\nu_3} z_5 &= -\gamma z_5 + \eta(z_1^2 + z_2^2),
\end{aligned} \quad (10)$$

where
$$\Omega_3(z_3) = -\kappa_{31}{}^C D_k^{\nu_2} z_3 - \kappa_{32} z_3$$

is the controller of fractional power and κ_{2i}, $i = 1, 2$, is the feedback connection. If
$$\gamma > 0, \quad \sigma > 0, \quad \kappa_{32} = \sigma + \kappa_{31}, \quad \sigma + \rho\kappa_{31} > 0,$$

then the unstable equilibrium point will be reached by asymptotically convergent in System (10).

3.2. *Computational outcome*

A numerical solution of the suggested system is investigated in this part. In many disciplines, including physics, engineering, economics and biology, computational investigations of dynamic systems incorporating complex variables are crucial. Variables in these systems, which make up the complex number space, have both real and fictitious components. Analyzing the performance and long-term evolution of such systems frequently involves the use of computational tools and computational models. Modeling chemical reactions, fluid dynamics, electronic circuits and population dynamics are only a few examples of the varied applications of computational studies of dynamic systems of complex variables. Other applications include the investigation of economic systems, neural networks and quantum systems. The individual system characteristics and the study goals influence the use of computational approaches.

Following are a few typical methods and methodologies for computational research of dynamic systems of complex variables:

Analysis of Stability: For many applications, knowledge of the stability of complicated dynamic systems is essential. The stability of fixed points, periodic orbits or chaotic behavior in complex systems can be evaluated using computational approaches.

Phase Portraits: Using the complex plane to visualize how complex systems behave can give us a better understanding of how they operate. Phase pictures show the system's trajectories and stability characteristics. In out investigation, we shall consider analytic computational technique using the definition of the K-symbol fractional power series.

Assume the partition and the initial conditions as follows:

$$h = T/M, \quad t_m = mh, \quad m = 1, \ldots, M\big(z_1(0), \ldots, z_5(0)\big).$$

Thus, System (3) can be viewed by the structure

$$z_1(m+1) = z_1(0) + \frac{h^{\nu_1/k}}{\Gamma_k(\nu_1+2)}\bigg[\sigma\big(-z_1^\ell(m+1) + z_3^\ell(m+1)\big)$$

$$+ \sum_{j=0}^{m} \Lambda_{1,j,m+1} \times \sigma\big(-z_1(j) + z_3(j)\big)\bigg],$$

$$z_2(m+1) = z_2(0) + \frac{h^{\nu_1/k}}{\Gamma_k(\nu_1+2)}\bigg[\sigma\big(-z_2^\ell(m+1) + z_4^\ell(m+1)\big)$$

$$+ \sum_{j=0}^{m} \Lambda_{2,j,m+1} \times \sigma\big(-z_2(j) + z_4(j)\big)\bigg],$$

$$z_3(m+1) = z_3(0) + \frac{h^{\nu_2/k}}{\Gamma_k(\nu_2+2)}\Big[\big(\rho z_1^\ell(m+1) - \delta z_1^\ell(m+1)z_5^\ell(m+1)\big)$$

$$+ \sum_{j=0}^{m} \Lambda_{3,j,m+1}\big(\rho z_1(j) - \delta z_1(j)z_5(j)\big)\Big],$$

$$z_4(m+1) = z_4(0) + \frac{h^{\nu_2/k}}{\Gamma_k(\nu_2+2)}\Big[\big(\rho z_2^\ell(m+1) - \delta z_2^\ell(m+1)z_5^\ell(m+1)\big)$$

$$+ \sum_{j=0}^{m} \Lambda_{4,j,m+1}\big(\rho z_2(j) - \delta z_2(j)z_5(j)\big)\Big],$$

$$z_5(m+1) = z_5(0) + \frac{h^{\nu_3/k}}{\Gamma_k(\nu_3+2)}\Big[-\gamma z_5^\ell(m+1)$$

$$+ \eta\big(z_1^{2\ell}(m+1) + z_2^{2\ell}(m+1)\big)$$

$$+ \sum_{j=0}^{m} \Lambda_{5,j,m+1}\big(-\gamma z_5(j) + \eta(z_1^2(j) + z_2^2(j))\big)\Big],$$

where

$$z_1^\ell(m+1) = z_1(0) + \sum_{j=0}^{m} \Pi_{1,j,m+1} \times \sigma\big(-z_1(j) + z_3(j)\big),$$

$$z_2^\ell(m+1) = z_2(0) + \sum_{j=0}^{m} \Pi_{2,j,m+1} \times \sigma\big(-z_2(j) + z_4(j)\big),$$

$$z_3^\ell(m+1) = z_3(0) + \sum_{j=0}^{m} \Pi_{3,j,m+1} \times \big(\rho z_1(j) - \delta z_1(j)z_5(j)\big),$$

$$z_4^\ell(m+1) = z_4(0) + \sum_{j=0}^{m} \Pi_{4,j,m+1} \times \big(\rho z_2(j) - \delta z_2(j)z_5(j)\big),$$

$$z_5^\ell(m+1) = z_5(0) + \sum_{j=0}^{m} \Pi_{5,j,m+1} \times \big(-\gamma z_5(j) + \eta(z_1^2(j) + z_2^2(j))\big),$$

$$\Lambda_{k,j,m+1} = \begin{cases} m^{\nu/k+1} - (m-\nu/k)(m+1)^{\nu/k}, & j = 0 \\ (m-j+2)^{\nu/k+1} + (m-j)^{\nu/k+1} \\ \quad -2(m-j+1)^{\nu/k+1}, & 1 \leq j \leq m \\ 1, & j = m+1, \end{cases}$$

and

$$\Pi_{k,j,m+1} = \frac{h^{\nu/k}}{\nu/k}[(m-j+1)^{\nu/k} - (m-j)^{\nu/k}], \quad 0 \leq j \leq m.$$

The error can be computed as follows:

$$|z_i(t_m) - z_i(m)| = o(h^{\nu/k}), \quad \ell = \min(2, 1 + \max \nu_{1,2,3}).$$

3.3. *Synchronizing at the origin*

When a factor is said to be "synchronizing at the origin", it usually refers to the process of aligning or coordinating many elements, systems or events such that they all begin or occur at the same time at a common place, which is frequently referred to as the "origin". Synchronizing at the origin, for instance, in the context of data analysis or signal processing could involve aligning several signals or data series so that their initial points meet at the same time or data index. To appropriately evaluate and interpret the data, this alignment is necessary. Synchronizing at the origin might be crucial for the efficient operation of systems or devices in other disciplines, such as engineering or communication. To ensure efficient and coordinated interpersonal interaction, for instance, endpoints in a network must synchronize their clocks to a common time reference (commonly referred to as network time protocol).

By using the fractional-order derivative, we get the system and implement a feedback controller of (3):

$$\begin{aligned}
{}^C D_k^{\nu_1} w_1 &= -\sigma w_1 + \sigma w_3, \\
{}^C D_k^{\nu_1} w_2 &= -\sigma w_2 + \sigma w_4, \\
{}^C D_k^{\nu_2} w_3 &= \rho w_1 - \delta w_1 w_5 + W, \\
{}^C D_k^{\nu_2} w_4 &= \rho w_2 - \delta w_2 w_5, \\
{}^C D_k^{\nu_3} w_5 &= -\gamma w_5 + \eta(w_1^2 + w_2^2),
\end{aligned} \tag{11}$$

where
$$W = \kappa_1[{}^C D_k^{\nu_1} w_1 - {}^C D_k^{\nu_1} z_1] + \kappa_2(w_1 - z_1) + \delta w_1 w_5 - \rho z_1$$

is the controller of fractional power and κ_i, $i = 1, 2$, is the feedback connect. We need the following result whose proof is similar to the case $k = 1$ [16].

Lemma 9. *The system*
$$^c D_k^\nu u = \Lambda u, \quad \nu \in (0, 1], \tag{12}$$

is asymptotically stable if and only if $|\arg[\lambda]| > \frac{\pi \nu}{2k}$ is fulfilled for all eigenvalues (λ) of Λ. Furthermore, it is stable if and only if $|\arg[\lambda]| \geq \frac{\pi \nu}{2k}$ achieves all eigenvalues (λ) of Λ and those critical eigenvalues which achieve $|\arg[\lambda]| = \frac{\pi \nu}{2k}, k \geq 1$, have geometric multiplicity one, where $\nu \in (0, 1), z \in \mathbb{R}^n$ and $\Lambda \in \mathbb{R}^n \times \mathbb{R}^n$.

Theorem 10. *If $\gamma > 0, \sigma > 0, \kappa_2 = -\rho + \kappa_1 \sigma$ and $\sigma(\kappa_1 + \rho) < 1$, then Systems (3) and (11) attained synchronization.*

Proof. Formulate the synchronization error variables as follows:
$$\varepsilon_i = w_i - z_i, \quad i = 1, \ldots, 5.$$

Therefore, we obtain the system
$$\begin{pmatrix} {}^C D_k^{\nu_1} \varepsilon_1 \\ {}^C D_k^{\nu_1} \varepsilon_2 \\ {}^C D_k^{\nu_2} \varepsilon_3 \\ {}^C D_k^{\nu_2} \varepsilon_4 \\ {}^C D_k^{\nu_3} \varepsilon_5 \end{pmatrix} = \Lambda \begin{pmatrix} \varepsilon_1 \\ \varepsilon_2 \\ \varepsilon_3 \\ \varepsilon_4 \\ \varepsilon_5 \end{pmatrix}, \tag{13}$$

where
$$\Lambda = \begin{pmatrix} -\sigma & 0 & \sigma & 0 & 0 \\ 0 & -\sigma & 0 & \sigma & 0 \\ \rho - \kappa_1 \sigma + \kappa_2 & 0 & k_1 \sigma & 0 & 0 \\ 0 & \rho - \delta z_5 & 0 & 0 & -\delta w_2 \\ \eta(w_1 + z_1) & \eta(w_2 + z_2) & 0 & 0 & -\gamma \end{pmatrix}.$$

Since $\kappa_2 = -\rho + \kappa_1\sigma$, then suitable values of z_i and w_i imply

$$\Lambda = \begin{pmatrix} -\sigma & 0 & \sigma & 0 & 0 \\ 0 & -\sigma & 0 & \sigma & 0 \\ 0 & 0 & \kappa_1\sigma & 0 & 0 \\ 0 & \rho & 0 & 0 & 0 \\ 0 & 0 & 0 & 0 & -\gamma \end{pmatrix}.$$

Thus, the characteristic equation is of the form

$$(\gamma + \lambda)(\sigma + \lambda)(\kappa_1\sigma - \lambda)[\lambda^2 + \sigma\lambda - \rho\sigma] = 0.$$

Since $\sigma > 0$, $\gamma > 0$, $\kappa_1 < 0$ yields that $|\arg[\lambda_i(\Lambda)]| > \frac{\pi}{2} > \frac{\pi\nu}{2k}$, $i = 1, \ldots, 5$. Since the equilibrium point of the error system, (13), is asymptotically stable due to Lemma 9, the Systems, (3) and (11), were synchronized by the fractional-order derivative of the K-symbol. The evidence is conclusive. □

3.4. *Solvability*

More investigation of System (3) is included in this part. This place is regarding the insolubility of the system. We aim to establish the existence and uniqueness of System (3). Let ξ_j, $j = 1, 2, 3$, be in the unit disk $U := \{z \in \mathbb{C} : |z| \leq 1\}$ and $t \in J = [0, T]$. In addition, we assume that $z_1, \ldots, z_5$ are continuous in J; we symbolize this set by $C(J)$. Let $\mathcal{S} = \{(z_1, \ldots, z_5)^T : z_1, \ldots, z_5 \in C(J)\}$ with the norm

$$\|\xi_1\|_{\mathcal{S}} = \|z_1\| + \|z_2\| = \sup_{t \in J}|z_1(t)| + \sup_{t \in J}|z_2(t)|,$$

$$\|\xi_2\|_{\mathcal{S}} = \|z_3\| + \|z_4\| = \sup_{t \in J}|z_3(t)| + \sup_{t \in J}|z_4(t)|$$

and

$$\|\xi_3\|_{\mathcal{S}} = \|z_5\| = \sup_{t \in J}|z_5(t)|.$$

Theorem 11. *Let $\rho = \delta$ and $\eta > -\gamma$ in System (3) with the initial condition $(z_1(0), \ldots, z_5(0))^T$. If*

$$\frac{|\sigma|T^{\nu_1/k}}{\Gamma_k(\nu_1+1)} + \frac{|\rho|T^{\nu_2/k}}{\Gamma_k(\nu_2+1)} + \frac{|\eta|T^{\nu_3/k}}{\Gamma_k(\nu_3+1)} < \frac{k}{4}, \quad T \in J,\ k \geq 1,$$

then (3) admits a single solution in $\mathcal{S}$.

Proof. System (3) can be formed as follows:

$$\left({}^C D_k^{\nu_1} z_1, {}^C D_k^{\nu_1} z_2\right)^T = \left(-\sigma z_1 + \sigma z_3, -\sigma z_2 + \sigma z_4\right)^T$$
$$\left({}^C D_k^{\nu_2} z_3, {}^C D_k^{\nu_2} z_4\right)^T = \left(\rho z_1 - \delta z_1 z_5, \rho z_2 - \delta z_2 z_5\right)^T \quad (14)$$
$${}^C D_k^{\nu_3} z_5 = -\gamma z_5 + \eta(z_1^2 + z_2^2).$$

Operating (14) by $I_k^{\nu_1}, I^{\mu_2}$ and $I_k^{\nu_3}$ respectively implies

$$(z_1, z_2)^T = \left(z_1(0) - \sigma I_k^{\nu_1} z_1 + \sigma I_k^{\nu_1} z_3, u_2(0) - \sigma I_k^{\nu_1} z_2 + \sigma I_k^{\nu_1} z_4\right)^T$$
$$(z_3, z_4)^T = \left(z_3(0) + \rho I_k^{\nu_2} z_1 - \delta I_k^{\nu_2} z_1 u_5, z_4(0) + \rho I_k^{\nu_2} z_2 - \delta I_k^{\nu_2} z_2 z_5\right)^T$$
$$z_5 = z_5(0) - \gamma I_k^{\nu_3} z_5 + \eta I_k^{\nu_3}(z_1^2 + z_2^2). \quad (15)$$

Define the operator $\mathcal{Z} : \mathcal{S} \to \mathcal{S}$ by

$$\mathcal{Z}(z_1, \ldots, z_5)^T = (z_1(0), \ldots, z_5(0))^T + \Big(-\sigma I_k^{\nu_1} z_1 + \sigma I_k^{\nu_1} z_3,$$
$$-\sigma I_k^{\nu_1} z_2 + \sigma I_k^{\nu_1} z_4, \rho I_k^{\nu_2} z_1 - \delta I_k^{\nu_2} z_1 z_5,$$
$$\rho I_k^{\nu_2} z_2 - \delta I_k^{\nu_2} z_2 z_5, -\gamma I_k^{\nu_3} z_5 + \eta I_k^{\nu_3}(z_1^2 + z_2^2)\Big)^T \quad (16)$$

Since $|z_i| < 1$, thus a computation yields

$$\|\mathcal{Z}(z_1, \ldots, z_5)^T - \mathcal{Z}(w_1, \ldots, w_5)^T\|_{\mathcal{S}}$$
$$= \Big\|\Big(-\sigma I_k^{\nu_1}(z_1 - w_1) + \sigma I_k^{\nu_1}(z_3 - w_3),$$
$$-\sigma I_k^{\nu_1}(z_2 - w_2) + \sigma I_k^{\nu_1}(z_4 - w_4),$$
$$\rho I_k^{\nu_2}(z_1 - w_1) - \delta I_k^{\nu_2}(z_1 z_5 - w_1 w_5),$$
$$\rho I_k^{\nu_2}(z_2 - w_2) - \delta I_k^{\nu_2}(z_2 z_5 - w_2 w_5),$$
$$-\gamma I^{\nu_3}(z_5 - w_5) + \eta I_k^{\nu_3}[(z_1^2 + z_2^2) - (w_1^2 + w_2^2)]\Big)^T\Big\|_{\mathcal{S}}$$

$$\leq \frac{4}{k}\left(\frac{|\sigma|T^{\nu_1/k}}{\Gamma_k(\nu_1+1)}+\frac{|\rho|T^{\nu_2/k}}{\Gamma_k(\nu_2+1)}+\frac{|\eta|T^{\nu_3/k}}{\Gamma_k(\nu_3+1)}\right)$$
$$\times \|(z_1,\ldots,z_5)^T-(w_1,\ldots,w_5)^T\|_{\mathcal{S}},$$

where $|v_i|<1$. Hence, by the Banach fixed point theorem, the system admits a single solution in $\mathcal{S}$. □

Remark 1. Since

$$\Gamma_k(v)=k^{\frac{v}{k}-1}\Gamma\left(\frac{v}{k}\right),$$

then the condition in Theorem 11 can be formulated as follows:

$$\frac{|\sigma|T^{\nu_1/k}}{k^{\frac{\nu_1+1}{k}}\Gamma\left(\frac{\nu_1+1}{k}\right)}+\frac{|\rho|T^{\nu_2/k}}{k^{\frac{\nu_2+1}{k}}\Gamma\left(\frac{\nu_2+1}{k}\right)}+\frac{|\eta|T^{\nu_3/k}}{k^{\frac{\nu_3+1}{k}}\Gamma\left(\frac{\nu_3+1}{k}\right)}<\frac{k}{4},\quad T\in J,\quad k\geq 1.$$

Example 4. Consider the system, with $t\in[0,1]$,

$$^CD_k^{14/17}z_1=0.05(-z_1+z_3),$$
$$^CD_k^{14/17}z_2=0.05(-z_2+z_4),$$
$$^CD_k^{15/17}z_3=0.06z_1-0.06z_1z_5,$$
$$^CD_k^{15/17}z_4=0.06z_2-0.06z_2z_5,$$
$$^CD_k^{16/17}z_5=-z_5+0.07(z_1^2+z_2^2). \tag{17}$$

Then

$$\frac{|\sigma|T^{\nu_1/k}}{k^{\frac{\nu_1+1}{k}}\Gamma\left(\frac{\nu_1+1}{k}\right)}+\frac{|\rho|T^{\nu_2/k}}{k^{\frac{\nu_2+1}{k}}\Gamma\left(\frac{\nu_2+1}{k}\right)}+\frac{|\eta|T^{\nu_3/k}}{k^{\frac{\nu_3+1}{k}}\Gamma\left(\frac{\nu_3+1}{k}\right)}$$
$$=\frac{0.05}{\Gamma(1.82)}+\frac{0.06}{\Gamma(1.88)}+\frac{0.07}{\Gamma(1.9)}$$
$$=0.1892<\frac{1}{4},\quad k=1.$$

Thus, in view of Theorem 11, System (17) has a unique solution for $k = 1$. Now, for $k = 2$, we have

$$\frac{|\sigma|T^{\nu_1/k}}{k^{\frac{\nu_1+1}{k}}\Gamma\left(\frac{\nu_1+1}{k}\right)} + \frac{|\rho|T^{\nu_2/k}}{k^{\frac{\nu_2+1}{k}}\Gamma\left(\frac{\nu_2+1}{k}\right)} + \frac{|\eta|T^{\nu_3/k}}{k^{\frac{\nu_3+1}{k}}\Gamma\left(\frac{\nu_3+1}{k}\right)}$$

$$= \frac{0.05}{2^{1.82/2}\Gamma(1.82/2)} + \frac{0.06}{2^{1.88/2}\Gamma(1.88/2)} + \frac{0.07}{2^{1.9/2}\Gamma(1.9/2)}$$

$$= 0.0903203\ldots < \frac{1}{2}, \quad k = 2.$$

As a consequence, Theorem 11 implies that System (17) has a unique solution for $k = 2$. Finally, for $k = 3$, we get

$$\frac{|\sigma|T^{\nu_1/k}}{k^{\frac{\nu_1+1}{k}}\Gamma\left(\frac{\nu_1+1}{k}\right)} + \frac{|\rho|T^{\nu_2/k}}{k^{\frac{\nu_2+1}{k}}\Gamma\left(\frac{\nu_2+1}{k}\right)} + \frac{|\eta|T^{\nu_3/k}}{k^{\frac{\nu_3+1}{k}}\Gamma\left(\frac{\nu_3+1}{k}\right)}$$

$$= \frac{0.05}{3^{1.82/3}\Gamma(1.82/3)} + \frac{0.06}{3^{1.88/3}\Gamma(1.88/3)} + \frac{0.07}{3^{1.9/3}\Gamma(1.9/3)}$$

$$= 0.0630574\ldots < \frac{3}{4}, \quad k = 3.$$

According to Theorem 11, System (17) has a unique solution for $k = 3$.

3.5. *Upper solution*

In this part, we investigate the upper bound of the solution, using the K-symbol Mittag–Leffler function (1). For a fractional power parameter $\nu \in (0,1)$, the set Δ can be defined as follows:

$$\Delta_k^\nu := \{\psi : J = [0,T] \to \mathbb{H}, \psi(\chi) = \Upsilon_k^\nu\varsigma(\chi), \varsigma \in L^\infty(J, \mathbb{H})\},$$

where $\mathbb{H}$ indicates the Hilbert space associated with the supremum norm.

Theorem 12. *Let $z_i, i = 1, \ldots, n \in \Delta_k^{\nu_i}$. Moreover, let*

$$^C D_k^{\nu_i} z_i \leq \lambda_i z_i(\chi) + b_i(\chi), \tag{18}$$

where $\lambda_i \in \mathbb{R}_+$ and $b_i : J \to \mathbb{R}_+$ is bounded and mensurable function. Then

$$z_i(\chi) \leq \Xi_{\nu_i,1}^k(\lambda_i \chi^{\nu_i}) z_i(0) + \int_0^\chi (\chi - x)^{\nu_i/k-1} \Xi_{\nu_i,\nu_i}^k$$
$$\left(\lambda_i(\chi - x)^{\nu_i/k}\right) b_i \, dx. \tag{19}$$

Proof. Suppose that there is a solution w_i which satisfies

$$^C D_k^{\nu_i} w_i(\chi) = \lambda_i w_i(\chi) + b_i(\chi)$$

with the initial condition $w_i(0) = z_i(0)$. Then in view of (18), we obtain

$$^C D_k^{\nu_i} (z_i(\chi) - w_i(\chi)) \leq \lambda_i (z_i(\chi) - w_i(\chi)).$$

Then a non-negative function u_i occurs on $[0, T]$, such that

$$^C D_k^{\nu_i} (z_i(\chi) - w_i(\chi)) = \lambda_i (z_i(\chi) - w_i(\chi)) - u_i(\chi).$$

The following non-negative function formulates the final equation's unique solution (see, for $k = 1$ [17], Theorem 2):

$$(w_i(\chi) - z_i(\chi)) = \int_0^\chi (\chi - x)^{\nu_i/k-1} \Xi_{\nu_i,\nu_i}^k \left(\lambda_i(\chi - x)^{\nu_i/k}\right) u_i(x) \, dx.$$

This yields that

$$z_i(\chi) \leq w_i(\chi), \quad i = 1, \ldots, n, \chi \in [0, T], \, w_i(0) = z_i(0).$$

Since w_i is a solution, then

$$w_i(\chi) = \Xi_{\nu_i,1}^k(\lambda_i \chi^{\nu_i}) w_i(0) + \int_0^\chi (\chi - x)^{\nu_i/k-1} \Xi_{\nu_i,\nu_i}^k$$
$$\times \left(\lambda_i(\chi - x)^{\nu_i/k}\right) b_i(x) \, dx.$$

Thus, we obtain the necessary inequality (19). □

As a consequence of Theorem 12, we have the following result:

Corollary 13. *Assume that in System (3), for $z_i, i = 1, \ldots, 5$, with non-negative connection parameters, the upper solution is given by the formula (19).*

Example 5.

- For $\nu_i = 16/17$, $\sigma = \gamma = 1$ and $\delta = 0, k = 1$, System (3) has a solution in terms of the K-symbol Mittag–Leffler function, under the initial conditions

$$(\zeta_1(0) = -3,\ \zeta_2(0) = 5,\ \zeta_3(0) = 1),$$

$$\zeta_1(\chi) \to \frac{1}{20}\left(25 - \sqrt{5}\right)\left(\sqrt{5} - 1\right) \Xi_{\frac{16}{17}}\left(\frac{1}{2}\left(\sqrt{5} - 1\right)\chi^{16/17}\right)$$

$$+ \frac{1}{20}\left(-\sqrt{5} - 1\right)\left(\sqrt{5} + 25\right) \Xi_{\frac{16}{17}}\left(\frac{1}{2}\left(-\sqrt{5} - 1\right)\chi^{16/17}\right),$$

$$\zeta_2(\chi) \to \frac{1}{10}\left(25 - \sqrt{5}\right) \Xi_{\frac{16}{17}}\left(\frac{1}{2}\left(\sqrt{5} - 1\right)\chi^{16/17}\right)$$

$$+ \frac{1}{10}\left(\sqrt{5} + 25\right) \Xi_{\frac{16}{17}}\left(\frac{1}{2}\left(-\sqrt{5} - 1\right)\chi^{16/17}\right),$$

$$\zeta_3(\chi) \to \Xi_{\frac{16}{17}}\left(-\chi^{16/17}\right).$$

- For $\nu_i = 16/17$, $\sigma = \gamma = 1$ and $\delta = 0, k = 2$, System (3) has a solution in terms of the K-symbol Mittag–Leffler function:

$$\zeta_1(\chi) \to \frac{1}{20}\left(25 - \sqrt{5}\right)\left(\sqrt{5} - 1\right) \Xi_{\frac{8}{17}}\left(\frac{1}{2}\left(\sqrt{5} - 1\right)\chi^{8/17}\right)$$

$$+ \frac{1}{20}\left(-\sqrt{5} - 1\right)\left(\sqrt{5} + 25\right) \Xi_{\frac{8}{17}}\left(\frac{1}{2}\left(-\sqrt{5} - 1\right)\chi^{8/17}\right),$$

$$\zeta_2(\chi) \to \frac{1}{10}\left(25 - \sqrt{5}\right) \Xi_{\frac{8}{17}}\left(\frac{1}{2}\left(\sqrt{5} - 1\right)\chi^{8/17}\right)$$

$$+ \frac{1}{10}\left(\sqrt{5} + 25\right) \Xi_{\frac{8}{17}}\left(\frac{1}{2}\left(-\sqrt{5} - 1\right)\chi^{8/17}\right),$$

$$\zeta_3(\chi) \to \Xi_{\frac{8}{17}}\left(-\chi^{8/17}\right).$$

4. Conclusion

In view of the K-symbol fractional calculus, the K-symbol fractional Liu system is defined and studied. We considered the system for complex variables. The analysis of the suggested system is presented to investigate the stability, stabilization, numerical solution and insolubility (the existence and uniqueness). Stability of the system is investigated theoretically. The method is based on the original case, when $K = 1$. Synchronizing at the origin is discussed by illustrating a set of conditions to obtain the controlling behavior of the system. Upper bound solution is determined by using the K-symbol Mittag–Leffler function. Examples are illustrated to show the movement of the findings. For future work, a modification of the suggested system can be held using the quantum calculus or mixed two-dimensional fractional calculus.

References

[1] C. Liu, T. Liu, L. Liu and K. Liu, A new chaotic attractor, *Chaos Solitons Fractals* **22**(5), 1031–1038 (2004).

[2] R. W. Ibrahim, Stability and stabilizing of fractional complex Lorenz systems, *Abstr. Appl. Anal.* (Hindawi Limited) **2013**, 1–13 (2013).

[3] R. Ibrahim and H. Jalab, Existence and uniqueness of a complex fractional system with delay, *Open Phys.* **11**(10), 1528–1535 (2013).

[4] C. Li, J. C. Sprott, X. Zhang, L. Chai and Z. Liu, Constructing conditional symmetry in symmetric chaotic systems, *Chaos Solitons Fract.* **155**, 111723 (2022).

[5] N. A. Saeed, H. A. Saleh, W. A. El-Ganaini, M. Kamel and M. S. Mohamed, On a new three-dimensional chaotic system with adaptive control and chaos synchronization, *Shock Vib.* **2023**, 1–19 (2023).

[6] A. Malik, M. Ali, F. S. Alsubaei, N. Ahmed and H. Kumar, A color image encryption scheme based on singular values and chaos, *CMES-Comput. Model. Eng. Sci.* **137**(1), 965–999 (2023).

[7] R. Diaz and E. Pariguan, On hypergeometric functions and Pochhammer K-symbol, *Divulg. Mat.* **15**, 179–192 (2007).

[8] C. Yildiz and L.-I. Cotirla, Examining the Hermite-Hadamard inequalities for k-fractional operators using the Green function, *Fractal Fract.* **7**(2), 161 (2023).

[9] J.-G. Liu, X.-J. Yang, Y.-Y. Feng and L.-L. Geng, A new fractional derivative for solving time fractional diffusion wave equation, *Math. Methods Appl. Sci.* **46**(1), 267–272 (2023).

[10] S. Rashid, M. Aslam Noor, K. Inayat Noor and Yu-Ming Chu, Ostrowski type inequalities in the sense of generalized K-fractional integral operator for exponentially convex functions, *AIMS Math.* **5**(3), 2629–2645 (2020).

[11] N. Samira and M. N. Naeem, On the generalization of k-fractional Hilfer-Katugampola derivative with Cauchy problem, *Turk. J. Math.* **45**(1), 110–124 (2021).

[12] S. B. Hadid and R. W. Ibrahim, Geometric study of 2D-wave equations in view of K-symbol airy functions, *Axioms* **11**(11), 590 (2022).

[13] R. W. Ibrahim, K-symbol fractional order discrete-time models of Lozi system, *J. Differ. Equ. Appl.* **12**, 1–20 (2022).

[14] D. Baleanu, A. S. Al-Shamayleh and R. W. Ibrahim, Image splicing detection using generalized Whittaker function descriptor, *Comput. Mater. Contin.* **75**(2), 3465–3477 (2023).

[15] A. K. Golmankhaneh, R. Arefi and D. Baleanu, The proposed modified Liu system with fractional order, *Adv. Math. Phys.* **2013**, 1–7 (2013).

[16] D. Matignon, Stability results for fractional differential equations with applications to control processing, *Comput. Eng. Syst. Appl.* **2**(1), 963–968 (1996).

[17] A. A. Kilbas, M. Rivero, L. Rodriguez-Germa and J. J. Trujillo, Caputo linear fractional differential equations, *IFAC Proc. Vol.* **39**(11), 52–57 (2006).

© 2024 World Scientific Publishing Company
https://doi.org/10.1142/9789811267048_0013

Chapter 13

Elastic Scattering by an Inhomogeneous Medium with Unknown Buried Obstacles

Angeliki K. Kaiafa[*], George Kanakoudis[†], and Vassilios Sevroglou[‡]

*Department of Statistics and Insurance Science,
School of Finance and Statistics,
University of Piraeus, Piraeus, Greece*

[*]*angeliki_kaiafa@unipi.gr*
[†]*gkanak@unipi.gr*
[‡]*bsevro@unipi.gr*

In this chapter, the direct and inverse scattering problem of time harmonic elastic waves by an inhomogeneous medium with buried objects inside is studied. Initially, the well-posedness of the direct scattering problem by the variational method in a suitable Sobolev space setting is presented and proved. Uniqueness, existence and continuity dependence of solution of the problem from the boundary data of the buried obstacles are established. Further, the corresponding inverse problem is studied and in particular the factorization method for shape reconstruction and location of the support of the inhomogeneous medium are exploited. In addition, an inversion algorithm for shape recovering of the medium is presented and proved as well. Last but not least, useful remarks and conclusions concerning the direct scattering problem and its linchpin with the corresponding inverse one in elastic media are given.

1. Introduction

In this chapter, a direct scattering problem and the corresponding inverse one in two-dimensional linear elasticity are studied. We consider the problem of scattering of time harmonic elastic waves by an inhomogeneous medium with unknown buried objects inside. Our inhomogeneous medium is assumed to be an open bounded domain having a smooth C^2-boundary. Our buried object(s) is a bounded rigid obstacle and its boundary which is assumed to be *Lipschitz* satisfies a Dirichlet boundary condition.

The applications on elastic materials or environments concerning buried obstacles are very extensive (see Refs. [1, 4, 5] and therein references). Some examples playing a very important role in the real world include problems that occur in various areas of applications, such as radar, remote sensing, geophysics, nondestructive testing, materials science, civil and mechanical engineering [2].

This chapter first deals with the well-posedness of the direct scattering problem. In Section 2, we present and establish the above issue by using a suitable variational method in the appropriate *Sobolev* space setting. In particular, we exploit the ideas of Qu [13] in the case of acoustic scattering and extend its results in the more complicated elastic case. The study of the solvability of the direct problem in elasticity, using a boundary integral equations approach, has been also established in Refs. [3, 11].

In this chapter, we also study the corresponding elastic inverse problem. In particular, the factorization method, first proposed by Kirsch [6] for inverse obstacle scattering problems in a homogeneous background medium, is applied. In Section 3, we are concerned with the recovering of shape and location of the boundary of the support of the inhomogeneous medium which has multiple buried objects in its interior. On each of them, different boundary conditions may occur, and for simplicity, we first consider the case of a Dirichlet boundary condition, the boundary of the buried obstacle. The proposed factorization method is applied to the inverse problem for the case when our buried obstacle has a Dirichlet boundary condition, with mass density in its exterior less than 1, i.e., $\rho(\mathbf{r}) < 1$. Finally, we end up with useful and fruitful conclusions as well as with remarks upon buried objects in elastic media.

2. The Direct Scattering Problem

In this section, the direct scattering problem is presented and its well-posedness is proved by studying its variational formulation. The problem is mathematically described by a Dirichlet boundary value problem. Let $D \subset \mathbb{R}^2$ denote an open connected bounded domain having a smooth C^2-boundary ∂D. In what follows, D is considered to be an inhomogeneous medium, which contains a finite number of impenetrable buried objects denoted by $D_0^k, k = 1, 2, \ldots, n$. We use the notation $D_0 = \cup_{k=1}^n D_0^k$, assuming that $D_0^{k_1} \cap D_0^{k_2} = \emptyset$, if $k_1 \neq k_2$ and let $D_1 := D \setminus \bar{D}_0$.

From the mathematical point of view, our problem is described by the following exterior boundary value problem: *Find a vector function* $\mathbf{u} \in [H_{loc}^1(\mathbb{R} \setminus D_0)]^2$ *such that*

$$\Delta^* \mathbf{u}(\mathbf{r}) + \rho(\mathbf{r})\omega^2 \mathbf{u}(\mathbf{r}) = \mathbf{0} \quad \text{in } \mathbb{R}^2 \setminus \bar{D}_0, \tag{1}$$

$$\mathbf{u}(\mathbf{r}) = \mathbf{0} \quad \text{on } \partial D_0, \tag{2}$$

$$\lim_{r \to +\infty} \sqrt{r} \left(\frac{\partial \mathbf{u}_\alpha^{sct}(\mathbf{r})}{\partial r} - ik_\alpha \mathbf{u}_\alpha^{sct}(\mathbf{r}) \right) = \mathbf{0}, \quad \alpha = p, s, \ r := |\mathbf{r}|. \tag{3}$$

The Kupradze radiation condition (3) holds uniformly in all directions $\hat{\mathbf{r}} = \frac{\mathbf{r}}{r}$ and k_p, k_s are the wave numbers for the longitudinal and the transverse waves, respectively. In the scattering problem (1)–(3), we use the notation $\mathbf{u}^{tot} \equiv \mathbf{u} = \mathbf{u}^{inc} + \mathbf{u}^{sct}$ where $\mathbf{u}$ is the total field in $\mathbb{R}^2 \setminus \bar{D}_0$, $\mathbf{u}^{inc}$ is the incident plane wave and $\mathbf{u}^{sct}$ denotes the corresponding scattered field. The explicit expression for Δ^* is given by

$$\Delta^* \mathbf{u}(\mathbf{r}) := \mu(\mathbf{r})\Delta \mathbf{u}(\mathbf{r}) + (\lambda(\mathbf{r}) + \mu(\mathbf{r})) \nabla \nabla \cdot \mathbf{u}(\mathbf{r}), \tag{4}$$

where $\omega \in \mathbb{R}$ is the so-called frequency parameter, the constants λ and μ are the Lamé constants and $\rho(\mathbf{r})$ is the density of the elastic medium. In order for the medium to sustain longitudinal as well as transverse waves, the following strong ellipticity conditions

$$\mu > 0, \quad \lambda + 2\mu > 0$$

are satisfied. Further, we assume $\rho \in L^\infty(\mathbb{R}^2 \setminus \bar{D}_0)$ having $\rho > 0$, with $\rho = 1$ in $\mathbb{R}^2 \setminus \bar{D}_1$ and $\rho \neq 1$ in D_1. Due to the linearity of the differential operator Δ^* and taking into account relation $\mathbf{u} = \mathbf{u}^{inc} + \mathbf{u}^{sct}$, equation (1) yields

$$\Delta^*\mathbf{u}^{inc}(\mathbf{r}) + \Delta^*\mathbf{u}^{sct}(\mathbf{r}) + \rho(\mathbf{r})\omega^2\mathbf{u}^{inc}(\mathbf{r}) + \rho(\mathbf{r})\omega^2\mathbf{u}^{sct}(\mathbf{r}) = \mathbf{0},$$

and therefore

$$\Delta^*\mathbf{u}^{sct}(\mathbf{r}) + \rho(\mathbf{r})\omega^2\mathbf{u}^{sct}(\mathbf{r}) = -(\Delta^*\mathbf{u}^{inc}(\mathbf{r}) + \rho(\mathbf{r})\omega^2\mathbf{u}^{inc}) \text{ in } \mathbb{R}^2 \setminus \bar{D}_0.$$

If we now consider $q := -(\Delta^* + \rho(\mathbf{r})\omega^2)$ and $\mathbf{f}_1 = \mathbf{u}^{inc}$ in $\mathbb{R}^2 \setminus \bar{D}_0$, the latter arrives at

$$\Delta^*\mathbf{u}^{sct} + \rho(\mathbf{r})\omega^2\mathbf{u}^{sct} = -q\mathbf{f}_1 \text{ in } \mathbb{R}^2 \setminus \bar{D}_0,$$

and since $\rho(\mathbf{r}) \neq 1$ in D_1 and $\rho(\mathbf{r}) = 1$ in $\mathbb{R}^2 \setminus \bar{D}_1$, we get $q\mathbf{f}_1 = (\rho(\mathbf{r}) - 1)\omega^2\mathbf{f}_1$ in $D_1 = D \setminus \bar{D}_0$. Furthermore, considering the boundary condition (2), i.e., $\mathbf{u}^{sct} = -\mathbf{u}^{inc}$ on ∂D_0, if we set the fact that $\mathbf{f}_2 := -\mathbf{u}^{inc}$ on ∂D_0, we get

$$\mathbf{u}^{sct} = -\mathbf{f}_2 \quad \text{on } \partial D_0.$$

Using the above relations, problem (1)–(3) can be written as follows:

$$\Delta^*\mathbf{w} + \rho(\mathbf{r})\omega^2\mathbf{w} = -q\mathbf{f}_1 \quad \text{in } \mathbb{R}^2 \setminus \bar{D}_0, \tag{5}$$

$$\mathbf{w} = -\mathbf{f}_2 \quad \text{on } \partial D_0, \tag{6}$$

$$\lim_{r \to +\infty} \sqrt{r}\left(\frac{\partial \mathbf{w}_\alpha \mathbf{r}}{\partial \mathbf{r}} - ik_\alpha \mathbf{w}_\alpha(\mathbf{r})\right) = \mathbf{0}, \quad \alpha = p, s, \quad r := |\mathbf{r}|. \tag{7}$$

We now consider a unit vector $\hat{\mathbf{d}} \in S^1$ being the direction of the wave propagation, where $S^1 := \{\mathbf{r} \in \mathbb{R}^2 : |\mathbf{r}| = 1\}$ and by $B_R := \{\mathbf{r} \in \mathbb{R}^2 : |\mathbf{r}| < R\}$ a circular disk centered at the origin of radius $R > 0$, such that $D \subset B_R$. Hence, in what follows, we consider the following domain Ω_R as

$$\Omega_R := B_R \setminus \bar{D}_0,$$

Elastic Scattering by an Inhomogeneous Medium 375

and moreover, we define the following Hilbert space:

$$X := \{\mathbf{p} \in H^1(\Omega_R) : \mathbf{p}|_{\partial D_0} = \mathbf{0}\}.$$

Now multiplying equation (5) with a vector test function $\bar{\boldsymbol{\phi}} \in X$, we get

$$\bar{\boldsymbol{\phi}} \cdot \Delta^* \mathbf{w} + \rho\omega^2 \mathbf{w} \cdot \bar{\boldsymbol{\phi}} = -q\mathbf{f}_1 \cdot \bar{\boldsymbol{\phi}} \quad \text{in } \Omega_R,$$

and therefore

$$\bar{\boldsymbol{\phi}} \cdot [\mu \Delta \mathbf{w} + (\lambda + \mu)\nabla(\nabla \cdot \mathbf{w})] + \rho\omega^2 \mathbf{w} \cdot \bar{\boldsymbol{\phi}} = -q\mathbf{f}_1 \cdot \bar{\boldsymbol{\phi}}. \qquad (8)$$

Integrating (8) in Ω_R, we get

$$\mu \int_{\Omega_R} \bar{\boldsymbol{\phi}} \cdot \Delta \mathbf{w} \, d\mathbf{r} + (\lambda + \mu) \int_{\Omega_R} \bar{\boldsymbol{\phi}} \cdot \nabla(\nabla \cdot \mathbf{w}) \, d\mathbf{r} + \int_{\Omega_R} \rho\omega^2 \mathbf{w} \cdot \bar{\boldsymbol{\phi}} \, d\mathbf{r}$$
$$= \int_{\Omega_R} -q\mathbf{f}_1 \cdot \bar{\boldsymbol{\phi}} \, d\mathbf{r}. \qquad (9)$$

If we take into account that $\bar{\boldsymbol{\phi}}|_{\partial D_0} = \mathbf{0}$, the following identity

$$\int_{\Omega_R} \nabla(\nabla \cdot \mathbf{w}) \cdot \bar{\boldsymbol{\phi}} \, ds = -\int_{\Omega_R} (\nabla \cdot \mathbf{w})(\nabla \cdot \bar{\boldsymbol{\phi}}) \, d\mathbf{r} - \int_{\partial \Omega_R} (\nabla \cdot \mathbf{w})\hat{\mathbf{n}} \cdot \bar{\boldsymbol{\phi}} \, ds \qquad (10)$$

can be written as

$$\int_{\Omega_R} \nabla(\nabla \cdot \mathbf{w}) \cdot \bar{\boldsymbol{\phi}} \, ds = -\int_{\Omega_R} (\nabla \cdot \mathbf{w})(\nabla \cdot \bar{\boldsymbol{\phi}}) \, ds - \int_{\partial B_R} (\nabla \cdot \mathbf{w})\hat{\mathbf{n}} \cdot \bar{\boldsymbol{\phi}} \, ds. \qquad (11)$$

Since now

$$\int_{\Omega_R} (\Delta \mathbf{w}) \cdot \bar{\boldsymbol{\phi}} \, d\mathbf{r} = -\int_{\Omega_R} (\nabla \mathbf{w}) : (\nabla \bar{\boldsymbol{\phi}}) \, d\mathbf{r} - \int_{\partial B_R} \hat{\mathbf{n}}(\nabla \cdot \mathbf{w}) \cdot \bar{\boldsymbol{\phi}} \, ds, \qquad (12)$$

where ":" denotes the *Frobenius product* between two dyadics and $q\mathbf{f}_1 = \mathbf{0}$ on $\mathbb{R}^2 \setminus \overline{D}_1$, i.e.,

$$\int_{\Omega_R} -q\mathbf{f}_1 \cdot \bar{\boldsymbol{\phi}} \, d\mathbf{r} = -\int_{D_1} q\mathbf{f}_1 \cdot \bar{\boldsymbol{\phi}} \, d\mathbf{r}, \qquad (13)$$

we can conclude that relation (9) via (11)–(13) arrives at

$$-\mu \int_{\Omega_R} (\nabla \mathbf{w}) : (\nabla \overline{\boldsymbol{\phi}}) \, \mathrm{d}r - \mu \int_{\partial B_R} \hat{\mathbf{n}} \cdot (\nabla \mathbf{w}) \cdot \overline{\boldsymbol{\phi}} \, \mathrm{d}s$$

$$-(\lambda + \mu) \int_{\Omega_R} (\nabla \cdot \mathbf{w})(\nabla \cdot \overline{\boldsymbol{\phi}}) \, \mathrm{d}r - (\lambda + \mu) \int_{\partial B_R} (\nabla \cdot \mathbf{w}) \hat{\mathbf{n}} \cdot \overline{\boldsymbol{\phi}} \, \mathrm{d}s$$

$$+ \int_{\Omega_R} \rho \omega^2 \mathbf{w} \cdot \overline{\boldsymbol{\phi}} \, \mathrm{d}r$$

$$= - \int_{D_1} \mathrm{q} \mathbf{f}_1 \cdot \overline{\boldsymbol{\phi}} \, \mathrm{d}r. \tag{14}$$

We now introduce the *Poincaré–Steklov* type operator

$$T_{dn} = \mu(\hat{\mathbf{n}} \cdot \nabla \mathbf{w}) + (\lambda + \mu)(\nabla \cdot \mathbf{w}) \, \hat{\mathbf{n}} \quad \text{on } \partial B_R, \tag{15}$$

where $\hat{\mathbf{n}}$ stands for the outward unit normal vector, defined a.e. on the boundary $\partial \Omega_R$ at the point $\mathbf{r}$. The operator T_{dn} maps $\tilde{\mathbf{w}}$ to $T_{dn}\mathbf{w}$, where $\tilde{\mathbf{w}}$ is a radiating solution of the exterior Dirichlet problem with boundary data $\tilde{\mathbf{w}}|_{\partial \Omega_R} = \mathbf{w}$. Due to (15), problems (5)–(7) are equivalent to the following:

$$\Delta^* \mathbf{w} + \rho \omega^2 \mathbf{w} = -\mathrm{q} \mathbf{f}_1 \quad \text{in } \Omega_R, \tag{16}$$

$$\mathbf{w} = -\mathbf{f}_2 \quad \text{on } \partial D_0, \tag{17}$$

$$\mu(\hat{\mathbf{n}} \cdot \nabla \mathbf{w}) + (\lambda + \mu)(\nabla \cdot \mathbf{w}) \, \hat{\mathbf{n}} = T_{dn}\mathbf{w} \quad \text{on } \partial B_R. \tag{18}$$

With the use of T_{dn}, relation (14) via relation (15) gives

$$\mu \int_{\Omega_R} (\nabla \mathbf{w}) : (\nabla \overline{\boldsymbol{\phi}}) \, \mathrm{d}r - (\lambda + \mu) \int_{\Omega_R} (\nabla \cdot \mathbf{w})(\nabla \cdot \overline{\boldsymbol{\phi}}) \, \mathrm{d}r$$

$$+ \omega^2 \int_{\Omega_R} \rho \mathbf{w} \cdot \overline{\boldsymbol{\phi}} \, \mathrm{d}r + <T_{dn}\mathbf{w}, \boldsymbol{\phi}>_{\partial B_R}$$

$$= - \int_{D_1} \mathrm{q} \mathbf{f}_1 \cdot \overline{\boldsymbol{\phi}} \, \mathrm{d}r. \tag{19}$$

Let us now assume $\mathbf{w} \in H^1(B_R \setminus D_0)$ such that $\mathbf{w}_0 = -\mathbf{f}_2$ on ∂D_0 and $\mathbf{w}_0 = \mathbf{0}$ on ∂B_R. Then for $\mathbf{u} = \mathbf{w} - \mathbf{w}_0$ in (19), we get the following variational formulation of (16)–(18): *Find a solution* $\mathbf{u} \in [H^1_{\partial D_0, \text{loc}}(\mathbb{R}^2 \setminus D_0)]^2$ *such that*

$$\alpha(\mathbf{u}, \boldsymbol{\phi}) = \ell(\boldsymbol{\phi}) - \alpha(\mathbf{w}_0, \boldsymbol{\phi}), \qquad (20)$$

where

$$\alpha(\mathbf{u}, \boldsymbol{\phi}) = \mu \int_{\Omega_R} (\nabla \boldsymbol{\phi}) : (\nabla \mathbf{u}) \, ds - (\lambda + \mu) \int_{\Omega_R} (\nabla \cdot \boldsymbol{\phi})(\nabla \cdot \mathbf{u}) \, dr$$
$$+ \int_{\Omega_R} \rho \omega^2 \mathbf{u} \cdot \boldsymbol{\phi} \, dr - <T_{dn}\mathbf{u}, \boldsymbol{\phi}>_{\partial B_R} \qquad (21)$$

and

$$\ell(\boldsymbol{\phi}) = \int_{D_1} q\mathbf{f}_1 \cdot \boldsymbol{\phi} \, dr \qquad (22)$$

for every $\boldsymbol{\phi} \in X$.

In order to establish existence and uniqueness for (20), we need first to prove the following three lemmas.

Lemma 1. *The linear functional $\ell(\boldsymbol{\phi})$ is bounded, i.e., there exists a positive constant c'_1 such that*

$$|\ell(\boldsymbol{\phi})| \leq c'_1 \|\boldsymbol{\phi}\|_{H^1(\Omega_R)}. \qquad (23)$$

Proof. Using the Cauchy–Schwarz inequality, relation (22) yields

$$|\ell(\boldsymbol{\phi})| \leq \|q\mathbf{f}_1\|_{L^2(D_1)} \|\boldsymbol{\phi}\|_{H^1(D_1)}. \qquad (24)$$

Considering that $q\mathbf{f}_1$ is bounded and $|\hat{\mathbf{n}}| = 1$, we have

$$|\ell(\boldsymbol{\phi})| \leq c_1 \|\boldsymbol{\phi}\|_{H^1(D_1)}, \qquad (25)$$

for some positive constant c_1 and therefore since $D_1 \subseteq \Omega_R$, there exists positive constant c'_1 such that

$$|\ell(\boldsymbol{\phi})| \leq c'_1 \|\boldsymbol{\phi}\|_{H^1(\Omega_R)}. \qquad (26)$$
□

Lemma 2. *The bilinear form $a(\mathbf{u}, \boldsymbol{\phi})$ is bounded, i.e.,*

$$|\alpha(\mathbf{u}, \boldsymbol{\phi})| \leq c_3 \|\mathbf{u}\|_{H^1(\Omega_R)} \|\boldsymbol{\phi}\|_{H^1(\Omega_R)}. \qquad (27)$$

Proof. By relation (21) and using the Cauchy–Schwarz inequality, we can arrive at

$$|\alpha(\mathbf{u}, \boldsymbol{\phi})| \leq \mu \|\nabla \mathbf{u}\|_{L^2(\Omega)} \|\nabla \boldsymbol{\phi}\|_{L^2(\Omega)} + (\lambda + \mu) \|\nabla \cdot \mathbf{u}\|_{L^2(\Omega)}$$
$$\times \|\nabla \cdot \boldsymbol{\phi}\|_{L^2(\Omega)} c_1 \omega^2 \|\mathbf{u}\|_{L^2(\Omega)} + \|\boldsymbol{\phi}\|_{L^2(\Omega)}$$
$$+ \|T_{dn}\mathbf{u}\|_{H^{-1/2}(\partial B_R)} \|\boldsymbol{\phi}\|_{H^{1/2}(\partial B_R)}, \qquad (28)$$

for some positive constant c_1. It has been proved in Ref. [8] that

$$\|T_{dn}\mathbf{u}\|_{H^{-\frac{1}{2}}(\partial\Omega_R)} \leq C \|\mathbf{u}\|_{H^{\frac{1}{2}}(\partial\Omega_R)}, \qquad (29)$$

and hence via (28), we get

$$|\alpha(\mathbf{u}, \boldsymbol{\phi})| \leq c_3 \|\mathbf{u}\|_{H^1(\Omega_R)} \|\boldsymbol{\phi}\|_{H^1(\Omega_R)}, \qquad (30)$$

for some positive constant c_3. □

Lemma 3. *The following coercivity property for $\alpha(\mathbf{u}, \mathbf{u})$ holds:*

$$\text{Re}\{\alpha(\mathbf{u}, \mathbf{u})\} \geq c' \|\mathbf{u}\|^2_{H^1(\Omega_R)}. \qquad (31)$$

Proof.

$$\text{Re}\{\alpha(\mathbf{u}, \mathbf{u})\} = \mu \|\nabla \mathbf{u}\|^2_{L^2(\Omega_R)} + (\lambda + \mu) \|\nabla \mathbf{u}\|^2_{L^2(\Omega_R)}$$
$$- \rho \omega^2 \|\mathbf{u}\|^2_{L^2(\Omega_R)} - \text{Re}\{<T_{dn}\mathbf{u}, \mathbf{u}>\}$$
$$\geq c_5 \|\mathbf{u}\|^2_{H^1(\Omega_R)} - c_6 \|\mathbf{u}\|^2_{L^2(\Omega_R)}, \qquad (32)$$

and therefore the assertion of the theorem is secured. □

We can now state the following.

Theorem 1. *Problems (16)–(18) have a unique solution $\mathbf{u} \in H^1(B_R \setminus \overline{D}_0)$, which satisfies the following inequality:*

$$\|\mathbf{u}\|_{H^1(\Omega_R)} \leq c \left(\|\mathbf{f}_2\|_{H^{1/2}(\partial D_0)} + \|\mathbf{f}_1\|_{L^2(D_1)} \right) \qquad (33)$$

for some positive constant c.

Proof. Let $\ell(\phi) = 0$. Then $q\mathbf{f}_1 = \mathbf{0}$ in D_1 and therefore $\mathbf{f}_1 = \mathbf{0}$ in D_1. It follows from (20) that $\mathbf{u} \in X$ solves problems (5)–(7) with $\mathbf{f}_1 = \mathbf{f}_2 = \mathbf{0}$, and by Rellich's lemma we have that $\mathbf{u} = \mathbf{0}$. Now, the homogeneous equation

$$\alpha(\mathbf{u}, \phi) = \mathbf{0} \tag{34}$$

admits only the trivial solution, hence equation (20) admits at most one solution. From this conclusion, along with Lemmas 1–3 and Lax–Milgram theorem, there is a unique solution of the variational problem satisfies inequality (33). □

Let the operator $G : [L^2(D_1)]^2 \times [H^{1/2}(\partial D_0)]^2 \to [L^2(S^1)]^2$, given by

$$G(\mathbf{f}_1, \mathbf{f}_2) = \mathbf{w}_\infty \tag{35}$$

with $\mathbf{w}_\infty$ being the far-field pattern of the solution $\mathbf{w}$ of (5)–(7) with given data $(\mathbf{f}_1, \mathbf{f}_2)^\top \in [L^2(D_1)]^2 \times [H^{1/2}(\partial D_0)]^2$.

Lemma 4. *The operator G is compact with a dense range in $L^2(S^1)$.*

Proof. We prove that $\text{Range}(G) = X$ and hence we show that the adjoint operator G^* has the property $Ker(G^*) = \{\mathbf{0}\}$. We consider that $\mathbf{w}$ is the scattering solution of (5)–(7) with data $(\mathbf{f}_1, \mathbf{f}_2)^\top \in [L^2(D_1)]^2 \times [H^{1/2}(\partial D_0)]^2$, whereas $\mathbf{u}$ is the solution of (1)–(3) which corresponds to the incident field $\mathbf{u}^{\text{inc}}$ (see later, relation (57) and Ref. [12]). Using the far-field pattern $\mathbf{w}_\infty$, relation (35) and the Kupradze radiation condition (3) satisfied by $\mathbf{w}$, we can arrive at

$$\begin{aligned}(G(\mathbf{f}_1, \mathbf{f}_2), \phi) &= \frac{1}{\sqrt{8\pi\omega}} \int_{\partial D} [\,T\mathbf{u}^{\text{inc}} \cdot \mathbf{w}\big|_+ - \mathbf{u}^{\text{inc}} \cdot T\mathbf{w}\big|_+]\,ds \\ &= \frac{1}{\sqrt{8\pi\omega}} \int_{\partial D} [\,T(\mathbf{u} - \mathbf{u}^{\text{sct}}) \cdot \mathbf{w}\big|_+ \\ &\quad + (\mathbf{u}^{\text{sct}} - \mathbf{u}) \cdot T\mathbf{w}\big|_+]\,ds \\ &= \frac{1}{\sqrt{8\pi\omega}} \int_{\partial D} [\,T\mathbf{u} \cdot \mathbf{w}\big|_+ - \mathbf{u} \cdot T\mathbf{w}\big|_+]\,ds, \end{aligned} \tag{36}$$

where relation $\mathbf{u}^{\text{inc}} = \mathbf{u} - \mathbf{u}^{\text{sct}}$ has been used, $\phi \in [L^2(S^1)]^2$ and the surface stress operator is given by

$$T^{(r)}\mathbf{u} := 2\mu\,\hat{\mathbf{n}}_\mathbf{r} \cdot \nabla + \lambda\,\hat{\mathbf{n}} + \mathbf{r}\,\nabla\cdot + \mu\,\hat{\mathbf{n}}_\mathbf{r} \times \nabla\times. \tag{37}$$

In equation (37), $\hat{\mathbf{n}}_\mathbf{r}$ stands for the outward unit normal vector on ∂D at the point $\mathbf{r}$, and the superscript on T (which is omitted from now on) denotes the action of the differential operator on the indicated variable. Taking now into account that $\partial D_1 = \partial D \cup \partial D_0$ as well as $\mathbf{u}|_{\partial D_0} = \mathbf{0}$, we get

$$(G(\mathbf{f}_1, \mathbf{f}_2), \phi) = \frac{1}{\sqrt{8\pi\omega}} \int_{\partial D_1} \left([T\mathbf{u} \cdot \mathbf{w}|_- - \mathbf{u} \cdot T\mathbf{w}|_-\right) ds$$
$$- \int_{\partial D_0} T\mathbf{u} \cdot \mathbf{w}|_- \, ds. \tag{38}$$

From third Betti's formula, we have

$$\int_{\partial D_1} \left(T\mathbf{u} \cdot \mathbf{w}|_- - \mathbf{u} \cdot T\mathbf{w}|_-\right) ds = \int_{D_1} (\mathbf{w} \cdot \Delta^*\mathbf{u} - \mathbf{u} \cdot \Delta^*\mathbf{w}) d\mathbf{r}, \tag{39}$$

hence, relation (38) via (39) can be written as

$$(G(\mathbf{f}_1, \mathbf{f}_2), \phi) = \frac{1}{\sqrt{8\pi\omega}} \left\{ \int_{D_1} (\mathbf{w} \cdot \Delta^*\mathbf{u} - \mathbf{u} \cdot \Delta^*\mathbf{w}) dv \right.$$
$$\left. - \int_{\partial D_0} T\mathbf{u} \cdot \mathbf{w}|_- \, ds \right\}. \tag{40}$$

In the domain D_1, we also have

$$\Delta^*\mathbf{w} = (1 - \rho(\mathbf{r}))\omega^2 \mathbf{f}_1 \tag{41}$$

and

$$\Delta^*\mathbf{u} = -\rho(\mathbf{r})\omega^2 \mathbf{u}, \tag{42}$$

and with the substitution of (41) and (42) in (40), we arrive at

$$(G(\mathbf{f}_1, \mathbf{f}_2), \phi) = \frac{1}{\sqrt{8\pi\omega}} \left\{ \int_{D_1} (\mathbf{w} \cdot (-\rho(\mathbf{r})\omega^2 \mathbf{u}) - \mathbf{u} \right.$$
$$\left. \cdot (1 - \rho(\mathbf{r}))\omega^2 \mathbf{f}_1) d\mathbf{r} - \int_{\partial D_0} T\mathbf{u} \cdot \mathbf{w} \, ds \right\}$$
$$= \frac{1}{\sqrt{8\pi\omega}} \left\{ \int_{\partial D_1} (\rho(\mathbf{r}) - 1)\omega^2 \mathbf{u} \cdot \mathbf{f}_1 \, d\mathbf{r} + \int_{\partial D_0} T\mathbf{u} \cdot \mathbf{f}_2 \, ds \right\}. \tag{43}$$

Therefore,

$$(G(\mathbf{f}_1, \mathbf{f}_2), \boldsymbol{\phi}) = \frac{1}{\sqrt{8\pi\omega}} \left\{ \int_{D_1} q\mathbf{f}_1 \cdot \mathbf{u}\,d\mathbf{r} - \int_{\partial D_0} T\mathbf{u} \cdot \mathbf{f}_2\,ds \right\} \quad (44)$$

or, equivalently

$$(G(\mathbf{f}_1, \mathbf{f}_2), \boldsymbol{\phi}) = \frac{1}{\sqrt{8\pi\omega}} \left\{ \int_{D_1} q\mathbf{f}_1 \cdot \overline{\overline{\mathbf{u}}}\,d\mathbf{r} - \int_{\partial D_0} \overline{T\mathbf{u}} \cdot \mathbf{f}_2 \right\} ds. \quad (45)$$

Combination of (36) and (45) yields

$$G^*\boldsymbol{\phi} = (\overline{q\mathbf{u}}|_{D_1}, -\overline{T\mathbf{u}}|_{\partial D_0}). \quad (46)$$

Let now for $\boldsymbol{\phi}$ assume that $G^*\boldsymbol{\phi} = \mathbf{0}$. It follows from (46) that $\mathbf{u} = \mathbf{0}$ in D_1, and from the unique continuation principle (see Ref. [10]), we get $\mathbf{u} = \mathbf{u}^{\text{inc}} + \mathbf{u}^{\text{sct}} = \mathbf{0}$ in $\mathbb{R}^2 \setminus \overline{D}$. Since $\mathbf{u}^{\text{inc}}$ does not satisfy the radiation condition, we have $\mathbf{u}^{\text{inc}} = \mathbf{0}$ in $\mathbb{R}^2 \setminus \overline{D}$. Then by Theorem 3.15 in Ref. [4], we obtain that $\boldsymbol{\phi} = \mathbf{0}$ and thus G^* is injective which completes the proof of the lemma. □

We now state the following theorem.

Theorem 2. *For* $\mathbf{a} \in \mathbb{R}^2$, *we define*

$$\boldsymbol{\varphi}_\mathbf{a}(\hat{\mathbf{r}}) := \left(\frac{1}{\lambda + 2\mu} \frac{i+1}{4\sqrt{\pi k_p}} e^{-ik_p\hat{\mathbf{r}}\cdot\mathbf{a}}\, \hat{\mathbf{r}} \cdot \mathbf{d},\ \frac{1}{\mu} \frac{i+1}{4\sqrt{\pi k_s}} e^{-ik_s\hat{\mathbf{r}}\cdot\mathbf{a}}\, \hat{\mathbf{r}}^\perp \cdot \mathbf{d} \right),$$

$$\hat{\mathbf{r}} \in S^1 \quad (47)$$

then

$$\mathbf{a} \in D \quad \Leftrightarrow \quad \boldsymbol{\varphi}_\mathbf{a} \in \text{Range}(G). \quad (48)$$

Proof. We let $\mathbf{a} \in D$ and define circle $B_\delta(\mathbf{a})$ of radius $\delta > 0$ centered at $\mathbf{a}$ such that $\overline{B}_\varepsilon$ is contained in D. We consider a cut-off function X such that $X(\mathbf{y}) = 1$ for every $\mathbf{y}$ with $|\mathbf{y}| \geq \delta$ and $X(\mathbf{y}) = 0$ for every $\mathbf{y}$ with $|\mathbf{y}| < \frac{\delta}{2}$. Further, we consider

$$\mathbf{w}(\mathbf{r}) = X(\mathbf{r} - \mathbf{a})\,\Gamma(\mathbf{r}, \mathbf{a}; \tau), \quad (49)$$

where $\boldsymbol{\Gamma}(\mathbf{r}, \mathbf{a}; \tau) = \tilde{\boldsymbol{\Gamma}}(\mathbf{r}, \mathbf{a}) \cdot \tau$, $\tilde{\boldsymbol{\Gamma}}(\mathbf{r}, \mathbf{a})$ is the fundamental solution of Navier equation when the point source is located on $\mathbf{a}$ and τ is the polarization vector. It is known that [3]

$$\boldsymbol{\Gamma}^\infty(\mathbf{r}, \mathbf{a}; \tau) = \Gamma_p^\infty(\hat{\mathbf{r}}, \mathbf{a}; \tau)\hat{\mathbf{r}}\frac{e^{ik_p r}}{\sqrt{r}} + \Gamma_s^\infty(\hat{\mathbf{r}}, \mathbf{a}; \tau)\hat{\mathbf{r}}^\perp \frac{e^{ik_s|r|}}{\sqrt{r}} + O(r^{-1/2}). \tag{50}$$

Then

$$\boldsymbol{\Gamma}^\infty(\mathbf{r}, \mathbf{a}; \tau) = \left(\frac{1}{\lambda + 2\mu}\frac{i+1}{4\sqrt{\pi k_p}}e^{-ik_p \hat{\mathbf{r}} \cdot \mathbf{a}}\hat{\mathbf{r}} \cdot \mathbf{d}, \ \frac{1}{\mu}\frac{i+1}{4\sqrt{\pi k_s}}e^{-ik_s \hat{\mathbf{r}} \cdot \mathbf{a}}\hat{\mathbf{r}}^\perp \cdot \mathbf{d}\right). \tag{51}$$

From (49), we have $\mathbf{w}(\mathbf{r}) = \boldsymbol{\Gamma}(\mathbf{r}, \mathbf{a}; \tau)$ for every $\mathbf{r}$ such that $|\mathbf{r} - \mathbf{a}| \geq \delta$. Using the identity

$$\Delta(X\boldsymbol{\Gamma}) = \boldsymbol{\Gamma}\Delta X + X\Delta\boldsymbol{\Gamma} + 2(\nabla X \cdot \nabla)\boldsymbol{\Gamma}, \tag{52}$$

and substituting the above result in Navier equation, we get

$$\Delta^* \mathbf{w} + \rho(\mathbf{r})\omega^2 \mathbf{w} = \mu\Delta\mathbf{w} + (\lambda + \mu)\nabla\nabla \cdot \mathbf{w} + \rho\omega^2\mathbf{w}$$
$$= \mu(\boldsymbol{\Gamma}\Delta X + X\Delta\boldsymbol{\Gamma} + 2(\nabla X \cdot \nabla)\boldsymbol{\Gamma})$$
$$+ (\lambda + \mu)\nabla\nabla \cdot X\boldsymbol{\Gamma} + \rho\omega^2 X\boldsymbol{\Gamma}. \tag{53}$$

Using the following identity

$$\nabla \cdot X\boldsymbol{\Gamma} = X\nabla \cdot \boldsymbol{\Gamma} + \boldsymbol{\Gamma} \cdot \nabla X \tag{54}$$

and after some calculations in (53), we get

$$\Delta^* \mathbf{w} + \rho\omega^2 \mathbf{w} = \mu\boldsymbol{\Gamma}\Delta X + X\Delta\boldsymbol{\Gamma} + 2(\nabla X \cdot \nabla)\boldsymbol{\Gamma}$$
$$+ (\lambda + \mu)\nabla(X\nabla \cdot \boldsymbol{\Gamma} + \boldsymbol{\Gamma} \cdot \nabla X)$$
$$+ \rho\omega^2 X\boldsymbol{\Gamma} =: -\mathbf{q}\,\mathbf{f}_1^0, \quad \text{in } D_1.$$

On the boundary ∂D_0, we get

$$\mathbf{w} = X\boldsymbol{\Gamma} =: \mathbf{f}_2^0, \tag{55}$$

where **w** is a solution to problems (5)–(7) with known data $\mathbf{f}_1^0$, $\mathbf{f}_2^0$ and therefore

$$G(\mathbf{f}_1^0, \mathbf{f}_2^0) = \mathbf{w}^\infty = \varphi_\mathbf{a}, \tag{56}$$

i.e., $\varphi_\mathbf{a} \in \text{Range}(G)$.

On the other hand, we consider $\mathbf{a} \notin D$ and that there exist $\mathbf{g}_1$, $\mathbf{g}_2$ such that $G(\mathbf{g}_1, \mathbf{g}_2) = \varphi_\mathbf{a}$. If **w** is the solution to the problem with data $\mathbf{g}_1$ and $\mathbf{g}_2$, by Rellich's lemma and unique continuation principle, we get that $\mathbf{w}_0(\mathbf{r}) = \mathbf{\Gamma}(\mathbf{r}, \mathbf{a}; \tau)$; that is a contradiction since $\|\mathbf{w}\| < \infty$ and $\|\mathbf{\Gamma}(\mathbf{r}, \mathbf{a}; \tau)\| = \infty$ in a small neighborhood of **a**. □

3. The Inverse Scattering Problem and the Factorization Method

In this section, we study the factorization method for reconstructing the shape and location ∂D of the support D of the inhomogeneous medium. In this study, we consider the case of Dirichlet boundary condition on the boundary ∂D_0 of the buried object. A similar reconstruction result also holds to other cases, such as a Neumann boundary condition or a mixed/impedance one.

Let us assume that our inhomogenous medium D is irradiated by an elastic incident plane wave of the form [12]

$$\mathbf{u}^{\text{inc}}(\mathbf{r}) = e^{-\frac{i\pi}{4}} \int_{S^1} \left[\sqrt{\frac{k^p}{\omega}} \mathbf{g}_p(\mathbf{d}) e^{ik^p \mathbf{r} \cdot \mathbf{d}} + \sqrt{\frac{k^s}{\omega}} \mathbf{g}_s(\mathbf{d}) e^{ik^s \mathbf{r} \cdot \mathbf{d}} \right] ds(\mathbf{d}) \tag{57}$$

and define the far-field operator $F : [L^2(S^1)]^2 \to [L^2(S^1)]^2$ given by Ref. [15]

$$(F\mathbf{g})(\mathbf{r}) = e^{-\frac{i\pi}{4}} \int_S \left[\sqrt{\frac{k^p}{\omega}} \mathbf{u}_\infty^p(\hat{\mathbf{r}}, \mathbf{d}, \mathbf{d}) \mathbf{g}_p(\mathbf{d}) \right.$$
$$\left. + \sqrt{\frac{k^s}{\omega}} \mathbf{u}_\infty^s(\hat{\mathbf{r}}, \mathbf{d}, \mathbf{d}^\top) \mathbf{g}_s(\mathbf{d}) \right] ds(\mathbf{d}), \tag{58}$$

where the kernel **g** has columns on the $L^2(S^1)$ vectors $e^{-\frac{i\pi}{4}} \left(\sqrt{\frac{k^p}{\omega}} \mathbf{g}_p, \sqrt{\frac{k^s}{\omega}} \mathbf{g}_s \right)$ whereas $\mathbf{d}, \hat{\mathbf{r}} \in \Omega$ denote the incident and observation directions, respectively. In addition,

$$\mathbf{g}_p(\mathbf{d}) = g_p(\mathbf{d})\,\mathbf{d}, \quad \mathbf{g}_s(\mathbf{d}) = g_s(\mathbf{d})\,\mathbf{d}^{\perp}, \tag{59}$$

$$\mathbf{u}_\infty(\cdot,\mathbf{d},\boldsymbol{\theta}) = (u_\infty^p(;,\mathbf{d},\boldsymbol{\theta}), u_\infty^s(;,\mathbf{d},\boldsymbol{\theta})), \tag{60}$$

with $\boldsymbol{\theta} = \mathbf{d}$ or $\mathbf{d}^{\perp}$ and the functions $\mathbf{u}_\infty$ are the far-field patterns of the scattered field $\mathbf{u}^{\text{sct}}$ of problems (1)–(3) due the incident plane wave $\mathbf{u}^{\text{inc}}$ and the Dirichlet boundary condition $\mathbf{u} = \mathbf{0}$ on ∂D_0.

At this point and for the convenience of the reader, we mention that relation (57) corresponds to an elastic Herglotz wave function, with operator F being the corresponding far-field operator (for more details one can see Ref. [14]). From the latter, we can obtain the following operator $H : [L^2(S^1)]^2 \to Y$ associated with problems (1)–(3):

$$(H_1\mathbf{g})(\mathbf{r}) := e^{-\frac{i\pi}{4}} \int_{S^1} \left[\sqrt{\frac{k^p}{\omega}} \mathbf{g}_p(\mathbf{d}) e^{ik^p \mathbf{r}\cdot\mathbf{d}} + \sqrt{\frac{k^s}{\omega}} \mathbf{g}_s(\mathbf{d}) e^{ik^s \mathbf{r}\cdot\mathbf{d}} \right] ds(\mathbf{d}),$$

$$\mathbf{r} \in D_1 \tag{61}$$

and

$$(H_2\mathbf{g})(\mathbf{r}) := e^{-\frac{i\pi}{4}} \int_{S^1} \left[\sqrt{\frac{k^p}{\omega}} \mathbf{g}_p(\mathbf{d}) e^{ik^p \mathbf{r}\cdot\mathbf{d}} + \sqrt{\frac{k^s}{\omega}} \mathbf{g}_s(\mathbf{d}) e^{ik^s \mathbf{r}\cdot\mathbf{d}} \right] ds(\mathbf{d}),$$

$$\mathbf{r} \in \partial D_0. \tag{62}$$

Due to the superposition principle and the definition of G (see relation (35)), it easily follows that

$$F = GH = G(H_1, H_2). \tag{63}$$

In order to prove a suitable factorization relation for the operator F, we show that the latter satisfies the Range identity given in Ref. [7] (Theorem 2.15, p. 57). Therefore, we move with our study by defining the following auxiliary integral operators $\mathbf{V}$ and $\mathbf{S}$ as:

$$(\mathbf{V}\boldsymbol{\phi}_1)(\mathbf{r}) = \int_{D_1} \tilde{\Gamma}(\mathbf{r}, \mathbf{r}_0)\boldsymbol{\phi}_1(\mathbf{r}_0)\,ds(\mathbf{r}_0), \quad \mathbf{r} \in D_1,\ \boldsymbol{\phi}_1 \in [L^2(D_1)]^2 \tag{64}$$

and

$$(\mathbf{S}\boldsymbol{\phi}_2)(\mathbf{r}) = \int_{\partial D_0} \tilde{\boldsymbol{\Gamma}}(\mathbf{r}, \mathbf{r}_0)\boldsymbol{\phi}_2(\mathbf{r})\,\mathrm{ds}(\mathbf{r}_0), \quad \mathbf{r} \in D_1,\ \boldsymbol{\phi}_2 \in [\mathrm{H}^{1/2}(\partial D_0)]^2 \tag{65}$$

and we also take into account their corresponding restrictions, given by

$$\mathbf{V}_{\mathrm{res}}\boldsymbol{\phi}_1 := (\mathbf{V}\boldsymbol{\phi}_1)|_{\mathbf{r} \in \partial D_0} \tag{66}$$

and

$$\mathbf{S}_{\mathrm{res}}\boldsymbol{\phi}_2 := (\mathbf{S}\boldsymbol{\phi}_2)|_{\mathbf{r} \in \partial D_0}. \tag{67}$$

The kernel $\tilde{\boldsymbol{\Gamma}}$ in (64) and (65) is the free-space fundamental solution of the Navier equation in $\mathbb{R}^2$, given by Ref. [2]:

$$\tilde{\boldsymbol{\Gamma}}(\mathbf{r}, \mathbf{r}_0) = \frac{i}{4}\left\{H_0^{(1)}(\mathrm{k}^\mathrm{s}|\mathbf{r} - \mathbf{r}_0|)\tilde{\mathrm{I}} - \frac{1}{\rho(\mathbf{r})\omega^2}\nabla_\mathbf{r} \otimes \nabla_\mathbf{r}[H_0^{(1)}(\mathrm{k}^\mathrm{p}|\mathbf{r} - \mathbf{r}_0|) \right.$$
$$\left. - H_0^{(1)}(\mathrm{k}^\mathrm{s}|\mathbf{r} - \mathbf{r}_0|)]\right\}, \tag{68}$$

where $H_0^{(1)}(z)$ is the Hankel function of first kind and zero order, $\tilde{\mathrm{I}}$ is the identity dyadic as usual and $\otimes$ is the juxtaposition between two vectors. Using Ref. [10] and the boundedness of the trace operator, we deduce that the operators

$$\mathbf{V} : [\mathrm{L}^2(D_1)]^2 \to [\mathrm{H}^2(D_1)]^2,$$
$$\mathbf{S} : [\mathrm{H}^{1/2}(\partial D_0)]^2 \to [\mathrm{H}^1(D_1)]^2 \tag{69}$$

as well as their restrictions

$$\mathbf{V}_{\mathrm{res}} : [\mathrm{L}^2(D_1)]^2 \to [\mathrm{H}^{3/2}(D_1)]^2,$$
$$\mathbf{S}_{\mathrm{res}} : [\mathrm{H}^{-1/2}(\partial D_0)]^2 \to [\mathrm{H}^{1/2}(\partial D_0)]^2 \tag{70}$$

are bounded.

Lemma 5. *The single-layer potential*

$$\mathbf{w}(\mathbf{r}) = \int_{D_1} \widetilde{\mathbf{\Gamma}}(\mathbf{r}, \mathbf{r}_0)\boldsymbol{\phi}_1(\mathbf{r}_0)\,d\mathbf{r}_0 + \int_{\partial D_0} \widetilde{\mathbf{\Gamma}}(\mathbf{r}, \mathbf{r}_0)\boldsymbol{\phi}_2(\mathbf{r})\,ds(\mathbf{r}_0),$$
$$\times\, \mathbf{r} \in \mathbb{R}^2 \setminus \overline{D}_0 \qquad (71)$$

has far-field pattern the conjugate operator $H^* : Y \to [L^2(S^1)]^2$ *of* H, *given by*

$$H^*(\alpha, \beta) = e^{\frac{i\pi}{4}} \int_{D_1} \alpha(\mathbf{r}) \left[\sqrt{\frac{k^p}{\omega}} e^{-ik^p \mathbf{r} \cdot \mathbf{d}} + \sqrt{\frac{k^s}{\omega}} e^{ik^s \mathbf{r} \cdot \mathbf{d}} \right] d\mathbf{r}$$
$$+ e^{\frac{i\pi}{4}} \int_{\partial D_0} \beta(\mathbf{r}) \left[\sqrt{\frac{k^p}{\omega}} e^{-ik^p \mathbf{r} \cdot \mathbf{d}} + \sqrt{\frac{k^s}{\omega}} e^{ik^s \mathbf{r} \cdot \mathbf{d}} \right] ds(\mathbf{r}_0),$$
$$(72)$$

for every $\alpha \in [L^2(D_1)]^2$ *and* $\beta \in [H^{1/2}(\partial D_0)]^2$.

Proof. We consider $\boldsymbol{\phi}_1 \in [L^2(D_1)]^2$, $\boldsymbol{\phi}_2 \in [H^{1/2}(\partial D_0)]^2$ and $\alpha \in [L^2(D_1)]^2$, $\beta \in [H^{1/2}(\partial D_0)]^2$. Then via relations (61) and (62), we have that

$$<H(\boldsymbol{\phi}_1, \boldsymbol{\phi}_2), (\alpha, \beta)> = <(H_1\boldsymbol{\phi}_1, H_2\boldsymbol{\phi}_2), (\alpha, \beta)>$$

$$= <H_1\boldsymbol{\phi}_1, \alpha> + <H_2\boldsymbol{\phi}_2, \beta>$$

$$= \int_{S^1} \boldsymbol{\phi}_1(\mathbf{d}) \overline{\left\{ e^{\frac{i\pi}{4}} \int_{D_1} \alpha(\mathbf{r}) \left[\sqrt{\frac{k^p}{\omega}} e^{-ik^p \mathbf{r} \cdot \mathbf{d}} + \sqrt{\frac{k^s}{\omega}} e^{ik^s \mathbf{r} \cdot \mathbf{d}} \right] d\mathbf{r} \right\}} ds(\mathbf{d})$$

$$+ \int_{S^1} \boldsymbol{\phi}_2(\mathbf{d}) \overline{\left\{ e^{\frac{i\pi}{4}} \int_{\partial D_0} \beta(\mathbf{r}) \left[\sqrt{\frac{k^p}{\omega}} e^{-ik^p \mathbf{r} \cdot \mathbf{d}} + \sqrt{\frac{k^s}{\omega}} e^{ik^s \mathbf{r} \cdot \mathbf{d}} \right] ds(\mathbf{r}) \right\}}$$

$$\times\, ds(\mathbf{d})$$

$$= <(\boldsymbol{\phi}_1, \boldsymbol{\phi}_2), H^*(\alpha, \beta)>,$$

and hence the proof is complete. □

Theorem 3. *The operator F has the factorization*

$$F = GP^*G^* \tag{73}$$

where $P : [L^2(D_1)]^2 \times [H^{-1/2}(\partial D_0)]^2 \to [L^2(D_1)]^2 \times [H^{1/2}(\partial D_0)]^2$ *is the operator matrix, defined by*

$$P = \begin{pmatrix} q_1 I - V & -S \\ -V_{res} & -S_{res} \end{pmatrix} \quad \text{with } q_1 \mathbf{f}_1 := \frac{1}{q}\mathbf{f}_1. \tag{74}$$

Proof. By the definition of $\mathbf{w}$ at (71) and the definitions of $\mathbf{V}$, $\mathbf{S}$ at (64) and (65) respectively, we get

$$\mathbf{w} = \mathbf{V}\phi_1 + \mathbf{S}\phi_2. \tag{75}$$

It is obvious that $\mathbf{w}$ solves problems (5)–(7) with data

$$\mathbf{f}_1 := q_1\phi_1 - (\mathbf{V}_1\phi_1 + \mathbf{S}\phi_2), \tag{76}$$
$$\mathbf{f}_2 := -\mathbf{V}_{res}\phi_1 - \mathbf{S}_{res}\phi_2, \tag{77}$$

and by relation (5), we can get

$$\Delta^*\mathbf{w} + \rho(\mathbf{r})\omega^2\mathbf{w} = -\phi_1, \quad \text{in } D_1. \tag{78}$$

We now consider the matrix

$$P = \begin{pmatrix} q_1 I - V & -S \\ -V_{res} & -S_{res} \end{pmatrix} \tag{79}$$

and since $G(\mathbf{f}_1, \mathbf{f}_2) = \mathbf{w}_\infty$ is the far-field pattern of the solution $\mathbf{w}$ of (5)–(7) with given data $(\mathbf{f}_1, \mathbf{f}_2)^\top \in [L^2(D_1)]^2 \times [H^{1/2}(\partial D_0)]^2$, we get $H^* = GP$ and therefore

$$H = P^*G^*. \tag{80}$$

Taking into account (58) and relations (61), (62), we arrive at

$$F = GH, \tag{81}$$

and hence (73) is secured. □

Theorem 4. *Let the operator* $P : [L^2(D_1)]^2 \times [H^{-1/2}(\partial D_0)]^2 \to Y$ *be defined by*

$$P = \begin{pmatrix} q_1 I - \mathbf{V} & -\mathbf{S} \\ -\mathbf{V}_{\text{res}} & -\mathbf{S}_{\text{res}} \end{pmatrix}, \qquad (82)$$

where $Y = [L^2(D_1)]^2 \times [H^{1/2}(\partial D_0)]^2$. *Then the following properties hold:*

(i) $P : Y^* \to Y$ *is an isomorphism if* ω^2 *is not a Dirichlet eigenvalue of* $-\Delta^*$ *in* D_0.
(ii) *If* $\rho(\mathbf{r}) < 1$, *there exists a coercive operator* $P^{(1)}$ *and a compact operator* $P^{(2)}$ *such that* $P = P^{(1)} + P^{(2)}$.
(iii) $Im(P)$ *is compact and nonpositive on* Y^*, *i.e.*, $Im < P\phi, \phi >_{Y^*} \leq 0$, *for all* $\phi \in Y^*$.

Proof. We begin our proof, by rewriting (82) as

$$P = \begin{pmatrix} q_1 I & 0 \\ 0 & -\mathbf{S}_{\text{res}}(\ell_p, \ell_s) \end{pmatrix} - \begin{pmatrix} \mathbf{V} & \mathbf{S} \\ \mathbf{V}_{\text{res}} & \mathbf{S}_{\text{res}}(k_p, k_s) - \mathbf{S}_{\text{res}}(\ell_p, \ell_s) \end{pmatrix}. \qquad (83)$$

If we define

$$P^{(1)} := \begin{pmatrix} q_1 I & 0 \\ 0 & -\mathbf{S}_{\text{res}}(\ell_p, \ell_s) \end{pmatrix} \qquad (84)$$

and

$$P^{(2)} := \begin{pmatrix} \mathbf{V} & \mathbf{S} \\ \mathbf{V}_{\text{res}} & \mathbf{S}_{\text{res}}(k_p, k_s) - \mathbf{S}_{\text{res}}(\ell_p, \ell_s) \end{pmatrix}, \qquad (85)$$

where $\mathbf{S}_{\text{res}}(k_p, k_s)$ and $\mathbf{S}_{\text{res}}(\ell_p, \ell_s)$ are the single-layer boundary operators defined on ∂D_0 and corresponding to the wave numbers k_p, k_s and $k_p = \ell_p, k_s = \ell_s$, respectively. It is obvious that $P^{(1)}$ is an isomorphism and $P^{(2)}$ is compact on Y^*. If we assume $\rho(\mathbf{r}) < 1$, we take $q < 0$ and by (74) we also get $q_1 < 0$. Then $\mathbf{S}(\ell_p, \ell_s)$ is coercive on $[H^{1/2}(\partial D_0)]^2$ and hence (ii) is proved, i.e., $P^{(1)}$ is coercive on Y^*.

For (i), we need to prove that P is injective on the space Y^*. Let $P\phi = \mathbf{0}$, for $(\phi_1, \phi_2)^\top \in Y^*$. Then $\mathbf{w}$ defined by (71) is a solution to the homogeneous problem of (5)–(7), i.e., $\mathbf{f}_1 = \mathbf{f}_2 = \mathbf{0}$, and by uniqueness of solution of the latter problem $\mathbf{w} = \mathbf{0}$ in $\mathbb{R}^2 \setminus \overline{D}_0$; the jump relation for (71) yields $\phi_1 = \mathbf{0}$. Since now $\Delta^*\mathbf{w} + \omega^2\mathbf{w} = \mathbf{0}$ in D_0 (recall here that $\rho(\mathbf{r}) = 1$ in D_0) with $\mathbf{w} = \mathbf{0}$ on ∂D_0, by assumption that ω^2 is not a Dirichlet eigenvalue of $-\Delta^*$ in D_0, we get that $\mathbf{w} = \mathbf{0}$ in D_0, and hence applying the jump relation for (71) once again yields to $\phi_2 = \mathbf{0}$.

Finally, concerning the proof of (iii), it follows analogous arguments as those in Ref. [13] and here is omitted for brevity. □

We end this chapter with the following result:

Theorem 5. *Assume that the Dirichlet boundary condition* $\mathbf{w} = -\mathbf{f}_2$ *holds and that* ω^2 *is not a Dirichlet eigenvalue of* $-\Delta^*$ *in* D_0 *with* $\rho(\mathbf{r}) < 1$ *in* D_1. *Then*

$$\mathbf{a} \in D \Leftrightarrow \boldsymbol{\varphi}_\mathbf{a} \in \mathrm{Range}(F_\sharp^{\frac{1}{2}}), \tag{86}$$

if and only if

$$W(\mathbf{a}) := \left[\frac{|<\boldsymbol{\varphi}_\mathbf{a}, \boldsymbol{\psi}_j>_{L^2(S^1)}|^2}{\lambda_j}\right]^{-1} > 0, \tag{87}$$

where $\boldsymbol{\varphi}_\mathbf{a}$ *is given by* (47) *and* $\{\lambda_i, \boldsymbol{\psi}_j\}_{j \in \mathbb{N}}$ *is a pair of eigensystem of the self-adjoint operator*

$$F_\sharp := |\Re(F)| + |\Im(F)|. \tag{88}$$

In conclusion, the direct and inverse scattering problem of elastic waves by an inhomogeneous medium with unknown buried obstacles in its interior was studied. Well-posedness of the direct scattering problem using a specific variational method in a suitable Sobolev space setting was established. The corresponding inverse elastic problem was also studied, and the issue of recovering the shape and location of the support of the inhomogeneous medium D was also addressed. The factorization method was the key ingredient for the above reconstruction (see Theorem 5), due to a Dirichlet boundary condition imposed on the buried object(s) D_0. The authors refer that the method is also valid for other boundary conditions such as *Neumann* or *impedance boundary condition* on ∂D_0, when $\rho(\mathbf{r}) < 1$, and they point out the complicated and more demanding case that elastic media meet.

References

[1] C. J. S. Alves and R. Kress, On the far field operator in elastic obstacle scattering, *IMA J. Appl. Math.* **67**, 1–21 (2002).

[2] C. E. Athanasiadis, D. Natroshvili, V. Sevroglou and I. G. Stratis, An application of the reciprocity gap functional to inverse mixed impedance problems in elasticity, *Inverse Probl.* **26**, 085011 (2010).

[3] C. E. Athanasiadis, D. Natroshvili, V. Sevroglou and I. G. Stratis, A boundary integral equations approach for mixed impedance problems in elasticity, *J. Integral Equ. Appl.* **23**, 183–222 (2011).

[4] D. Colton and R. Kress, *Inverse Acoustic and Electromagnetic Scattering Theory*, 3rd edn. (Springer, Berlin, 2013).

[5] G. Dassios and R. Kleinman, *Low Frequency Scattering*, 1st edn. (Oxford University Press, Oxford, 2000).

[6] A. Kirsch, Characterization of the shape of a scattering obstacle using the spectral data of the far field operator, *Inverse Probl.* **14**, 1489–1512 (1998).

[7] A. Kirsch and N. Grinberg, *The Factorization Method for Inverse Problems* (Oxford University Press, Oxford, 2008).

[8] P. Li, Y. Wang, Z. Wang and Y. Zhao, Inverse obstacle scattering for elastic waves, *Inverse Probl.* **32**, 115–218 (2016).

[9] X. Liu, B. Zhang and G. Hu, Uniqueness in the inverse scattering problem in a piecewise homogeneous medium, *Inverse Probl.* **26**, 015002 (2009).

[10] W. McLean, *Strongly Elliptic Systems and Boundary Integral Equations* (Cambridge University Press, Cambridge, 2000).

[11] D. Natrosvilli and Z. Tediashvili, Mixed type direct and inverse scattering problems. In J. Elschner, I. Gohberg and B. Silbermann (eds.) *Problems and Methods in Mathematical Physics*. Operator Theory: Advances and Applications, Vol. 121 (Birkhäuser, Basel, 2001), pp. 366–389.

[12] G. Pelekanos and V. Sevroglou, The (F*F) 1/4-method for the transmission problem in two-dimensional linear elasticity. *Appl. Anal.* **84**(3), 311–328 (2006).

[13] F. Qu, J. Yang and B. Zhang, An approximate factorization method for inverse medium scattering with unknown buried objects, *Inverse Probl.* **33**, 035007 (2017), 24 p.

[14] V. Sevroglou and G. Pelekanos, Two dimensional elastic Herglotz functions and their applications in inverse scattering, *J. Elast.* **68**, 123–144 (2002).

[15] V. Sevroglou, The far-field operator for penetrable and absorbing obstacles in 2D inverse elastic scattering, *Inverse Probl.* **17**, 717–738 (2005).

© 2024 World Scientific Publishing Company
https://doi.org/10.1142/9789811267048_0014

Chapter 14

Some New Bounds of Gauss–Jacobi and Hermite–Hadamard-Type Integral Inequalities

Artion Kashuri

*Department of Mathematical Engineering,
Polytechnic University of Tirana, Tirana, Albania*

a.kashuri@fimif.edu.al

In this chapter, we first introduced a new class of functions called generalized-$\mathbf{m}-(((h_1 \circ g)^p, (h_2 \circ g)^q); (\eta_1, \eta_2))$-convex defined on $(\mathbf{m}, g; \sigma)$-invex set and also discover two interesting identities regarding Gauss–Jacobi and Hermite–Hadamard-type integral inequalities. By using the first lemma as an auxiliary result, some new bounds with respect to Gauss–Jacobi-type integral inequalities are established. Also, using the second lemma, some new estimates with respect to Hermite–Hadamard-type integral inequalities via k-fractional integrals for generalized-$\mathbf{m}-(((h_1 \circ g)^p, (h_2 \circ g)^q); (\eta_1, \eta_2))$-convex mappings are obtained. It is pointed out that some new special cases can be deduced from main results. At the end, some applications to special means for different positive real numbers are provided as well.

1. Introduction

The following notations are used throughout this chapter. We use $\mathcal{I}$ to denote an interval on the real line $\Re = (-\infty, +\infty)$. For any subset $\mathcal{K} \subseteq \Re^n$, $\mathcal{K}^\circ$ is the interior of $\mathcal{K}$. The set of integrable functions on the interval $[\varrho_1, \varrho_2]$ is denoted by $\mathcal{L}[\varrho_1, \varrho_2]$.

The following inequality, named Hermite–Hadamard inequality, is one of the most famous inequalities in the literature for convex functions.

Theorem 1. *Let $f : \mathcal{I} \subseteq \Re \to \Re$ be a convex function on $\mathcal{I}$ and $\varrho_1, \varrho_2 \in I$ with $\varrho_1 < \varrho_2$. Then the following inequality holds:*

$$f\left(\frac{\varrho_1 + \varrho_2}{2}\right) \leq \frac{1}{\varrho_2 - \varrho_1} \int_{\varrho_1}^{\varrho_2} f(x)\mathrm{d}x \leq \frac{f(\varrho_1) + f(\varrho_2)}{2}. \qquad (1)$$

This inequality (1) is also known as trapezium inequality.

The trapezium-type inequality has remained an area of great interest due to its wide applications in the field of mathematical analysis. For other recent results which generalize, improve and extend the inequality (1) through various classes of convex functions, interested readers are referred to Refs. [1–20, 22, 23, 25–31, 35, 38, 39].

Let us recall some special functions and evoke some basic definitions as follows.

Definition 1. *For $k \in \Re^+$ and $x \in \mathbb{C}$, the k-gamma function is defined by*

$$\Gamma_k(x) = \lim_{n \to \infty} \frac{n! k^n (nk)^{\frac{x}{k}-1}}{(x)_{n,k}}. \qquad (2)$$

Its integral representation is given by

$$\Gamma_k(\alpha) = \int_0^\infty t^{\alpha-1} e^{-\frac{t^k}{k}} \mathrm{d}t. \qquad (3)$$

One can note that

$$\Gamma_k(\alpha + k) = \alpha \Gamma_k(\alpha). \qquad (4)$$

For $k = 1$, (3) gives integral representation of gamma function.

Definition 2 ([25]). Let $f \in \mathcal{L}[\varrho_1, \varrho_2]$. Then k-fractional integrals of order $\alpha, k > 0$ with $\varrho_1 \geq 0$ are defined as

$$I_{\varrho_1^+}^{\alpha,k} f(x) = \frac{1}{k\Gamma_k(\alpha)} \int_{\varrho_1}^{x} (x-t)^{\frac{\alpha}{k}-1} f(t) dt, \quad x > \varrho_1$$

and

$$I_{\varrho_2^-}^{\alpha,k} f(x) = \frac{1}{k\Gamma_k(\alpha)} \int_{x}^{\varrho_2} (t-x)^{\frac{\alpha}{k}-1} f(t) dt, \quad \varrho_2 > x. \tag{5}$$

For $k = 1$, k-fractional integrals give Riemann–Liouville integrals.

Definition 3 ([37]). A set $\mathcal{S} \subseteq \Re^n$ is said to be an invex set with respect to the mapping $\eta : \mathcal{S} \times \mathcal{S} \to \Re^n$, if $x + t\eta(y,x) \in \mathcal{S}$ for every $x, y \in \mathcal{S}$ and $t \in [0,1]$.

The invex set $\mathcal{S}$ is also termed an η-connected set.

Definition 4 ([24]). Let $h : [0,1] \to \Re$ be a non-negative function and $h \neq 0$. The function f on the invex set $\mathcal{K}$ is said to be h-preinvex with respect to η, if

$$f(x + t\eta(y,x)) \leq h(1-t)f(x) + h(t)f(y) \tag{6}$$

for each $x, y \in \mathcal{K}$ and $t \in [0,1]$, where $f(\cdot) > 0$.

If the mapping $\eta(y,x) = y - x$ in Definition 4, then the non-negative function f reduces to h-convex mappings [34].

Definition 5 ([36]). Let $\mathcal{S} \subseteq \Re^n$ be an invex set with respect to $\eta : \mathcal{S} \times \mathcal{S} \to \Re^n$. A function $f : \mathcal{S} \to [0, +\infty)$ is said to be s-preinvex (or s-Breckner-preinvex) with respect to η and $s \in (0,1]$, if for every $x, y \in \mathcal{S}$ and $t \in [0,1]$,

$$f(x + t\eta(y,x)) \leq (1-t)^s f(x) + t^s f(y). \tag{7}$$

Definition 6 ([26]). A function $f : \mathcal{K} \to \Re$ is said to be s-Godunova–Levin–Dragomir-preinvex of second kind, if

$$f(x + t\eta(y,x)) \leq (1-t)^{-s} f(x) + t^{-s} f(y), \tag{8}$$

for each $x, y \in \mathcal{K}, t \in (0,1)$ and $s \in (0,1]$.

Definition 7 ([33]). A non-negative function $f : \mathcal{K} \subseteq \Re \to \Re$ is said to be tgs-convex on $\mathcal{K}$ if the inequality

$$f((1-t)x + ty) \leq t(1-t)[f(x) + f(y)] \tag{9}$$

grips for all $x, y \in \mathcal{K}$ and $t \in (0, 1)$.

Definition 8 ([21]). A function $f : \mathcal{I} \subseteq \Re \to \Re$ is said to be MT-convex, if it is non-negative and $\forall\, x, y \in \mathcal{I}$ and $t \in (0, 1)$ satisfies the subsequent inequality

$$f(tx + (1-t)y) \leq \frac{\sqrt{t}}{2\sqrt{1-t}} f(x) + \frac{\sqrt{1-t}}{2\sqrt{t}} f(y). \tag{10}$$

The concept of η-convex functions (at the beginning was named by φ-convex functions), considered in Ref. [11], has been introduced as follows.

Definition 9. Consider a convex set $\mathcal{I} \subseteq \Re$ and a bifunction $\eta : f(\mathcal{I}) \times f(\mathcal{I}) \to \Re$. A function $f : \mathcal{I} \to \Re$ is called convex with respect to η (briefly η-convex), if

$$f(\lambda x + (1 - \lambda)y) \leq f(y) + \lambda \eta(f(x), f(y)) \tag{11}$$

is valid for all $x, y \in \mathcal{I}$ and $\lambda \in [0, 1]$.

Geometrically, it says that if a function is η-convex on $\mathcal{I}$, then for any $x, y \in \mathcal{I}$, its graph is on or under the path starting from $(y, f(y))$ and ending at $(x, f(y) + \eta(f(x), f(y)))$. If $f(x)$ should be the end point of the path for every $x, y \in \mathcal{I}$, then we have $\eta(x, y) = x - y$ and the function reduces to a convex one. For more results about η-convex functions, see Refs. [6, 7, 10, 11].

Definition 10 ([1]). Let $\mathcal{I} \subseteq \Re$ be an invex set with respect to $\eta_1 : \mathcal{I} \times \mathcal{I} \to \Re$. Consider $f : \mathcal{I} \to \Re$ and $\eta_2 : f(\mathcal{I}) \times f(\mathcal{I}) \to \Re$. The function f is said to be (η_1, η_2)-convex if

$$f(x + \lambda \eta_1(y, x)) \leq f(x) + \lambda \eta_2(f(y), f(x)) \tag{12}$$

is valid for all $x, y \in \mathcal{I}$ and $\lambda \in [0, 1]$.

The Gauss–Jacobi-type quadrature formula is as follows:

$$\int_{\varrho_1}^{\varrho_2} (x-\varrho_1)^p (\varrho_2 - x)^q f(x) \mathrm{d}x = \sum_{k=0}^{+\infty} B_{m,k} f(\gamma_k) + R_m^\star |f|, \qquad (13)$$

for certain $B_{m,k}, \gamma_k$ and rest $R_m^\star |f|$, see Ref. [32].

Recently, in Liu, Ref. [19] obtained several integral inequalities for the left-hand side of (13). Also in Özdemir et al. [28] established several integral inequalities concerning the left-hand side of (13) via some kind of convexity.

Motivated by the above literatures, the main objective of this chapter is to discover in Section 2 and in Section 3 two interesting identities regarding Gauss–Jacobi and Hermite–Hadamard-type integral inequalities and to established new bounds associated with generalized-$\mathbf{m} - (((h_1 \circ g)^p, (h_2 \circ g)^q); (\eta_1, \eta_2))$-convex mappings. By using the first lemma as an auxiliary result in Section 3, some new bounds with respect to Gauss–Jacobi-type integral inequalities will be given. Also, using the second lemma in Section 3, some new estimates with respect to Hermite–Hadamard-type integral inequalities via k-fractional integrals are obtained. It is pointed out that some new special cases are deduced from main results. In Section 4, some applications to special means for different positive real numbers are provided as well.

2. Some New Bounds of the Quadrature Formula of Gauss–Jacobi Type

The following definitions are used in this section.

Definition 11. Let $\sigma : \mathcal{I} \to \mathcal{K}$, $g : [0,1] \to [0,1]$ be continuous and $\mathbf{m} : [0,1] \to (0,1]$. A set $\mathcal{K} \subseteq \Re^n$ is named as $(\mathbf{m}, g; \sigma)$-invex with respect to the mapping $\eta : \mathcal{K} \times \mathcal{K} \to \Re^n$, if $\mathbf{m}(t)\sigma(x) + g(\xi)\eta(\sigma(y), \mathbf{m}(t)\sigma(x)) \in \mathcal{K}$ holds for each $x, y \in \mathcal{I}$ and any $t, \xi \in [0,1]$.

Remark 1. In Definition 11, under certain conditions, the mapping $\eta(\sigma(y), \mathbf{m}(t)\sigma(x))$ for any $t, \xi \in [0,1]$ could reduce to $\eta(y, mx)$. For example, when $\mathbf{m}(t) = m$ for all $t \in [0,1]$, $g(\xi) = \xi$, $\forall \xi \in [0,1]$ and $\sigma(x) = x$ for all $x \in \mathcal{I}$, then the $(\mathbf{m}, g; \sigma)$-invex set degenerates an

m-invex set on $\mathcal{K}$. Also, taking $\mathbf{m}(t) \equiv 1$ for all $t \in [0,1]$, $g(\xi) = \xi$, $\forall \xi \in [0,1]$ and $\sigma(x) = x$ for all $x \in \mathcal{I}$, in Definition 11, we get Definition 3.

We next introduce the concept of generalized-$\mathbf{m} - (((h_1 \circ g)^p, (h_2 \circ g)^q); (\eta_1, \eta_2))$-convex mappings.

Definition 12. Let $\mathcal{K} \subseteq \Re$ be an open $(\mathbf{m}, g; \sigma)$-invex set with respect to the mapping $\eta_1 : \mathcal{K} \times \mathcal{K} \to \Re$, where $\sigma : \mathcal{I} \to \mathcal{K}$, $g : [0,1] \to [0,1]$ are continuous and $\mathbf{m} : [0,1] \to (0,1]$. Consider $f : \mathcal{K} \to (0, +\infty)$, where $h_1, h_2 : [0,1] \to [0, +\infty)$ are continuous and $\eta_2 : f(\mathcal{K}) \times f(\mathcal{K}) \to \Re$. The mapping f is said to be generalized-$\mathbf{m} - (((h_1 \circ g)^p, (h_2 \circ g)^q); (\eta_1, \eta_2))$-convex if

$$f\big(\mathbf{m}(t)\sigma(x) + g(\xi)\eta_1(\sigma(y), \mathbf{m}(t)\sigma(x))\big)$$
$$\leq \Big[\mathbf{m}(\xi)(h_1 \circ g)^p(\xi) f^r(x) + (h_2 \circ g)^q(\xi)\eta_2(f^r(y), f^r(x))\Big]^{\frac{1}{r}} \quad (14)$$

holds for all $x, y \in \mathcal{I}$, $r \neq 0$, $t, \xi \in [0,1]$ and any fixed $p, q > -1$.

Remark 2. In Definition 12, if we choose $p = q = 1$, $\mathbf{m}(t) = m$, $\forall t \in [0,1]$ and $g(\xi) = \xi$, $\forall \xi \in [0,1]$, then we get Definition 2.3 in Ref. [15]. Also, in Definition 12, if we choose $\mathbf{m} = p = q = r = 1$, $g(\xi) = \xi$ and $\sigma(x) = x$, then we obtain Definition 10.

Remark 3. In Definition 12, if we choose $\mathbf{m} = p = q = r = 1$, $h_1(t) = 1$, $h_2(t) = t$, $\eta_1(\sigma(y), \mathbf{m}(t)\sigma(x)) = \sigma(y) - \mathbf{m}(t)\sigma(x)$, $\eta_2(f^r(y), f^r(x)) = \eta(f^r(y), f^r(x))$, $g(\xi) = \xi$ and $\sigma(x) = x$, then we get Definition 9. Also, in Definition 12, if we choose $\mathbf{m} = p = q = r = 1$, $h_1(t) = 1$, $h_2(t) = t$, $g(\xi) = \xi$ and $\sigma(x) = x$, then we obtain Definition 10. Under some suitable choices as we have done above, we can have also Definitions 5 and 6.

Remark 4. Let us discuss some special cases where $g(\xi) = \xi$, in Definition 12 as follows:

(i) Taking $h_1(t) = h(1-t)$ and $h_2(t) = h(t)$, we have generalized-$\mathbf{m} - ((h^p(1-t), h^q(t)); (\eta_1, \eta_2))$-convex mappings.
(ii) Choosing $h_1(t) = (1-t)^s$ and $h_2(t) = t^s$ for $s \in (0,1]$, we get generalized-$\mathbf{m} - (((1-t)^{sp}, t^{sq}); (\eta_1, \eta_2))$-Breckner-convex mappings.

(iii) Taking $h_1(t) = (1-t)^{-s}$ and $h_2(t) = t^{-s}$ for $s \in (0,1]$, we obtain generalized-**m** $- (((1-t)^{-sp}, t^{-sq}); (\eta_1, \eta_2))$-Godunova–Levin–Dragomir-convex mappings.

(iv) Choosing $h_1(t) = h_2(t) = t(1-t)$, we have generalized-**m** $- ((t(1-t))^{sp}, (t(1-t))^{sq}); (\eta_1, \eta_2))$-convex mappings.

(v) Taking $h_1(t) = \frac{\sqrt{1-t}}{2\sqrt{t}}$ and $h_2(t) = \frac{\sqrt{t}}{2\sqrt{1-t}}$, we get generalized-**m** $- \left(\left(\left(\frac{\sqrt{1-t}}{2\sqrt{t}}\right)^p, \left(\frac{\sqrt{t}}{2\sqrt{1-t}}\right)^q\right); (\eta_1, \eta_2)\right)$-convex mappings.

It is worth mentioning here that to the best of our knowledge all the special cases discussed above are new in the literature.

Let see the following example of a generalized-**m** $- (((h_1 \circ g)^p, (h_2 \circ g)^q); (\eta_1, \eta_2))$-convex mapping which is not convex.

Example 1. Let **m** $= r = \frac{1}{2}$, $h_1(t) = t^l$, $h_2(t) = t^s$ for all $l, s \in [0,1]$, any fixed $p, q \geq 1$, where $g(t) = t^r$, $\forall r > 0$ and σ be an identity function. Consider the function $f : [0, +\infty) \to [0, +\infty)$ by

$$f(x) = \begin{cases} x, & 0 \leq x \leq 2, \\ 4, & x > 2. \end{cases}$$

Define two bifunctions $\eta_1 : [0, +\infty) \times [0, +\infty) \to \Re$ and $\eta_2 : [0, +\infty) \times [0, +\infty) \to [0, +\infty)$ by

$$\eta_1(x, y) = \begin{cases} -y, & 0 \leq y \leq 2, \\ x + y, & y > 2 \end{cases}$$

and

$$\eta_2(x, y) = \begin{cases} x + y, & x \leq y, \\ 4(x + y), & x > y. \end{cases}$$

Then f is generalized $\frac{1}{2} - ((t^{rlp}, t^{rsq}); (\eta_1, \eta_2))$-convex mapping. But f is not preinvex with respect to η_1 and also it is not convex (consider $x = 0, y = 3$ and $t \in (0,1]$).

For establishing our results regarding new bounds' integral inequalities for Gauss–Jacobi type, we need the following lemma.

Lemma 1. *Let $\sigma : \mathcal{I} \to \mathcal{K}$ be a continuous function, $g : [0,1] \to [0,1]$ be a strictly increasing function on $(0,1)$ and $\mathbf{m} : [0,1] \to (0,1]$.*

Suppose $\mathcal{K} = (\boldsymbol{m}(t)\sigma(\varrho_1), \boldsymbol{m}(t)\sigma(\varrho_1) + \Xi_t) \subseteq \Re$ is an open $(\boldsymbol{m}, g; \sigma)$-invex subset with respect to $\Psi : \mathcal{K} \times \mathcal{K} \to \Re$ for $\Xi_t = \Psi(\sigma(\varrho_2), \boldsymbol{m}(t)\sigma(\varrho_1)) > 0$ and $\forall t \in [0,1]$. Assume that $f : \mathcal{K} \to \Re$ is a continuous mapping on $\mathcal{K}^\circ$. Then for some fixed $\boldsymbol{m}$ and any fixed $p, q > 0$, we have

$$\int_{\boldsymbol{m}(t)\sigma(\varrho_1)}^{\boldsymbol{m}(t)\sigma(\varrho_1)+\Xi_t} (x - \boldsymbol{m}(t)\sigma(\varrho_1))^p (\boldsymbol{m}(t)\sigma(\varrho_1) + \Xi_t - x)^q f(x) \mathrm{d}x$$

$$= \Xi_t^{p+q+1} \times \int_0^1 g^p(\xi)(1-g(\xi))^q f(\boldsymbol{m}(t)\sigma(\varrho_1) + g(\xi)\Xi_t) \mathrm{d}g(\xi). \tag{15}$$

We denote

$$\mathcal{T}_{f,g}^{p,q}(\Psi, \sigma, \boldsymbol{m}; \varrho_1, \varrho_2)$$

$$:= \Xi_t^{p+q+1} \times \int_0^1 g^p(\xi)(1-g(\xi))^q f(\boldsymbol{m}(t)\sigma(\varrho_1) + g(\xi)\Xi_t) \mathrm{d}g(\xi). \tag{16}$$

Proof. Changing the variable $x = \boldsymbol{m}(t)\sigma(\varrho_1) + g(\xi)\Xi_t$, we get

$$\mathcal{T}_{f,g}^{p,q}(\Psi, \sigma, \mathbf{m}; \varrho_1, \varrho_2) = \Xi_t \times \int_0^1 (\boldsymbol{m}(t)\sigma(\varrho_1) + g(\xi)\Xi_t - \boldsymbol{m}(t)\sigma(\varrho_1))^p$$

$$\times (\boldsymbol{m}(t)\sigma(\varrho_1) + \Xi_t - \boldsymbol{m}(t)\sigma(\varrho_1) - g(\xi)\Xi_t)^q f(\boldsymbol{m}(t)\sigma(\varrho_1)$$

$$+ g(\xi)\Xi_t) \mathrm{d}g(\xi)$$

$$= \int_{\boldsymbol{m}(t)\sigma(\varrho_1)}^{\boldsymbol{m}(t)\sigma(\varrho_1)+\Xi_t} (x - \boldsymbol{m}(t)\sigma(\varrho_1))^p (\boldsymbol{m}(t)\sigma(\varrho_1) + \Xi_t - x)^q f(x) \mathrm{d}x.$$

This completes the proof of the lemma. □

Remark 5. Under the conditions of Lemma 1 for $g(\xi) = \xi$, $\sigma(x) = x$ and $\Xi_t = \sigma(\varrho_2) - \boldsymbol{m}(t)\sigma(\varrho_1)$, where $\boldsymbol{m}(t) \equiv 1$, we get the left-hand side of equation (13).

Using Lemma 1, we now state the following theorems.

Theorem 2. Let $\sigma : \mathcal{I} \to \mathcal{K}$ be a continuous function, $g : [0,1] \to [0,1]$ be a strictly increasing function on $(0,1)$ and $\boldsymbol{m} : [0,1] \to (0,1]$. Suppose $\mathcal{K} = (\boldsymbol{m}(t)\sigma(\varrho_1), \boldsymbol{m}(t)\sigma(\varrho_1) + \Xi_{t,1}) \subseteq \Re$ be an open $(\boldsymbol{m}, g; \sigma)$-invex subset with respect to $\Psi : \mathcal{K} \times \mathcal{K} \to \Re$ for $\Xi_{t,1} = \Psi_1(\sigma(\varrho_2), \boldsymbol{m}(t)\sigma(\varrho_1)) > 0$ for all $t \in [0,1]$ and $\Psi_2 : f(\mathcal{K}) \times f(\mathcal{K}) \to \Re$. Assume that $f : \mathcal{K} \to (0, +\infty)$ is a mapping on $\mathcal{K}^\circ$ such that $f \in \mathcal{L}(\mathcal{K})$. If $f^{\frac{k}{k-1}}$ is generalized-$\boldsymbol{m}-(((h_1 \circ g)^{p_1}, (h_2 \circ g)^{p_2}); (\Psi_1, \Psi_2))$-convex mapping, $k > 1$, $0 < r \le 1$ and $p_1, p_2 > -1$, then for any fixed $p, q > 0$, the following inequality holds:

$$\left| T_{f,g}^{p,q}(\Psi_1, \sigma, \boldsymbol{m}; \varrho_1, \varrho_2) \right| \le \Xi_{t,1}^{p+q+1} \times \sqrt[k]{B(g(\xi))}$$

$$\times \left[f^{\frac{rk}{k-1}}(\varrho_1) I^r((h_1 \circ g)(\xi)) + \Psi_2 \left(f^{\frac{rk}{k-1}}(\varrho_2), f^{\frac{rk}{k-1}}(\varrho_1) \right) \right.$$

$$\left. \times I^r((h_2 \circ g)(\xi)) \right]^{\frac{k-1}{rk}}, \tag{17}$$

where

$$I((h_1 \circ g)(\xi)) := \int_0^1 \boldsymbol{m}^{\frac{1}{r}}(\xi)(h_1 \circ g)^{\frac{p_1}{r}}(\xi) dg(\xi); \quad I((h_2 \circ g)(\xi))$$

$$:= \int_0^1 (h_2 \circ g)^{\frac{p_2}{r}}(\xi) dg(\xi)$$

and

$$B(g(\xi)) := \int_0^1 g^{kp}(\xi)(1 - g(\xi))^{kq} dg(\xi).$$

Proof. From Lemma 1, generalized-$\boldsymbol{m} - (((h_1 \circ g)^{p_1}, (h_2 \circ g)^{p_2}); (\Psi_1, \Psi_2))$-convexity of $f^{\frac{k}{k-1}}$, Hölder inequality, Minkowski inequality and properties of the modulus, we have

$$\left| T_{f,g}^{p,q}(\Psi_1, \sigma, \boldsymbol{m}; \varrho_1, \varrho_2) \right| \le |\Xi_{t,1}|^{p+q+1} \left[\int_0^1 g^{kp}(\xi)(1 - g(\xi))^{kq} dg(\xi) \right]^{\frac{1}{k}}$$

$$\times \left[\int_0^1 \left| f(\boldsymbol{m}(t)\sigma(\varrho_1) + g(\xi)\Xi_{t,1}) \right|^{\frac{k}{k-1}} dg(\xi) \right]^{\frac{k-1}{k}}$$

$$\leq \Xi_{t,1}^{p+q+1} \times \sqrt[k]{B(g(\xi))}$$

$$\times \left[\int_0^1 \left[\mathbf{m}(\xi)(h_1 \circ g)^{p_1}(\xi) f^{\frac{rk}{k-1}}(\varrho_1) + (h_2 \circ g)^{p_2}(\xi) \Psi_2 \right. \right.$$

$$\left. \left. \times \left(f^{\frac{rk}{k-1}}(\varrho_2), f^{\frac{rk}{k-1}}(\varrho_1) \right) \right]^{\frac{1}{r}} \mathrm{d}g(\xi) \right]^{\frac{k-1}{k}}$$

$$\leq \Xi_{t,1}^{p+q+1} \times \sqrt[k]{B(g(\xi))}$$

$$\times \left[\left(\int_0^1 \mathbf{m}^{\frac{1}{r}}(\xi) f^{\frac{k}{k-1}}(\varrho_1)(h_1 \circ g)^{\frac{p_1}{r}}(\xi) \mathrm{d}g(\xi) \right)^r \right.$$

$$\left. + \left(\int_0^1 \Psi_2^{\frac{1}{r}} \left(f^{\frac{rk}{k-1}}(\varrho_2), f^{\frac{rk}{k-1}}(\varrho_1) \right) (h_2 \circ g)^{\frac{p_2}{r}}(\xi) \mathrm{d}g(\xi) \right)^r \right]^{\frac{k-1}{rk}}$$

$$= \Xi_{t,1}^{p+q+1} \times \sqrt[k]{B(g(\xi))}$$

$$\times \left[f^{\frac{rk}{k-1}}(\varrho_1) I^r((h_1 \circ g)(\xi)) + \Psi_2 \left(f^{\frac{rk}{k-1}}(\varrho_2), f^{\frac{rk}{k-1}}(\varrho_1) \right) I^r \right.$$

$$\left. \times ((h_2 \circ g)(\xi)) \right]^{\frac{k-1}{rk}}.$$

So, the proof of this theorem is completed. □

We point out some special cases of Theorem 2.

Corollary 3. *In Theorem 2 for* $k = 2$, *we get the following inequality:*

$$\left| \mathcal{T}_{f,g}^{p,q}(\Psi_1, \sigma, \mathbf{m}; \varrho_1, \varrho_2) \right| \leq \Xi_{t,1}^{p+q+1} \times \sqrt{B(g(\xi))}$$
$$\times \sqrt[2r]{f^{2r}(a) I^r((h_1 \circ g)(\xi)) + \Psi_2 \left(f^{2r}(b), f^{2r}(a) \right) I^r((h_2 \circ g)(\xi))}.$$
(18)

Corollary 4. In Theorem 2 for $g(\xi) = \xi$, we get the following inequality:

$$\left|T_f^{p,q}(\Psi_1, \sigma, m; \varrho_1, \varrho_2)\right| \leq \Xi_{t,1}^{p+q+1} \times \sqrt[k]{\beta(kp+1, kq+1)}$$

$$\times \left[f^{\frac{rk}{k-1}}(\varrho_1)I^r(h_1(\xi)) + \Psi_2\left(f^{\frac{rk}{k-1}}(\varrho_2), f^{\frac{rk}{k-1}}(\varrho_1)\right)I^r(h_2(\xi))\right]^{\frac{k-1}{rk}}.$$

(19)

Corollary 5. Under the conditions of Remark 5 using Theorem 2, we get the following inequality:

$$\int_{\varrho_1}^{\varrho_2} (x-\varrho_1)^p(\varrho_2-x)^q f(x)\mathrm{d}x \leq (\varrho_2-\varrho_1)^{p+q+1}$$

$$\times \sqrt[k]{\beta(kp+1, kq+1)} \times \left[f^{\frac{rk}{k-1}}(\varrho_1)I^r(h_1(\xi))\right.$$

$$\left. + \Psi_2\left(f^{\frac{rk}{k-1}}(\varrho_2), f^{\frac{rk}{k-1}}(\varrho_1)\right)I^r(h_2(\xi))\right]^{\frac{k-1}{rk}}. \quad (20)$$

Corollary 6. In Corollary 5 for $h_1(t) = h(1-t)$ and $h_2(t) = h(t)$, we get the following inequality for generalized-$1 - ((h^{p_1}(1-t), h^{p_2}(t)); (\Psi_1, \Psi_2))$-convex mappings:

$$\int_{\varrho_1}^{\varrho_2} (x-\varrho_1)^p(\varrho_2-x)^q f(x)\mathrm{d}x \leq (\varrho_2-\varrho_1)^{p+q+1}$$

$$\times \sqrt[k]{\beta(kp+1, kq+1)} \times \left[f^{\frac{rk}{k-1}}(\varrho_1)I^r(h(1-\xi))\right.$$

$$\left. + \Psi_2\left(f^{\frac{rk}{k-1}}(\varrho_2), f^{\frac{rk}{k-1}}(\varrho_1)\right)I^r(h(\xi))\right]^{\frac{k-1}{rk}}. \quad (21)$$

Corollary 7. In Corollary 6 for $h_1(t) = (1-t)^s$ and $h_2(t) = t^s$, we get the following inequality for generalized-$1 - (((1-t)^{sp_1}, t^{sp_2});$

$(\Psi_1, \Psi_2))$-*Breckner-convex mappings:*

$$\int_{\varrho_1}^{\varrho_2} (x - \varrho_1)^p (\varrho_2 - x)^q f(x) \mathrm{d}x \leq (\varrho_2 - \varrho_1)^{p+q+1}$$

$$\times \sqrt[k]{\beta(kp+1, kq+1)} \left[f^{\frac{rk}{k-1}}(\varrho_1) \left(\frac{r}{r + sp_1} \right)^r \right.$$

$$\left. + \Psi_2 \left(f^{\frac{rk}{k-1}}(\varrho_2), f^{\frac{rk}{k-1}}(\varrho_1) \right) \left(\frac{r}{r + sp_2} \right)^r \right]^{\frac{k-1}{rk}}. \qquad (22)$$

Corollary 8. *In Corollary 6 for* $h_1(t) = (1-t)^{-s}$, $h_2(t) = t^{-s}$ *and* $r > s \cdot \max\{p_1, p_2\}$, *we get the following inequality for generalized-*$1 - (((1-t)^{-sp_1}, t^{-sp_2})$;
$(\Psi_1, \Psi_2))$-*Godunova–Levin–Dragomir-convex mappings:*

$$\int_{\varrho_1}^{\varrho_2} (x - \varrho_1)^p (\varrho_2 - x)^q f(x) dx \leq (\varrho_2 - \varrho_1)^{p+q+1}$$

$$\times \sqrt[k]{\beta(kp+1, kq+1)} \times \left[f^{\frac{rk}{k-1}}(\varrho_1) \left(\frac{r}{r - sp_1} \right)^r \right.$$

$$\left. + \Psi_2 \left(f^{\frac{rk}{k-1}}(\varrho_2), f^{\frac{rk}{k-1}}(\varrho_1) \right) \left(\frac{r}{r - sp_2} \right)^r \right]^{\frac{k-1}{rk}}. \qquad (23)$$

Corollary 9. *In Corollary 6 for* $h_1(t) = h_2(t) = t(1-t)$, *we get the following inequality for generalized-*$1 - ((t(1-t))^{sp_1}, (t(1-t))^{sp_2}); (\Psi_1, \Psi_2))$-*convex mappings:*

$$\int_{\varrho_1}^{\varrho_2} (x - \varrho_1)^p (\varrho_2 - x)^q f(x) \mathrm{d}x \leq (\varrho_2 - \varrho_1)^{p+q+1}$$

$$\times \sqrt[k]{\beta(kp+1, kq+1)} \times \left[f^{\frac{rk}{k-1}}(\varrho_1) \beta^r \left(1 + \frac{p_1}{r}, 1 + \frac{p_1}{r} \right) \right.$$

$$\left. + \Psi_2 \left(f^{\frac{rk}{k-1}}(\varrho_2), f^{\frac{rk}{k-1}}(\varrho_1) \right) \beta^r \left(1 + \frac{p_2}{r}, 1 + \frac{p_2}{r} \right) \right]^{\frac{k-1}{rk}}. \qquad (24)$$

Corollary 10. *In Corollary 6 for* $h_1(t) = \frac{\sqrt{1-t}}{2\sqrt{t}}$, $h_2(t) = \frac{\sqrt{t}}{2\sqrt{1-t}}$ *and* $r > \frac{1}{2} \cdot \max\{p_1, p_2\}$, *we get the following inequality for generalized*-$1 - \left(\left(\left(\frac{\sqrt{1-t}}{2\sqrt{t}}\right)^{p_1}, \left(\frac{\sqrt{t}}{2\sqrt{1-t}}\right)^{p_2}\right); (\Psi_1, \Psi_2)\right)$-*convex mappings:*

$$\int_{\varrho_1}^{\varrho_2} (x - \varrho_1)^p (\varrho_2 - x)^q f(x) dx \leq (\varrho_2 - \varrho_1)^{p+q+1}$$

$$\times \sqrt[k]{\beta(kp+1, kq+1)} \times \left[f^{\frac{rk}{k-1}}(\varrho_1) \frac{\beta^r \left(1 - \frac{p_1}{2r}, 1 + \frac{p_1}{2r}\right)}{2^{p_1}} \right.$$

$$\left. + \Psi_2 \left(f^{\frac{rk}{k-1}}(\varrho_2), f^{\frac{rk}{k-1}}(\varrho_1) \right) \frac{\beta^r \left(1 - \frac{p_2}{2r}, 1 + \frac{p_2}{2r}\right)}{2^{p_2}} \right]^{\frac{k-1}{rk}}. \quad (25)$$

Theorem 11. *Let* $\sigma : \mathcal{I} \to \mathcal{K}$ *be a continuous function*, $g : [0,1] \to [0,1]$ *be a strictly increasing function on* $(0,1)$ *and* $\boldsymbol{m} : [0,1] \to (0,1]$. *Suppose* $\mathcal{K} = (\boldsymbol{m}(t)\sigma(\varrho_1), \boldsymbol{m}(t)\sigma(\varrho_1) + \Xi_{t,1}) \subseteq \Re$ *is an open* $(\boldsymbol{m}, g; \sigma)$-*invex subset with respect to* $\Psi : \mathcal{K} \times \mathcal{K} \to \Re$ *for* $\Xi_{t,1} = \Psi_1(\sigma(\varrho_2), \boldsymbol{m}(t)\sigma(\varrho_1)) > 0$ *for all* $t \in [0,1]$ *and* $\Psi_2 : f(\mathcal{K}) \times f(\mathcal{K}) \to \Re$. *Assume that* $f : \mathcal{K} \to (0, +\infty)$ *is a mapping on* $\mathcal{K}^\circ$ *such that* $f \in \mathcal{L}(\mathcal{K})$. *If* f^l *is generalized-*$\boldsymbol{m} - (((h_1 \circ g)^{p_1}, (h_2 \circ g)^{p_2}); (\Psi_1, \Psi_2))$-*convex mapping*, $l \geq 1$, $0 < r \leq 1$ *and* $p_1, p_2 > -1$, *then for any fixed* $p, q > 0$, *the following inequality holds:*

$$\left| T^{p,q}_{f,g}(\Psi_1, \sigma, \boldsymbol{m}; \varrho_1, \varrho_2) \right| \leq \Xi^{p+q+1}_{t,1} \times C^{\frac{l-1}{l}}(g(\xi))$$

$$\times \sqrt[rl]{f^{rl}(\varrho_1) F^r((h_1 \circ g)(\xi)) + \Psi_2 \left(f^{rl}(\varrho_2), f^{rl}(\varrho_1) \right) F^r((h_2 \circ g)(\xi))}, \quad (26)$$

where

$$F((h_1 \circ g)(\xi)) := \int_0^1 \boldsymbol{m}^{\frac{1}{r}}(\xi) g^p(\xi)(1 - g(\xi))^q (h_1 \circ g)^{\frac{p_1}{r}}(\xi) dg(\xi),$$

$$F((h_2 \circ g)(\xi)) := \int_0^1 g^p(\xi)(1 - g(\xi))^q (h_2 \circ g)^{\frac{p_2}{r}}(\xi) dg(\xi)$$

and

$$C(g(\xi)) := \int_0^1 g^p(\xi)(1 - g(\xi))^q dg(\xi).$$

Proof. From Lemma 1, generalized-$\mathbf{m} - (((h_1 \circ g)^{p_1}, (h_2 \circ g)^{p_2}); (\Psi_1, \Psi_2))$-convexity of f^l, the well-known power mean inequality, Minkowski inequality and properties of the modulus, we have

$$\left|\mathcal{T}_{f,g}^{p,q}(\Psi_1, \sigma, \mathbf{m}; \varrho_1, \varrho_2)\right| = |\Xi_{t,1}|^{p+q+1}$$

$$\times \left| \int_0^1 \left[g^p(\xi)(1-g(\xi))^q\right]^{\frac{l-1}{l}} \left[g^p(\xi)(1-g(\xi))^q\right]^{\frac{1}{l}} f(\mathbf{m}(t)\sigma(\varrho_1) \right.$$

$$\left. + g(\xi)\Xi_{t,1})\mathrm{d}g(\xi) \right|$$

$$\leq \Xi_{t,1}^{p+q+1} \times \left[\int_0^1 g^p(\xi)(1-g(\xi))^q \mathrm{d}g(\xi)\right]^{\frac{l-1}{l}}$$

$$\times \left[\int_0^1 g^p(\xi)(1-g(\xi))^q f^l(\mathbf{m}(t)\sigma(\varrho_1) + g(\xi)\Xi_{t,1})\mathrm{d}g(\xi)\right]^{\frac{1}{l}}$$

$$\leq \Xi_{t,1}^{p+q+1} \times C^{\frac{l-1}{l}}(g(\xi))$$

$$\times \left[\int_0^1 g^p(\xi)(1-g(\xi))^q \times \left[\mathbf{m}(\xi)(h_1 \circ g)^{p_1}(\xi) f^{rl}(\varrho_1) \right.\right.$$

$$\left.\left. + (h_2 \circ g)^{p_2}(\xi)\Psi_2 \times \left(f^{rl}(\varrho_2), f^{rl}(\varrho_1)\right)\right]^{\frac{1}{r}} \mathrm{d}g(\xi)\right]^{\frac{1}{l}}$$

$$\leq \Xi_{t,1}^{p+q+1} \times C^{\frac{l-1}{l}}(g(\xi)) \times \left[\left(\int_0^1 \mathbf{m}^{\frac{1}{r}}(\xi) f^l(\varrho_1) g^p(\xi) \right.\right.$$

$$\times (1-g(\xi))^q (h_1 \circ g)^{\frac{p_1}{r}}(\xi) \mathrm{d}g(\xi)\bigg)^r$$

$$+ \left(\int_0^1 \Psi_2^{\frac{1}{2}} \left(f^{rl}(\varrho_2), f^{rl}(\varrho_1)\right) g^p(\xi)(1-g(\xi))^q \right.$$

$$\left.\left. \times (h_2 \circ g)^{\frac{p_2}{r}}(\xi)\mathrm{d}g(\xi)\right)^r \right]^{\frac{1}{rl}}$$

$$= \Xi_{t,1}^{p+q+1} \times C^{\frac{l-1}{l}}(g(\xi))$$

$$\times \sqrt[rl]{\frac{f^{rl}(\varrho_1) F^r((h_1 \circ g)(\xi)) + \Psi_2\left(f^{rl}(\varrho_2), f^{rl}(\varrho_1)\right)}{F^r((h_2 \circ g)(\xi))}}.$$

So, the proof of this theorem is completed. □

We point out some special cases of Theorem 11.

Corollary 12. *In Theorem 11 for $l = 1$, we get the following inequality:*

$$\left| T_{f,g}^{p,q}(\Psi_1, \sigma, \boldsymbol{m}; \varrho_1, \varrho_2) \right| \leq \Xi_{t,1}^{p+q+1}$$

$$\times \sqrt[r]{f^r(\varrho_1) F^r((h_1 \circ g)(\xi)) + \Psi_2 (f^r(\varrho_2), f^r(\varrho_1)) F^r((h_2 \circ g)(\xi))}. \tag{27}$$

Corollary 13. *In Theorem 11 for $g(\xi) = \xi$, we get the following inequality:*

$$\left| T_f^{p,q}(\Psi_1, \sigma, \boldsymbol{m}; \varrho_1, \varrho_2) \right| \leq \Xi_{t,1}^{p+q+1} \times \beta^{\frac{l-1}{l}}(p+1, q+1)$$

$$\times \sqrt[rl]{f^{rl}(\varrho_1) F^r(h_1(\xi)) + \Psi_2 (f^{rl}(\varrho_2), f^{rl}(\varrho_1)) F^r(h_2(\xi))}. \tag{28}$$

Corollary 14. *Under the conditions of Remark 5 using Theorem 11, we get the following inequality:*

$$\int_{\varrho_1}^{\varrho_2} (x - \varrho_1)^p (\varrho_2 - x)^q f(x) dx \leq (\varrho_2 - \varrho_1)^{p+q+1} \times \beta^{\frac{l-1}{l}}(p+1, q+1)$$

$$\times \sqrt[rl]{f^{rl}(\varrho_1) F^r(h_1(\xi)) + \Psi_2 (f^{rl}(\varrho_2), f^{rl}(\varrho_1)) F^r(h_2(\xi))}. \tag{29}$$

Corollary 15. *In Corollary 14 for $h_1(t) = h(1 - t)$ and $h_2(t) = h(t)$, we get the following inequality for generalized-$1 - ((h^{p_1}(1-t), h^{p_2}(t)); (\Psi_1, \Psi_2))$-convex mappings:*

$$\int_{\varrho_1}^{\varrho_2} (x - \varrho_1)^p (\varrho_2 - x)^q f(x) \mathrm{d}x \leq (\varrho_2 - \varrho_1)^{p+q+1} \times \beta^{\frac{l-1}{l}}(p+1, q+1)$$

$$\times \sqrt[rl]{f^{rl}(\varrho_1) F^r(h(1 - \xi)) + \Psi_2 (f^{rl}(\varrho_2), f^{rl}(\varrho_1)) F^r(h(\xi))}. \tag{30}$$

Corollary 16. *In Corollary 15 for $h_1(t) = (1-t)^s$ and $h_2(t) = t^s$, we get the following inequality for generalized-$1 - (((1-t)^{sp_1}, t^{sp_2}); (\Psi_1, \Psi_2))$-Breckner-convex mappings:*

$$\int_{\varrho_1}^{\varrho_2} (x-\varrho_1)^p (\varrho_2-x)^q f(x) dx \leq (\varrho_2-\varrho_1)^{p+q+1} \times \beta^{\frac{l-1}{l}}(p+1, q+1)$$

$$\times \sqrt[rl]{\begin{array}{l} f^{rl}(\varrho_1) \beta^r \left(p+1, \frac{sp_1}{r}+q+1\right) \\ + \Psi_2 \left(f^{rl}(\varrho_2), f^{rl}(\varrho_1)\right) \beta^r \left(q+1, \frac{sp_2}{r}+p+1\right) \end{array}}. \quad (31)$$

Corollary 17. *In Corollary 15 for $h_1(t) = (1-t)^{-s}$, $h_2(t) = t^{-s}$ where $p > \frac{sp_2}{r} - 1$ and $q > \frac{sp_1}{r} - 1$, we get the following inequality for generalized-$1 - (((1-t)^{-sp_1}, t^{-sp_2}); (\Psi_1, \Psi_2))$-Godunova–Levin–Dragomir-convex mappings:*

$$\int_{\varrho_1}^{\varrho_2} (x-\varrho_1)^p (\varrho_2-x)^q f(x) dx \leq (\varrho_2-\varrho_1)^{p+q+1} \times \beta^{\frac{l-1}{l}}(p+1, q+1)$$

$$\times \sqrt[rl]{\begin{array}{l} f^{rl}(\varrho_1) \beta^r \left(p+1, q-\frac{sp_1}{r}+1\right) \\ + \Psi_2 \left(f^{rl}(\varrho_2), f^{rl}(\varrho_1)\right) \beta^r \left(q+1, p-\frac{sp_2}{r}+1\right) \end{array}}. \quad (32)$$

Corollary 18. *In Corollary 15 for $h_1(t) = h_2(t) = t(1-t)$, we get the following inequality for generalized-$1 - ((t(1-t))^{sp_1}, (t(1-t))^{sp_2}); (\Psi_1, \Psi_2))$-convex mappings:*

$$\int_{\varrho_1}^{\varrho_2} (x-\varrho_1)^p (\varrho_2-x)^q f(x) dx \leq (\varrho_2-\varrho_1)^{p+q+1} \times \beta^{\frac{l-1}{l}}(p+1, q+1)$$

$$\times \sqrt[rl]{\begin{array}{l} f^{rl}(\varrho_1) \beta^r \left(\frac{p_1}{r}+p+1, \frac{p_1}{r}+q+1\right) \\ + \Psi_2 \left(f^{rl}(\varrho_2), f^{rl}(\varrho_1)\right) \beta^r \left(\frac{p_2}{r}+p+1, \frac{p_2}{r}+q+1\right) \end{array}}. \quad (33)$$

Corollary 19. *In Corollary 15 for $h_1(t) = \frac{\sqrt{1-t}}{2\sqrt{t}}$, $h_2(t) = \frac{\sqrt{t}}{2\sqrt{1-t}}$ and $r > \frac{1}{2} \cdot \max\{p_1, p_2\}$, we get the following inequality for generalized-$1 - \left(\left(\left(\frac{\sqrt{1-t}}{2\sqrt{t}}\right)^{p_1}, \left(\frac{\sqrt{t}}{2\sqrt{1-t}}\right)^{p_2}\right); (\Psi_1, \Psi_2)\right)$-convex mappings:*

$$\int_{\varrho_1}^{\varrho_2} (x-\varrho_1)^p (\varrho_2-x)^q f(x) dx \leq (\varrho_2-\varrho_1)^{p+q+1} \times \beta^{\frac{l-1}{l}}(p+1, q+1)$$

$$\times \sqrt[rl]{\begin{array}{l} f^{rl}(\varrho_1) \frac{\beta^r \left(p-\frac{p_1}{2r}+1, \frac{p_1}{2r}+q+1\right)}{2^{p_1}} + \Psi_2 \left(f^{rl}(\varrho_2), f^{rl}(\varrho_1)\right) \\ \times \frac{\beta^r \left(\frac{p_2}{2r}+p+1, q-\frac{p_2}{2r}+1\right)}{2^{p_2}} \end{array}}. \quad (34)$$

Remark 6. By taking particular values of parameters p, q, r, p_1 and p_2 in Theorems 2 and 11, several new bounds integral inequalities for Gauss–Jacobi-type associated with generalized-$\mathbf{m} - (((h_1 \circ g)^{p_1}, (h_2 \circ g)^{p_2}); (\Psi_1, \Psi_2))$-convex mappings can be obtained.

3. Some New Bounds of Hermite–Hadamard Type via k-Fractional Integral Inequalities

For establishing our next results, we need to prove the following lemma.

Lemma 2. *Let $\sigma : \mathcal{I} \to \mathcal{K}$ be a continuous function, $g : [0,1] \to [0,1]$ be a strictly increasing function on $(0,1)$ and $\mathbf{m} : [0,1] \to (0,1]$.*
Suppose $\mathcal{K} = (\mathbf{m}(t)\sigma(\varrho_1), \mathbf{m}(t)\sigma(\varrho_1) + g(1)\Xi_t) \subseteq \Re$ is an open $(\mathbf{m}, g; \sigma)$-invex subset with respect to $\Psi : \mathcal{K} \times \mathcal{K} \to \Re$ for $\Xi_t = \Psi(\sigma(\varrho_2), \mathbf{m}(t)\sigma(\varrho_1)) > 0$ and $\forall t \in [0,1]$. Assume that $f : \mathcal{K} \to \Re$ is a twice differentiable mapping on $\mathcal{K}^\circ$ such that $f'' \in \mathcal{L}(\mathcal{K})$. Then for $\alpha, k > 0$, the following equality holds:

$$\frac{\Xi_t^2}{2} \times \left\{ \frac{f'(\mathbf{m}(t)\sigma(\varrho_1) + g(1)\Xi_t) - f'(\mathbf{m}(t)\sigma(\varrho_1) + g(0)\Xi_t)}{\Xi_t} \right.$$

$$-\frac{(1-g(1))^{\frac{\alpha}{k}+1}f'(\mathbf{m}(t)\sigma(\varrho_1) + g(1)\Xi_t)}{\Xi_t}$$
$$-\frac{(1-g(0))^{\frac{\alpha}{k}+1}f'(\mathbf{m}(t)\sigma(\varrho_1) + g(0)\Xi_t)}{\Xi_t} - \frac{\left(\frac{\alpha}{k}+1\right)}{\Xi_t}$$

$$\times \left[\frac{(1-g(1))^{\frac{\alpha}{k}}f(\mathbf{m}(t)\sigma(\varrho_1) + g(1)\Xi_t)}{-(1-g(0))^{\frac{\alpha}{k}}f(\mathbf{m}(t)\sigma(\varrho_1) + g(0)\Xi_t)} \right]$$

$$-\frac{g^{\frac{\alpha}{k}+1}(1)f'(\mathbf{m}(t)\sigma(\varrho_1) + g(1)\Xi_t)}{\Xi_t}$$
$$\left. -\frac{g^{\frac{\alpha}{k}+1}(0)f'(\mathbf{m}(t)\sigma(\varrho_1) + g(0)\Xi_t)}{\Xi_t} + \frac{\left(\frac{\alpha}{k}+1\right)}{\Xi_t} \right.$$

$$\times \left[\frac{g^{\frac{\alpha}{k}}(1)f(\boldsymbol{m}(t)\sigma(\varrho_1)+g(1)\Xi_t) - g^{\frac{\alpha}{k}}(0)f(\boldsymbol{m}(t)\sigma(\varrho_1)+g(0)\Xi_t)}{\Xi_t} \right] \right\} - \frac{\alpha\left(\frac{\alpha}{k}+1\right)}{2k\Xi_t^{\frac{\alpha}{k}}}$$

$$\times \int_{\boldsymbol{m}(t)\sigma(\varrho_1)+g(0)\Xi_t}^{\boldsymbol{m}(t)\sigma(\varrho_1)+g(1)\Xi_t} \left[(w-\boldsymbol{m}(t)\sigma(\varrho_1))^{\frac{\alpha}{k}-1} \right.$$

$$\left. + (\boldsymbol{m}(t)\sigma(\varrho_1)+\Xi_t-w)^{\frac{\alpha}{k}-1} \right] f(w)\mathrm{d}w$$

$$= \frac{\Xi_t^2}{2} \times \int_0^1 \left[1-(1-g(\xi))^{\frac{\alpha}{k}+1} - g^{\frac{\alpha}{k}+1}(\xi) \right] f''(\boldsymbol{m}(t)\sigma(\varrho_1)$$

$$+ g(\xi)\Xi_t)\mathrm{d}g(\xi). \tag{35}$$

We denote

$$\mathcal{T}_{f,g}^{\alpha,k}(\Psi,\sigma,\boldsymbol{m};\varrho_1,\varrho_2) := \frac{\Xi_t^2}{2}$$

$$\times \int_0^1 \left[1-(1-g(\xi))^{\frac{\alpha}{k}+1} - g^{\frac{\alpha}{k}+1}(\xi) \right] f''(\boldsymbol{m}(t)\sigma(\varrho_1)+g(\xi)\Xi_t)\mathrm{d}g(\xi). \tag{36}$$

Proof. Integrating by parts twice equation (36) and changing the variable $w = \boldsymbol{m}(t)\sigma(\varrho_1) + g(\xi)\Xi_t$, we get

$$\mathcal{T}_{f,g}^{\alpha,k}(\Psi,\sigma,\mathbf{m};\varrho_1,\varrho_2) = \frac{\Xi_t^2}{2}$$

$$\times \left[\int_0^1 f''(\mathbf{m}(t)\sigma(\varrho_1)+g(\xi)\Xi_t)\mathrm{d}g(\xi) \right.$$

$$- \int_0^1 (1-g(\xi))^{\frac{\alpha}{k}+1} f''(\mathbf{m}(t)\sigma(\varrho_1)+g(\xi)\Xi_t)\mathrm{d}g(\xi)$$

$$\left. - \int_0^1 g^{\frac{\alpha}{k}+1}(\xi) f''(\mathbf{m}(t)\sigma(\varrho_1)+g(\xi)\Xi_t)\mathrm{d}g(\xi) \right]$$

$$= \frac{\Xi_t^2}{2} \times \left\{ \frac{f'(\mathbf{m}(t)\sigma(\varrho_1) + g(1)\Xi_t) - f'(\mathbf{m}(t)\sigma(\varrho_1) + g(0)\Xi_t)}{\Xi_t} \right.$$

$$- \left. \frac{(1-g(\xi))^{\frac{\alpha}{k}+1} f'(\mathbf{m}(t)\sigma(\varrho_1) + g(\xi)\Xi_t)}{\Xi_t} \right|_0^1$$

$$- \frac{\left(\frac{\alpha}{k}+1\right)}{\Xi_t} \times \int_0^1 (1-g(\xi))^{\frac{\alpha}{k}} f'(\mathbf{m}(t)\sigma(\varrho_1) + g(\xi)\Xi_t) \mathrm{d}g(\xi)$$

$$- \left. \frac{g^{\frac{\alpha}{k}+1}(\xi) f'(\mathbf{m}(t)\sigma(\varrho_1) + g(\xi)\Xi_t)}{\Xi_t} \right|_0^1 + \frac{\left(\frac{\alpha}{k}+1\right)}{\Xi_t}$$

$$\left. \times \int_0^1 g^{\frac{\alpha}{k}}(\xi) f'(\mathbf{m}(t)\sigma(\varrho_1) + g(\xi)\Xi_t) \mathrm{d}g(\xi) \right\}$$

$$= \frac{\Xi_t^2}{2} \times \left\{ \frac{f'(\mathbf{m}(t)\sigma(\varrho_1) + g(1)\Xi_t) - f'(\mathbf{m}(t)\sigma(\varrho_1) + g(0)\Xi_t)}{\Xi_t} \right.$$

$$- \frac{(1-g(1))^{\frac{\alpha}{k}+1} f'(\mathbf{m}(t)\sigma(\varrho_1) + g(1)\Xi_t)}{\Xi_t} - \frac{\left(\frac{\alpha}{k}+1\right)}{\Xi_t}$$

$$\times \left[\left. \frac{(1-g(\xi))^{\frac{\alpha}{k}} f(\mathbf{m}(t)\sigma(\varrho_1) + g(\xi)\Xi_t)}{\Xi_t} \right|_0^1 \right.$$

$$\left. + \frac{\alpha}{k\Xi_t} \times \int_0^1 (1-g(\xi))^{\frac{\alpha}{k}-1} f(\mathbf{m}(t)\sigma(\varrho_1) + g(\xi)\Xi_t) \mathrm{d}g(\xi) \right]$$

$$- \frac{g^{\frac{\alpha}{k}+1}(1) f'(\mathbf{m}(t)\sigma(\varrho_1) + g(1)\Xi_t)}{\Xi_t} + \frac{\left(\frac{\alpha}{k}+1\right)}{\Xi_t}$$

$$\times \left[\left. \frac{g^{\frac{\alpha}{k}}(\xi) f(\mathbf{m}(t)\sigma(\varrho_1) + g(\xi)\Xi_t)}{\Xi_t} \right|_0^1 - \frac{\alpha}{k\Xi_t} \right.$$

$$\left. \left. \times \int_0^1 g^{\frac{\alpha}{k}-1}(\xi) f(\mathbf{m}(t)\sigma(\varrho_1) + g(\xi)\Xi_t) \mathrm{d}g(\xi) \right] \right\}$$

$$= \frac{\Xi_t^2}{2} \times \left\{ \frac{f'(\mathbf{m}(t)\sigma(\varrho_1) + g(1)\Xi_t) - f'(\mathbf{m}(t)\sigma(\varrho_1) + g(0)\Xi_t)}{\Xi_t} \right.$$

$$- \frac{(1-g(1))^{\frac{\alpha}{k}+1} f'(\mathbf{m}(t)\sigma(\varrho_1) + g(1)\Xi_t) - (1-g(0))^{\frac{\alpha}{k}+1} f'(\mathbf{m}(t)\sigma(\varrho_1) + g(0)\Xi_t)}{\Xi_t} - \frac{\left(\frac{\alpha}{k}+1\right)}{\Xi_t}$$

$$\times \left[\frac{(1-g(1))^{\frac{\alpha}{k}} f(\mathbf{m}(t)\sigma(\varrho_1) + g(1)\Xi_t) - (1-g(0))^{\frac{\alpha}{k}} f(\mathbf{m}(t)\sigma(\varrho_1) + g(0)\Xi_t)}{\Xi_t} + \frac{\alpha}{k\Xi_t^{\frac{\alpha}{k}+1}} \right.$$

$$\left. \times \int_{\mathbf{m}(t)\sigma(\varrho_1) + g(0)\Xi_t}^{\mathbf{m}(t)\sigma(\varrho_1) + g(1)\Xi_t} (\mathbf{m}(t)\sigma(\varrho_1) + \Xi_t - w)^{\frac{\alpha}{k}-1} f(w)\mathrm{d}w \Xi_t \right]$$

$$- \frac{g^{\frac{\alpha}{k}+1}(1) f'(\mathbf{m}(t)\sigma(\varrho_1) + g(1)\Xi_t) - g^{\frac{\alpha}{k}+1}(0) f'(\mathbf{m}(t)\sigma(\varrho_1) + g(0)\Xi_t)}{\Xi_t} + \frac{\left(\frac{\alpha}{k}+1\right)}{\Xi_t}$$

$$\times \left[\frac{g^{\frac{\alpha}{k}}(1) f(\mathbf{m}(t)\sigma(\varrho_1) + g(1)\Xi_t) - g^{\frac{\alpha}{k}}(0) f(\mathbf{m}(t)\sigma(\varrho_1) + g(0)\Xi_t)}{\Xi_t} - \frac{\alpha}{k\Xi_t^{\frac{\alpha}{k}+1}} \right.$$

$$\left. \left. \times \int_{\mathbf{m}(t)\sigma(\varrho_1) + g(0)\Xi_t}^{\mathbf{m}(t)\sigma(\varrho_1) + g(1)\Xi_t} (w - \mathbf{m}(t)\sigma(\varrho_1))^{\frac{\alpha}{k}-1} f(w)\mathrm{d}w \right] \right\}$$

$$= \frac{\Xi_t^2}{2} \times \left\{ \frac{f'(\mathbf{m}(t)\sigma(\varrho_1) + g(1)\Xi_t) - f'(\mathbf{m}(t)\sigma(\varrho_1) + g(0)\Xi_t)}{\Xi_t} \right.$$

$$- \frac{(1-g(1))^{\frac{\alpha}{k}+1} f'(\mathbf{m}(t)\sigma(\varrho_1) + g(1)\Xi_t)}{\Xi_t} - \frac{(1-g(0))^{\frac{\alpha}{k}+1} f'(\mathbf{m}(t)\sigma(\varrho_1) + g(0)\Xi_t)}{\Xi_t} - \frac{\left(\frac{\alpha}{k}+1\right)}{\Xi_t}$$

$$\times \left[\frac{(1-g(1))^{\frac{\alpha}{k}} f(\mathbf{m}(t)\sigma(\varrho_1) + g(1)\Xi_t)}{\Xi_t} - (1-g(0))^{\frac{\alpha}{k}} f(\mathbf{m}(t)\sigma(\varrho_1) + g(0)\Xi_t)}{\Xi_t} \right]$$

$$- \frac{g^{\frac{\alpha}{k}+1}(1) f'(\mathbf{m}(t)\sigma(\varrho_1) + g(1)\Xi_t)}{\Xi_t} - g^{\frac{\alpha}{k}+1}(0) f'(\mathbf{m}(t)\sigma(\varrho_1) + g(0)\Xi_t)}{\Xi_t} + \frac{\left(\frac{\alpha}{k}+1\right)}{\Xi_t}$$

$$\times \left[\frac{g^{\frac{\alpha}{k}}(1) f(\mathbf{m}(t)\sigma(\varrho_1) + g(1)\Xi_t)}{\Xi_t} - g^{\frac{\alpha}{k}}(0) f(\mathbf{m}(t)\sigma(\varrho_1) + g(0)\Xi_t)}{\Xi_t} \right] \right\} - \frac{\alpha\left(\frac{\alpha}{k}+1\right)}{2k\Xi_t^{\frac{\alpha}{k}}}$$

$$\times \int_{\mathbf{m}(t)\sigma(\varrho_1)+g(0)\Xi_t}^{\mathbf{m}(t)\sigma(\varrho_1)+g(1)\Xi_t} \left[(w - \mathbf{m}(t)\sigma(\varrho_1))^{\frac{\alpha}{k}-1} + (\mathbf{m}(t)\sigma(\varrho_1) + \Xi_t - w)^{\frac{\alpha}{k}-1} \right] f(w) dw.$$

This completes the proof of our lemma. □

Corollary 20. *Under the conditions of Lemma 2 for $g(\xi) = \xi$, the following identity for k-fractional integrals holds:*

$$\mathcal{T}_f^{\alpha,k}(\Psi, \sigma, \mathbf{m}; \varrho_1, \varrho_2) := \frac{\Xi_t^2}{2}$$

$$\times \int_0^1 \left[1 - (1-\xi)^{\frac{\alpha}{k}+1} - \xi^{\frac{\alpha}{k}+1} \right] f''(\mathbf{m}(t)\sigma(\varrho_1) + \xi\Xi_t) d\xi.$$

$$= \left(\frac{\alpha}{k}+1\right) \times \left\{ \frac{f(\mathbf{m}(t)\sigma(\varrho_1)) + f(\mathbf{m}(t)\sigma(\varrho_1) + \Xi_t)}{2} \right.$$

$$-\frac{\Gamma_k(\alpha+k)}{2\Xi_t^{\frac{\alpha}{k}}} \times \left[I_{(m(t)\sigma(\varrho_1))^+}^{\alpha,k} f(m(t)\sigma(\varrho_1) + \Xi_t)\right.$$

$$\left. + I_{(m(t)\sigma(\varrho_1)+\Xi_t)^-}^{\alpha,k} f(m(t)\sigma(\varrho_1))\right]\bigg\}. \tag{37}$$

Remark 7. Using Corollary 20, for $\Xi_t = \sigma(\varrho_2) - m(t)\sigma(\varrho_1)$, where $m(t) \equiv 1$ for all $t \in [0,1]$, we get the following Hermite–Hadamard k-fractional integral identity:

$$\frac{f(\sigma(\varrho_1)) + f(\sigma(\varrho_2))}{2} - \frac{\Gamma_k(\alpha+k)}{2(\sigma(\varrho_2) - \sigma(\varrho_1))^{\frac{\alpha}{k}}}$$

$$\times \left[I_{(\sigma(\varrho_1))^+}^{\alpha,k} f(\sigma(\varrho_2)) + I_{(\sigma(\varrho_2))^-}^{\alpha,k} f(\sigma(\varrho_1))\right]$$

$$= \frac{(\sigma(\varrho_2) - \sigma(\varrho_1))^2}{2\left(\frac{\alpha}{k}+1\right)} \int_0^1 \left[1 - (1-\xi)^{\frac{\alpha}{k}+1} - \xi^{\frac{\alpha}{k}+1}\right] f''(\sigma(\varrho_1)$$

$$+ \xi(\sigma(\varrho_2) - \sigma(\varrho_1)))\mathrm{d}\xi. \tag{38}$$

Using Lemma 2, we now state the following theorems for the corresponding version for power of second derivative.

Theorem 21. *Let $h_1, h_2 : [0,1] \to [0,+\infty)$, $\sigma : \mathcal{I} \to \mathcal{K}$ be continuous functions, $g : [0,1] \to [0,1]$ be a strictly increasing function on $(0,1)$ and $m : [0,1] \to (0,1]$. Suppose $\mathcal{K} = (m(t)\sigma(\varrho_1), m(t)\sigma(\varrho_1) + g(1)\Xi_{t,1}) \subseteq \Re$ is an open $(m, g; \sigma)$-invex subset with respect to $\Psi_1 : \mathcal{K} \times \mathcal{K} \to \Re$ for $\Xi_{t,1} = \Psi_1(\sigma(\varrho_2), m(t)\sigma(\varrho_1)) > 0, \forall t \in [0,1]$ and $\Psi_2 : f(\mathcal{K}) \times f(\mathcal{K}) \to \Re$. Assume that $f : \mathcal{K} \to (0,+\infty)$ is a twice differentiable mapping on K° such that $f'' \in \mathcal{L}(\mathcal{K})$. If $(f''(x))^q$ is positive generalized-$m - (((h_1 \circ g)^{p_1}, (h_2 \circ g)^{p_2}); (\Psi_1, \Psi_2))$-convex mapping, $0 < r \le 1$, $p_1, p_2 > -1$, $q > 1$, $p^{-1} + q^{-1} = 1$, then the following inequality for $\alpha, k > 0$ holds:*

$$\left|\mathcal{T}_{f,g}^{\alpha,k}(\Psi_1, \sigma, m; \varrho_1, \varrho_2)\right| \le \frac{\Xi_{t,1}^2}{2}$$

$$\times \sqrt[p]{(g(1)-g(0)) - \frac{g^{p(\frac{\alpha}{k}+1)+1}(0) - g^{p(\frac{\alpha}{k}+1)+1}(1)}{p\left(\frac{\alpha}{k}+1\right)+1}}$$

$$\times \sqrt[rq]{\frac{(f''(\varrho_1))^{rq} I^r((h_1 \circ g)(\xi))}{+\Psi_2\left((f''(\varrho_2))^{rq}, (f''(\varrho_1))^{rq}\right) I^r((h_2 \circ g)(\xi))}}, \tag{39}$$

where $I((h_1 \circ g)(\xi))$ and $I((h_2 \circ g)(\xi))$ are defined as in Theorem 2.

Proof. From Lemma 2, positive generalized-$\mathbf{m} - (((h_1 \circ g)^{p_1}, (h_2 \circ g)^{p_2}); (\Psi_1, \Psi_2))$-convexity of $(f''(x))^q$, Hölder inequality, Minkowski inequality, properties of the modulus and changing the variable $u = \mathbf{m}(t)\sigma(\varrho_1) + g(\xi)\Xi_{t,1}, \forall t, \xi \in [0,1]$, we have

$$\left|\mathcal{T}_{f,g}^{\alpha,k}(\Psi_1, \sigma, \mathbf{m}; \varrho_1, \varrho_2)\right| \leq \frac{\Xi_{t,1}^2}{2}$$

$$\times \int_0^1 \left|1 - (1 - g(\xi))^{\frac{\alpha}{k}+1} - g^{\frac{\alpha}{k}+1}(\xi)\right|$$

$$\times \left|f''(\mathbf{m}(t)\sigma(\varrho_1) + g(\xi)\Xi_{t,1})\right| dg(\xi)$$

$$\leq \frac{\Xi_{t,1}^2}{2} \times \left(\int_0^1 \left(1 - (1 - g(\xi))^{\frac{\alpha}{k}+1} - g^{\frac{\alpha}{k}+1}(\xi)\right)^p dg(\xi)\right)^{\frac{1}{p}}$$

$$\times \left(\int_0^1 (f''(\mathbf{m}(t)\sigma(\varrho_1) + g(\xi)\Xi_{t,1}))^q dg(\xi)\right)^{\frac{1}{q}}$$

$$\leq \frac{\Xi_{t,1}^2}{2} \times \left(\int_0^1 \left(1 - (1 - g(\xi))^{p(\frac{\alpha}{k}+1)} - g^{p(\frac{\alpha}{k}+1)}(\xi)\right) dg(\xi)\right)^{\frac{1}{p}}$$

$$\times \left(\int_0^1 \left[\mathbf{m}(\xi)(h_1 \circ g)^{p_1}(\xi)(f''(\varrho_1))^{rq}\right.\right.$$

$$\left.\left. + (h_2 \circ g)^{p_2}(\xi)\Psi_2\left((f''(\varrho_2))^{rq}, (f''(\varrho_1))^{rq}\right)\right]^{\frac{1}{r}} dg(\xi)\right)^{\frac{1}{q}} \leq \frac{\Xi_{t,1}^2}{2}$$

$$\times \sqrt[p]{(g(1) - g(0)) - \frac{g^{p(\frac{\alpha}{k}+1)+1}(0) - g^{p(\frac{\alpha}{k}+1)+1}(1) + (1 - g(0))^{p(\frac{\alpha}{k}+1)+1} - (1 - g(1))^{p(\frac{\alpha}{k}+1)+1}}{p(\frac{\alpha}{k}+1) + 1}}$$

$$\times \left[\left(\int_0^1 \mathbf{m}^{\frac{1}{r}}(\xi)(f''(\varrho_1))^q(h_1 \circ g)^{\frac{p_1}{r}}(\xi)dg(\xi)\right)^r\right.$$

$$\left. + \left(\int_0^1 \Psi_2^{\frac{1}{r}}\left((f''(\varrho_2))^{rq}, (f''(\varrho_1))^{rq}\right)(h_2 \circ g)^{\frac{p_2}{r}}(\xi)dg(\xi)\right)^r\right]^{\frac{1}{rq}}$$

$$= \frac{\Xi_{t,1}^2}{2}$$

$$\times \sqrt[p]{(g(1)-g(0)) - \frac{g^{p(\frac{\alpha}{k}+1)+1}(0) - g^{p(\frac{\alpha}{k}+1)+1}(1) + (1-g(0))^{p(\frac{\alpha}{k}+1)+1} - (1-g(1))^{p(\frac{\alpha}{k}+1)+1}}{p(\frac{\alpha}{k}+1)+1}}$$

$$\times \sqrt[rq]{(f''(\varrho_1))^{rq} I^r((h_1 \circ g)(\xi)) + \Psi_2\left((f''(\varrho_2))^{rq}, (f''(\varrho_1))^{rq}\right) I^r((h_2 \circ g)(\xi))}.$$

So, the proof of this theorem is completed. $\square$

We point out some special cases of Theorem 21.

Corollary 22. *In Theorem 21 for $p = q = 2$, we get the following inequality for k-fractional integrals:*

$$\left| \mathcal{T}_{f,g}^{\alpha,k}(\Psi_1, \sigma, \boldsymbol{m}; \varrho_1, \varrho_2) \right| \le \frac{\Xi_{t,1}^2}{2}$$

$$\times \sqrt{(g(1)-g(0)) - \frac{g^{\frac{2\alpha}{k}+3}(0) - g^{\frac{2\alpha}{k}+3}(1) + (1-g(0))^{\frac{2\alpha}{k}+3} - (1-g(1))^{\frac{2\alpha}{k}+3}}{\frac{2\alpha}{k}+3}}$$

$$\times \sqrt[2r]{\frac{(f''(a))^{2r} I^r((h_1 \circ g)(\xi))}{+ \Psi_2\left((f''(b))^{2r}, (f''(a))^{2r}\right) I^r((h_2 \circ g)(\xi))}}. \quad (40)$$

Corollary 23. *In Theorem 21 for $g(\xi) = \xi$, we get the following inequality for k-fractional integrals:*

$$\left| \mathcal{T}_{f}^{\alpha,k}(\Psi_1, \sigma, \boldsymbol{m}; \varrho_1, \varrho_2) \right| \le \frac{\Xi_{t,1}^2}{2}$$

$$\times \sqrt[rq]{(f''(\varrho_1))^{rq} I^r(h_1(\xi)) + \Psi_2\left((f''(\varrho_2))^{rq}, (f''(\varrho_1))^{rq}\right) I^r(h_2(\xi))}. \quad (41)$$

Corollary 24. *Under the conditions of Remark 7 using Corollary 23, we get the following Hermite–Hadamard k-fractional integral* *inequality:*

$$\left| \frac{f(\sigma(\varrho_1)) + f(\sigma(\varrho_2))}{2} - \frac{\Gamma_k(\alpha+k)}{2(\sigma(\varrho_2) - \sigma(\varrho_1))^{\frac{\alpha}{k}}} \right.$$
$$\left. \times \left[I^{\alpha,k}_{(\sigma(\varrho_1))^+} f(\sigma(\varrho_2)) + I^{\alpha,k}_{(\sigma(\varrho_2))^-} f(\sigma(\varrho_1)) \right] \right|$$
$$\leq \frac{(\sigma(\varrho_2) - \sigma(\varrho_1))^2}{2\left(\frac{\alpha}{k}+1\right)}$$
$$\times \sqrt[rq]{(f''(\varrho_1))^{rq} I^r(h_1(\xi)) + \Psi_2 \left((f''(\varrho_2))^{rq}, (f''(\varrho_1))^{rq}\right) I^r(h_2(\xi))}. \tag{42}$$

Corollary 25. *In Corollary 23 for* $\Xi_1 = \Psi_1(\sigma(\varrho_2), m\sigma(\varrho_1))$, $h_1(t) = h(1-t)$, $h_2(t) = h(t)$ *and* $\boldsymbol{m}(t) = m \in (0,1]$ *for all* $t \in [0,1]$, *we get the following k-fractional integral inequality for generalized-m $-$ $((h^{p_1}(1-t), h^{p_2}(t)); (\Psi_1, \Psi_2))$-convex mappings:*

$$\left| \mathcal{T}_f^{\alpha,k}(\Psi_1, \sigma, m; \varrho_1, \varrho_2) \right| \leq \frac{\Xi_1^2}{2}$$
$$\times \sqrt[rq]{m(f''(\varrho_1))^{rq} I^r(h(1-\xi)) + \Psi_2 \left((f''(\varrho_2))^{rq}, (f''(\varrho_1))^{rq}\right) I^r(h(\xi))}. \tag{43}$$

Corollary 26. *In Corollary 25 for* $h_1(t) = (1-t)^s$ *and* $h_2(t) = t^s$, *we get the following k-fractional integral inequality for generalized-$m - (((1-t)^{sp_1}, t^{sp_2}); (\Psi_1, \Psi_2))$-Breckner-convex mappings:*

$$\left| \mathcal{T}_f^{\alpha,k}(\Psi_1, \sigma, m; \varrho_1, \varrho_2) \right| \leq \frac{\Xi_1^2}{2}$$
$$\times \sqrt[rq]{m(f''(\varrho_1))^{rq} \left(\frac{r}{r+sp_1}\right)^r + \Psi_2 \left((f''(\varrho_2))^{rq}, (f''(\varrho_1))^{rq}\right) \left(\frac{r}{r+sp_2}\right)^r}. \tag{44}$$

Corollary 27. *In Corollary 25 for* $h_1(t) = (1-t)^{-s}$, $h_2(t) = t^{-s}$ *and* $r > s \cdot \max\{p_1, p_2\}$, *we get the following k-fractional* integral inequality for generalized-$m - (((1-t)^{-sp_1}, t^{-sp_2}); (\Psi_1, \Psi_2))$-Godunova–Levin–Dragomir-convex mappings:

$$\left|\mathcal{T}_f^{\alpha,k}(\Psi_1, \sigma, m; \varrho_1, \varrho_2)\right| \leq \frac{\Xi_1^2}{2}$$
$$\times \sqrt[rq]{m(f''(\varrho_1))^{rq}\left(\frac{r}{r-sp_1}\right)^r + \Psi_2\left((f''(\varrho_2))^{rq}, (f''(\varrho_1))^{rq}\right)\left(\frac{r}{r-sp_2}\right)^r}. \tag{45}$$

Corollary 28. *In Corollary 25 for* $h_1(t) = h_2(t) = t(1-t)$ *and* $\boldsymbol{m}(t) = m \in (0,1]$ *for all* $t \in [0,1]$, *we get the following k-fractional integral inequality for generalized-$m - ((t(1-t))^{sp_1}, (t(1-t))^{sp_2}); (\Psi_1, \Psi_2))$-convex mappings:*

$$\left|\mathcal{T}_f^{\alpha,k}(\Psi_1, \sigma, m; \varrho_1, \varrho_2)\right| \leq \frac{\Xi_1^2}{2}$$
$$\times \sqrt[rq]{\begin{array}{l} m(f''(\varrho_1))^{rq}\beta^r\left(1+\frac{p_1}{r}, 1+\frac{p_1}{r}\right) \\ + \Psi_2\left((f''(\varrho_2))^{rq}, (f''(\varrho_1))^{rq}\right)\beta^r\left(1+\frac{p_2}{r}, 1+\frac{p_2}{r}\right) \end{array}}. \tag{46}$$

Corollary 29. *In Corollary 25 for* $h_1(t) = \frac{\sqrt{1-t}}{2\sqrt{t}}, h_2(t) = \frac{\sqrt{t}}{2\sqrt{1-t}}$ *and* $r > \frac{1}{2} \cdot \max\{p_1, p_2\}$, *we get the following k-fractional integral inequality for generalized-$m - \left(\left(\left(\frac{\sqrt{1-t}}{2\sqrt{t}}\right)^{p_1}, \left(\frac{\sqrt{t}}{2\sqrt{1-t}}\right)^{p_2}\right); (\Psi_1, \Psi_2)\right)$-convex mappings:*

$$\left|\mathcal{T}_f^{\alpha,k}(\Psi_1, \sigma, m; \varrho_1, \varrho_2)\right| \leq \frac{\Xi_1^2}{2}$$
$$\times \left[m(f''(\varrho_1))^{rq}\frac{\beta^r\left(1-\frac{p_1}{2r}, 1+\frac{p_1}{2r}\right)}{2^{p_1}}\right.$$
$$\left. + \Psi_2\left((f''(\varrho_2))^{rq}, (f''(\varrho_1))^{rq}\right)\frac{\beta^r\left(1-\frac{p_2}{2r}, 1+\frac{p_2}{2r}\right)}{2^{p_2}}\right]^{\frac{1}{rq}}. \tag{47}$$

Theorem 30. *Let* $h_1, h_2 : [0,1] \to [0,+\infty)$, $\sigma : \mathcal{I} \to \mathcal{K}$ *be continuous functions,* $g : [0,1] \to [0,1]$ *be a strictly increasing function on* $(0,1)$ *and* $\boldsymbol{m} : [0,1] \to (0,1]$. *Suppose* $\mathcal{K} = (\boldsymbol{m}(t)\sigma(\varrho_1), \boldsymbol{m}(t)\sigma(\varrho_1) + g(1)\Xi_{t,1}) \subseteq \Re$ *is an open* $(\boldsymbol{m}, g; \sigma)$*-invex subset with respect to* $\Psi_1 : \mathcal{K} \times \mathcal{K} \to \Re$ *for* $\Xi_{t,1} = \Psi_1(\sigma(\varrho_2), \boldsymbol{m}(t)\sigma(\varrho_1)) > 0, \forall t \in [0,1]$

and $\Psi_2 : f(\mathcal{K}) \times f(\mathcal{K}) \to \Re$. Assume that $f : \mathcal{K} \to (0, +\infty)$ is a twice differentiable mapping on $\mathcal{K}^\circ$ such that $f'' \in \mathcal{L}(\mathcal{K})$. If $(f''(x))^q$ is positive generalized-$m - (((h_1 \circ g)^{p_1}, (h_2 \circ g)^{p_2}); (\Psi_1, \Psi_2))$-convex mapping, $0 < r \leq 1$, $p_1, p_2 > -1$ and $q \geq 1$, then the following inequality for $\alpha, k > 0$ holds:

$$\left| \mathcal{T}_{f,g}^{\alpha,k}(\Psi_1, \sigma, \boldsymbol{m}; \varrho_1, \varrho_2) \right|$$

$$\leq \frac{\Xi_{t,1}^2}{2} \left((g(1) - g(0)) - \frac{g^{\frac{\alpha}{k}+2}(0) - g^{\frac{\alpha}{k}+2}(1) + (1-g(0))^{\frac{\alpha}{k}+2} - (1-g(1))^{\frac{\alpha}{k}+2}}{\frac{\alpha}{k}+2} \right)^{1-\frac{1}{q}}$$

$$\times \sqrt[rq]{(f''(\varrho_1))^{rq} G^r((h_1 \circ g)(\xi)) + \Psi_2\left((f''(\varrho_2))^{rq}, (f''(\varrho_1))^{rq}\right) G^r((h_2 \circ g)(\xi))}, \quad (48)$$

where

$$G((h_1 \circ g)(\xi)) := \int_0^1 m^{\frac{1}{r}}(\xi) \left(1 - (1-g(\xi))^{\frac{\alpha}{k}+1} - g^{\frac{\alpha}{k}+1}(\xi)\right)$$

$$\times (h_1 \circ g)^{\frac{p_1}{r}}(\xi) \mathrm{d}g(\xi),$$

$$G((h_2 \circ g)(\xi)) := \int_0^1 \left(1 - (1-g(\xi))^{\frac{\alpha}{k}+1} - g^{\frac{\alpha}{k}+1}(\xi)\right)$$

$$\times (h_2 \circ g)^{\frac{p_2}{r}}(\xi) \mathrm{d}g(\xi).$$

Proof. From Lemma 2, positive generalized-$m - (((h_1 \circ g)^{p_1}, (h_2 \circ g)^{p_2}); (\Psi_1, \Psi_2))$-convexity of $(f''(x))^q$, the well-known power mean inequality, Minkowski inequality, properties of the modulus and changing the variable $u = \boldsymbol{m}(t)\sigma(\varrho_1) + g(\xi)\Xi_{t,1}, \forall t, \xi \in [0,1]$, we have

$$\left| \mathcal{T}_{f,g}^{\alpha,k}(\Psi_1, \sigma, \boldsymbol{m}; \varrho_1, \varrho_2) \right|$$

$$\leq \frac{\Xi_{t,1}^2}{2} \int_0^1 \left| 1 - (1-g(\xi))^{\frac{\alpha}{k}+1} - g^{\frac{\alpha}{k}+1}(\xi) \right|$$

$$\times \left| f''(\boldsymbol{m}(t)\sigma(\varrho_1) + g(\xi)\Xi_{t,1}) \right| \mathrm{d}g(\xi)$$

$$\leq \frac{\Xi_{t,1}^2}{2} \times \left(\int_0^1 \left(1 - (1-g(\xi))^{\frac{\alpha}{k}+1} - g^{\frac{\alpha}{k}+1}(\xi)\right) \mathrm{d}g(\xi) \right)^{1-\frac{1}{q}}$$

$$\times \left(\int_0^1 \left(1 - (1-g(\xi))^{\frac{\alpha}{k}+1} - g^{\frac{\alpha}{k}+1}(\xi)\right) \right.$$

$$\left. \times (f''(\mathbf{m}(t)\sigma(\varrho_1) + g(\xi)\Xi_{t,1}))^q \, \mathrm{d}g(\xi) \right)^{\frac{1}{q}}$$

$$\leq \frac{\Xi_{t,1}^2}{2} \times \left((g(1) - g(0)) - \frac{g^{\frac{\alpha}{k}+2}(0) - g^{\frac{\alpha}{k}+2}(1) + (1-g(0))^{\frac{\alpha}{k}+2} - (1-g(1))^{\frac{\alpha}{k}+2}}{\frac{\alpha}{k}+2} \right)^{1-\frac{1}{q}}$$

$$\times \left[\int_0^1 \left(1 - (1-g(\xi))^{\frac{\alpha}{k}+1} - g^{\frac{\alpha}{k}+1}(\xi)\right) \right.$$

$$\times \left[\mathbf{m}(\xi)(h_1 \circ g)^{p_1}(\xi)(f''(\varrho_1))^{rq} \right.$$

$$\left. \left. + (h_2 \circ g)^{p_2}(\xi)\Psi_2 \left((f''(\varrho_2))^{rq}, (f''(\varrho_1))^{rq}\right) \right]^{\frac{1}{r}} \mathrm{d}g(\xi) \right]^{\frac{1}{q}}$$

$$\leq \frac{\Xi_{t,1}^2}{2} \times \left((g(1) - g(0)) - \frac{g^{\frac{\alpha}{k}+2}(0) - g^{\frac{\alpha}{k}+2}(1) + (1-g(0))^{\frac{\alpha}{k}+2} - (1-g(1))^{\frac{\alpha}{k}+2}}{\frac{\alpha}{k}+2} \right)^{1-\frac{1}{q}}$$

$$\times \left[\left(\int_0^1 \mathbf{m}^{\frac{1}{r}}(\xi)(f''(\varrho_1))^q \left(1 - (1-g(\xi))^{\frac{\alpha}{k}+1} - g^{\frac{\alpha}{k}+1}(\xi)\right) \right. \right.$$

$$\left. \times (h_1 \circ g)^{\frac{p_1}{r}}(\xi)\mathrm{d}g(\xi) \right)^r$$

$$+ \left(\int_0^1 \Psi_2^{\frac{1}{r}} \left((f''(\varrho_2))^{rq}, (f''(\varrho_1))^{rq}\right) \left(1 - (1-g(\xi))^{\frac{\alpha}{k}+1} \right. \right.$$

$$\left. \left. \left. - g^{\frac{\alpha}{k}+1}(\xi)\right) (h_2 \circ g)^{\frac{p_2}{r}}(\xi)\mathrm{d}g(\xi) \right)^r \right]^{\frac{1}{rq}}$$

$$= \frac{\Xi_{t,1}^2}{2} \times \left((g(1) - g(0)) - \frac{g^{\frac{\alpha}{k}+2}(0) - g^{\frac{\alpha}{k}+2}(1) + (1-g(0))^{\frac{\alpha}{k}+2} - (1-g(1))^{\frac{\alpha}{k}+2}}{\frac{\alpha}{k}+2} \right)^{1-\frac{1}{q}}$$

$$\times \sqrt[rq]{(f''(\varrho_1))^{rq} G^r((h_1 \circ g)(\xi)) + \Psi_2\left((f''(\varrho_2))^{rq}, (f''(\varrho_1))^{rq}\right) G^r((h_2 \circ g)(\xi))}.$$

So, the proof of this theorem is completed. □

We point out some special cases of Theorem 30.

Corollary 31. *In Theorem 30 for $q = 1$, we get the following inequality for k-fractional integrals:*

$$\left| \mathcal{T}_{f,g}^{\alpha,k}(\Psi_1, \sigma, \mathbf{m}; \varrho_1, \varrho_2) \right| \leq \frac{\Xi_{t,1}^2}{2}$$
$$\times \sqrt[r]{(f''(a))^r G^r((h_1 \circ g)(\xi)) + \Psi_2\left((f''(b))^r, (f''(a))^r\right) G^r((h_2 \circ g)(\xi))}.$$
(49)

Corollary 32. *In Theorem 30 for $g(\xi) = \xi$, we get the following inequality for k-fractional integrals:*

$$\left| \mathcal{T}_{f}^{\alpha,k}(\Psi_1, \sigma, \mathbf{m}; \varrho_1, \varrho_2) \right| \leq \frac{\Xi_{t,1}^2}{2}$$
$$\times \sqrt[rq]{(f''(\varrho_1))^{rq} G^r(h_1(\xi)) + \Psi_2\left((f''(\varrho_2))^{rq}, (f''(\varrho_1))^{rq}\right) G^r(h_2(\xi))}.$$
(50)

Corollary 33. *Under the conditions of Remark 7 using Corollary 32, we get the following Hermite–Hadamard k-fractional integrals:*

$$\left| \frac{f(\sigma(\varrho_1)) + f(\sigma(\varrho_2))}{2} - \frac{\Gamma_k(\alpha+k)}{2(\sigma(\varrho_2) - \sigma(\varrho_1))^{\frac{\alpha}{k}}} \right.$$
$$\left. \times \left[I_{(\sigma(\varrho_1))^+}^{\alpha,k} f(\sigma(\varrho_2)) + I_{(\sigma(\varrho_2))^-}^{\alpha,k} f(\sigma(\varrho_1)) \right] \right|$$
$$\leq \frac{(\sigma(\varrho_2) - \sigma(\varrho_1))^2}{2\left(\frac{\alpha}{k}+1\right)} \times \sqrt[rq]{(f''(\varrho_1))^{rq} G^r(h_1(\xi)) + \Psi_2\left((f''(\varrho_2))^{rq}, (f''(\varrho_1))^{rq}\right) G^r(h_2(\xi))}.$$
(51)

Corollary 34. *In Corollary 32 for $\Xi_1 = \Psi_1(\sigma(\varrho_2), m\sigma(\varrho_1))$, $h_1(t) = h(1-t)$, $h_2(t) = h(t)$ and $\boldsymbol{m}(t) = m \in (0,1]$ for all $t \in [0,1]$, we get the following k-fractional integral inequality for generalized-m $-$ $((h^{p_1}(1-t), h^{p_2}(t)); (\Psi_1, \Psi_2))$-convex mappings:*

$$\left|\mathcal{T}_f^{\alpha,k}(\Psi_1, \sigma, m; \varrho_1, \varrho_2)\right| \leq \frac{\Xi_1^2}{2}$$

$$\times \sqrt[rq]{m(f''(\varrho_1))^{rq} G^r(h(1-\xi)) + \Psi_2\left((f''(\varrho_2))^{rq}, (f''(\varrho_1))^{rq}\right) G^r(h(\xi))}.$$
(52)

Corollary 35. *In Corollary 34 for $h_1(t) = (1-t)^s$ and $h_2(t) = t^s$, we get the following k-fractional integral inequality for generalized-m $-$ $(((1-t)^{sp_1}, t^{sp_2}); (\Psi_1, \Psi_2))$-Breckner-convex mappings:*

$$\left|\mathcal{T}_f^{\alpha,k}(\Psi_1, \sigma, m; \varrho_1, \varrho_2)\right| \leq \frac{\Xi_1^2}{2}$$

$$\times \left[m(f''(\varrho_1))^{rq} \left(\frac{r}{r+sp_1} - \frac{1}{\frac{sp_1}{r} + \frac{\alpha}{k} + 2} - \beta\left(\frac{sp_1}{r} + 1, \frac{\alpha}{k} + 2\right) \right)^r \right.$$

$$+ \Psi_2\left((f''(\varrho_2))^{rq}, (f''(\varrho_1))^{rq}\right) \left(\frac{r}{r+sp_2} - \frac{1}{\frac{sp_2}{r} + \frac{\alpha}{k} + 2} \right.$$

$$\left. \left. - \beta\left(\frac{sp_2}{r} + 1, \frac{\alpha}{k} + 2\right) \right)^r \right]^{\frac{1}{rq}}.$$
(53)

Corollary 36. *In Corollary 34 for $h_1(t) = (1-t)^{-s}$, $h_2(t) = t^{-s}$ and $r > s \cdot \max\{p_1, p_2\}$, we get the following k-fractional integral inequality for generalized-m $-$ $(((1-t)^{-sp_1}, t^{-sp_2}); (\Psi_1, \Psi_2))$-Godunova–Levin–Dragomir-convex mappings:*

$$\left|\mathcal{T}_f^{\alpha,k}(\Psi_1, \sigma, m; \varrho_1, \varrho_2)\right| \leq \frac{\Xi_1^2}{2}$$

$$\times \left[m(f''(\varrho_1))^{rq} \left(\frac{r}{r-sp_1} - \frac{1}{\frac{\alpha}{k} - \frac{sp_1}{r} + 2} - \beta\left(1 - \frac{sp_1}{r}, \frac{\alpha}{k} + 2\right) \right)^r \right.$$

$$+ \Psi_2\left((f''(\varrho_2))^{rq}, (f''(\varrho_1))^{rq}\right) \left(\frac{r}{r-sp_2} - \frac{1}{\frac{\alpha}{k} - \frac{sp_2}{r} + 2} \right.$$

$$\left. \left. - \beta\left(1 - \frac{sp_2}{r}, \frac{\alpha}{k} + 2\right) \right)^r \right]^{\frac{1}{rq}}.$$
(54)

Corollary 37. *In Corollary 34 for* $h_1(t) = h_2(t) = t(1-t)$ *and* $\boldsymbol{m}(t) = m \in (0,1]$ *for all* $t \in [0,1]$, *we get the following k-fractional integral inequality for generalized-m* $- ((t(1-t))^{sp_1}, (t(1-t))^{sp_2}); (\Psi_1, \Psi_2))$-*convex mappings:*

$$\left| T_f^{\alpha,k}(\Psi_1, \sigma, m; \varrho_1, \varrho_2) \right| \leq \frac{\Xi_1^2}{2}$$

$$\times \left[m(f''(\varrho_1))^{rq} \left(\beta \left(\frac{p_1}{r} + 1, \frac{p_1}{r} + 1 \right) \right. \right.$$

$$\left. - 2\beta \left(\frac{p_1}{r} + \frac{\alpha}{k} + 2, \frac{p_1}{r} + 1 \right) \right)^r$$

$$+ \Psi_2 \left((f''(\varrho_2))^{rq}, (f''(\varrho_1))^{rq} \right) \left(\beta \left(\frac{p_2}{r} + 1, \frac{p_2}{r} + 1 \right) \right.$$

$$\left. \left. - 2\beta \left(\frac{p_2}{r} + \frac{\alpha}{k} + 2, \frac{p_2}{r} + 1 \right) \right)^r \right]^{\frac{1}{rq}}. \tag{55}$$

Corollary 38. *In Corollary 34 for* $h_1(t) = \frac{\sqrt{1-t}}{2\sqrt{t}}$, $h_2(t) = \frac{\sqrt{t}}{2\sqrt{1-t}}$ *and* $r > \frac{1}{2} \cdot \max\{p_1, p_2\}$, *we get the following k-fractional integral inequality for generalized-m* $- \left(\left(\left(\frac{\sqrt{1-t}}{2\sqrt{t}} \right)^{p_1}, \left(\frac{\sqrt{t}}{2\sqrt{1-t}} \right)^{p_2} \right); (\Psi_1, \Psi_2) \right)$-*convex mappings:*

$$\left| T_f^{\alpha,k}(\Psi_1, \sigma, m; \varrho_1, \varrho_2) \right| \leq \frac{\Xi_1^2}{2} \left[\frac{m(f''(\varrho_1))^{rq}}{2^{p_1}} \right.$$

$$\times \left(\beta \left(1 - \frac{p_1}{2r}, \frac{p_1}{2r} + 1 \right) - \beta \left(1 - \frac{p_1}{2r}, \frac{p_1}{2r} + \frac{\alpha}{k} + 2 \right) \right.$$

$$\left. - \beta \left(\frac{p_1}{2r} + 1, \frac{\alpha}{k} - \frac{p_1}{2r} + 2 \right) \right)^r + \frac{\Psi_2 \left((f''(\varrho_2))^{rq}, (f''(\varrho_1))^{rq} \right)}{2^{p_2}}$$

$$\times \left(\beta \left(1 - \frac{p_2}{2r}, \frac{p_2}{2r} + 1 \right) - \beta \left(\frac{p_2}{2r} + 1, \frac{\alpha}{k} - \frac{p_2}{2r} + 2 \right) \right.$$

$$\left. \left. - \beta \left(1 - \frac{p_2}{2r}, \frac{p_2}{2r} + \frac{\alpha}{k} + 2 \right) \right)^r \right]^{\frac{1}{rq}}. \tag{56}$$

Remark 8. By taking particular values of parameters α, k, r, p_1, p_2 and different appropriate choices of function $\boldsymbol{m} \neq 1$, m in

Theorems 21 and 30, several k-fractional integral inequalities associated with generalized-$\mathbf{m} - (((h_1 \circ g)^{p_1}, (h_2 \circ g)^{p_2}); (\Psi_1, \Psi_2))$-convex mappings can be obtained. In particular, for $k = 1$, by our Theorems 21 and 30, we can get some new special Hermite–Hadamard-type inequalities via fractional integrals of order $\alpha > 0$. Also, for $\alpha = k = 1$, we can get some new special Hermite–Hadamard-type inequalities via classical integrals.

Remark 9. Also, applying our Theorems 21 and 30 for appropriate choices of functions g (see examples: $g(x) = x^\alpha$, where $\alpha > 1$ and $\forall\, x \in (0,1]$; $g(x) = -\frac{1}{x}, \forall\, x \in (0,1]$; $g(x) = e^x, \forall\, x \in [0,1]$; $g(x) = \ln x, \forall\, x \in (0,1]$; $g(x) = \tan x, \forall\, x \in [0,1]$; $g(x) = \arctan x, \forall\, x \in [0,1]$; etc.), several k-fractional integral inequalities can be obtained.

Remark 10. Finally, applying our Theorems 21 and 30, for $0 < f''(x) \leq K, \forall\, x \in I$, we can get some new k-fractional integral inequalities.

4. Applications to Special Means

Definition 13. A function $\mathcal{M} : \Re_+^2 \to \Re_+$ is called a mean function if it has the following properties:

(1) Homogeneity: $\mathcal{M}(ax, ay) = a\mathcal{M}(x, y)$, for all $a > 0$.
(2) Symmetry: $\mathcal{M}(x, y) = \mathcal{M}(y, x)$.
(3) Reflexivity: $\mathcal{M}(x, x) = x$.
(4) Monotonicity: If $x \leq x'$ and $y \leq y'$, then $\mathcal{M}(x, y) \leq \mathcal{M}(x', y')$.
(5) Internality: $\min\{x, y\} \leq \mathcal{M}(x, y) \leq \max\{x, y\}$.

Let us consider some special means for arbitrary positive real numbers $\alpha \neq \beta$ as follows: the arithmetic mean $\mathcal{A} := \mathcal{A}(\alpha, \beta)$, the geometric mean $\mathcal{G} := \mathcal{G}(\alpha, \beta)$, the harmonic mean $\mathcal{H} := \mathcal{H}(\alpha, \beta)$, the power mean $\mathcal{P}_r := \mathcal{P}_r(\alpha, \beta)$, the identric mean $\mathcal{I}^* := \mathcal{I}^*(\alpha, \beta)$, the logarithmic mean $\mathcal{L}^* := \mathcal{L}^*(\alpha, \beta)$, the generalized log-mean $\mathcal{L}_p := \mathcal{L}_p(\alpha, \beta)$ and the weighted p-power mean $\mathcal{M}^* = \mathcal{M}_p^*$. Now, let ϱ_1 and ϱ_2 be positive real numbers such that $\varrho_1 < \varrho_2$. Let us consider continuous functions $h_1, h_2 : [0,1] \to [0, +\infty)$, $\sigma : \mathcal{I} \to \mathcal{K}$; $g : [0,1] \to [0,1]$ as a strictly increasing function on $(0,1)$ and $\Psi_1 : \mathcal{K} \times \mathcal{K} \to \Re$, $\Psi_2 : f(\mathcal{K}) \times f(\mathcal{K}) \to \Re$, where

$\overline{\mathcal{M}} := \mathcal{M}(\sigma(\varrho_1), \sigma(\varrho_2)) : (\sigma(\varrho_1), \sigma(\varrho_1) + g(1)\Psi_1(\sigma(\varrho_2), \sigma(\varrho_1))) \times (\sigma(\varrho_1), \sigma(\varrho_1) + g(1)\Psi_1(\sigma(\varrho_2), \sigma(\varrho_1))) \to \Re_+$, which is one of the abovementioned means. Replace $\Psi_1(\sigma(y), \mathbf{m}(t)\sigma(x))$ with $\Psi_1(\sigma(y), \sigma(x))$, where $\mathbf{m}(t) \equiv 1$, for all $t \in [0,1]$ and setting $\Psi_1(\sigma(y), \sigma(x)) = \mathcal{M}(\sigma(x), \sigma(y))$ for all $x, y \in \mathcal{I}$, in (39) and (48), one can obtain the following inequalities involving means:

$$\left| T^{\alpha,k}_{f,g}(\mathcal{M}(\cdot,\cdot), \sigma, 1; \varrho_1, \varrho_2) \right| \leq \frac{\overline{\mathcal{M}}^2}{2}$$

$$\times \sqrt[p]{(g(1) - g(0)) - \frac{g^{p(\frac{\alpha}{k}+1)+1}(0) - g^{p(\frac{\alpha}{k}+1)+1}(1) + (1-g(0))^{p(\frac{\alpha}{k}+1)+1} - (1-g(1))^{p(\frac{\alpha}{k}+1)+1}}{p\left(\frac{\alpha}{k}+1\right)+1}}$$

$$\times \sqrt[rq]{\begin{array}{c}(f''(\varrho_1))^{rq} I^r((h_1 \circ g)(\xi)) \\ + \Psi_2\left((f''(\varrho_2))^{rq}, (f''(\varrho_1))^{rq}\right) I^r((h_2 \circ g)(\xi))\end{array}}, \quad (57)$$

$$\left| T^{\alpha,k}_{f,g}(\mathcal{M}(\cdot,\cdot), \sigma, 1; \varrho_1, \varrho_2) \right| \leq \frac{\overline{\mathcal{M}}^2}{2}$$

$$\times \left((g(1) - g(0)) - \frac{g^{\frac{\alpha}{k}+2}(0) - g^{\frac{\alpha}{k}+2}(1) + (1-g(0))^{\frac{\alpha}{k}+2} - (1-g(1))^{\frac{\alpha}{k}+2}}{\frac{\alpha}{k}+2}\right)^{1-\frac{1}{q}}$$

$$\times \sqrt[rq]{\begin{array}{c}(f''(\varrho_1))^{rq} G^r((h_1 \circ g)(\xi)) \\ + \Psi_2\left((f''(\varrho_2))^{rq}, (f''(\varrho_1))^{rq}\right) G^r((h_2 \circ g)(\xi))\end{array}}. \quad (58)$$

Letting $\overline{\mathcal{M}} := \mathcal{A}, \mathcal{G}, \mathcal{H}, \mathcal{P}_r, \mathcal{I}^*, \mathcal{L}^*, \mathcal{L}_p, \mathcal{M}_p^*$ in (57) and (58), we get the inequalities involving means for a particular choices of positive $(f''(x))^q$ that are generalized-$1 - (((h_1 \circ g)^{p_1}, (h_2 \circ g)^{p_2}); (\Psi_1, \Psi_2))$-convex mappings.

Remark 11. Also, applying our Theorems 21 and 30 for appropriate choices of functions h_1 and h_2 (see Remark 4) such that $(f''(x))^q$ to be positive generalized-$1 - (((h_1 \circ g)^{p_1}, (h_2 \circ g)^{p_2}); (\Psi_1, \Psi_2))$-convex mappings (see examples: $f(x) = x^\alpha$, where $\alpha > 1$ and $\forall\, x > 0$; $f(x) = \dfrac{1}{x}$, $\forall\, x > 0$; $f(x) = e^x$, $\forall\, x \in \Re$; $f(x) = -\ln x$, $\forall\, x > 0$; $f(x) = -\arctan x$, $\forall\, x \in \Re$; etc.), we can deduce some new k-fractional integral inequalities using above special means. The details are left to the interested reader.

Acknowledgements

We thank anonymous referee for his/her valuable suggestion regarding this manuscript.

References

[1] S. M. Aslani, M. R. Delavar and S. M. Vaezpour, Inequalities of Fejér type related to generalized convex functions with applications, *Int. J. Anal. Appl.* **16**(1), 38–49 (2018).

[2] F. X. Chen and S. H. Wu, Several complementary inequalities to inequalities of Hermite–Hadamard type for s–convex functions, *J. Nonlinear Sci. Appl.* **9**(2), 705–716 (2016).

[3] Y.-M. Chu, M. A. Khan, T. U. Khan and T. Ali, Generalizations of Hermite–Hadamard type inequalities for MT-convex functions, *J. Nonlinear Sci. Appl.* **9**(5), 4305–4316 (2016).

[4] T. S. Du, J. G. Liao and Y. J. Li, Properties and integral inequalities of Hadamard–Simpson type for the generalized (s, m)-preinvex functions, *J. Nonlinear Sci. Appl.* **9**, 3112–3126 (2016).

[5] Z. Dahmani, On Minkowski and Hermite–Hadamard integral inequalities via fractional integration, *Ann. Funct. Anal.* **1**(1), 51–58 (2010).

[6] M. R. Delavar and S. S. Dragomir, On η-convexity, *Math. Inequal. Appl.* **20**, 203–216 (2017).

[7] M. R. Delavar and M. De La Sen, Some generalizations of Hermite–Hadamard type inequalities, *SpringerPlus*, **5**(1661), 1–9 (2016).

[8] S. S. Dragomir and R. P. Agarwal, Two inequalities for differentiable mappings and applications to special means of real numbers and trapezoidal formula, *Appl. Math. Lett.* **11**(5), 91–95 (1998).

[9] G. Farid and A. U. Rehman, Generalizations of some integral inequalities for fractional integrals, *Ann. Math. Sil.* **32**(1), 201–214 (2018).

[10] M. E. Gordji, S. S. Dragomir and M. R. Delavar, An inequality related to η–convex functions (II), *Int. J. Nonlinear Anal. Appl.* **6**(2), 26–32 (2016).

[11] M. E. Gordji, M. R. Delavar and M. De La Sen, On φ–convex functions, *J. Math. Inequal. Wiss* **10**(1), 173–183 (2016).
[12] A. Kashuri and R. Liko, Hermite–Hadamard type fractional integral inequalities for generalized $(r; s, m, \varphi)$–preinvex functions, *Eur. J. Pure Appl. Math.* **10**(3), 495–505 (2017).
[13] A. Kashuri and R. Liko, Hermite–Hadamard type inequalities for generalized (s, m, φ)–preinvex functions via k-fractional integrals, *Tbil. Math. J.* **10**(4), 73–82 (2017).
[14] A. Kashuri and R. Liko, Hermite–Hadamard type fractional integral inequalities for $MT_{(m,\varphi)}$–preinvex functions, *Stud. Univ. Babeş-Bolyai Math.* **62**(4), 439–450 (2017).
[15] A. Kashuri, R. Liko and S. S. Dragomir, Some new Gauss-Jacobi and Hermite–Hadamard type inequalities concerning $(n+1)$-differentiable generalized $((h_1^p, h_2^q); (\eta_1, \eta_2))$–convex mappings, *Tamkang J. Math.* **49**(4), 317–337 (2018).
[16] M. A. Khan, Y.-M. Chu, A. Kashuri and R. Liko, Hermite–Hadamard type fractional integral inequalities for $MT_{(r;g,m,\phi)}$–preinvex functions, *J. Comput. Anal. Appl.* **26**(8), 1487–1503 (2019).
[17] M. A. Khan, Y. Khurshid and T. Ali, Hermite–Hadamard inequality for fractional integrals via η–convex functions, *Acta Math. Univ. Comenianae* **79**(1), 153–164 (2017).
[18] M. A. Khan, Y.-M. Chu, A. Kashuri, R. Liko and G. Ali, Conformable fractional integrals versions of Hermite–Hadamard inequalities and their generalizations, *J. Funct. Spaces* **2018**, 1–9 (2018). Article ID 6928130.
[19] W. Liu, New integral inequalities involving beta function via P–convexity, *Miskolc Math. Notes* **15**(2), 585–591 (2014).
[20] W. J. Liu, Some Simpson type inequalities for h–convex and (α, m)–convex functions, *J. Comput. Anal. Appl.* **16**(5), 1005–1012 (2014).
[21] W. Liu, W. Wen and J. Park, Ostrowski type fractional integral inequalities for MT–convex functions, *Miskolc Math. Notes* **16**(1), 249–256 (2015).
[22] W. Liu, W. Wen and J. Park, Hermite–Hadamard type inequalities for MT–convex functions via classical integrals and fractional integrals, *J. Nonlinear Sci. Appl.* **9**, 766–777 (2016).
[23] C. Luo, T. S. Du, M. A. Khan, A. Kashuri and Y. Shen, Some k-fractional integrals inequalities through generalized $\lambda_{\phi m}$-MT-preinvexity, *J. Comput. Anal. Appl.* **27**(4), 690–705 (2019).
[24] M. Matloka, Inequalities for h-preinvex functions, *Appl. Math. Comput.* **234**, 52–57 (2014).

[25] S. Mubeen and G. M. Habibullah, k-Fractional integrals and applications, *Int. J. Contemp. Math. Sci.* **7**, 89–94 (2012).
[26] M. A. Noor, K. I. Noor, M. U. Awan and S. Khan, Hermite–Hadamard inequalities for s–Godunova–Levin preinvex functions, *J. Adv. Math. Stud.* **7**(2), 12–19 (2014).
[27] O. Omotoyinbo and A. Mogbodemu, Some new Hermite–Hadamard integral inequalities for convex functions, *Int. J. Sci. Innov. Tech.* **1**(1), 1–12 (2014).
[28] M. E. Özdemir, E. Set and M. Alomari, Integral inequalities via several kinds of convexity, *Creat. Math. Inform.* **20**(1), 62–73 (2011).
[29] C. Peng, C. Zhou and T. S. Du, Riemann-Liouville fractional Simpson's inequalities through generalized (m, h_1, h_2)–preinvexity, *Ital. J. Pure Appl. Math.* **38**, 345–367 (2017).
[30] E. Set, M. A. Noor, M. U. Awan and A. Gözpinar, Generalized Hermite–Hadamard type inequalities involving fractional integral operators, *J. Inequal. Appl.* **169**, 1–10 (2017).
[31] H. N. Shi, Two Schur-convex functions related to Hadamard-type integral inequalities, *Publ. Math. Debrecen* **78**(2), 393–403 (2011).
[32] D. D. Stancu, G. Coman and P. Blaga, Analiză numerică şi teoria aproximării, *Cluj-Napoca: Presa Universitară Clujeană* **2**, 1–434 (2002).
[33] M. Tunç, E. Göv and Ü. Şanal, On tgs-convex function and their inequalities, *Facta Univ. Ser. Math. Inform.* **30**(5), 679–691 (2015).
[34] S. Varošanec, On h-convexity, *J. Math. Anal. Appl.* **326**(1), 303–311 (2007).
[35] H. Wang, T. S. Du and Y. Zhang, k-fractional integral trapezium-like inequalities through (h, m)–convex and (α, m)–convex mappings, *J. Inequal. Appl.* **2017**(311), 20 (2017).
[36] Y. Wang, S. H. Wang and F. Qi, Simpson type integral inequalities in which the power of the absolute value of the first derivative of the integrand is s-preinvex, *Facta Univ. Ser. Math. Inform.* **28**(2), 151–159 (2013).
[37] T. Weir and B. Mond, Preinvex functions in multiple objective optimization, *J. Math. Anal. Appl.* **136**, 29–38 (1988).
[38] X. M. Zhang, Y.-M. Chu and X. H. Zhang, The Hermite–Hadamard type inequality of GA–convex functions and its applications, *J. Inequal. Appl.* **2010**, 1–11 (2010). Article ID 507560.
[39] Y. Zhang, T. S. Du, H. Wang, Y. J. Shen and A. Kashuri, Extensions of different type parameterized inequalities for generalized (m, h)-preinvex mappings via k-fractional integrals, *J. Inequal. Appl.* **2018**(49), 30 (2018).

© 2024 World Scientific Publishing Company
https://doi.org/10.1142/9789811267048_0015

Chapter 15

Payoff-Independent Action Update for Continuous Action Social Dilemmas: A Preliminary Investigation

Ath. Kehagias

Department of Electrical and Computer Engineering, Faculty of Engineering, Aristotle University, Thessaloniki, Greece

kehagiat@ece.auth.gr

In this chapter, we introduce an *action update* model for two-player *continuous action social dilemma* games. The model applies to continuous action versions of social dilemma games such as *Prisoner's Dilemma, Hawk and Dove*, and *Stag Hunt*. In our formulation, action updates depend on the current *cooperation levels* of the two players, *independent of specific payoffs*. This means that the *same* update equations can be applied to each of the abovementioned games. We present a preliminary investigation, limited to a particular action update model. As we explain in the sequel, this model admits a large number of modifications and extensions and in fact it belongs to a more general family which is further explored in future publications.

1. Introduction

In this chapter, we introduce an *action update* model for two-player *continuous action social dilemma* games as defined by Dawes and Lange [6, 7, 14]. The model applies to continuous action versions of social dilemma games, such as *Prisoner's Dilemma* [21], *Hawk and Dove* [28] and *Stag Hunt* [27]. In our formulation, action updates depend on the current *cooperation levels* of the two players, *independent of specific payoffs*. This means that the *same* update equations can be applied to each of the abovementioned games. We present a preliminary investigation, limited to a particular action update model. As we explain in the sequel, this model admits a large number of modifications and extensions and in fact it belongs to a more general family which is further explored in future publications.

The seminal papers on social dilemmas have been authored by Dawes [6] and Liebrand [16], which have been followed by an extensive literature; for example, see Dawes [7] and, for a recent list of references, the book by Lange [15]. The study of continuous action games forms a basic branch of game theory [3, 18, 20] and provides the tools to study continuous action social dilemmas. For instance, several authors [8, 9, 31] have used differential equations to model the evolution of actions over time. These equations are actually quite similar to the ones used to model the evolution of *probabilities of discrete actions* [4, 21, 26]; in addition, differential equations (the *replicator* equations) are also used in *evolutionary* game theory [10, 11, 25, 30] to model the evolution of action *frequencies* in a population.

This chapter is organized as follows. In Section 2, we present mathematical preliminaries regarding differential equations and game theory. In Section 3, we discuss social dilemma games with both discrete and continuous actions. Section 4 is devoted to the introduction and analysis of our basic action update model for a continuous action social dilemma game. In Section 5, we briefly present various modifications and generalizations of the basic model. In Section 6, we study an inverse problem related to the selection of update coefficients such that the update equations have a prescribed equilibrium. Finally, in Section 7, we summarize and present future research directions.

2. Preliminaries

2.1. *Differential equations*

Our model involves continuous actions evolving in continuous time. Consequently, our main tool is vector differential equations of the form

$$\frac{d\mathbf{x}}{dt} = \mathbf{f}(\mathbf{x}).$$

Here $\mathbf{x}, \mathbf{f}(\mathbf{x}) \in \mathbb{R}^N$ (for some $N \in \mathbb{N}$, usually $N = 2$); in other words,

$$\mathbf{x} = (x_1, \ldots, x_N) \quad \text{and} \quad \mathbf{f}(\mathbf{x}) = (f_1(\mathbf{x}), \ldots, f_N(\mathbf{x})).$$

Notation 1. We denote the norm of any $\mathbf{z} \in \mathbb{R}^N$ by $|\mathbf{z}| = \left(\sum_{n=1}^{N} z_n^2 \right)^{1/2}$.

We always assume that $\mathbf{f}$ satisfies appropriate conditions to ensure the existence of a unique solution of the problem

$$\frac{d\mathbf{x}}{dt} = \mathbf{f}(x), \quad \mathbf{x}(0) = \mathbf{x}_0. \tag{1}$$

Hence the following notation is meaningful.

Notation 2. We denote the unique solution of (1) by $\mathbf{x}(t|\mathbf{x}_0)$.

Definition 1. We say that the set $S \subseteq \mathbb{R}^n$ is an *invariant set* of the DE $\frac{d\mathbf{x}}{dt} = \mathbf{f}(\mathbf{x})$ (with $\mathbf{x} \in \mathbb{R}^N$) iff we have

$$\mathbf{x}_0 \in S \Rightarrow (\forall t \geq 0 : \mathbf{x}(t|\mathbf{x}_0) \in S).$$

Definition 2. We say that $\bar{\mathbf{x}} \in \mathbb{R}^n$ is an *invariant point* or an *equilibrium* of the DE $\frac{d\mathbf{x}}{dt} = \mathbf{f}(\mathbf{x})$ (with $\mathbf{x} \in \mathbb{R}^N$) iff we have

$$\mathbf{x}_0 = \bar{\mathbf{x}} \Rightarrow (\forall t \geq 0 : \mathbf{x}(t|\mathbf{x}_0) = \bar{\mathbf{x}}).$$

Remark 1. Clearly, $\bar{\mathbf{x}}$ is an equilibrium of $\frac{d\mathbf{x}}{dt} = f(\mathbf{x})$ iff $\{\bar{\mathbf{x}}\}$ is an invariant set of $\frac{d\mathbf{x}}{dt} = \mathbf{f}(\mathbf{x})$.

Proposition 1. *The point $\overline{\mathbf{x}}$ is an equilibrium of $\frac{d\mathbf{x}}{dt} = \mathbf{f}(\mathbf{x})$ iff $\overline{\mathbf{x}}$ is a solution of the algebraic system $\mathbf{f}(\mathbf{x}) = 0$.*

Definition 3. We say that $\overline{\mathbf{x}} \in \mathbb{R}^n$ is a *stable equilibrium* of the DE $\frac{d\mathbf{x}}{dt} = \mathbf{f}(\mathbf{x})$ iff

$$\forall \varepsilon > 0 : \exists \delta_\varepsilon > 0 : |x_0 - \overline{x}| < \delta_\varepsilon \Rightarrow (\forall t \geq 0 : |\mathbf{x}(t|\mathbf{x}_0) - \overline{\mathbf{x}}| < \varepsilon).$$

Otherwise, we say that $\overline{\mathbf{x}} \in \mathbb{R}^n$ is an *unstable equilibrium*.

Definition 4. Given the DE system $\frac{d\mathbf{x}}{dt} = \mathbf{f}(\mathbf{x})$, where $\mathbf{x} \in \mathbb{R}^N$, the *Jacobian* matrix $J(\mathbf{x})$ is defined by $J_{ij}(\mathbf{x}) = \frac{\partial f_i}{\partial x_j}(\mathbf{x})$ for $i, j \in \{1, \ldots, N\}$.

Proposition 2. *Given the DE system $\frac{d\mathbf{x}}{dt} = \mathbf{f}(\mathbf{x})$, where $\mathbf{x} \in \mathbb{R}^N$, assume it has an equilibrium $\overline{\mathbf{x}}$ and $\mathbf{f}$ has continuous first derivatives with respect to $x_1, \ldots, x_N$ in a neighborhood of $\overline{\mathbf{x}}$. Let $\lambda_1, \ldots, \lambda_N$ be the eigenvalues of $J(\overline{\mathbf{x}})$. If $\mathrm{Re}(\lambda_n) < 0$ for all $n \in \{1, \ldots, N\}$, then $\overline{\mathbf{x}}$ is a stable equilibrium; otherwise (i.e., if there exists some n such that $\mathrm{Re}(\lambda_n) \geq 0$) $\overline{\mathbf{x}}$ is an unstable equilibrium.*

Definition 5. Given the DE system $\frac{d\mathbf{x}}{dt} = f(\mathbf{x})$, assume it has an equilibrium $\overline{\mathbf{x}}$. We define the *basin of attraction of $\overline{\mathbf{x}}$* by

$$U(\overline{\mathbf{x}}) = \left\{ \mathbf{z} : \lim_{t \to \infty} \mathbf{x}(t|\mathbf{z}) = \overline{\mathbf{x}} \right\}.$$

Theorem 1 (Petrovitsch [19]). *If the scalar functions $y(t)$ and $z(t)$ satisfy*

$$\frac{dy}{dt} = f(t, y(t)), \quad \frac{dz}{dt} > f(t, z(t)),$$

then the following inequality holds:

$$\forall c, \forall t \in (0, t_1] : z(t|c) > y(t|c).$$

2.2. Bimatrix games

A *bimatrix game* [3, 18] is a two-player game in which each player has a finite number of possible actions and the players choose their action simultaneously. The name comes from the fact that the "*normal form*" of such a game can be described by two matrices: matrix A^1 (resp. A^2) describes the *payoffs* of the first player P_1 (resp. the second player P_2). If P_1 has M_1 possible actions and P_2 has M_2 possible actions (in which case we will speak of a "*finite actions game*"), then A^1 and A^2 are $M_1 \times M_2$ matrices. When P_1 selects his mth action and P_2 selects his nth action, the payoff to P_1 (resp. P_2) is A^1_{mn} (resp. A^2_{mn}). These payoffs can also be combined into a *bimatrix*

$$A = (A^1, A^2) = \begin{bmatrix} (A^1_{11}, A^2_{11}) & (A^1_{12}, A^2_{21}) \\ (A^1_{21}, A^2_{21}) & (A^1_{22}, A^2_{22}) \end{bmatrix}.$$

Definition 6. A *(mixed) strategy* for P_1 (resp. for P_2) is a nonnegative vector $\mathbf{p}^1 = (p^1_1, \ldots, p^1_{M_1})$ (resp. $\mathbf{p}^2 = (p^2_1, \ldots, p^2_{M_2})$) such that $\sum_{m=1}^{M_1} p^1_m = 1$ (resp. $\sum_{n=1}^{M_2} p^2_n = 1$) where

$$\forall m : p^1_m = \Pr(P_1 \text{ plays his } m\text{th action}),$$
$$\forall n : p^2_n = \Pr(P_2 \text{ plays his } n\text{th action}).$$

If $\mathbf{p}^i$ ($i \in \{1, 2\}$) is such that there exists some $\overline{m}$ satisfying $p^i_{\overline{m}} = 1$ (and hence, for all $m \neq \overline{m}$, $p^i_m = 0$), then we say that $\mathbf{p}^i$ is a *pure strategy*.

Notation 3. When the players play mixed strategies $\mathbf{p}^1$ and $\mathbf{p}^2$, the expected payoffs for P_1 and P_2 are

$$Q_n(\mathbf{p}^1, \mathbf{p}^2) = \sum_{m=1}^{M_1} \sum_{n=1}^{M_2} p^1_m A^n_{mn} p^2_n. \qquad (2)$$

When P_1 (resp. P_2) plays a pure strategy $\mathbf{p}^1$ (resp. $\mathbf{p}^2$) concentrating all probability on action m_1 (resp. m_2), we also write $Q(m_1, m_2)$ instead of $Q_n(\mathbf{p}^1, \mathbf{p}^2)$.

Definition 7. A *Nash equilibrium* (NE) of the bimatrix game (A^1, A^2) is a pair of mixed strategies $(\widehat{\mathbf{p}}^1, \widehat{\mathbf{p}}^2)$ such that

$$\forall \mathbf{p}^1 \in \Sigma_1 : Q_1(\widehat{\mathbf{p}}^1, \widehat{\mathbf{p}}^2) \geq Q_1(\mathbf{p}^1, \widehat{\mathbf{p}}^2) \quad \text{and}$$
$$\forall \mathbf{p}^2 \in \Sigma_2 : Q_2(\widehat{\mathbf{p}}^1, \widehat{\mathbf{p}}^2) \geq Q_2(\widehat{\mathbf{p}}^1, \mathbf{p}^2).$$

The interpretation is that when both players have committed to an NE $(\widehat{\mathbf{p}}^1, \widehat{\mathbf{p}}^2)$, neither player gains an advantage by *unilaterally* changing his strategy.

Proposition 3 (Nash [18]). *Every bimatrix game has a Nash equilibrium in (possibly) mixed strategies.*

It is worth noting that in a bimatrix game the interests of the two players are not, in general, diametrically opposed. In other words, one player's gain is not necessarily the other player's loss. The special case of completely opposed interests occurs in so-called *zero-sum* games, i.e., when $A^2 = -A^1$. But in a general (i.e., nonzero-sum) bimatrix game, there is scope for *cooperation* between the players.

3. Social Dilemmas

In this chapter, we follow Liebrand [16]: a social dilemma is *"defined as a [game] in which (1) there is a strategy that yields the person the best payoff in at least one configuration of strategy choices and that has a negative impact on the interests of the other persons involved, and (2) the choice of that particular strategy by all persons results in a deficient outcome"*. We now introduce some very simple social dilemma games which are the focus of our investigation in this chapter.

3.1. Two players, two actions

The simplest form of social dilemma games emerges when we consider two-action bimatrix games. Since there exists an infinite number of bimatrices, there also exists an infinite number of bimatrix games. However, the essential characteristics of such games have been elucidated by Rapoport [22, 23] and Liebrand [16], by concentrating on the *ordering* rather than the *magnitude* of the payoffs. In particular, a family of 78 "prototypical" games is defined by Rapoport [22] and, as explained by Liebrand [16], exactly three of these games are *symmetric*[a] *social dilemmas*. These are the games

---

[a]That is, A^2 is the transpose of A^1, resulting in symmetric payoffs for the two players.

determined by the following bimatrices[b]:

Prisoner's Dilemma (PD)   **Hawk and Dove (HD)**   **Stag Hunt (SH)**

$$A = \begin{bmatrix} (3,3) & (1,4) \\ (4,1) & (2,2) \end{bmatrix} \quad A = \begin{bmatrix} (3,3) & (2,4) \\ (4,2) & (1,1) \end{bmatrix} \quad A = \begin{bmatrix} (4,4) & (1,3) \\ (3,1) & (2,2) \end{bmatrix}$$

The names of the above games are obtained from accompanying "back-stories" giving illustrative examples of situations which can be modeled by the respective games. We omit these stories (the interested reader can consult Rapoport [21] for Prisoner's Dilemma, Smith [28] for Hawk and Dove, and Skyrms [27] for Stag Hunt) and proceed to a discussion of the similarities and differences between the three games.

For reasons which will soon become apparent, we denote the first action of each player by C (for cooperation) and the second by D (for defection). For example, supposing that P_1 chooses C and P_2 chooses D, the payoff pair is $(1,4)$ in PD, $(2,4)$ in HD and $(1,3)$ in SH. Now let us briefly discuss the dilemma involved in each game.

(1) **Prisoner's Dilemma:** In this game, each player should always play D irrespective of the other player's action. To see this, suppose P_2 plays C: if P_1 plays C, his payoff is 3, and if he plays D, his payoff is $4 > 3$; similarly, suppose P_2 plays D: if P_1 plays C, his payoff is 1, and if he plays D, his payoff is $2 > 1$. Hence, in every case, C is better than D for P_1; by symmetry, the same holds for P_2. More generally, it is easy to prove that neither player gains anything by playing C with positive probability. In short, $(\widehat{\mathbf{p}}^1, \widehat{\mathbf{p}}^2) = ((0,1), (0,1))$ is the *unique* NE of PD. The "dilemma"

---

[b]These are the simplest versions of the above games. We can obtain variants by changing the payoff values, as long as the ordering of the payoffs is preserved. For example, the following bimatrix also describes a Prisoner's dilemma

$$\widehat{A} = \begin{bmatrix} (7,7) & (0,8) \\ (8,0) & (4,4) \end{bmatrix}$$

because the inequalities $A^1_{1,2} < A^1_{2,2} < A^1_{1,1} < A^1_{2,1}$ are preserved as $\widehat{A}^1_{1,2} < \widehat{A}^1_{2,2} < \widehat{A}^1_{1,1} < \widehat{A}^1_{2,1}$.

or paradox is that both players would be better off if they played C (cooperation); in this case, they would get

$$Q_1((1,0),(1,0)) = Q_2((1,0),(1,0)) = 3,$$
$$Q_1((0,1),(0,1)) = Q_2((0,1),(0,1)) = 2$$

and $3 > 2$; however, $(\widetilde{\mathbf{p}}^1, \widetilde{\mathbf{p}}^2) = ((1,0),(1,0))$ is *not an* NE.

(2) **Hawk and Dove:** In this game, each player's best payoff is 4, obtained when he plays D and the other player plays C; however, if both players play D, each ends up with the worst possible payoff, namely 1. It turns out that the game has three NEs: $((1,0),(0,1))$ with payoffs $(2,4)$, $((0,1),(1,0))$ with payoffs $(4,2)$ and the mixed strategies pair $\left(\left(\frac{1}{2},\frac{1}{2}\right),\left(\frac{1}{2},\frac{1}{2}\right)\right)$ with payoffs $\left(\frac{5}{2},\frac{5}{2}\right)$.

(3) **Stag Hunt:** In this game, each player's best payoff is 4, when he plays C and the other player also plays C; this is an NE. However, if both players play D, this is also an NE, in which each player plays D and receives a payoff of 2. So, we have two pure NEs: $((1,0),(1,0))$ with payoffs $(4,4)$ and $((0,1),(0,1))$ with payoffs $(2,2)$. It turns out there exists also an NE in mixed strategies: $\left(\left(\frac{1}{2},\frac{1}{2}\right),\left(\frac{1}{2},\frac{1}{2}\right)\right)$ with payoffs $\left(\frac{5}{2},\frac{5}{2}\right)$.

In each of the above games, both players would be better off if they both played C (i.e., if they *cooperated*). The "social dilemma" is that this fact is not supported by the observed NE: in both PD and HD, (C,C) is not an NE at all; in SH, (C,C) is an NE but there exist additional "deficient" NE[c]. In simpler terms, while both players would be better off under mutual cooperation, each is "tempted" to defect. This raises serious questions regarding the foundations of game theoretic analysis, in particular the concept of Nash equilibrium. Consequently, a vast literature exists on social dilemmas (especially Prisoner's Dilemma) and various approaches have been introduced to justify the *emergence of cooperation* [1, 2, 17].

An important observation, which is also supported by experimental study of actual games played between humans [12, 32, 33], is that, in many cases, what determines the outcome of a game is not so much the exact payoff values but the "attitude" of the players.

---

[c] By "deficient" we mean that both players are worse off (in terms of their payoffs) than if they played (C,C).

For example, mutual cooperation tends to reinforce itself: in a game of PD, when both players play C once, they are more likely to play C again in future replays of the game (despite the fact that equilibrium analysis suggests the D is better). The same has been observed for mutual defection: players who repeatedly and mutually defect may in time recognize that defection generates further *mutual* defection resulting in mutual disadvantage and hence may switch to mutual cooperation. We will return to these considerations a little later.

3.2. Two players, continuous actions

A *continuous action* game is one in which each player can choose from a continuum of actions. In the context of social dilemmas, this approach is quite appropriate to model situations which are characterized by "degrees of cooperation" (e.g., how much private income to contribute to a common cause) [9, 13, 24, 31]. For example, P_n's action can be a number $x_n \in [0,1]$; a larger value indicates higher cooperation (hence 0 indicates full defection and 1 indicates full cooperation).

"Continuous social dilemmas" can be obtained from the corresponding bimatrix games by interpolation. Recall that in a bimatrix game we have

$$Q_n(1,1) = A_{11}^n, \quad Q_n(1,2) = A_{12}^n, \quad Q_n(2,1) = A_{21}^n, \quad Q_n(2,2) = A_{22}^n.$$

Now consider the functions (for $n \in \{1,2\}$) $\overline{Q}_n = [0,1] \times [0,1] \to \mathbb{R}$ defined as follows:

$$\forall n : \overline{Q}_n(x_1, x_2) = A_{11}^n x_1 x_2 + A_{12}^n x_1 (1-x_2) + A_{21}^n (1-x_1) x_2 \\ + A_{22}^n (1-x_1)(1-x_2). \tag{3}$$

Obviously,

$$Q_n(1,1) = \overline{Q}_n(1,1), \quad Q_n(1,2) = \overline{Q}_n(1,0),$$
$$Q_n(2,1) = \overline{Q}_n(0,1), \quad Q_n(2,2) = \overline{Q}_n(0,0).$$

What we have done is first to map C to 1 (i.e., full cooperation) and D to 0 (i.e., null cooperation) and then to interpolate the "interior" $\overline{Q}_n$ values from the "corner" Q_n values. In this way, a bimatrix game has been extended to a "*game on the unit square*" [20] with polynomial payoff functions. We have the following.

Proposition 4 (Raghavan [20]). *Every two-person game on the unit square, with the payoffs $\overline{Q}_1(x_1, x_2)$ and $\overline{Q}_2(x_1, x_2)$ being polynomials in x_1 and x_2, has at least one NE.*

The previously discussed questions ("dilemmas") regarding the emergence of cooperation in bimatrix social dilemmas also appear in connection to continuous action social dilemmas. In fact, one reason for the introduction of continuous action social dilemmas has been the hope that these questions can be answered satisfactorily in the continuous action context. However, even in continuous social dilemmas, emergence of cooperation is not guaranteed [9, 31].

As a final remark, it is worth noting that the discrete action payoff function (2) and the continuous action payoff function (3) are remarkably similar. To see this, let us rewrite (2) as follows: for $n \in \{1, 2\}$, let $p_n^1 = p_n$, $p_n^2 = 1 - p_n^1 = 1 - p_n$. Then (2) becomes

$$\forall n : Q_n(p_1, p_2) = A_{11}^n p_1 p_2 + A_{12}^n p_1 (1 - p_2) + A_{21}^n (1 - p_1) p_2 \\ + A_{22}^n (1 - p_1)(1 - p_2) \quad (4)$$

and the similarity of (3) and (4) becomes obvious from the correspondence $x_1 \leftrightarrow p_1$ and $x_2 \leftrightarrow p_2$. However, an important distinction must be made: under the (usual) assumption of *fully observable actions*, in the continuous action game, both P_1 and P_2 know the values of x_1, x_2, while in the finite action game, P_1 does not know p_2 and P_2 does not know p_1.

4. The Basic Action Update Model

In this section, we present and study a model for the evolution of actions in a two-player game on the unit square. The model is intended to capture the drift toward or away from cooperation in repeated plays of the game. The *update equations* have the form

$$\frac{dx_1}{dt} = f_1(x_1, x_2), \quad \frac{dx_2}{dt} = f_2(x_1, x_2), \quad (5)$$

where $x_n(t)$ is P_n's action at time t. The above equations imply that $x_1(t), x_2(t)$ are functions of a continuous time variable t; in other words, we assume that the replays of the game occur in *continuous* time. This is a common approach and can be understood as an approximation of replays at *discrete* times $t_i = i \cdot \Delta t$, with $i \in \mathbb{N}_0$ and $\Delta t \to 0$. Furthermore, (5) implies that, at each time t, each player knows not only his own but also the opponent's action (this is the "fully observable actions" hypothesis; in game theoretic terms, we assume a game of *perfect information*).

4.1.   The update equations

Our first task is to choose a concrete form for equations (5). To make our arguments specific, let us initially assume that the game being played is a continuous action Prisoner's Dilemma.[d]

We assume that each player is continuously modifying his cooperation level $x_n(t)$ in response to the other player's cooperation level. We also assume that both players use the same (more accurately, symmetric) update equations. Specifically, we assume that the nth player's action x_n (for $n \in \{1, 2\}$) is updated according to

$$\frac{dx_1}{dt} = (a_{11}x_1x_2 + a_{10}x_1(1-x_2) + a_{01}(1-x_1)x_2$$
$$+ a_{00}(1-x_1)(1-x_2))\,x_1(1-x_1), \qquad (6)$$

$$\frac{dx_2}{dt} = (a_{11}x_1x_2 + a_{10}(1-x_1)x_2 + a_{01}x_1(1-x_2)$$
$$+ a_{00}(1-x_1)(1-x_2))\,x_2(1-x_2). \qquad (7)$$

To justify the specific form of the above equations, let us look at (6) and consider separately each term in the right hand of the equation; a similar interpretation can be given for (7). It is easier to understand the following interpretations by first considering the "extremal" cases, i.e., when $(x_1, x_2) \in \{0, 1\} \times \{0, 1\}$:

(1) $a_{11}x_1x_2$ is P_1's tendency to change x_1 in the face of reciprocated cooperation.
(2) $a_{10}x_1(1-x_2)$ is P_1's tendency to change x_1 in the face of unreciprocated cooperation.

---

[d] We later argue that the exact same equations can be used for updates in a continuous action HD or SH game and in fact for any social dilemma in the unit square.

(3) $a_{01}(1-x_1)x_2$ is P_1's tendency to change x_1 in the face of unreciprocated defection.
(4) $a_{00}x_1x_2$ is P_1's tendency to change x_1 in the face of reciprocated defection.

It remains to choose the values of the parameters a_{mn} ($m,n \in \{1,2\}$). For the sake of simplicity, we assume $|a_{mn}| = 1$ and it remains to specify the *sign* of each a_{mn} coefficient. We make the following choices:

(1) $a_{11} = 1$ because reciprocated cooperation reinforces itself.
(2) $a_{10} = -1$ because unreciprocated cooperation makes one spiteful and uncooperative.
(3) $a_{01} = -1$ because unreciprocated defection is rewarded and hence discourages cooperation.
(4) $a_{00} = 1$ because reciprocated defection makes a player understand that cooperation is better.

Hence, equations (6)–(7) reduce to

$$\frac{dx_1}{dt} = (x_1x_2 - x_1(1-x_2) - (1-x_1)x_2 + (1-x_1)(1-x_2))$$
$$\times x_1(1-x_1),$$

$$\frac{dx_2}{dt} = (x_1x_2 - x_1(1-x_2) - (1-x_1)x_2 + (1-x_1)(1-x_2))$$
$$\times x_2(1-x_2).$$

Simplifying the above equations, we obtain

$$\frac{dx_1}{dt} = 4\left(\frac{1}{2} - x_1\right)\left(\frac{1}{2} - x_2\right)x_1(1-x_1), \tag{8}$$

$$\frac{dx_2}{dt} = 4\left(\frac{1}{2} - x_1\right)\left(\frac{1}{2} - x_2\right)x_2(1-x_2), \tag{9}$$

which are the forms we use in our subsequent study. Before proceeding any further, we note that, from standard existence and uniqueness theorems [5, 29], we have the following.

Proposition 5. *For any initial conditions* $\mathbf{x}(0) = \mathbf{x}_0 \in [0,1] \times [0,1]$, *(8)–(9) have a unique solution* $\mathbf{x}(t|\mathbf{x}_0)$.

4.2. Equilibria

Let us find and characterize the equilibria of (8)–(9). By definition, these are the solutions of the system of algebraic equations:

$$0 = 4\left(\frac{1}{2} - x_1\right)\left(\frac{1}{2} - x_2\right)x_1(1-x_1), \tag{10}$$

$$0 = 4\left(\frac{1}{2} - x_1\right)\left(\frac{1}{2} - x_2\right)x_2(1-x_2). \tag{11}$$

We easily find that (10)–(11) has exactly the following solutions in $[0,1] \times [0,1]$:

$(x_1, x_2) = (0,0)$, $(x_1, x_2) = (1,1)$, $(x_1, x_2) = (0,1)$, $(x_1, x_2) = (1,0)$,

$(x_1, x_2) \in \left\{\left(\frac{1}{2}, z\right) : z \in [0,1]\right\}$, $(x_1, x_2) \in \left\{\left(z, \frac{1}{2}\right) : z \in [0,1]\right\}$.

Furthermore, after simplification, the Jacobian matrix is

$J(x_1, x_2)$

$$= \begin{bmatrix} -(6x_1^2 - 6x_1 + 1)(2x_2 - 1) & -2(2x_1 - 1)x_1(x_1 - 1) \\ -2(2x_2 - 1)x_2(x_2 - 1) & -(6x_2^2 - 6x_2 + 1)(2x_1 - 1) \end{bmatrix}$$

and, from Proposition 2, we have the following cases:

(1) $(0,0)$ is unstable because $J(0,0) = \begin{bmatrix} 1 & 0 \\ 0 & 1 \end{bmatrix}$, eigenvalues are $\lambda_1 = 1$, $\lambda_2 = 1$.

(2) $(1,1)$ is stable because $J(1,1) = \begin{bmatrix} -1 & 0 \\ 0 & -1 \end{bmatrix}$, eigenvalues are $\lambda_1 = -1$, $\lambda_2 = -1$.

(3) $(1,0)$ is unstable because $J(1,0) = \begin{bmatrix} 1 & 0 \\ 0 & -1 \end{bmatrix}$, eigenvalues are $\lambda_1 = 1$, $\lambda_2 = -1$.

(4) $(0,1)$ is unstable because $J(0,1) = \begin{bmatrix} -1 & 0 \\ 0 & 1 \end{bmatrix}$, eigenvalues are $\lambda_1 = -1$, $\lambda_2 = 1$.

(5) $\left(\frac{1}{2}, z\right)$ is unstable because $J\left(\frac{1}{2}, z\right) = \begin{bmatrix} z - \frac{1}{2} & 0 \\ 0 & 0 \end{bmatrix}$, eigenvalues are $\lambda_1 = z - \frac{1}{2}$, $\lambda_2 = 0$.

(6) $\left(z, \frac{1}{2}\right)$ is unstable because $J\left(z, \frac{1}{2}\right) = \begin{bmatrix} 0 & 0 \\ 0 & z - \frac{1}{2} \end{bmatrix}$, eigenvalues are $\lambda_1 = 0$, $\lambda_2 = z - \frac{1}{2}$.

However, the above stability analysis does not tell the full story. We can get a better idea of the behavior of (8)–(9) by studying its *phase plot*, illustrated in Figure 1.

We see in Figure 1 that $[0, 1] \times [0, 1]$ can be partitioned into several regions; furthermore, the value of $\lim_{t \to \infty} (x_1(t), x_2(t))$ is determined by the region to which the initial condition $(x_1(0), x_2(0))$ belongs. For example, it appears from Figure 1 that

$$\forall (x_1(0), x_2(0)) \in \left(\frac{1}{2}, 1\right) \times \left(\frac{1}{2}, 1\right) : \lim_{t \to \infty} (x_1(t), x_2(t)) = (1, 1).$$

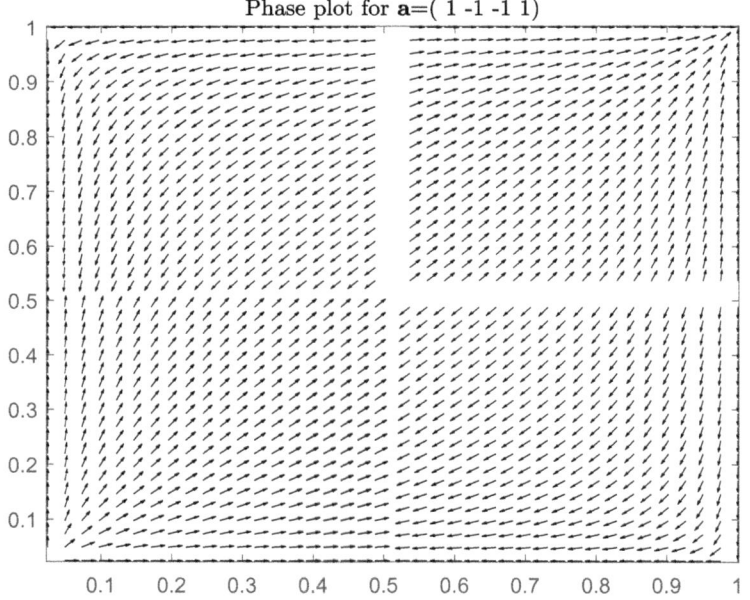

Figure 1. Phase plot of the system (8)–(9).

Similar remarks can be made when $(x_1(0), x_2(0))$ to some other region. Hence, we next embark upon a study of the *attraction basins* of (8)–(9).

We see in Figure 1 that $[0,1] \times [0,1]$ can be partitioned into several regions; furthermore, the value of $\lim_{t\to\infty} (x_1(t), x_2(t))$ is determined by the region to which the initial condition $(x_1(0), x_2(0))$ belongs. For example, it appears from Figure 1 that

$$\forall (x_1(0), x_2(0)) \in \left(\frac{1}{2}, 1\right) \times \left(\frac{1}{2}, 1\right) : \lim_{t\to\infty} (x_1(t), x_2(t)) = (1, 1).$$

Similar remarks can be made when $(x_1(0), x_2(0))$ to some other region. Hence, we next embark upon a study of the *attraction basins* of (8)–(9).

4.3. Attraction basins

Looking at (8)–(9), we see that, for all $m \in \{1, 2\}$, $x_m(1 - x_m) > 0$. Hence, the sign of $\frac{dx_m}{dt}$ is determined by $4\left(\frac{1}{2} - x_1\right)\left(\frac{1}{2} - x_2\right)$, which is positive on $\left(0, \frac{1}{2}\right) \times \left(0, \frac{1}{2}\right)$ and $\left(\frac{1}{2}, 1\right) \times \left(\frac{1}{2}, 1\right)$, and negative on $\left(0, \frac{1}{2}\right) \times \left(\frac{1}{2}, 1\right)$ and $\left(\frac{1}{2}, 1\right) \times \left(0, \frac{1}{2}\right)$. Accordingly, in our study of the basins of attraction, we start by considering these four "subsquares" of $[0, 1] \times [0, 1]$.

Before proceeding with a detailed analysis of each subsquare (as well as of their boundaries), we remind the reader that, by virtue of Proposition 5, systems (8)–(9) have a unique solution for any initial condition $(x_1(0), x_2(0)) \in [0, 1] \times [0, 1]$.

4.3.1. Initial conditions $(x_1(0), x_2(0)) \in \left(\frac{1}{2}, 1\right) \times \left(\frac{1}{2}, 1\right)$

Proposition 6. *The system*

$$\frac{dx_1}{dt} = 4\left(\frac{1}{2} - x_1\right)\left(\frac{1}{2} - x_2\right) x_1(1 - x_1), \quad x_1(0) = c_1 \in \left(\frac{1}{2}, 1\right),$$

$$\frac{dx_2}{dt} = 4\left(\frac{1}{2} - x_1\right)\left(\frac{1}{2} - x_2\right) x_2(1 - x_2), \quad x_2(0) = c_2 \in \left(\frac{1}{2}, 1\right)$$

has a unique solution $(x_1(t), x_2(t))$ *which satisfies* $\lim_{t\to\infty} x_1(t) = \lim_{t\to\infty} x_2(t) = 1$.

Proof. Existence and uniqueness follow from Proposition 5. For any $k \geq 3$, consider

$$\frac{\mathrm{d}x_1}{\mathrm{d}t} = 4\left(x_2 - \frac{1}{2}\right)x_1\left(x_1 - \frac{1}{2}\right)(1-x_1), \quad x_1(0) = c_1 \in \left(\frac{1}{2} + \frac{1}{k}, 1\right), \tag{12}$$

$$\frac{\mathrm{d}x_2}{\mathrm{d}t} = 4\left(x_2 - \frac{1}{2}\right)x_2\left(x_1 - \frac{1}{2}\right)(1-x_2), \quad x_2(0) = c_2 \in \left(\frac{1}{2} + \frac{1}{k}, 1\right). \tag{13}$$

The solution is $(x_1(t), x_2(t))$ where both $x_1(t)$ and $x_2(t)$ are strictly increasing functions. Hence,

$$\forall m \in \{1,2\}, \quad \forall t : x_m(t) \in \left(\frac{1}{2} + \frac{1}{k}, 1\right] \tag{14}$$

Consequently, we have

$$4\left(x_2 - \frac{1}{2}\right)x_1 \geq 4 \cdot \frac{1}{k} \cdot \frac{1}{2} = \frac{2}{k}$$

and

$$\frac{\mathrm{d}x_1}{\mathrm{d}t} \geq \frac{2}{k}\left(x_1 - \frac{1}{2}\right)(1-x_1). \tag{15}$$

Now consider the DE

$$\frac{\mathrm{d}y_k}{\mathrm{d}t} = \frac{2}{k}\left(y_k - \frac{1}{2}\right)(1-y_k), \quad y_k(0) = c_1 \in \left(\frac{1}{2} + \frac{1}{k}, 1\right). \tag{16}$$

This has the solution

$$y_k(t) = \frac{\frac{2c_1-1}{c_1-1}e^{t/k} - 1}{\frac{2c_1-1}{c_1-1}e^{t/k} - 2} = 1 + \frac{1}{\frac{2c_1-1}{c_1-1}e^{t/k} - 2}$$

We see that $y_k(t)$ is increasing and $\lim_{t\to\infty} y_k(t) = 1$. On the other hand, from (12)–(16) we have

$$\forall t : \frac{\mathrm{d}x_1}{\mathrm{d}t} \geq \frac{\mathrm{d}y_k}{\mathrm{d}t}.$$

Hence, from Petrovich's theorem we have: $\forall t : x_1(t) \geq y_k(t)$. Hence,

$$1 = \lim_{t\to\infty} y_k(t) \leq \lim_{t\to\infty} x_1(t) \leq 1 \Rightarrow \lim_{t\to\infty} x_1(t) = 1.$$

In the same manner, we prove $\lim_{t\to\infty} x_2(t) = 1$. Since these hold when $(x_1(0), x_2(0)) \in (\frac{1}{2} + \frac{1}{k}, 1) \times (\frac{1}{2} + \frac{1}{k}, 1)$ and for any $k \geq 3$, we conclude that

$$\forall (x_1(0), x_2(0)) \in \left(\frac{1}{2}, 1\right) \times \left(\frac{1}{2}, 1\right) : \lim_{t\to\infty} x_1(t) = \lim_{t\to\infty} x_2(t) = 1$$

and we have proved the proposition. $\square$

4.3.2. *Initial conditions* $(x_1(0), x_2(0)) \in (\frac{1}{2}, 1) \times (0, \frac{1}{2})$

Proposition 7. *The system*

$$\frac{dx_1}{dt} = 4\left(\frac{1}{2} - x_1\right)\left(\frac{1}{2} - x_2\right) x_1(1 - x_1), \quad x_1(0) = c_1 \in \left(\frac{1}{2}, 1\right),$$

$$\frac{dx_2}{dt} = 4\left(\frac{1}{2} - x_1\right)\left(\frac{1}{2} - x_2\right) x_2(1 - x_2), \quad x_2(0) = c_2 \in \left(0, \frac{1}{2}\right)$$

has a unique solution $(x_1(t), x_2(t))$ *which satisfies* $\lim_{t\to\infty} x_1(t) = \frac{1}{2}$ *and* $\lim_{t\to\infty} x_2(t) \in [0, \frac{1}{2})$.

Proof. Existence and uniqueness follow from Proposition 5. Now fix some $k \geq 3$ and consider the system

$$\frac{dx_1}{dt} = -4\left(\frac{1}{2} - x_2\right) x_1 \left(x_1 - \frac{1}{2}\right)(1 - x_1),$$

$$x_1(0) = c_1 \in \left(\frac{1}{2}, 1 - \frac{1}{k}\right),$$

$$\frac{dx_2}{dt} = -4\left(\frac{1}{2} - x_2\right)\left(x_1 - \frac{1}{2}\right) x_2(1 - x_2),$$

$$x_2(0) = c_2 \in \left(0, \frac{1}{2} - \frac{1}{k}\right)$$

(note that we have rewritten the DEs (8)–(9) in slightly different but equivalent form). The problem has as unique solution $(x_1(t), x_2(t))$ where both $x_1(t)$ and $x_2(t)$ are strictly decreasing functions, with

$x_1(t) \in [\frac{1}{2}, 1 - \frac{1}{k})$ and $x_2(t) \in [0, \frac{1}{2} - \frac{1}{k})$. The following limits exist:

$$\lim_{t \to \infty} x_1(t) = \bar{c}_1 \in \left[\frac{1}{2}, 1\right), \quad \lim_{t \to \infty} x_2(t) = \bar{c}_2 \in \left[0, \frac{1}{2}\right).$$

Furthermore, we have

$$4\left(\frac{1}{2} - x_2\right) x_1 \geq \frac{2}{k}.$$

Hence,

$$-4\left(\frac{1}{2} - x_2\right) x_1 \left(x_1 - \frac{1}{2}\right)(1 - x_1) \leq -\frac{2}{k}\left(x_1 - \frac{1}{2}\right)(1 - x_1)$$

and

$$\frac{dx_1}{dt} \leq -\frac{2}{k}\left(x_1 - \frac{1}{2}\right)(1 - x_1), \quad x_1(0) = c_1 \in \left(\frac{1}{2}, 1\right).$$

Now consider the DE

$$\frac{dy_k}{dt} = -\frac{2}{k}\left(y_k - \frac{1}{2}\right)(1 - y_k), \quad y(0) = c_1 \in \left(\frac{1}{2}, 1 - \frac{1}{k}\right).$$

This has a unique solution $y_k(t)$ which is strictly decreasing and satisfies $\lim_{t \to \infty} y_k(t) = \frac{1}{2}$. Hence, $\forall t : y_k(t) \geq x_1(t) \geq \frac{1}{2}$ and so we have

$$\frac{1}{2} = \lim_{t \to \infty} y_k(t) \geq \lim_{t \to \infty} x_1(t) \geq \frac{1}{2} \Rightarrow \lim_{t \to \infty} x_1(t) = \frac{1}{2}.$$

In short, $\lim_{t \to \infty} x_1(t) = \frac{1}{2}$ and $\lim_{t \to \infty} x_2(t) \in [0, \frac{1}{2})$. Since these hold when $(x_1(0), x_2(0)) \in (\frac{1}{2} + \frac{1}{k}, 1) \times (0, \frac{1}{2} - \frac{1}{k})$ and for any $k \geq 3$, we conclude that

$$\forall (x_1(0), x_2(0)) \in \left(\frac{1}{2}, 1\right) \times \left(0, \frac{1}{2}\right) : \lim_{t \to \infty} x_1(t) = \frac{1}{2}$$

$$\text{and} \lim_{t \to \infty} x_2(t) \in \left[0, \frac{1}{2}\right)$$

and we have proved the proposition. $\square$

4.3.3. Initial conditions $(x_1(0), x_2(0)) \in \left(0, \frac{1}{2}\right) \times \left(\frac{1}{2}, 1\right)$

In the same manner as in the previous case, but with the roles of $x_1(t)$ and $x_2(t)$ interchanged, we prove the following.

Proposition 8. *The system*

$$\frac{dx_1}{dt} = 4\left(\frac{1}{2} - x_1\right)\left(\frac{1}{2} - x_2\right)x_1(1-x_1), \quad x_1(0) = c_1 \in \left(0, \frac{1}{2}\right),$$

$$\frac{dx_2}{dt} = 4\left(\frac{1}{2} - x_1\right)\left(\frac{1}{2} - x_2\right)x_2(1-x_2), \quad x_2(0) = c_2 \in \left(\frac{1}{2}, 1\right)$$

has a unique solution $(x_1(t), x_2(t))$ which satisfies $\lim_{t \to \infty} x_1(t) \in \left[0, \frac{1}{2}\right)$ and $\lim_{t \to \infty} x_2(t) = \frac{1}{2}$.

4.3.4. Initial conditions $(x_1(0), x_2(0)) \in \left(0, \frac{1}{2}\right) \times \left(0, \frac{1}{2}\right)$

The proof of the following proposition is a little more involved than the previous ones.

Proposition 9. *The system*

$$\frac{dx_1}{dt} = 4\left(\frac{1}{2} - x_1\right)\left(\frac{1}{2} - x_2\right)x_1(1-x_1), \quad x_1(0) = c_1 \in \left(0, \frac{1}{2}\right),$$

$$\frac{dx_2}{dt} = 4\left(\frac{1}{2} - x_1\right)\left(\frac{1}{2} - x_2\right)x_2(1-x_2), \quad x_2(0) = c_2 \in \left(0, \frac{1}{2}\right)$$

has a unique solution $(x_1(t), x_2(t))$ which satisfies $\lim_{t \to \infty} x_1(t) = \bar{c}_1$ and $\lim_{t \to \infty} x_2(t) = \bar{c}_2$ and either $\bar{c}_1 = \frac{1}{2}$ or $\bar{c}_2 = \frac{1}{2}$ or both.

Proof. Fix some $k_0 \geq 3$ and consider the system

$$\frac{dx_1}{dt} = 4\left(\frac{1}{2} - x_1\right)\left(\frac{1}{2} - x_2\right)x_1(1-x_1), \quad x_1(0) = c_1 \in \left(0, \frac{1}{2} - \frac{1}{k_0}\right),$$

$$\frac{dx_2}{dt} = 4\left(\frac{1}{2} - x_1\right)\left(\frac{1}{2} - x_2\right)x_2(1-x_2), \quad x_2(0) = c_2 \in \left(0, \frac{1}{2} - \frac{1}{k_0}\right).$$

This system has a unique solution $(x_1(t), x_2(t))$ where both $x_1(t)$ and $x_2(t)$ are strictly increasing functions which take values inside $(0, \frac{1}{2}]$. Since $x_1(t)$, $x_2(t)$ are bounded and increasing, the following limits exist: $\lim_{t\to\infty} x_1(t) = \bar{c}_1 \in (0, \frac{1}{2}]$ and $\lim_{t\to\infty} x_2(t) = \bar{c}_2 \in (0, \frac{1}{2}]$. Furthermore, we have

$$\forall k \geq 3 : \forall (x_1(t), x_2(t)) \in \left(0, \frac{1}{2} - \frac{1}{k}\right)$$
$$\times \left(0, \frac{1}{2} - \frac{1}{k}\right) : 4\left(\frac{1}{2} - x_2(t)\right)(1 - x_1(t)) \geq \frac{2}{k} \quad (17)$$

and

$$\frac{dx_1}{dt} \geq \frac{2}{k}\left(\frac{1}{2} - x_1\right)x_1, \quad x_1(0) = c_1 \in \left(0, \frac{1}{2} - \frac{1}{k_0}\right).$$

Now take any $k \geq k_0$ and consider the DE

$$\frac{dy_k}{dt} = \frac{2}{k}\left(\frac{1}{2} - y_k\right)y_k, \quad y_k(0) = c_1 \in \left(0, \frac{1}{2} - \frac{1}{k_0}\right).$$

This has the unique solution

$$y_k(t) = \frac{1}{2 + \frac{1-2c_1}{c_1}e^{-\frac{1}{k}t}}$$

which is strictly increasing. From $y_k(0) = c_1 \in \left(0, \frac{1}{2} - \frac{1}{k_0}\right)$, $k_0 \leq k$ and $\lim_{t\to\infty} y_k(t) = \frac{1}{2}$, we conclude that there exists a t_k at which we have $y_k(t_k) = \frac{1}{2} - \frac{1}{k}$. In fact, solving

$$\frac{1}{2 + \frac{1-2c_1}{c_1}e^{-\frac{1}{k}t}} = \frac{1}{2} - \frac{1}{k}$$

we get

$$t_k = k \ln \frac{(1 - 2c_1)k - 2 + 4c_1}{4c_1}$$

Hence, the sequence $t_k, t_{k+1}, \ldots$ is strictly increasing. Now we have two possibilities:

(1) The inequality (17) holds for every $t \in [0, t_k]$. Then, by Petrovitsch's theorem, at every t_k we have $x_1(t_k) \geq y_k(t_k) = \frac{1}{2} - \frac{1}{k} > \frac{1}{2} - \sqrt{\frac{2}{k}}$.

(2) The inequality (17) does not hold for some $\widetilde{t}_k \leq t_k$, i.e., $\left(\frac{1}{2} - x_2\left(\widetilde{t}_k\right)\right)\left(1 - x_1\left(\widetilde{t}_k\right)\right) < \frac{2}{k}$. In this case, we must have

(a) either $1 - x_1\left(\widetilde{t}_k\right) < \sqrt{\frac{2}{k}}$ which implies $x_1(t_k) \geq x_1\left(\widetilde{t}_k\right) > 1 - \sqrt{\frac{2}{k}} > \frac{1}{2} - \sqrt{\frac{2}{k}}$

(b) or $\frac{1}{2} - x_2\left(\widetilde{t}_k\right) < \sqrt{\frac{2}{k}}$ which implies $x_2(t_k) \geq x_2\left(\widetilde{t}_k\right) \geq \frac{1}{2} - \sqrt{\frac{2}{k}}$.

In any case, for every large enough k, we have that either $x_1(t_k) > \frac{1}{2} - \sqrt{\frac{2}{k}}$ or $x_2(t_k) > \frac{1}{2} - \sqrt{\frac{2}{k}}$; hence one of the two inequalities must hold infinitely often. Consequently, there exist strictly increasing sequences $k_1, k_2, \ldots$ and $t_{k_1}, t_{k_2}, \ldots$ such that at least one of the following holds for all n and all $t \geq t_{k_n}$:

(1) either $x_1(t) \geq x_1(t_{k_n}) > \frac{1}{2} - \sqrt{\frac{2}{k_n}}$, then, since also $\frac{1}{2} \geq x_1(t)$, we have $\lim_{t \to \infty} x_1(t) = \frac{1}{2}$,

(2) or $x_2(t) \geq x_2(t_{k_n}) \geq \frac{1}{2} - \sqrt{\frac{2}{k_n}}$, then, since also $\frac{1}{2} \geq x_2(t)$, we have that $\lim_{t \to \infty} x_2(t) = \frac{1}{2}$.

Finally, the above holds for any $(x_1(0), x_2(0)) \in \left(0, \frac{1}{2} - \frac{1}{k_0}\right) \times \left(0, \frac{1}{2} - \frac{1}{k_0}\right)$ and for any $k_0 \geq 3$; hence, it also holds for any $(x_1(0), x_2(0)) \in \left(0, \frac{1}{2}\right) \times \left(0, \frac{1}{2}\right)$ and the proof is complete. □

Remark 2. With a little more effort, we can prove the following stronger result.

Proposition 10. *The system*

$$\frac{dx_1}{dt} = 4\left(\frac{1}{2} - x_1\right)\left(\frac{1}{2} - x_2\right)x_1(1 - x_1), \quad x_1(0) = c_1 \in \left(0, \frac{1}{2}\right),$$

$$\frac{dx_2}{dt} = 4\left(\frac{1}{2} - x_1\right)\left(\frac{1}{2} - x_2\right)x_2(1 - x_2), \quad x_2(0) = c_2 \in \left(0, \frac{1}{2}\right)$$

has a unique solution $(x_1(t), x_2(t))$ which satisfies $\lim_{t\to\infty} x_1(t) = \bar{c}_1$ and $\lim_{t\to\infty} x_2(t) = \bar{c}_2$. Now define the sets

$$A = \left\{ (z_1, z_2) \in \left(0, \frac{1}{2}\right) \times \left(0, \frac{1}{2}\right) \text{ and } z_1 < z_2 \right\},$$

$$B = \left\{ (z_1, z_2) \in \left(0, \frac{1}{2}\right) \times \left(0, \frac{1}{2}\right) \text{ and } z_1 > z_2 \right\},$$

$$C = \left\{ (z_1, z_2) \in \left(0, \frac{1}{2}\right) \times \left(0, \frac{1}{2}\right) \text{ and } z_1 = z_2 \right\}.$$

Then we have the following:

$$(x_1(0), x_2(0)) \in A \Rightarrow \lim_{t\to\infty} x_1(t) = \frac{1}{2} \text{ and } \lim_{t\to\infty} x_2(t) < \frac{1}{2},$$

$$(x_1(0), x_2(0)) \in B \Rightarrow \lim_{t\to\infty} x_1(t) < \frac{1}{2} \text{ and } \lim_{t\to\infty} x_2(t) = \frac{1}{2},$$

$$(x_1(0), x_2(0)) \in C \Rightarrow \lim_{t\to\infty} x_1(t) = \frac{1}{2} \text{ and } \lim_{t\to\infty} x_2(t) = \frac{1}{2}.$$

4.3.5. *Some initial conditions belonging to the set $\{0, \frac{1}{2}, 1\}$*

Now we present a proposition which applies to cases in which $x_1(0) \in \{0, \frac{1}{2}, 1\}$.

Proposition 11. *For the problem*

$$\frac{dx_1}{dt} = 4\left(\frac{1}{2} - x_1\right)\left(\frac{1}{2} - x_2\right) x_1(1 - x_1), \quad x_1(0) = c_1,$$

$$\frac{dx_2}{dt} = 4\left(\frac{1}{2} - x_1\right)\left(\frac{1}{2} - x_2\right) x_2(1 - x_2), \quad x_2(0) = c_2,$$

the following hold (the indicated limits always exist):

(1) *When $c_1 = 0$ and $c_2 \in (0, \frac{1}{2})$, we have $\lim_{t\to\infty} x_1(t) = 0$ and $\lim_{t\to\infty} x_2(t) = \frac{1}{2}$.*
(2) *When $c_1 = \frac{1}{2}$ and $c_2 \in (0, \frac{1}{2})$, we have $\lim_{t\to\infty} x_1(t) = \frac{1}{2}$ and $\lim_{t\to\infty} x_2(t) = c_2$.*
(3) *When $c_1 = 1$ and $c_2 \in (0, \frac{1}{2})$, we have $\lim_{t\to\infty} x_1(t) = 1$ and $\lim_{t\to\infty} x_2(t) = 0$.*

(4) When $c_1 = 0$ and $c_2 \in \left(\frac{1}{2}, 1\right)$, we have $\lim_{t \to \infty} x_1(t) = 0$ and $\lim_{t \to \infty} x_2(t) = \frac{1}{2}$.
(5) When $c_1 = \frac{1}{2}$ and $c_2 \in \left(\frac{1}{2}, 1\right)$, we have $\lim_{t \to \infty} x_1(t) = \frac{1}{2}$ and $\lim_{t \to \infty} x_2(t) = c_2$.
(6) When $c_1 = 1$ and $c_2 \in \left(\frac{1}{2}, 1\right)$, we have $\lim_{t \to \infty} x_1(t) = 1$ and $\lim_{t \to \infty} x_2(t) = 1$.

Proof. The proofs of all cases are straightforward and so only an outline is given. For example, when $c_1 = 0$ and $c_2 \in \left(0, \frac{1}{2}\right)$, the system reduces to

$$\frac{dx_1}{dt} = 0, \quad \frac{dx_2}{dt} = 2\left(\frac{1}{2} - x_2\right) x_2 (1 - x_2). \tag{18}$$

From the first DE of (18), we see that $x_1(t) = 0$ for all t. The second DE of (18) involves only x_2 and can be easily solved; inspection of the solution readily shows that $\lim_{t \to \infty} x_2(t) = \frac{1}{2}$.

The remaining cases are similar; we only give the reduced form of the DEs and leave verification of the claimed results to the reader:

(1) When $c_1 = \frac{1}{2}$ and $c_2 \in \left(0, \frac{1}{2}\right)$, the DEs reduce to

$$\frac{dx_1}{dt} = 0, \quad \frac{dx_2}{dt} = 0.$$

(2) When $c_1 = 1$ and $c_2 \in \left(0, \frac{1}{2}\right)$, the DEs reduce to

$$\frac{dx_1}{dt} = 0, \quad \frac{dx_2}{dt} = -2\left(\frac{1}{2} - x_2\right) x_2 (1 - x_2).$$

(3) When $c_1 = 0$ and $c_2 \in \left(\frac{1}{2}, 1\right)$, the DEs reduce to

$$\frac{dx_1}{dt} = 0, \quad \frac{dx_2}{dt} = 2\left(\frac{1}{2} - x_2\right) x_2 (1 - x_2).$$

(4) When $c_1 = \frac{1}{2}$ and $c_2 \in \left(\frac{1}{2}, 1\right)$, the DEs reduce to

$$\frac{dx_1}{dt} = 0, \quad \frac{dx_2}{dt} = 0.$$

(5) When $c_1 = 1$ and $c_2 \in \left(\frac{1}{2}, 1\right)$, the DEs reduce to

$$\frac{dx_1}{dt} = 0, \quad \frac{dx_2}{dt} = -2\left(\frac{1}{2} - x_2\right)x_2(1 - x_2).$$

This completes the proof. □

We also have a complementary proposition for the cases in which $x_2(0) \in \{0, \frac{1}{2}, 1\}$. We present it without proof, since it is identical to that of Proposition 11.

Proposition 12. *For the problem*

$$\frac{dx_1}{dt} = 4\left(\frac{1}{2} - x_1\right)\left(\frac{1}{2} - x_2\right)x_1(1 - x_1), \quad x_1(0) = c_1, \tag{19}$$

$$\frac{dx_2}{dt} = 4\left(\frac{1}{2} - x_1\right)\left(\frac{1}{2} - x_2\right)x_2(1 - x_2), \quad x_2(0) = c_2, \tag{20}$$

the following hold (the indicated limits always exist):

(1) When $c_1 \in \left(0, \frac{1}{2}\right)$ and $c_2 = 0$, we have $\lim_{t \to \infty} x_1(t) = \frac{1}{2}$ and $\lim_{t \to \infty} x_2(t) = 0$.

(2) When $c_1 \in \left(0, \frac{1}{2}\right)$ and $c_2 = \frac{1}{2}$ we have $\lim_{t \to \infty} x_1(t) = c_1$ and $\lim_{t \to \infty} x_2(t) = \frac{1}{2}$.

(3) When $c_1 \in \left(0, \frac{1}{2}\right)$ and $c_2 = 1$, we have $\lim_{t \to \infty} x_1(t) = 0$ and $\lim_{t \to \infty} x_2(t) = 1$.

(4) When $c_1 \in \left(\frac{1}{2}, 1\right)$ and $c_2 = 0$, we have $\lim_{t \to \infty} x_1(t) = \frac{1}{2}$ and $\lim_{t \to \infty} x_2(t) = 0$.

(5) When $c_1 \in \left(\frac{1}{2}, 1\right)$ and $c_2 = \frac{1}{2}$ we have $\lim_{t \to \infty} x_1(t) = c_1$ and $\lim_{t \to \infty} x_2(t) = \frac{1}{2}$.

(6) When $c_1 \in \left(\frac{1}{2}, 1\right)$ and $c_2 = 1$, we have $\lim_{t \to \infty} x_1(t) = 1$ and $\lim_{t \to \infty} x_2(t) = 1$.

4.3.6. Summary and discussion

We collect our results in the following table. In each row, we indicate the following: the sets which contain $\mathbf{x}(0) = (x_1(0), x_2(0))$ and $\lim_{t \to \infty} \mathbf{x}(t) = (\lim_{t \to \infty} x_1(t), \lim_{t \to \infty} x_2(t))$ (the indicated limits always exist).

Several conclusions can be reached from the above table and they can be arranged in analogy to the shape of the attraction basins:

(1) Suppose $(x_1(0), x_2(0)) \in \left(\frac{1}{2}, 1\right) \times \left(\frac{1}{2}, 1\right)$. Then, as seen in Table 1 and also by Proposition 6, we have $\lim_{t \to \infty} (x_1(t), x_2(t)) = (1, 1)$. This can be quite reasonably interpreted as follows: if both players start with a "better-than-average" tendency to cooperate, then they will end up with full cooperation.

(2) The case $(x_1(0), x_2(0)) \in \left(0, \frac{1}{2}\right) \times \left(0, \frac{1}{2}\right)$ is more interesting. As seen in Table 1 and also by Proposition 9, we now have $\lim_{t \to \infty} (x_1(t), x_2(t)) = (\bar{c}_1, \bar{c}_2)$, where at least one of $\bar{c}_1, \bar{c}_2$ will equal $\frac{1}{2}$. In fact, by looking at either Figure 1 or at the proof of Proposition 9, we can say more: both $x_1(t)$ and $x_2(t)$ are increasing functions of time (both players' levels of cooperation increase). This can be interpreted as follows: both players realize

Table 1. Attraction basins of the systems (8)–(9).

$\mathbf{x}(0)$ belongs to	$\lim_{t \to \infty} \mathbf{x}(t)$ belongs to
$\left(\frac{1}{2}, 1\right) \times \left(\frac{1}{2}, 1\right)$	$\{(1, 1)\}$
$\left(\frac{1}{2}, 1\right) \times \left(0, \frac{1}{2}\right)$	$\left\{\frac{1}{2}\right\} \times \left[0, \frac{1}{2}\right)$
$\left(0, \frac{1}{2}\right) \times \left(\frac{1}{2}, 1\right)$	$\left[0, \frac{1}{2}\right) \times \left\{\frac{1}{2}\right\}$
$\left(0, \frac{1}{2}\right) \times \left(0, \frac{1}{2}\right)$	$\left[\left(0, \frac{1}{2}\right) \times \left\{\frac{1}{2}\right\}\right] \cup \left[\left\{\frac{1}{2}\right\} \times \left(0, \frac{1}{2}\right)\right]$
$\{0\} \times \left(0, \frac{1}{2}\right)$	$\left\{\left(0, \frac{1}{2}\right)\right\}$
$\left\{\frac{1}{2}\right\} \times \left(0, \frac{1}{2}\right)$	$\left\{\left(\frac{1}{2}, x_2(0)\right)\right\}$
$\{1\} \times \left(0, \frac{1}{2}\right)$	$\{(1, 0)\}$
$\{0\} \times \left(\frac{1}{2}, 1\right)$	$\left\{\left(0, \frac{1}{2}\right)\right\}$
$\left\{\frac{1}{2}\right\} \times \left(\frac{1}{2}, 1\right)$	$\left\{\left(\frac{1}{2}, x_2(0)\right)\right\}$
$\{1\} \times \left(\frac{1}{2}, 1\right)$	$\{(1, 1)\}$
$\left(0, \frac{1}{2}\right) \times \{0\}$	$\left\{\left(\frac{1}{2}, 0\right)\right\}$
$\left(0, \frac{1}{2}\right) \times \left\{\frac{1}{2}\right\}$	$\left\{\left(x_1(0), \frac{1}{2}\right)\right\}$
$\left(0, \frac{1}{2}\right) \times \{1\}$	$\{(0, 1)\}$
$\left(\frac{1}{2}, 1\right) \times \{0\}$	$\left\{\left(\frac{1}{2}, 0\right)\right\}$
$\left(\frac{1}{2}, 1\right) \times \left\{\frac{1}{2}\right\}$	$\left\{\left(x_1(0), \frac{1}{2}\right)\right\}$
$\left(\frac{1}{2}, 1\right) \times \{1\}$	$\{(1, 1)\}$

that defection does not payoff and increase their cooperation levels until at least one of them reaches "average" cooperation (i.e., at least one of $\lim_{t\to\infty} x_1(t)$ $\lim_{t\to\infty} x_2(t)$ equals $\frac{1}{2}$). At this point, the players are trapped into a situation of "average" cooperation, namely one of the equilibrium sets $\{\frac{1}{2}\} \times (0, \frac{1}{2})$, $(0, \frac{1}{2}) \times \{\frac{1}{2}\}$.

(3) The players will also end up into one of the above invariant sets if they start in either $(0, \frac{1}{2}) \times (\frac{1}{2}, 1)$ or $(\frac{1}{2}, 1) \times (0, \frac{1}{2})$. In both of these cases, cooperation is decreasing over time; a "psychological" interpretation could be the following:

(a) The initially more cooperative player decreases his cooperation level because he observes it is not reciprocated.

(b) The initially less cooperative player increases his cooperation level because he observes that his defections have a negative outcome.

Eventually, the two levels of cooperation equilibrate at one of the above equilibrium sets.

Overall, the results are rather encouraging, in the following sense: if two continuous PD players actually update their strategies according to the update equations (8)–(9), then they will indeed reach some level of nonzero (and possibly full) cooperation. This is certainly more hopeful than the mutual defection predicted by the NE analysis of the discrete action PD.

5. Extensions

In this section, we introduce several possible extensions of the basic model presented in Section 4. The discussion is brief and further study of the proposed extensions is relegated to future publications.

5.1. *Applicability to general continuous action social dilemmas*

An important part of the basic model is the determination of the signs of the a_{mn} coefficients appearing in the action update equations (6)–(7). These signs were determined by some arguments (for example, reinforcement of mutual cooperation) regarding the effect of combinations of pure cooperation/defection levels in $\{0, 1\}$. While we initially presented these arguments in the context of PD, they also hold (for the same reasons) for the other social dilemmas, i.e., HD and SH. Hence, the update equations (8)–(9) can be applied to any continuous action social dilemma extrapolated from the discrete action PD, HD or SH game; in fact, the same arguments apply even to continuous action social dilemmas obtained independent of the discrete action ones.

5.2. Symmetric update equations with unitary coefficients

We now argue that *any* combination of unitary action update coefficients, i.e., any

$$\mathbf{a} = (a_{11}, a_{10}, a_{01}, a_{00}) \in \{-1, 1\}^4$$

can be used in the action update equations (6)–(7). Our argument is as follows:

(1) We can justify $a_{11} = 1$ because mutual cooperation reinforces itself; we can justify $a_{11} = -1$ because the opponent's cooperation increases temptation to take advantage by defection.
(2) We can justify $a_{10} = -1$ because unreciprocated cooperation makes one spiteful and uncooperative; we can justify $a_{10} = 1$ because when the opponent defects, one may attempt to lure him back to cooperation by showing "good faith" (this argument holds when the outcome of defection is worse than that of cooperation, e.g., in SH).
(3) We can justify $a_{01} = -1$ because unreciprocated defection is rewarded and hence obstructs cooperation; we can justify $a_{01} = 1$ because when a player defects, he may become afraid of causing the opponent to also defect and hence incur the high cost of mutual defection.
(4) We can justify $a_{00} = -1$ because reciprocated defection makes one spiteful and unwilling to cooperate; we can justify $a_{00} = 1$ because reciprocated defection makes one understand that cooperation is better.

Under the assumption of *symmetric* update equations, since there exist $2^4 = 16$ possible coefficient vectors $\mathbf{a}$, there also exist 16 possible update equation systems of the form (6)–(7). The system corresponding to $\mathbf{a} = (1, -1, -1, 1)$ is the one we have studied in Section 4.

An equally detailed analysis of all 16 systems requires more space than is available in this chapter; hence, we simply present the phase plots of these 16 systems in Figures 2 and 3. The reader is invited to interpret the observed behaviors for each system, with special attention to the conditions under which either full defection or full cooperation is obtained asymptotically.

5.3. *Asymmetric update equations with unitary coefficients*

If we relax the symmetry condition, we get update equations of the form

$$\frac{dx_1}{dt} = (a_{11}x_1x_2 + a_{10}x_1(1-x_2) + a_{01}(1-x_1)x_2$$
$$+ a_{00}(1-x_1)(1-x_2))x_1(1-x_1), \qquad (21)$$

$$\frac{dx_2}{dt} = (b_{11}x_1x_2 + b_{10}(1-x_1)x_2 + b_{01}x_1(1-x_2)$$
$$+ b_{00}(1-x_1)(1-x_2))x_2(1-x_2), \qquad (22)$$

where each of $\mathbf{a} = (a_{11}, a_{10}, a_{01}, a_{00})$ and $\mathbf{b} = (b_{11}, b_{10}, b_{01}, b_{00})$ can take any value in $\{-1, 1\}^4$. We now have a total of $16 \times 16 = 256$ possible combinations. We plot some of these in Figure 4 and observe the appearance of rather interesting dynamics. The interpretation of these in terms of social dilemma behaviors is left to the reader.

5.4. *General update equations*

We can obtain even more general action update equations using the forms (21)–(22) but removing the constraint that $|a_{mn}| = |b_{mn}| = 1$. In this manner, we can obtain even more involved dynamics. For example, using

$$\mathbf{a} = (2, -2, 8, 4), \quad \mathbf{b} = (3, -8, -2, 4),$$

Action Update for Continuous Action Social Dilemmas 455

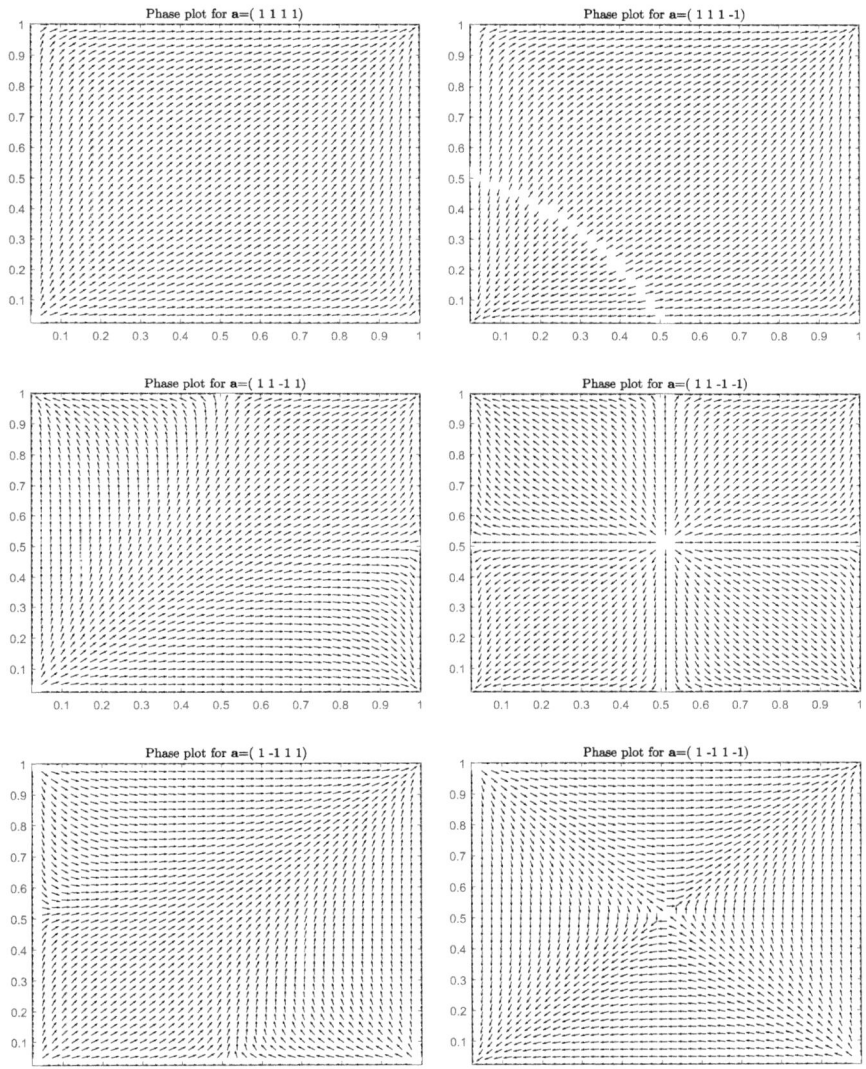

Figure 2. Phase plots for systems of the forms (6)–(7) and various values of the **a** coefficients.

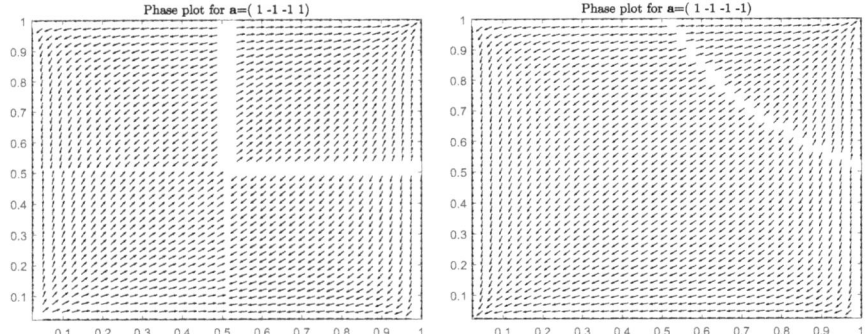

Figure 2. (*Continued*)

we get the update equations

$$\frac{dx_1}{dt} = (2x_1x_2 - 2x_1(1-x_2) + 8(1-x_1)x_2 + 4(1-x_1)(1-x_2))$$
$$\times x_1(1-x_1), \tag{23}$$

$$\frac{dx_2}{dt} = (3x_1x_2 - 8(1-x_1)x_2 - 2x_1(1-x_2) + 4(1-x_1)(1-x_2))$$
$$\times x_2(1-x_2), \tag{24}$$

which have the phase plot of Figure 5.

5.5. *More than two players*

All of the models presented up to this point involved two players. A natural extension is to consider a game played between N players. There are several ways to generalize the update equations (21)–(22). We propose the general form

$$\forall n \in \{1, \ldots, N\} : \frac{dx_n}{dt} = F_n(\mathbf{x}) x_n (1 - x_n)$$

and it remains to choose the form $F_n(\mathbf{x})$. We propose two forms, both of which can be understood as generalizations of the basic two-player model. In both cases, we only give the form of $F_1(\mathbf{x})$;

Action Update for Continuous Action Social Dilemmas 457

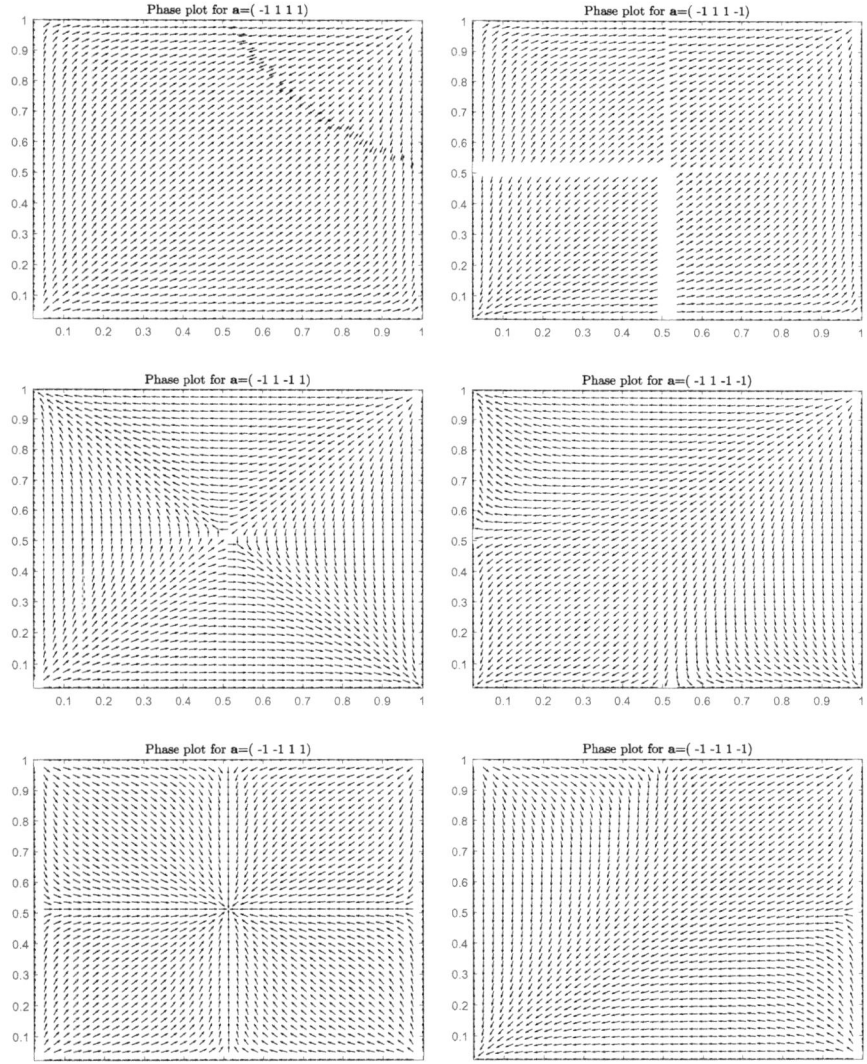

Figure 3. Phase plots for systems of the forms (6)–(7) and various values of the **a** coefficients.

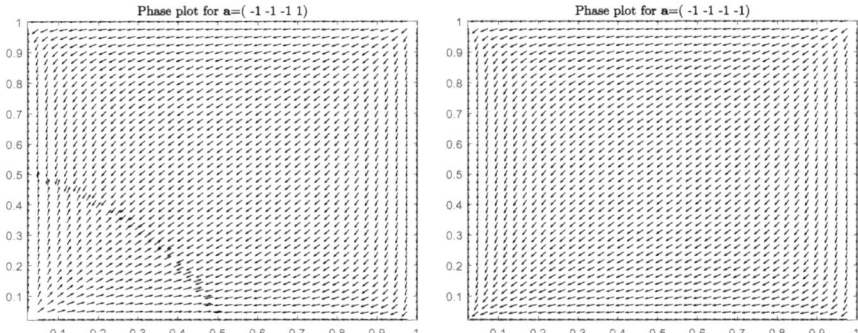

Figure 3. (*Continued*)

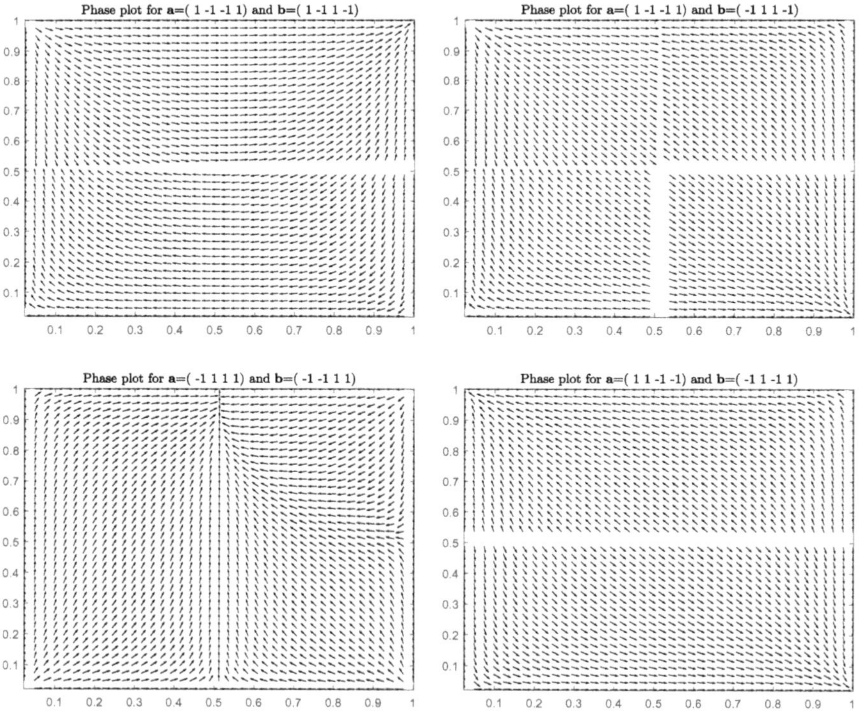

Figure 4. Phase plots for systems of the forms (6)–(7) and various values of the **a** and **b** coefficients.

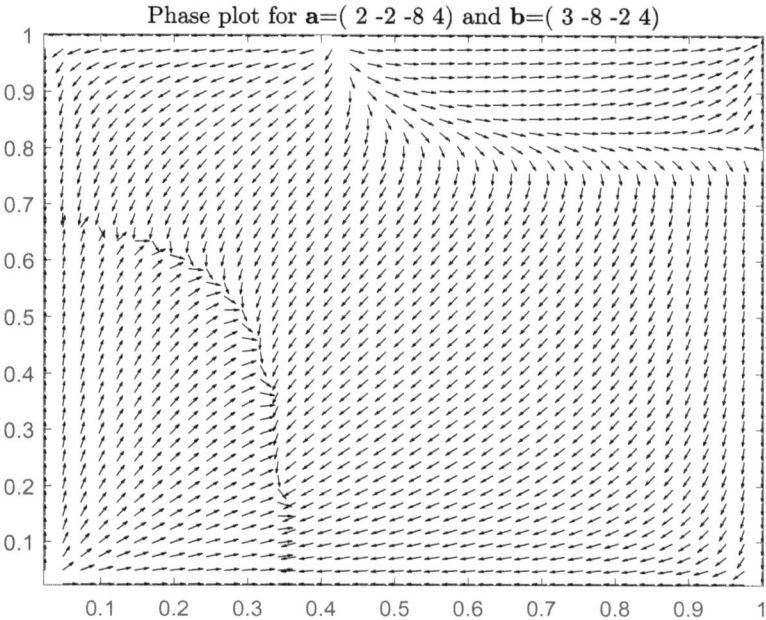

Figure 5. Phase plots for the systems of (23)–(24).

the form of the remaining F_n's is obtained similarly:

(1) The first model has the form

$$\forall n \in \{1,\ldots,N\} : F_1(\mathbf{x}) = a_{11}x_1 \sum_{m \neq n} x_m + a_{10}x_n \left(1 - \sum_{m \neq n} x_m\right)$$

$$+ a_{01}(1 - x_n) \sum_{m \neq n} x_m + a_{00}(1 - x_n) 1 - \sum_{m \neq n} x_m.$$

In this case, the change in P_n's cooperation level $\frac{dx_n}{dt}$ depends on x_n itself and on $\sum_{m \neq n} x_m$, the *aggregate* cooperation levels of the other players. Hence, $\sum_{m \neq n} x_m$ now plays the role of the single "other" player's cooperation level in equations (21)–(22). For the sake of concreteness, let us consider an example with $N = 3$ and using the symmetric selections $a_{11} = a_{00} = 1$, $a_{10} = a_{01} = -1$.

Then, after simplification, the first equation of the first model is

$$\frac{dx_1}{dt} = (4x_1x_2 + 4x_1x_3 - 2x_1 - 2x_2 - 2x_3 + 1)\, x_1\,(1 - x_1).$$

By symmetry, the full system is

$$\frac{dx_1}{dt} = (4x_1(x_2 + x_3) - 2(x_1 + x_2 + x_3) + 1)\, x_1\,(1 - x_1),$$

$$\frac{dx_2}{dt} = (4x_2(x_3 + x_1) - 2(x_1 + x_2 + x_3) + 1)\, x_2\,(1 - x_2),$$

$$\frac{dx_3}{dt} = (4x_3(x_1 + x_2) - 2(x_1 + x_2 + x_3) + 1)\, x_3\,(1 - x_3).$$

(2) The second model has the form

$$\forall n \in \{1, \ldots, N\} : F(\mathbf{x}) = a_{11} x_1 \prod_{m \neq n} x_m + a_{10} x_n \prod_{m \neq n} (1 - x_m)$$

$$+ a_{01}(1 - x_n) \prod_{m \neq n} x_m + a_{00}(1 - x_n) \prod_{m \neq n} (1 - x_m).$$

In this case, P_1 treats each other player's level of cooperation "individually" rather than "aggregately". For the sake of concreteness, let us consider an example with $N = 3$ and using the symmetric selections $a_{11} = a_{00} = 1$, $a_{10} = a_{01} = -1$. Then the first equation of the second model is

$$\frac{dx_1}{dt} = (2x_1x_2 + 2x_1x_3 + 1 - 2x_1 - x_2 - x_3)\, x_1\,(1 - x_1).$$

By symmetry, the full system is

$$\frac{dx_1}{dt} = (2x_1x_2 + 2x_1x_3 + 1 - 2x_1 - x_2 - x_3)\, x_1\,(1 - x_1),$$

$$\frac{dx_2}{dt} = (2x_2x_3 + 2x_2x_1 + 1 - 2x_2 - x_3 - x_1)\, x_2\,(1 - x_2),$$

$$\frac{dx_3}{dt} = (2x_3x_1 + 2x_3x_2 + 1 - 2x_3 - x_1 - x_2)\, x_3\,(1 - x_3).$$

6. Achieving a Prescribed NE

In our study of the basic model, we have seen that, for every $z \in (0, 1)$, the points $(z, \frac{1}{2})$ and $(\frac{1}{2}, z)$ are (unstable) equilibria. It is worth noting that these points are equilibria for *any* payoff values. Now we are going to consider an "inverse" problem: suppose we know that, for a given symmetric continuous social dilemma, we can compute an interior symmetric NE $(b, b) \in [0, 1] \times [0, 1]$. Is there a system of update equations of the form (6)–(7) such that one of its equilibria is (b, b)?

To achieve this goal, we must find appropriate values of the coefficients $a_{11}, a_{10}, a_{01}, a_{00}$. Rather than assuming all coefficients take arbitrary values, we first consider the case $a_{11} = 1$, $a_{10} = -1$, $a_{01} = -1$ and let the only free parameter be $a_{00} = a$. Then, we have the following.

Proposition 13. *For any $b \in (0, 1)$, there is a value of a such that the system*

$$\frac{dx_1}{dt} = x_1 x_2 - x_1(1 - x_2) - (1 - x_1)x_2 + a(1 - x_1)(1 - x_2), \quad (25)$$

$$\frac{dx_2}{dt} = x_1 x_2 - x_2(1 - x_1) - (1 - x_2)x_1 + a(1 - x_1)(1 - x_2) \quad (26)$$

has an equilibrium $\widehat{\mathbf{x}} = (b, b)$.

Proof. At any *symmetric* equilibrium (x, x) of (25)–(26), we have

$$0 = xx - x(1 - x) - (1 - x)x + a(1 - x)(1 - x).$$

This equation has solutions

$$r_1(a) = \frac{1 + a + \sqrt{1 - a}}{3 + a}, \quad r_2(a) = \frac{1 + a - \sqrt{1 - a}}{3 + a}.$$

Hence, for any value of a, $(r_1(a), r_1(a))$ and $(r_2(a), r_2(a))$ are equilibria of (25)–(26).

Now, $r_1(a)$ is continuous in $(-\infty, 1)$, with $\lim_{a \to -\infty} r_1(a) = 1$ and $r_1(1) = \frac{1}{2}$. Hence, for any $\widehat{b} \in [\frac{1}{2}, 1)$, there exists some $\widehat{a}$ such that $r_1(\widehat{a}) = \widehat{b}$. Similarly, $r_2(a)$ is continuous in $[0, \frac{1}{2}]$, with $r_2(0) = 0$

and $r_2(1) = \frac{1}{2}$. Hence, for any $\widehat{b} \in \left[0, \frac{1}{2}\right]$, there exists some $\widehat{a}$ such that $r_2(\widehat{a}) = \widehat{b}$. □

7. Conclusion

We have studied an action update model for two-player continuous action social dilemma games, such as Prisoner's Dilemma, Hawk and Dove, and Stag Hunt. The proposed action updates depend on the cooperation levels of the two players but are independent of specific payoffs; consequently, the update equations have the same form for each of the abovementioned games. We have presented a particular action update model, which can be generalized and extended in various ways, as indicated in Section 5; these extensions are further explored in future publications.

References

[1] R. Axelrod and W. D. Hamilton, The evolution of cooperation, *Science* **211**(4489), 1390–1396 (1981).
[2] R. Axelrod, The emergence of cooperation among egoists. *Am. Polit. Sci. Rev.* **75**(2), 306–318 (1981).
[3] E. N. Barron, *Game Theory: An Introduction* (John Wiley & Sons, Hoboken, New Jersey, 2013).
[4] M. Bowling and M. Veloso, Multiagent learning using a variable learning rate, *Artif. Intell.* **136**(2), 215–250 (2002).
[5] C. Chicone, *Ordinary Differential Equations with Applications* (Springer, New York, 2006).
[6] R. M. Dawes, Social dilemmas, *Annu. Rev. Psychol.* **31**(1), 169–193 (1980).
[7] R. M. Dawes and D. M. Messick, Social dilemmas, *Int. J. Psychol.* **35**(2), 111–116 (2000).
[8] M. Doebeli and N. Knowlton, The evolution of interspecific mutualisms, *Proc. Natl. Acad. Sci.* **95**(15), 8676–8680 (1998).
[9] M. Frean, The evolution of degrees of cooperation, *J. Theor. Biol.* **182**(4), 549–559 (1996).
[10] J. Hofbauer and K. Sigmund, *Evolutionary Games and Population Dynamics* (Cambridge University Press, Cambridge, United Kingdom, 1998).

[11] J. Hofbauer and K. Sigmund, Evolutionary game dynamics, *Bull. Am. Math. Soc.* **40**(4), 479–519 (2003).
[12] P. S. Gallo Jr. and C. G. McClintock, Cooperative and competitive behavior in mixed-motive games, *J. Conflict Resol.* **9**(1), 68–78 (1965).
[13] T. Killingback and M. Doebeli, Spatial evolutionary game theory: Hawks and Doves revisited, *Proc. R. Soc. Lond. Ser. B Biol. Sci.* **263**(1374), 1135–1144 (1996).
[14] P. A. M. Van Lange, *et al.* The psychology of social dilemmas: A review. *Organ. Behav. Hum. Decis. Process.* **120**(2), 125–141 (2013).
[15] P. A. M. Van Lange, B. Rockenbach and T. Yamagishi (eds.) *Trust in Social Dilemmas* (Oxford University Press, New York, 2017).
[16] W. B. G. Liebrand, A classification of social dilemma games, *Simul. Games* **14**(2), 123–138 (1983).
[17] M. A. Nowak, *et al.* Emergence of cooperation and evolutionary stability in finite populations, *Nature* **428**(6983), 646–650 (2004).
[18] G. Owen, *Game Theory* (Academic Press, San Diego, California, 1968).
[19] M. Petrovitsch, Sur une manière d'étendre le théorème de la moyence aux équations différentielles du premier ordre, *Math. Ann.* **54**(3), 417–436 (1901).
[20] T. E. S. Raghavan, Non-zero-sum two-person games, *Handbook of Game Theory with Economic Applications*, Vol. 3 (2002), pp. 1687–1721.
[21] A. Rapoport, A. M. Chammah and C. J. Orwant, *Prisoner's Dilemma: A Study in Conflict and Cooperation* (University of Michigan Press, Ann Arbor, Michigan, 1965).
[22] A. Rapoport and M. Guyer, A taxonomy of 2 × 2 games, *Gen. Syst.* **11** (1966).
[23] A. Rapoport, Exploiter, leader, hero, and martyr: The four archetypes of the 2 × 2 game, *Behav. Sci.* **12**(2), 81–84 (1967).
[24] G. Roberts and T. N. Sherratt, Development of cooperative relationships through increasing investment, *Nature* **394**(6689), 175–179 (1998).
[25] W. H. Sandholm, *Population Games and Evolutionary Dynamics. Economic Learning and Social Evolution* (The MIT Press, Boston, Massachusets, 2010).
[26] S. P. Singh, M. J. Kearns and Y. Mansour, *Nash Convergence of Gradient Dynamics in General-Sum Games* (UAI, 2000).
[27] B. Skyrms, *The Stag Hunt and the Evolution of Social Structure* (Cambridge University Press, Cambridge, United Kingdom, 2004).
[28] J. M. Smith, *Evolution and the Theory of Games* (Cambridge University Press, Cambridge, United Kingdom, 1982).

[29] S. H. Strogatz, *Nonlinear Dynamics and Chaos: With Applications to Physics, Biology, Chemistry, and Engineering* (CRC Press, 2018).
[30] J. Tanimoto, *Fundamentals of Evolutionary Game Theory and Its Applications* (Springer, Tokyo, Japan, 2015).
[31] T. Verhoeff, The trader's dilemma: A continuous version of the prisoner's dilemma, *Comput. Sci. Notes* **93**(2) (1998); *Computer Science* (2005).
[32] A. P. Wit and H. A. M. Wilke, The effect of social categorization on cooperation in three types of social dilemmas, *J. Econ. Psychol.* **13**(1), 135–151 (1992).
[33] M. Wubben, Social functions of emotions in social dilemmas. No. EPS-2010-187-ORG (2010).

© 2024 World Scientific Publishing Company
https://doi.org/10.1142/9789811267048_0016

Chapter 16

Applying Logarithm Sobolev Inequalities to Probability and Statistics

Christos P. Kitsos

Department of Informatics, University of West Attica, Egaleo, Athens, Greece

xkitsos@uniwa.gr

The target of this chapter is to investigate (and try to add more results to) the statistical results based on logarithm Sobolev inequalities. Heisenberg's uncertainty principle can be considered as such a problem. Moreover, the γ-order generalized normal distribution is emerged from logarithm Sobolev inequality and provides a number of extensions to statistical results, such as measures on information theory and the heat equation.

1. Introduction

The normal distribution plays an important role in the statistical world: either as a limiting distribution to the Binomial distribution or as the Gaussian error function. The weak law of large numbers, by Bernoulli in 1713, was the first step toward probability theory. Later on, in 1733, normal distribution appeared as an approximation to the probability for sums of binomial distributed quantities, to be in between two given values, by de Moivre, as the spiritual research of Ref. [1] refers. It was Gauss in 1809 in his "Theoria Motus Corponum Coelestim" stating the least squares theory. known to him since 1795, and declaring that the model was appropriate when the "errors" were coming from a normal distribution (in current terminology) [2]. The hypotheses of errors were established and adapted by the astronomers especially [3], who appreciated the normal distributions as the early astronomers the sphere. It is in Ref. [4] where the limiting form for the binomial for large n is extensively discussed, as well as the particular case that the normal distribution is an error distribution.

The first who used the bell-shaped normal curve, outside the field of astronomy, was Quetelet and the first English who adopted this work who first noticed that the logarithms of the observations might be assumed that are coming from the normal distribution was Francis Galton. The mathematical foundation of probability came later [5]. There are also some technical attempts to present as a generalization of the normal distribution the one with pdf

$$f(x) = \frac{\beta}{2\alpha \Gamma\left(\frac{1}{\beta}\right)} \exp\left\{-\left(\frac{|x-\mu|}{\alpha}\right)^\beta\right\}$$

but there is no theoretical insight for it.

In this chapter, the γ-order generalized normal distribution is discussed as an extension of the normal distribution with mean μ and variance σ^2, introducing an extra shape parameter. This distribution has emerged as an external from the Euclidean Logarithm Sobolev Inequality (LSI) [6] and provides food for thought for the use of LSI in statistics. Their pioneering work [7, 8] contributed to the relation of Fisher's information measure and the entropy measure. Therefore, under the LSI, various extensions are discussed in this chapter.

2. Sobolev Inequalities

In his early paper [9], Sobolev presented an integral inequality and the set of functions which turn inequality to quality. Under this light of thought some years later, while Sobolev was working on the problem of evaluation of the relation between the lower order of the derivatives of a given function, with their upper order, he came across the Sobolev inequalities, see Ref. [10]. The impressive Sobolev inequality is

$$\left(\int_{\mathbb{R}^p}|f(x)|^q\mathrm{d}x\right)^{1/q}\leq C\left(\int_{\mathbb{R}^p}|\nabla f(x)|^2\mathrm{d}x\right)^{1/2} \tag{1}$$

with $q=\frac{2p}{p-2}$ or in a compact form

$$\|f(x)\|_q \leq C\|\nabla f\|_2 \tag{2}$$

with p being the number of the involved variables (in analysis, the usual notation is n, but in statistics, it is p). Inequality (2) is valid for differentiable function with compact support, ∇f is the gradient of f and the Sobolev constant C reads

$$C=\left(\frac{1}{\pi p(p-2)}\right)^{1/2}\left[\frac{\Gamma(p)}{\Gamma(p/2)}\right]^{2/p} > 0. \tag{3}$$

Note that in (1) equality holds if and only if $f(x) = \kappa(\lambda^2 + |x-r|^2)^{-q/2}$ with $\kappa \in \mathbb{R}, \lambda > 0, r \in \mathbb{R}^p$. Moreover, working on the unit sphere $S_p := \{x \in \mathbb{R}^{p+1} : \|x\| = 1\}$ with $\mathrm{d}x$ being the surface measure on S_p and $|S_p| = \frac{2\pi^{\frac{p+1}{2}}}{\Gamma(\frac{p+1}{2})}$ with $\mathrm{d}s(x) = \frac{1}{\|S_p\|}\mathrm{d}x$ the normalized measure, the LSI can be defined as a sphere, see Ref. [11].

The exponent q is crucial in (1), as only with such q it holds, where the dimension $p > 2$. Moreover, Sobolev proved that there exists a function embedding the Banach space $W^{m,p}(X)$ of the functions of $L_p(X)$ into $L_p(X)$ or to the space of continuous function $C(X)$, for particular m, p, q.

As a first attempt to relate integration, derivatives and a constant term can be considered the known result from calculus:

$$(b-a) \leq \int_a^b (f'(x))^2 \mathrm{d}x, \tag{4}$$

for the function $f : [a,b] \mapsto \mathbb{R}$ with $f(a) = a$ and $f(b) = b$ and f differentiable.

An important application of LSI is to Markov chains by Ref. [12].

Moreover, working on measure space $(\mathbb{R}^p, \mu)$ with the standard Gaussian measure $\mathrm{d}\mu(x) = \mathrm{d}\varphi_p(x) = (2\pi)^{-p/2}\exp\left(-\frac{|x|^2}{2}\right)$, while

dx is the Lebesgue measure. Then from (10), the "classical Gross inequality" is

$$\|f\|_2 = 1 \Rightarrow$$
$$E_{nt}(f^2) \leq C \int_{\mathbb{R}^p} |\nabla f|^2 \mathrm{d}\varphi_p. \tag{5}$$

Note that $E_{nt}(f) = -\int f \ln f \, dx$ is the entropy of a (non-negative) measurable function. The particular feature in relation (5) is that the constant C does not depend on p, the space dimension. Reference [13] proved that the following similar form inequality holds:

$$\|f\|_2^{1+2/p} \leq C \|\nabla f\|_2^2 \|f\|_1^{4/p}, \tag{6}$$

where the value of C was evaluated as in Ref. [14], which is equivalent to (1) as above. Moreover, it is equivalent to evaluate the function $h(t, x, y)$ such that

$$\sup h(t, x, y) \leq C t^{-p/2}, \quad \forall t > 0, \tag{7}$$

with $h(t, x, y)$ being the fundamental solution of the Cauchy problem, namely

$$\frac{\partial u}{\partial t} = \Delta u,$$
$$u(0, x) = \delta(x - y), \tag{8}$$

with Δ being the Laplace operator, δ the Dirac function and $u(x, y)$ defined on $D = (0, \infty) \times \mathbb{R}^p$. The function $h(t, x, y)$ is the solution of the heat equation, see Refs. [15, 16], and equals

$$h(t, x, y) = \frac{1}{(4\pi t)^{p/2}} \exp\left(-\frac{|x-y|^2}{4t}\right). \tag{9}$$

The improvement of relation (1) came from Refs. [17, 18], at the same year, independent of one from the other. The Logarithm Sobolev Inequality (LSI) was due to Ref. [8], and we shall recall the Gross [19] logarithm inequality, with respect to the Gaussian weight of the form

$$\int_{\mathbb{R}^p} |g|^2 \log |g|^2 \mathrm{d}m \leq \frac{1}{\pi} \int_{\mathbb{R}^p} |\nabla g|^2 \mathrm{d}m, \tag{10}$$

where
$$\|g\|_2 = 1, \ dm = \exp\{-\pi x^2\} dx,$$
$$\|g\|_2 \in L^2(\mathbb{R}^p, dm).$$

Inequality (10) is equivalent [20] to the Euclidean LSI as in (11):

$$\int_{\mathbb{R}^p} |g|^2 \log |g|^2 dm \leq \frac{p}{2} \log \left[\frac{2}{\pi pe} \int_{\mathbb{R}^p} |\nabla g|^2 dx \right] \quad (11)$$

with

$$g \in W^{1,2}(\mathbb{R}^p), \ \int_{\mathbb{R}^p} |g|^2 dx = 1.$$

Relation (11) is very crucial, being in the border with the normal distribution, as the extremals of (11) with $g(x) = f(x, \mu, \sigma^2), \sigma > 0$, $\mu \in \mathbb{R}^p$, see Refs. [6, 21], are normal distributions. From (11), the generalized normal distribution was emerged, see (30). The inequality (11) is optimal in the sense that

$$\frac{2}{\pi pe} = \inf \left\{ \frac{\int_{\mathbb{R}^n} |\nabla g|^2 dx}{\exp\{\int_{\mathbb{R}^n} |g|^2 \log |g|^2 dx\}} : g \in W^{1,2}(\mathbb{R}^n), \|g\|_2 = 1 \right\}$$

while external functions for (11) are precisely the Gaussians of the form

$$g(x) = f(x; \mu, \sigma^2) = \left(\frac{\pi \sigma^2}{2} \right)^{-p/4} \exp \left\{ -\frac{1}{\sigma^2} |x - \mu|^2 \right\}.$$

We remind that a function $f : \mathbb{R}^n \mapsto \mathbb{R}$ is said to be in a Sobolev space $W^{1,\alpha}, \alpha > 1$ if $f \in L^\alpha(\mathbb{R}^n)$ and its gradient, ∇f, is a function in the Lebesgue space $L^\alpha(\mathbb{R}^n)$, i.e.,

$$W^{1,\alpha} := \left\{ f : \int_{\mathbb{R}^n} |f(x)|^\alpha dx < \infty, \int_{\mathbb{R}^n} |\nabla f(x)|^\alpha dx < \infty \right\}.$$

Proposition 1. *The constant* $C = C_p$ *as in* (3) *is a function of* w_{p-1} *the volume of the unit sphere* S_p *in* $\mathbb{R}^{p+1}$, *namely*

$$C_p = \frac{4}{p(p-2)} w^{-2/p}, \quad \text{with } w = w_{p-1}. \quad (12)$$

Proof. See in Appendix A.1. □

3. Adopting LSI to Statistics

Heisenberg's uncertainly principle (HUP) is fundamental in Quantum Mechanics (QM) and not only that, analysis, QM and probability theory are joined and tackling the subject. The work of Heisenberg in 1927 was based on physical consideration, while Kennard also published in 1927 at Zeit. Physik as well as Weyl published in 1928 (with a revised English edition). The HUP states that for a given particle at position x, with a standard deviation σ_x, with momentum M and standard deviation of σ_M, both σ_x and σ_M satisfy the inequality

$$\sigma_x \sigma_M \geq \frac{\hbar}{2}, \qquad (13)$$

where $\hbar = \frac{h}{2\pi}$, and h is the Planck constant.

Example 1. Let us assume that the accuracy of the position of an object provides that the uncertainty in the position is $\sigma_x = 1.5 \times 10^{-11}$ m. Then the minimum uncertainty in the momentum of the object is due to (13)

$$\sigma_M = \frac{h}{2\pi\sigma_x} = \frac{6.63 \times 10^{-34} \text{ J} \cdot \text{s}}{2\pi(1.5 \times 10^{-11}) \text{ m}} = 7.0 \times 10^{-24} \text{ kg} \cdot \text{m/s}.$$

Note that moment M is defined at the direction x. Moreover, the corresponding minimum uncertainty in the speed of the object depends on the mass of the object as $v_x = \frac{M_x}{m}$. If the object is an electron, then $m_e = 9.1 \times 10^{-31}$ kg and $\sigma(v_x) = 7.7 \times 10^6$ m/s. If the object is a ping-pong ball, then $m_{pp} = 2.2 \times 10^{-3}$ kg and the corresponding uncertainty of the speed of the ping-pong ball, v_{pp}, is $\sigma(v_{pp}) = 3.2 \times 10^{-21}$ m/s. Therefore, it is clear that in "real situations", it is getting very close to the neighborhood of zero.

In (13) is included ideas from statistics, due to the involved standard deviation, analysis, due to the fact that the standard deviations are evaluated through the wave functions for position and momentum — note that these functions are Fourier transforms of each other — therefore, Fourier analysis is also involved, and eventually from quantum mechanics: not only through to $\hbar$ but also through the brilliant presentation of Heisenberg, mainly through quantum mechanics. We keep to this scenario for (13), and the existence of Fourier transform as the quantum mechanics part is not included in our target but only as the source of our discussion. It is essential to note that due to Ref. [22], the HUP can be also stated as a nonzero function and its Fourier transform cannot be sharply located.

Recall that the Fourier transform of a given function $f(x)$ is defined as

$$\hat{f}(\xi) = \int_{-\infty}^{\infty} \exp[-2\pi i \xi x] f(x) \mathrm{d}x. \qquad (14)$$

Reference [22] worked adopting the Fourier transform and provided certain bounds, the most important being as in the following paragraph, we believe.

For $f \in L^2(\mathbb{R}^p)$ with $\|f\|_2 = 1$, both $|f|^2$ and $|\hat{f}|^2$ are probability density functions on $\mathbb{R}^p$ and for their corresponding variances $\mathrm{Var}(\cdot)$ it holds

$$\frac{1}{16\pi^2} \leq \mathrm{Var}(|f|^2) \mathrm{Var}(|\hat{f}|^2). \qquad (15)$$

Considering the determinants of the variance–covariance matrices, it holds

$$\frac{1}{(16\pi^2)^p} \leq \det\ \mathrm{Var}(|f|^2) \det \mathrm{Var}(|\hat{f}|^2). \qquad (16)$$

It is very clearly stated by Ref. [23] that the LSI can be interpreted as sharping the uncertainty principle. It is clear in practical problems that there is not a unique distribution approaching a given dataset. There is not only practical considerations and discussions but also tests to investigate "how close" to the real one is the decided theoretical distribution. The following theorem offers to this discussion.

Theorem 1. *Let f be the density function describing the underlying phenomenon. There is an uncertainty on the choice of (the right one) distribution.*

Proof. For $p = 1$, consider $f, |f|^2$ and the Fourier transform of $f, \hat{f}$ which is the characteristic function of the distribution f. Then, due to (15), with all the components being positive, it holds

$$\frac{1}{4\pi} \leq [\mathrm{Var}(|f|^2)]^{1/2} [\mathrm{Var}(|\hat{f}|^2)]^{1/2}.$$

If we denote by $\sigma(g)$ the standard deviation of the distribution g, then

$$\frac{1}{2}\frac{1}{2\pi} \leq \sigma(|f|^2)\sigma(|\hat{f}|^2). \tag{17}$$

Relation (17) obeys to the notation of (13) and the spirit of it for statistical distribution, being well defined from f and its characteristic function.

For $p > 1$, we can work either through (16) and obtain

$$\left(\frac{1}{4\pi}\right)^p \leq [\det \text{ Var}(|f|^2)]^{1/2} \det \text{ Var}(|\hat{f}|^2)]^{1/2}$$

or thinking in statistical terms uncertainty to one direction provides uncertainty to the choice of f. □

Moreover, for the typical LSI as in (10), where the Gaussian measure has been considered, one can reach the inequality

$$\frac{p}{4}\ln\left(\frac{p\pi e}{2}\right) + \int_{\mathbb{R}^p} |f|^2 \ln|f| \mathrm{d}x \leq \frac{p}{4}\ln \int_{\mathbb{R}^p} |\nabla f|^2 \mathrm{d}x. \tag{18}$$

Recall that for f being a probability density function, the entropy, $E_{nt}(\cdot)$, and the entropy-type information, $I_E(\cdot)$, are defined as

$$E_{nt}(f) := -\int f \ln f \mathrm{d}x, \tag{19}$$

$$I_E(f) := \int_{\mathbb{R}^p} |\nabla f|^2 \frac{1}{f} \mathrm{d}x. \tag{20}$$

Thus, from (18), see Ref. [24], we can arrive to

$$E_{nt}(f) + \frac{p}{2}\ln\left[\frac{1}{2\pi p e}I_E(f)\right] \geq 0. \tag{21}$$

Applying the entropy inequality for a probability density f, we have

$$E_{nt}(f) \leq \frac{p}{2}\ln\left[\frac{2\pi e}{p}\text{Var}(f)\right]. \tag{22}$$

Therefore, it is clear that we can arrive to the Statistical Cramer–Rao inequality

$$I_E(f)\,\text{Var}(f) \geq p^2 \qquad (23)$$

starting from the physical HUP.

4. Statistical Extensions

Following the extension of Ref. [25] for LSI, as defined in (10) for $1 < \gamma < p$ and $f \in W^{1,\gamma}(\mathbb{R}^p)$ with $\|f\|_p = 1$ of the form

$$I(f, \gamma) = \int_{\mathbb{R}^p} \|f\|^\gamma \log \|f\| \mathrm{d}x,$$

with

$$J(f, \gamma, \Lambda_\gamma) = \frac{p}{\gamma^2} \log\left[\Lambda_\gamma \int_{\mathbb{R}^p} |\nabla f|^\gamma \mathrm{d}x\right].$$

It holds that

$$I(f, \gamma) \leq J(f, \gamma, \Lambda_\gamma)$$

and the optimal constant Λ_γ to be

$$\Lambda_\gamma = \frac{\gamma}{p}\left(\frac{\gamma-1}{e}\right)^{\gamma-1} \pi^{-\gamma/2} A^{\gamma/p},$$

with

$$A = A(p, \gamma) = \frac{\Gamma(\frac{p}{2}+1)}{\Gamma(p\frac{\gamma-1}{\gamma}+1)}.$$

Reference [6] came across as the external to the hyper-multivariate distribution with mean vector μ, covariance matrix $\Sigma = (\sigma^2/\gamma)^{\frac{\gamma-1}{\gamma}} I$ and density function $\phi_\gamma(x; \mu, \Sigma)$ equal to

$$\phi_\gamma(x; \mu, \Sigma) = C \exp\left\{-\frac{\gamma-1}{\gamma}[Q(x)]^{\frac{\gamma}{2(\gamma-1)}}\right\} \qquad (24)$$

with the normalizing factor C to be

$$C = C(p, \gamma, \Sigma) = \frac{(\frac{\gamma-1}{\gamma})^{p\frac{\gamma-1}{\gamma}}}{\pi^{p/2}|\Sigma|^{1/2}} A(p, \gamma) \qquad (25)$$

and the quadratic form

$$Q(x) = (x - \mu)^T \Sigma^{-1} (x - \mu). \qquad (26)$$

This generalized γ-order (multivariate) normal distribution is a Kotz-type distribution [6].

As for the MLE, let us consider a random sample $X_1, X_2, \ldots, X_n$ coming from (24). Then the log-likelihood function $l_\gamma(\mu, \Sigma)$ is

$$l_\gamma(\mu, \Sigma) = \sum_{i=1}^n \log \phi_\gamma(x_i; \mu, \Sigma)$$

$$= \sum_{i=1}^n \left(\log C - \frac{1}{2} \log |\Sigma| - \frac{\gamma-1}{\gamma} (Q(x_i))^{\frac{\gamma}{2(\gamma-1)}} \right)$$

$$= n \log C - \frac{n}{2} \log |\Sigma|$$

$$- \frac{\gamma-1}{\gamma} \sum_{i=1}^n \left((x_i - \mu)^T \Sigma^{-1} (x_i - \mu) \right)^{\frac{\gamma}{2(\gamma-1)}}.$$

Interest has been focused for the multivariate case, while the partial derivatives are in Appendix A.2, and comments for the multivariate case are also placed here. The inverse of Fisher's information matrix is

$$I_E^{-1}(\phi_\gamma) = \begin{bmatrix} \sigma^2(\gamma-1) T_\gamma & 0 \\ 0 & \sigma^2 \frac{\gamma-1}{\gamma} \end{bmatrix}$$

with $T_\gamma = \frac{\Gamma(\frac{\gamma-1}{\gamma})}{\Gamma(\frac{1}{\gamma})} \left(\frac{\gamma}{\gamma-1} \right)^{\frac{\gamma-2}{\gamma}}$.

Moreover, for $\gamma = 2$, the well-known normal is achieved, see Figure 1. Moreover, the classical entropy inequality [26] can be extended to

$$\left[\frac{2\pi e}{p} \operatorname{Var}(X) \right]^{1/2} \left[\Lambda_\gamma \frac{1}{\gamma^\gamma} J_\gamma(X) \right]^{1/\gamma} \geq 1. \qquad (27)$$

Actually, $J_\gamma(X)$ represents an extension of Fisher's entropy-type information measure

$$J_\gamma(X) = \int_{\mathbb{R}^p} |\nabla \log f|^\gamma \mathrm{d}x$$

$$= \int_{\mathbb{R}^p} |\nabla f|^\gamma f^{1-\gamma} \mathrm{d}x \tag{28}$$

as with $\gamma = 2$

$$J_2(X) = \int_{\mathbb{R}^p} |\nabla \log f|^2 \mathrm{d}x$$

$$= \int_{\mathbb{R}^p} (\nabla f)(\nabla \log f) \mathrm{d}x = J(X), \tag{29}$$

see Ref. [6], which proved that the Blachman–Stam inequality holds also for J. Moreover, it might be that the "dimension"/the "order" of Fisher's entropy type is different than γ. In such cases, it holds.

Theorem 2 ([27]). *The generalized Fisher's information J_α for the random variable $X_\gamma \sim N_\gamma(0, I_p)$ consisted of orthogonal columns with the same norm given by*

$$J_\alpha(X_\gamma) = p \left(\frac{\gamma - 1}{\gamma}\right)^{\frac{\gamma-\alpha}{\gamma}} \frac{\Gamma(\frac{\alpha+p(\gamma-1)}{\gamma})}{\Gamma(1 + p\frac{(\gamma-1)}{\gamma})}.$$

Corollary 1. *When $\alpha = \gamma$, i.e., the order of the generalized normal coincides with the order of the generalized Fisher's entropy-type information, then*

$$J_\gamma(X_\gamma) = p$$

with $X_\gamma \sim N_\gamma(0, I_p)$.

Corollary 2. *For the corresponding extension of Shannon exponential entropy, $N_\gamma(X)$ holds:*

$$N_\gamma(X_\gamma) = p J_\gamma^{-1}(X_\gamma)$$

with J_γ as in (28).

Proof. Due to the inequality,
$$J_\gamma(X_\gamma) \cdot N_\gamma(X_\gamma) \geq p.$$
Reference [6] is where Shannon exponential entropy has been extended. □

Corollary 3. *It holds*
$$N_\gamma(X_\gamma) = 1.$$

Proof. From Corollaries 1 and 2. □

As far as (8) concerns, the heat equation and relation (9) can be generalized: Consider a standard Brownian motion $\{X(t); t > 0\}$ coming from $N_\gamma(0, t)$, i.e., from the γ-order generalized normal distribution with density function

$$\phi_\gamma(x; 0, t) = \frac{\lambda_\gamma}{\sqrt{\pi t}} \exp\left\{-\frac{\gamma-1}{\gamma}\left(\frac{x}{\sqrt{t}}\right)^{\frac{\gamma}{\gamma-1}}\right\} \tag{30}$$

with

$$\lambda_\gamma = \frac{\Gamma(\frac{1}{2}+1)}{\Gamma(\frac{\gamma-1}{\gamma}+1)}\left(\frac{\gamma-1}{\gamma}\right)^{\frac{\gamma-1}{\gamma}}, \tag{31}$$

see Ref. [28] for details. Note the simplification for $\gamma = 2$ to the standard normal, see Figure 1. For the different values of γ, see Table 1. Then, it can be proved.

Theorem 3 (Kitsos, 2023). *There exists a well-defined function* $K = K(x; t, \gamma)$ *such that*

$$\frac{\partial^2 \phi_\gamma}{\partial x^2} = K \frac{\partial \phi_\gamma}{\partial t} \tag{32}$$

with $K = K(x; t, \gamma) = \frac{N(x;t,\gamma)}{D(x;t,\gamma)}$, where

$$N(\cdot) = t^{-\gamma_1} x^{\frac{2}{\gamma-1}} - \frac{1}{\gamma-1} t^{-\frac{1}{2}\gamma_1} x^{\frac{2-\gamma}{\gamma-1}},$$

$$D(\cdot) = \frac{1}{2}\left(-\frac{1}{t} + \frac{x^{\gamma_1}}{t^{\frac{3\gamma-2}{2(\gamma-1)}}}\right), \quad \gamma_1 = \frac{\gamma}{\gamma-1}, \quad \gamma_0 = \frac{\gamma-1}{\gamma}, \quad \gamma_2 = \gamma_0^{\gamma_0}. \tag{33}$$

Applying Logarithm Sobolev Inequalities to Probability and Statistics

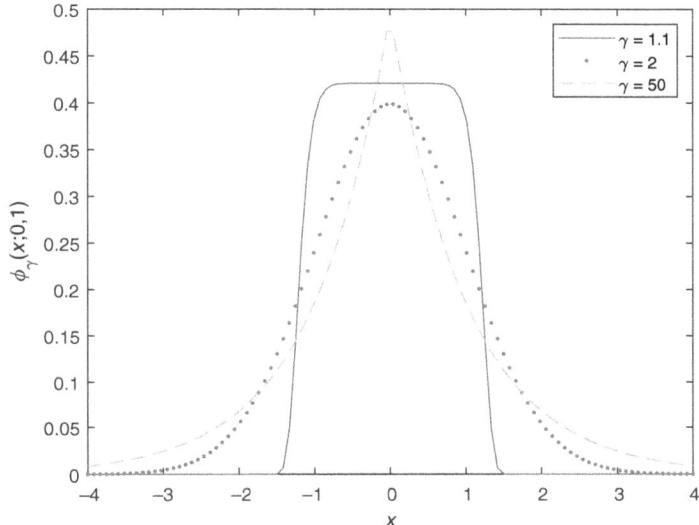

Figure 1. Plots of $\phi_\gamma(x; 0, 1)$ for different values of $\gamma = 1.1, 2, 50$.

Table 1. Values of λ_γ for given values of *gamma*.

γ	-20	-10	-1.9	-0.9	2	10	20
λ_γ	0.9126	0.9405	1.2477	1.9321	0.7071	0.8381	0.8614

A sketch of the proof is in Appendix A.3. Now, for $t = 1$, we find that $D(x; 1, \gamma) = \frac{1}{2}(-1 + x^{\gamma_1})$ and therefore $K(x; 1, \gamma)$ is defined for $x^{\gamma_1} \neq 1$ and in principle, $x \neq 1$. See Figure 2 for a graphical representation of $K(x; 1, \gamma)$ for various values of γ, where a special consideration is needed for the x-values at MATLAB. Note that, see Ref. [28, Corollary 4.1], for $\gamma = 2$,

$$K = K(x; t, 2) = 2, \tag{34}$$

and therefore, (26) is reduced to the classical heat equation [15]:

$$\frac{\partial^2 \phi_2}{\partial x^2} = 2 \frac{\partial \phi_2}{\partial t}. \tag{35}$$

For the case $p > 1$ and different values of γ, see Ref. [29] for a number of graphs concerning $N_\gamma(\mu, \sigma^2 I)$. Recall that for $\gamma \to 1$

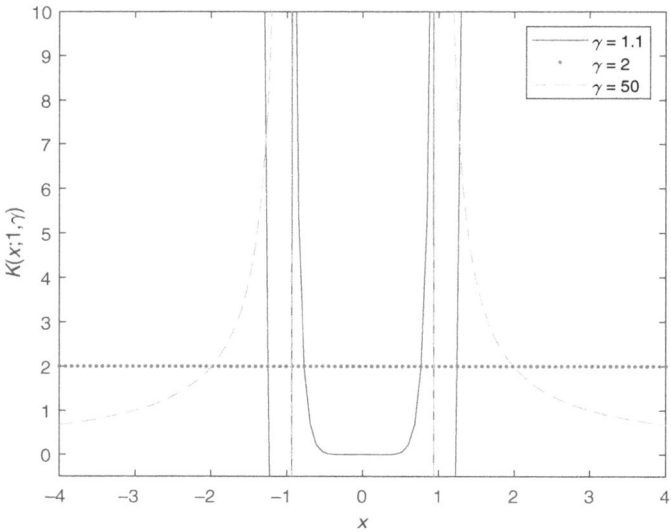

Figure 2. Plots of $K(x; 1, \gamma)$ for different values of $\gamma = 1.1, 2, 50$.

and $p = 1, \mu = 0, t = 1 = \sigma_t$, the defined $\phi_\gamma(x)$ distribution as in (24) approaches the uniform distribution, while for $\gamma \to \infty$, the $\phi_\gamma(x)$ distribution approaches the Laplace distribution [27], see Figure 1. As far as the corresponding values of $K(x; 1, \gamma)$ concerns, for $\gamma = 2$, the γ-order generalized normal is reduced to normal and $\gamma = 2$, while for $\gamma \to 1, K(0; 1, 1) = 0$ and for $x \in (0 - \epsilon, 0 + \epsilon)$, $K(x; 1, 1)$ is constant for the corresponding uniform distribution for this particular γ, see Figure 2.

5. Discussion

The LSI appears, recently, to be applied in a number of applications related with uncertainty and statistical information theory [27]. We introduced LSI in a compact form and investigated some essential, to our consideration, areas of application. More extensions can be obtained such as the Blachman–Stam inequality, see Ref. [6], while a number of nice theoretical results can be obtained, see Ref. [30] among others.

If we call by $E_{ner}(f)$ the energy of a given function f, under a given measure μ and $\text{Var}_\mu(f) = \inf_{\tau \in \mathbb{R}} E_\mu[(f-\tau)^2]$

$$E_{ner}(f) = E_\mu(|\nabla f|^2)$$

with E the expected values (under μ), then Poincare's inequality is

$$\text{Var}_\mu(f) \leq c_p E_{ner}(f),$$

while the logarithm Sobolev inequality is

$$E_{ent}(f^2) \leq c_s E_{ner}(f).$$

Both c_p and c_s constants are very essential as they connect mathematics, statistics and physics and provide strong evidence for the existence of interrelation, especially in measuring uncertainty. Therefore, there is a scientific area covering elements from physics, analysis and statistics that needs a particular investigation under the involved inequalities, especially the LSI [31]. Therefore, more attempts are needed starting from statistics to approach this common area. Theorem 1 establishes a mathematical background for the existent uncertainty for the practitioners in statistics, who really do not know if the assumed distribution for the study under consideration is the appropriate one. Moreover, the introduced generalized normal distribution covers a range of "fat tailed distributions", usually needed in economical problems. Therefore, due to LSI and the corresponding influence in the probability theory, there are statistical benefits on the way to analyze the dataset in hand.

A.1. Proof of Proposition 1

It is $C = C_p = \frac{1}{\pi p(p-2)} \left[\frac{\Gamma(p)}{\Gamma(p/2)} \right]^{2/p}$. Recall that it holds

$$\frac{\Gamma(p)}{\Gamma(p/2)} = \frac{2^{p-1}}{\pi^{1/2}} \Gamma\left(\frac{p+1}{2}\right).$$

Thus,

$$C = \frac{1}{\pi p(p-2)} \left[\frac{2^{p-1}}{\pi^{1/2}}\right]^{2/p} \left[\Gamma\left(\frac{p+1}{2}\right)\right]^{2/p}$$

$$= \frac{1}{\pi} \left[\frac{2^{p-1}}{\pi^{1/2}}\right]^{2/p} \frac{1}{p(p-2)} \left[\Gamma\left(\frac{p+1}{2}\right)\right]^{2/p}$$

$$= 4 \frac{1}{\pi} \frac{2^{-2/p}}{\pi^{1/p}} \frac{1}{p(p-2)} \left[\Gamma\left(\frac{p+1}{2}\right)\right]^{2/n}$$

$$= \frac{4}{p(p-2)} \left[\frac{\Gamma(\frac{p+1}{2})}{2\pi^{\frac{p+1}{2}}}\right]^{2/p}$$

$$= \frac{4}{p(p-2)} w^{-2/p},$$

where $w = w_p = V(S_p) = \frac{2\pi^{\frac{p+1}{2}}}{\Gamma(\frac{p+1}{2})}$ with S_p the sphere in $\mathbb{R}^{p+1}$.

A.2. The "Normal Equations" for $N_\gamma(\mu, \sigma^2 I)$

The partial derivatives of the log-likelihood with

$$Q_i = |x_i - \mu|^{\frac{2-\gamma}{\gamma-1}}, \quad R_i = |x_i - \mu|^{\frac{\gamma}{\gamma-1}}$$

are

$$\frac{\partial l_\gamma}{\partial \mu} = \sigma^{-\frac{\gamma}{\gamma-1}} \sum_{i=1}^{n} Q_i(x_i - \mu), \quad \frac{\partial^2 l_\gamma}{\partial \mu^2} = -\frac{1}{\gamma-1} \sigma^{-\frac{\gamma}{\gamma-1}} \sum_{i=1}^{n} Q_i,$$

$$\frac{\partial l_\gamma}{\partial \sigma^2} = -\frac{n}{2\sigma^2} = \frac{1}{2} \sigma^{\frac{2-3\gamma}{\gamma-1}} \sum_{i=1}^{n} R_i,$$

$$\frac{\partial^2 l_\gamma}{\partial (\sigma^2)^2} = \frac{n}{2\sigma^4} + \frac{2-3\gamma}{4(\gamma-1)} \sigma^{\frac{4-5\gamma}{\gamma-1}} \sum_{i=1}^{n} R_i,$$

$$\frac{\partial^2 l_\gamma}{\partial \mu \partial \sigma^2} = -\frac{\gamma}{2(\gamma-1)} \sigma^{\frac{2-3\gamma}{\gamma-1}} \sum_{i=1}^{n} Q_i(x_i - \mu),$$

see for details Ref. [32].

A.3. Sketch of the Proof

Let $\phi_\gamma = \phi_\gamma(x;0,t)$ as in (30). Then, it holds

$$\frac{\partial \phi_\gamma}{\partial t} = \phi_\gamma(x;0,t)\left[-\frac{1}{2}t^{-1} + \frac{1}{2}x^{\gamma_1}t^{-\frac{\gamma_1+2}{2}}\right],$$

$$\frac{\partial \phi_\gamma}{\partial x} = \phi_\gamma(x;0,t)B_1(x;t,\gamma),$$

$$\frac{\partial^2 \phi_\gamma}{\partial x^2} = \phi_\gamma(x;0,t)B_2(x;t,\gamma)$$

with

$$B_1(x;t,\gamma) = -t^{-\frac{1}{2}\gamma_1}x^{\gamma_1-1} \quad \text{and}$$

$$B_2(x;t,\gamma) = \left[B_1^2(x;t,\gamma) + \left(-t^{-\frac{1}{2}\gamma_1}(\gamma_1-1)x^{\gamma_1-2}\right)\right],$$

see for details Ref. [28].

References

[1] R. H. Daw and E. S. Pearson, Studies in the history of probability and statistics, xxx. Abraham de Moivre's 1733 derivation of the normal curve: A bibliographical note, *Biometrika* **59**(3), 677–680 (1972). doi: 10.2307/2334818.

[2] H. L. Seal, Studies in the history of probability and statistics, xv: The historical development of the Gauss linear model, *Biometrika* **54**(1/2), 1–24 (1967).

[3] G. Airy, *On the Algebraical and Numerical Theory of Errors of Observations and the Combination of Observations* (MacMillan, London, 1861).

[4] G. U. Yule and M. G. Kendall, *An Introduction to the Theory of Statistics*, 14th edn. (Universal Book Stall, New Delhi, 1993).

[5] A. Kolmogorov, *Grundbegriffe der Wahrscheinlichkeitsrechnung* (in German) [Translation: *Foundations of the Theory of Probability*], 2nd edn. (Springer, New York, 1956).

[6] C. P. Kitsos and N. K. Tavoularis, Logarithmic Sobolev inequalities for information measures, *IEEE Trans. Inform. Theory* **55**(6), 2554–2561 (2009). doi: 10.1109/tit.2009.2018179.

[7] C. E. Shannon, A mathematical theory of communication, *Bell Syst. Tech. J.* **27**(3–4), 379–423, 623–656 (1948).

[8] A. J. Stam, Some inequalities satisfied by the quantities of information of fisher and Shannon, *Inform. Control* **2**, 255–269 (1959).
[9] G. A. Bliss, An integral inequality, *J. Lond. Math. Soc.* **s1–5**(1), 40–46 (1930). doi: 10.1112/jlms/s1-5.1.40.
[10] S. L. Sobolev, On a theorem of functional analysis, *Mat. Sb.* (translated by the American Mathematical Society Translations) **4**(46), 471–497 (1938).
[11] W. Beckner, Sharp Sobolev inequalities on the sphere and the Moser–Trudinger inequality, *Ann. Math.* **138**(1), 213–242 (July, 1993). doi: 10.2307/2946638.
[12] P. Diaconis and L. Saloff-Coste, Logarithmic Sobolev inequalities for finite Markov chains, *Ann. Appl. Probab.* **6**(3), 695–750 (1996). doi: 10.1214/aoap/1034968224.
[13] J. Nash, Continuity of solutions of parabolic and elliptic equations, *Am. J. Math.* **80**(4), 931–954 (1958).
[14] E. A. Carlen and M. Loss, Sharp constant in Nashs inequality, *Int. Math. Res. Not.* **1993**(7), 213–215 (1993). doi: 10.1155/S107379289 3000224.
[15] S. Karlin and M. H. Taylor, *A First Course in Stochastic Processes*, 2nd edn. (Academic Press, New York, 1975).
[16] A. M. Wazwaz, *Partial Differential Equations, Methods and Applications* (A. A. Balkema, Tokyo, 2002).
[17] T. Aubin, Problemes isometriques et espaces de Sobolev, *J. Differ. Geom.* **11**, 573–598 (1976).
[18] G. Talenti, Best constants in Sobolev inequality, *Ann. Mat. Pura Appl.* **110**, 353–372 (1976).
[19] L. Gross, Logarithmic Sobolev inequalities, *Am. J. Math.* **97**(761), 1061–1083 (1975).
[20] F. B. Weissler, Logarithmic Sobolev inequalities for the heat-diffusion semigroup, *Trans. Am. Math. Soc.* **237**, 225–269 (1978).
[21] E. A. Carlen, Superadditivity of Fishers information and logarithmic Sobolev inequalities, *J. Funct. Anal.* **101**(1), 194–211 (1991). doi: 10.1016/0022-1236(91)90155-x.
[22] B. Folland and A. Sitaram, The uncertainty principle: A mathematical survey, *J. Fourier Anal. Appl.* **3**(3), 207–238 (1997).
[23] W. Beckner, Inequalities in Fourier analysis, *Ann. Math.* **102**(1), 159–182 (1975). doi: 10.2307/1970980.
[24] W. Beckner, Pitt's inequality and the uncertainty principle, *Proc. Am. Math. Soc.* **123**(6), 1897–1905 (1995). doi: 10.2307/2161009.
[25] J. D. Pino, J. Dolbeault and I. Gentil, Nonlinear diffusions, hypercontractivity and the optimal Lp-Euclidean logarithmic Sobolev inequality, *J. Math. Anal. Appl.* **293**(2), 375–388 (2004).

[26] T. M. Cover and J. A. Thomas, *Elements of Information Theory*, 2nd edn. (Wiley, New York, 2006). doi: 10.1002/047174882x.
[27] C. Kitsos and T. Toulias, New information measures for the generalized normal distribution, *Information* **1**, 13–27 (2010).
[28] C. P. Kitsos, Generalizing the heat equation, *REVSTAT — Stat. J.* (2023). https://revstat.ine.pt/index.php/REVSTAT/article/view/517.
[29] G. Halkos and C. P. Kitsos, Uncertainty in environmental economics: The problem of entropy and model choice, *Econ. Anal. Policy* **60**, 128–140 (2018).
[30] W. Beckner and M. Pearson, On sharp Sobolev embedding and the logarithmic Sobolev inequality, *Bull. Lond. Math. Soc.* **30**(1), 80–84 (1998). doi: 10.1112/s0024609397003901.
[31] S. G. Bobkov and M. Ledoux, On modified logarithmic Sobolev inequalities for Bernoulli and Poisson measures, *J. Funct. Anal.* **156**(2), 347–365 (July, 1998). doi: 10.1006/jfan.1997.3187.
[32] C. P. Kitsos, V. G. Vassiliadis and T. L. Toulias, MLE for the γ-order generalized normal distribution, *Discuss. Math. Probab. Stat.* **34**(1–2), 143–158 (2014).

© 2024 World Scientific Publishing Company
https://doi.org/10.1142/9789811267048_0017

Chapter 17

Some Certain Families of Catalan-Type Numbers and Polynomials: Analysis of Their Generating Function with Their Functional Equations

Irem Kucukoglu[*,‡] and Yilmaz Simsek[†,§]

[*]*Department of Engineering Fundamental Sciences,
Faculty of Engineering,
Alanya Alaaddin Keykubat University, Antalya, Turkey*
[†]*Department of Mathematics, Faculty of Science,
University of Akdeniz, Antalya, Turkey*
[‡]*irem.kucukoglu@alanya.edu.tr*
[§]*ysimsek@akdeniz.edu.tr*

In order to present known and new identities and relations for Catalan-type numbers and polynomials, the method used in this chapter covers the techniques used for p-adic integral equations and special functions, especially generating functions with their functional equations. The following briefly summarizes the results to be given in this chapter. The aim of this chapter is to survey some certain families of Catalan-type numbers and polynomials by generating function approach. In this chapter, in the sequel of reminding the Catalan numbers and some features of these numbers and their relations with some special numbers, we focus on some Catalan-type numbers and analyze some of their fundamental properties. Moreover, by applying p-adic integrals to a function arising from the Catalan–Qi function, we get some p-adic integral formulas involving the Catalan–Qi function and other certain families of special numbers. Finally, with the implementation of our algorithms in Wolfram Mathematica Online (namely, in the Wolfram Cloud (Wolfram Research Inc., 2022)) by using the Wolfram programming language, we demonstrate some numerical values of Catalan-type numbers inside tables.

1. Preliminaries

Before giving the main concepts of this chapter, we set the notations and definitions briefly as follows:

Let $\mathbb{N}$, $\mathbb{Z}$, $\mathbb{Q}$, $\mathbb{R}$ and $\mathbb{C}$ denote the set of natural numbers, the set of integers, the set of rational numbers, the set of real numbers and the set of complex numbers, respectively. Furthermore, $\mathbb{N}_0 = \mathbb{N} \cup \{0\}$. It is assumed that

$$0^n = \begin{cases} 1, & (n=0) \\ 0, & (n \in \mathbb{N}) \end{cases}$$

and, for $z \in \mathbb{C}$, we also assume that $\ln z$ denotes the principle branch of the many-valued function:

$$\ln z = \ln |z| + i \arg z; \quad (|z| > 0),$$

where $-\pi < \arg z \leq \pi$.

1.1. *Catalan numbers and their generating functions*

The Catalan numbers, C_n, are defined by the following ordinary generating function:

$$g(t) := \frac{1 - \sqrt{1 - 4t}}{2t} = \sum_{n=0}^{\infty} C_n t^n,$$

where $0 < |t| \leq \frac{1}{4}$ (cf. Refs. [7, Theorem 3.2], [11], [19, pp. 96–106], [60, pp. 109–110], [96], [42]).

The Catalan numbers inherently emerge in the solution of many kinds of combinatorial enumeration problems, such as Euler's polygon problem, the Ballot problems and the Dyck path (see, for details, Refs. [11], [19, pp. 96–106], [42], [60, pp. 109–110], [96]). In Ref. [21], Cruz et al. have studied on the Catalan numbers. They showed that the Catalan numbers play a significant role in the context of hypercomplex function theory. As for the studies that used the techniques of generating function in order to derive identities for the Catalan numbers, the interested readers may refer to Refs. [20, 26, 42, 49, 54, 58–60, 66, 82, 96, 102].

The Catalan numbers can be calculated with many different formulas explicitly or with combinatorial means. Some of these formulas (among others) are given as follows:

$$C_n = \frac{1}{n+1}\binom{2n}{n}, \quad (n \in \mathbb{N}_0), \tag{1}$$

$$C_n = \binom{2n}{n} - \binom{2n}{n+1}, \quad (n \in \mathbb{N}_0), \tag{2}$$

$$C_n = \frac{1}{n+1}\sum_{k=0}^{n}\binom{n}{k}^2, \quad (n \in \mathbb{N}_0), \tag{3}$$

$$C_n = \frac{1}{(n+1)!}\prod_{j=1}^{n}(4j-2), \quad (n \in \mathbb{N}), \tag{4}$$

$$C_n = \prod_{j=2}^{n}\frac{n+j}{j},$$

and these numbers have a recurrence relation given by

$$\frac{C_{n+1}}{C_n} = \frac{2(2n+1)}{n+2} \tag{5}$$

(cf. Refs. [10, 11, 18, 19, 23, 24, 26, 27, 35, 36, 49, 58–61, 96, 102, 109, 125]).

By using the above formulas, first eight values of the numbers C_n are computed as follows:

$$C_0 = 1, \quad C_1 = 1, \quad C_2 = 2, \quad C_3 = 5,$$
$$C_4 = 14, \quad C_5 = 42, \quad C_6 = 132, \quad C_7 = 429,$$

and so on (cf. Refs. [11, 18, 19, 23, 24, 26, 27, 35, 36, 49, 58–61, 96, 102, 109, 125]).

Let $r \in \mathbb{N}$ and $C_n^{(r)}$ denote the Catalan numbers of order r. Let we set

$$(g(t))^r = \sum_{n=0}^{\infty} C_n^{(r)} t^n, \qquad (6)$$

where

$$(g(t))^r = \underbrace{g(t)g(t)\ldots g(t)}_{r\text{-times}} \qquad (7)$$

(cf. Ref. [102]).

Using the well-known Cauchy product in (6), one can easily compute the Catalan numbers of order r. For example, considering the order as $r = 2$ in (6) and then using the Cauchy product yields

$$C_n^{(2)} = \sum_{j=0}^{n} C_j C_{n-j}.$$

By the above formula, the second author [102] gave few values of the Catalan numbers of order 2 as follows:

$$C_0^{(2)} = 1 = C_1,$$
$$C_1^{(2)} = 2 = C_2,$$
$$C_2^{(2)} = 5 = C_3$$

(cf. Ref. [102]). By the following commands written in Wolfram Cloud (by Wolfram programming language), we extend this table (in Figure 1) which contains the values of $C_n^{(2)}$ from $C_0^{(2)}$ to $C_{10}^{(2)}$:

```
Unprotect[Power];
Power[0,0]=1;
Protect[Power];
   CatalanTwoHigherNumber[n_] :=Sum[CatalanNumber[j]*
CatalanNumber[n−j], {j,0,n}]
   TableForm[Table [StringJoin[ToString[
CatalanTwoHigherNumber[n]]," ⌣=⌣",ToString[Subscript[Style["
C", Italic],n+1], StandardForm]] , {n,0,10}],
   TableHeadings−>{Table [ToString[Superscript[Subscript[
Style[" C", Italic], k], "(2)" ], StandardForm],{k,0,10}] }]
```

Some Certain Families of Catalan-Type Numbers and Polynomials 489

$$\begin{array}{c|c}
C_0^{(2)} & 1 = C_1 \\
C_1^{(2)} & 2 = C_2 \\
C_2^{(2)} & 5 = C_3 \\
C_3^{(2)} & 14 = C_4 \\
C_4^{(2)} & 42 = C_5 \\
C_5^{(2)} & 132 = C_6 \\
C_6^{(2)} & 429 = C_7 \\
C_7^{(2)} & 1430 = C_8 \\
C_8^{(2)} & 4862 = C_9 \\
C_9^{(2)} & 16796 = C_{10} \\
C_{10}^{(2)} & 58786 = C_{11}
\end{array}$$

Figure 1. Some of the Catalan numbers of order 2 and their relation with the Catalan numbers.

Remark 1. As indicated in the table (in Figure 1), the Catalan numbers $C_n^{(2)}$ of order 2 are respectively equal to the Catalan numbers C_{n+1} in a shifting way by one index.

In order to give some interesting formulas for the Catalan-type numbers, we need to recall some properties of the beta and gamma functions.

1.2. Some notations and formulas for the beta and gamma functions and factorial polynomials

Here, we give some well-known formulas and definitions regarding the beta and gamma functions and factorial polynomials, and also Vandermonde identities.

The beta function $B(\alpha, \beta)$ is defined by

$$B(\alpha, \beta) = \int_0^1 t^{\alpha-1}(1-t)^{\beta-1}dt = B(\beta, \alpha), \quad (\operatorname{Re}(\alpha) > 0, \operatorname{Re}(\beta) > 0),$$
(8)

where α and β are complex variables (cf. Refs. [91], [124, p. 9, Equation (62)]).

For $z \in \mathbb{C}$, the Pochhammer symbol $(z)_v$, which is also known as the rising factorial, is defined by

$$(z)_v := \frac{\Gamma(z+v)}{\Gamma(z)} = \begin{cases} z(z+1)\ldots(z+v-1) & \text{if } v \in \mathbb{N}, \\ 1 & \text{if } v = 0, \end{cases}$$

where $\Gamma(z)$ denotes Euler's gamma function defined by

$$\Gamma(z) = \int_0^\infty t^{z-1} e^{-t} dt; \quad \text{Re}(z) > 0 \tag{9}$$

which is related to the beta function due to the following relation:

$$B(\alpha, \beta) = \frac{\Gamma(\alpha)\Gamma(\beta)}{\Gamma(\alpha + \beta)}$$

such that replacing α and β respectively by $n \in \mathbb{N}$ and $m \in \mathbb{N}$ yields

$$B(n, m) = \frac{\Gamma(n)\Gamma(m)}{\Gamma(n+m)} = \frac{(n-1)!(m-1)!}{(n+m-1)!} \tag{10}$$

due to the fact that

$$\Gamma(n) = (n-1)! \quad (n \in \mathbb{N})$$

(cf. Refs. [91], [124, p. 9, Equation (62)]).

Note that

$$\Gamma(z) = \frac{\Gamma(z+n+1)}{z(z+1)\ldots(z+n)}, \tag{11}$$

$$\Gamma\left(\frac{1}{2}\right) = \sqrt{\pi}, \tag{12}$$

$$\Gamma\left(n + \frac{1}{2}\right) = \frac{(2n)!}{4^n n!} \sqrt{\pi} \tag{13}$$

(cf. Refs. [91, 124]).

For $z \in \mathbb{C}$, the falling factorial is given by

$$(z)^{\underline{v}} := \begin{cases} z(z-1)\ldots(z-v+1) & \text{if } v \in \mathbb{N}, \\ 1 & \text{if } v = 0, \end{cases}$$

whose relation with the Pochhammer symbol is given by

$$(z)^{\underline{v}} = (-1)^v (-z)_v$$

(cf. Refs. [91, 124]).

Thus, the binomial coefficient is given in terms of the falling factorial as follows:

$$\binom{z}{v} := \begin{cases} \dfrac{(z)^{\underline{v}}}{v!} & \text{if } v \in \mathbb{N},\ z \in \mathbb{C} \\ 1 & \text{if } v = 0,\ z \in \mathbb{C}. \end{cases}$$

The generalized Vandermonde's convolution is given by

$$\binom{x + \sum_{k=1}^{v} y_k}{n} = \sum_{k_0 + k_1 + \cdots + k_v = n} \binom{x}{k_0}\binom{y_1}{k_1}\cdots\binom{y_v}{k_v} \qquad (14)$$

(cf. Ref. [34, Exercise 62, p. 248]). Note that in the special case when $v = 1$ and $y_1 = y$, (14) is reduced to the following well-known Chu–Vandermonde identity in terms of the falling factorial:

$$\binom{x+y}{n} = \frac{1}{n!}\sum_{k=0}^{n}\binom{n}{k}(x)^{\underline{k}}(y)^{\underline{n-k}}, \qquad (15)$$

(cf. Refs. [18, 44, 65, 111, 114]).

1.3. *Some certain families of special numbers and polynomials*

Here, we recall some certain families of special numbers and polynomials with their generating functions and properties.

The Stirling numbers of the first kind, $s(n, k)$, are defined by the following generating functions:

$$\frac{(\ln(1+t))^k}{k!} = \sum_{k=0}^{\infty} s(n,k)\frac{t^n}{n!} \qquad (16)$$

and

$$(x)^{\underline{n}} = \sum_{k=0}^{n} s(n,k)\, x^k \qquad (17)$$

(cf. Refs. [11, 25, 103, 125]).

The Stirling numbers of the second kind, $S(n,k)$, are defined by the following generating functions:

$$\frac{(e^t - 1)^k}{k!} = \sum_{n=0}^{\infty} S(n,k) \frac{t^n}{n!} \qquad (18)$$

and

$$x^n = \sum_{k=0}^{n} S(n,k) (x)_{\underline{k}} \qquad (19)$$

(cf. Refs. [9, 18, 27, 55, 73, 101, 121, 125]).

Equation (18) yields

$$S(n,k) = \frac{1}{k!} \sum_{j=0}^{k} (-1)^{k-j} \binom{k}{j} j^n \qquad (20)$$

so that if $k > n$ or $k < 0$, then $S(n,k) = 0$ (cf. Refs. [9, 18, 27, 55, 73, 87, 101, 121, 125]).

The Apostol–Bernoulli numbers, $\mathcal{B}_k(\omega)$, are defined by the following generating function:

$$\frac{t}{\omega e^t - 1} = \sum_{k=0}^{\infty} \mathcal{B}_k(\omega) \frac{t^k}{k!}, \quad (\omega \in \mathbb{C}) \qquad (21)$$

which converges when $|t| < 2\pi$ if $\omega = 1$ and $|t| < |\ln(\omega)|$ if $\omega \neq 1$. By using (21), the first few values of these numbers are computed as follows:

$$\mathcal{B}_0(\omega) = 0,$$
$$\mathcal{B}_1(\omega) = (\omega - 1)^{-1},$$
$$\mathcal{B}_2(\omega) = -2\omega(\omega - 1)^{-2},$$
$$\mathcal{B}_3(\omega) = 3\omega(\omega + 1)(\omega - 1)^{-3},$$
$$\mathcal{B}_4(\omega) = -4\omega(\omega^2 + 4\omega + 1)(\omega - 1)^{-4},$$
$$\mathcal{B}_5(\omega) = 5\omega(\omega^3 + 11\omega^2 + 11\omega + 1)(\omega - 1)^{-5},$$

and so on (cf. Refs. [4, 12, 62–64, 73, 121, 123, 125, 131] and the references cited therein).

The positive-order Bernoulli numbers, $B_k^{(n)}$, are defined by the following generating function:

$$\left(\frac{t}{e^t - 1}\right)^n = \sum_{k=0}^{\infty} B_k^{(n)} \frac{t^k}{k!} \qquad (22)$$

where $n \in \mathbb{N}_0$ (cf. Refs. [9, 13, 16, 18, 27, 31, 55, 73, 101, 121, 125]).

Equation (22) yields the following recurrence relation for the numbers $B_k^{(n)}$, for $n, k \in \mathbb{N} \setminus \{1\}$:

$$B_k^{(n)} = \left(1 - \frac{k}{n-1}\right) B_k^{(n-1)} - k B_{k-1}^{(n-1)}, \qquad (23)$$

which implies

$$B_k^{(k+1)} = (-1)^k k! \qquad (24)$$

(cf. Refs. [13, 14]).

By (24), we have

$$B_0^{(1)} = 1, \quad B_1^{(2)} = -1, \quad B_2^{(3)} = 2, \quad B_3^{(4)} = -6,$$
$$B_4^{(5)} = 24, \quad B_5^{(6)} = -120, \quad B_6^{(7)} = 720,$$

and so on. More generally, by (23), few values of the numbers $B_k^{(n)}$ are computed as follows:

$$B_0^{(n)} = 1,$$
$$B_1^{(n)} = -\frac{1}{2}n,$$
$$B_2^{(n)} = \frac{1}{12}(3n^2 - n),$$
$$B_3^{(n)} = -\frac{1}{8}(n^3 - n^2),$$
$$B_4^{(n)} = \frac{1}{240}(15n^4 - 30n^3 + 5n^2 + 2n),$$
$$B_5^{(n)} = -\frac{1}{96}(n^3 - n^2)(3n^2 - 7n - 2),$$
$$B_6^{(n)} = \frac{1}{4032}(63n^6 - 315n^5 + 315n^4 + 91n^3 - 42n^2 - 16n),$$

and so on (cf. Refs. [75, 116] and see also the references cited therein).

For example, by choosing the upper and lower indices exactly the same in the above formulas, we have

$$B_1^{(1)} = -\frac{1}{2}, \quad B_2^{(2)} = \frac{5}{6}, \quad B_3^{(3)} = -\frac{9}{4}, \quad B_4^{(4)} = \frac{251}{30},$$

$$B_5^{(5)} = -\frac{475}{12}, \quad B_6^{(6)} = \frac{19087}{84}, \quad B_7^{(7)} = -\frac{36799}{24},$$

$$B_8^{(8)} = \frac{1070017}{90}, \quad B_9^{(9)} = -\frac{2082753}{20}, \quad B_{10}^{(10)} = \frac{134211265}{132},$$

and so on (cf. Refs. [75, 116] and see also the references cited therein).

Note that the special case of (21) when $\omega = 1$ and the special case of (22) when $n = 1$ both give the generating function for the Bernoulli numbers B_k of the first kind, namely:

$$B_k = \mathcal{B}_k(1)$$

and

$$B_k = B_k^{(1)}$$

whose recurrence relation is given as

$$\begin{cases} B_0 = 1 \quad \text{(initial condition)} \\ B_k = \sum_{j=0}^{k} \binom{k}{j} B_j \quad \text{for } k \in \mathbb{N} \setminus \{1\} \end{cases}$$

by which, first few values are given as follows:

$$B_0 = 1, \quad B_1 = -\frac{1}{2}, \quad B_2 = \frac{1}{6}, \quad B_4 = -\frac{1}{30}, \quad B_6 = \frac{1}{42},$$

$$B_8 = -\frac{1}{30}, \quad B_{10} = \frac{5}{66},$$

such that $B_{2k+1} = 0$ ($k \in \mathbb{N}$) (cf. Refs. [1, 4, 9, 73, 78, 80, 81, 120–123, 125] and the references cited therein).

The negative-order Bernoulli numbers, $B_k^{(-m)}$, are defined by the following generating function:

$$\left(\frac{e^t - 1}{t}\right)^m = \sum_{k=0}^{\infty} B_k^{(-m)} \frac{t^k}{k!} \quad (25)$$

where $m \in \mathbb{N}$ (cf. Refs. [9, 18, 27, 55, 73, 101, 121, 125]).

Some Certain Families of Catalan-Type Numbers and Polynomials   495

A formula for computing the numbers $B_k^{(-n)}$ is given as follows:

$$B_k^{(-n)} = \frac{\binom{k+n}{n}^{-1}}{n!} \sum_{j=0}^{k+n-1} \sum_{v=0}^{j} (-1)^v \binom{j}{k} \binom{k+n+1}{v} (j+1-v)^{n+k}$$

(cf. Ref. [36]).

The relationship between the positive-order Bernoulli numbers and the negative-order Bernoulli numbers is as follows:

$$\sum_{j=0}^{k} \binom{k}{j} B_j^{(-n)} B_{k-j}^{(n)} = 0, \qquad (26)$$

where $n, k \in \mathbb{N}$ (cf. Refs. [36, 73, 101, 121, 125]).

The Apostol–Euler numbers, $\mathcal{E}_k(\omega)$, are defined by the following generating function:

$$\frac{2}{\omega e^t + 1} = \sum_{k=0}^{\infty} \mathcal{E}_k(\omega) \frac{t^k}{k!}, \qquad (27)$$

which converges when $|t| < \pi$ if $\omega = 1$ and $|t| < |\ln(-\omega)|$ if $\omega \neq 1$. By using (27), first few values of these numbers are computed as follows:

$$\mathcal{E}_0(\omega) = 2(\omega + 1)^{-1},$$
$$\mathcal{E}_1(\omega) = -2\omega(\omega + 1)^{-2},$$
$$\mathcal{E}_2(\omega) = 2\omega(\omega - 1)(\omega + 1)^{-3},$$
$$\mathcal{E}_3(\omega) = -2\omega(\omega^2 - 4\omega + 1)(\omega + 1)^{-4},$$
$$\mathcal{E}_4(\omega) = 2\omega(\omega^3 - 11\omega^2 + 11\omega - 1)(\omega + 1)^{-5},$$
$$\mathcal{E}_5(\omega) = -2\omega(\omega^4 - 26\omega^3 + 66\omega^2 - 26\omega + 1)(\omega + 1)^{-6},$$

and so on (cf. Refs. [15, 25, 71–73, 120, 121, 123, 125, 131] and the references cited therein).

The special case of (27) when $\omega = 1$ gives the generating function of the Euler numbers E_k of the first kind, namely

$$E_k = \mathcal{E}_k(1), \qquad (28)$$

whose recurrence relation is given as

$$\begin{cases} E_0 = 1 & \text{(initial condition)} \\ E_k = -\sum_{j=0}^{k} \binom{k}{j} E_j & \text{for } k \in \mathbb{N} \end{cases}$$

by which the first few values of these numbers are given as follows:

$$E_0 = 1, \quad E_1 = -\frac{1}{2}, \quad E_3 = \frac{1}{4}, \quad E_5 = -\frac{1}{2}, \quad E_7 = \frac{17}{8}, \quad E_9 = -\frac{31}{2}$$

and so on; $E_{2k} = 0$ ($k \in \mathbb{N}$) (cf. Refs. [1, 9, 15, 18, 25, 73, 78, 80, 81, 120, 121, 123, 125] and the references cited therein).

The Apostol–Genocchi numbers, $\mathcal{G}_k(\omega)$, are defined by the following generating function:

$$\frac{2t}{\omega e^t + 1} = \sum_{k=0}^{\infty} \mathcal{G}_k(\omega) \frac{t^k}{k!} \qquad (29)$$

which converges when $|t| < \pi$ if $\omega = 1$ and $|t| < |\ln(-\omega)|$ if $\omega \neq 1$ (cf. Refs. [73, 121, 123, 125] and the references cited therein). By using (29), the first few values of these numbers are computed as follows:

$$\mathcal{G}_0(\omega) = 0,$$
$$\mathcal{G}_1(\omega) = 2(\omega + 1)^{-1},$$
$$\mathcal{G}_2(\omega) = -4\omega(\omega + 1)^{-2},$$
$$\mathcal{G}_3(\omega) = 6\omega(\omega - 1)(\omega + 1)^{-3},$$
$$\mathcal{G}_4(\omega) = -8\omega(\omega^2 - 4\omega + 1)(\omega + 1)^{-4},$$
$$\mathcal{G}_5(\omega) = 10\omega(\omega^3 - 11\omega^2 + 11\omega - 1)(\omega + 1)^{-5}$$

and so on (cf. Refs. [73, 121, 123, 125, 131] and the references cited therein).

The special case of (29) when $\omega = 1$ gives the generating function of the Genocchi numbers G_k, namely

$$G_k = \mathcal{G}_k(1), \qquad (30)$$

whose recurrence relation is given as

$$\begin{cases} G_0 = 0, \ G_1 = 1 \quad \text{(initial conditions)} \\ G_k = -\sum_{j=0}^{k} \binom{k}{j} G_j \quad \text{for } k \in \mathbb{N} \setminus \{1\} \end{cases}$$

by which the first few values of these numbers are given as follows:

$$G_0 = 0, \quad G_1 = 1, \quad G_2 = -1, \quad G_4 = 1, \quad G_6 = -3,$$
$$G_8 = 17, \quad G_{10} = -155$$

and so on; $G_{2k+1} = 0$ ($k \in \mathbb{N}$) (cf. Refs. [73, 121, 123, 125, 134] and the references cited therein).

The equations showing the relationships among the Apostol–Bernoulli numbers, the Apostol–Euler numbers and the Apostol–Genocchi numbers are as follows, for $k \in \mathbb{N}$:

$$\mathcal{B}_k(\omega) = -\frac{k\mathcal{E}_{k-1}(-\omega)}{2}, \tag{31}$$

$$\mathcal{E}_k(\omega) = \frac{\mathcal{G}_{k+1}(\omega)}{k+1}, \tag{32}$$

$$\mathcal{G}_k(\omega) = -2\mathcal{B}_k(-\omega) \tag{33}$$

(cf. Refs. [73, 121, 123, 125] and the references cited therein).

In addition, the following equation shows the relationships among the Bernoulli numbers of the first kind, the Euler numbers of the first kind and the Genocchi numbers, for $k \in \mathbb{N}_0$:

$$2\left(1 - 2^{2k}\right) B_{2k} = 2k E_{2k-1} = G_{2k} \tag{34}$$

(cf. Refs. [43, 126, 127] and the references cited therein).

The equation showing the relationship between the Apostol–Bernoulli numbers and the Fubini numbers is given as follows:

$$\mathcal{B}_k\left(\frac{1}{2}\right) = -2k w_g(k-1) \tag{35}$$

in which $w_g(k)$ denotes the Fubini numbers defined by the following generating function:

$$\frac{1}{2-e^t} = \sum_{k=0}^{\infty} w_g(k) \frac{t^k}{k!} \tag{36}$$

where $w_g(0) = 1$ and $|t| < \ln 2$ (cf. Refs. [32, 45] and the references cited therein).

It is well known that the Fubini numbers have a relationship with the Stirling numbers of the second kind as follows:

$$w_g(k) = \sum_{j=0}^{k} j! S(k,j) \tag{37}$$

(cf. Refs. [18, 45] and the references cited therein).

The equation showing the relationship between the Bernoulli numbers of the first kind and the tangent numbers is given as follows:

$$\mathcal{T}_{2k-1} = (-1)^{k-1} \frac{2^{2k}(2^{2k}-1)B_{2k}}{2k} \tag{38}$$

in which $\mathcal{T}_k$ denotes the tangent numbers defined by the following generating function:

$$\tan(t) = \sum_{k=1}^{\infty} \mathcal{T}_k \frac{t^{2k-1}}{(2k-1)!} \tag{39}$$

(cf. Ref. [90, p. 20, (11.2)–(11.3)]; see also Refs. [18, 100]).

The Bernoulli numbers of the second kind (also known as Cauchy numbers), b_n, are defined by

$$b_n = \int_0^1 (u)^{\underline{n}} \, du \tag{40}$$

whose exponential generating function is given by

$$F_C(t) = \frac{t}{\ln(t+1)} = \sum_{n=0}^{\infty} b_n \frac{t^n}{n!} \tag{41}$$

(cf. Refs. [56, 74], [95, p. 116]).

The combination of (16) and (41) yields

$$b_n = \sum_{j=0}^{n} \frac{s(n,j)}{j+1}, \qquad (42)$$

and by either of (40) and (42), the first few values of these numbers are computed as follows:

$$b_0 = 1, \quad b_1 = \frac{1}{2}, \quad b_2 = -\frac{1}{6}, \quad b_3 = \frac{1}{4}, \quad b_4 = -\frac{19}{30},$$

$$b_5 = \frac{9}{4}, \quad b_6 = -\frac{863}{84}$$

and so on (cf. Refs. [18, 74, 95, 98]).

Let $k \in \mathbb{N}$ and ω be a real or complex number. Then, the numbers $Y_m^{(k)}(\omega)$ are defined by the following generating function (cf. Ref. [65]):

$$\frac{2^k}{(\omega(1+\omega t) - 1)^k} = \sum_{n=0}^{\infty} Y_n^{(k)}(\omega) \frac{t^n}{n!} \qquad (43)$$

whose explicit formula is given by

$$Y_n^{(k)}(\omega) = (-1)^n \binom{n+k-1}{n} \frac{2^k n! \omega^{2n}}{(\omega-1)^{k+n}} \qquad (44)$$

(cf. Ref. [65]).

Setting $k = 1$ in (43) yields

$$Y_n(\omega) = Y_n^{(1)}(\omega),$$

whose explicit formula is given by

$$Y_n(\omega) = (-1)^n \frac{2n!}{\omega - 1} \left(\frac{\omega^2}{\omega - 1}\right)^n$$

(cf. Ref. [113]).

Note that the numbers $Y_n^{(k)}(\omega)$ and the numbers $Y_n(\omega)$ are associated with numerous number families, such as the Bernoulli-type numbers, the Euler-type numbers, the Stirling numbers and the Fibonacci numbers. Besides, these numbers are of many applications in some problems from probability and enumerative combinatorics (see, for details, Refs. [65, 113, 128]).

The Bernstein basis functions, $B_k^n(x)$, are of relationship with the Catalan numbers, and these functions are explicitly defined by

$$B_k^n(x) = \binom{n}{k} x^k (1-x)^{n-k}, \quad (k = 0, 1, \ldots, n; \; n \in \mathbb{N}_0), \qquad (45)$$

which is defined by means of the following exponential generating function:

$$\frac{(xt)^k e^{(1-x)t}}{k!} = \sum_{n=0}^{\infty} B_k^n(x) \frac{t^n}{n!} \qquad (46)$$

(cf. Refs. [2, 70, 102, 104, 107, 118] and also the references cited therein).

The integral formula for the Bernstein basis functions is given by

$$\int_0^1 B_k^n(x) \, \mathrm{d}x = \binom{n}{k} B(k+1, n-k+1) \qquad (47)$$

$$= \frac{1}{n+1} \qquad (48)$$

(cf. Refs. [33, 70, 102]).

2. Some Formulas for the Catalan Numbers

2.1. Some formulas for the Catalan numbers derived from the Euler's Gamma function and Beta function

By combining identities of the Bernstein basis function, $B_k^n(x)$ with the beta function $B(\alpha, \beta)$, Simsek [102] gave the following formulas for the Catalan numbers:

$$C_n = \frac{1}{2n+1} \sum_{k=n}^{2n} \binom{k}{n},$$

$$C_n = \frac{1}{(2n+1)!} \sum_{l=n}^{2n} \binom{2n}{l} \binom{l}{n} (3n-l)! \, (l-n)!,$$

$$C_n = \frac{1}{(n+1)^2} \sum_{l=0}^{n} \frac{\binom{2n}{l}\binom{2n-l}{n}}{\binom{n}{l}},$$

$$C_n = \frac{2n+1}{n+1} \sum_{j=n}^{2n} (-1)^{n+j} \binom{2n}{j} \frac{1}{j+1},$$

$$C_n = \sum_{l=n}^{2n} \sum_{j=0}^{l-n} (-1)^{l-n-j} \frac{(2n)!}{j!n!\,(l-n-j)!\,(2n-l)!\,(2n-j+1)}$$

and

$$\frac{1}{C_n} = (n+1)(2n+1) \sum_{l=0}^{n} (-1)^{n-l} \binom{n}{l} \frac{1}{2n-l+1}.$$

The Catalan numbers can be written in terms of Euler's gamma function as in the following forms:

$$C_n = \frac{4^{n+1}}{2\pi} \frac{\Gamma\left(n+\frac{1}{2}\right)\Gamma\left(\frac{3}{2}\right)}{\Gamma(n+2)} \tag{49}$$

$$= \frac{4^n \Gamma\left(n+\frac{1}{2}\right)}{\sqrt{\pi}\,\Gamma(n+2)} \tag{50}$$

(cf. Refs. [60, Equation (5.5)], [86, 96] and see also cited references therein).

2.2. Riemann integral representations for the Catalan numbers

An integral representation of the Catalan numbers is given by

$$C_n = \frac{1}{2\pi} \int_0^4 x^{n-\frac{1}{2}}(4-x)^{\frac{1}{2}} dx; \quad (n \in \mathbb{N} \cup \{0\}) \tag{51}$$

(cf. Ref. [96, Theorem 3.2]).

Other integral representations of the Catalan numbers are given by
$$C_n = \frac{4^{n+1}}{\pi} \int_0^1 u^{2n} \sqrt{1-u^2}\, du$$

(cf. Ref. [23]),
$$C_n = \frac{4^{n+1}}{2\pi} \int_0^1 t^{n-\frac{1}{2}} (1-t)^{\frac{1}{2}}\, dt$$

and
$$C_n = \frac{4^{n+1}}{\pi} \int_0^\infty \frac{u^2}{(u^2+1)^{n+2}}\, du$$

(cf. Ref. [36]).

For the reciprocals of the Catalan numbers, the integral representation is given by
$$\frac{1}{C_n} = \frac{(2n+3)(2n+2)(2n+1)}{2^{4n+4}} \int_0^2 x^{2n+1}(4-x^2)^{\frac{1}{2}}\, dx \qquad (52)$$

(cf. Ref. [22]).

For a variety of Riemann integral representations of the Catalan numbers, the reader may also refer to Ref. [84] and the references cited therein.

2.3. *p-adic integral representations for the Catalan numbers*

Let $\mathbb{Z}_p$ stand for the set of p-adic integers, and let $\mu_1(x)$ be the Haar distribution defined by
$$\mu_1(x) = \mu_1\left(x + p^N \mathbb{Z}_p\right) = \frac{1}{p^N}.$$

Then, the Volkenborn (p-adic bosonic) integral of $h(x)$, which is a uniformly differentiable function on $\mathbb{Z}_p$, is defined by
$$\int_{\mathbb{Z}_p} h(x)\, d\mu_1(x) = \lim_{M \to \infty} p^{-M} \sum_{j=0}^{p^M - 1} h(j) \qquad (53)$$

(cf. Ref. [97]; see also Refs. [51–53, 111]).

Let $\mu_{-1}(x) = (-1)^x$. The p-adic fermionic integral of $h(x)$, which is uniformly differentiable function on $\mathbb{Z}_p$, is defined by

$$\int_{\mathbb{Z}_p} h(x) \, d\mu_{-1}(x) = \lim_{M \to \infty} \sum_{l=0}^{p^M-1} (-1)^j h(j) \qquad (54)$$

(cf. Refs. [51, 52]).

The Volkenborn integral of the falling factorial $(x)^{\underline{n}}$ is given as

$$\int_{\mathbb{Z}_p} (x)^{\underline{n}} \, d\mu_1(x) = \frac{(-1)^n n!}{n+1} \qquad (55)$$

(cf. Ref. [97] and see also Refs. [46, 51–53, 111, 114]).

The fermionic p-adic integral of the falling factorial $(x)^{\underline{n}}$ is given as

$$\int_{\mathbb{Z}_p} (x)^{\underline{n}} \, d\mu_{-1}(x) = \frac{(-1)^n n!}{2^n} \qquad (56)$$

(cf. Ref. [47] and see also Refs. [52, 53, 111, 114]).

In Ref. [54, Theorem 3, p. 497], Kim gave the following p-adic integral representation of the Catalan numbers:

$$C_n = \frac{(-1)^n 2^{2n}}{n!} \int_{\mathbb{Z}_p} \left(\frac{x}{2}\right)^{\underline{n}} d\mu_{-1}(x) \qquad (57)$$

and also

$$\int_{\mathbb{Z}_p} \left(\frac{x}{2}\right)^{\underline{n}} d\mu_{-1}(x) = \sum_{k=0}^{n} 2^{-k} s(n, k) E_k. \qquad (58)$$

Thus, the combination of (57) and (58) implies

$$C_n = \frac{(-1)^n 2^{2n}}{n!} \sum_{k=0}^{n} 2^{-k} s(n, k) E_k \qquad (59)$$

(cf. Refs. [54, 59, 110]).

2.4. A formula covering the Catalan numbers, the Stirling numbers of the second kind and the Bernoulli numbers of negative order

Recently, Gun and Simsek [36] gave formulas and integral representations. For instance, Gun and Simsek [36] gave a relation among the Catalan numbers, the Stirling numbers of the second kind and the Bernoulli numbers of negative order as in the following form:

$$S(2n,n) = (n+1)C_n B_n^{(-n)}, \quad (n \in \mathbb{N}). \tag{60}$$

2.5. Inequalities and asymptotic expansions including the Catalan numbers

Some inequalities for the Catalan numbers are given as follows:

$$C_n \geq \frac{2^{2n-1} n^{-\frac{1}{2}}}{(n+1)} \tag{61}$$

(cf. Ref. [109, Corollary 6.3, p. 1154]), and also the Catalan numbers satisfy the following inequalities:

$$C_n \geq \frac{2^{2n}}{2n+1}$$

and

$$C_n \geq \frac{2^{2n}}{(n+1)^2}. \tag{62}$$

where $n \in \mathbb{N}_0$ (cf. Ref. [37]).

2.6. Approximation for the Catalan numbers

It is well known that Stirling's approximation for factorials is given by (cf. Refs. [60, 125])

$$n! \approx \sqrt{2\pi} n^{n+\frac{1}{2}} e^{-n}. \tag{63}$$

Using (63), Koshy [60, p. 110] gave the following approximation for the Catalan numbers:

$$C_n \approx \pi^{-\frac{1}{2}} 4^n n^{-\frac{3}{2}} \tag{64}$$

(cf. Ref. [60, p. 110] and see also Refs. [18, 27, 36, 125]).

3. Some Catalan-Type Numbers

The Catalan numbers and their generalizations are emerging in almost everywhere and every fields of natural sciences. Some associated numbers with the Catalan numbers are the Fuss–Catalan numbers, the Catalan–Qi numbers, Catalan–Daehee numbers, the Motzkin numbers, the little Schroder numbers (also known as the Schroder–Hipparchus numbers or the super-Catalan numbers), Narayana numbers, q-Catalan numbers, (q,t)-Catalan numbers and so on. In the next, we survey some of these numbers mentioned above.

Now, we start with the so-called Fuss–Catalan numbers.

Swiss mathematician Nikolai Fuss (1755–1826) and French and Belgian mathematician Eugène Charles Catalan (1814–1894) defined a family of numbers which are so-called two-parameter Fuss–Catalan numbers (or Raney numbers) as in following forms:

$$\mathcal{A}_n(p,r) = \frac{r}{np+r} \binom{np+r}{n} \tag{65}$$

$$= \frac{r}{n!} \prod_{j=1}^{n-1} (np+r-j) \tag{66}$$

$$= \frac{r\Gamma(np+r)}{\Gamma(n+1)\Gamma(n(p-1)+r+1)} \tag{67}$$

where $n \geq 0$, $p > 1$ and $r > 0$ (cf. Ref. [30]).

The ordinary generating function for the numbers $\mathcal{A}_n(p,r)$ is given by

$$G_\mathcal{A}(t;p,r) := \sum_{n=0}^{\infty} \mathcal{A}_n(p,r) t^n \tag{68}$$

which satisfies the following functional equation:

$$G_\mathcal{A}(t;p,r) = \left(1 + t\left(G_\mathcal{A}(t;p,r)\right)^{\frac{p}{r}}\right)^r \tag{69}$$

(cf. Refs. [34, 77, 132]).

By using densities of the Ranney distribution, Mlotkowski et al. [77] investigated some special cases of the functions $G_\mathcal{A}(t;p,r)$.

The order-p Fuss–Catalan numbers are defined by the following formula:

$$\mathcal{A}_{n-1}(p,p) = {}_p d_n = C_p(n) = \frac{1}{(p-1)n+1}\binom{pn}{n}, \quad (p, n \in \mathbb{N}) \tag{70}$$

which are related to some combinatorial problems, such as the Ballot problem, and some combinatorial objects, such as the Dyck paths and ternary trees. Especially, the numbers $C_p(n)$ enumerate p-ary trees, or alternatively p-Dyck paths (cf. Refs. [6, 42, 85] and see also the references cited therein).

For $p = 2$ and $r = 1$, the two-parameter Fuss–Catalan numbers reduce to the Catalan numbers, i.e.,

$$\mathcal{A}_n(2,1) = C_n; \quad (n \in \mathbb{N}) \tag{71}$$

(cf. Ref. [86] and see the references cited therein).

It is time to give another Catalan-type number which is so-called Catalan–Qi number:

For Re(a), Re(b) > 0, Re(z) ≥ 0, the Catalan–Qi function $C(a, b; z)$ is defined by

$$C(a, b; z) = \frac{\Gamma(b)}{\Gamma(a)}\left(\frac{b}{a}\right)^z \frac{\Gamma(z+a)}{\Gamma(z+b)} \tag{72}$$

which, in the special case when $z = n \in \mathbb{N} \cup \{0\}$, the Catalan–Qi function reduces to the Catalan–Qi numbers $C(a, b; n)$:

$$C(a, b; n) = \left(\frac{b}{a}\right)^n \frac{(a)_n}{(b)_n} \tag{73}$$

(cf. Ref. [86] and see also the references cited therein).

For $a = \frac{1}{2}$ and $b = 2$, the numbers $C(a, b; n)$ reduce to the Catalan numbers, i.e.,

$$C\left(\frac{1}{2}, 2; n\right) = C_n \tag{74}$$

(cf. Ref. [86] and see also the references cited therein).

Recall that the generalized hypergeometric function $_pF_p$ is defined by

$$_pF_q\begin{bmatrix}\alpha_1,\ldots,\alpha_p\\\beta_1,\ldots,\beta_q\end{bmatrix};z = {}_pF_q(\alpha_1,\alpha_2,\ldots,\alpha_p;\beta_1,\beta_2,\ldots,\beta_q;z)$$

$$= \sum_{k=0}^{\infty}\left(\frac{\prod_{j=1}^{p}(\alpha_j)_k}{\prod_{j=1}^{q}(\beta_j)_k}\right)\frac{z^k}{k!}.$$

The above series converges for all z if $p < q+1$, and for $|z| < 1$ if $p = q+1$. For this series, one can assumed that all parameters have general values, real or complex, except for the β_j, $j = 1, 2, \ldots, q$, none of which is equal to zero or to a negative integer. For instance, Gauss's hypergeometric function, which is a special case of the generalized hypergeometric function $_pF_q$ when $p = 2$ and $q = 1$, is given by the following series:

$$_2F_1(\alpha_1;\alpha_2;\beta_1;z) = \sum_{k=0}^{\infty}\frac{(\alpha_1)_k(\alpha_2)_k}{(\beta_1)_k}\frac{z^k}{k!},$$

which converges when $|z| < 1$ (cf. Refs. [3, 17, 40, 41, 76, 86, 88, 89, 91, 112, 125, 129, 130], and see the references cited therein).

The ordinary generating function for the numbers $C(a, b; n)$ is given by the following Gauss's hypergeometric function:

$$_2F_1\left(a,1;b;\frac{bt}{a}\right) = \sum_{n=0}^{\infty}C(a,b;n)t^n, \tag{75}$$

where $a, b > 0$ and $n \geq 0$ (cf. Ref. [83, Theorem 10, Equation (33)]).

A relation between the Fuss–Catalan numbers and the numbers $C(a, b; n)$ is given by

$$A_n(p,r) = r^n \frac{\prod_{k=0}^{p-1}C(\frac{k+r}{p},1;n)}{\prod_{k=0}^{p-2}C(\frac{k+r+1}{p-1},1;n)} \tag{76}$$

(cf. Refs. [85, 86]).

For more information on the Catalan–Qi function and the Catalan–Qi numbers, the reader may refer to Ref. [86] and the references cited therein.

We assume that a and b are positive integers. Then, by using Euler's gamma function with the case of $b = ka$, (72) reduces to

$$C(a, ka; z) = \frac{(ka-1)!}{(a-1)!} k^z \frac{(z)_a}{(z)_{ka}}. \tag{77}$$

Setting $k = 1$, we have

$$C(a, a; z) = 1. \tag{78}$$

By (77), we also have

$$(z)_{ka} C(a, ka; z) = \frac{(ka-1)!}{(a-1)!} k^z (z)_a. \tag{79}$$

By applying the Volkenborn (p-adic bosonic) integral to (79) on $\mathbb{Z}_p$, we get

$$\int_{\mathbb{Z}_p} (z)_{ka} C(a, ka; z) \mathrm{d}\mu_1(z) = \frac{(ka-1)!}{(a-1)!} \int_{\mathbb{Z}_p} k^z (z)_a \mathrm{d}\mu_1(z). \tag{80}$$

It is well known that

$$(z)_a = \sum_{j=0}^{a} |s(a,j)| z^j \tag{81}$$

(cf. Ref. [111]).

Thus, combining (79) with (81) yields

$$\int_{\mathbb{Z}_p} (z)_{ka} C(a, ka; z) \mathrm{d}\mu_1(z) = \frac{(ka-1)!}{(a-1)!} \sum_{j=0}^{a} |s(a,j)| \int_{\mathbb{Z}_p} k^z z^j \mathrm{d}\mu_1(z). \tag{82}$$

Since

$$\mathfrak{B}_n(\lambda) = \int_{\mathbb{Z}_p} \lambda^z z^n \mathrm{d}\mu_1(z), \quad (n \geq 0), \tag{83}$$

where $\mathfrak{B}_n(\lambda)$ denotes a family of Apostol-type Bernoulli numbers (see, for details, Refs. [55, 99, 111, 112]), combining the above equation with (82), we arrive at the following theorem:

Some Certain Families of Catalan-Type Numbers and Polynomials

Theorem 1. *Let a be positive integer. Then we have*

$$\int_{\mathbb{Z}_p} (z)_{ka} C(a, ka; z) \mathrm{d}\mu_1(z) = \frac{(ka-1)!}{(a-1)!} \sum_{j=0}^{a} |s(a,j)| \mathfrak{B}_j(k). \tag{84}$$

By applying the *p*-adic fermionic integral to (79) on $\mathbb{Z}_p$, we get

$$\int_{\mathbb{Z}_p} (z)_{ka} C(a, ka; z) \mathrm{d}\mu_{-1}(z) = \frac{(ka-1)!}{(a-1)!} \int_{\mathbb{Z}_p} k^z (z)_a \mathrm{d}\mu_{-1}(z). \tag{85}$$

Combination of (79) with (81) yields

$$\int_{\mathbb{Z}_p} (z)_{ka} C(a, ka; z) \mathrm{d}\mu_{-1}(z) = \frac{(ka-1)!}{(a-1)!} \sum_{j=0}^{a} |s(a,j)| \int_{\mathbb{Z}_p} k^z z^j \mathrm{d}\mu_{-1}(z). \tag{86}$$

Since

$$\mathcal{E}_n(\lambda) = \int_{\mathbb{Z}_p} \lambda^z z^n \mathrm{d}\mu_{-1}(z) \quad (n \geq 0), \tag{87}$$

where $\mathcal{E}_n(\lambda)$ denotes a family of Apostol-type Euler numbers (cf. Ref. [111, Equation (3.45)]), combining the above equation with (86), we arrive at the following theorem:

Theorem 2. *Let a be positive integer. Then we have*

$$\int_{\mathbb{Z}_p} (z)_{ka} C(a, ka; z) \mathrm{d}\mu_{-1}(z) = \frac{(ka-1)!}{(a-1)!} \sum_{j=0}^{a} |s(a,j)| \mathcal{E}_j(k). \tag{88}$$

Next, we give another Catalan-type number which is the so-called *w*-Catalan number.

As a generalization of Catalan numbers, Kim and Kim [50] introduced so-called *w*-Catalan numbers $C_{n,w}$ by the following ordinary generating functions:

$$\frac{2}{1 + (1-4t)^{\frac{w}{2}}} = \sum_{n=0}^{\infty} C_{n,w} t^n, \tag{89}$$

where *w* is any positive integer.

In the special case when $w = 1$, the numbers $C_{n,w}$ reduce to the usual Catalan numbers, i.e.,

$$C_{n,1} = C_n$$

(cf. Ref. [50]).

Next, we give another Catalan-type number which is the so-called Catalan–Daehee number.

In Ref. [58], Kim and Kim introduced the so-called Catalan–Daehee numbers of higher order, $d_n^{(k)}$, by the following ordinary generating functions:

$$\left(\frac{\frac{1}{2} \ln (1 - 4t)}{\sqrt{1 - 4t} - 1} \right)^k = \sum_{n=0}^{\infty} d_n^{(k)} t^n \quad (k \in \mathbb{N}) \tag{90}$$

which, for $k = 1$, yields the generating functions of the numbers $d_n = d_n^{(1)}$ called the Catalan–Daehee numbers (cf. Refs. [57, 58]).

In Ref. [26, Corollary 3, p. 5], Dolgy et al. gave the following p-adic integral representation of the Catalan–Daehee numbers:

$$d_n = \frac{(-1)^n \, 2^{2n}}{n!} \int_{\mathbb{Z}_p} \left(\frac{x}{2}\right)^n d\mu_1(x) \tag{91}$$

and also

$$\int_{\mathbb{Z}_p} \left(\frac{x}{2}\right)^n d\mu_1(x) = \sum_{k=0}^{n} 2^{-k} s(n,k) B_k. \tag{92}$$

Thus, the combination of (91) and (92) implies

$$d_n = \frac{(-1)^n \, 2^{2n}}{n!} \sum_{k=0}^{n} 2^{-k} s(n,k) B_k \tag{93}$$

(cf. Refs. [26, 59]).

In the sequel, we present another type of Catalan number which is the so-called w-Catalan–Daehee number.

In Ref. [59], Kim et al. introduced the w-Catalan–Daehee numbers, $d_{n,w}$, by the following ordinary generating function:

$$\frac{\frac{1}{2}\ln(1-4t)}{(1-4t)^{\frac{w}{2}}-1} = \sum_{n=0}^{\infty} d_{n,w} t^n \qquad (94)$$

such that the p-adic integral representation of the w-Catalan–Daehee numbers is given by

$$d_{n,w} = (-1)^n 2^{2n} \int_{\mathbb{Z}_p} \left(\frac{wx}{2}\right)^n d\mu_1(x) \qquad (95)$$

(cf. Ref. [59]).

4. Further Classes of Catalan-Type Numbers and Polynomials

Recently, Kucukoglu and Simsek [69] introduced another Catalan-type number and polynomial of higher order, denoted respectively by $V_n^{(k)}(\omega)$ and $V_n^{(k)}(z;\omega)$, by the following ordinary generating functions:

$$\mathcal{F}_V(t,k;\omega) = \left(\frac{1-\omega+\sqrt{(\omega-1)^2+8\omega^2 t}}{2\omega^2 t}\right)^k$$

$$= \sum_{n=0}^{\infty} V_n^{(k)}(\omega) t^n \qquad (96)$$

and

$$\mathcal{F}_V(t,x;k;\omega) = \mathcal{F}_V(t,k;\omega)(1+t)^{\frac{z}{2}}$$

$$= \sum_{n=0}^{\infty} V_n^{(k)}(z;\omega) t^n \qquad (97)$$

where $k \in \mathbb{N}_0$, $\omega \in \mathbb{R}$ such that $0 < \left|\frac{\omega^2 t}{(\omega-1)^2}\right| \leq \frac{1}{8}$.

A relation between the numbers $V_n^{(k)}(\omega)$ and the polynomials $V_n^{(k)}(z;\omega)$ is by

$$V_n^{(k)}(z;\omega) = \sum_{j=0}^{n} \frac{\left(\frac{z}{2}\right)^j}{j!} V_{n-j}^{(k)}(\omega), \qquad (98)$$

where $n \in \mathbb{N}_0$ (cf. Ref. [69]).

For $k = 1$, (96) and (97) yield

$$F_V(t,\omega) = \frac{1 - \omega + \sqrt{(\omega - 1)^2 + 8\omega^2 t}}{2\omega^2 t} = \sum_{n=0}^{\infty} V_n(\omega) t^n \qquad (99)$$

and

$$F_V(t,z;\omega) = F_V(t,\omega)(1+t)^{\frac{z}{2}} = \sum_{n=0}^{\infty} V_n(z;\omega) t^n, \qquad (100)$$

which mean

$$V_n(\omega) = V_n^{(1)}(\omega)$$

and

$$V_n(z;\omega) = V_n^{(1)}(z;\omega)$$

(cf. Ref. [66]).

It should be noted here that due to their following relation with the Catalan numbers, the numbers $V_n(\omega)$ and the polynomials $V_n(z;\omega)$ are called "Catalan-type":

$$V_n(\omega) = (-1)^n \frac{2^{n+1} \omega^{2n} C_n}{(\omega - 1)^{2n+1}}, \quad (n \in \mathbb{N}_0) \qquad (101)$$

(cf. Ref. [66]).

Thanks to equation (101), the Catalan numbers can also be expressed in terms of the numbers $V_n(\omega)$ as follows:

$$C_n = (-1)^n \frac{(\omega-1)^{2n+1} V_n(\omega)}{2^{n+1} \omega^{2n}}, \quad (n \in \mathbb{N}_0)$$

(cf. Ref. [66]).

We now present an investigation on the properties of the function $F_V(t,\omega)$ on the following set:

$$S = \left\{ t \in \mathbb{R} : |t| < \frac{(\omega-1)^2}{8\omega^2} \right\}.$$

For $t \in S$, we have the properties regarding the function $F_V(t,\omega)$ on the sets S and $\bar{S}$, respectively, as follows:

- $F_V(t,\omega)$ is analytic in S.
- $F_V(t,\omega)$ is continuous on $\bar{S}$.
- $F_V(t,\omega)$ is a generating function for the numbers $V_n(\omega)$ (see, for details, Ref. [66]).

Another feature provided by the function $F_V(t,\omega)$ is that it satisfies the following algebraic equation:

$$\omega^2 t \left(F_V(t,\omega)\right)^2 + (\omega-1) F_V(t,\omega) = 2, \tag{102}$$

which yields the following recurrence relation for the numbers $V_n(\omega)$:

$$V_n(\omega) = \frac{n\omega^2}{1-\omega} \sum_{j=0}^{n-1} V_j(\omega) V_{n-j-1}(\omega), \quad (n \in \mathbb{N}) \tag{103}$$

with the initial condition

$$V_0(\omega) = \frac{2}{\omega-1}$$

(cf. Ref. [66]).

By using the recurrence relation given by (103), a computation algorithm (Algorithm 1) was given by Kucukoglu et al. [66].

Algorithm 1 Let n be a non-negative integer and $\omega \in \mathbb{C}$. This algorithm will return the numbers $V_n(\omega)$, recursively (cf. Ref. [66]).

procedure V_CATALAN_TYPE_NUM(n: nonnegative integer, ω)
   Begin
   Local variable j : nonnegative integer
   if $n = 0$ then
      return $2/(\omega - 1)$
   else
      return $\big((n * \text{power}\,(\omega, 2))/(1 - \omega)\big) *$
          $\hookrightarrow \text{sum}\Big(\text{V\_CATALAN\_TYPE\_NUM}(j, \omega) *$
              $\hookrightarrow \text{V\_CATALAN\_TYPE\_NUM}\,(n - j - 1, \omega), j, 0, n - 1\Big)$
   end if
end procedure

Algorithm 1 consists of the procedure V_CATALAN_TYPE_NUM which returns the values of the numbers $V_n(\omega)$.

By implementing Algorithm 1 with the following commands in Wolfram Cloud (by Wolfram programming language), one can easily make the table (in Figure 2):

Unprotect[Power];
Power[0,0]=1;
Protect[Power];
VCatalanTypeNum[0, \[Omega]_] := 2 / (\[Omega] − 1);
VCatalanTypeNum[n_, \[Omega]_] := ((n ∗ (\[Omega]^2)) / (1 − \[Omega])) ∗ **Sum**[VCatalanTypeNum[j, \[Omega]] ∗ VCatalanTypeNum[n − j − 1, \[Omega]], {j, 0, n − 1}]
TableForm[**Evaluate**[**Table**[**Simplify**[VCatalanTypeNum[n, \ [Omega]]], {n, 0, 5}]],
TableHeadings−>{**Table** [**StringJoin**[**ToString**[**Subscript**[Style["V", Italic], k],**StandardForm**],**ToString**["(\[Omega])", **StandardForm**]],{k,0,5}]}]

For the algorithm which computes the values of the polynomials $V_n(z;\omega)$ (by the procedure entitled V_CATALAN_TYPE_POLY), the readers may refer to Ref. [66].

$V_0(\omega)$	$\frac{2}{-1+\omega}$
$V_1(\omega)$	$-\frac{4\omega^2}{(-1+\omega)^3}$
$V_2(\omega)$	$\frac{32\omega^4}{(-1+\omega)^5}$
$V_3(\omega)$	$-\frac{432\omega^6}{(-1+\omega)^7}$
$V_4(\omega)$	$\frac{7936\omega^8}{(-1+\omega)^9}$
$V_5(\omega)$	$-\frac{181120\omega^{10}}{(-1+\omega)^{11}}$

Figure 2. First few values of the Catalan-type numbers $V_n(\omega)$.

4.1. Further properties of the Catalan-type numbers $V_n(\omega)$

Here, we give further properties of the Catalan-type numbers $V_n(\omega)$.

In Ref. [67], Kucukoglu and Simsek also obtain the following formulas for the Catalan-type numbers $V_n(\omega)$, for $n \in \mathbb{N}_0$ and $\omega \neq 1$:

$$V_n(\omega) = (-1)^n \left[\binom{2n}{n} - \binom{2n}{n+1}\right] \frac{2^{n+1}\omega^{2n}}{(\omega-1)^{2n+1}}, \tag{104}$$

$$V_n(\omega) = \frac{(-1)^n 2^{n+1}}{(n+1)} \frac{\omega^{2n}}{(\omega-1)^{2n+1}} \sum_{k=0}^{n} \binom{n}{k}^2 \tag{105}$$

and

$$V_n(\omega) = \frac{(-1)^n 2^{n+1}}{(n+1)!} \frac{\omega^{2n}}{(\omega-1)^{2n+1}} \prod_{j=1}^{n}(4j-2) \tag{106}$$

(cf. Ref. [67]).

4.2. Asymptotic behaviour of the Catalan-type numbers $V_n(\omega)$

Here, we give asymptotic behavior of the Catalan-type numbers $V_n(\omega)$.

For $n \in \mathbb{N}_0$ and $\omega \neq 1$, the ratio of consecutive Catalan-type numbers $V_n(\omega)$ is given by

$$\frac{V_{n+1}(\omega)}{V_n(\omega)} = -\frac{8n+4}{n+2}\left(\frac{\omega}{\omega-1}\right)^2 \tag{107}$$

(cf. Ref. [67]).

When n goes to infinity, (107) reduces to the following known result:

$$\lim_{n \to \infty} \frac{V_{n+1}(\omega)}{V_n(\omega)} = -8\left(\frac{\omega}{\omega-1}\right)^2 \tag{108}$$

(cf. Ref. [67]).

The above two facts yield that we have the following approximation:

$$V_{n+1}(\omega) \approx -8\left(\frac{\omega}{\omega-1}\right)^2 V_n(\omega), \tag{109}$$

when n is sufficiently large and $\omega \neq 1$ (cf. Ref. [67]).

With the aid of Stirling's approximation for factorials, namely applying (63) and (64) to equation (101), Kucukoglu and Simsek [67] gave an approximation for the numbers $V_n(\omega)$ as in the following form:

$$V_n(\omega) \approx V_n^\star(\omega), \tag{110}$$

where n is sufficiently large and $\omega \neq 1$ such that

$$V_n^\star(\omega) = (-1)^n \pi^{-\frac{1}{2}} 2^{3n+1} n^{-\frac{3}{2}} \frac{\omega^{2n}}{(\omega-1)^{2n+1}}$$

(cf. Ref. [67]).

Due to (110), it can be inferred that the numbers $V_n(\omega)$ grow asymptotically as (110). That is, the quotient

$$\frac{V_n^\star(\omega)}{V_n(\omega)}$$

tends toward 1 as n goes to infinity (cf. Ref. [67]).

4.3. Derivative formulas for the Catalan-type polynomials $V_n(z;\omega)$

The formulas given in the previous sections show that the numbers $V_n(\omega)$ have close relationships with some kinds of special numbers and polynomials. Now, it is time to give some properties of the Catalan-type polynomials $V_n(z;\omega)$ including derivative formulas of them and their generating functions.

Taking k-times derivative of (100) with respect to z yields

$$\frac{\partial^k}{\partial z^k}\{F_V(t,z;\omega)\} = \frac{(\ln(t+1))^k}{2^k} F_V(t,z;\omega) \tag{111}$$

which, by the Cauchy product, implies the following theorem:

Theorem 3 (cf. Ref. [66]). Let $n, k \in \mathbb{N}$. Then we have

$$\frac{\partial^k}{\partial z^k}\{V_n(z;\omega)\} = \frac{k!}{2^k} \sum_{j=k}^{n} \frac{s(j,k)V_{n-j}(z;\omega)}{j!}. \tag{112}$$

In the special case when $k = 1$, (112) yields

$$\frac{\partial}{\partial z}\{V_n(z;\omega)\} = \sum_{j=1}^{n} \frac{s(j,1)V_{n-j}(z;\omega)}{2(j)!} \tag{113}$$

and

$$\frac{\partial}{\partial z}\{V_n(z;\omega)\} = \sum_{j=1}^{n} (-1)^{j+1} \frac{V_{n-j}(z;\omega)}{2j}. \tag{114}$$

For the above formulas, there are many applications given in Ref. [66].

4.4. Integral formulas for the Catalan-type numbers $V_n(\omega)$ and Catalan-type polynomials $V_n(z;\omega)$

Here, we give some integral formulas for the Catalan-type numbers $V_n(\omega)$ and Catalan-type polynomials $V_n(z;\omega)$.

The following three sections involve representations of Riemann integral, p-adic integral and also contour integral for the Catalan-type polynomials $V_n(z;w)$:

4.4.1. Formulas for Riemann integral of the polynomials $V_n(z;w)$

Riemann integral of the polynomials $V_n(z;w)$ is given by the following theorem, which was proved by Kucukoglu et al. [66]:

Theorem 4. *Let $n \in \mathbb{N}$. Then we have*

$$\int_0^1 V_{n-1}(z;w)\,dz = \sum_{j=0}^{n} \frac{2b_{n-j}}{(n-j)!}(V_j(1;w) - V_j(w)). \qquad (115)$$

Proof. We shall give just a brief sketch of the proof as the details are similar to those in Ref. [66]: Taking integral of both sides of (100) yields the following functional equation for the function $F_V(t,z;w)$:

$$\int_0^1 F_V(t,z;w)\,dz = \frac{2}{t} F_C(t)(F_V(t,1,w) - F_V(t,w)), \qquad (116)$$

which, by combining with (41) and (100) and also by applying the Cauchy product, yields (115). □

Integrating the case $k=1$ of equation (98) yields the following another Riemann integral of the polynomials $V_n(z;w)$:

$$\int_0^1 V_n(z;w)\,dz = \sum_{j=0}^{n} V_{n-j}(w)k_j, \qquad (117)$$

in which the numbers k_j stand for

$$k_j = \int_0^1 \binom{\frac{z}{2}}{j} dz, \qquad (118)$$

and the numbers k_j are so-called Bernoulli-type numbers of the second kind (or Cauchy-type numbers). By (118), one can easily compute first few values of the numbers k_j as follows:

$$k_0 = 1, \quad k_1 = \frac{1}{4}, \quad k_2 = -\frac{1}{3}, \quad k_3 = \frac{27}{16}$$

and so on (see, for details, Ref. [66]).

Applications of the Riemann integral to the polynomials $V_n(z;\omega)$ with multi-variables are given as follows:

Theorem 5 (cf. Ref. [67]). *Let $n \in \mathbb{N}_0$ and $\omega \neq 1$. Then we have*

$$\underbrace{\int_0^1 \int_0^1 \cdots \int_0^1}_{(v+1)\text{-times}} V_n\left(2\left(x + \sum_{k=1}^v y_k\right); \omega\right) dx\, dy_1 \ldots dy_v$$

$$= \sum_{j=0}^n \sum_{k_0+k_1+\cdots+k_v=j} \prod_{m=0}^v \frac{b_{k_m}}{k_m!} V_{n-j}(\omega). \tag{119}$$

Proof. We shall give just a brief sketch of the proof as the details are similar to those in Ref. [67]: Replacing z by $2(x + \sum_{k=1}^v y_k)$ in the case $k = 1$ of the equation (98), we have

$$V_n\left(2\left(x + \sum_{k=1}^v y_k\right); \omega\right) = \sum_{j=0}^n \frac{V_{n-j}(\omega)}{j!}\left(x + \sum_{k=1}^v y_k\right)^j. \tag{120}$$

By combining (14) with (120), we get

$$V_n\left(2\left(x + \sum_{k=1}^v y_k\right); \omega\right)$$

$$= \sum_{j=0}^n \sum_{k_0+k_1+\cdots+k_v=j} \binom{x}{k_0}\binom{y_1}{k_1}\cdots\binom{y_v}{k_v} V_{n-j}(\omega). \tag{121}$$

By applying the Riemann integral, $(v + 1)$-times, to (121), we get

$$\underbrace{\int_0^1 \int_0^1 \cdots \int_0^1}_{(v+1)\text{-times}} V_n\left(2\left(x + \sum_{k=1}^v y_k\right); \omega\right) dx\, dy_1 \ldots dy_v$$

$$= \sum_{j=0}^n \sum_{k_0+k_1+\cdots+k_v=j} \underbrace{\int_0^1 \int_0^1 \cdots \int_0^1}_{(v+1)\text{-times}} \binom{x}{k_0}\binom{y_1}{k_1}\cdots\binom{y_v}{k_v}$$

$$\times V_{n-j}(\omega)\, dx\, dy_1 \ldots dy_v.$$

Combining (40) with the above equation implies the assertion of Theorem 5. □

Example 1 (cf. Ref. [67]). For the purpose of presenting an example, setting $v = 1$ and $y_1 = y$ in (119) implies the following interesting formula:

$$\int_0^1 \int_0^1 V_n(2x + 2y; \omega) \, \mathrm{d}x \, \mathrm{d}y = \sum_{j=0}^n \sum_{k_0+k_1=j} \frac{b_{k_0} b_{k_1} V_{n-j}(\omega)}{k_0! k_1!}.$$

4.4.2. Formulas for p-adic integrals of the polynomials $V_n(z; \omega)$

Kucukoglu et al. [66] gave the Volkenborn and the fermionic p-adic integrals of the polynomials $V_n(z; \omega)$, respectively, by the following formulas:

$$\int_{\mathbb{Z}_p} V_n(z; \omega) \, \mathrm{d}\mu_1(z) = \sum_{j=0}^n \sum_{k=0}^j \frac{V_{n-j}(\omega) s(j, k) B_k}{j! 2^k} \quad (122)$$

and

$$\int_{\mathbb{Z}_p} V_n(z; \omega) \, \mathrm{d}\mu_{-1}(z) = \sum_{j=0}^n \sum_{k=0}^j \frac{V_{n-j}(\omega) s(j, k) E_k}{j! 2^k} \quad (123)$$

(see, for details, Ref. [66]).

Applications of the Volkenborn integral to the polynomials $V_n(z; \omega)$ with multi-variables are given as follows:

Theorem 6 (cf. Ref. [67]). Let $n \in \mathbb{N}_0$ and $\omega \neq 1$. Then we have

$$\underbrace{\int_{\mathbb{Z}_p} \int_{\mathbb{Z}_p} \cdots \int_{\mathbb{Z}_p}}_{(v+1)\text{-times}} V_n\left(2\left(x + \sum_{k=1}^v y_k\right); \omega\right) \mathrm{d}\mu_1(x) \, \mathrm{d}\mu_1(y_1) \ldots \mathrm{d}\mu_1(y_v)$$

$$= \sum_{j=0}^n \sum_{k_0+k_1+\cdots+k_v=j} \frac{(-1)^{\sum_{m=0}^v k_m}}{\prod_{m=0}^v (k_m + 1)} V_{n-j}(\omega). \quad (124)$$

Proof. We shall give just a brief sketch of the proof as the details are similar to those in Ref. [67]: By applying $(v+1)$-times the Volkenborn integral to (121), we get

$$\underbrace{\int_{\mathbb{Z}_p}\int_{\mathbb{Z}_p}\cdots\int_{\mathbb{Z}_p}}_{(v+1)\text{-times}} V_n\left(2\left(x+\sum_{k=1}^{v}y_k\right);\omega\right)d\mu_1(x)\,d\mu_1(y_1)\ldots d\mu_1(y_v)$$

$$=\sum_{j=0}^{n} V_{n-j}(\omega) \sum_{k_0+k_1+\cdots+k_v=j}$$

$$\times \underbrace{\int_{\mathbb{Z}_p}\int_{\mathbb{Z}_p}\cdots\int_{\mathbb{Z}_p}}_{(v+1)\text{-times}} \binom{x}{k_0}\binom{y_1}{k_1}\cdots\binom{y_v}{k_v} d\mu_1(x)\,d\mu_1(y_1)\ldots d\mu_1(y_v).$$

Combining (55) with the above equation implies the assertion of Theorem 6. □

Example 2 (cf. Ref. [67]). For the purpose of presenting an example, putting $v=1$ and $y_1=y$ in (124) implies the following interesting formula:

$$\int_{\mathbb{Z}_p}\int_{\mathbb{Z}_p} V_n(2x+2y;\omega)\,d\mu_1(x)\,d\mu_1(y)$$

$$=\sum_{j=0}^{n}\sum_{k_0+k_1=j}\frac{(-1)^{k_0+k_1} V_{n-j}(\omega)}{(k_0+1)(k_1+1)}.$$

Combining (101) with (124) implies the following corollary:

Corollary 1 (cf. Ref. [67]). Let $n \in \mathbb{N}_0$ and $\omega \neq 1$. Then we have

$$\underbrace{\int_{\mathbb{Z}_p}\int_{\mathbb{Z}_p}\cdots\int_{\mathbb{Z}_p}}_{(v+1)\text{-times}} V_n\left(2\left(x+\sum_{k=1}^{v}y_k\right);\omega\right)d\mu_1(x)\,d\mu_1(y_1)\ldots d\mu_1(y_v)$$

$$=\sum_{j=0}^{n}\sum_{k_0+k_1+\cdots+k_v=j}\frac{(-1)^{n-j+\sum_{m=0}^{v}k_m}\,2^{n-j+1}\omega^{2(n-j)}}{(\omega-1)^{2(n-j)+1}\prod_{m=0}^{v}(k_m+1)}C_{n-j}. \quad (125)$$

Theorem 6 can also be written in terms of the Daehee numbers D_n as follows (cf. Ref. [67]), for $\omega \neq 1$:

$$\underbrace{\int_{\mathbb{Z}_p} \int_{\mathbb{Z}_p} \cdots \int_{\mathbb{Z}_p}}_{(v+1)\text{-times}} V_n \left(2 \left(x + \sum_{k=1}^{v} y_k \right); \omega \right) d\mu_1(x) d\mu_1(y_1) \ldots d\mu_1(y_v)$$

$$= \sum_{j=0}^{n} \sum_{k_0+k_1+\cdots+k_v=j} \prod_{m=0}^{v} \frac{D_{k_m} V_{n-j}(\omega)}{k_m!}, \tag{126}$$

where

$$D_n = \frac{(-1)^n n!}{n+1}$$

(cf. Ref. [46]).

Applications of the fermionic integral to the polynomials $V_n(z; \omega)$ with multi-variables are given as follows:

Theorem 7 (cf. Ref. [67]). Let $n \in \mathbb{N}_0$ and $\omega \neq 1$. Then we have

$$\underbrace{\int_{\mathbb{Z}_p} \int_{\mathbb{Z}_p} \cdots \int_{\mathbb{Z}_p}}_{(v+1)\text{-times}} V_n \left(2 \left(x + \sum_{k=1}^{v} y_k \right); \omega \right) d\mu_{-1}(x) d\mu_{-1}(y_1) \ldots d\mu_{-1}(y_v)$$

$$= \sum_{j=0}^{n} \sum_{k_0+k_1+\cdots+k_v=j} \left(-\frac{1}{2} \right)^{\sum_{m=0}^{v} k_m} V_{n-j}(\omega). \tag{127}$$

Proof. We shall give just a brief sketch of the proof as the details are similar to those in Ref. [67]: By applying $(v+1)$-times the fermionic p-adic integral to (121), we get

$$\underbrace{\int_{\mathbb{Z}_p} \int_{\mathbb{Z}_p} \cdots \int_{\mathbb{Z}_p}}_{(v+1)\text{-times}} V_n \left(2 \left(x + \sum_{k=1}^{v} y_k \right); \omega \right) d\mu_{-1}(x) d\mu_{-1}(y_1) \ldots d\mu_{-1}(y_v)$$

$$= \sum_{j=0}^{n} V_{n-j}(\omega) \sum_{k_0+k_1+\cdots+k_v=j}$$

$$\times \underbrace{\int_{\mathbb{Z}_p} \int_{\mathbb{Z}_p} \cdots \int_{\mathbb{Z}_p}}_{(v+1)\text{-times}} \binom{x}{k_0} \binom{y_1}{k_1} \cdots \binom{y_v}{k_v} d\mu_{-1}(x) d\mu_{-1}(y_1) \ldots d\mu_{-1}(y_v).$$

Combining (56) with the above equation implies the assertion of Theorem 7. □

Example 3 (cf. Ref. [67]). To present an example, substituting $v = 1$ and $y_1 = y$ into (127) gives the following formula:

$$\int_{\mathbb{Z}_p} \int_{\mathbb{Z}_p} V_n(2x + 2y; \omega) \, d\mu_{-1}(x) \, d\mu_{-1}(y)$$

$$= \sum_{j=0}^{n} V_{n-j}(\omega) \sum_{k_0+k_1=j} \left(-\frac{1}{2}\right)^{k_0+k_1}.$$

Combination of (101) with (127) implies the following corollary:

Corollary 2 (cf. Ref. [67]). Let $n \in \mathbb{N}_0$ and $\omega \neq 1$. Then we have

$$\underbrace{\int_{\mathbb{Z}_p} \int_{\mathbb{Z}_p} \cdots \int_{\mathbb{Z}_p}}_{(v+1)\text{-times}} V_n\left(2\left(x + \sum_{k=1}^{v} y_k\right); \omega\right) d\mu_{-1}(x) \, d\mu_{-1}(y_1) \ldots d\mu_{-1}(y_v)$$

$$= \sum_{j=0}^{n} \sum_{k_0+k_1+\cdots+k_v=j} \frac{(-1)^{n-j+\sum_{m=0}^{v} k_m} \omega^{2(n-j)}}{2^{j-n-1+\sum_{m=0}^{v} k_m} (\omega - 1)^{2(n-j)+1}} C_{n-j}. \quad (128)$$

It is known from Ref. [47] that the so-called Changhee numbers Ch_n satisfy

$$\text{Ch}_n = \frac{(-1)^n n!}{2^n}. \quad (129)$$

Therefore, Theorem 7 can be written as

$$\underbrace{\int_{\mathbb{Z}_p} \int_{\mathbb{Z}_p} \cdots \int_{\mathbb{Z}_p}}_{(v+1)\text{-times}} V_n\left(2\left(x + \sum_{k=1}^{v} y_k\right); \omega\right) d\mu_{-1}(x) \, d\mu_{-1}(y_1) \ldots d\mu_{-1}(y_v)$$

$$= \sum_{j=0}^{n} \sum_{k_0+k_1+\cdots+k_v=j} \prod_{m=0}^{v} \frac{\text{Ch}_{k_m}}{k_m!} V_{n-j}(\omega) \quad (130)$$

(see, for details, Ref. [67]).

4.4.3. Contour integral formulas for the numbers $V_n(\omega)$

Thanks to the Cauchy residue theorem, it is well known that the binomial coefficients can be computed by following the Cauchy integral formula:

$$\int_\Omega \frac{(1+z)^k}{z^{j+1}} \mathrm{d}z = 2\pi i \binom{k}{j}, \qquad (131)$$

where $z \in \mathbb{C}$ and Ω is a simple closed contour surrounding the origin positive oriented (cf. Ref. [8]).

Replacing k by $2k$ and j by k in (131), and then combining the final equation with (1) and (101), implies a contour integral representation of the numbers $V_n(\omega)$ given by the following theorem:

Theorem 8 (cf. Ref. [66]). Let $k \in \mathbb{N}_0$. Then, we have

$$V_k(\omega) = \frac{i^{2k-1} (2\omega^2)^k}{\pi (\omega - 1)^{2k+1}} \int_\Omega \frac{(1+z)^{2k}}{z^{k+1}} \mathrm{d}z, \qquad (132)$$

where $z \in \mathbb{C}$ and Ω is a simple closed contour surrounding the origin positive oriented.

Applying contour integration to the following fraction:

$$\frac{C_n}{V_n(-\omega)} = (-1)^{n+1} \frac{(\omega+1)^{2n+1}}{2^{n+1}\omega^{2n}},$$

we obtain a contour integral representation for the reciprocal of the Catalan-type number as in the following theorem:

Theorem 9. Let $n \in \mathbb{N}_0$. Then we have

$$\int_\Omega \frac{\mathrm{d}\omega}{V_n(-\omega)} = \frac{(-1)^{n+1} \pi i}{2^n C_n} \binom{2n+1}{2n-1}, \qquad (133)$$

where $\omega \in \mathbb{C}$ and Ω is a simple closed contour surrounding the origin positive oriented.

4.5. Infinite series representations for the Catalan-type numbers

Here, information about new generating functions involving the Catalan-type numbers and how these numbers are represented by infinite series are given. More information can be consulted from the appropriate references provided.

An infinite series involving a quotient of the numbers $V_n(\omega)$ and the numbers C_n is given by the following theorem:

Theorem 10 (cf. Ref. [66]). *The following series holds true*

$$h(\omega) := \frac{2(\omega-1)}{3\omega^2 - 2\omega + 1} = \sum_{n=0}^{\infty} \frac{V_n(\omega)}{C_n}$$

where

$$\left| \frac{2\omega^2}{(\omega-1)^2} \right| < 1.$$

It is clear that the function $h(\omega)$ satisfies the following relation:

$$h(\omega) = 2(\omega-1)H(\omega; 2, -3; 1, 1, 1),$$

where

$$H(\omega; x, y; a, b, c) = \frac{1}{1 - x^a\omega - y^b\omega^{b+c}} \quad (a, b, c \in \mathbb{N}_0)$$

(cf. Ref. [66]). The function $H(\omega; x, y; a, b, c)$ is a generating function of the Fibonacci-type polynomials in two variables (for details, see Ref. [79]).

Setting $t = \frac{(\omega-1)^2}{8\omega^2}$ in (99) with $\omega > 1$ implies an infinite series containing the numbers $V_n(\omega)$ by the following theorem:

Theorem 11 (cf. Ref. [67]). *Let $\omega > 1$. Then we have*

$$\sum_{n=0}^{\infty} \frac{V_n(\omega)}{2^n} \left(\frac{\omega-1}{2\omega} \right)^{2n} = \frac{4(\sqrt{2}-1)}{\omega-1}. \tag{134}$$

The combination of (101) with (134) yields the following example:

Example 4 (cf. Ref. [67]).

$$\sum_{n=0}^{\infty}(-1)^n \frac{C_n}{2^{2n-1}} = 4\left(\sqrt{2}-1\right). \tag{135}$$

4.6. *Some inequalities for the Catalan-type numbers*

Here, we present some inequalities for the Catalan-type numbers.

By combining (1), (61) and (101) with the following well-known inequality

$$\binom{2n}{n} \leq 2^{2n} \tag{136}$$

(cf. Ref. [8]), we get upper and lower bounds for the Catalan-type numbers as in the following two theorems:

Theorem 12 (cf. Ref. [67]). *Let $n \in \mathbb{N}_0$ and $\omega \neq 1$. Then we have*

$$V_n(\omega) \leq \frac{(-1)^n 2^{3n+1}\omega^{2n}}{(n+1)(\omega-1)^{2n+1}}. \tag{137}$$

Theorem 13 (cf. Ref. [67]). *Let $n \in \mathbb{N}$ and $\omega \neq 1$. Then we have*

$$V_n(\omega) \geq \frac{(-1)^n 2^{3n}\omega^{2n}}{(n+1)\sqrt{n}(\omega-1)^{2n+1}}. \tag{138}$$

4.7. *Relations of the Catalan numbers with other special numbers*

By applying generating function and functional equations, here we give some relations and formulas involving the numbers $V_n(\omega)$, the numbers $Y_n^{(k)}(\omega)$, the numbers B_n, the polynomials $B_n(x)$, the numbers $s(n,k)$, the numbers $S(n,k)$ and the numbers C_n.

Theorem 14 (cf. Ref. [66]). *Let $n \in \mathbb{N}_0$. Then we have*

$$Y_n^{(n+1)}(\omega) = (n+1)!V_n(\omega). \tag{139}$$

Theorem 15 (cf. Ref. [66]). Let $n \in \mathbb{N}$. Then we have

$$\sum_{j=0}^{n-1} j! \left(2^{-3}\omega^{-2}(1-\omega)^2\right)^j V_j(\omega) S(n-1,j) = \frac{4\left(B_n\left(\frac{1}{2}\right) - B_n\right)}{(\omega-1)n}. \tag{140}$$

By using the following well-known identity (cf. Refs. [25, 125])

$$B_n\left(\frac{1}{2}\right) = \left(2^{1-n} - 1\right) B_n$$

with (140), we get the following theorem:

Theorem 16 (cf. Ref. [66]). Let $n \in \mathbb{N}$. Then we have

$$B_n = \frac{n}{2^3 - 2^{3-n}} \sum_{j=0}^{n-1} \frac{(1-\omega)^{2j+1} j! V_j(\omega) S(n-1,j)}{(8\omega^2)^j}. \tag{141}$$

The combination of (141) with (34) implies the following two formulas, for $n \in \mathbb{N}$:

$$E_{2n-1} = -2^{2n-2} \sum_{j=0}^{n-1} \frac{(1-\omega)^{2j+1} j! V_j(\omega) S(n-1,j)}{(8\omega^2)^j} \tag{142}$$

and

$$G_{2n} = -n 2^{2n-1} \sum_{j=0}^{n-1} \frac{(1-\omega)^{2j+1} j! V_j(\omega) S(n-1,j)}{(8\omega^2)^j}. \tag{143}$$

The combination of (38) with the case $n = 2k$ (141), we get a formula for the tangent numbers in terms of the Stirling numbers of the second kind and the Catalan-type numbers, given as in the following corollary:

Corollary 3. Let $k \in \mathbb{N}$. Then, we have

$$T_{2k-1} = (-1)^{k-1} 2^{4k-3} \sum_{j=0}^{2k-1} \frac{(1-\omega)^{2j+1} j! V_j(\omega) S(2k-1,j)}{(8\omega^2)^j}. \tag{144}$$

Theorem 17 (cf. Ref. [67]). Let $n \in \mathbb{N}$ and $\omega \neq 1$. With the initial condition

$$V_0(\omega) = \frac{2}{\omega - 1},$$

we have

$$V_n(\omega) = -\sum_{j=0}^{n} \frac{\left(\frac{1}{2}\right)_j 8^j}{j!} \left(\frac{\omega}{\omega - 1}\right)^{2j} V_{n-j}(\omega). \tag{145}$$

Example 5 (cf. Ref. [67]). Combining the following well-known relation

$$\frac{\left(\frac{1}{2}\right)_n}{n!} = \frac{(-1)^{n+1}}{2^{2n}(2n-1)} \binom{2n}{n} \quad (n \in \mathbb{N}_0)$$

(cf. Ref. [60]), with equation (145), we have

$$V_n(\omega) = \sum_{j=0}^{n} (-1)^j \frac{2^j}{2j-1} \binom{2j}{j} \left(\frac{\omega}{\omega - 1}\right)^{2j} V_{n-j}(\omega).$$

By (1), we also have

$$V_n(\omega) = \sum_{j=0}^{n} (-1)^j \frac{2^j(j+1)}{2j-1} \left(\frac{\omega}{\omega - 1}\right)^{2j} C_j V_{n-j}(\omega), \tag{146}$$

where $n \in \mathbb{N}$ and $\omega \neq 1$.

Corollary 4 (cf. Ref. [67]). Let $n \in \mathbb{N}_0$ and $\omega \neq 1$. Then we have

$$Y_n(\omega) = \frac{(n+1)!\,(n!)^2\,(\omega - 1)^n}{(2n)!\,2^n} V_n(\omega). \tag{147}$$

The combination of (1) with (147) implies the following identity:

$$Y_n(\omega) = n! \left(\frac{\omega - 1}{2}\right)^n \frac{V_n(\omega)}{C_n} \tag{148}$$

(cf. Ref. [67]).

Theorem 18 (cf. Ref. [67]). Let $m \in \mathbb{N}$ and $w \neq 0, 1$. Then we have

$$\mathcal{B}_m(w) = \frac{m}{2} \sum_{n=0}^{m-1} n! \left(\frac{w-1}{2w}\right)^n \frac{S(m-1,n) V_n(w)}{C_n}. \tag{149}$$

Setting $w = \frac{1}{2}$ in (18) gives

$$\mathcal{B}_m\left(\frac{1}{2}\right) = \frac{m}{2} \sum_{n=0}^{m-1} (-1)^n \frac{n! S(m-1,n) V_n\left(\frac{1}{2}\right)}{2^n C_n}. \tag{150}$$

Combining the above equation with (35), we get a formula for the Fubini numbers in terms of the Stirling numbers of the second kind and the Catalan-type numbers, given as in the following corollary:

Corollary 5. Let $m \in \mathbb{N}$. Then we have

$$w_g(m) = \sum_{n=0}^{m} (-1)^{n+1} \frac{n! S(m,n) V_n\left(\frac{1}{2}\right)}{2^{n+2} C_n}. \tag{151}$$

Remark 2. By using the case $w = \frac{1}{2}$ of (101) in (151), the identity, given by (37), can also be obtained.

The combination of (31) and (33) with (149) gives the following two corollaries:

Corollary 6 (cf. Ref. [67]). Let $m \in \mathbb{N}_0$ and $w \neq 0, -1$. Then we have

$$\mathcal{E}_m(w) = -\sum_{n=0}^{m} n! \left(\frac{w+1}{2w}\right)^n \frac{S(m,n) V_n(-w)}{C_n}. \tag{152}$$

Corollary 7 (cf. Ref. [67]). Let $m \in \mathbb{N}$ and $w \neq 0, -1$. Then we have

$$\mathcal{G}_m(w) = -m \sum_{n=0}^{m-1} n! \left(\frac{w+1}{2w}\right)^n \frac{S(m-1,n) V_n(-w)}{C_n}. \tag{153}$$

Theorem 19 (cf. Ref. [67]). Let $m \in \mathbb{N}_0$ and $\omega \neq -1$. Then we have

$$V_m(-\omega) = -\frac{C_m}{m!} \left(\frac{2\omega}{\omega+1}\right)^m \sum_{n=0}^{m} \mathcal{E}_n(\omega) s(m,n). \tag{154}$$

The combination of (31) and (33) with (154) gives the following two corollaries:

Corollary 8 (cf. Ref. [67]). Let $m \in \mathbb{N}_0$ and $\omega \neq 1$. Then we have

$$V_m(\omega) = \frac{2C_m}{m!} \left(\frac{2\omega}{\omega-1}\right)^m \sum_{n=0}^{m} \frac{\mathcal{B}_{n+1}(\omega) s(m,n)}{n+1}. \tag{155}$$

Corollary 9 (cf. Ref. [67]). Let $m \in \mathbb{N}_0$ and $\omega \neq 1$. Then we have

$$V_m(\omega) = -\frac{C_m}{m!} \left(\frac{2\omega}{\omega-1}\right)^m \sum_{n=0}^{m} \frac{\mathcal{G}_{n+1}(-\omega) s(m,n)}{n+1}. \tag{156}$$

Example 6 (cf. Ref. [67]). Setting $\omega = 1$ in (154), we have

$$V_m(-1) = -\frac{C_m}{m!} \sum_{n=0}^{m} \mathcal{E}_n(1) s(m,n). \tag{157}$$

Combining the above equation with

$$Y_m(-1) = (-1)^{m+1} \sum_{n=0}^{m} \mathcal{E}_n(1) s(m,n) \tag{158}$$

(cf. Ref. [128, Remark 1]), we have

$$V_m(-1) = (-1)^m \frac{C_m}{m!} Y_m(-1). \tag{159}$$

On the other hand, it is known that

$$Y_m(-1) = (-1)^{m+1} \mathrm{Ch}_m, \tag{160}$$

where Ch_m denotes the Changhee numbers (cf. Ref. [128, Equation (31)]). Thus, using the above equality in (159) yields

$$V_m(-1) = -\frac{C_m}{m!}\text{Ch}_m. \qquad (161)$$

References

[1] M. Abramowitz and I. A. Stegun, *Handbook of Mathematical Functions with Formulas, Graphs and Mathematical Tables* (Dover Publication, New York, 1970).

[2] M. Acikgoz and S. Araci, On generating function of the Bernstein polynomials, *AIP Conf. Proc.* **1281**(1), 1141 (2010). doi: 10.1063/1.3497855.

[3] M. Ali, M. Ghayasuddin and T. K. Pogany, Integrals with two-variable generating function in the integrand, *Montes Taurus J. Pure Appl. Math.* **3**(3), 95–103 (2021). Article ID: MTJPAM-D-20-00048.

[4] T. M. Apostol, On the Lerch zeta function, *Pac. J. Math.* **1**, 161–167 (1951).

[5] E. Artin, *The Gamma Function* (Holt, Rinehart & Winston, Inc., New York, 1964).

[6] J.-C. Aval, Multivariate Fuss–Catalan numbers, *Discrete Math.* **308**, 4660–4669 (2008).

[7] I. Bajunaid, J. M. Cohen, F. Colonna and D. Singman, Function series, Catalan numbers, and random walks on trees, *Am. Math. Mon.* **112**, 765–785 (2005).

[8] J. Bak and D. J. Newman, *Complex Analysis*, 3rd edn. (Springer, New York, 2010).

[9] H. Bateman and A. Erdélyi, *Higher Transcendental Functions, Volumes I–III: Bateman Manuscript Project* (McGraw-Hill Book Company, New York, 1953).

[10] R. J. Betts, A uniform convergent series for $\zeta(s)$ and closed formulas that including Catalan numbers (2010). Preprint arXiv:1008.0387v3.

[11] M. Bona, *Introduction to enumerative combinatorics* (The McGraw-Hill Companies Inc., New York, 2007).

[12] K. N. Boyadzhiev, Apostol-Bernoulli functions, derivative polynomials and Eulerian polynomials (2007). Preprint arXiv:0710.1124.

[13] L. Carlitz, Some theorems on Bernoulli numbers of higher order, *Pac. J. Math.* **2**(2), 127–139 (1952).

[14] L. Carlitz, D. P. Roselle and R. Scoville, Permutations and sequences with repetitions by number of increase, *J. Combin. Theory* **1**, 350–374 (1966).

[15] C. A. Charalambides, *Enumerative Combinatorics* (Chapman and Hall/CRC Press Company, London, 2002).
[16] J. Choi, Explicit formulas for Bernoulli polynomials of order n, *Indian J. Pure Appl. Math.* **27**, 667–674 (1996).
[17] J. Choi and A. K. Rathie, General summation formulas for the Kampe de Feriet function, *Montes Taurus J. Pure Appl. Math.* **1**(1), 107–128 (2019). Article ID: MTJPAM-D-19-00004.
[18] L. Comtet, *Advanced Combinatorics* (D. Reidel Publication Company, Dordrecht, 1974).
[19] J. H. Conway and R. K. Guy, *In the Book of Numbers* (Springer-Verlag, New York, 1996).
[20] G. E. Cossali, A common generating function for Catalan numbers and other integer sequences, *J. Integer Seq.* **6** (2003). Article ID: 03.1.8.
[21] C. Cruz, M. I. Falcao and H. R. Malonek, Monogenic pseudo-complex power functions and their applications, *Math. Methods Appl. Sci.* **37**(12), 1723–1735 (2014).
[22] T. Dana-Picard, Parametric integrals and Catalan numbers, *Int. J. Math. Educ. Sci. Tech.* **36**, 410–414 (2005).
[23] T. Dana-Picard, Integral presentations of Catalan numbers, *Int. J. Math. Educ. Sci. Tech.* **41**(1), 63–69 (2009).
[24] T. Dana-Picard, Integral presentations of Catalan numbers and Wallis formula, *Int. J. Math. Educ. Sci. Tech.* **42**(1), 122–129 (2011).
[25] G. B. Djordjevic and G. V. Milovanović, *Special Classes of Polynomials* (University of Nis, Faculty of Technology, Leskovac, 2014).
[26] D. V. Dolgy, G.-W. Jang, D. S. Kim and T. Kim, Explicit expressions for Catalan-Daehee numbers, *Proc. Jangjeon Math. Soc.* **20**(1), 1–9 (2017). doi: 10.23001/pjms2017.20.1.1.
[27] N. Elezovic, Asymptotic expansions of gamma and related functions, binomial coefficients, inequalities and means, *J. Math. Inequ.* **9**(4), 1001–1054 (2015).
[28] P. Flajolet and R. Sedgewick, *Analytic Combinatorics* (Cambridge University Press, Cambridge, 2009).
[29] P. Flajolet and A. Odlyzko, Singularity analysis of generating functions, *SIAM J. Discrete Math.* **3**(2), 216–240 (1990).
[30] N. I. Fuss, Solutio quaestionis, quot modis polygonum n laterum in polygona m laterum, per diagonales resolvi queat, *Nova Acta Acad. Sci. Petropolitanae* **9**, 243–251 (1791).
[31] J. W. L. Glaisher, On the residues of the sums of products of the first $p-1$ numbers, and their powers, to modulus $p-2$ or $p-3$, *Q. J. Pure Appl. Math.* **31**, 321–353 (1900).

[32] I. J. Good, The number of ordering of n candidates when ties are permitted, *Fibonacci Q.* **13**, 11–18 (1975).
[33] R. Goldman, *Pyramid Algorithms: A Dynamic Programming Approach to Curves and Surfaces for Geometric Modeling* (Morgan Kaufmann Publishers, R. Academic Press, San Diego, 2002).
[34] R. L. Graham, D. E. Knuth and O. Patashnik, *Concrete Mathematics: A Foundation for Computer Science*, 2nd edn. (Addison-Wesley, Reading, Massachussets, 1994).
[35] R. P. Grimaldi, *Fibonacci and Catalan numbers and Introduction* (John Wiley & Sons, Inc., Hoboken, 2012).
[36] D. Gun and Y. Simsek, Some new identities and inequalities for Bernoulli polynomials and numbers of higher order related to the Stirling and Catalan numbers, *RACSAM Rev. R. Acad. A.* **114**(167), 1–12 (2020).
[37] D. Gun and Y. Simsek, Combinatorial sums involving Stirling, Fubini, Bernoulli numbers and approximate values of Catalan numbers, *Adv. Stud. Contemp. Math. (Kyungshang)* **30**(4), 503–513 (2020).
[38] D. Gun and Y. Simsek, Inequalities and formulas associated with Bernoulli numbers of higher order and special numbers, 2849, 060005 (2023). https://doi.org/10.1063/5.0162294.
[39] D. Gun and Y. Simsek, Some series representations for special numbers involving Fibonacci polynomials and combinatorial numbers, *AIP Conf. Proc.* 2849, 060006 (2023). https://doi.org/10.1063/5.0162295.
[40] A. Hasanov and T. G. Ergashev, New decomposition formulas associated with the Lauricella multivariable hypergeometric functions, *Montes Taurus J. Pure Appl. Math.* **3**(3), 317–326 (2021). Article ID: MTJPAM-D-20-00049.
[41] A. Hasanov, A. Ryskan and J. Choi, Decomposition formulas for second-order quadruple Gaussian hypergeometric series by means of operators $H(\alpha, \beta)$ and $\overline{H}(\alpha, \beta)$, *Montes Taurus J. Pure Appl. Math.* **4**(3), 41–60 (2022). Article ID: MTJPAM-D-21-00019.
[42] P. Hilton and J. Pedersen, Catalan numbers, their generalization, and their uses, *Math. Intell.* **13**(2), 64–75 (1991). doi: 10.1007/BF03024089.
[43] A. F. Horadam, Genocchi polynomials. In G. E. Bergum, A. N. Philippou, A. F. Horadam (ed.) *Applications of Fibonacci Numbers*, Volume 4 (Kluwer Academic Publishers Group, Dordrecht, 1991), pp. 145–166.
[44] C. Jordan, *Calculus of Finite Differences*, 2nd edn. (Chelsea Publishing Company, New York, 1950).

[45] N. Kilar and Y. Simsek, A new family of Fubini type numbers and polynomials associated with Apostol-Bernoulli numbers and polynomials, *J. Korean Math. Soc.* **54**(5), 1605–1621 (2017).

[46] D. S. Kim and T. Kim, Daehee numbers and polynomials, *Appl. Math. Sci. (Ruse)* **7**(120), 5969–5976 (2013). doi: 10.12988/ams.2013.39535

[47] D. S. Kim, T. Kim and J. Seo, A note on Changhee numbers and polynomials, *Adv. Stud. Theor. Phys.* **7**, 993–1003 (2013).

[48] D. S. Kim, T. Kim and H. I. Kwon, Identities of some special mixed-type polynomials (2014). *Preprint* arXiv:1406.2124v1.

[49] D. S. Kim and T. Kim, A new approach to Catalan numbers using differential equations, *Russ. J. Math. Phys.* **24**(4), 465–475 (2017).

[50] D. S. Kim and T. Kim, Triple symmetric identities for w-Catalan polynomials, *J. Korean Math. Soc.* **54**(4), 1243–1264 (2017).

[51] T. Kim, q-Volkenborn integration, *Russ. J. Math. Phys.* 19, 288–299 (2002).

[52] T. Kim, q-Euler numbers and polynomials associated with p-adic q-integral and basic q-zeta function, *Trends Int. Math. Sci. Study* **9**, 7–12 (2006).

[53] T. Kim, An invariant p-adic q-integral on $\mathbb{Z}_p$, *Appl. Math. Lett.* **21**, 105–108 (2008).

[54] T. Kim, A note on Catalan numbers associated with p-adic integral on $\mathbb{Z}_p$, *Proc. Jangjeon Math. Soc.* **19**(3), 493–501 (2016).

[55] T. Kim, S. H. Rim, Y. Simsek and D. Kim, On the analogs of Bernoulli and Euler numbers, related identities and zeta and l-functions, *J. Korean Math. Soc.* **45**(2), 435–453 (2008).

[56] T. Kim, D. S. Kim, D. V. Dolgy and J.-J. Seo, Bernoulli polynomials of the second kind and their identities arising from Umbral calculus, *J. Nonlinear Sci. Appl.* **9**, 860–869 (2016). doi: 10.22436/jnsa.009.03.14.

[57] T. Kim and D. S. Kim, Some identities of Catalan-Daehee polynomials arising from Umbral calculus, *Appl. Comput. Math.* **16**(2), 177–189 (2017).

[58] T. Kim and D. S. Kim, Differential equations associated with Catalan-Daehee numbers and their applications, *RACSAM Rev. R. Acad. A* **111**, 1071–1081 (2017).

[59] T. Kim, S.-H. Rim, D. V. Dolgy and S.-S. Pyo, Explicit expression for symmetric identities of w-Catalan-Daehee polynomials, *Notes Number Theory Discrete Math.* **24**(4), 99–111 (2018). doi: 10.7546/nntdm.2018.24.4.99-111.

[60] T. Koshy, *Catalan Numbers with Applications* (Oxford University Press, Oxford, 2009).

[61] I. Kucukoglu and Y. Simsek, Combinatorial identities associated with new families of the numbers and polynomials and their approximation values (2017). *Preprint* arXiv:1711.00850.

[62] I. Kucukoglu, Implementation of computation formulas for certain classes of Apostol-type polynomials and some properties associated with these polynomials, *Commun. Fac. Sci. Univ. Ank. Ser. A1 Math. Stat.* **70**(1), 426–442 (2021).

[63] I. Kucukoglu and Y. Simsek, On a family of special numbers and polynomials associated with Apostol-type numbers and poynomials and combinatorial numbers, *Appl. Anal. Discrete Math.* **13**, 478–494 (2019).

[64] I. Kucukoglu, Y. Simsek and H. M. Srivastava, A new family of Lerch-type zeta functions interpolating a certain class of higher-order Apostol-type numbers and Apostol-type polynomials, *Quaest. Math.* **42**(4), 465–478 (2019).

[65] I. Kucukoglu, B. Simsek and Y. Simsek, An approach to negative hypergeometric distribution by generating function for special numbers and polynomials, *Turk. J. Math.* **43**, 2337–2353 (2019).

[66] I. Kucukoglu, B. Simsek and Y. Simsek, New classes of Catalan-type numbers and polynomials with their applications related to p-adic integrals and computational algorithms, *Turk. J. Math.* **44**, 2337–2355 (2020).

[67] I. Kucukoglu and Y. Simsek, Computational identities for extensions of some families of special numbers and polynomials, *Turk. J. Math.* **45**(5), 2341–2365 (2021).

[68] I. Kucukoglu and Y. Simsek, New formulas and numbers arising from analyzing combinatorial numbers and polynomials, *Montes Taurus J. Pure Appl. Math.* **3**(3), 238–259 (2021). Article ID: MTJPAM-D-20-00038.

[69] I. Kucukoglu and Y. Simsek, Formulas and combinatorial identities for Catalan-type numbers and polynomials: Their analysis with computational algorithms, *Appl. Comput. Math.* **21**(2), 158–177 (2022).

[70] G. G. Lorentz, *Bernstein Polynomials* (Chelsea Publishing Company, New York, 1986).

[71] Q.-M. Luo and H. M. Srivastava, Some generalizations of the Apostol-Bernoulli and Apostol-Euler polynomials, *J. Math. Anal. Appl.* **308**, 290–302 (2005).

[72] Q.-M. Luo, Apostol-Euler polynomials of higher order and Gaussian hypergeometric functions, *Taiwan. J. Math.* **10**, 917–925 (2006).

[73] Q.-M. Luo and H. M. Srivastava, Some generalizations of the Apostol-Genocchi polynomials and the Stirling numbers of the second kind, *Appl. Math. Comput.* **217**, 5702–5728 (2011).

[74] D. Merlini, R. Sprugnoli and M. C. Verri, The Cauchy numbers, *Discrete Math.* **306**(16), 1906–1920 (2006).
[75] L. M. Milne-Thomson, *Calculus of Finite Differences* (Macmillan and Co. Ltd., London, 1933).
[76] G. V. Milovanovic and A. K. Rathie, Four unified results for reducibility of Srivastava's triple hypergeometric series H_B, *Montes Taurus J. Pure Appl. Math.* **3**(3), 155–164 (2021). Article ID: MTJPAM-D-20-00062.
[77] W. Mlotkowski, A. K. Penson and K. Zyczkowski, Densities of the Raney distributions, *Doc. Math.* **18**, 1593–1596 (2013). arXiv:1211.7259.
[78] M. A. Özarslan, Unified Apostol–Bernoulli, Euler and Genocchi polynomials, *Comput. Math. Appl.* **62**, 2452–2462 (2011).
[79] G. Ozdemir, Y. Simsek and G. V. Milovanović, Generating functions for special polynomials and numbers including Apostol-type and Humbert-type polynomials, *Mediterr. J. Math.* **14**, 1–17 (2017). Article ID: 117.
[80] H. Ozden, Unification of generating function of the Bernoulli, Euler and Genocchi numbers and polynomials, *AIP Conf. Proc.* **1281**, 1125 (2010). https://doi.org/10.1063/1.3497848.
[81] H. Ozden, Y. Simsek and H. M. Srivastava, A unified presentation of the generating functions of the generalized Bernoulli, Euler and Genocchi polynomials, *Comput. Math. Appl.* **60**, 2779–2787 (2010).
[82] F. Qi, Explicit formulas for computing Bernoulli numbers of the second kind and Stirling numbers of the first kind, *Filomat* **28**, 319–327 (2014).
[83] F. Qi, M. Mahmoud, X.-T. Shi and F.-F. Liu, Some properties of the Catalan–Qi function related to the Catalan numbers, *SpringerPlus* **5**, 1126 (2016). doi: 10.1186/s40064-016-2793-1.
[84] F. Qi and B.-N. Guo, Integral representations of the Catalan numbers and their applications, *Mathematics* **5**, 40 (2017). doi: 10.3390/math5030040.
[85] F. Qi and P. Cerone, Some properties of the Fuss–Catalan numbers, *Mathematics* **6**(12), 277 (2018). doi: 10.3390/math6120277.
[86] F. Qi and B.-N. Guo, Sums of infinite power series whose coefficients involve products of the Catalan–Qi numbers, *Montes Taurus J. Pure Appl. Math.* **1**(2), 1–12 (2019). Article ID: MTJPAM-D-19-00007.
[87] J. Quaintance, *Combinatorial Identities for Stirling Numbers* (The Unpublished Notes of H. W. Gould) (World Scientific Publishing Co. Pte. Ltd., Singapore, 2015).

[88] M. I. Qureshi and S. Ahmad Dar, Generalizations and applications of Srinivasa Ramanujan's integral $R_S(m,n)$ via hypergeometric approach and integral transforms, *Montes Taurus J. Pure Appl. Math.* **3**(3), 216–226 (2021).

[89] M. I. Qureshi and S. A. Dar, Some hypergeometric summation theorems and reduction formulas via Laplace transform method, *Montes Taurus J. Pure Appl. Math.* **3**(3), 182–199 (2021). Article ID: MTJPAM-D-20-00016.

[90] H. Rademacher, *Topics in Analytic Number Theory*. Grundlehren der mathematischen Wissenschaften, Vol. 169 (Springer-Verlag, Berlin, 1973).

[91] E. D. Rainville, *Special Functions* (The Macmillan Company, New York, 1960).

[92] R. P. Stanley, *Enumerative Combinatorics, Volume 1* (Wadsworth and Brooks/Cole, Monterey, 1986).

[93] R. P. Stanley, *Enumerative Combinatorics, Volume 1* (Cambridge University Press, Cambridge, 1999).

[94] R. P. Stanley, *Catalan numbers* (Cambridge University Press, Cambridge, 2015).

[95] S. Roman, *The Umbral Calculus* (Dover Publications Inc., New York, 2005).

[96] S. Roman, *An Introduction to Catalan Numbers* (Birkhäuser, Cham, 2015).

[97] W. H. Schikhof, *Ultrametric Calculus: An Introduction to p-Adic Analysis*. Cambridge Studies in Advanced Mathematics, Vol. 4 (Cambridge University Press, Cambridge, 1984).

[98] N. J. A. Sloane, *The On-Line Encyclopedia of Integer Sequences*. OEIS Sequence: A006232 and A006233.

[99] Y. Simsek, Twisted (h,q)-Bernoulli numbers and polynomials related to twisted (h,q)-zeta function and L-function, *J. Math. Anal. Appl.* **324**, 790–804 (2006).

[100] Y. Simsek, Special functions related to Dedekind-type DC-sums and their applications, *Russ. J. Math. Phys.* **17**(4), 495–508 (2010).

[101] Y. Simsek, On q-deformed Stirling numbers, *Int. J. Math. Comput.* **15**(2), 70–80 (2010).

[102] Y. Simsek, Analysis of the Bernstein basis functions: An approach to combinatorial sums involving binomial coefficients and Catalan numbers, *Math. Methods Appl. Sci.* **38**(14), 3007–3021 (2015).

[103] Y. Simsek, Generating functions for generalized Stirling type numbers, array type polynomials, Eulerian type polynomials and their applications, *Fixed Point Theory Appl.* **87**, 1–28 (2013).

[104] Y. Simsek, Functional equations from generating functions: A novel approach to deriving identities for the Bernstein basis functions, *Fixed Point Theory Appl.* **2013**, 1–13 (2013).

[105] Y. Simsek, Unification of the Bernstein-type polynomials and their applications, *Bound. Value Probl.*, **2013**, 56 (2013). doi: 10.1186/1687-2770-2013-56.

[106] Y. Simsek, Deriving novel formulas and identities for the Bernstein basis functions and their generating functions. Lecture Notes in Computer Science, Vol. 8177 (Springer-Verlag, Berlin, 2014), pp. 471–490.

[107] Y. Simsek, Generating functions for the Bernstein type polynomials: A new approach to deriving identities and applications for the polynomials, *Hacet. J. Math. Stat.* **43**(1), 1–14 (2014).

[108] Y. Simsek, *Families of twisted Bernoulli numbers, twisted Bernoulli polynomials and their applications.* In G. V. Milovanovic and M. T. Rassias (eds.) *Analytic Number Theory, Approximation Theory, and Special Functions* (Springer, Berlin, 2014), pp. 149–214.

[109] Y. Simsek, Combinatorial sums and binomial identities associated with the beta-type polynomials, *Hacet. J. Math. Stat.* **47**(5), 1144–1155 (2018).

[110] Y. Simsek, Identities on the Changhee numbers and Apostol-type Daehee polynomials, *Adv. Stud. Contemp. Math. (Kyungshang)* **27**(2), 199–212 (2017).

[111] Y. Simsek, Explicit formulas for p-adic integrals: Approach to p-adic distributions and some families of special numbers and polynomials, *Montes Taurus J. Pure Appl. Math.* **1**(1), 1–76 (2019). Article ID: MTJPAM-D-19-00005.

[112] Y. Simsek, Generating functions for finite sums involving higher powers of binomial coefficients: Analysis of hypergeometric functions including new families of polynomials and numbers, *J. Math. Anal. Appl.* **477**, 1328–1352 (2019).

[113] Y. Simsek, Construction of some new families of Apostol-type numbers and polynomials via Dirichlet character and p-adic q-integrals, *Turkish J. Math.* **42**, 557–577 (2018).

[114] Y. Simsek, Interpolation functions for new classes special numbers and polynomials via applications of p-adic integrals and derivative operator, *Montes Taurus J. Pure Appl. Math.* **3**(1), 38–61 (2021). Article ID: MTJPAM-D-20-00000.

[115] Y. Simsek, Analysis of Apostol-type numbers and polynomials with their approximations and asymptotic behavior. In T. M. Rassias (ed.) *Approximation Theory and Analytic Inequalities* (Springer, Cham., Springer Nature Switzerland AG, 2021), pp. 435–486.

[116] Y. Simsek, Applications of Apostol-type numbers and polynomials: Approach to techniques of computation algorithms in approximation and interpolation functions. In N. J. Daras and T. M. Rassias (eds.) *Approximation and Computation in Science and Engineering* (Springer, Cham., Springer Nature Switzerland AG, 2022), pp. 783–860.

[117] Y. Simsek, Some classes of finite sums related to the generalized harmonic functions and special numbers and polynomials, *Montes Taurus J. Pure Appl. Math.* **4**(3), 61–79 (2022).

[118] Y. Simsek and M. Acikgoz, A new generating function of (q-) Bernstein-type polynomials and their interpolation function, *Abstr. Appl. Anal.* **2010**, 769095 (2010). doi: 10.1155/2010/769095.

[119] Y. Simsek and A. Yardimci, Applications on the Apostol-Daehee numbers and polynomials associated with special numbers, polynomials, and p-adic integrals, *Adv. Differ. Equ.* **308** (2016). doi: 10.1186/s13662-016-1041-x.

[120] H. M. Srivastava, Some formulas for the Bernoulli and Euler polynomials at rational arguments, *Math. Proc. Camb. Philos. Soc.* **129**, 77–84 (2000). doi: 10.1017/S0305004100004412.

[121] H. M. Srivastava, Some generalizations and basic (or q-) extensions of the Bernoulli, Euler and Genocchi polynomials, *Appl. Math. Inf. Sci.* **5**, 390–444 (2011).

[122] H. M. Srivastava and A. Pinter, Remarks on some relationships between the Bernoulli and Euler polynomials, *Appl. Math. Lett.* **17**, 375–380 (2004).

[123] H. M. Srivastava, T. Kim and Y. Simsek, q-Bernoulli numbers and polynomials associated with multiple q-zeta functions and basic L-series, *Russ. J. Math. Phys.* **12**, 241–268 (2005).

[124] H. M. Srivastava and J. Choi, *Series Associated with the Zeta and Related Functions* (Kluwer Academic Publishers, Dordrecht, 2001).

[125] H. M. Srivastava and J. Choi, *Zeta and q-Zeta Functions and Associated Series and Integrals* (Elsevier Science Publishers, Amsterdam, 2012).

[126] H. M. Srivastava, B. Kurt and Y. Simsek, Some families of Genocchi type polynomials and their interpolation functions, *Integral Transforms Spec. Funct.* **23**(12), 919–938 (2012).

[127] H. M. Srivastava, B. Kurt and Y. Simsek, Corrigendum: Some families of Genocchi type polynomials and their interpolation functions, *Integral Transforms Spec. Funct.* **23**(12), 939–940 (2012). doi: 10.1080/10652469.2012.690950.

[128] H. M. Srivastava, I. Kucukoglu and Y. Simsek, Partial differential equations for a new family of numbers and polynomials unifying the Apostol-type numbers and the Apostol-type polynomials, *J. Number Theory* **181**, 117–146 (2017).

[129] R. Tremblay, New quadratic transformations of hypergeometric functions and associated summation formulas obtained with the well-poised fractional calculus operator, *Montes Taurus J. Pure Appl. Math.* **2**(1), 36–62 (2020). Article ID: MTJPAM-D-20-00005.

[130] R. Tremblay, Using the well-poised fractional calculus operator $_{g(z)}O_\beta^\alpha$ to obtain transformations of the Gauss hypergeometric function with higher level arguments, *Montes Taurus J. Pure Appl. Math.* **3**(3), 260–283 (2021). Article ID: MTJPAM-D-20-00057.

[131] T. Usman, N. Khan, M. Saif and J. Choi, A unified family of Apostol-Bernoulli based poly-Daehee polynomials, *Montes Taurus J. Pure Appl. Math.* **3**(3), 1–11 (2021). Article ID: MTJPAM-D-20-00009.

[132] https://en.wikipedia.org/wiki/Fuss-Catalan_number.

[133] Wolfram Research Inc. Mathematica Online (Wolfram Cloud) (Champaign, IL, USA, 2022). https://www.wolframcloud.com.

[134] Q. Zou, Identities on Genocchi polynomials and Genocchi numbers concerning binomial coefficients, *Int. J. Anal. Appl.* **14**(2), 140–146 (2017).

© 2024 World Scientific Publishing Company
https://doi.org/10.1142/9789811267048_0018

Chapter 18

On the Generalizations of the Cauchy–Schwarz–Bunyakovsky Inequality with Applications to Elasticity

D. Labropoulou[*,§], T. Labropoulos[†,¶], P. Vafeas[*,∥], and D. M. Manias[‡,**]

[*]*Department of Chemical Engineering,*
University of Patras, Patras, Greece
[†]*Department of Mathematics,*
School of Applied Mathematics and Physical Sciences,
National Technical University of Athens, Athens, Greece
[‡]*Department of Mathematics,*
Khalifa University of Science and Technology,
Abu Dhabi, UAE

[§] *dlabropoulou@chemeng.upatras.gr*
[¶] *tlabropoulos@mail.ntua.gr*
[∥] *vafeas@chemeng.upatras.gr*
[**] *dimitris.manias@ku.ac.ae*

In this article, we present both the discrete and the integral form of Cauchy–Bunyakovsky–Schwarz (CBS) inequality, some important generalizations in the n-dimensional Euclidean space and in linear subspaces of it, as well as the strengthened CBS. The last CBS inequality plays an important role in elasticity problems. A geometrical interpretation and a collection of the most important proofs of it are, also, presented.

1. Introduction

Cauchy [1] in his monograph published in 1821 presented an inequality (for finite discrete sums), which subsequently played and continues to play an important role in mathematics. In 1859, Bunyakovsky [2] extended this inequality to its integral version (that is, for the case where we have continuous summation). Some years later, in 1888, Schwarz [3] established the general form of this inequality, valid in vector spaces endowed with dot product, the so-called "Cauchy–Bunyakovsky–Schwarz inequality" (abbreviated CBS inequality).

The CBS inequality is one of the most famous inequalities in mathematics, on the one hand, because of its close relations with a multitude of other important inequalities of mathematics and, on the other hand, because of its utility in a particularly wide and varied range of applications in almost all branches of classical and modern mathematics, such as real and complex analysis, numerical analysis, Hilbert space theory, differential equations, probability theory, statistics and number theory. More details on this topic are in the works of Dragomir [4] and Steele [5].

Additionally, new applications of various branches of mathematics are constantly appearing, where the CBS inequality itself or CBS-type inequalities play a significant role. Bîrsan and Bîrsan [6] presented a CBS-type inequality in both the discrete and the integral forms. The integral version of this inequality appears in the study of mechanical properties of thin elastic rods. Also, Achchab and Maitre [7] presented the strengthened Cauchy–Bunyakovsky–Schwarz inequality scheme, which plays a fundamental role in the convergence rate of multilevel iterative methods. Its optimal constant plays an important role in elasticity problems (see Margenov [8], Achchab et al. [9], Achchab and Maitre [7], Dragomir [4], Jung and Maitre [10]). Besides, this handy theory can be readily applied to more recent developments in anisotropic elasticity, wherein even nowadays novel methods have been introduced in the work of Labropoulou et al. [11].

In this chapter, we attempt a presentation of the basic versions of the CBS inequality, an inequality of CBS type, the integral version of which appears in the study of mechanical properties of thin elastic rods [6, 12], demonstrating that the strengthened CBS inequality is very useful in elasticity problems. Finally, in the Appendix we present an anthology of clever and fairly short proofs of the CBS inequality (see also the work of Wu and Wu [13]).

2. Cauchy–Bunyakovsky–Schwarz Inequality

2.1. *Cauchy inequality*

Cauchy's inequality for non-negative real numbers is expressed as follows:

Proposition 1. *If $\alpha_1, \alpha_2, \ldots, \alpha_n$ are non-negative real numbers, then*

$$\frac{\alpha_1 + \alpha_2 + \cdots + \alpha_\nu}{\nu} \geqslant \sqrt[\nu]{\alpha_1 \cdot \alpha_2 \ldots \cdot \alpha_\nu}. \qquad (1)$$

If in addition the numbers $\alpha_1, \alpha_2, \ldots, \alpha_n$ are different from zero, then

$$\frac{\alpha_1 + \alpha_2 + \cdots + \alpha_\nu}{\nu} \geqslant \sqrt[\nu]{\alpha_1 \cdot \alpha_2 \ldots \cdot \alpha_\nu} \geqslant \frac{\nu}{\dfrac{1}{\alpha_1} + \dfrac{1}{\alpha_2} + \cdots + \dfrac{1}{\alpha_\nu}}. \qquad (2)$$

Equality is true only when $\alpha_1 = \alpha_2 = \cdots = \alpha_\nu$.

For a nice proof, we propose the book of Rassias [14].

Historical Notation 1.[a] *This inequality first appeared in Augustin–Louis Cauchy's book (1789–1857) entitled "Coursd' Analysis of the Polycole Polytechnique" (1821) for students at the Ecole Polytechnique in Paris. The method of proving inequality is a "kind of mathematical induction". A very beautiful proof of this inequality is presented in the book (see Ref. [14, p. 271]) and is due to the great Hungarian mathematician George Polya (1887–1985). It is worth noting that this inequality has been given a particularly large amount of evidence.*

---

[a] All Historical Notations are found in the Wikipedia biographies of Cauchy, Schwarz, Bunyakovsky, Holder and Minkowski.

Historical Notation 2. *Baron Augustin–Louis Cauchy, French mathematician and pioneer of Analysis, was born on August 21, 1789 and died on May 23, 1857. He graduated as a civil engineer in 1807 at the age of 18 from the Polytechnic School of Paris, where he was a professor. He began to formulate and prove the theorems of infinite calculus in a rigorous way, rejecting any heuristic principle of the generality of algebra used by older writers. He defined the sequence in infinite terms, gave several important theorems in complex analysis, and began the study of transpositional groups in abstract algebra. Cauchy, being a thoughtful mathematician, greatly influenced modern mathematicians as well as later ones.*

Before formulating the Cauchy–Schwarz–Bunyakovsky inequality, we recall the concept of the inner product in $\mathbb{R}^n$, $n \geqslant 1$, and its basic properties. In calculus, we had to define the concept of absolute value in order to study distance in real line. The same can happen in the vector space $\mathbb{R}^n$. We know that the n vectors $e_1 = (1, 0, \ldots, 0)$, $e_2 = (0, 1, \ldots, 0)$, $\ldots, e_n = (0, 0, \ldots, 1)$ of $\mathbb{R}^n$ consist a basis of it and in fact the normal basis (that is, it is unit and perpendicular to two) and they generalize its usual orthogonal basis $B = \{i, j, k\}$ $(i = e_1, j = e_2, k = e_3)$ of $\mathbb{R}^3$.

We know that for two possible vectors $x = (x_1, x_2, x_3)$ and $y = (y_1, y_2, y_3)$ of $\mathbb{R}^3$ their inner product is defined as

$$x \cdot y = x_1 y_1 + x_2 y_2 + x_3 y_3 \in \mathbb{R}.$$

We can now define in a similar way the inner product of two vectors of $\mathbb{R}^n$ for any n. So if $x = (x_1, \ldots, x_n) \in \mathbb{R}^n$ and $y = (y_1, \ldots, y_n) \in \mathbb{R}^n$, extending the definition we gave for the inner product of two vectors of $\mathbb{R}^3$ for any n, we can define the inner product of the vectors of $\mathbb{R}^n$ in a similar way via

$$x \cdot y = x_1 y_1 + x_2 y_2 + \cdots + x_n y_n = \sum_{i=1}^{n} x_i y_i.$$

The length or Euclidean norm of $x = (x_1, \ldots, x_n)$ is defined to be the non-negative real

$$\|x\| = \sqrt{x \cdot x} = \sqrt{x_1^2 + \cdots + x_n^2}.$$

The pair $(\mathbb{R}^n, \|\cdot\|)$ is called the Euclidean space of dimension n. We often use the notation $\langle x, y\rangle$ instead of $x \cdot y$ and the notation $|x|$ instead of $\|x\|$.

The basic properties of the inner product are described in the following proposition:

Proposition 2. *For all $x, y, z \in \mathbb{R}^n$, the following properties hold:*

(i) $\langle x, y\rangle = \langle y, x\rangle$,
(ii) $\langle x + z, y\rangle = \langle x, y\rangle + \langle z, y\rangle$,
(iii) $\langle x, x\rangle \geq 0$,
(iv) $\langle x, x\rangle = 0 \Leftrightarrow x = 0 \Leftrightarrow x_1 = x_2 = \cdots = x_n = 0$.

The proof of properties is achieved by simply applying the definition of the inner product.

2.2. Cauchy–Bunyakovsky–Schwarz inequality

The CBS inequality is expressed as follows:

Theorem 1. *Let $x, y \in \mathbb{R}^n$. Then*

$$|x \cdot y| \leq \|x\| \cdot \|y\|. \tag{3}$$

The equality in CBS inequality holds if and only if the vectors x and y are collinear.

Proof. If $x = 0$ or $y = 0$, then inequality applies trivially as equality. Assume that $x \neq 0$ and $y \neq 0$.

(i) Let's say that $\|x\| = 1 = \|y\|$. Then from the obvious inequality $(|x_i| - |y_i|)^2 \geq 0$, we get that

$$x_i^2 + y_i^2 - 2|x_i y_i| \geq 0$$

from which it follows that

$$|x_i y_i| \leq \frac{x_i^2 + y_i^2}{2}, \quad i = 1, 2, \ldots, n.$$

Therefore,

$$\left|\sum_{i=1}^{n} x_i y_i\right| \leqslant \sum_{i=1}^{n} |x_i y_i| \leqslant \frac{1}{2} \sum_{i=1}^{n} (x_i^2 + y_i^2)$$

$$= \frac{1}{2} \sum_{i=1}^{n} x_i^2 + \frac{1}{2} \sum_{i=1}^{n} y_i^2 = \frac{1}{2} + \frac{1}{2} = 1,$$

that is,

$$|x \cdot y| = \left|\sum_{i=1}^{n} x_i y_i\right| \leqslant 1 = 1 \cdot 1 = |x| \, |y|$$

Thus, in this case, inequality applies.

(ii) **General case:** Let $x = (x_1, \ldots, x_n) \in \mathbb{R}^n$ and $y = (y_1, \ldots, y_n) \in \mathbb{R}^n$ be two possible vectors. We set

$$X_i = \frac{x_i}{\|x\|}, \quad Y_i = \frac{y_i}{\|y\|}, \quad i = 1, 2, \ldots, n.$$

It follows from case (i) that

$$\left|\sum_{i=1}^{n} X_i Y_i\right| = \left|\sum_{i=1}^{n} \left(\frac{x_i}{\|x\|} \cdot \frac{y_i}{\|y\|}\right)\right| \leqslant \sum_{i=1}^{n} \frac{|x_i|}{\|x\|} \cdot \frac{|y_i|}{\|y\|} \leqslant 1.$$

Since

$$\sum_{i=1}^{n} X_i^2 = \sum_{i=1}^{n} \left(\frac{x_i}{\|x\|}\right)^2 = \sum_{i=1}^{n} \frac{x_i^2}{\|x\|^2} = \frac{1}{\|x\|^2} \sum_{i=1}^{n} x_i^2 = \frac{1}{\|x\|^2} \|x\|^2 = 1$$

and similarly

$$\sum_{i=1}^{n} Y_i^{'2} = 1.$$

It follows that

$$|x \cdot y| = \left|\sum_{i=1}^{n} x_i y_i\right| \leqslant 1 \cdot 1 = \|x\| \, \|y\| \, .$$

If the vectors x and y are linear, that is, if we have for someone, then we easily find the equality in the CBS inequality.

We now assume that $|x \cdot y| = \|x\| \|y\|$. If one of x and y is 0, let x be, then we set $\lambda = 0$ and observe that $x = \lambda y$. We now assume that $x = (x_1, \ldots, x_n) \neq 0$ and $y = (y_1, \ldots, y_n) \neq 0$. Then the equality $|x \cdot y| = \|x\| \|y\|$ is written as

$$\left| \sum_{i=1}^n \frac{|x_i|}{\|x\|} \cdot \frac{|y_i|}{\|y\|} \right| = 1$$

and posing $X_i = \dfrac{x_i}{\|x\|}$ and $Y_i = \dfrac{y_i}{\|y\|}$ yields

$$\left| \sum_{i=1}^n X_i \cdot Y_i \right| = 1,$$

where as we proved above $\sum_{i=1}^n X_i^2 = 1 = \sum_{i=1}^n Y_i^2$.

Therefore,

$$1 = \left| \sum_{i=1}^n X_i \cdot Y_i \right| \leqslant \sum_{i=1}^n |X_i \cdot Y| \leqslant \frac{1}{2} \sum_{i=1}^n X_i^2 + \frac{1}{2} \sum_{i=1}^n Y_i^2 = 1. \quad (4)$$

We conclude that the numbers X_i, Y_i, $i = 1, 2, \ldots, n$, are identical, since the equation holds in the triangular inequality, that is,

$$\left| \sum_{i=1}^n X_i \cdot Y_i \right| = \sum_{i=1}^n |X_i \cdot Y_i|.$$

By (4), also, it follows that necessary

$$|X_i Y_i| = \frac{1}{2} X_i^2 + \frac{1}{2} Y_i^2, \quad i = 1, 2, \ldots, n. \quad (5)$$

(a) If the common sign of the numbers X_i, Y_i, $i = 1, 2, \ldots, n$, is positive, then from equalities (5) we conclude that

$$X_i Y_i = \frac{1}{2}(X_i^2 + Y_i^2) \Leftrightarrow (X_i - Y_i)^2 = 0 \Leftrightarrow X_i = Y_i, \quad \forall\, i = 1, 2, \ldots, n$$

or

$$\frac{x_i}{\|x\|} = \frac{y_i}{\|y\|} \Leftrightarrow x_i = \frac{\|x\|}{\|y\|} \cdot y_i \Leftrightarrow x = \frac{\|x\|}{\|y\|} \cdot y$$

and

$$\lambda = \frac{\|x\|}{\|y\|}.$$

(b) If the common sign of the numbers X_i, Y_i, $i = 1, 2, \ldots, n$, is negative, then from equalities (5) we find that

$$-X_i Y_i = \frac{1}{2}\left(X_i^2 + Y_i^2\right) \Leftrightarrow (X_i + Y_i)^2 = 0$$

$$\Leftrightarrow X_i = -Y_i, \quad \forall\, i = 1, 2, \ldots, n$$

and

$$\lambda = -\frac{\|x\|}{\|y\|}. \qquad \square$$

Historical Notation 3. *Karl Hermann Amandus Schwarz, a German mathematician known for his work on complex analysis, was born on January 25, 1843 in Hermsnstorf, Silesia (southern Poland) and died on November 30, 1921 in Berlin. Schwarz initially studied chemistry in Berlin, but Kummer and Karl Weierstrass encouraged him to switch to mathematics. He received his Ph.D. from the University of Berlin in 1864 under the supervision of Kummer and Weierstrass. Between 1867 and 1869, he worked at the University of Halle and then at the Swiss Federal Polytechnic. From 1875 he worked at the University of Gottingen and was involved in complex analysis, differential geometry and calculus of change.*

Historical Notation 4. *Viktor Yakovlevich Bunyakovsky, a Russian mathematician known for his work in theoretical engineering and number theory, was born on December 16, 1804 in Bar, Russia, and died on November 12, 1889 in St. Petersburg. He was a member and later vice-president of the Russian Academy of Sciences. He graduated from the Sorbonne University in 1824 and received his doctorate under Cauchy. He proved the CBS inequality for the infinite case in 1859, long before Schwarz dealt with it.*

2.3. Geometrical interpretation of the CBS inequality

From the CBS inequality, it follows that for nonzero vectors x and y in $\mathbb{R}^n$, the quotient

$$\frac{x \cdot y}{\|x\| \|y\|}$$

belongs to the interval $[-1, 1]$, i.e.,

$$-1 \leqslant \frac{x \cdot y}{\|x\| \|y\|} \leqslant 1.$$

This defines the angle θ between x and y by the formula

$$\cos \theta = \frac{x \cdot y}{\|x\| \cdot \|y\|}, \quad \theta \in [0, \pi]. \tag{6}$$

Moreover, if the vectors x and y belong to $\mathbb{R}^3$, then CBS inequality follows from the law of the cosine of trigonometry.

Applying the law of cosines to the triangle with vertices $O(0,0,0)$ and sides the vectors x and y, it will result that

$$\|x - y\|^2 = \|x\|^2 + \|y\|^2 - 2\|x\| \|y\| \cos \theta, \tag{7}$$

where θ is the angle of x and y. From (7), using the properties of the inner product, we get successively

$$\|x - y\|^2 = \|x\|^2 + \|y\|^2 - 2\|x\| \cdot \|y\| \cdot \cos \theta$$
$$\Leftrightarrow (x - y) \cdot (x - y) = \|x\|^2 + \|y\|^2 - 2\|x\| \|y\| \cos \theta$$
$$\Leftrightarrow x^2 - 2x \cdot y + y^2 = x^2 + y^2 - 2\|x\| \|y\| \cos \theta$$
$$\Leftrightarrow -2x \cdot y = -2\|x\| \|y\| \cos \theta$$
$$\Leftrightarrow x \cdot y = \|x\| \|y\| \cos \theta.$$

Therefore,

$$|\cos \theta| = \frac{|x \cdot y|}{\|x\| \|y\|} \leqslant 1 \Leftrightarrow |x \cdot y| \leqslant \|x\| \|y\|.$$

If the vectors x and y are collinear, that is, $x = \lambda y$, for some $\lambda \in \mathbb{R}$, then $\theta = 0$ or $\theta = \pi$, therefore $\cos \theta = 1$ and so $|x \cdot y| = \|x\| \|y\|$. If the vectors x and y are not collinear, that is, linearly independent, then $\theta \in (0, \pi)$, therefore $|\cos \theta| < 1$ and so $|x \cdot y| < \|x\| \|y\|$. So, we come to the part where we talk about the inner product of nonzeros vectors

$$x \cdot y = \|x\| \|y\| \cos \theta, \quad x, y \in \mathbb{R}^n,$$

where $\theta \in [0, \pi]$ is the unique angle between x and y. Two vectors $x, y \in \mathbb{R}^n \backslash \{0\}$ are called *orthogonal* if $x \cdot y = 0 \Leftrightarrow \theta = \dfrac{\pi}{2}$.

2.4. *Triangular inequality*

We can prove our well-known triangular inequality in $\mathbb{R}^n$ on the basis of the CBS inequality. That is, we prove that if $x, y \in \mathbb{R}^n$, then we have

$$\|x + y\| \leqslant \|x\| + \|y\|. \tag{8}$$

Indeed using the CBS inequality (3), we have

$$\begin{aligned}
\|x + y\|^2 &= (x + y) \cdot (x + y) \\
&= x \cdot x + x \cdot y + y \cdot x + y \cdot y \\
&= \|x\|^2 + 2x \cdot y + \|y\|^2 \\
&\leqslant \|x\|^2 + 2 \|x\| \|y\| + \|y\|^2 \\
&= (\|x\| + \|y\|)^2.
\end{aligned}$$

Taking square roots in each member results in the required inequality.

Based on the CBS inequality, we can get the following very useful proposition:

Proposition 3. *If $(x_1, \ldots, x_n)$ and $(y_1, \ldots, y_n)$ are two sequences of real numbers, then the following inequality holds*

$$\left(\sum_{i=1}^n x_i^2 \right) \left(\sum_{i=1}^n y_i^2 \right) \geqslant \left(\sum_{i=1}^n x_i y_i \right)^2, \tag{9}$$

with the equality to be valid if and only if the sequences $(x_1, \ldots, x_n)$ and $(y_1, \ldots, y_n)$ are proportional term by term, that is, if there exists a fixed number λ such that $x_k = \lambda y_\kappa \, \forall \, k \in \{1, 2, \ldots, n, \ldots\}$.

CBS inequality is known to be one of the most basic and useful inequalities in algebra, and because of its particular weight, it has greatly preoccupied mathematicians. This has resulted in a very large number of proofs of great interest. Then we present some of the ones we have chosen based on the diversity, the brevity and the beauty that they present, and of course, these criteria are subjective.

In the Appendix, we present some of the most beautiful and useful proofs of the CBS inequality.

CBS inequality, as mentioned above, without any doubt, is one of the most widely used and important inequalities in all mathematics. If, for example, we write it in the form

$$x_1 y_1 + x_2 y_2 + \cdots + x_n y_n \leqslant \sqrt{x_1^2 + x_2^2 + \cdots + x_n^2} \sqrt{y_1^2 + y_2^2 + \cdots + y_n^2},$$

one of the most important and immediate conclusions that emerges from this is that from $\sum_{i=1}^{\infty} x_i^2 < \infty$ and $\sum_{i=1}^{\infty} y_i^2 < \infty$ it implies that $\sum_{i=1}^{\infty} |x_i y_i| < \infty$.

We can generalize the above CBS inequality in arbitrary linear spaces. Let, for example, V be an arbitrary real vector space. If we can define a function in $V \times V$ which is defined by the map $(x, y) \mapsto \langle x, y \rangle$ in a such way that the $\langle x, y \rangle$ satisfies the properties in Proposition 2, then the $\langle x, y \rangle$ is an inner product and $(V, \langle \cdot, \cdot \rangle)$ is a real space with inner product.

If we go one step further and think that the integration of a function in a space is nothing more than a sum in the continuous case, the question arises whether the CBS inequality holds in this continuous case. That is, if we replace the sums with integrals, then the following inequality arises:

$$\int_a^b f(x) g(x) \, \mathrm{d}x \leqslant \left(\int_a^b f^2(x) \, \mathrm{d}x \right)^{\frac{1}{2}} \left(\int_a^b g^2(x) \, \mathrm{d}x \right)^{\frac{1}{2}}.$$

The question therefore is whether such inequality applies. The answer to this question is affirmative, and in fact, we can get a more general inequality, the so-called Hölder inequality. In other words, the following proposition is valid:

Proposition 4 (Hölder's inequality). *Let p, q be two real numbers such that $1 < p < +\infty$ and $\frac{1}{p} + \frac{1}{q} = 1$. Let, also, the integrable* functions $f, g : [\alpha, \beta] \to [0, +\infty]$. Then the following inequality holds:

$$\int_a^b f(x) g(x) \, dx \leqslant \left(\int_a^b f^p(x) \, dx\right)^{\frac{1}{p}} \left(\int_a^b g^q(x) \, dx\right)^{\frac{1}{q}}.$$

A special case of the latter is taken for $p = q = \frac{1}{2}$. The positive exponents p and q are called conjugate exponents if $\frac{1}{p} + \frac{1}{q} = 1$ or equivalent $p + q = pq$.

Historical Notation 5. *We mention that this inequality first appeared in print in Victor Yacovlevich Bunyakovsky's Memoire which was published by the Imperial Academy of Sciences of St. Petersburg in 1859.*

In aim to prove Hölder's inequality, we assume that

$$0 < \left(\int_a^b f^p(x) \, dx\right)^{\frac{1}{p}} < +\infty, \ 0 < \left(\int_a^b g^q(x) \, dx\right)^{\frac{1}{q}} < +\infty$$

and that

$$\left(\int_a^b f^p(x) \, dx\right)^{\frac{1}{p}} \neq 0, \ \left(\int_a^b g^q(x) \, dx\right)^{\frac{1}{q}} \neq 0,$$

that is, that the functions f^p and g^q are integrable with nonzero integrals because these cases are trivial.

For the proof, we need the following proposition:

Proposition 5. *Let p, q be two real numbers such that $1 < p < +\infty$ and $\dfrac{1}{p} + \dfrac{1}{q} = 1$. Then, for all $x, y \geq 0$, the following inequality holds:*

$$xy \leqslant \frac{x^p}{p} + \frac{y^q}{q}.$$

Proof. We observe that if $x = 0$ or $y = 0$, the inequality holds, then we assume that $x > 0$ and $y > 0$. Dividing inequality by members with y^q, we obtain the inequality

$$\frac{x}{y^{q-1}} \leqslant \frac{x^p}{py^q} + \frac{1}{q}$$

and if we put in the latter $t = \dfrac{x^p}{y^q}$, because of $\dfrac{1}{p} + \dfrac{1}{q} = 1$, we have $(q-1)p = q$ and so it turns out that

$$\dfrac{x}{y^{q-1}} = \left(\dfrac{x^p}{y^{(q-1)p}}\right)^{\frac{1}{p}} = \left(\dfrac{x^p}{y^{pq-p}}\right)^{\frac{1}{p}} = \left(\dfrac{x^p}{y^q}\right)^{\frac{1}{p}} = t^{\frac{1}{p}},$$

so we get the inequality

$$t^{\frac{1}{p}} \leqslant \dfrac{1}{p}t + \dfrac{1}{q}, \quad t > 0.$$

The function $f(t) = t^{\frac{1}{p}} - \dfrac{1}{p}t - \dfrac{1}{q}$, $t > 0$, is differentiable with $f'(t) = \dfrac{1}{p}\left(t^{-\frac{1}{q}} - 1\right)$.

We observe that $f'(1) = 0$, $f'(t) < 0$, $\forall\, t > 1$ and $f'(t) > 0$, $\forall\, t \in (0,1)$, so $f(t) \leqslant f(1) = 0$ $\forall\, t > 0$, and then

$$t^{\frac{1}{p}} - \dfrac{1}{p}t - \dfrac{1}{q} \leqslant 0, \quad \forall\, t > 0,$$

where the equality holds only when $t = 1$. □

Proof of Hölder's Inequality. We set $A = \left(\int_a^b f^p(x)\,\mathrm{d}x\right)^{\frac{1}{p}}$ and $B = \left(\int_a^b g^q(x)\,\mathrm{d}x\right)^{\frac{1}{q}}$. So from the above proposition for the numbers $f(x)/A$ and $g(x)/B$, we get

$$\dfrac{f(x)}{A}\dfrac{g(x)}{B} \leqslant \dfrac{(f(x)/A)^p}{p} + \dfrac{(g(x)/B)^q}{q} = \dfrac{(f(x))^p}{pA^p} + \dfrac{(g(x))^q}{qB^q}.$$

By integration by parts from the last, we obtain

$$\dfrac{1}{AB}\int_\alpha^\beta f(x)g(x)\,\mathrm{d}x \leqslant \dfrac{1}{pA^p}\int_\alpha^\beta (f(x))^p\,\mathrm{d}x + \dfrac{1}{qB^q}\int_\alpha^\beta (g(x))^q\,\mathrm{d}x$$

$$= \dfrac{1}{pA^p}A^p + \dfrac{1}{qB^q}B^q = \dfrac{1}{p} + \dfrac{1}{q} = 1,$$

so

$$\int_\alpha^\beta f(x)g(x)\,dx \leqslant AB \Leftrightarrow \int_\alpha^\beta f(x)g(x)\,dx$$
$$\leqslant \left(\int_\alpha^\beta f^p(x)\right)^{\frac{1}{p}} \left(\int_\alpha^\beta g^q(x)\right)^{\frac{1}{q}}.$$

If we use the classic notation for the norm in L^p spaces, that is, if we set

$$\|f\|_p = \left(\int_\alpha^\beta f^p(x)\right)^{\frac{1}{p}}, \quad 1 \leqslant p < +\infty,$$

the last inequality takes the form

$$\|fg\|_1 = \|f\|_p \|g\|_q. \qquad \square$$

Historical Notation 6. *Ludwig Otto Holder, German mathematician, was born on December 22, 1859 in Stuttgart and died on August 29, 1937. He studied at the present University of Stuttgart where he was a student of Kronecker, Weirstrass and Kummer. He received his doctorate from the University of Tubingen in 1882. He is known for many theorems such as the Holder Inequality, the Jordan–Holder theorem and the abnormal external automorphism of the symmetric group.*

Completing this generalization in the continuous case, that is, in spaces, we can get the generalization of the triangular inequality called Minkowski inequality.

Proposition 6. *Let $1 < p < +\infty$ and the integrable functions $f, g : [\alpha, \beta] \to [0, +\infty]$. Then,*

$$\left(\int_a^b (f+g)^p(x)\,dx\right)^{\frac{1}{p}} \leqslant \left(\int_a^b f^p(x)\,dx\right)^{\frac{1}{p}} + \left(\int_a^b g^p(x)\,dx\right)^{\frac{1}{p}}$$

Proof. We note that

$$[f(x) + g(x)]^p \leq [2\max\{f(x), g(x)\}]^p$$
$$= 2^p \max\{f^p(x), g^p(x)\}$$
$$\leq 2^p [f^p(x) + g^p(x)]$$

or

$$(f(x) + g(x))^p \leq 2^p f^p(x) + 2^p g^p(x),$$

so with integration by parts, we get this inequality

$$\int_a^b (f+g)^p(x)\,dx \leq 2^p \int_a^b f^p(x)\,dx + 2^p \int_a^b g^p(x)\,dx.$$

We can assume that $\int_a^b f^p(x)\,dx < +\infty$, $\int_a^b g^p(x)\,dx \leq +\infty$, and then due to the last inequality arises $\int_a^b (f+g)^p(x)\,dx < +\infty$ and so in this case, the inequality we want to prove is obvious.

Let now be the conjugate exponent q of p. Then, $(p-1)q = p$, so from Hölder's inequality, we get

$$\int_a^b f(f+g)^{p-1}(x)\,dx \leq \left(\int_a^b f^p(x)\,dx\right)^{\frac{1}{p}} \left(\int_a^b (f+g)^p(x)\,dx\right)^{\frac{1}{q}}$$

and

$$\int_a^b g(f+g)^{p-1}(x)\,dx \leq \left(\int_a^b g^p(x)\,dx\right)^{\frac{1}{p}} \left(\int_a^b (f+g)^p(x)\,dx\right)^{\frac{1}{q}}.$$

Adding by members the last two inequalities, we reach

$$\int_a^b (f+g)^p(x)\,dx \leq \left[\left(\int_a^b f^p(x)\,dx\right)^{\frac{1}{p}} + \left(\int_a^b g^p(x)\,dx\right)^{\frac{1}{p}}\right]$$
$$\times \left(\int_a^b (f+g)^p(x)\,dx\right)^{\frac{1}{q}}$$

and from the latter, by division by members, we obtain the inequality

$$\left(\int_a^b (f+g)^p(x)\,\mathrm{d}x\right)^{1-\frac{1}{q}} \leqslant \left(\int_a^b f^p(x)\,\mathrm{d}x\right)^{\frac{1}{p}} + \left(\int_a^b g^p(x)\,\mathrm{d}x\right)^{\frac{1}{p}}$$

and because $1 - \dfrac{1}{q} = \dfrac{1}{p}$ we finally arrive to

$$\left(\int_a^b (f+g)^p(x)\,\mathrm{d}x\right)^{\frac{1}{p}} \leqslant \left(\int_a^b f^p(x)\,\mathrm{d}x\right)^{\frac{1}{p}} + \left(\int_a^b g^p(x)\,\mathrm{d}x\right)^{\frac{1}{p}}.$$

The last inequality, again using the classic notation of L^p-norm, can be written in the form

$$\|f+g\|_p \leqslant \|f\|_p + \|g\|_p,$$

which is the generalization of triangular inequality in spaces. □

Historical Notation 7. *Herman Minkowski, German mathematician, was born on June 22, 1864 in Lithuania and died on January 12, 1909 in Gottingen. He received his PhD from the University of Konigsbgerg at the age of 21. He taught at the Federal Institute of Technology (ETH) in Zurich and in the last years of his life at the University of Gottingen, where he had the mathematics course created for him by David Hilbert. But Schwarz needed a two-dimensional form of Cauchy inequality. In particular, he had to show that if $S \subset \mathbb{R}^2$ and $f,g : S \to \mathbb{R}$, then the double integrals $A = \iint_S f^2 \mathrm{d}x\mathrm{d}y$, $B = \iint_S fg \mathrm{d}x\mathrm{d}y$, and $A = \iint_S g^2 \mathrm{d}x\mathrm{d}y$ must satisfy the inequality $|B| \leqslant \sqrt{A} \cdot \sqrt{C}$ and he had to know that the inequality is strict unless the functions are proportional. Schwarz would face a problem for a number of reasons, including the fact that the severity of the inequality can be lost using integrals. So he had to find a different solution, which he did as he discovered a proof that went down in history. He based his proof on an impressive observation. He observed that the real polynomial $p(t) = \iint_S (t\,f(x,y) + g(x,y))^2 \mathrm{d}x\mathrm{d}y = At^2 + 2Bt + C$ is always non-negative and that $p(t)$ it is strictly positive unless f and g are proportional. The binomial formula tells us that the coefficients must satisfy $B^2 \leqslant AC$ and unless f and g are proportional then we* have strict inequality $B^2 < AC$. So from a simple algebraic observation Schwarz discovered everything he needed to know.

3. A Cauchy–Bunyakovsky–Schwarz-Type Inequality

In this section, we present a Cauchy–Schwarz-type inequality, which arose from the study of mechanical properties of elastic rod (see Ref. [6]). First, we present the discreet case and second, its integral version.

Theorem 2. *Let $x = (x_1, \ldots, x_n)$ and $y = (y_1, \ldots, y_n)$ be arbitrary elements of $\mathbb{R}^n$. Then the following holds:*

$$\left(\sum_{i=1}^n p_i\right)\left(\sum_{i=1}^n p_i x_i y_i\right) - \left(\sum_{i=1}^n p_i x_i\right)\left(\sum_{i=1}^n p_i y_i\right)$$
$$\leq \sqrt{\left(\sum_{i=1}^n p_i\right)\left(\sum_{i=1}^n p_i x_i^2\right) - \left(\sum_{i=1}^n p_i x_i\right)^2}$$
$$\times \sqrt{\left(\sum_{i=1}^n p_i\right)\left(\sum_{i=1}^n p_i y_i^2\right) - \left(\sum_{i=1}^n p_i y_i\right)^2}, \quad (10)$$

for any positive real numbers $p_1, p_2, \ldots, p_n$ (weights). The above relation (10) becomes an equality if and only if there exist $a, b \in \mathbb{R}$ such that $ax + by = c$, a constant vector, i.e., if there exists a linear combination of the vectors x and y which is a constant vector.

Sketch of the Proof. We present the basic steps of the proof. (A proof in detail and some important remarks are available in Ref. [6].)

We define the function of two variables $\langle .,. \rangle : \mathbb{R}^n \times \mathbb{R}^n \to \mathbb{R}$ defined by

$$\langle x, y \rangle = \left(\sum_{i=1}^n p_i\right)\left(\sum_{i=1}^n p_i x_i y_i\right) - \left(\sum_{i=1}^n p_i x_i\right)\left(\sum_{i=1}^n p_i y_i\right). \quad (11)$$

It is easy to prove that the function $\langle .,. \rangle : \mathbb{R}^n \times \mathbb{R}^n \to \mathbb{R}$ is a bilinear form, i.e., it satisfies the properties of Proposition 2. These properties can be easy verified by taking into account definition (11). Concerning the first property, it can be proved using the CBS inequality (3), that is,

$$\langle x, x \rangle = \left(\sum_{i=1}^{n} p_i\right)\left(\sum_{i=1}^{n} p_i x_i^2\right) - \left(\sum_{i=1}^{n} p_i x_i\right)\left(\sum_{i=1}^{n} p_i x_i\right)$$

$$= \left[\sum_{i=1}^{n} (\sqrt{p_i})^2\right]\left[\sum_{i=1}^{n} (\sqrt{p_i} x_i)^2\right] - \left(\sum_{i=1}^{n} \sqrt{p_i}\sqrt{p_i} x_i\right)^2 \geqslant 0.$$

The inequality becomes equality, i.e., $\langle x, x \rangle = 0$, if and only if the vectors $(\sqrt{p_1}, \sqrt{p_2}, \ldots, \sqrt{p_n})$ and $(\sqrt{p_1} x_1, \sqrt{p_2} x_2, \ldots, \sqrt{p_n} x_n)$ are linear dependent, that is, $x = (x_1, x_2, \ldots, x_n)$ is a constant vector. With these considerations, inequality (11) takes the form

$$\langle x, y \rangle \leqslant \sqrt{\langle x, x \rangle}\sqrt{\langle y, y \rangle}, \quad \forall\, x, y \in \mathbb{R}^n,$$

which is inequality (2).

The integral version of the CBS-type inequality is presented by the following proposition. (We relax the notation $\int_\Omega f(x)\mathrm{d}x$ and simply write $\int_\Omega f$.)

Proposition 7. *Let the functions $f, g, p : \Omega \subset \mathbb{R}^n \to \mathbb{R}$ be continuous on a compact domain Ω, and let p be positive. The following inequality holds:*

$$\int_\Omega p \int_\Omega pfg - \int_\Omega pf \int_\Omega pg \leqslant \sqrt{\int_\Omega p \int_\Omega pf^2 - \left(\int_\Omega pf\right)^2}$$

$$\times \sqrt{\int_\Omega p \int_\Omega pg^2 - \left(\int_\Omega pg\right)^2}, \qquad (12)$$

where the inequality becomes equality if and only if there exist $\alpha, \beta \in \mathbb{R}$ such that $\alpha f + \beta g = c$, a constant.

Notation. We can imagine that inequality (12) is obtained directly from (10) by replacing the discrete "sums" by continuous ones, i.e., by integrals, and what remains to be shown is that the steps in the proof of the continuous case are exactly the same as those of the discrete one.

Sketch of the Proof. We set

$$\langle f,g\rangle = \int_\Omega p \int_\Omega pfg - \int_\Omega pf - \int_\Omega pg. \tag{13}$$

It easy to verify that the properties (i)–(v) in Proposition 2 are satisfied and inequality (12) can be written as

$$\langle f,g\rangle \leqslant \sqrt{\langle f,f\rangle}\sqrt{\langle g,g\rangle}, \quad \forall f,g \in C(\Omega), \tag{14}$$

which leads to the proof of the result.

4. Strengthened Cauchy–Bunyakovsky–Schwarz Inequality for Elasticity

In this section, we present the strengthened CBS inequality:

$$|\langle u,v\rangle| \leq \gamma\sqrt{\langle u,u\rangle}\sqrt{\langle v,v\rangle}, \quad u \in U, \ v \in V, \ U \cap V = \{0\}, \tag{15}$$

where U and V are two linear subspaces, $\langle .,.\rangle$ is the bilinear form corresponding to the variational formulation of the problem and γ is a constant between 0 and 1 depending only on the spaces U and V. (In fact, the constant γ could take the values 0 and 1, but if $\gamma = 0$, then the vectors u and v are orthogonal and for $\gamma = 1$ arises the classical CBS inequality.) The estimate of the constant that appears in the second term of this inequality and in particular its optimal value (that is, the smallest value it can taken to hold for all $u \in U$ and $v \in V$) is of fundamental practical importance in solving the problems that appear, especially in solving 2D and 3D elasticity problems, which are the ones that interest us. For example, Margenov [8] gave estimates on the 2D elasticity problem on a triangular mesh and proved that for a uniform mesh of right isosceles triangles, the constant γ^2 is bounded above by 3/4 uniformly on the mesh. Achchab and Maitre [7] proved that this result remains true for every triangular mesh. In the case of a 3D problem, Jung and Maire [10] reported that numerical experiments on a particular triangulation and specifically in the refinement of a standard tetrahedron showed that the constant γ^2 is bounded by 9/10. Achchab et al. [9] generalized this result and proved that it remains valid for an arbitrary tetrahedron mesh. In addition, Achchab *et al.* proved that the constant γ plays a significant role *in posteriori* error estimates for elasticity problems.

In this section, we present the basic theorem needed for the establishment of the strengthened CBS inequality, which is needed for the estimation of the constant γ. (For more details and for the proofs, see also the work of Achchab *et al.* [9].)

Theorem 3. *Given a finite-dimensional Hilbert space H, an inner product $\langle .,.\rangle$ on it and the subspaces U and V of H with $U \cap V = \{0\}$, there exist a constant $\gamma \in (0,1)$ depending only on U and V such that for all $u \in U$ and $v \in V$ the following strengthened CBS inequality holds:*

$$|\langle u, v \rangle| \leq \gamma \|u\| \|v\|, \tag{16}$$

where the norm is induced by the inner product.

Theorem 4. *Let A be a symmetric, positive semidefine 2×2 block matrix*

$$A = \begin{pmatrix} A_{11} & A_{12} \\ A_{21} & A_{22} \end{pmatrix},$$

where A_{11} is invertible and where the partitioning components to U, V, which are the spaces of vectors with only nonzero first, and second components $u \in U$, $v \in V$ of the form $u = \begin{pmatrix} u_1 \\ 0 \end{pmatrix}$ and $v = \begin{pmatrix} 0 \\ v_2 \end{pmatrix}$, respectively. Assuming that the kernel of A, $N(A)$, is included in V, the optimal value of the constant γ is given by

$$\gamma^2 = \sup_{\substack{u \in U \\ v \in V \setminus N(A)}} \frac{\left(u^T A v\right)^2}{\left(u^T A u\right)\left(v^T A v\right)}.$$

Let Ω be a bounded open connected subset of $\mathbb{R}^3$ with a Lipschitz continuous boundary Γ. Let, also, Γ_0 be a measurable subset of Γ and Γ_1 its complement on Γ. We consider the following three-dimensional boundary value problem of linear elasticity:

$$\sum_{j=1}^{3} \frac{\partial \sigma_{ij}(u)}{\partial x_j} + f_i \quad \text{in } \Omega, \quad i = 1, 2, 3,$$

$$u_i = 0 \quad \text{on } \Gamma_0, \quad i = 1, 2, 3,$$

$$\sum_{j=1}^{3} \sigma_{ij}(u) v_j = g_i \quad \text{on } \Gamma_1, \quad i = 1, 2, 3.$$

$v = (v_1, v_2, v_3)$ is the external normal to Ω, with the classical constitutive law for isotropic material with Láme modula λ and μ

$$c_{ij}(u) = \lambda \left(\sum_{k=1}^{3} \varepsilon_{kk}(u) \right) \delta_{ij} + 2\mu \varepsilon_{ij}(u),$$

where

$$\varepsilon_{ij}(u) = \frac{1}{2} \left(\frac{\partial u_i}{\partial x_j} + \frac{\partial u_j}{\partial x_i} \right).$$

We introduce the space

$$V(\Omega) = \left\{ v \in \left[H^1(\Omega) \right]^3 ; \ v = 0 \text{ on } \Gamma_0 \right\}.$$

For $f \in \left[L^2(\Omega) \right]^3$ and $g \in \left[L^2(\Gamma_0) \right]^3$, the variational formulation of the problem is given by

$$\text{Find } u \in V(\Omega) \text{ such that } a(u, v) = L(v), \quad \forall v \in \Omega,$$

where

$$a(u, v) = \lambda \int_\Omega \operatorname{div}(u) \operatorname{div}(v) \, \mathrm{d}x + 2\mu \sum_{i,j=1}^{3} \varepsilon_{ij}(u) \varepsilon_{ij}(v) \mathrm{d}x$$

and

$$L(v) = \int_\Omega fv \, \mathrm{d}x + \int_{\Gamma_1} gv \, \mathrm{d}\sigma.$$

Finally, we define the two forms $a_1(,.,)$ and $a_2(,.,)$ by

$$a_1(u,v) = \int_\Omega \text{div}(u)\text{div}(v)dx$$

and

$$a_2(u,v) = \sum_{i,j=1}^{3} \int_\Omega c_{ij}(u)\, c_{ij}(v)\, dx.$$

Under the above considerations, the following theorem holds:

Theorem 5. *The constants γ_1, γ_2 of the strengthened CBS inequality associated with the forms $a_1(u,v)$ and $a_2(u,v)$ satisfy*

$$\gamma_i^2 \leq \frac{9}{10}, \quad i = 1, 2.$$

Appendix

Proof 1. Consider the double sum $\sum_{i=1}^{n}\sum_{j=1}^{n}(x_i y_j - x_j y_i)^2$. By deleting the parentheses and grouping the same terms, we successively get

$$\sum_{i=1}^{n}\sum_{j=1}^{n}(x_i y_j - x_j y_i)^2 = \sum_{i=1}^{n}\sum_{j=1}^{n}(x_i^2 y_j^2 + x_j^2 y_i^2 - 2x_i x_j y_i y_j)$$

$$= \sum_{i=1}^{n}\left(\sum_{j=1}^{n} x_i^2 y_j^2 + \sum_{j=1}^{n} x_j^2 y_i^2 - \sum_{j=1}^{n} 2x_i x_j y_i y_j\right)$$

$$= \sum_{i=1}^{n}\left(x_i^2 \sum_{j=1}^{n} y_j^2 + y_i^2 \sum_{j=1}^{n} x_j^2 - 2x_i y_i \sum_{j=1}^{n} x_j y_j\right)$$

$$= \sum_{i=1}^{n}\left(x_i^2 \sum_{j=1}^{n} y_j^2\right) + \sum_{i=1}^{n}\left(y_i^2 \sum_{j=1}^{n} x_j^2\right)$$

$$- 2\sum_{i=1}^{n}\left(x_i y_i \sum_{j=1}^{n} x_j y_j\right)$$

$$= \sum_{j=1}^{n} y_j^2 \sum_{i=1}^{n} x_i^2 + \sum_{j=1}^{n} x_j^2 \sum_{i=1}^{n} y_i^2 - 2 \sum_{j=1}^{n} x_j y_j \sum_{i=1}^{n} x_i y_i$$

$$= \sum_{i=1}^{n} x_i^2 \sum_{j=1}^{n} y_j^2 + \sum_{i=1}^{n} y_i^2 \sum_{j=1}^{n} x_j^2 - 2 \sum_{i=1}^{n} x_i y_i \sum_{j=1}^{n} x_j y_j$$

$$= 2 \left(\sum_{i=1}^{n} x_i^2 \right) \left(\sum_{i=1}^{n} y_i^2 \right) - 2 \left(\sum_{j=1}^{n} x_i y_i \right)^2.$$

Since the left-hand side of the equation is a sum of squares of real numbers greater or equal to zero,

$$\left(\sum_{i=1}^{n} x_i^2 \right) \left(\sum_{i=1}^{n} y_i^2 \right) \geqslant \left(\sum_{i=1}^{n} x_i y_i \right)^2,$$

that is, the inequality (9). □

Proof 2. Consider the next trinomial with respect to t:

$$f(t) = \left(\sum_{i=1}^{n} x_i^2 \right) t^2 - 2 \left(\sum_{i=1}^{n} x_i y_i \right) t + \sum_{i=1}^{n} y_i^2.$$

Doing the operations in the second member, we note that

$$f(t) = \left(\sum_{i=1}^{n} x_i^2 \right) t^2 - 2 \left(\sum_{i=1}^{n} x_i y_i \right) t + \sum_{i=1}^{n} y_i^2$$

$$= \sum_{i=1}^{n} x_i^2 t^2 - 2 \sum_{i=1}^{n} x_i y_i t + \sum_{i=1}^{n} y_i^2$$

$$= \sum_{i=1}^{n} (x_i t)^2 - 2 \sum_{i=1}^{n} (x_i t) y_i + \sum_{i=1}^{n} y_i^2$$

$$= \sum_{i=1}^{n} \left[(x_i t)^2 - 2 (x_i t) y_i + y_i^2 \right]$$

$$= \sum_{i=1}^{n} (x_i t - y_i)^2 \geqslant 0.$$

Since $f(t) \geq 0$ for every $t \in \mathbb{R}$, the discriminant of $f(t)$ is negative, so it holds that

$$\left(\sum_{i=1}^{n} x_i y_i\right)^2 - \left(\sum_{i=1}^{n} x_i^2\right)\left(\sum_{i=1}^{n} y_i^2\right) \leq 0 \Leftrightarrow \left(\sum_{i=1}^{n} x_i y_i\right)^2$$

$$\leq \left(\sum_{i=1}^{n} x_i^2\right)\left(\sum_{i=1}^{n} y_i^2\right),$$

that is, inequality (9) holds. □

Proof 3. We assume that the CBS inequality (4) is true for every $(x_1, \ldots, x_n)$ and $(y_1, \ldots, y_n)$. Based on this hypothesis, we come to the true. If $\sum_{i=1}^{n} x_i^2 = 0$ or $\sum_{i=1}^{n} y_i^2 = 0$, then $x_1 = x_2 = \cdots = x_n = 0$ or $y_1 = y_2 = \cdots = y_n = 0$ and then the conclusion is obvious. Now we assume that $X_n = \sum_{i=1}^{n} x_i^2 \neq 0$ and $Y_n = \sum_{i=1}^{n} y_i^2 \neq 0$ and we set $a_i = \frac{x_i}{\sqrt{X_n}}$ and $b_i = \frac{y_i}{\sqrt{Y_n}}$, for each $i = 1, 2, \ldots, n$. Then we have

$$\sum_{i=1}^{n} a_i^2 = \sum_{i=1}^{n}\left(\frac{x_i}{\sqrt{X_n}}\right)^2 = \sum_{i=1}^{n}\frac{x_i^2}{X_n} = \frac{1}{X_n}\sum_{i=1}^{n} x_i^2 = \frac{1}{X_n}X_n = 1$$

and

$$\sum_{i=1}^{n} b_i^2 = 1.$$

Since we have assumed that the initial inequality (4) is true for all n-groups, we apply it to $(a_1, \ldots, a_n)$ and $(b_1, \ldots, b_n)$ and then we get

$$1 \cdot 1 \geq \left(\sum_{i=1}^{n} a_i b_i\right)^2$$

from which inequality arises

$$a_1 b_1 + a_2 b_2 + \cdots + a_n b_n \leq 1.$$

From the latter, we take successively

$$2(a_1b_1 + a_2b_2 + \cdots + a_nb_n) \leqslant 2 \Leftrightarrow$$
$$2(a_1b_1 + a_2b_2 + \cdots + a_nb_n) \leqslant 1 + 1 \Leftrightarrow$$
$$2(a_1b_1 + a_2b_2 + \cdots + a_nb_n) \leqslant (a_1^2 + a_2^2 + \cdots + a_n^2)$$
$$+ (b_1^2 + b_2^2 + \cdots + b_n^2),$$

thus

$$(a_1 - b_1)^2 + (a_2 - b_2)^2 + \cdots + (a_n - b_n)^2 \geqslant 0,$$

which is obviously true. □

Proof 4. Let

$$X = \sqrt{x_1^2 + x_2^2 + \cdots + x_n^2}, \quad Y = \sqrt{y_1^2 + y_2^2 + \cdots + y_n^2}.$$

From the inequality of the arithmetic-geometric mean, we have

$$\sum_{i=1}^{n} \frac{x_i y_i}{XY} \leqslant \frac{1}{2} \sum_{i=1}^{n} \left(\frac{x_i^2}{X^2} + \frac{y_i^2}{Y^2} \right)$$

$$= \frac{1}{2} \left(\sum_{i=1}^{n} \frac{x_i^2}{X^2} + \sum_{i=1}^{n} \frac{y_i^2}{Y^2} \right)$$

$$= \frac{1}{2} \left(\frac{1}{X^2} \sum_{i=1}^{n} x_i^2 + \frac{1}{Y^2} \sum_{i=1}^{n} y_i^2 \right)$$

$$= 1,$$

so it will be

$$\sum_{i=1}^{n} x_i y_i \leqslant XY = \sqrt{x_1^2 + x_2^2 + \cdots + x_n^2} \sqrt{y_1^2 + y_2^2 + \cdots + y_n^2}.$$

Therefore,
$$\left(\sum_{i=1}^n x_i y_i\right)^2 \leqslant \left(\sum_{i=1}^n x_i^2\right)\left(\sum_{i=1}^n y_i^2\right). \qquad \square$$

Proof 5. Let
$$A_n = x_1^2 + x_2^2 + \cdots + x_n^2,$$
$$B_n = y_1^2 + y_2^2 + \cdots + y_n^2$$
and
$$C_n = x_1 y_1 + x_2 y_2 + \cdots + x_n y_n.$$

It follows from the inequality of the arithmetic-geometric mean that
$$\frac{\sum_{i=1}^n x_i^2 \sum_{i=1}^n y_i^2}{\left(\sum_{i=1}^n x_i y_i\right)^2} + 1 = \frac{A_n B_n}{C_n^2} + 1 = \frac{\sum_{i=1}^n x_i^2 B_n}{C_n^2} + \frac{\sum_{i=1}^n y_i^2}{B_n}$$
$$= \sum_{i=1}^n \left(\frac{x_i^2 B_n}{C_n^2} + \frac{y_i^2}{B_n}\right) a$$
$$\geqslant \sum_{i=1}^n \left(2 \frac{x_i^2 B_n}{C_n^2} \cdot \frac{y_i^2}{B_n}\right)$$
$$= 2 \frac{1}{C_n} \sum_{i=1}^n x_i y_i = 2 \frac{1}{C_n} C_n = 2.$$

Therefore,
$$A_n B_n \geqslant C_n^2,$$
that is,
$$\left(x_1^2 + x_2^2 + \cdots + x_n^2\right)\left(y_1^2 + y_2^2 + \cdots + y_n^2\right)$$
$$\geqslant (x_1 y_1 + x_2 y_2 + \cdots + x_n y_n)^2. \qquad \square$$

Proof 6. We prove the CBS inequality (9) by mathematical induction. Starting the induction for 1, the hypothesis is trivial. Note that

$$(x_1y_1 + x_2y_2)^2 = x_1^2y_1^2 + 2x_1y_1x_2y_2 + x_2^2y_2^2$$
$$\leqslant x_1^2y_1^2 + x_1^2y_2^2 + x_2^2y_1^2 + x_2^2y_2^2$$
$$= (x_1^2 + x_2^2)(y_1^2 + y_2^2),$$

which shows that inequality (9) holds for $n = 2$. Assume that inequality (9) holds for $n = k$, that is,

$$\left(\sum_{i=1}^{k} x_i y_i\right)^2 \leqslant \left(\sum_{i=1}^{k} x_i^2\right)\left(\sum_{i=1}^{k} y_i^2\right),$$

and we prove that it is true for $n = k + 1$. We have

$$\sqrt{\sum_{i=1}^{k+1} x_i^2} \cdot \sqrt{\sum_{i=1}^{k+1} y_i^2} = \sqrt{\sum_{i=1}^{k} x_i^2 + x_{k+1}^2} \cdot \sqrt{\sum_{i=1}^{k} y_i^2 + y_{k+1}^2}$$

$$\geqslant \sqrt{\sum_{i=1}^{k} x_i^2} \cdot \sqrt{\sum_{i=1}^{k} y_i^2} + |x_{k+1}y_{k+1}|$$

$$\geqslant \sum_{i=1}^{k} |x_i y_i| + |x_{k+1}y_{k+1}| = \sum_{i=1}^{k+1} |x_i y_i|.$$

This means that inequality (9) holds for $n = k + 1$, and so we conclude that CBS holds for all natural numbers. This completes the proof. □

Proof 7. Let

$$X = \{x_1y_1, \ldots, x_1y_n, x_2y_1, \ldots, x_2y_n, \ldots, x_ny_1, \ldots, x_ny_n\},$$
$$Y = \{x_1y_1, \ldots, x_1y_n, x_2y_1, \ldots, x_2y_n, \ldots, x_ny_1, \ldots, x_ny_n\},$$
$$Z = \{x_1y_1, \ldots, x_1y_n, x_2y_1, \ldots, x_2y_n, \ldots, x_ny_1, \ldots, x_ny_n\},$$
$$W = \{x_1y_1, \ldots, x_ny_1, x_1y_2, \ldots, x_ny_2, \ldots, x_1y_n, \ldots, x_ny_n\}.$$

It is easy to observe that the X and Y have the same classification, while the Z and W have a mixed classification. Applying the rearrangement inequality, we have

$$(x_1y_1)(x_1y_1) + \cdots + (x_1y_n)(x_1y_n) + (x_2y_1)(x_2y_1) + \cdots$$
$$+ (x_2y_n)(x_2y_n) + \cdots$$
$$+ (x_ny_1)(x_ny_1) + \cdots + (x_ny_n)(x_ny_n)$$
$$\geqslant (x_1y_1)(x_1y_1) + \cdots + (x_1y_n)(x_ny_1)$$
$$+ (x_2y_1)(x_1y_2) + \cdots + (x_2y_n)(x_ny_2) + \cdots$$
$$+ (x_ny_1)(x_ny_1) + \cdots + (x_ny_n)(x_ny_n),$$

which can be simplified and written as follows:

$$\left(x_1^2 + x_2^2 + \cdots + x_n^2\right)\left(y_1^2 + y_2^2 + \cdots + y_n^2\right)$$
$$\geqslant (x_1y_1 + x_2y_2 + \cdots + x_ny_n)^2.$$

The last inequality is the desired one. □

Proof 8. Since if $x = (0, 0, \ldots, 0)$ or $y = (0, 0, \ldots, 0)$ inequality (4) is always valid, we consider the nonzero vectors $x = (x_1, x_2, \ldots, x_n)$ and $y = (y_1, y_2, \ldots, y_n)$. Then for any real number t, the following holds:

$$|x + ty|^2 = (x + ty) \cdot (x + ty) = x \cdot x + 2(x \cdot y)t + (y \cdot y)t^2 = |x|^2 + 2(x \cdot y)t + |y|^2 t^2,$$

so inequality

$$|x|^2 + 2(x \cdot y)t + |y|^2 t^2 \geqslant 0$$

holds for every $t \in \mathbb{R}$. But the first term of the last inequality is a trinomial with respect to t and because it is greater than or equal to 0 for each $t \in \mathbb{R}$ and its coefficient is $|y|^2 > 0$ (because we have assumed that $y = (y_1, y_2, \ldots, y_n) \neq 0$), its discriminant will be less

than or equal to zero. Therefore,
$$4(x \cdot y)^2 - 4|x|^2|y|^2 \leqslant 0,$$
so it will be
$$(x \cdot y)^2 \leqslant |x|^2|y|^2.$$
Using the notations
$$x \cdot y = x_1y_1 + x_2y_2 + \cdots + x_ny_n, \ |x|^2 = \sum_{i=1}^{k} x_i^2, \ |y|^2 = \sum_{i=1}^{k} y_i^2,$$
we get the following:
$$\left(\sum_{i=1}^{n} x_iy_i\right)^2 \leqslant \left(\sum_{i=1}^{n} x_i^2\right)\left(\sum_{i=1}^{n} y_i^2\right),$$
which is inequality (9). □

Proof 9. Consider the vectors
$$x = (x_1, x_2, \ldots, x_n), \quad y = (y_1, y_2, \ldots, y_n).$$
From the formula of inner product $x \cdot y = |x||y|\cos(x, y)$, we conclude that $x \cdot y \leqslant |x||y|$. Using the notations
$$x \cdot y = x_1y_1 + x_2y_2 + \cdots + x_ny_n, \ |x|^2 = \sum_{i=1}^{k} x_i^2, \ |y|^2 = \sum_{i=1}^{k} y_i^2,$$
the CBS inequality arises. □

References

[1] A. L. Cauchy, *Cours d'analyse de l'Ecole royale polytechnique; par m. Augustin-Louis Cauchy... 1. re partie. Analyse algébrique* (de l'Imprimerie royale, 1821).

[2] V. t. Buniakovskiĭ, *Sur quelques inégalités concernant les intégrales ordinaires et les intégrales aux différences finies*, Vol. 1 (Imperatorskoj Akademīi Nauk, 1859).

[3] H. A. Schwarz, Über ein die flächen kleinsten flächeninhalts betreffendes problem der variationsrechnung. In *Gesammelte Mathematische Abhandlungen* (Springer, Berlin, 1890), pp. 223–269.

[4] S. S. Dragomir, A survey on Cauchy-Bunyakovsky-Schwarz type discrete inequalities, *J. Inequal. Pure Appl. Math.* **4**(3), 1–142 (2003).

[5] J. M. Steele, *The Cauchy-Schwarz Master Class: An Introduction to the Art of Mathematical Inequalities* (Cambridge University Press, Cambridge, 2004).

[6] M. Birsan and T. Birsan, An inequality of Cauchy Schwarz type with application in the theory of elastic rods, *Lib. Math.* (Vols. I–XXXI) **31**, 123–126 (2011).

[7] B. Achchab and J. Maitre, Estimate of the constant in two strengthened CBS inequalities for fem systems of 2d elasticity: Application to multilevel methods and a posteriori error estimators, *Numer. Linear Algebra Appl.* **3**(2), 147–159 (1996).

[8] S. D. Margenov, Upper bound of the constant in the strengthened CBS inequality for FEM 2D elasticity equations, *Numer. Linear Algebra Appl.* **1**(1), 65–74 (1994).

[9] B. Achchab, O. Axelsson, L. Laayouni and A. Souissi, Strengthened Cauchy–Bunyakowski–Schwarz inequality for a three-dimensional elasticity system, *Numer. Linear Algebra Appl.* **8**(3), 191–205 (2001).

[10] M. Jung and J.-F. Maitre, Some remarks on the constant in the strengthened CBS inequality: Estimate for hierarchical finite element discretizations of elasticity problems, *Numer. Methods Partial Differ. Equ. Int. J.* **15**(4), 469–487 (1999).

[11] D. Labropoulou, P. Vafeas and G. Dassios, Anisotropic elasticity and harmonic functions in Cartesian geometry. In *Mathematical Analysis in Interdisciplinary Research* (Springer, Cham, Switzerland, 2021), pp. 523–553.

[12] M. Bîrsan and H. Altenbach, On the theory of porous elastic rods, *Int. J. Solids Struct.* **48**(6), 910–924 (2011).

[13] H.-H. Wu and S. Wu, Various proofs of the Cauchy-Schwarz inequality, *Octogon Math. Mag.* **17**(1), 221–229 (2009).

[14] T. Rassias, *Mathematics I*, 2nd edn. (Greek) (Tsotras Publications, Athens, Greece, 2017).

© 2024 World Scientific Publishing Company
https://doi.org/10.1142/9789811267048_0019

Chapter 19

Lie Bracket Derivations in Banach Lie Algebras

Jung Rye Lee[*,§], Choonkil Park[†,¶], and Michael Th. Rassias[‡,||]

[*]*Department of Data Science, Daejin University, Kyunggi, Korea*
[†]*Department of Mathematics, Hanyang University, Seoul, Korea*
[‡]*Department of Mathematics and Engineering Sciences, Hellenic Military Academy, Vari Attikis, Greece*
[§]*jrlee@daejin.ac.kr*
[¶]*baak@hanyang.ac.kr*
[||]*mthrassias@yahoo.com*

In this chapter, we introduce Lie bracket derivations in complex Banach Lie algebras. Using the direct method and the fixed point method, we prove the Hyers–Ulam stability of Lie bracket derivations in complex Banach Lie algebras.

1. Introduction and Preliminaries

Let A be a complex Banach Lie algebra with Lie bracket $[\cdot,\cdot]$ and $\text{Der}(A)$ be the set of $\mathbb{C}$-linear (bounded) Lie derivations in A.

For $\delta_1, \delta_2 \in \mathrm{Der}(A)$,

$$\delta_1 \circ \delta_2([a,b]) = [\delta_1 \circ \delta_2(a), b] + [\delta_2(a), \delta_1(b)] + [\delta_1(a), \delta_2(b)] + [a, \delta_1 \circ \delta_2(b)],$$
$$\delta_2 \circ \delta_1([a,b]) = [\delta_2 \circ \delta_1(a), b] + [\delta_1(a), \delta_2(b)] + [\delta_2(a), \delta_1(b)] + [a, \delta_2 \circ \delta_1(b)]$$

for all $a, b \in A$. Let $[\delta_1, \delta_2] = \delta_1 \circ \delta_2 - \delta_2 \circ \delta_1$. Then,

$$[\delta_1, \delta_2]([a,b]) = [[\delta_1, \delta_2](a), b] + [a, [\delta_1, \delta_2](b)]$$

for all $a, b \in A$. Since $[\delta_1, \delta_2] : A \to A$ is $\mathbb{C}$-linear, $[\delta_1, \delta_2] \in \mathrm{Der}(A)$ for all $\delta_1, \delta_2 \in \mathrm{Der}(A)$. Thus, $\mathrm{Der}(A)$ is a Lie algebra with Lie bracket $[\delta_1, \delta_2]$, since $\delta_1 + \delta_2$ and $\alpha \delta_1$ are $\mathbb{C}$-linear derivations in A for all $\delta_1, \delta_2 \in \mathrm{Der}(A)$ and all $\alpha \in \mathbb{C}$. One can easily show that $\mathrm{Der}(A)$ is a Banach space, since A is complete.

In this chapter, we introduce and investigate Lie bracket derivations in a complex Banach Lie algebra.

Definition 1. Let A be a complex Banach Lie algebra and $G, H : A \to A$ be $\mathbb{C}$-linear mappings. Let $[G, H](a) = G(H(a)) - H(G(a))$ for all $a \in A$. A $\mathbb{C}$-linear mapping $[G, H] : A \to A$ is called a *Lie bracket derivation* in A if $[G, H]$ is a Lie derivation in A, i.e.,

$$[G, H]([a, b]) = [[G, H](a), b] + [a, [G, H](b)]$$

for all $a, b \in A$.

Since $[\delta_1, \delta_2] \in \mathrm{Der}(A)$ for $\delta_1, \delta_2 \in \mathrm{Der}(A)$, $[\delta_1, \delta_2]$ is a Lie bracket derivation.

The stability problem of functional equations originated from a question of Ulam [23] concerning the stability of group homomorphisms. Hyers [11] gave a first affirmative partial answer to the question of Ulam for Banach spaces. Hyers' theorem was generalized by Aoki [1] for additive mappings and by Rassias [20] for linear mappings by considering an unbounded Cauchy difference. A generalization of the Rassias theorem was obtained by Găvruta [10] by replacing the unbounded Cauchy difference by a general control function in the spirit of Rassias' approach. The stability of quadratic functional equation was proved by Skof [21] for mappings $f : E_1 \to E_2$, where E_1 is a normed space and E_2 is a Banach space. Cholewa [6] noted that the theorem of Skof is still true if the relevant domain E_1 is replaced by an abelian group.

Park [15, 16] defined additive ρ-functional inequalities and proved the Hyers–Ulam stability of the additive ρ-functional inequalities in Banach spaces and non-Archimedean Banach spaces. The stability problems of various functional equations have been extensively investigated by a number of authors [2, 9, 22, 24].

We recall a fundamental result in fixed point theory.

Theorem 1 ([3, 7]). *Let (X, d) be a complete generalized metric space and let $J : X \to X$ be a strictly contractive mapping with Lipschitz constant $\alpha < 1$. Then for each given element $x \in X$, either*

$$d(J^n x, J^{n+1} x) = \infty$$

for all non-negative integers n or there exists a positive integer n_0 such that

(1) $d(J^n x, J^{n+1} x) < \infty, \quad \forall n \geq n_0,$
(2) *the sequence $\{J^n x\}$ converges to a fixed point y^* of J,*
(3) *y^* is the unique fixed point of J in the set $Y = \{y \in X \mid d(J^{n_0} x, y) < \infty\}$,*
(4) $d(y, y^*) \leq \frac{1}{1-\alpha} d(y, Jy)$ *for all $y \in Y$.*

In 1996, Isac and Rassias [12] were the first to provide applications of stability theory of functional equations for the proof of new fixed point theorems with applications. By using fixed point methods, the stability problems of several functional equations have been extensively investigated by a number of authors [4, 5, 17–19].

This chapter is organized as follows: In Section 2, we prove the Hyers–Ulam stability of Lie bracket derivations in a complex Banach Lie algebra by using the fixed point method. In Section 3, we prove the Hyers–Ulam stability of Lie bracket derivations in a Banach Lie algebra by using the direct method.

Throughout this chapter, let A be a complex Banach Lie algebra and p be a nonzero complex number with $|p| < 1$.

2. Hyers–Ulam Stability of Lie Bracket Derivations in Complex Banach Lie Algebras: Fixed Point Method

In this section, we prove the Hyers–Ulam stability of Lie bracket derivations in complex Banach Lie algebras by using the fixed point method.

Lemma 1 ([14, Theorem 2.1]). *Let X be a complex normed space and Y be a complex Banach space. Let $f : X \to Y$ be a mapping such that*

$$f(\lambda(a+b)) = \lambda f(a) + \lambda f(b)$$

for all $\lambda \in \mathbb{T}^1 := \{\xi \in \mathbb{C} : |\xi| = 1\}$ and all $a, b \in X$. Then $f : X \to Y$ is $\mathbb{C}$-linear.

For given mappings $\phi, \psi : A \to A$, we define

$$E_\lambda \phi(x,y) := \phi(\lambda(x+y)) - \lambda\phi(x) - \lambda\phi(y),$$
$$F_\lambda \psi(x,y) := 2\psi\left(\lambda\frac{x+y}{2}\right) - \lambda\psi(x) - \lambda\psi(y)$$

for all $\lambda \in \mathbb{T}^1$ and all $x, y \in A$.

Lemma 2. *Let $g, h : X \to Y$ be mappings satisfying*

$$\|E_\lambda g(a,b)\| + \|E_\lambda h(a,b)\| \leq \|pF_\lambda g(a,b)\| + \|pF_\lambda h(a,b)\| \quad (1)$$

for all $\lambda \in \mathbb{T}^1$ and all $a, b \in X$. Then $g, h : X \to Y$ are $\mathbb{C}$-linear.

Proof. Letting $b = a$ and $\lambda = 1$ in (1), $g(2a) = 2g(a)$ and $h(2a) = 2h(a)$ for all $a \in X$. So

$$\|E_\lambda g(a,b)\| + \|E_\lambda h(a,b)\| \leq \|pF_\lambda g(a,b)\| + \|pF_\lambda h(a,b)\|$$
$$= \|pE_\lambda g(a,b)\| + \|pE_\lambda h(a,b)\|$$

for all $\lambda \in \mathbb{T}^1$ and all $a, b \in X$. Thus,

$$g(\lambda(a+b)) = \lambda g(a) + \lambda g(b),$$
$$h(\lambda(a+b)) = \lambda h(a) + \lambda h(b)$$

for all $\lambda \in \mathbb{T}^1$ and all $a, b \in X$, since $|p| < 1$. By Lemma 1, $g, h : X \to Y$ are $\mathbb{C}$-linear. □

Theorem 2. *Let $\varphi : A^2 \to [0, \infty)$ be a function such that there exists an $L < 1$ with*

$$\varphi\left(\frac{x}{2}, \frac{y}{2}\right) \le \frac{L}{4}\varphi(x, y) \le \frac{L}{2}\varphi(x, y) \qquad (2)$$

for all $x, y \in A$. Let $g, h : A \to A$ be mappings satisfying

$$\|E_\lambda g(x, y)\| + \|E_\lambda h(x, y)\| \le \|pF_\lambda g(x, y)\| + \|pF_\lambda h(x, y)\|$$
$$+ \varphi(x, y), \qquad (3)$$
$$\|[g, h]([x, y]) - [[g, h](x), y] - [x, [g, h](y)]\| \le \varphi(x, y) \qquad (4)$$

for all $\lambda \in \mathbb{T}^1$ and all $x, y \in A$. Then there exist unique $\mathbb{C}$-linear mappings $G, H : A \to A$ such that

$$\|g(x) - G(x)\| + \|h(x) - H(x)\| \le \frac{L}{2(1-L)}\varphi(x, x) \qquad (5)$$

for all $x \in A$. Furthermore, the $\mathbb{C}$-linear mapping $[G, H] : A \to A$ is a Lie bracket derivation in A.

Proof. Letting $y = x$ and $\lambda = 1$ in (3), we get

$$\|g(2x) - 2g(x)\| + \|h(2x) - 2h(x)\| \le \varphi(x, x) \qquad (6)$$

for all $x \in A$.

Consider the set

$$S := \{(\phi, \psi) : \phi, \psi : A \to A\}$$

and introduce the generalized metric on S:

$$d((\phi_1, \psi_1), (\phi_2, \psi_2))$$
$$= \inf\{\mu \in \mathbb{R}_+ : \|\phi_1(x) - \phi_2(x)\| + \|\psi_1(x) - \psi_2(x)\| \le \mu\varphi(x, x),$$
$$\times \forall x \in A\},$$

where, as usual, $\inf \phi = +\infty$. It is easy to show that (S, d) is complete [13].

Now we consider the linear mapping $J : S \to S$ such that

$$J(\phi, \psi)(x) := \left(2\phi\left(\frac{x}{2}\right), 2\psi\left(\frac{x}{2}\right)\right)$$

for all $x \in A$.

Let $(\phi_1, \psi_1), (\phi_2, \psi_2) \in S$ be given such that $d((\phi_1, \psi_1), (\phi_2, \psi_2)) = \varepsilon$. Then,

$$\|\phi_1(x) - \phi_2(x)\| + \|\psi_1(x) - \psi_2(x)\| \le \varepsilon \varphi(x, x)$$

for all $x \in A$. Since

$$\left\|2\phi_1\left(\frac{x}{2}\right) - 2\phi_2\left(\frac{x}{2}\right)\right\| + \left\|2\psi_1\left(\frac{x}{2}\right) - 2\psi_2\left(\frac{x}{2}\right)\right\| \le 2\varepsilon\varphi\left(\frac{x}{2}, \frac{x}{2}\right)$$
$$\le 2\varepsilon \frac{L}{2}\varphi(x, x)$$
$$= L\varepsilon\varphi(x, x)$$

for all $x \in A$, $d(J(\phi_1, \psi_1), J(\phi_2, \psi_2)) \le L\varepsilon$. This means that

$$d(J(\phi_1, \psi_1), J(\phi_2, \psi_2)) \le L d((\phi_1, \psi_1), (\phi_2, \psi_2))$$

for all $(\phi_1, \psi_1), (\phi_2, \psi_2) \in S$.

It follows from (6) that

$$\left\|g(x) - 2g\left(\frac{x}{2}\right)\right\| + \left\|h(x) - 2h\left(\frac{x}{2}\right)\right\| \le \varphi\left(\frac{x}{2}, \frac{x}{2}\right)$$
$$\le \frac{L}{2}\varphi(x, x)$$

for all $x \in A$. So $d((g, h), J(g, h)) \le \frac{L}{2}$.

By Theorem 1, there exist mappings $G, H : A \to A$ satisfying the following:

(1) (G, H) is a fixed point of J, i.e.,

$$(G(x), H(x)) = \left(2G\left(\frac{x}{2}\right), 2H\left(\frac{x}{2}\right)\right), \qquad (7)$$

for all $x \in A$. The pair (G, H) is a unique fixed point of J. This implies that the pair (G, H) is a unique pair satisfying (7) such that there exists a $\mu \in (0, \infty)$ satisfying

$$\|g(x) - G(x)\| + \|h(x) - H(x)\| \le \mu \varphi(x, x)$$

for all $x \in A$.
(2) $d(J^l(g, h), (G, H)) \to 0$ as $l \to \infty$. This implies the equality

$$\lim_{n \to \infty} 2^n g\left(\frac{x}{2^n}\right) = G(x), \quad \lim_{n \to \infty} 2^n h\left(\frac{x}{2^n}\right) = H(x)$$

for all $x \in A$.
(3) $d((g, h), (G, H)) \le \frac{1}{1-L} d((g, h), J(g, h))$, which implies

$$\|g(x) - G(x)\| + \|h(x) - H(x)\| \le \frac{L}{2(1-L)} \varphi(x, x)$$

for all $x \in A$.

It follows from (3) and (7) that

$$\|E_\lambda G(x, y)\| + \|E_\lambda H(x, y)\|$$
$$= \lim_{n \to \infty} 2^n \left(\left\| E_\lambda g\left(\frac{x}{2^n}, \frac{y}{2^n}\right) \right\| + \left\| E_\lambda h\left(\frac{x}{2^n}, \frac{y}{2^n}\right) \right\| \right)$$
$$\le \lim_{n \to \infty} 2^n \left(\left\| pF_\lambda g\left(\frac{x}{2^n}, \frac{y}{2^n}\right) \right\| + \left\| pF_\lambda h\left(\frac{x}{2^n}, \frac{y}{2^n}\right) \right\| \right)$$
$$+ \lim_{n \to \infty} 2^n \varphi\left(\frac{x}{2^n}, \frac{y}{2^n}\right)$$
$$= \|pF_\lambda G(x, y)\| + \|pF_\lambda H(x, y)\|$$

for all $x, y \in A$, since $\lim_{n \to \infty} 2^n \varphi\left(\frac{x}{2^n}, \frac{y}{2^n}\right) \le \lim_{n \to \infty} \frac{2^n L^n}{2^n} \varphi(x, y) = 0$. So,

$$\|E_\lambda G(x, y)\| + \|E_\lambda H(x, y)\| \le \|pF_\lambda G(x, y)\| + \|pF_\lambda H(x, y)\|$$

for all $x, y \in A$. By Lemma 2, the mappings $G, H : A \to A$ are $\mathbb{C}$-linear. So there exist unique $\mathbb{C}$-linear mappings $G, H : A \to A$ satisfying (5).

It follows from (4) that

$$\|[G,H]([x,y]) - [[G,H](x),y] - [x,[G,H](y)]\|$$
$$= \lim_{n\to\infty} 4^n \left\| [g,h]\left(\frac{[x,y]}{4^n}\right) - \left[[g,h]\left(\frac{x}{2^n}\right), \frac{y}{2^n}\right] \right.$$
$$\left. - \left[\frac{x}{2^n}, [g,h]\left(\frac{y}{2^n}\right)\right] \right\|$$
$$\leq \lim_{n\to\infty} 4^n \varphi\left(\frac{x}{2^n}, \frac{y}{2^n}\right) \leq \lim_{n\to\infty} \frac{4^n L^n}{4^n} \varphi(x,y) = 0$$

for all $x, y \in A$. So,

$$[G,H]([x,y]) = [[G,H](x),y] + [x,[G,H](y)]$$

for all $x, y \in A$.

Therefore, the $\mathbb{C}$-linear mapping $[G,H]: A \to A$ is a Lie bracket derivation in A. □

Corollary 1. *Let $r > 2$ and θ be non-negative real numbers and $g, h : A \to A$ be mappings satisfying*

$$\|E_\lambda g(x,y)\| + \|E_\lambda h(x,y)\| \leq \|pF_\lambda g(x,y)\| + \|pF_\lambda h(x,y)\|$$
$$+ \theta(\|x\|^r + \|y\|^r), \tag{8}$$
$$\|[g,h]([x,y]) - [[g,h](x),y] - [x,[g,h](y)]\| \leq \theta(\|x\|^r + \|y\|^r) \tag{9}$$

for all $\lambda \in \mathbb{T}^1$ and all $x, y \in A$. Then there exist unique $\mathbb{C}$-linear mappings $G, H : A \to A$ such that

$$\|g(x) - G(x)\| + \|h(x) - H(x)\| \leq \frac{4\theta}{2^r - 4}\|x\|^r \tag{10}$$

for all $x \in A$. Furthermore, the $\mathbb{C}$-linear mapping $[G,H] : A \to A$ is a Lie bracket derivation in A.

Proof. The proof follows from Theorem 2 by taking $L = 2^{2-r}$ and $\varphi(x,y) = \theta(\|x\|^r + \|y\|^r)$ for all $x, y \in A$. □

Theorem 3. Let $\varphi : A^2 \to [0,\infty)$ be a function such that there exists an $L < 1$ with

$$\varphi(x,y) \leq 2L\varphi\left(\frac{x}{2}, \frac{y}{2}\right) \tag{11}$$

for all $x, y \in A$. Let $g, h : A \to A$ be mappings satisfying (3) and (4). Then there exist unique $\mathbb{C}$-linear mappings $G, H : A \to A$ such that

$$\|g(x) - G(x)\| + \|h(x) - H(x)\| \leq \frac{1}{2(1-L)}\varphi(x,x)$$

for all $x \in A$. Furthermore, the $\mathbb{C}$-linear mapping $[G, H] : A \to A$ is a Lie bracket derivation in A.

Proof. Let (S, d) be the generalized metric space defined in the proof of Theorem 2.

Now we consider the linear mapping $J : S \to S$ such that

$$J(\phi, \psi)(x) := \left(\frac{1}{2}\phi(2x), \frac{1}{2}\psi(2x)\right)$$

for all $x \in A$.

It follows from (6) that

$$\left\|g(x) - \frac{1}{2}g(2x)\right\| + \left\|h(x) - \frac{1}{2}h(2x)\right\| \leq \frac{1}{2}\varphi(x,x)$$

for all $x \in A$.

The rest of the proof is similar to the proof of Theorem 2. □

Corollary 2. Let $r < 1$ and θ be non-negative real numbers and $g, h : A \to A$ be mappings satisfying (8) and (9). Then there exist unique $\mathbb{C}$-linear mappings $G, H : A \to A$ such that

$$\|g(x) - G(x)\| + \|h(x) - H(x)\| \leq \frac{4\theta}{4 - 2^r}\|x\|^r \tag{12}$$

for all $x \in A$. Furthermore, the $\mathbb{C}$-linear mapping $[G, H] : A \to A$ is a Lie bracket derivation in A.

Proof. The proof follows from Theorem 3 by taking $L = 2^{r-2}$ and $\varphi(x,y) = \theta(\|x\|^r + \|y\|^r)$ for all $x, y \in A$. □

3. Hyers–Ulam Stability of Lie Bracket Derivations in Banach Lie Algebras: Direct Method

In this section, we prove the Hyers–Ulam stability of Lie bracket derivations in complex Banach Lie algebras by using the direct method.

Theorem 4. *Let $\varphi : A^2 \to [0, \infty)$ be a function such that*

$$\Psi(x, y) := \sum_{j=0}^{\infty} \frac{1}{2^j} \varphi(2^j x, 2^j y) < \infty \tag{13}$$

and $g, h : A \to A$ be mappings satisfying (3) and (4). Then there exist unique $\mathbb{C}$-linear mappings $G, H : A \to A$ such that

$$\|g(x) - G(x)\| + \|h(x) - H(x)\| \leq \frac{1}{2} \Psi(x, x) \tag{14}$$

for all $x \in A$. Furthermore, the $\mathbb{C}$-linear mapping $[G, H] : A \to A$ is a Lie bracket derivation in A.

Proof. Letting $y = x$ and $\lambda = 1$ in (3), we get

$$\|g(2x) - 2g(x)\| + \|h(2x) - 2h(x)\| \leq \varphi(x, x) \tag{15}$$

for all $x \in A$. Thus,

$$\left\|g(x) - \frac{1}{2}g(2x)\right\| + \left\|h(x) - \frac{1}{2}h(2x)\right\| \leq \frac{1}{2}\varphi(x, x)$$

for all $x \in A$. So,

$$\left\|\frac{1}{2^l}g\left(2^l x\right) - \frac{1}{2^m}g\left(2^m x\right)\right\| + \left\|\frac{1}{2^l}h\left(2^l x\right) - \frac{1}{2^m}h\left(2^m x\right)\right\|$$

$$\leq \sum_{j=l}^{m-1}\left(\left\|\frac{1}{2^j}g\left(2^j x\right) - \frac{1}{2^{j+1}}g\left(2^{j+1} x\right)\right\|\right.$$

$$\left. + \left\|\frac{1}{2^j}h\left(2^j x\right) - \frac{1}{2^{j+1}}h\left(2^{j+1} x\right)\right\|\right)$$

$$\leq \sum_{j=l}^{m-1} \frac{1}{2} \cdot \frac{1}{2^j} \varphi\left(2^j x, 2^j x\right) \tag{16}$$

for all non-negative integers m and l with $m > l$ and all $x \in A$. It follows from (16) that the sequences $\{\frac{1}{2^k}g(2^k x)\}$ and $\{\frac{1}{2^k}h(2^k x)\}$ are Cauchy sequences for all $x \in A$. Since A is complete, the sequences $\{\frac{1}{2^k}g(2^k x)\}$ and $\{\frac{1}{2^k}h(2^k x)\}$ converge. So one can define the mappings $G, H : A \to A$ by

$$G(x) := \lim_{k\to\infty} \frac{1}{2^k} g\left(2^k x\right), \qquad H(x) := \lim_{k\to\infty} \frac{1}{2^k} h\left(2^k x\right)$$

for all $x \in A$. Moreover, letting $l = 0$ and passing the limit $m \to \infty$ in (16), we get (14).

It follows from (3) that

$$\|E_\lambda G(x,y)\| + \|E_\lambda H(x,y)\|$$
$$= \lim_{n\to\infty} 2^n \left(\left\| E_\lambda g\left(\frac{x}{2^n}, \frac{y}{2^n}\right) \right\| + \left\| E_\lambda h\left(\frac{x}{2^n}, \frac{y}{2^n}\right) \right\| \right)$$
$$\leq \lim_{n\to\infty} 2^n \left(\left\| pF_\lambda g\left(\frac{x}{2^n}, \frac{y}{2^n}\right) \right\| + \left\| pF_\lambda h\left(\frac{x}{2^n}, \frac{y}{2^n}\right) \right\| \right)$$
$$+ \lim_{n\to\infty} 2^n \varphi\left(\frac{x}{2^n}, \frac{y}{2^n}\right)$$
$$= \|pF_\lambda G(x,y)\| + \|pF_\lambda H(x,y)\|$$

for all $x, y \in A$, since $\lim_{n\to\infty} 2^n \varphi\left(\frac{x}{2^n}, \frac{y}{2^n}\right) \leq \lim_{n\to\infty} \frac{2^n L^n}{2^n}$ $\varphi(x,y) = 0$. So,

$$\|E_\lambda G(x,y)\| + \|E_\lambda H(x,y)\| \leq \|pF_\lambda G(x,y)\| + \|pF_\lambda H(x,y)\|$$

for all $x, y \in A$. By Lemma 2, the mappings $G, H : A \to A$ are $\mathbb{C}$-linear.

Now, let $T, L : A \to A$ be another $\mathbb{C}$-linear mappings satisfying (14). Then we have

$$\|G(x) - T(x)\| + \|H(x) - L(x)\|$$
$$= \left\| \frac{1}{2^q} G(2^q x) - \frac{1}{2^q} T(2^q x) \right\| + \left\| \frac{1}{2^q} H(2^q x) - \frac{1}{2^q} L(2^q x) \right\|$$

$$\leq \left\| \frac{1}{2^q} G(2^q x) - \frac{1}{2^q} g(2^q x) \right\| + \left\| \frac{1}{2^q} T(2^q x) - \frac{1}{2^q} g(2^q x) \right\|$$
$$+ \left\| \frac{1}{2^q} H(2^q x) - \frac{1}{2^q} h(2^q x) \right\| + \left\| \frac{1}{2^q} L(2^q x) - \frac{1}{2^q} h(2^q x) \right\|$$
$$\leq \frac{1}{2^q} \Psi(2^q x, 2^q x),$$

which tends to zero as $q \to \infty$ for all $x \in A$. So we can conclude that $G(x) = T(x)$ and $H(x) = L(x)$ for all $x \in A$. This proves the uniqueness of (G, H).

It follows from (4) that

$$\|[G,H]([x,y]) - [[G,H](x),y] - [x,[G,H](y)]\|$$
$$= \lim_{n\to\infty} \frac{1}{4^n} \|[g,h](4^n[x,y]) - [[g,h](2^n x), 2^n y] - [2^n x, [g,h](2^n y)]\|$$
$$\leq \lim_{n\to\infty} \frac{1}{4^n} \varphi(2^n x, 2^n y) \leq \lim_{n\to\infty} \frac{1}{2^n} \varphi(2^n x, 2^n y) = 0$$

for all $x, y \in A$. So,

$$[G,H]([x,y]) = [[G,H](x),y] + [x,[G,H](y)]$$

for all $x, y \in A$.

Therefore, the $\mathbb{C}$-linear mapping $[G, H] : A \to A$ is a Lie bracket derivation in A. □

Corollary 3. *Let $r < 1$ and θ be non-negative real numbers, and $g, h : A \to A$ be mappings satisfying (8) and (9). Then there exist unique $\mathbb{C}$-linear mappings $G, H : A \to A$ satisfying (12). Furthermore, the $\mathbb{C}$-linear mapping $[G, H] : A \to A$ is a Lie bracket derivation in A.*

Similarly, we can obtain the following:

Theorem 5. *Let $\varphi : A^2 \to [0, \infty)$ be a function such that*

$$\sum_{j=1}^{\infty} 4^j \varphi\left(\frac{x}{2^j}, \frac{y}{2^j}\right) < \infty \tag{17}$$

for all $x, y \in A$ and $g, h : A \to A$ be mappings satisfying (3) and (4). Then there exist unique $\mathbb{C}$-linear mappings $G, H : A \to A$ such that

$$\|g(x) - G(x)\| + \|h(x) - H(x)\| \leq \frac{1}{2}\Psi(x, x) \tag{18}$$

for all $x \in A$, where

$$\Psi(x, y) := \sum_{j=1}^{\infty} 2^j \varphi\left(\frac{x}{2^j}, \frac{y}{2^j}\right)$$

for all $x, y \in A$. Furthermore, the $\mathbb{C}$-linear mapping $[G, H] : A \to A$ is a Lie bracket derivation in A.

Proof. It follows from (15) that

$$\left\|g(x) - 2g\left(\frac{x}{2}\right)\right\| + \left\|h(x) - 2h\left(\frac{x}{2}\right)\right\| \leq \varphi\left(\frac{x}{2}, \frac{x}{2}\right)$$

for all $x \in A$.

By the same reasoning as in the proof of Theorem 4, there exist unique $\mathbb{C}$-mappings $G, H : A \to A$ satisfying (18). The $\mathbb{C}$-linear mappings $G, H : A \to A$ are defined by

$$G(x) = \lim_{n \to \infty} 2^n g\left(\frac{x}{2^n}\right), \quad H(x) = \lim_{n \to \infty} 2^n h\left(\frac{x}{2^n}\right)$$

for all $x \in A$.

It follows from (4) that

$$\|[G, H]([x, y]) - [[G, H](x), y] - [x, [G, H](y)]\|$$

$$= \lim_{n \to \infty} 4^n \left\| [g, h]\left(\frac{[x, y]}{4^n}\right) - \left[[g, h]\left(\frac{x}{2^n}\right), \frac{y}{2^n}\right] \right.$$

$$\left. - \left[\frac{x}{2^n}, [g, h]\left(\frac{y}{2^n}\right)\right] \right\|$$

$$\leq \lim_{n \to \infty} 4^n \varphi\left(\frac{x}{2^n}, \frac{y}{2^n}\right) = 0$$

for all $x, y \in A$. So,

$$[G, H]([x, y]) = [[G, H](x), y] + [x, [G, H](y)]$$

for all $x, y \in A$.

Therefore, the $\mathbb{C}$-linear mapping $[G, H] : A \to A$ is a Lie bracket derivation in A. $\square$

Corollary 4. Let $r > 2$ and θ be non-negative real numbers and $g, h : A \to A$ be mappings satisfying (8) and (9). Then there exist unique $\mathbb{C}$-linear mappings $G, H : A \to A$ satisfying (10). Furthermore, the $\mathbb{C}$-linear mapping $[G, H] : A \to A$ is a Lie bracket derivation in A.

References

[1] T. Aoki, On the stability of the linear transformation in Banach spaces, *J. Math. Soc. Jpn.* **2**, 64–66 (1950).

[2] L. Cădariu, L. Găvruta and P. Găvruta, On the stability of an affine functional equation, *J. Nonlinear Sci. Appl.* **6**, 60–67 (2013).

[3] L. Cădariu and V. Radu, Fixed points and the stability of Jensen's functional equation, *J. Inequal. Pure Appl. Math.* **4**(1) (2003). Article ID 4.

[4] L. Cădariu and V. Radu, On the stability of the Cauchy functional equation: A fixed point approach, *Grazer Math. Ber.* **346**, 43–52 (2004).

[5] L. Cădariu and V. Radu, Fixed point methods for the generalized stability of functional equations in a single variable, *Fixed Point Theory Appl.* **2008** (2008). Article ID 749392.

[6] P. W. Cholewa, Remarks on the stability of functional equations, *Aequationes Math.* **27**, 76–86 (1984).

[7] J. Diaz and B. Margolis, A fixed point theorem of the alternative for contractions on a generalized complete metric space, *Bull. Am. Math. Soc.* **74**, 305–309 (1968).

[8] N. Eghbali, J. M. Rassias and M. Taheri, On the stability of a k-cubic functional equation in intuitionistic fuzzy n-normed spaces, *Results Math.* **70**, 233–248 (2016).

[9] I. EL-Fassi, Solution and approximation of radical quintic functional equation related to quintic mapping in quasi-β-Banach spaces, *Rev. R. Acad. Cienc. Exactas Fís. Nat. Ser. A Mat. RACSAM* **113**(2), 675–687 (2019).

[10] P. Găvruta, A generalization of the Hyers-Ulam-Rassias stability of approximately additive mappings, *J. Math. Anal. Appl.* **184**, 431–436 (1994).

[11] D. H. Hyers, On the stability of the linear functional equation, *Proc. Nat. Acad. Sci. USA* **27**, 222–224 (1941).

[12] G. Isac and T. M. Rassias, Stability of ψ-additive mappings: Applications to nonlinear analysis, *Int. J. Math. Math. Sci.* **19**, 219–228 (1996).

[13] D. Miheţ and V. Radu, On the stability of the additive Cauchy functional equation in random normed spaces, *J. Math. Anal. Appl.* **343**, 567–572 (2008).
[14] C. Park, Homomorphisms between Poisson JC^*-algebras, *Bull. Braz. Math. Soc.* **36**, 79–97 (2005).
[15] C. Park, Additive ρ-functional inequalities and equations, *J. Math. Inequal.* **9**, 17–26 (2015).
[16] C. Park, Additive ρ-functional inequalities in non-Archimedean normed spaces, *J. Math. Inequal.* **9**, 397–407 (2015).
[17] C. Park, Fixed point method for set-valued functional equations, *J. Fixed Point Theory Appl.* **19**, 2297–2308 (2017).
[18] C. Park, Set-valued additive ρ-functional inequalities, *J. Fixed Point Theory Appl.* **20**(2) (2018). Paper No. 70.
[19] V. Radu, The fixed point alternative and the stability of functional equations, *Fixed Point Theory* **4**, 91–96 (2003).
[20] T. M. Rassias, On the stability of the linear mapping in Banach spaces, *Proc. Am. Math. Soc.* **72**, 297–300 (1978).
[21] F. Skof, Propriet locali e approssimazione di operatori, *Rend. Sem. Mat. Fis. Milano* **53**, 113–129 (1983).
[22] L. Székelyhidi, Superstability of functional equations related to spherical functions, *Open Math.* **15**(1), 427–432 (2017).
[23] S. M. Ulam, *A Collection of the Mathematical Problems* (Interscience Publishers, New York, 1960).
[24] Z. Wang, Stability of two types of cubic fuzzy set-valued functional equations, *Results Math.* **70**, 1–14 (2016).

© 2024 World Scientific Publishing Company
https://doi.org/10.1142/9789811267048_0020

Chapter 20

Aboodh Transform and Ulam Stability of Second-Order Linear Differential Equations

Ramdoss Murali[*,∥], Arumugam Ponmana Selvan[†,**],
Sanmugam Baskaran[*,††], Choonkil Park[§,‡‡],
and Michael Th. Rassias[¶,§§]

[*]*PG and Research Department of Mathematics,
Sacred Heart College (Autonomous), Tirupattur,
Tirupattur Dt., Tamil Nadu, India*
[†]*Department of Mathematics,
Sri Sai Ram Institute of Technology,
Sai Leo Nagar, West Tambaram,
Chennai, Tamil Nadu, India*
[§]*Department of Mathematics, Hanyang University,
Seoul, Korea*
[¶]*Department of Mathematics and Engineering Sciences,
Hellenic Military Academy, Vari Attikis, Greece*

[∥]*shcrmurali@yahoo.co.in*
[**]*selvaharry@yahoo.com*
[††]*sps.baskaran@gmail.com*
[‡‡]*baak@hanyang.ac.kr*
[§§]*mthrassias@yahoo.com*

In this chapter, we establish approximate solutions of linear differential equations of second order with constant coefficients in the sense of Ulam and Hyers by using Aboodh integral transformation. By applying Aboodh integral transform, we prove the different types of Ulam stabilities such as Hyers–Ulam stability, Hyers–Ulam–Rassias stability, Mittag–Leffler–Hyers–Ulam stability and Mittag–Leffler–Hyers–Ulam–Rassias stability of linear differential differential equations of second order with constant coefficients for homogeneous and nonhomogeneous cases. We also obtain the Hyers–Ulam stability constants of these differential equations by using the Aboodh integral transform and give some examples to illustrate our main results.

1. Introduction

In 1940, Ulam gave a wide ranging talk before the Mathematics Club of the University of Wisconsin, in which he discussed a number of important unsolved problems (see Ref. [58]). Among those was the question concerning the stability of homomorphisms: let G_1 be a group and let G_2 be a metric group with a metric $d(.,.)$. Given any $\delta > 0$, does there exist an $\epsilon > 0$ such that if a function $h : G_1 \to G_2$ satisfies the inequality $d(h(xy), h(x)h(y)) < \epsilon$ for all $x, y \in G_1$, then there exists a homomorphism $H : G_1 \to G_2$ with $d(h(x), H(x)) < \delta$ for all $x \in G_1$? In 1941, Hyers [17] partially solved the Ulam problem for the case where G_1 and G_2 are Banach spaces. Furthermore, the result of Hyers has been generalized by Rassias [56]. Since then, the stability problems of various functional equations have been investigated by many authors [1, 2, 4, 8, 10–13, 15, 18–20, 25, 27, 31, 38, 49, 54, 59].

Consider the Hyers–Ulam stability problem for the differential equation: assume that Y is a normed space over a scalar field $\mathbb{K}$ and that I is an open interval, where $\mathbb{K}$ denotes either $\mathbb{R}$ or $\mathbb{C}$. Assume that $a_0, a_1, \ldots, a_n : I \to \mathbb{K}$ and $g : I \to X$ are given continuous functions and that $y : I \to X$ is an n times continuously differentiable function satisfying the inequality

$$\left\| a_n(t) y^{(n)}(t) + a_{n-1}(t) y^{(n-1)}(t) + \cdots + a_1(t) y'(t) \right.$$
$$\left. + a_0(t) y(t) + g(t) \right\| \leq \epsilon$$

for all $t \in I$ and for a given $\epsilon > 0$. If there exists an n times continuously differentiable function $y_0 : I \to X$ satisfying the differential equation

$$a_n(t)y_0^{(n)}(t) + a_{n-1}(t)y_0^{(n-1)}(t) + \cdots$$
$$+ a_1(t)y_0'(t) + a_0(t)y_0(t) + g(t) = 0$$

and

$$\|y(t) - y_0(t)\| \leq K(\epsilon)$$

for any $t \in I$, where $K(\epsilon)$ is an expression of ϵ with $\lim_{\epsilon \to 0} K(\epsilon) = 0$, then we say that the corresponding differential equation has the Hyers–Ulam stability. If the preceding statement is also true when we replace ϵ and $K(\epsilon)$ by $\phi(t)$ and $\varphi(t)$, where ϕ, φ are appropriate functions not depending on x and x_a explicitly, then we say that the corresponding differential equation has the Hyers–Ulam–Rassias stability.

Now, we introduce a result of Alsina and Ger [6]: if $f : I \to R$ is a differentiable function, which is a solution of the following differential inequality $\|x'(t) - x(t)\| \leq \epsilon$, where I is an open subinterval of $\mathbb{R}$, then there is a solution $g : I \to R$ of $x'(t) = x(t)$ such that for any $t \in I$, we have $\|f(t) - g(t)\| \leq 3\epsilon$. This result has been generalized by Takahasi et al. [57]. They proved that the Hyers–Ulam stability holds true for the Banach space-valued differential equation $y'(t) = \lambda y(t)$ [37]. Indeed, the Hyers–Ulam stability has been proved for the first-order linear differential equations in more general settings [21–24, 26, 36]. In 2007, Wang et al. [61] established the Hyers–Ulam stability of a class of first-order linear differential equations.

In 2014, Alqifiary and Jung [5] investigated the Hyers–Ulam–Rassias stability of

$$x^{(n)}(t) + \sum_{k=0}^{n-1} \alpha_k\, x^{(k)}(t) = f(t)$$

by using the Laplace transform method. Nowadays, the Hyers–Ulam stability of differential equations has been given attention. For more recent results about this subject, we can refer to Refs. [9, 14, 16, 34, 35, 39–42, 44, 45, 47, 48, 60].

Very recently, Murali et al. [46] investigated the Ulam stability of various differential equations by using Fourier transform method [43, 46, 55].

In this chapter, our main intention is to apply the Aboodh integral transform method to investigate the Hyers–Ulam stability and the Mittag–Leffler–Hyers–Ulam stability of the following general homogeneous-type second-order linear differential equation:

$$u''(t) + \mu u'(t) + \zeta u(t) = 0 \tag{1}$$

and the nonhomogeneous-type second-order linear differential equation of the form

$$u''(t) + \mu u'(t) + \zeta u(t) = q(t) \tag{2}$$

for all $t \in I$, $u(t) \in C^2(I)$ and $q(t) \in C(I)$, $I = [a,b]$, $-\infty < a < b < \infty$, where μ and ζ are constants.

Throughout this chapter, $\mathbb{F}$ denotes the real field $\mathbb{R}$ or the complex field $\mathbb{C}$. Assume that a function $f : (0, \infty) \to \mathbb{F}$ is of exponential order, i.e., there exist constants $A, B \in \mathbb{R}$ such that $|f(t)| \leq Ae^{tB}$ for all $t > 0$.

Consider the set S, which is defined by

$$S = \left\{ f(t) : \exists M,\ k_1, k_2 > 0,\ |f(t)| < M\ e^{-\xi t} \right\}.$$

For a given function $f(t)$ in the set S, M must be a finite number and k_1 and k_2 may be finite or infinite [7]. The Ulam stability for partial differential equations was studied for the first time by Prástaro and Rassias (cf. Refs. [50–53]).

2. Preliminaries

In this section, we introduce some standard notations and definitions which are useful to prove our main results.

Definition 1 ([3, 7]). The Aboodh (integral) transform is defined by, for a function $f(t)$ of exponential order,

$$\mathcal{A}\{f(t)\} = \frac{1}{\xi} \int_0^\infty f(t)\ e^{-\xi t}\ dt = F(\xi), \quad t \geq 0,$$

provided that the integral exists for some ξ, where $\xi \in (k_1, k_2)$. Here $\mathcal{A}$ is called the Aboodh (integral) transform operator.

For two functions f and g, both Lebesgue integrable on $(-\infty, +\infty)$, let S denote the set of x for which the Lebesgue integral

$$h(x) = \int_{-\infty}^{\infty} f(t)\, g(x-t)\, \mathrm{d}t$$

exists. This integral defines a function h on S called the *convolution* of f and g. We also write $h = f * g$ to denote this function.

Definition 2 ([30]). The Mittag–Leffler function of one parameter is denoted by $E_\alpha(z)$ and defined as

$$E_\alpha(z) = \sum_{k=0}^{\infty} \frac{1}{\Gamma(\alpha k + 1)} z^k,$$

where $z, \alpha \in \mathbb{C}$ and $\mathrm{Re}(\alpha) > 0$. If we put $\alpha = 1$, then the above equation becomes

$$E_1(z) = \sum_{k=0}^{\infty} \frac{1}{\Gamma(k+1)} z^k = \sum_{k=0}^{\infty} \frac{z^k}{k} = e^z.$$

Definition 3 ([30]). A generalization of $E_\alpha(z)$ is defined as a function

$$E_{\alpha,\beta}(z) = \sum_{k=0}^{\infty} \frac{1}{\Gamma(\alpha k + \beta)} z^k,$$

where $z, \alpha, \beta \in \mathbb{C}$, $\mathrm{Re}(\alpha) > 0$ and $\mathrm{Re}(\beta) > 0$.

Let $I, J \subseteq \mathbb{R}$. Throughout this chapter, we denote the space of k continuously differentiable functions from I to J by $C^k(I, J)$ and denote $C^k(I, I)$ by $C^k(I)$. Furthermore, $C(I, J) = C^0(I, J)$ denotes the space of continuous functions from I to J. In addition, $\mathbb{R}_+ := [0, \infty)$. From now on, we assume that $I = [a, b]$, where $-\infty < a < b < \infty$.

Now, we give some definitions of various forms of Hyers–Ulam stability and Mittag–Leffler–Hyers–Ulam stability of the differential equations (1) and (2).

Definition 4. We say that the differential equation (1) has the Hyers–Ulam stability if there exists a constant $L > 0$ satisfying the following condition: if, for every $\epsilon > 0$, there exists $u(t) \in C^2(I)$ satisfying the inequality

$$\left|u''(t) + \mu u'(t) + \zeta u(t)\right| \leq \epsilon,$$

for all $t \in I$, then there exists a solution $v \in C^2(I)$ satisfying the differential equation $v''(t) + \mu v'(t) + \zeta v(t) = 0$ and $|u(t) - v(t)| \leq L\epsilon$ for all $t \in I$. We call such L the Hyers–Ulam stability constant for (1).

Definition 5. We say that the differential equation (1) has the Hyers–Ulam–Rassias stability with respect to $\phi \in C(\mathbb{R}_+, \mathbb{R}_+)$ if there exists a constant $L_\phi > 0$ with the following property: if, for every $\epsilon > 0$, there exists $u(t) \in C^2(I)$ satisfying the inequality

$$\left|u''(t) + \mu u'(t) + \zeta u(t)\right| \leq \epsilon \phi(t)$$

for all $t \in I$, then there exists a solution $v \in C^2(I)$ satisfying the differential equation $v''(t) + \mu v'(t) + \zeta v(t) = 0$ such that

$$|u(t) - v(t)| \leq L_\phi \epsilon \phi(t)$$

for all $t \in I$. We call such L the Hyers–Ulam–Rassias stability constant for (1).

Definition 6. We say that the differential equation (2) has the Hyers–Ulam stability if there exists a constant $L > 0$ satisfying the following condition: if, for every $\epsilon > 0$, there exists $u(t) \in C^2(I)$ satisfying the inequality

$$\left|u''(t) + \mu u'(t) + \zeta u(t) - q(t)\right| \leq \epsilon$$

for all $t \in I$, then there exists some $v \in C^2(I)$ satisfying $v''(t) + \mu v'(t) + \zeta v(t) = q(t)$ and

$$|u(t) - v(t)| \leq L\epsilon$$

for all $t \in I$. We call such L the Hyers–Ulam stability constant for (2).

Definition 7. We say that the differential equation (2) has the Hyers–Ulam–Rassias stability with respect to $\phi \in C(\mathbb{R}_+, \mathbb{R}_+)$ if there exists a constant $L_\phi > 0$ such that for every $\epsilon > 0$ and each $u(t) \in C^2(I)$ satisfying the inequality

$$\left|u''(t) + \mu u'(t) + \zeta u(t) - q(t)\right| \leq \epsilon \phi(t)$$

for all $t \in I$, there exists some $v \in C^2(I)$ satisfying $v''(t) + \mu v'(t) + \zeta v(t) = q(t)$ and

$$|u(t) - v(t)| \leq L_\phi \epsilon \phi(t)$$

for all $t \in I$. We call such L the Hyers–Ulam–Rassias stability constant for (2).

Definition 8. We say that the differential equation (1) has the Mittag–Leffler–Hyers–Ulam stability if there exists a positive constant L satisfying the following condition: if, for every $\epsilon > 0$, there exists $u(t) \in C^2(I)$ satisfying the inequality

$$\left|u''(t) + \mu u'(t) + \zeta u(t)\right| \leq \epsilon E_\alpha(t)$$

for all $t \in I$, then there exists a solution $v \in C^2(I)$ satisfying $v''(t) + \mu v'(t) + \zeta v(t) = 0$ and

$$|u(t) - v(t)| \leq L \epsilon E_\alpha(t)$$

for all $t \in I$. We call such L the Mittag–Leffler–Hyers–Ulam stability constant for (1).

Definition 9. We say that the differential equation (1) has the Mittag–Leffler–Hyers–Ulam–Rassias stability with respect to $\phi : (0, \infty) \to (0, \infty)$ if there exists a positive constant L_ϕ satisfying the following condition: if, for every $\epsilon > 0$, there exists $u(t) \in C^2(I)$ satisfying the inequality

$$\left|u''(t) + \mu u'(t) + \zeta u(t)\right| \leq \phi(t) \epsilon E_\alpha(t)$$

for all $t \in I$, then there exists a solution $v \in C^2(I)$ satisfying $v''(t) + \mu v'(t) + \zeta v(t) = 0$ and

$$|u(t) - v(t)| \leq L_\phi \phi(t) \epsilon E_\alpha(t)$$

for all $t \in I$. We call such L_ϕ the Mittag–Leffler–Hyers–Ulam–Rassias stability constant for (1).

Definition 10. We say that the differential equation (2) has the Mittag–Leffler–Hyers–Ulam stability if there exists a positive constant L satisfying the following condition: if, for every $\epsilon > 0$, there exists $u(t) \in C^2(I)$ satisfying the inequality

$$\left|u''(t) + \mu u'(t) + \zeta u(t) - q(t)\right| \leq \epsilon E_\alpha(t)$$

for all $t \in I$, then there exists a solution $v \in C^2(I)$ satisfying the linear differential equation $v''(t) + \mu v'(t) + \zeta v(t) = q(t)$ and

$$|u(t) - v(t)| \leq L\epsilon E_\alpha(t)$$

for all $t \in I$. We call such L the Mittag–Leffler–Hyers–Ulam stability constant for (2).

Definition 11. We say that the differential equation (2) has the Mittag–Leffler–Hyers–Ulam–Rassias stability with respect to $\phi : (0, \infty) \to (0, \infty)$ if there exists a positive constant L_ϕ satisfying the following condition: if, for every $\epsilon > 0$, there exists $u(t) \in C^2(I)$ satisfying the inequality

$$\left|u''(t) + \mu u'(t) + \zeta u(t) - q(t)\right| \leq \phi(t)\epsilon E_\alpha(t)$$

for all $t \in I$, then there exists a solution $v \in C^2(I)$ satisfying the linear differential equation $v''(t) + \mu v'(t) + \zeta v(t) = q(t)$ and $|u(t) - v(t)| \leq L_\phi \phi(t)\epsilon E_\alpha(t)$ for all $t \in I$. We call such L_ϕ the Mittag–Leffler–Hyers–Ulam–Rassias stability constant for (2).

3. Ulam Stabilities for (1)

In this section, we prove the Hyers–Ulam stability, Hyers–Ulam–Rassias stability, Mittag–Leffler–Hyers–Ulam stability and Mittag–Leffler–Hyers–Ulam–Rassias stability of the differential equation (1) by using the Aboodh transform.

Theorem 1. *The homogeneous linear differential equation* (1) *has Hyers–Ulam stability.*

Proof. Let $\epsilon > 0$. Suppose that $u(t) \in C^2(I)$ satisfies the inequality

$$\left|u''(t) + \mu u'(t) + \zeta u(t)\right| \leq \epsilon \qquad (3)$$

for all $t \in I$. We show that there exists a real number $L > 0$ which is independent of ϵ and u such that $|u(t) - v(t)| \le L\epsilon$ for some $v \in C^2(I)$ satisfying the differential equation $v''(t) + \mu v'(t) + \zeta v(t) = 0$ for all $t \in I$.

Define a function $p : (0, \infty) \longrightarrow \mathbb{F}$ such that $p(t) =: u''(t) + \mu u'(t) + \zeta u(t)$ for all $t > 0$. In view of (3), we have $|p(t)| \le \epsilon$. Taking the Aboodh transform to $p(t)$, we have

$$\mathcal{A}\{p\} = \xi^2 \mathcal{A}\{u\} - \frac{u'(0)}{\xi} - u(0) + \mu \left[\xi \mathcal{A}\{u\} - \frac{u'(0)}{\xi}\right] + \zeta \frac{u'(0)}{\xi}$$

$$= (\xi^2 + \mu\xi + \zeta)\mathcal{A}\{u\} - \frac{u'(0)}{\xi} - u(0)\left[1 + \frac{\mu}{\xi}\right].$$

It can be written as

$$\mathcal{A}\{p\} = P(\xi) = (\xi^2 + \mu\xi + \zeta)U(\xi) - \frac{u'(0)}{\xi} - u(0)\left[\frac{\xi + \mu}{\xi}\right] \quad (4)$$

and thus

$$\mathcal{A}\{u\} = U(\xi) = \frac{P(\xi) + u(0)\left[\frac{\xi + \mu}{\xi}\right] + \frac{u'(0)}{\xi}}{\xi^2 + \mu\xi + \zeta}.$$

By (4), a function $u_0 : (0, \infty) \longrightarrow \mathbb{F}$ is a solution of (1) if and only if

$$(\xi^2 + \mu\xi + \zeta)\mathcal{A}\{u_0\} - u_0(0)\left[\frac{\xi + \mu}{\xi}\right] - \frac{u_0'(0)}{\xi} = 0.$$

If there exist constants l and m in $\mathbb{F}$ such that $\xi^2 + \mu\xi + \zeta = (\xi - l)(\xi - m)$ with $l + m = -\mu$ and $lm = \zeta$, then (4) becomes

$$\mathcal{A}\{u\} = \frac{P(\xi) + u(0)\left[\frac{\xi + \mu}{\xi}\right] + \frac{u'(0)}{\xi}}{(\xi - l)(\xi - m)}. \quad (5)$$

Set

$$v(t) = u(0)\left(\frac{l\, e^{lt} - m\, e^{mt}}{l - m}\right) + [\mu\, u(0) + u'(0)]\, G_{l,m}(t),$$

where $G_{l,m}(t) = \left(\frac{e^{lt}-e^{mt}}{l-m}\right)$. Then we have $v(0) = u(0)$ and $u'(0) = v'(0)$. Taking the Aboodh transform to $v(t)$, we obtain

$$\mathcal{A}\{v\} = \frac{u(0)}{(\xi-l)(\xi-m)} + \frac{\mu\, u(0) + u'(0)}{\xi(\xi-l)(\xi-m)}. \tag{6}$$

On the other hand, one can obtain that

$$\mathcal{A}\{v''(t) + \mu v'(t) + \zeta v(t)\} = (\xi^2 + \mu\xi + \zeta)\mathcal{A}\{v\}$$
$$- v(0)\left[\frac{\xi+\mu}{\xi}\right] - \frac{v'(0)}{\xi}. \tag{7}$$

Using (6) in (7), we get $\mathcal{A}\{v''(t) + \mu v'(t) + \zeta v(t)\} = 0$. Since $\mathcal{A}$ is a one-to-one operator and linear, $v''(t) + \mu v'(t) + \zeta v(t) = 0$. This means that $v(t)$ is a solution of (1). It follows from (5) and (6) that

$$\mathcal{A}\{u\} - \mathcal{A}\{v\} = \frac{\mathcal{A}\{p\} + u(0)\left[\frac{\xi+\mu}{\xi}\right] + \frac{u'(0)}{\xi}}{(\xi-l)(\xi-m)} - \frac{u(0)}{(\xi-l)(\xi-m)}$$
$$- \frac{\mu\, u(0) + u'(0)}{\xi(\xi-l)(\xi-m)}$$
$$= \frac{\mathcal{A}\{p\}}{(\xi-l)(\xi-m)} = \frac{\xi\mathcal{A}\{p\}}{\xi(\xi-l)(\xi-m)},$$
$$\mathcal{A}\{u(t) - v(t)\} = \mathcal{A}\{p(t) * G_{l,m}(t)\}.$$

The above equalities show that

$$u(t) - v(t) = p(t) * G_{l,m}(t).$$

Taking the modulus on both sides and using $|p(t)| \leq \epsilon$, we get

$$|u(t) - v(t)| = |p(t) * G_{l,m}(t)| = \left|p(t) * \left(\frac{e^{lt} - e^{mt}}{l-m}\right)\right|$$
$$\leq \left|\int_0^t p(x)\left(\frac{e^{l(t-x)} - e^{m(t-x)}}{l-m}\right)\mathrm{d}x\right|$$
$$\leq \epsilon \left|\int_0^t \left(\frac{e^{l(t-x)} - e^{m(t-x)}}{l-m}\right)\mathrm{d}x\right|$$

for all $t > 0$, where

$$L = \left| \int_0^t \left(\frac{e^{l(t-x)} - e^{m(t-x)}}{l - m} \right) dx \right|$$

$$\leq \frac{1}{|l - m|} \left\{ e^{\mathcal{R}(l)t} \int_0^t e^{-\mathcal{R}(l)x} dx + e^{\mathcal{R}(m)t} \int_0^t e^{-\mathcal{R}(m)x} dx \right\} \leq \frac{\mathcal{K}}{|l - m|}$$

and $\int_0^t e^{-\mathcal{R}(l)x} dx$ and $\int_0^t e^{-\mathcal{R}(m)x} dx$ exist. Hence, $|u(t) - v(t)| \leq \frac{\mathcal{K}}{|l-m|} \epsilon = L \epsilon$. By the virtue of Definition 4, the linear differential equation (1) has the Hyers–Ulam stability. This completes the proof. □

By using the same technique in Theorem 1, we can also prove the following theorem which shows the Hyers–Ulam–Rassias stability of the differential equation (1).

Theorem 2. *Assume that $u(t) \in C^2(I)$ satisfies the following inequality*

$$|u''(t) + \mu u'(t) + \zeta u(t)| \leq \epsilon \phi(t) \tag{8}$$

for $\epsilon > 0$ and $\phi \in C(\mathbb{R}_+, \mathbb{R}_+)$ is an integrable function for all $t \in I$. Then there exist a real number $L_\phi > 0$ and some $v \in C^2(I)$ satisfying the differential equation $v''(t) + \mu v'(t) + \zeta v(t) = 0$ for all $t \in I$ and

$$|u(t) - v(t)| \leq L_\phi \epsilon \phi(t).$$

Proof. Consider the function $p : (0, \infty) \longrightarrow \mathbb{F}$ defined by

$$p(t) =: u''(t) + \mu u'(t) + \zeta u(t)$$

for all $t > 0$. By (8), we have $|p(t)| \leq \epsilon \phi(t)$. Now, taking the Aboodh transform to $p(t)$, we have

$$\mathcal{A}\{p\} = (\xi^2 + \mu \xi + \zeta)\mathcal{A}\{u\} - \frac{u'(0)}{\xi} - u(0)\left[1 + \frac{\mu}{\xi}\right]. \tag{9}$$

Then we know that a function $u_0 : (0, \infty) \to \mathbb{F}$ is a solution of (1) if and only if

$$(\xi^2 + \mu \xi + \zeta)\mathcal{A}\{u_0\} - u_0(0)\left[\frac{\xi + \mu}{\xi}\right] - \frac{u_0'(0)}{\xi} = 0.$$

If there exist constants l and m in $\mathbb{F}$ such that $\xi^2 + \mu\xi + \zeta = (\xi - l)(\xi - m)$ with $l + m = -\mu$ and $lm = \zeta$, then (9) becomes

$$U(\xi) = \frac{P(\xi) + u(0)\left[\frac{\xi+\mu}{\xi}\right] + \frac{u'(0)}{\xi}}{(\xi - l)(\xi - m)}. \tag{10}$$

Let us take the solution function

$$v(t) = u(0)\left(\frac{l\, e^{lt} - m\, e^{mt}}{l - m}\right) + \left[\mu\, u(0) + u'(0)\right] G_{l,m}(t),$$

where $G_{l,m}(t) = \left(\frac{e^{lt} - e^{mt}}{l - m}\right)$. Then $v(0) = u(0)$ and $u'(0) = v'(0)$. Taking again the Aboodh transform to $v(t)$, we have

$$\mathcal{A}\{v\} = \frac{u(0)}{(\xi - l)(\xi - m)} + \frac{\mu\, u(0) + u'(0)}{\xi(\xi - l)(\xi - m)}. \tag{11}$$

Furthermore, one can have

$$\mathcal{A}\{v''(t) + \mu v'(t) + \zeta v(t)\} = (\xi^2 + \mu\xi + \zeta)\mathcal{A}\{v\} - v(0)\left[\frac{\xi + \mu}{\xi}\right] - \frac{v'(0)}{\xi}.$$

Thus, using (11) in the above equation, we get $\mathcal{A}\{v''(t) + \mu v'(t) + \zeta v(t)\} = 0$, and so $v''(t) + \mu v'(t) + \zeta v(t) = 0$. By (10) and (11), we get

$$\mathcal{A}\{u\} - \mathcal{A}\{v\} = \frac{\mathcal{A}\{p\}\xi}{\xi(\xi - l)(\xi - m)},$$

$$\mathcal{A}\{u(t) - v(t)\} = \mathcal{A}\{p(t) * G_{l,m}(t)\}.$$

So $u(t) - v(t) = p(t) * G_{l,m}(t)$. By $|p(t)| \leq \epsilon\phi(t)$, we get

$$|u(t) - v(t)| = \left|p(t) * \left(\frac{e^{lt} - e^{mt}}{l - m}\right)\right|$$

$$\leq \left|\int_0^t p(x)\left(\frac{e^{l(t-x)} - e^{m(t-x)}}{l - m}\right)\mathrm{d}x\right|$$

$$\leq \epsilon\left|\int_0^t \left(\frac{e^{l(t-x)} - e^{m(t-x)}}{l - m}\right)\phi(x)\mathrm{d}x\right|$$

for all $t > 0$, where

$$L_\phi = \left| \int_0^t \left(\frac{e^{l(t-x)} - e^{m(t-x)}}{l-m} \right) \phi(x) \mathrm{d}x \right|$$

$$\leq \frac{1}{|l-m|} \left\{ e^{\mathcal{R}(l)t} \int_0^t e^{-\mathcal{R}(l)x} \phi(x) \mathrm{d}x \right.$$

$$\left. + e^{\mathcal{R}(m)t} \int_0^t e^{-\mathcal{R}(m)x} \phi(x) \mathrm{d}x \right\} \leq \frac{\mathcal{K}_\phi \, \phi(t)}{|l-m|}$$

and the integrals $\int_0^t e^{-\mathcal{R}(l)x} \phi(x) \, \mathrm{d}x$ and $\int_0^t e^{-\mathcal{R}(m)x} \phi(x) \mathrm{d}x$ exist for all $t > 0$ and ϕ is an integrable function. Hence,

$$|u(t) - v(t)| \leq \frac{\mathcal{K}_\phi \, \phi(t)}{|l-m|} \, \epsilon = L_\phi \, \epsilon \phi(t).$$

Then, by Definition 5, we confirm that the homogeneous linear differential equation (1) has Hyers–Ulam–Rassias stability. □

Now, we prove the Mittag–Leffler–Hyers–Ulam stability of the homogeneous differential equation (1). The proof is similar to the proof of Theorem 1, but we include it for the sake of completeness.

Theorem 3. *Suppose that $u(t)$ is a twice continuously differentiable function on I, which satisfies the differential equation (1), and $E_\alpha(t)$ is a Mittag–Leffler function. Then the homogeneous linear differential equation (1) has Mittag–Leffler–Hyers–Ulam stability.*

Proof. Let $\epsilon > 0$ and $u(t) \in C^2(I)$ be a function satisfying the inequality

$$\left| u''(t) + \mu u'(t) + \zeta u(t) \right| \leq \epsilon E_\alpha(t) \tag{12}$$

for all $t \in I$. We prove that there exists a real number $L > 0$ which is independent of ϵ and u such that $|u(t) - v(t)| \leq L \epsilon E_\alpha(t)$ for some $v \in C^2(I)$ satisfying $v''(t) + \mu v'(t) + \zeta v(t) = 0$ for all $t \in I$.

Define a function $p : (0, \infty) \longrightarrow \mathbb{F}$ such that $p(t) :=: u''(t) + \mu u'(t) + \zeta u(t)$ for all $t > 0$. In view of (12), we have $|p(t)| \leq \epsilon E_\alpha(t)$. Taking the Aboodh transform to $p(t)$, we have

$$\mathcal{A}\{p\} = (\xi^2 + \mu \xi + \zeta) \mathcal{A}\{u\} - \frac{u'(0)}{\xi} - u(0) \left[1 + \frac{\mu}{\xi} \right] \tag{13}$$

and thus

$$\mathcal{A}\{u\} = \frac{\mathcal{A}\{p\} + u(0)\left[\frac{\xi+\mu}{\xi}\right] + \frac{u'(0)}{\xi}}{\xi^2 + \mu\xi + \zeta}.$$

By (13), a function $u_0 : (0, \infty) \longrightarrow \mathbb{F}$ is a solution of (1) if and only if

$$(\xi^2 + \mu\xi + \zeta)\mathcal{A}\{u_0\} - u_0(0)\left[\frac{\xi+\mu}{\xi}\right] - \frac{u'_0(0)}{\xi} = 0.$$

If there exist constants l and m in $\mathbb{F}$ such that $\xi^2 + \mu\xi + \zeta = (\xi - l)(\xi - m)$ with $l + m = -\mu$ and $lm = \zeta$, then (13) becomes

$$\mathcal{A}\{u\} = \frac{P(\xi) + u(0)\left[\frac{\xi+\mu}{\xi}\right] + \frac{u'(0)}{\xi}}{(\xi - l)(\xi - m)}. \tag{14}$$

Set

$$v(t) = u(0)\left(\frac{l\, e^{lt} - m\, e^{mt}}{l - m}\right) + [\mu\, u(0) + u'(0)]\, G_{l,m}(t),$$

where $G_{l,m}(t) = \left(\frac{e^{lt} - e^{mt}}{l-m}\right)$. Then we have $v(0) = u(0)$ and $v'(0) = u'(0)$. Taking the Aboodh transform to $v(t)$, we obtain

$$\mathcal{A}\{v\} = \frac{u(0)}{(\xi - l)(\xi - m)} + \frac{[\mu\, u(0) + u'(0)]}{\xi(\xi - l)(\xi - m)}. \tag{15}$$

On the other hand, we can have

$$\mathcal{A}\{v''(t) + \mu v'(t) + \zeta v(t)\} = (\xi^2 + \mu\xi + \zeta)\mathcal{A}\{v\}$$
$$- v(0)\left[\frac{\xi+\mu}{\xi}\right] - \frac{v'(0)}{\xi}.$$

Using (15) in the above equality, we get $\mathcal{A}\{v''(t) + \mu v'(t) + \zeta v(t)\} = 0$. Since $\mathcal{A}$ is a one-to-one operator and linear, $v''(t) + \mu v'(t) + \zeta v(t) = 0$. This means that $v(t)$ is a solution of (1). It follows from (14) and (15) that

$$\mathcal{A}\{u\} - \mathcal{A}\{v\} = \frac{P(\xi) + u(0)\left[\frac{\xi+\mu}{\xi}\right] + \frac{u'(0)}{\xi}}{(\xi - l)(\xi - m)} - \frac{u(0)}{(\xi - l)(\xi - m)}$$
$$- \frac{[\mu\, u(0) + u'(0)]}{\xi(\xi - l)(\xi - m)} = \frac{\mathcal{A}\{p\}}{(\xi - l)(\xi - m)},$$

$$\mathcal{A}\{u(t) - v(t)\} = \mathcal{A}\left\{p(t) * \left(\frac{e^{lt} - e^{mt}}{l - m}\right)\right\}.$$

The above equalities show that

$$u(t) - v(t) = p(t) * \left(\frac{e^{lt} - e^{mt}}{l - m}\right)$$

and by $|p(t)| \leq \epsilon E_\alpha(t)$, we get

$$|u(t) - v(t)| \leq \epsilon E_\alpha(t) \left| \int_0^t \left(\frac{e^{l(t-x)} - e^{m(t-x)}}{l - m}\right) dx \right|$$

for all $t > 0$. Choose a constant

$$L = \left| \int_0^t \left(\frac{e^{l(t-x)} - e^{m(t-x)}}{l - m}\right) dx \right|$$
$$\leq \frac{1}{|l - m|} \left\{ e^{\mathcal{R}(l)t} \int_0^t e^{-\mathcal{R}(l)x} dx + e^{\mathcal{R}(m)t} \int_0^t e^{-\mathcal{R}(m)x} dx \right\} \leq \frac{\mathcal{K}}{|l - m|},$$

where the integrals $\int_0^t e^{-\mathcal{R}(l)x} dx$ and $\int_0^t e^{-\mathcal{R}(m)x} dx$ exist. Hence, $|u(t) - v(t)| \leq L\, \epsilon E_\alpha(t)$. By the virtue of Definition 8, the linear differential equation (1) has the Mittag–Leffler–Hyers–Ulam stability. This finishes the proof. □

The following corollary shows the Mittag–Leffler–Hyers–Ulam–Rassias stability of the differential equation (1). The proof is similar to the proof of Theorem 3, but we include some part of the proof for completion.

Corollary 1. *Suppose that $u(t)$ is a twice continuously differentiable function on I, which satisfies the second-order differential equation (1) and $E_\alpha(t)$ is a Mittag–Leffler function. Then the differential equation (1) has Mittag–Leffler–Hyers–Ulam–Rassias stability.*

Proof. For every $\epsilon > 0$, let $u(t)$ be a twice continuously differentiable function on I which satisfies the inequality

$$\left| u''(t) + \mu u'(t) + \zeta u(t) \right| \leq \epsilon \phi(t) E_\alpha(t) \tag{16}$$

for all $t \in I$.

We prove that there exists a real number $L > 0$ which is independent of ϵ and u, for some $v \in C^2(I)$ satisfying the differential equation $v''(t) + \mu v'(t) + \zeta v(t) = 0$ and

$$|u(t) - v(t)| \leq L_\phi \epsilon \phi(t) E_\alpha(t)$$

for all $t \in I$. Now, let us choose the function $p : (0, \infty) \longrightarrow \mathbb{F}$ such that

$$p(t) =: u''(t) + \mu u'(t) + \zeta u(t)$$

for all $t > 0$. In view of (16), we have $|p(t)| \leq \epsilon \phi(t) E_\alpha(t)$. Then by the same reasoning as in the proof of Theorem 3, we have

$$|u(t) - v(t)| \leq \epsilon E_\alpha(t) \left| \int_0^t \left(\frac{e^{l(t-x)} - e^{m(t-x)}}{l - m} \right) \phi(x) \mathrm{d}x \right|$$

for all $t > 0$, by choosing

$$L_\phi = \left| \int_0^t \left(\frac{e^{l(t-x)} - e^{m(t-x)}}{l - m} \right) \phi(x) \mathrm{d}x \right|$$

$$\leq \frac{1}{|l - m|} \left\{ e^{\mathcal{R}(l)t} \int_0^t e^{-\mathcal{R}(l)x} \phi(x) \mathrm{d}x \right.$$

$$\left. + e^{\mathcal{R}(m)t} \int_0^t e^{-\mathcal{R}(m)x} \phi(x) \mathrm{d}x \right\} \leq \frac{\mathcal{K}_\phi \, \phi(t)}{|l - m|},$$

where the integrals $\int_0^t e^{-\mathcal{R}(l)x}\phi(x)dx$ and $\int_0^t e^{-\mathcal{R}(m)x}\phi(x)dx$ exist for all $t>0$ and ϕ is an integrable function. Hence, $|u(t)-v(t)| \leq L_\phi \ \epsilon\phi(t)E_\alpha(t)$. Then, by Definition 9, the linear differential equation (1) has Mittag–Leffler–Hyers–Ulam–Rassias stability. $\square$

4. Ulam Stabilities for (2)

In this section, we investigate the Hyers–Ulam stability, Hyers–Ulam–Rassias stability, Mittag–Leffler–Hyers–Ulam stability and Mittag–Leffler–Hyers–Ulam–Rassias stability of the differential equation (2). First, we prove the Hyers–Ulam stability of the nonhomogeneous linear differential equation (2).

Theorem 4. *The nonhomogeneous linear second-order differential equation (2) has Hyers–Ulam stability.*

Proof. Let $\epsilon > 0$ and $u(t) \in C^2(I)$ satisfy

$$|u''(t) + \mu u'(t) + \zeta u(t) - q(t)| \leq \epsilon \qquad (17)$$

for all $t \in I$. In what follows, we shall prove that there exists a real number $L > 0$ which is independent of ϵ and u such that $|u(t) - v(t)| \leq L\epsilon$ for some $v \in C^2(I)$ satisfying $v''(t) + \mu v'(t) + \zeta v(t) = q(t)$ for all $t \in I$.

Let $p : (0, \infty) \to \mathbb{F}$ be a function defined by

$$p(t) =: u''(t) + \mu u'(t) + \zeta u(t) - q(t),$$

which satisfies $|p(t)| \leq \epsilon$. Taking the Aboodh transform to $p(t)$, we have

$$\mathcal{A}\{u\} = \frac{\mathcal{A}\{p\} + u(0)\left[\dfrac{\xi+\mu}{\xi}\right] + \dfrac{u'(0)}{\xi} + \mathcal{A}\{q\}}{\xi^2 + \mu\xi + \zeta}. \qquad (18)$$

The equality (18) shows that a function $u_0 : (0, \infty) \longrightarrow \mathbb{F}$ is a solution of (2) if and only if

$$(\xi^2 + \mu\xi + \zeta)\mathcal{A}\{u_0\} - u_0(0)\left[\dfrac{\xi+\mu}{\xi}\right] - \dfrac{u'_0(0)}{\xi} = \mathcal{A}\{q\}.$$

If there exist constants l and m in $\mathbb{F}$ such that $\xi^2 + \mu\xi + \zeta = (\xi - l)(\xi - m)$ with $l + m = -\mu$ and $lm = \zeta$, then (18) converts to

$$\mathcal{A}\{u\} = \frac{P(\xi) + u(0)\left[\frac{\xi + \mu}{\xi}\right] + \frac{u'(0)}{\xi} + \mathcal{A}\{q\}}{(\xi - l)(\xi - m)}. \tag{19}$$

Set $G_{l,m}(t) = \dfrac{e^{lt} - e^{mt}}{l - m}$ and

$$v(t) = u(0)\left(\frac{l\, e^{lt} - m\, e^{mt}}{l - m}\right) + [\mu\, u(0) + u'(0)]\, G_{l,m}(t)$$
$$+ [(G_{l,m} * q)(t)].$$

Then, $v(0) = u(0)$ and $u'(0) = v'(0)$. Taking the Aboodh transform to $v(t)$, we get

$$\mathcal{A}\{v\} = \frac{u(0) + \mathcal{A}\{q\}}{(\xi - l)(\xi - m)} + \frac{\mu\, u(0) + u'(0)}{\xi(\xi - l)(\xi - m)}. \tag{20}$$

On the other hand,

$$\mathcal{A}\{v''(t) + \mu v'(t) + \zeta v(t)\} = (\xi^2 + \mu\xi + \zeta)\mathcal{A}\{v\} - v(0)\left[\frac{\xi + \mu}{\xi}\right] - \frac{v'(0)}{\xi}.$$

Using (20), the last equality becomes $\mathcal{A}\{v''(t) + \mu v'(t) + \zeta v(t)\} = \mathcal{A}\{q\}$. Since $\mathcal{A}$ is a one-to-one operator and linear, $v''(t) + \mu v'(t) + \zeta v(t) = q(t)$. This shows that $v(t)$ is a solution of (2). Now, by (19) and (20),

$$\mathcal{A}\{u(t) - v(t)\} = \mathcal{A}\{u\} - \mathcal{A}\{v\} = \frac{\mathcal{A}\{p\}}{(\xi - l)(\xi - m)} = \mathcal{A}\{p(t) * r(t)\}$$

and hence $u(t) - v(t) = p(t) * G_{l,m}(t)$. Taking modulus on both sides of the last equality and using $|p(t)| \leq \epsilon$, we get

$$|u(t) - v(t)| = |p(t) * G_{l,m}(t)|$$
$$\leq \epsilon \left|\int_0^t \left(\frac{e^{l(t-x)} - e^{m(t-x)}}{l - m}\right) dt\right| \leq L\,\epsilon,$$

where

$$L = \left| \int_0^t \left(\frac{e^{l(t-x)} - e^{m(t-x)}}{l-m} \right) dx \right|$$

$$\leq \frac{1}{|l-m|} \left\{ e^{\mathcal{R}(l)t} \int_0^t e^{-\mathcal{R}(l)x} dx + e^{\mathcal{R}(m)t} \int_0^t e^{-\mathcal{R}(m)x} dx \right\} \leq \frac{\mathcal{K}}{|l-m|},$$

and the integrals $\int_0^t e^{-\mathcal{R}(l)x} dx$ and $\int_0^t e^{-\mathcal{R}(m)x} dx$ exist for all $t > 0$. Therefore, the linear differential equation (2) has the Hyers–Ulam stability, by the virtue of Definition 6. □

In analogous to Theorem 4, we have the following result which shows the Hyers–Ulam–Rassias stability of the differential equation (2).

Theorem 5. *Let $\epsilon > 0$ and $\phi \in C(\mathbb{R}_+, \mathbb{R}_+)$. Suppose that $u(t) \in C^2(I)$ satisfies*

$$\left| u''(t) + \mu u'(t) + \zeta u(t) - q(t) \right| \leq \epsilon \phi(t) \tag{21}$$

for all $t \in I$. Then there exist a real number $L_\phi > 0$ and $v(t) \in C^2(I)$ satisfying s the differential equation $v''(t) + \mu v'(t) + \zeta v(t) = q(t)$ such that for all $t > 0$

$$|u(t) - v(t)| \leq L_\phi \epsilon \phi(t).$$

Proof. Define $p : (0, \infty) \longrightarrow \mathbb{F}$ by $p(t) =: u''(t) + \mu u'(t) + \zeta u(t) - q(t)$ for all $t > 0$. In view of (21), we have $|p(t)| \leq \epsilon \phi(t)$. Now, taking the Aboodh transform to $p(t)$, we get

$$\mathcal{A}\{u\} = \frac{\mathcal{A}\{p\} + u(0) \left[\frac{\xi + \mu}{\xi} \right] + \frac{u'(0)}{\xi} + \mathcal{A}\{q\}}{\xi^2 + \mu \xi + \zeta}. \tag{22}$$

In addition, by (22), a function $u_0 : (0, \infty) \longrightarrow \mathbb{F}$ is a solution of (2) if and only if

$$(\xi^2 + \mu \xi + \zeta)\mathcal{A}\{u_0\} - u_0(0) \left[\frac{\xi + \mu}{\xi} \right] - \frac{u_0'(0)}{\xi} = \mathcal{A}\{q\}.$$

However, (22) becomes

$$\mathcal{A}\{u\} = \frac{P(\xi) + u(0)\left[\frac{\xi+\mu}{\xi}\right] + \frac{u'(0)}{\xi} + Q(\xi)}{(\xi - l)(\xi - m)}. \tag{23}$$

Assume that there exist constants l and m in $\mathbb{F}$ such that $\xi^2 + \mu\xi + \zeta = (\xi-l)(\xi-m)$ with $l+m = -\mu$ and $lm = \zeta$. Put $G_{l,m}(t) = \frac{e^{lt}-e^{mt}}{l-m}$ and

$$v(t) = u(0)\left(\frac{l\,e^{lt} - m\,e^{mt}}{l-m}\right)$$
$$+ \left[\mu\, u(0) + u'(0)\right] G_{l,m}(t) + \left[(G_{l,m} * q)(t)\right].$$

Then one can easily obtain $v(0) = u(0)$ and $u'(0) = v'(0)$. Taking the Aboodh transform to $v(t)$, we have

$$\mathcal{A}\{v\} = \frac{u(0) + \mathcal{A}\{q\}}{(\xi-l)(\xi-m)} + \frac{\mu\, u(0) + u'(0)}{\xi(\xi-l)(\xi-m)}. \tag{24}$$

Furthermore, we can have

$$\mathcal{A}\{v'' + \mu v'(t) + \zeta v(t)\} = (\xi^2 + \mu\xi + \zeta)\mathcal{A}\{v\} - v(0)\left[\frac{\xi+\mu}{\xi}\right] - \frac{v'(0)}{\xi}. \tag{25}$$

By (24) in (25), we can show that $\mathcal{A}\{v''(t) + \mu v'(t) + \zeta v(t)\} = \mathcal{A}\{q\}$. The last equality implies that $v''(t) + \mu v'(t) + \zeta v(t) = q(t)$. This means that $v(t)$ is a solution of (2). By (23) and (24), we obtain

$$\mathcal{A}\{u(t) - v(t)\} = \mathcal{A}\{u\} - \mathcal{A}\{v\} = \frac{\mathcal{A}\{p\}}{(\xi-l)(\xi-m)}$$
$$= \mathcal{A}\{p(t) * G_{l,m}(t)\}.$$

Thus, $u(t) - v(t) = p(t) * G_{l,m}(t)$. By $|p(t)| \leq \epsilon\phi(t)$, we get

$$|u(t) - v(t)| \leq \epsilon \left|\int_0^t \left(\frac{e^{l(t-x)} - e^{m(t-x)}}{l-m}\right)\phi(t)\mathrm{d}x\right| \leq L_\phi\, \epsilon\phi(t),$$

where

$$L_\phi = \left| \int_0^t \left(\frac{e^{l(t-x)} - e^{m(t-x)}}{l - m} \right) \phi(x) \mathrm{d}x \right|$$

$$\leq \frac{1}{|l - m|} \left\{ e^{\mathcal{R}(l)t} \int_0^t e^{-\mathcal{R}(l)x} \phi(x) \mathrm{d}x \right.$$

$$\left. + e^{\mathcal{R}(m)t} \int_0^t e^{-\mathcal{R}(m)x} \phi(x) \mathrm{d}x \right\} \leq \frac{K_\phi \, \phi(t)}{|l - m|}$$

and the integrals $\int_0^t e^{-\mathcal{R}(l)x} \phi(x) \, \mathrm{d}x$ and $\int_0^t e^{-\mathcal{R}(m)x} \phi(x) \, \mathrm{d}x$ exist for all $t > 0$ and ϕ is an integrable function. This concludes the proof by the virtue of Definition 7. □

By using the same technique as in Theorem 4, we can also prove the following theorem which shows the Mittag–Leffler–Hyers–Ulam stability of the differential equation (2). The proof is similar, but we include it for the sake of completeness.

Theorem 6. *Let $\epsilon > 0$ and $u(t) \in C^2(I)$ satisfy*

$$\left| u''(t) + \mu u'(t) + \zeta u(t) - q(t) \right| \leq \epsilon E_\alpha(t) \qquad (26)$$

for all $t \in I$. Then there exist a real number $L > 0$ which is independent of ϵ and u and some $v \in C^2(I)$ satisfying the differential equation $v''(t) + \mu v'(t) + \zeta v(t) = q(t)$ and $|u(t) - v(t)| \leq L\epsilon E_\alpha(t)$ for all $t \in I$.

Proof. Choose $p : (0, \infty) \to \mathbb{F}$ defined by

$$p(t) =: u''(t) + \mu u'(t) + \zeta u(t) - q(t),$$

which satisfies $|p(t)| \leq \epsilon E_\alpha(t)$. Taking the Aboodh transform to $p(t)$, we have

$$\mathcal{A}\{u\} = \frac{\mathcal{A}\{p\} + u(0)\left[\frac{\xi + \mu}{\xi}\right] + \frac{u'(0)}{\xi} + \mathcal{A}\{q\}}{\xi^2 + \mu\xi + \zeta}. \qquad (27)$$

The equality (27) shows that a function $u_0 : (0, \infty) \to \mathbb{F}$ is a solution of (2) if and only if

$$(\xi^2 + \mu\xi + \zeta)\mathcal{A}\{u_0\} - u_0(0)\left[\frac{\xi + \mu}{\xi}\right] - \frac{u_0'(0)}{\xi} = \mathcal{A}\{q\}.$$

If there exist constants l and m in $\mathbb{F}$ such that $\xi^2 + \mu\xi + \zeta = (\xi - l)(\xi - m)$ with $l + m = -\mu$ and $lm = \zeta$, then (27) converts to

$$U(\xi) = \frac{P(\xi) + u(0)\left[\frac{\xi + \mu}{\xi}\right] + \frac{u'(0)}{\xi} + Q(\xi)}{(\xi - l)(\xi - m)}. \tag{28}$$

Set $G_{l,m}(t) = \dfrac{e^{lt} - e^{mt}}{l - m}$ and

$$v(t) = u(0)\left(\frac{l\, e^{lt} - m\, e^{mt}}{l - m}\right) + [\mu\, u(0) + u'(0)]\, G_{l,m}(t)$$
$$+ [(G_{l,m} * q)(t)].$$

Then $v(0) = u(0)$ and $v'(0) = u'(0)$. Taking the Aboodh transform to $v(t)$, we obtain

$$V(\xi) = \frac{u(0) + Q(\xi)}{(\xi - l)(\xi - m)} + \frac{\mu\, u(0) + u'(0)}{\xi(\xi - l)(\xi - m)}. \tag{29}$$

On the other hand, one can easily obtain that

$$\mathcal{A}\{v''(t) + \mu v'(t) + \zeta v(t)\}$$
$$= (\xi^2 + \mu\xi + \zeta)\mathcal{A}\{v\} - v(0)\left[\frac{\xi + \mu}{\xi}\right] - \frac{v'(0)}{\xi}.$$

By (29), the last equality becomes $\mathcal{A}\{v''(t) + \mu v'(t) + \zeta v(t)\} = \mathcal{A}\{q\}$. Since $\mathcal{A}$ is a one-to-one operator and linear, $v''(t) + \mu v'(t) + \zeta v(t) = q(t)$. This shows that $v(t)$ is a solution of (2). Now, by (28) and (29), we have

$$\mathcal{A}\{u(t) - v(t)\} = \mathcal{A}\{u\} - \mathcal{A}\{v\} = \frac{\xi \mathcal{A}\{p\}}{\xi(\xi - l)(\xi - m)}$$
$$= \mathcal{A}\{p(t) * G_{l,m}(t)\}$$

and hence $u(t) - v(t) = p(t) * G_{l,m}(t)$. Taking modulus on both sides of the last equality and using $|p(t)| \leq \epsilon E_\alpha(t)$, we get

$$|u(t) - v(t)| \leq L \, \epsilon E_\alpha(t),$$

where

$$L = \left| \int_0^t \left(\frac{e^{l(t-x)} - e^{m(t-x)}}{l - m} \right) dx \right|$$

$$\leq \frac{1}{|l-m|} \left\{ e^{\mathcal{R}(l)t} \int_0^t e^{-\mathcal{R}(l)x} dx + e^{\mathcal{R}(m)t} \int_0^t e^{-\mathcal{R}(m)x} dx \right\} \leq \frac{K}{|l-m|}$$

and the integrals $\int_0^t e^{-\mathcal{R}(l)x} dx$ and $\int_0^t e^{-\mathcal{R}(m)x} dx$ exist for all $t > 0$. Hence, the linear differential equation (2) has the Mittag–Leffler–Hyers–Ulam stability by the virtue of Definition 10. □

In analogous to Theorem 5, we have the following corollary which shows the Mittag–Leffler–Hyers–Ulam–Rassias stability of the differential equation (2). The proof is similar to the proof of Theorem 5, but we include some parts of the proof for completeness of the result.

Corollary 2. *Assume that $u(t)$ is a twice continuously differentiable function on I, which satisfies the second-order differential equation* (2) *and $E_\alpha(t)$ is a Mittag–Leffler function. Then the nonhomogeneous linear differential equation* (2) *has Mittag–Leffler–Hyers–Ulam–Rassias stability.*

Proof. Let $\epsilon > 0$ and $u(t) \in C^2(I)$ satisfy the following inequality

$$|u''(t) + \mu u'(t) + \zeta u(t) - q(t)| \leq \epsilon \phi(t) E_\alpha(t) \tag{30}$$

for all $t \in I$. In what follows, we shall prove that there exist real number $L_\phi > 0$ which is independent of ϵ and u and some $v \in C^2(I)$ satisfying the nonhomogeneous differential equation $v''(t) + \mu v'(t) + \zeta v(t) = q(t)$ such that

$$|u(t) - v(t)| \leq L_\phi \epsilon \phi(t) E_\alpha(t)$$

for all $t \in I$. Define $p : (0, \infty) \to \mathbb{F}$ by $p(t) =: u''(t) + \mu u'(t) + \zeta u(t) - q(t)$ for all $t > 0$. In view of (30), we have $|p(t)| \leq \epsilon \phi(t) E_\alpha(t)$.

Hence by the same reasoning as in the proof of Theorem 5, we can easily show that

$$|u(t) - v(t)| \leq \epsilon\, E_\alpha(t) \left| \int_0^t \left(\frac{e^{l(t-x)} - e^{m(t-x)}}{l-m} \right) \phi(t)\mathrm{d}x \right|$$

$$\leq L_\phi\, \epsilon E_\alpha(t)\phi(t),$$

where

$$L_\phi = \left| \int_0^t \left(\frac{e^{l(t-x)} - e^{m(t-x)}}{l-m} \right) \phi(x)\mathrm{d}x \right|$$

$$\leq \frac{1}{|l-m|} \left\{ e^{\mathcal{R}(l)t} \int_0^t e^{-\mathcal{R}(l)x}\, \phi(x)\mathrm{d}x \right.$$

$$\left. + e^{\mathcal{R}(m)t} \int_0^t e^{-\mathcal{R}(m)x}\, \phi(x)\mathrm{d}x \right\} \leq \frac{\mathcal{K}_\phi\, \phi(t)}{|l-m|}$$

and the integrals $\int_0^t e^{-\mathcal{R}(l)x}\, \phi(x)\,\mathrm{d}x$ and $\int_0^t e^{-\mathcal{R}(m)x}\, \phi(x)\,\mathrm{d}x$ exist for all $t > 0$ and ϕ is an integrable function. Hence by Definition 11, the differential equation (2) has the Mittag–Leffler–Hyers–Ulam–Rassias stability. This concludes the proof. □

5. Examples and Remarks

In this section, we introduce some examples to make it easier to understand the main results of this chapter.

Example 1. Consider the following homogeneous linear differential equation of second order

$$u''(t) + u(t) = 0, \tag{31}$$

where $\mu = 0$ and $\zeta = 1$, with initial conditions $u(0) = u'(0) = 1$. By Theorem 1, for $p(t) = u''(t) + u(t)$, by taking the Aboodh transform, we have

$$P(\xi) = \xi^2 U(\xi) - \frac{u'(0)}{\xi} - u(0) + U(\xi).$$

By using initial conditions, we have $\mathcal{A}\{u\} = \frac{\xi P(\xi)+\xi+1}{\xi(\xi^2+1)}$. If a continuously differentiable function $v : [0,\infty) \to \mathbb{F}$ of exponential order satisfies

$$|u''(t) + u(t)| \leq \epsilon$$

for all $t \geq 0$ and some $\epsilon > 0$, then Theorem 1 implies that there exists a solution $v : [0,\infty) \to \mathbb{F}$ of the differential equation (31) such that

$$|u(t) - v(t)| \leq K\epsilon$$

for all $t \geq 0$, where $\frac{1}{\Re(\zeta)} = 1$.

Example 2. Consider the following nonhomogeneous linear differential equation

$$u''(t) - 3u'(t) + 2u(t) = 4e^{3t} \tag{32}$$

with initial conditions

$$y(0) = -3 \quad \text{and} \quad y'(0) = 5,$$

where $q(t) = 4e^{3t}$ is a function of exponential order with $\mu = -3$ and $\zeta = 2$.

By Theorem 1, for $p(t) = u''(t) - 3u'(t) + 2u(t) - 4e^{3t}$, by taking the Aboodh transform, we have

$$P(\xi) = \xi^2 U(\xi) - \frac{u'(0)}{\xi} - u(0) - 3\left(\xi U(\xi) - \frac{u(0)}{\xi}\right)$$
$$+ 2U(\xi) - \frac{4}{\xi(\xi-3)}.$$

By using initial conditions, we have

$$U(\xi) = \mathcal{A}\{u\} = \frac{\xi P(\xi) - 3\xi + 14}{\xi(\xi^2 - 3\xi + 2)} + \frac{4}{\xi(\xi-3)(\xi^2 - 3\xi + 2)}.$$

It can be also written as

$$\mathcal{A}\{u\} = \frac{\xi \mathcal{A}\{p\} - 3\xi + 14}{\xi(\xi-1)(\xi-2)} + \frac{4}{\xi(\xi-1)(\xi-2)(\xi-3)}.$$

If a continuously differentiable function $z : [0, \infty) \to \mathbb{F}$ of exponential order satisfies

$$|u''(t) - 3u'(t) + 2u(t) - 4e^{3t}| \leq \epsilon$$

for all $t \geq 0$ and some $\epsilon > 0$, then Theorem 4 implies that there exists a solution $y : [0, \infty) \to \mathbb{F}$ of the differential equation (32) such that

$$|u(t) - v(t)| \leq K\epsilon$$

for all $t \geq 0$, where $\frac{1}{\Re(\mu)} = \frac{1}{3}$ and $\frac{1}{\Re(\zeta)} = \frac{1}{5}$.

Remark 1. The above examples are also true when we replace ε and $K\varepsilon$ with $\phi(t)\varepsilon$ and $K\phi(t)\varepsilon$, respectively, where $\phi(t)$ is an increasing function. In this case, we see that the corresponding differential equations have the Hyers–Ulam–Rassias stability.

Remark 2. The differential equations (31) and (32) have the Mittag–Leffler–Hyers–Ulam stability if $\nu > 0$. In particular, they also have the Mittag–Leffler–Hyers–Ulam–Rassias stability when $\phi(t)$ is an increasing function and $\nu > 0$.

References

[1] M. R. Abdollahpour, R. Aghayaria and M. T. Rassias, Hyers–Ulam stability of associated Laguerre differential equations in a subclass of analytic functions, *J. Math. Anal. Appl.* **437**, 605–612 (2016).

[2] M. R. Abdollahpour and M. T. Rassias, Hyers–Ulam stability of hypergeometric differential equations, *Aequationes Math.* **93**, 691–698 (2019).

[3] K. S. Aboodh, Solving porous medium equation using Aboodh transform homotopy perturbation method, *Pure Appl. Math. J.* **4(6)**, 271–276 (2016).

[4] J. Aczel and J. Dhombres, *Functional Equations in Several Variables* (Cambridge University Press, Cambridge, 1989).

[5] Q. H. Alqifiary and S. Jung, Laplace transform and generalized Hyers–Ulam stability of differential equations, *Electron. J. Differ. Equ.* **2014** (2014). Paper No. 80.

[6] C. Alsina and R. Ger, On some inequalities and stability results related to the exponential function, *J. Inequal. Appl.* **2**, 373–380 (1998).

[7] A. A. Alsshikh and M. M. A. Mahgob, A comparative study between Laplace transform and two new integrals "ELzaki" transform and "Aboodh" transform, *Pure Appl. Math. J.* **5**(5), 145–150 (2016).

[8] T. Aoki, On the stability of the linear transformation in Banach spaces, *J. Math. Soc. Jpn.* **2**, 64–66 (1950).

[9] A. Buakird and S. Saejung, Ulam stability with respect to a directed graph for some fixed point equations, *Carpathian J. Math.* **35**, 23–30 (2019).

[10] D. G. Bourgin, Classes of transformations and bordering transformations, *Bull. Am. Math. Soc.* **57**, 223–237 (1951).

[11] S. Czerwik, *Functional Equations and Inequalities in Several Variables* (World Scientific, Singapore, 2002).

[12] M. Eshaghi Gordji, A. Fazeli and C. Park, 3-Lie multipliers on Banach 3-Lie algebras, *Int. J. Geom. Meth. Mod. Phys.* **9**(7), (2012). Art. ID 1250052.

[13] M. Eshaghi Gordji, M. B. Ghaemi and B. Alizadeh, A fixed point method for perturbation of higher ring derivations in non-Archimedean Banach algebras, *Int. J. Geom. Methods Mod. Phys.* **8**(7), 1611–1625 (2011).

[14] R. Fukutaka and M. Onitsuka, Best constant in Hyers-Ulam stability of first-order homogeneous linear differential equations with a periodic coefficient, *J. Math. Anal. Appl.* **473**, 1432–1446 (2019).

[15] P. Găvruta, A generalization of the Hyers-Ulam-Rassias stability of approximately additive mappings, *J. Math. Anal. Appl.* **184**, 431–436 (1994).

[16] J. Huang, S. Jung and Y. Li, On Hyers-Ulam stability of nonlinear differential equations, *Bull. Korean Math. Soc.* **52**, 685–697 (2015).

[17] D. H. Hyers, On the stability of the linear functional equation, *Proc. Nat. Acad. Sci. USA* **27**, 222–224 (1941).

[18] D. H. Hyers, G. Isac and T. M. Rassias, *Stability of Functional Equations in Several Variables* (Birkhäuser, New York, 1998).

[19] D. H. Hyers and T. M. Rassias, Approximate homomorphisms, *Aequationes Math.* **44**, 125–153 (1992).

[20] G. Isac and T. M. Rassias, Stability of ψ-additive mappings: Applications to nonlinear analysis, *Int. J. Math. Math. Sci.* **19**, 219–228 (1996).

[21] S. Jung, Hyers-Ulam stability of linear differential equation of first order, *Appl. Math. Lett.* **17**, 1135–1140 (2004).

[22] S. Jung, Hyers-Ulam stability of linear differential equations of first order (III), *J. Math. Anal. Appl.* **311**, 139–146 (2005).

[23] S. Jung, Hyers-Ulam stability of linear differential equations of first order (II), *Appl. Math. Lett.* **19**, 854–858 (2006).

[24] S. Jung, Hyers-Ulam stability of a system of first order linear differential equations with constant coefficients, *J. Math. Anal. Appl.* **320**, 549–561 (2006).

[25] S. Jung, *Hyers-Ulam-Rassias, Stability of Functional Equations in Nonlinear Analysis* (Springer, New York, 2011).

[26] S. Jung, Approximate solution of a linear differential equation of third order, *Bull. Malay. Math. Sci. Soc.* **35**(4), 1063–1073 (2012).

[27] S. Jung, C. Mortici and M. T. Rassias, On a functional equation of trigonometric type, *Appl. Math. Comput.* **252**, 294–303 (2015).

[28] S. Jung, D. Popa and M. T. Rassias, On the stability of the linear functional equation in a single variable on complete metric groups, *J. Global Optim.* **59**, 165–171 (2014).

[29] S. Jung and M. T. Rassias, A linear functional equation of third order associated to the Fibonacci numbers, *Abstr. Appl. Anal.* **2014**, (2014). Article ID 137468.

[30] V. Kalvandi, N. Eghbali and J. M. Rassias, Mittag-Leffler-Hyers-Ulam stability of fractional differential equations of second order, *J. Math. Ext.* **13**(1), 1–15 (2019).

[31] P. Kannappan, *Functional Equations and Inequalities with Applications* (Springer, New York, 2009).

[32] Y. Lee, S. Jung and M. T. Rassias, On an n-dimensional mixed type additive and quadratic functional equation, *Appl. Math. Comput.* **228**, 13–16 (2014).

[33] Y. Lee, S. Jung and M. T. Rassias, Uniqueness theorems on functional inequalities concerning cubic-quadratic-additive equation, *J. Math. Inequal.* **12**, 43–61 (2018).

[34] T. Li, A. Zada and S. Faisal, Hyers-Ulam stability of nth order linear differential equations, *J. Nonlinear Sci. Appl.* **9**, 2070–2075 (2016).

[35] Y. Li and Y. Shen, Hyers-Ulam stability of linear differential equations of second order, *Appl. Math. Lett.* **23**, 306–309 (2010).

[36] T. Miura, On the Hyers-Ulam stability of a differentiable map, *Sci. Math. Jpn.* **55**, 17–24 (2002).

[37] T. Miura, S. Jung and S. E. Takahasi, Hyers-Ulam-Rassias stability of the Banach space valued linear differential equations $y' = \lambda y$, *J. Korean Math. Soc.* **41**, 995–1005 (2004).

[38] C. Mortici, S. Jung and M. T. Rassias, On the stability of a functional equation associated with the Fibonacci numbers, *Abstr. Appl. Anal.* **2014**, (2014). Article ID 546046.

[39] R. Murali, A. Bodaghi and A. Ponmana Selvan, Stability for the third order linear ordinary differential equation, *Int. J. Math. Comput.* **30**, 87–92 (2019).

[40] R. Murali and A. Ponmana Selvan, On the generalized Hyers-Ulam stability of linear ordinary differential equations of higher order, *Int. J. Pure. Appl. Math.* **117**(12), 317–326 (2017).

[41] R. Murali and A. Ponmana Selvan, Hyers-Ulam-Rassias stability for the linear ordinary differential equation of third order, *Kragujevac J. Math.* **42**(4), 579–590 (2018).

[42] R. Murali and A. Ponmana Selvan, Ulam stability of third order linear differential equations, *Int. J. Pure Appl. Math.* **120**(9), 217–225 (2018).

[43] R. Murali and A. Ponmana Selvan, Fourier transforms and Ulam stabilities of linear differential equations. In *Frontiers in Functional Equations and Analytic Inequalities* (Springer, Switzerland, 2019), pp. 195–217.

[44] R. Murali and A. Ponmana Selvan, Hyers-Ulam stability of nth order linear differential equation, *Proyecciones* **38**(3), 553–566 (2019).

[45] R. Murali and A. Ponmana Selvan, Hyers-Ulam stability of a free and forced vibrations, *Kragujevac J. Math.* **44**(2), 299–312 (2020).

[46] R. Murali, A. Ponmana Selvan and C. Park, Ulam stability of linear differential equations using Fourier transform, *AIMS Math.* **5**, 766–780 (2019).

[47] M. Onitsuka, Hyers-Ulam stability of first order linear differential equations of Caratheodory type and its application, *Appl. Math. Lett.* **90**, 61–68 (2019).

[48] M. Onitsuka and T. Shoji, Hyers-Ulam stability of first order homogeneous linear differential equations with a real valued coefficients, *Appl. Math. Lett.* **63**, 102–108 (2017).

[49] C. Park and M. T. Rassias, Additive functional equations and partial multipliers in C^*-algebras, *Rev. R. Acad. Cien. Exactas Fís. Nat. Ser. A Mat. RACSAM* **113**, 2261–2275 (2019).

[50] A. Prástaro and T. M. Rassias, *Geometry in Partial Differential Equations* (World Scientific Publishing Co., Singapore, 1994).

[51] A. Prástaro and T. M. Rassias, On the set of solutions of the generalized d'Alembert equation, *C. R. Acad. Sci. Sér. I* (Paris) **328**, 389–394 (1999).

[52] A. Prástaro and T. M. Rassias, Ulam stability in geometry of PDE's, *Nonlinear Funct. Anal. Appl.* **8**(2), 259–278 (2003).

[53] A. Prástaro and T. M. Rassias, On Ulam stability in the geometry of PDE's. In *Functional Equations, Inequalities and Applications* (Kluwer Academic Publishers, Dordrecht, 2003), pp. 139–147.

[54] J. M. Rassias, On approximately of approximately linear mappings by linear mappings, *J. Funct. Anal.* **46**, 126–130 (1982).

[55] J. M. Rassias, R. Murali and A. Ponmana Selvan, Mittag-Leffler-Hyers-Ulam stability of linear differential equations using Fourier transforms, *J. Comput. Anal. Appl.* **29**, 68–85 (2021).

[56] T. M. Rassias, On the stability of the linear mapping in Banach spaces, *Proc. Am. Math. Soc.* **72**, 297–300 (1978).

[57] S. E. Takahasi, T. Miura and S. Miyajima, On the Hyers-Ulam stability of the Banach space-valued differential equation $y' = \alpha y$, *Bull. Korean Math. Soc.* **39**, 309–315 (2002).

[58] S. M. Ulam, *A Collection of the Mathematical Problems* (Interscience Publishers, New York, 1960).

[59] H. Vaezi, Hyers-Ulam stability of weighted composition operators on disc algebra, *Int. J. Math. Comput.* **10**(M11), 150–154 (2011).

[60] D. Venturi and A. Dektor, Spectral methods for nonlinear functionals and functional differential equations, *Res. Math. Sci.* **8**(2), (2021). Paper No. 27.

[61] G. Wang, M. Zhou and L. Sun, Hyers-Ulam stability of linear differential equations of first order, *Appl. Math. Lett.* **21**, 1024–1028 (2008).

© 2024 World Scientific Publishing Company
https://doi.org/10.1142/9789811267048_0021

Chapter 21

Some Classes of Extended General Variational Inequalities

Muhammad Aslam Noor* and Khalida Inayat Noor[†]

*Department of Mathematics,
COMSATS University Islamabad,
Park Road, Islamabad, Pakistan*

*noormaslam@gmail.com
[†]khalidan@gmail.com

It is well known that extended general variational inequalities provide us with a unified, natural, novel and simple framework to study a wide class of unrelated problems, which arise in pure and applied sciences. In this chapter, we present a number of new and known numerical techniques for solving general variational inequalities and equilibrium problems using various techniques including projection, Wiener–Hopf equations, dynamical systems and auxiliary principle. Our results present a significant improvement of previously known methods for solving variational inequalities and related optimization problems. Since the extended general variational inequalities include variational inequalities and implicit complementarity problems as special cases, results presented continue to hold for these problems. Several open problems have been suggested for further research in these areas.

1. Introduction

Variational inequality theory, which was introduced in Stampacchia [103], has emerged as an interesting and fascinating branch of applicable mathematics with a wide range of applications in industry, finance, economics, social, pure and applied sciences. Variational inequalities may be viewed as novel generalization of the variational principles, which have placed a crucial and important in the development of various fields of sciences and have appeared as a unifying force. Variational inequalities have been extended and generalized in several directions using novel and new techniques. To be more precise, the minimum of a differentiable convex function F on the convex set K is equivalent to finding $u \in K$ such that

$$\langle F'(u), v - u \rangle \geq 0, \quad \forall v \in K,$$

which is known as the variational inequality. Stampacchia [103] proved that potential problems associated with elliptic equations can be studied by the variational inequality. This simple fact inspired a great interest in variational inequalities. Lions and Stampacchia [39] studied the existence of a solution of variational inequalities using essentially the auxiliary principle technique coupled with projection idea.

Motivated and inspired by the ongoing research in these fields, Noor [75–77] introduced and investigated a new class of variational inequalities involving three operators. For given nonlinear operators T, g, h, consider the problem of finding $u \in H : h(u) \in K$, such that

$$\langle Tu, g(v) - h(u) \rangle \geq 0, \quad \forall v \in H : g(v) \in K, \tag{1}$$

which is known as the **extended general variational inequalities**. It turned out that odd-order and nonsymmetric obstacle, free, unilateral and moving boundary value problems arising in pure and applied sciences can be studied via the general variational inequalities, see Refs. [49–53, 66, 70, 72, 74–77].

If K is a convex cone, then the implicit complementarity problem (1) is equivalent to finding $u \in H$ such that

$$g(u) \geq 0, \quad Tu \in K^*, \quad \langle Tu, g(u) \rangle = 0, \tag{2}$$

which is known as the general complementarity problem, where K^* is the dual (polar) cone.

During the years which have been elapsed since its discovery, a number of numerical methods including projection method and its variant forms, Wiener–Hopf equations, auxiliary principle and dynamical systems have been developed for solving the variational inequalities. Projection method and its variants forms including the Wiener–Hopf equations represent important tools for finding the approximate solution of variational inequalities, the origin of which can be traced back to Lions and Stampacchia [39]. The main idea in this technique is to establish the equivalence between the variational inequalities and the fixed-point problem by using the concept of projection. This alternative formulation has played a significant part in developing various projection-type methods for solving variational inequalities. It is well known that the convergence of the projection methods requires that the operator must be strongly monotone and Lipschitz continuous. Unfortunately, these strict conditions rule out many applications of this method. This fact motivated to modify the projection method or to develop other methods. The extragradient-type methods overcome this difficulty by performing an additional forward step and a projection at each iteration according to the double projection. These methods can be viewed as predictor–corrector methods. Their convergence requires only that a solution exists and the monotone operator is Lipschitz continuous. Recently, several modified projection and extragradient-type methods have been suggested and developed for solving variational inequalities, see Refs. [1–10, 17, 18, 24, 36, 37, 39, 42–44, 49–53, 56, 66, 70, 72, 74–79, 82–84, 91, 93, 96, 101–105, 107, 108, 111–115, 117–119] and the references therein.

In Section 4, we have the concept of the extended general Wiener–Hopf equations, which was introduced by Noor [52]. As a special case, we obtain the original Wiener–Hopf equations, which were considered and studied by Shi [99] and Robinson [97] in conjunction with variational inequalities from different point of views. Using the projection technique, one usually establishes the equivalence between the variational inequalities and the Wiener–Hopf equations. It turned out that the Wiener–Hopf equations are more general and flexible. This approach has played an important part not only in developing various efficient projection-type methods but also in studying the sensitivity analysis, dynamical systems as well as other concepts of variational inequalities. Noor, Wang and Xiu 90 have suggested and analyzed some predictor–corrector-type projection methods by modifying the Wiener–Hopf equations. These methods are also known as forward–backward methods. It has been shown that these predictor–corrector-type methods are efficient and robust.

Section 5 is devoted to the concept of projected dynamical system in the context of variational inequalities, which was introduced by Dupuis and Nagurney [12] by using the fixed-point formulation of the variational inequalities. For the recent development and applications of the dynamical systems, see Refs. [11, 12, 17, 18, 42, 43, 65, 68, 89, 110, 116, 118]. In this technique, we reformulate the variational inequality problem as an initial value problem. Using the discretizing of the dynamical systems, we suggest some new iterative methods for solving the general variational inequalities.

It is a well-known fact that to implement the projection-type methods, one has to evaluate the projection, which is itself a difficult problem. Second, the projection and Wiener–Hopf equations techniques can't be extended and generalized for some classes of variational inequalities involving the nonlinear (non)differentiable functions, see Refs. [92, 94, 108]. These facts motivated to use the auxiliary principle technique. This technique deals with finding the auxiliary variational inequality and proving that the solution of the auxiliary problem is the solution of the original problem by using the fixed-point approach. Glowinski et al. [24] used this technique to study the existence of a solution of mixed variational inequalities. Noor [54–56, 59, 60, 71, 72, 77] has used this technique to suggest some predictor–corrector methods for solving various classes of variational inequalities. It is well known that a substantial number of numerical methods can be obtained as special cases from this technique. We use this technique to suggest and analyze some explicit predictor–corrector methods for extended general variational inequalities. In this chapter, we give the basic idea of the inertial proximal methods. It is an open problem to compare the efficiency of the inertial methods with other methods and this is another direction for future research.

Theory of variational inequalities is quite broad, so we shall content ourselves here to give the flavor of the ideas and techniques involved. The techniques used to analyze the iterative methods and other results for general variational inequalities are a beautiful blend of ideas of pure and applied mathematical sciences. In this chapter, we present the some results regarding the development of various algorithms and their convergence analysis. Although this chapter is expository in nature, our choice has been rather to consider some interesting aspects of general variational inequalities. The framework chosen should be seen as a model setting for more general results for other classes of variational inequalities and variational inclusions. One of the main purposes of this expository chapter is to demonstrate the close connection among various classes of algorithms for the solution of the general variational inequalities and to point out that researchers in different fields of variational inequalities and optimization have been considering parallel paths. We would like to emphasize that the results obtained and discussed in this chapter may motivate and bring a large number of novel, innovate and potential applications, extensions and interesting topics in these areas. For some other aspects of the general variational inequalities, readers are referred to the state-of-art articles of Noor [49–53, 66, 70, 72, 74–77] and the references therein. The interested reader is advised to explore this field further and discover novel and fascinating applications of this theory in Banach and topological spaces.

The general theory is quite technical, so we shall content ourselves here to give the flavor of the main ideas involved. The framework chosen should be seen as a model setting for more general results. However, by just relying on these special results, interesting problems arising in the applications can be dealt with easily. Our main motivation of this chapter is to give a summary account of the basic theory of extended general variational inequalities set in the framework of nonlinear operators defined on convex sets in a real Hilbert space. We focus our attention on the iterative methods for solving variational inequalities.

2. Preliminaries and Basic Concepts

Let H be a real Hilbert space, whose inner product and norm are denoted by $\langle \cdot, \cdot \rangle$ and $\|\cdot\|$, respectively.

Definition 1. The set K in H is said to be a convex set, if

$$u + t(v - u) \in K, \quad \forall u, v \in K, t \in [0, 1].$$

Definition 2. A function F is said to be a convex function, if
$$F((1-t)u + tv) \leq (1-t)F(u) + tF(v), \quad \forall u, v \in K, t \in [0,1].$$

It is well known that a function F is a convex functions, if and only if it satisfies the inequality

$$F\left(\frac{a+b}{2}\right) \leq \frac{2}{b-a} \int_a^b F(x)dx \leq \frac{F(a)+F(b)}{2}, \quad \forall a, b \in I = [a,b],$$

which is known as the Hermite–Hadamard-type inequality. Such types of the inequalities provide us with the upper and lower bounds for the mean value integral.

If the convex function F is differentiable, then $u \in K$ is the minimum of the F, if and only if $u \in K$ satisfies the inequality

$$\langle F'(u), v - u \rangle \geq 0, \quad \forall v \in K,$$

which is called the variational inequality, introduced and studied by Stampacchia [103] in 1964. For the applications, formulation, sensitivity, dynamical systems, generalizations and other aspects of the variational inequalities, see Refs. [1–10, 17, 18, 24, 36, 37, 39, 42–44, 49–53, 56, 66, 70, 72, 74–79, 82, 83, 91, 93, 96, 101–105, 107, 108, 111–115, 117–119] and the references therein.

It is known that a set may not be a convex set. However, a set may be made convex set with respect to some arbitrary functions. Motivated by this fact, Noor introduced the concept of an extended general convex set involving arbitrary two functions.

Definition 3. The set K in H is said to be a general convex set, if there exists arbitrary two functions g, h such that

$$h(u) + t(g(v) - h(u)) \in K_{hg}, \quad \forall u, v \in H : h(u), g(v) \in K_{hg}, t \in [0,1].$$

Note that, if $g = h$, then the extended general convex set reduces to the general convex set, which was introduced by Youness [115].

Definition 4. The set K in H is said to be a general convex set, if there exists arbitrary two functions g, h, such that

$$g(u) + t(g(v) - g(u)) \in K_g, \quad \forall u, v \in H : g(u), g(v) \in K_g, t \in [0,1].$$

If $g = I$, the identity operator, then the general convex set reduces to the classical convex set.

Clearly, every convex set is a general convex set and the general convex set is an extended general convex set, but the converse is not true.

For the sake of simplicity, we always assume that $\forall u, v \in H : g(u), h(v) \in K_{hg}$, unless otherwise.

Definition 5. A function F is said to an extended general convex function, if there exists arbitrary two functions g, h such that

$$F((1-t)h(u) + tg(v)) \leq (1-t)F(h(u)) + tF(g(v)),$$
$$\forall u, v \in H : h(u), g(v) \in K_{hg}, t \in [0,1].$$

It is known that every convex function is an extended general convex function, but the converse is not true.

We now define the general convex functions on $I_g = [h(a), g(b)]$.

Definition 6. Let $I_g = [g(a)h(b)]$. Then F is a general convex function, if and only if

$$\begin{vmatrix} 1 & 1 & 1 \\ h(a) & g(x) & g(b) \\ F(h(a)) & F(g(x)) & F(g(b)) \end{vmatrix} \geq 0; \quad h(a) \leq g(x) \leq g(b).$$

One can easily show that the following are equivalent:

(1) F is a general convex function.

(2) $F(g(x)) \leq F(g(a)) + \frac{F(g(b)) - F(h(a))}{g(b) - h(a)} (g(x) - h(a))$.

(3) $\frac{F(g(x)) - F(h(a))}{g(x) - h(a)} \leq \frac{F(g(b)) - F(h(a))}{g(b) - h(a)}$.

(4) $(g(x) - h(a))F(h(a)) + (g(b) - h(a))F(g(x)) + (h(a) - g(x))F(g(b)) \geq 0$.

(5) $\frac{F(h(a))}{(g(b)-h(a))(h(a)-g(x))} + \frac{F(g(x))}{(g(x)-g(b))(h(a)-g(x))} + \frac{F(g(b))}{(g(b)-h(a))(g(x)-g(b))} \geq 0$,

where $g(x) = (1-t)h(a) + tg(b), \in [0,1]$.

We now show that the minimum of a differentiable extended general convex function on K_{hg} in H can be characterized by the extended general variational inequality. This result is mainly due to Noor [75–77].

Theorem 1. *Let $F : K_{hg} \longrightarrow H$ be a differentiable extended general convex function. Then $u \in H : h(u) \in K$ is the minimum of a differentiable general convex function F on K_{hg}, if and only if $u \in : h(u) \in K_{hg}$ satisfies the inequality*

$$\langle F'(h(u)), g(v) - h(u) \rangle \geq 0, \quad \forall g(v) \in K_{hg}, \tag{3}$$

where F' is the differential of F at $h(u) \in K_{hg}$ in the direction $g(v) - h(u)$.

Proof. Let $u \in H : h(u) \in K_{hg}$ be a minimum of general convex function F on K_{hg}. Then

$$F(h(u)) \leq F(g(v)), \quad \forall g(v) \in K_{hg}. \tag{4}$$

Since K_{hg} is a general convex set, so, for all $u, v \in K_{hg}, t \in [0,1], g(v_t) = h(u) + t(g(v) - h(u)) \in K_{hg}$. Setting $g(v) = g(v_t)$ in (4), we have

$$F(h(u)) \leq F(h(u) + t(g(v) - h(u))) \leq F(h(u)) + t(F(g(v)) - (h(u))).$$

Dividing the above inequality by t and taking $t \longrightarrow 0$, we have

$$\langle F'(h(u)), g(v) - h(u) \rangle \geq 0,$$

which is the required result (3).

Conversely, let $u \in H, h(u) \in K_{hg}$ satisfy the inequality (3). Since F is a general convex function, so $\forall g(u), h(v) \in K_{hg}, \quad t \in [0,1], \quad h(u) + t(g(v) - h(u)) \in K_{hg}$ and

$$F(h(u) + t(g(v) - h(u))) \leq (1-t)F(h(u)) + tF(g(v)),$$

which implies that

$$F(g(v)) - F(h(u)) \geq \frac{F(h(u) + t(g(v) - h(u))) - F(h(u))}{t}.$$

Letting $t \longrightarrow 0$, we have
$$F(g(v)) - F(h(u)) \geq \langle F'(h(u)), g(v) - h(u) \rangle \geq 0, \quad \text{using (3)},$$
which implies that
$$F(h(u)) \leq F(g(v)), \quad \forall g(v) \in K_{hg},$$
showing that $u \in H : h(u) \in K_{hg}$ is the minimum of F on K_{hg} in H. □

Theorem 1 implies that extended general convex programming problem can be studied via the extended general variational inequality (5) with $Tu = F'(h(u))$. In a similar way, one can show that the general variational inequality is the Fritz–John condition of the inequality constrained optimization problem.

In many applications, the general variational inequalities do not arise as the minimization of the differentiable general convex functions. Also, it is known that the variational inequality introduced by Stampacchia [103] can only be used to study the even-order boundary value problems. These facts motivated Noor [75–77] to introduce extended general variational inequality involving three distinct operators. Extended general variational inequalities provide us a unified framework to study a wide class of problems, which arise in various fields of pure and applied sciences.

Let K be a closed convex set in H and $T, g, h : H \longrightarrow H$ be nonlinear operators. We now consider the problem of finding $u \in H, h(u) \in K$ such that
$$\langle \rho Tu + h(u) - g(u), g(v) - h(u) \rangle \geq 0, \quad \forall v \in H; h(v) \in K, \quad (5)$$
where $\rho > 0$ is a constant. Problem (5) is also called the **extended general variational inequality.**

We now discuss some special classes of extended general variational inequalities (5):

(I) If $g(u) = h(u)$, then (5) reduces to finding $u \in H, h(u) \in K$ such that
$$\langle Tu, g(v) - h(u) \rangle \geq 0, \quad \forall v \in H; g(v) \in K,$$
which is exactly *the general variational inequality* (1), introduced and studied by Noor [49] in 1988.

(II) If $h = I$, then problem (5) reduces to finding $u \in H, g(u) \in K$ such that

$$\langle \rho Tu + u - g(u), g(v) - u \rangle \geq 0, \quad \forall v \in H; g(v) \in K, \qquad (6)$$

which appears to be a new ones.

(III) If $g = h$, then (5) reduces to finding $u \in H, g(u) \in K$ such that

$$\langle Tu, g(v) - g(u) \rangle \geq 0, \quad \forall v \in H; g(v) \in K, \qquad (7)$$

which is called the *general variational inequality*, introduced and studied by Noor [49] in 1988.

(IV) If $g = I$, then (5) reduces to finding $u \in H, g(u) \in K$ such that

$$\langle \rho Tu, g(v) - u \rangle \geq 0, \quad \forall v \in H; g(v) \in K, \qquad (8)$$

which was introduced and studied by Noor [50] in 1988.

(V) For $g = I = h$, where I is the identity operator, problem (5) is equivalent to finding $u \in K$ such that

$$\langle Tu, v - u \rangle \geq 0 \quad \forall v \in K, \qquad (9)$$

which is known as the classical variational inequality introduced and studied by Stampacchia [103] and Fichera [15] independently in 1964. For recent state-of-the-art in this field, see Refs. [3–8, 10, 13, 16, 21–24, 26, 32, 33, 45, 53, 66, 72, 82, 85, 94, 119] and the references therein.

(VII) If $K^* = \{u \in H : \langle u, v \rangle \geq 0, \forall v \in K\}$ is a polar (dual) cone of a convex cone K in H, then problem (7) is equivalent to finding $u \in H$ such that

$$g(u) \in K, \quad Tu \in K^* \quad \text{and} \quad \langle Tu, g(u) \rangle = 0, \qquad (10)$$

which is known as the general complementarity problem, see Noor [49]. $g(u) = m(u) + K$, where m is a point-to-point mapping, is called the implicit (quasi) complementarity problem. If $g \equiv I$, then problem (10) is known as the generalized complementarity problems. Such problems have been studied extensively in recent years.

(VIII) If $K = H$, then the general variational inequality (7) is equivalent to finding $u \in H : g(u) \in H$ such that

$$\langle Tu, g(v) \rangle = 0, \quad \forall v \in H : g(v) \in H,$$

which is called the weak formulation of the odd-order and nonsymmetric boundary value problems.

It has been shown that a large class of unrelated odd-order and nonsymmetric obstacle, unilateral, contact, free, moving and equilibrium problems arising in regional, physical, mathematical, engineering and applied sciences can be studied in the unified and general framework of the general variational inequalities (5). Luc and Noor [40] have studied the local uniqueness of the solution of the general variational inequality (5) by using the concept of Frechet approximate Jacobian. For a suitable and appropriate choice of the operators and spaces, one can obtain several classes of variational inequalities and related optimization problems as special cases of the general variational inequalities (5).

We also need the following result, which plays a key role in the studies of variational inequalities and optimization theory.

Lemma 1 ([33]). For a given $z \in H$, $u \in K$ satisfies the inequality

$$\langle u - z, v - u \rangle \geq 0, \quad \forall v \in K, \tag{11}$$

if and only if

$$u = P_K z,$$

where P_K is the projection of H onto K.

Also, the projection operator P_K is nonexpansive, that is,

$$\|P_K(u) - P_K(v)\| \leq \|u - v\|, \quad \forall u, v \in H,$$

and satisfies the inequality

$$\|P_K z - u\|^2 \leq \|z - u\|^2 - \|z - P_K z\|^2, \quad \forall z, u \in H.$$

We now recall the well known concepts.

Definition 7. For all $u, v \in H$, the operator $T : H \longrightarrow H$ is said to be **(i)** *g-monotone*, if

$$\langle Tu - Tv, g(u) - g(v) \rangle \geq 0,$$

(ii) *g-pseudomonotone*, if

$$\langle Tu, g(v) - g(u) \rangle \geq 0 \quad \text{implies} \quad \langle Tv, g(v) - g(u) \rangle \geq 0.$$

For $g \equiv I$, Definition 7 reduces to the usual definition of monotonicity and pseudomonotonicity of the operator T. Note that monotonicity implies pseudomonotonicity but the converse is not true, see Ref. [35].

Definition 8. A function F is said to be strongly general hg-convex on the general convex set K with modulus $\mu > 0$, if $\forall \ h(u), g(v) \in K_g, t \in [0, 1]$,

$$F(h(u) + t(g(v) - h(u))) \leq (1-t)F(h(u)) + tF(g(v))$$
$$- t(1-t)\mu \|g(v) - h(u)\|^2.$$

For differentiable strongly general convex function F, the following statements are equivalent:

(1) $F(g(v)) - F(h(u)) \geq \langle F'(h(u)), g(v) - h(u) \rangle + \mu \|g(v) - h(u)\|^2$,
(2) $F'(h(u)) - F'(g(v)), h(u) - g(v) \rangle \geq \mu \|g(v) - h(u)\|^2$.

It is well known [9, 115] that the general convex functions are not convex function, but they have some nice properties which the convex functions have. Note that, for $g = I$, the general convex functions are convex functions and definition (8) is the well-known result in convex analysis.

3. Projection Methods

In this section, we use the fixed-point formulation to suggest and analyze some new implicit methods for solving the variational inequalities. Using Lemma 1, one can show that the extended general variational inequalities are equivalent to the fixed-point problems.

Lemma 2. *The function $u \in H : h(u) \in K$ is a solution of the extended general variational inequalities (5), if and only if $u \in H$: $h(u) \in K$ satisfies the relation*

$$h(u) = P_K[g(u) - \rho Tu], \tag{12}$$

where P_K is the projection operator and $\rho > 0$ is a constant.

Lemma 2 implies that the general variational inequality (5) is equivalent to the fixed-point problem (12). This equivalent fixed-point formulation was used to suggest some implicit iterative methods for solving the general variational inequalities. One uses the equivalent fixed-point formulation (12) to suggest the following iterative methods for solving general variational inequalities (5).

Algorithm 1. For a given $u_0 \in H$, compute u_{n+1} by the iterative scheme

$$u_{n+1} = u_n - h(u_n) + P_K[g(u_n) - \rho Tu_n], \quad n = 0, 1, 2, \ldots$$

which is known as the projection method and has been studied extensively.

Algorithm 2. For a given $u_0 \in H$, compute u_{n+1} by the iterative scheme

$$u_{n+1} = u_n - h(u_n) + P_K[g(u_n) - \rho Tu_{n+1}], \quad n = 0, 1, 2, \ldots$$

which is known as the extragradient method, which was suggested and analyzed by Korperlevech [34] and has been studied extensively. Noor [66, 72] has proved the convergence of the extragradient for pseudomonotone operators.

Algorithm 3. For a given $u_0 \in H$, compute u_{n+1} by the iterative scheme

$$u_{n+1} = u_n - h(u_n) + P_K[g(u_{n+1}) - \rho Tu_{n+1}], \quad n = 0, 1, 2, \ldots$$

which is known as the modified projection method and has been studied extensively, see Noor [72].
We can rewrite equation (12) as:

$$h(u) = P_K\left[g\left(\frac{u+u}{2}\right) - \rho Tu\right].$$

This fixed-point formulation was used to suggest the following implicit method for solving variational inequalities, which is due to Noor et al. [72]. We used this equivalent formulation to suggest implicit methods for general variational inequality (5).

Algorithm 4. For a given $u_0 \in H$, compute u_{n+1} by the iterative scheme

$$u_{n+1} = u_n - h(u_n) + P_K\left[g\left(\frac{u_n + u_{n+1}}{2}\right) - \rho T u_{n+1}\right], \quad n = 0, 1, 2, \ldots$$

For the implementation of Algorithm 4, one can use the predictor–corrector technique to suggest the following two-step iterative method for solving general variational inequalities.

Algorithm 5. For a given $u_0 \in H$, compute u_{n+1} by the iterative scheme

$$h(y_n) = P_K[g(u_n) - \rho T u_n]$$

$$u_{n+1} = u_n - h(u_n) + P_K\left[g\left(\frac{y_n + u_n}{2}\right) - \rho T y_n\right],$$

$$\lambda \in [0,1], \quad n = 0, 1, 2, \ldots$$

which is a two-step iterative method:
From equation (12), we have

$$h(u) = P_K\left[g(u) - \rho T\left(\frac{u+u}{2}\right)\right].$$

This fixed-point formulation is used to suggest the implicit method for solving the variational inequalities.

Algorithm 6. For a given $u_0 \in H$, compute u_{n+1} by the iterative scheme

$$u_{n+1} = u_n - h(u_n) + P_K\left[g(u_n) - \rho T\left(\frac{u_n + u_{n+1}}{2}\right)\right], \quad n = 0, 1, 2, \ldots$$

which is another implicit method, see Noor et al. [86, 88]. To implement this implicit method, one can use the predictor–corrector technique to rewrite Algorithm 6 as equivalent two-step iterative method.

Algorithm 7. For a given $u_0 \in H$, compute u_{n+1} by the iterative scheme

$$h(y_n) = P_K[g(u_n) - \rho T u_n],$$

$$u_{n+1} = u_n - h(u_n) + P_K\left[g(u_n) - \rho T\left(\frac{u_n + y_n}{2}\right)\right], \quad n = 0, 1, 2, \ldots$$

which is known as the mid-point implicit method for solving general variational inequalities. For the convergence analysis and other aspects of algorithm [88], see Noor *et al.* [86].

It is obvious that Algorithms 4 and 6 have been suggested using different variants of the fixed-point formulations (12). It is natural to combine these fixed-point formulation to suggest a hybrid implicit method for solving the general variational inequalities and related optimization problems, which is the main motivation of this chapter.

One can rewrite equation (12) as

$$h(u) = P_K\left[g\left(\frac{u+u}{2}\right) - \rho T\left(\frac{u+u}{2}\right)\right].$$

This equivalent fixed-point formulation enables the suggestion of the following method for solving the general variational inequalities.

Algorithm 8. For a given $u_0 \in H$, compute u_{n+1} by the iterative scheme

$$u_{n+1} = u_n - h(u_n) + P_K\left[g\left(\frac{u_n + u_{n+1}}{2}\right) - \rho T\left(\frac{u_n + u_{n+1}}{2}\right)\right],$$

$$n = 0, 1, 2, \ldots$$

which is an implicit method.

We would like to emphasize that Algorithm 8 is an implicit method. To implement the implicit method, one uses the predictor–corrector technique. We use Algorithm 1 as the predictor and Algorithm 9 as the corrector. Thus, we obtain a new two-step method for solving general variational inequalities.

Algorithm 9. For a given $u_0 \in H$, compute u_{n+1} by the iterative scheme

$$h(y_n) = P_K[g(u_n) - \rho T u_n]$$
$$u_{n+1} = u_n - h(u_n) + P_K\left[g\left(\frac{y_n + u_n}{2}\right) - \rho T\left(\frac{y_n + u_n}{2}\right)\right],$$
$$n = 0, 1, 2, \ldots$$

which is the two-step method introduced.

For constants $\lambda, \xi \in [0.1]$, we can rewrite equation (12) as

$$h(u) = P_K\left[(1-\lambda)g(u) + \lambda g(u)) - \rho T((1-\xi)u + \xi u)\right].$$

This equivalent fixed-point formulation enables the suggestion of the following method for solving the general variational inequalities.

Algorithm 10. For a given $u_0 \in H$, compute u_{n+1} by the iterative scheme

$$h(u_{n+1}) = P_K\left[(1-\lambda)g(u_n) + \lambda g(u_{n+1})) - \rho T((1-\xi)u_n + \xi u_{n+1})\right].$$
$$n = 0, 1, 2, \ldots$$

which is an implicit method.

Using the prediction–correction technique, Algorithm 12 can be written in the following form.

Algorithm 11. For a given $u_0 \in H$, compute u_{n+1} by the iterative scheme:

$$h(y_n) = P_K[g(u_n) - \rho T u_n]$$
$$h(u_{n+1}) = P_K\left[(1-\lambda)g(u_n) + \lambda g(y_n)) - \rho T((1-\xi)u_n + \xi y_n)\right],$$
$$n = 0, 1, 2, \ldots$$

which is the two-step method.

Remark 1. It is worth mentioning that Algorithm 12 is a unified ones. For suitable and appropriate choice of the constant λ and ξ, one can obtain a wide class of iterative methods for solving general variational inequalities and related optimization problems.

4. Wiener–Hopf Equations Technique

In this section, we consider the problem of the general Wiener–Hopf equations. To be more precise, let $Q_K = I - P_K$, where I is the identity operator and P_K is the projection of H onto K. For given nonlinear operators $T, g, h : H \to H$, consider the problem of finding $z \in H$ such that

$$\rho T g(h^{-1}) P_K z + Q_K z = 0, \qquad (13)$$

provided h^{-1} exists. Equations of type (13) are called the *extended general Wiener–Hopf equations*, which were introduced and studied by Noor. For $h = I$, we obtain the original general Wiener–Hopf equations [58], and for $g = I = h$, the classical Wiener–Hopf equations, introduced and studied by Shi [99] and Robinson [97] in different settings independently. Using the projection operator technique, one can show that the general variational inequalities are equivalent to the general Wiener–Hopf equations. This equivalent alternative formulation has played a fundamental and important role in studying various aspects of variational inequalities. It has been shown that Wiener–Hopf equations are more flexible and provide a unified framework to develop some efficient and powerful numerical techniques for solving variational inequalities and related optimization problems.

Lemma 3 ([79]). *The element $u \in H : h(u) \in K$ is a solution of the general variational inequality (5), if and only if $z \in H$ satisfies the extended general Wiener–Hopf equation (13), where*

$$h(u) = P_K z, \qquad (14)$$
$$z = g(u) - \rho T u, \qquad (15)$$

where $\rho > 0$ is a constant.

From Lemma 3, it follows that the variational inequalities (5) and the Wiener–Hopf equations (13) are equivalent. This alternative equivalent formulation is used to suggest and analyze a wide class of efficient and robust iterative methods for solving general variational inequalities and related optimization problems, see Refs. [66, 72] and the references therein.

We use the Wiener–Hopf equations (13) to suggest some new iterative methods for solving the extended general variational inequalities. From (14) and (15),

$$z = g(h^{-1})P_K z - \rho T h^{-1} P_K z$$
$$= g(h^{-1})P_K[g(u) - \rho Tu] - \rho T h^{-1} P_K[g(u) - \rho Tu].$$

Thus, we have

$$h(u) = \rho Tu + [g(h^{-1})P_K[g(u) - \rho Tu] - \rho T h^{-1} P_K[g(u) - \rho Tu$$
$$+ P_K[g(u) - \rho Tu] - h(u)].$$

Consequently, for a constant $\alpha_n > 0$, we have

$$h(u) = (1 - \alpha_n)h(u) + \alpha_n\{g(h^{-1})P_K[P_K[g(u) - \rho Tu] + \rho Tu$$
$$- \rho T h^{-1} P_K[g(u) - \rho Tu] + g(h^{-1})P_K[g(u) - \rho Tu] - g(u)]\}$$
$$= (1 - \alpha_n)h(u) + \alpha_n\{g(h^{-1})P_K[g(y) - \rho Ty]$$
$$+ g(h^{-1})P_K[g(y) - \rho Tu] - h(u)]\}, \tag{16}$$

where

$$h(y) = P_K[g(u) - \rho Tu]. \tag{17}$$

Using (16) and (17), we can suggest the following new predictor–corrector method for solving variational inequalities.

Algorithm 12. For a given $u_0 \in H$, compute u_{n+1} by the iterative scheme

$$h(y_n) = P_K[g(u_n) - \rho Tu_n],$$
$$h(u_{n+1}) = (1 - \alpha_n)h(u_n)$$
$$+ \alpha_n\left\{P_K[g(y_n) - \rho Ty_n + g(y_n) - (g(u_n) - \rho Tu_n)]\right\}.$$

Algorithm 12 can be rewritten in the following equivalent form:

Algorithm 13. For a given $u_0 \in H$, compute u_{n+1} by the iterative scheme

$$u_{n+1} = (1 - \alpha_n)u_n \\ + \alpha_n\{g(h^{-1})P_K[P_K[g(u_n) - \rho T u_n] - \rho T h^{-1} P_K[g(u_n) - \rho T u_n] \\ + P_K[g(u_n) - \rho T u_n) - (g(u_n) - \rho T u_n])\},$$

which is an explicit iterative method and appears to be a new one.

If $\alpha_n = 1$, then Algorithm 13 reduces to the following:

Algorithm 14. For a given $u_0 \in H$, compute u_{n+1} by the iterative scheme

$$g(y_n) = P_K[g(u_n) - \rho T u_n],$$
$$u_{n+1} = P_K[g(y_n) - \rho T y_n + g(y_n) - (g(u_n) - \rho T u_n])\},$$
$$n = 0, 1, 2, \ldots,$$

which appears to be a new one.

This shows that our Algorithms 13 and 14 can be considered as practical alternative to the extragradient and other modified projection methods. The comparison of new methods developed in this chapter with the recent methods is an interesting problem for future research.

5. Dynamical System Technique

In this section, we consider the projected dynamical systems associated with variational inequalities. We investigate the convergence analysis of these new methods involving only the monotonicity of the operator.

We now define the residue vector $R(u)$ by the relation

$$R(u) = h(u) - P_K[g(u) - \rho T u]. \tag{18}$$

Invoking Lemma 3, one can easily conclude that $u \in H : h(u) \in K$ is a solution of (5), if and only if $u \in H : h(u) \in K$ is a zero of the equation

$$R(u) = 0. \tag{19}$$

We now consider a projected dynamical system associated with the variational inequalities. Using the equivalent formulation (19), we suggest a class of projected dynamical systems as

$$\frac{dh(u)}{dt} = \lambda P_K[g(u) - \rho Tu] - h(u)\}, \quad u(t_0) = u_0 \in K, \quad (20)$$

where λ is a parameter. The system of type (20) is called the projected dynamical system associated with variational inequalities (5). Here the right hand is related to the resolvent and is discontinuous on the boundary. From the definition, it is clear that the solution of the dynamical system always stays in H. This implies that the qualitative results such as the existence, uniqueness and continuous dependence of the solution of (20) can be studied. These projected dynamical systems are associated with the general variational inequalities (5), which have been studied extensively.

We use the projected dynamical system (20) to suggest some iterative for solving variational inequalities (5). These methods can be viewed in the sense of Koperlevich [34] and Noor [66, 72] involving the double projection operator.

For simplicity, we consider the dynamical system

$$\frac{dh(u)}{dt} + h(u) = P_K[g(u) - \rho Tu], \quad u(t_0) = \alpha. \quad (21)$$

We construct the implicit iterative method using the forward difference scheme. Discretizing equation (21), we have

$$\frac{h(u_{n+1}) - h(u_n)}{h_1} + h(u_{n+1}) = P_K[g(u_n) - \rho Tu_{n+1}], \quad (22)$$

where h_1 is the step size. Now, we can suggest the following implicit iterative method for solving the variational inequality (5).

Algorithm 15. For a given $u_0 \in H$, compute u_{n+1} by the iterative scheme

$$h(u_{n+1}) = P_K\left[g(u_n) - \rho Tu_{n+1} - \frac{h(u_{n+1}) - h(u_n)}{h_1}\right],$$
$$n = 0, 1, 2, \ldots.$$

This is an implicit method and is quite different from the implicit method [16, 32]. Using Lemma 1, Algorithm 15 can be rewritten in the equivalent form as follows:

Algorithm 16. For a given $u_0 \in H$, compute u_{n+1} by the iterative scheme

$$\left\langle \rho T u_{n+1} + \frac{1+h_1}{h_1}(h(u_{n+1}) - h(u_n)), g(v) - h(u_{n+1}) \right\rangle \geq 0,$$

$$\forall g(v) \in K. \quad (23)$$

We now study the convergence analysis of Algorithm 16 under some mild conditions.

Theorem 2. Let $u \in H : g(v) \in K$ be a solution of the general variational inequality (5). Let u_{n+1} be the approximate solution obtained from (23). If T is hg-monotone, then

$$\|h(u) - h(u_{n+1})\|^2 \leq \|h(u) - h(u_n)\|^2 - \|h(u_n) - h(u_{n+1})\|^2. \quad (24)$$

Proof. Let $u \in H : h(v) \in K$ be a solution of (5). Then

$$\langle Tv, g(v) - h(u) \rangle \geq 0, \quad \forall v \in H : g(v) \in K, \quad (25)$$

since T is a g-monotone operator.

Set $v = u_{n+1}$ in (25), to have

$$\langle T u_{n+1}, g(u_{n+1}) - h(u) \rangle \geq 0. \quad (26)$$

Taking $v = u$ in (23), we have

$$\left\langle \rho T u_{n+1} + \left\{ \frac{(1+h_1)h(u_{n+1}) - (1+h_1)h(u_n)}{h_1} \right\}, \right.$$

$$\left. h(u) - h(u_{n+1}) \right\rangle \geq 0. \quad (27)$$

From (26) and (27), we have

$$\langle (1+h_1)(h(u_{n+1}) - h(u_n)), h(u) - h(u_{n+1}) \rangle \geq 0. \quad (28)$$

From (28) and using $2\langle a,b\rangle = \|a+b\|^2 - \|a\|^2 - \|b\|^2$, $\forall a,b \in H$, we obtain

$$\|h(u_{n+1}) - h(u)\|^2 \leq \|h(u) - h(u_n)\|^2 - \|h(u_{n+1}) - h(u_n)\|^2, \tag{29}$$

which is the required result. □

Theorem 3. *Let $u \in K$ be the solution of variational inequality (5). Let u_{n+1} be the approximate solution obtained from (23). If T is an hg-monotone operator and h^{-1} exists, then u_{n+1} converges to $u \in H$ satisfying (5).*

Proof. Let T be an hg-monotone operator. Then, from (24), it follows that the sequence $\{u_i\}_{i=1}^{\infty}$ is a bounded sequence and

$$\sum_{i=1}^{\infty} \|h(u_n) - h(u_{n+1})\|^2 \leq \|h(u) - h(u_0)\|^2,$$

which implies that

$$\lim_{n \to \infty} \|u_{n+1} - u_n\|^2 = 0, \tag{30}$$

since h^{-1} exists.

Since sequence $\{u_i\}_{i=1}^{\infty}$ is bounded, there exists a cluster point $\hat{u}$ to which the subsequence $\{u_{ik}\}_{k=1}^{\infty}$ converges. Taking limit in (23) and using (30), it follows that $\hat{u} \in K$ satisfies

$$\langle T\hat{u}, g(v) - h(\hat{u})\rangle \geq 0, \quad \forall v \in H : g(v) \in K,$$

and

$$\|h(u_{n+1}) - h(u)\|^2 \leq \|h(u) - h(u)_n\|^2.$$

Using this inequality, one can show that the cluster point $\hat{u}$ is unique and

$$\lim_{n \to \infty} u_{n+1} = \hat{u}.$$

□

We now suggest an other implicit iterative method for solving (5). Discretizing (23), we have

$$\frac{h(u_{n+1}) - h(u_n)}{h_1} + h(u_{n+1}) = P_K\left[g(u_{n+1}) - \rho T u_{n+1}\right], \quad (31)$$

where h is the step size.

This formulation enables us to suggest the following iterative method.

Algorithm 17. For a given $u_0 \in K$, compute u_{n+1} by the iterative scheme

$$h(u_{n+1}) = P_K\left[g(u_{n+1}) - \rho T u_{n+1} - \frac{h(u_{n+1}) - h(u_n)}{h_1}\right].$$

Using Lemma 1, Algorithm 20 can be rewritten in the equivalent form as follows:

Algorithm 18. For a given $u_0 \in K$, compute u_{n+1} by the iterative scheme

$$\left\langle \rho T u_{n+1} + \left\{\frac{h(u_{n+1}) - h(u_n)}{h_1}\right\}, g(v) - h(u_{n+1}) \right\rangle \geq 0,$$
$$\forall v \in H : g(v) \in K. \quad (32)$$

Again using the dynamical systems, we suggested some iterative methods for solving the variational inequalities and related optimization problems.

Algorithm 19. For a given $u_0 \in K$, compute u_{n+1} by the iterative scheme

$$u_{n+1} = P_K\left[\frac{(h_1 + 1)(h(u_n) - h(u_{n+1}))}{h_1} - \rho T u_n\right], \quad n = 0, 1, 2, \ldots,$$

which can be written in the equivalent form as follows:

Algorithm 20. For a given $u_0 \in K$, compute u_{n+1} by the iterative scheme

$$\left\langle \rho T u_n + \left\{\frac{h_1+1}{h_1}(h(u_{n+1}) - h(u_n))\right\}, g(v) - h(u_{n+1}) \right\rangle \geq 0,$$
$$\forall g(v) \in K. \quad (33)$$

In a similar way, one can suggest a wide class of implicit iterative methods for solving variational inequalities and related optimization problems. The comparison of these methods with other methods is an interesting problem for future research.

6. Auxiliary Principle Technique

In the previous sections, we have considered and analyzed several projection-type methods for solving variational inequalities. It is well known that to implement such types of the methods, one has to evaluate the projection, which is itself a difficult problem. Second, one can't extend the technique of projection for solving some other classes of variational inequalities. These facts are motivated to consider other methods. One of these techniques is known as the auxiliary principle. This technique is basically due to Lions and Stampacchia [39]. Glowinski et al. [24] used this technique to study the existence of a solution of mixed variational inequalities. Noor [45, 54, 55, 71, 72, 77] has used this technique to develop some predictor–corrector methods for solving variational inequalities. It have been be shown that various classes of methods including projection, Wiener–Hopf, decomposition and descent can be obtained from this technique as special cases.

For a given $u \in H, h(u) \in K$ satisfying (1), consider the problem of finding a unique $w \in H, h(w) \in K$ such that

$$\langle \rho T u + h(w) - g(u), g(v) - h(w) \rangle \geq 0, \quad \forall g(v) \in K, \qquad (34)$$

where $\rho > 0$ is a constant.

Note that, if $w = u$, then w is clearly a solution of the general variational inequality (5). This simple observation enables us to suggest and analyze the following predictor–corrector method.

Algorithm 21. For a given $u_0 \in H$, compute the approximate solution u_{n+1} by the iterative schemes:

$$\langle \mu T u_n + h(y_n) - g(u_n), g(v) - h(y_n) \rangle \geq 0, \quad \forall g(v) \in K,$$
$$\langle \beta T y_n + h(w_n) - g(y_n), g(v) - h(w_n) \rangle \geq 0, \quad \forall g(v) \in K,$$
$$\langle \rho T w_n + h(u_{n+1}) - g(w_n), g(v) - h(u_{n+1}) \rangle \geq 0, \quad \forall g(v) \in K,$$

where $\rho > 0$, $\beta > 0$ and $\mu > 0$ are constants.

Algorithm 21 can be considered as a three-step predictor–corrector method, which was suggested and studied by Noor [66, 72].

If $\mu = 0$, then Algorithm 21 reduces to the following:

Algorithm 22. For a given $u_0 \in H$, compute the approximate solution u_{n+1} by the iterative schemes:

$$\langle \beta T u_n + h(w_n) - g(u_n), g(v) - h(w_n) \rangle \geq 0, \quad \forall g(v) \in K,$$
$$\langle \rho T w_n + h(u_{n+1}) - g(w_n), g(v) - h(u_{n+1}) \rangle \geq 0, \quad \forall g(v) \in K,$$

which is known as the two-step predictor–corrector method.

If $\mu = 0$, $\beta = 0$, then Algorithm 21 becomes as follows:

Algorithm 23. For a given $u_0 \in H$, compute u_{n+1} by the iterative scheme

$$\langle \rho T u_n + h(u_{n+1}) - g(u_n), g(v) - h(u_{n+1}) \rangle \geq 0, \quad \forall g(v) \in K.$$

Using the projection technique, Algorithm 21 can be written as follows:

Algorithm 24. For a given $u_0 \in H$, compute u_{n+1} by the iterative schemes

$$h(y_n) = P_K[g(u_n) - \mu T u_n],$$
$$h(w_n) = P_K[g(y_n) - \beta T y_n],$$
$$h(u_{n+1}) = P_K[g(w_n) - \rho T w_n], \quad n = 0, 1, 2, \ldots$$

or

$$h(u_{n+1}) = P_K[I - \mu T h^{-1}] P_K[I - \beta T h^{-1}] P_K[I - \rho T h^{-1}] g(u_n),$$
$$n = 0, 1, 2, \ldots$$

or

$$h(u_{n+1}) = (I + \rho T h^{-1})^{-1} \{ P_K[I - \rho T h^{-1}] P_K[I - \rho T h^{-1}]$$
$$P_K[I - \rho T h^{-1}] + \rho T g^{-1} \} g(u_n), \quad n = 0, 1, 2, \ldots,$$

which is the three-step forward–backward method. Algorithm 24 can be considered as a generalization of a two-step forward–backward splitting method of Tseng [106, 107].

7. General Equilibrium Problems

In this section, we introduce and consider a class of equilibrium problems known as general equilibrium problems. It is known that equilibrium problems include variational and complementarity problems as special cases. We note that the projection and its variant forms including the Wiener–Hopf equations cannot be extended to equilibrium problems, since it is not possible to find the projection of the bifunction $F(.,.)$. In this chapter, we show that the auxiliary principle technique can be used to suggest and analyze some iterative methods for solving general equilibrium problems. We also study the convergence analysis of these iterative methods and discuss some special cases.

For a given nonlinear function $F(.,.) : H \times H \longrightarrow H$ and operator $g, h : H \longrightarrow H$, we consider the problem of finding $u \in H$, $h(u) \in K$ such that

$$F(u, g(v) - h(u)) \geq 0, \quad \forall g(v) \in K, \tag{35}$$

which is known as the *extended general equilibrium problem.*

We now discuss some special cases of the extended general equilibrium problem (35),

(I) For $g \equiv I$, the identity operator, problem (35) is equivalent to finding $u \in H : h(u) \in K$ such that

$$F(u, v - h(u)) \geq 0, \quad \forall\, v \in K, \tag{36}$$

which is called the equilibrium problem. For the recent applications and development, see Refs. [6, 21, 22, 55] and the reference therein.

(II) If $F(u, g(v) - h(u)) = \langle Tu, \eta(g(v), h(u)) \rangle$ and the set K is an invex set in H, then problem (35) is equivalent to finding $u \in H : h(u) \in K$ such that

$$\langle Tu, \eta(g(v), h(u)) \rangle \geq 0, \quad \forall g(v) \in K. \tag{37}$$

Inequality of type (37) is known as the the extended general variational-like inequality, which arises as a minimum of general preinvex functions on the invex set K.

(III) We note that for $F(u, g(v)) \equiv \langle Tu, g(v) - g(u) \rangle$, problem (35) reduces to problem (1), that is, find $u \in K$, $h(u) \in K$ such that

$$\langle Tu, g(v) - h(u) \rangle \geq 0, \quad \forall g(v) \in K,$$

Equation (5), which is exactly the the extended general variational inequality (1). Thus, we conclude that extended general equilibrium problems (35) are more general than general variational inequalities.

We now use the auxiliary principle technique as developed in Section 6 to suggest and analyze some iterative methods for solving general equilibrium problems (35).

For a given $u \in H$, $g(u) \in K$ satisfying (35), consider the auxiliary equilibrium problem of finding $w \in H, g(w) \in K$ such that

$$\rho F(u, g(v) - h(w)) + \langle h(w) - h(u), g(v) - h(w)\rangle \geq 0, \quad \forall g(v) \in K, \tag{38}$$

where $\rho > 0$ is constant.

Obviously, if $w = u$, then w is a solution of the general equilibrium problem (35). This fact allows us to suggest and analyze the following iterative method for solving (35).

Algorithm 25. For a given $u_0 \in H$, compute the approximate solution u_{n+1} by the iterative scheme:

$$\rho F(w_n, g(v) - h(w_n)) + \langle h(u_{n+1}) - g(w_n), g(v) - h(u_{n+1})\rangle \geq 0,$$
$$\forall g(v) \in K. \tag{39}$$
$$\beta F(u_n, g(v) - h(w_n)) + \langle h(w_n) - g(u_n), g(v) - h(w_n)\rangle \geq 0,$$
$$\forall g(v) \in K, \tag{40}$$

where $\rho > 0$ and $\beta > 0$ are constants.

Algorithm 25 is called the predictor–corrector method for solving general equilibrium problem (35).

For $g = I$, where I is the identity operator, Algorithm 25 reduces to the following:

Algorithm 26. For a given $u_0 \in H$, compute the approximate solution u_{n+1} by the iterative schemes:

$$\rho F(w_n, v - h(w_n)) + \langle u_{n+1} - h(w_n), v - u_{n+1}\rangle \geq 0, \quad \forall v \in K,$$
$$\beta F(u_n, v - h(u_n)) + \langle w_n - h(u_n), v - w_n\rangle \geq 0, \quad \forall v \in K.$$

Algorithm 26 is also a predictor–corrector method for solving equilibrium problem and appears to be a new one.

In brief, for suitable and appropriate choice of the functions $F(.,.)$ and the operators T, g, h, one can obtain various algorithms developed in the previous sections.

For the convergence analysis of Algorithm 26, we need the following concepts:

Definition 9. The function $F(.,.) : H \times H \longrightarrow H$ is said to be:

(i) *hg-monotone*, if
$$F(u, g(v) - h(u)) + F(v, g(u) - h(v)) \leq 0, \quad \forall u, v \in H,$$

(ii) *hg-pseudomonotone*, if
$$F(u, g(v) - h(u)) \leq 0 \quad \text{implies} \quad F(v, g(u) - h(v)) \leq 0, \quad \forall u, v \in H,$$

(iii) *hg-partially relaxed strongly monotone*, if there exists a constant $\alpha > 0$ such that
$$F(u, g(v) - h(u)) + F(v, h(z) - g(u)) \leq \alpha \|h(z) - g(u)\|^2,$$
$$\forall u, v, z \in H.$$

Note that for $u = z$, g-partially relaxed strongly monotonicity reduces to g-monotonicity of $F(.,.)$.

We now consider the convergence analysis of Algorithm 26.

Theorem 4. *Let $\bar{u} \in H : h(\bar{u})$ be a solution of (35) and let u_{n+1} be an approximate solution obtained from Algorithm 26. If the bifunction $F(.,.)$ is hg-partially relaxed strongly monotone with constant $\alpha > 0$, then*
$$\|g(\bar{u}) - g(u_{n+1})\|^2 \leq \|g(\bar{u}) - g(w_n)\|^2 - (1 - 2\alpha\rho)$$
$$\|g(w_n) - g(u_{n+1})\|^2, \tag{41}$$
$$\|g(\bar{u}) - g(w_n)\|^2 \leq \|g(\bar{u}) - g(u_n)\|^2 - (1 - 2\beta\rho)$$
$$\|g(w_n) - g(u_n)\|^2. \tag{42}$$

Proof. Let $\bar{u} \in H$, $h(\bar{u}) \in K$ be a solution of (35). Then,
$$\rho F(\bar{u}, g(v) - h(\bar{u})) \geq 0, \quad \forall g(v) \in K, \tag{43}$$
$$\beta F(\bar{u}, g(v)) - h(\bar{u}) \geq 0, \quad \forall g(v) \in K, \tag{44}$$

where $\rho > 0$ and $\beta > 0$ are constants.

Now taking $v = u_{n+1}$ in (39) and $v = \bar{u}$ in (43), we have

$$\rho F(\bar{u}, g(u_{n+1}) - - h(\bar{u})) \geq 0 \qquad (45)$$

and

$$\rho F(w_n, g(\bar{u} - h(w_n))) + \langle g(u_{n+1}) - h(w_n), h(\bar{u}) - g(u_{n+1}) \rangle \geq 0. \qquad (46)$$

Adding (45) and (46), we have

$$\langle h(u_{n+1}) - h(w_n), h(\bar{u}) - h(u_{n+1}) \rangle \geq -\alpha\rho \|h(u_{n+1}) - h(w_n)\|^2, \qquad (47)$$

where we have used the fact that $F(.,.)$ is hg-partially relaxed strongly monotone with constant $\alpha > 0$.

Using the inequality

$$2\langle a, b \rangle - \|a + b\|^2 - \|a\|^2 - \|b\|^2, \quad \forall a, b \in H,$$

we obtain

$$2\langle h(u_{n+1}) - h(w_n), h(\bar{u}) - h(u_{n+1}) \rangle$$
$$= \|h(\bar{u}) - h(w_n)\|^2 - \|h(\bar{u}) - h(u_{n+1})\|^2 - \|h(u_{n+1}) - h(w_n)\|^2. \qquad (48)$$

Combining (47) and (48), we have

$$\|h(\bar{u}) - h(u_{n+1})\|^2 \leq \|h(\bar{u}) - h(w_n)\|^2$$
$$- (1 - 2\rho\alpha)\|h(w_n) - h(u_{n+1})\|^2, \qquad (49)$$

the required (41).

Taking $v = \bar{u}$ in (40) and $v = w_n$ in (44), we obtain

$$\beta F(\bar{u}, g(w_n) - h(\bar{u})) \geq 0 \qquad (50)$$

and

$$\beta F(u_n, g(\bar{u}) - h(w_n)) + \langle h(w_n) - g(u_n), g(\bar{u}) - h(w_n) \rangle \geq 0. \qquad (51)$$

Adding (50), (51) and rearranging the terms, we have

$$\langle h(w_n) - h(u_n), h(\bar{u}) - h(w_n) \rangle \geq -\alpha\beta \|h(u_n) - h(w_n)\|^2, \qquad (52)$$

since $F(.,.)$ is hg-partially strongly monotone with constant $\alpha > 0$.
Consequently, from (52), we have

$$\|h(\bar{u}) - h(w_n)\|^2 \leq \|h(\bar{u}) - h(u_n)\|^2 - (1 - 2\alpha\beta)\|h(u_n) - h(w_n)\|^2,$$

the required (42). □

Theorem 5. *Let H be a finite dimension subspace and let $0 < \rho < \frac{1}{2\alpha}$ and $0 < \beta < \frac{1}{2\alpha}$. If $\bar{u} \in H : g(\bar{u}) \in K$ is a solution of (35) and u_{n+1} is an approximate solution obtained from Algorithm 26, then $\lim_{n \to \infty} u_n = \bar{u}$.*

Proof. Its proof is very much similar to that of Noor [72]. □

We again use the auxiliary principle technique to suggest an inertial proximal method for solving general equilibrium problem (35). It is noted that inertial proximal method includes the proximal method as a special case.

For a given $u \in H$, $g(u) \in K$ satisfying (35), consider the auxiliary general equilibrium problem of finding $w \in H$, $h(w) \in K$ such that

$$\rho F(w, g(v) - h(w)) + \langle h(w) - g(u) - \alpha_n(h(u) - h(u)),$$
$$g(v) - h(w)\rangle \geq 0, \forall g(v) \in K, \qquad (53)$$

where $\rho > 0$ and $\alpha_n > 0$ are constants.

It is clear that if $w = u$, then w is a solution of the general equilibrium problem (35). This fact enables us to suggest an iterative method for solving (35) as:

Algorithm 27. *For a given $u_0 \in H$, compute the approximate solution u_{n+1} by the iterative scheme*

$$(\rho F(u_{n+1}, g(v) - h(u_{n+1})))$$
$$+ \langle h(u_{n+1}) - g(u_n) - \alpha_n(g(u_n) - h(u_{n-1})), \quad \forall g(v) \in K,$$

where $\rho > 0$ and $\alpha_n > 0$ are constants.

Algorithm 27 is called the inertial proximal point method.

For $\alpha_n = 0$, Algorithm 27 reduces to the following:

Algorithm 28. For a given $u_0 \in H$, find the approximate solution u_{n+1} by the iterative schemes

$$(\rho F(u_{n+1}), g(v) - h(u_{n+1})))$$
$$+ \langle h(u_{n+1}) - g(u_n), g(v) - h(u_{n+1}) \rangle \geq 0, \quad \forall g(v) \in K,$$

which is known as the proximal method and appears to be a new one. Note that for $g = I = h$, the identity operator, one can obtain inertial proximal method for solving equilibrium problems (36).

Using the technique of Noor [72], one can easily analyze the convergence of Algorithm 27 under some suitable conditions.

8. Generalizations and Extensions

We would like to mention that some of the results obtained and presented in this chapter can be extended for more general strongly variational inequalities. To be more precise, for a given nonlinear operator T, A, g, h, consider the problem of finding $u \in H : h(u) \in K$ such that

$$\langle Tu, g(v) - h(u) \rangle \geq \langle A(u), g(v) - h(u) \rangle, \quad \forall v \in H : g(v) \in K, \quad (54)$$

which is called the generalized extended general variational inequality. We would like to mention that one can obtain various classes of general variational inequalities for appropriate and suitable choices of the operators T, A, g, h. Using Lemma 1, one can show that problem (54) is equivalent to finding $u \in H : g(u) \in K$ such that

$$h(u) = P_K[g(u) - \rho(Tu - A(u))]. \quad (55)$$

These alternative formulations can be used to suggest and analyze similar techniques for solving generalized extended general variational inequalities (54) as considered in this chapter under certain extra conditions. A complete study of these algorithms for problem (54) is the subject of subsequent research. Development of efficient and implementable algorithms for problems (54) need further research efforts.

Some special cases are as follows:

(I) For a given nonlinear operator T, A, g, h, consider the problem of finding $\in H : h(u) \in K$ such that

$$\langle Tu, v - h(u) \rangle \geq \langle A(u), v - h(u) \rangle, \quad \forall v \in K, \tag{56}$$

which is also called the strongly general variational inequalities.

(II) For a given nonlinear operator T, A, g, h, consider the problem of finding $\in H : h(u) \in K$ such that

$$\langle Tu, g(v) - u \rangle \geq \langle A(u), g(v) - u \rangle, \quad \forall v \in H : g(v) \in K, \tag{57}$$

is also known as the strongly general variational inequality.

Remark 2. We would like to point out that problems (54), (56) and (57) are quite distinct and different from each other and have significant applications in various branches of pure and applied sciences. It is an open and interesting problem for future research. We would like to emphasize that problems (54), (56) and (57) are equivalent in many respect and share the basic and fundamental properties. In particular, they have the same equivalent fixed-point formulations. Consequently, most of the results obtained in this chapter continue to hold for these problems with minor modifications.

(III) If $H = K$, then problem (54) is equivalent to finding $u \in H : h(u) \in H$, such that

$$\langle Tu, g(v) - h(u) \rangle = \langle A(u), g(v) - h(u) \rangle, \quad \forall v \in H : g(v) \in H, \tag{58}$$

which can be viewed as a representation theorem for the nonlinear functions involving an arbitrary function g and h. For more details, see Noor and Noor [82].

(IV) If $H = K$ and $A(u) = A|u|$, then problem (54) is equivalent to finding $u \in H : h(u) \in H$, such that

$$\langle Tu, g(v) - h(u) \rangle = \langle A|u|, g(v) - h(u) \rangle, \quad \forall v \in H : g(v) \in H, \tag{59}$$

which is called the system of absolute value equations involving an arbitrary function g and h. Batool et al. [5] have discussed some cases of problem (59). This is another direction for future research.

The theory of extended general variational inequalities is a new one and does not appear to have developed to an extent that it provides a complete framework for studying these problems. Much more research is needed in all these areas to develop a sound basis for applications. Results proved in this chapter can be extended for multivalued and system of extended general variational inequalities using the technique of this chapter. The comparison of the iterative method for solving extended general variational inequalities is an interesting problem for future research. One can study the sensitivity analysis and the properties of the associated dynamical system related to the extended general variational inequalities. In fact, this field has been continuing and will continue to foster new, innovative and novel applications in various branches of pure and applied sciences. We have given only a brief introduction of this fast growing field. We hope that the ideas and techniques of this chapter may stimulate further research in this field. The interested reader is advised to explore this field further. It is our hope that this brief introduction may inspire and motivate the reader to discover new and interesting applications of this theory in other areas of sciences.

Acknowledgements

We wish to express our deepest gratitude to our colleagues, collaborators and friends, who have direct or indirect contributions in the process of this chapter. We are also grateful to Rector, COMSATS University Islamabad, Islamabad, Pakistan, for the research facilities and support in our research endeavors.

References

[1] E. A. Al-Said, M. A. Noor and T. M. Rassias, Numerical solutions of third-order obstacle problems, *Int. J. Comput. Math.* **69**, 75–84 (1998).
[2] F. Alvarez, On the minimization property of a second order dissipative system in Hilbert space, *SIAM J. Control Optim.* **38**, 1102–1119 (2000).
[3] C. Baiocchi and A. Capelo, *Variational and Quasi-Variational Inequalities* (J. Wiley and Sons, New York, 1984).

[4] S. Batool, M. A. Noor and K. I. Noor, Absolute value variational inequalities and dynamical systems, *Int. J. Math. Anal.* **18**(3) (2020).
[5] M. I. Bloach and M. A. Noor, Perturbed mixed variational-like inequalities, *AIMS Math.* **5**(3), 2153–2162 (2019).
[6] E. Blum and W. Oettli, From optimization adn variational inequalities to equilibrium problems, *Math. Stud.* **63**, 123–145 (1994).
[7] R. W. Cottle, J. S. Pang and R. E. Stone, *The Linear Complementarity Problem* (Academic Press, New York, 1992).
[8] J. Crank, *Free and Moving Boundary Problems* (Clarendon Press, Oxford, 1984).
[9] G. Cristescu and L. Lupsa, *Non-Connected Convexities and Applications* (Kluwer Academic Publishers, Dordrecht, 2002).
[10] V. F. Demyanov, G. E. Stavroulakis, L. N. Polyakova and P. D. Panagiotoulos, *Quasidifferentiability and Nonsmooth Modeling in Mechanics, Engineering and Economics* (Kluwer Academic Publishers, Boston, 1996).
[11] J. Dong, D. Zhang and A. Nagurney, A projected dynamical systems model of general financial equilibrium with stability analysis, *Math. Comput. Model.* **24**(2), 35–44 (1996).
[12] P. Dupuis and A. Nagurney, Dynamical systems and variational inequalities, *Ann. Oper. Res.* **44**, 19–42 (1993).
[13] G. Duvaut and J. L. Lions, *Inequalities in Mechanics and Physics* (Springer-Verlag, Berlin, 1976).
[14] I. Ekland and R. Temam, *Convex Analysis and Variational Problems* (North-Holland, Amsterdam, 1976).
[15] G. Fichera, Problemi elastostatici con vincoli unilaterali: il problema di Signorini con ambique condizione al contorno, *Atti. Acad. Naz. Lincei. Mem. Cl. Sci. Nat. Sez. Ia* **7**(8), 91–140 (1963–64).
[16] V. M. Filippov, *Variational Principles for Nonpotential Operators*, Vol. 77 (American Mathematical Society, USA, 1989).
[17] T. L. Friesz, D. H. Bernstein, N. J. Mehta and S. Ganjliazadeh, Day-to-day dynamic network disequilibria and idealized traveler information systems, *Oper. Res.* **42**, 1120–1136 (1994).
[18] T. L. Friesz, D. H. Bernstein and R. Stough, Dynamic systems, variational inequalities and control theoretic models for predicting time-varying urban network flows, *Trans. Sci.* **30**, 14–31 (1996).
[19] M. Fukushima, Equivalent differentiable optimization problems and descent methods for asymmetric variational inequality problems, *Math. Program.* **53**, 99–110 (1992).
[20] C. Fulga and V. Preda, Nonlinear programming with φ-preinvex and local φ-preinvex functions, *Eur. J. Oper. Res.* **192**, 737–743 (2009).

[21] F. Giannessi and A. Maugeri, *Variational Inequalities and Network Equilibrium Problems* (Plenum Press, New York, 1995).
[22] F. Giannessi, A. Maugeri and P. M. Pardalos, *Equilibrium Problems: Nonsmooth Optimization and Variational Inequality Models* (Kluwer Academic Publishers, Dordrecht, 2001).
[23] G. Glowinski, *Numerical Methods for Nonlinear Variational Problems* (Springer-Verlag, Berlin, 1984).
[24] R. Glowinski, J. J. Lions and R. Tremolieres, *Numerical Analysis of Variational Inequalities* (North-Holland, Amsterdam, 1981).
[25] D. Han and H. K. Lo, Two new self-adaptive projection methods for variational inequality problems, *Comput. Math. Appl.* **43**, 1529–1537 (2002).
[26] P. T. Harker and J. S. Pang, Finite dimensional variational inequalities and nonlinear complementarity problems: A survey of theory, algorithms and applications, *Math. Program.* **48**, 161–220 (1990).
[27] B. S. He, A class of projection and contraction methods for variational inequalities, *Appl. Math. Optim.* **35**, 69–76 (1997).
[28] B. S. He, Inexact implicit methods for monotone general variational inequalities, *Math. Program.* **86**, 199–217 (1999).
[29] B. S. He and L. Z. Liao, Improvement of some projection methods for monotone nonlinear variational inequalities, *J. Optim. Theory Appl.* **112**, 111–128 (2002).
[30] S. Jabeen, M. A. Noor and K. I. Noor, Inertial iterative methods for general quasi variational inequalities and dynamical systems. *J. Math. Anal.* **11**(3) (2020).
[31] S. Karamardian, Generalized complementarity problems, *J. Optim. Theory Appl.* **8**, 161–168 (1971).
[32] N. Kikuchi and J. T. Oden, *Contact Problems in Elasticity* (SIAM Publishing Co., Philadelphia, 1988).
[33] D. Kinderlehrer and G. Stampacchia, *An Introduction to Variational Inequalities and Their Applications* (SIAM, Philadelphia, 2000).
[34] G. M. Korpelevich, The extragradient method for finding saddle points and other problems, *Matecon.* **12**, 747–756 (1976).
[35] T. Larsson and M. Patriksson, A class of gap functions for variational inequalities, *Math. Program.* **64**, 53–79 (1994).
[36] C. E. Lemke, Bimatrix equilibrium points and mathematical programming, *Manag. Sci.* **11**, 681–689 (1965).
[37] H. Lewy and G. Stampacchia, On the regularity of the solutions of the variational inequalities, *Commun. Pure Appl. Math.* **22**, 153–188 (1969).

[38] G. H. Lin and M. Fukushima, Some exact penalty results for nonlinear programs and mathematical programs with equilibrium constraints, *J. Optim. Theory Appl.* **118**(1), 67–80 (2003).

[39] J. L. Lions and G. Stampacchia, Variational inequalities, *Commun. Pure Appl. Math.* **20**, 493–512 (1967).

[40] D. T. Luc and M. A. Noor, Local uniqueness of solutions of general variational inequalities, *J. Optim. Theory Appl.* **117**, 103–119 (2003).

[41] B. Martinet, Regularization d'inequations variationnelles par approximations successive, *Revue Fran. d'Informat. Rech. Oper.* **4**, 154–159 (1970).

[42] B. B. Mohsin, M. A. Noor, K. I. Noor and R. Latif, Resolvent dynamical systems and mixed variational inequalities, *J. Nonlinear Sci. Appl.* **10**, 2925–2933 (2017).

[43] A. Nagurney and D. Zhang, *Projected Dynamical Systems and Variational Inequalities with Applications* (Kluwer Academic Publishers, Dordrecht, 1996).

[44] M. A. Noor, On Variational Inequalities, Ph.D. Thesis, Brunel University, London, UK (1975).

[45] M. A. Noor, An iterative scheme for a class of quasi variational inequalities, *J. Math. Anal. Appl.* **110**(2), 463–468 (1985).

[46] M. A. Noor, Generalized quasi complemetarity problems, *J. Math. Anal. Appl.* **120**, 321–327 (1986).

[47] M. A. Noor, Fixed-point approach for complementarity problems, *J. Math. Anal. Appl.* **133**, 437–448 (1988).

[48] M. A. Noor, General variational inequalities, *Appl. Math. Lett.* **1**, 119–121 (1988).

[49] M. A. Noor, Quasi variational inequalities, *Appl. Math. Lett.* **1**, 367–370 (1988).

[50] M. A. Noor, An iterative algorithm for variational inequalities, *J. Math. Anal. Appl.* **158**, 448–455 (1991).

[51] M. A. Noor, Wiener-Hopf equations and variational inequalities, *J. Optim. Theory Appl.* **79**, 197–206 (1993).

[52] M. A. Noor, Variational inequalities in physical oceanography. In M. Rahman (ed.) *Ocean Wave Engineering* (Computational Mechanics Publications, Southampton, 1994), pp. 201–226.

[53] M. A. Noor, Variational-like inequalities, *Optimization* **30**, 323–330 (1994).

[54] M. A. Noor, Invex equilibrium problems, *J. Math. Anal. Appl.* **302**, 463–475 (2005).

[55] M. A. Noor, Fundamentals of equilibrium problems, *Math. Inequal. Appl.* **9**(3), 529–566 (2006).

[56] M. A. Noor, General variational inequalities and nonexpansive mappings, *J. Math. Anal. Appl.* **331**(2), 810–822 (2007).
[57] M. A. Noor, Sensitivity analysis for quasi variational inequalities, *J. Optim. Theory Appl.* **95**, 399–407 (1997).
[58] M. A. Noor, Wiener-Hopf equations techniques for variational inequalities, *Korean J. Comput. Appl. Math.* **7**, 581–599 (2000).
[59] M. A. Noor, Some recent advances in variational inequalities, Part I, basic concepts, *New Zealand J. Math.* **26**, 53–80 (1997).
[60] M. A. Noor, Some recent advances in variational inequalities, Part II, other concepts, *New Zealand J. Math.* **26**, 229–255 (1997).
[61] M. A. Noor, Generalized quasi variational inequalities and implicit Wiener-Hopf equations, *Optimization* **45**, 197–222 (1999).
[62] M. A. Noor, Some algorithms for general monotone mixed variational inequalities, *Math. Comput. Model.* **29**, 1–9 (1999).
[63] M. A. Noor, Merit functions for variational-like inequalities, *Math. Inequal. Appl.* **3**, 117–128 (2000).
[64] M. A. Noor, A predictor-corrector method for general variational inequalities, *Appl. Math. Lett.* **14**, 53–87 (2001).
[65] M. A. Noor, A Wiener-Hopf dynamical system for variational inequalities, *New Zealand J. Math.* **31**, 173–182 (2002).
[66] M. A. Noor, New approximation schemes for general variational inequalities, *J. Math. Anal. Appl.* **251**, 217–229 (2000).
[67] M. A. Noor, Proximal Methods for mixed variational inequalities, *J. Optim. Theory Appl.* **115**, 447–451 (2002).
[68] M. A. Noor, Implicit dynamical systems and quasi variational inequalities, *Appl. Math. Comput.* **134**, 69–81 (2002).
[69] M. A. Noor, Extragradient method for pseudomonotone variational inequalities, *J. Optim. Theory Appl.* **117**(3), 475–488 (2003).
[70] M. A. Noor, New extragradient-type methods for general variational inequalities, *J. Math. Anal. Appl.* **277**, 379–395 (2003).
[71] M. A. Noor, Auxiliary principle technique for equilibrium problems, *J. Optim. Theory Appl.* **122**(2), 371–386 (2004).
[72] M. A. Noor, Some devlopments in general variational inequalites, *Appl. Math. Comput.* **152**, 199–277 (2004).
[73] M. A. Noor, Merit functions for general variational inequalities, *J. Math. Anal. Appl.* **316**(2), 736–752 (2006).
[74] M. A. Noor, Differentiable non-convex functions and general variational inequalities, *Appl. Math. Comput.* **99**, 623–630 (2008).
[75] M. A. Noor, Extended general variational inequalities, *Appl. Math. Lett.* **22**(2), 182–185 (2009).
[76] M. A. Noor, Extended general quasi-variational inequalities, *Nonlinear Anal. Forum* **15**, 33–39 (2010).

[77] M. A. Noor, Some aspects of extended general variational inequalities, *Abst. Appl. Anal.* **2012** (2012), 16 p. ID: 303569.

[78] M. A. Noor and K. I. Noor, Sensitivity analysis for quasi variational inclusions, *J. Math. Anal. Appl.* **236**, 290–299 (1999).

[79] M. A. Noor and K. I. Noor, Some characterization of strongly preinvex functions, *J. Math. Anal. Appl.* **316**(2), 697–706 (2006).

[80] M. A. Noor and K. I. Noor, On strongly exponentially preinvex functions, *UPB Sci. Bull. Ser. A* **81**(4), 75–84 (2019).

[81] M. A. Noor and K. I. Noor, New classes of exponentially preinvex functions, *AIMS Math.* **4**(6), 1554–1568 (2019).

[82] M. A. Noor and K. I. Noor, From representation theorems to variational inequalities. In N. J. Daras and T. M. Rassias (eds.) *Computational Mathematics and Variational Analysis* (Springer Verlag, 2020).

[83] M. A. Noor and K. I. Noor, Higher order strongly general convex functions and variational inequalities, *AIMS Math.* **5**(4), 3646–3663 (2020).

[84] M. A. Noor, K. I. Noor and M. T. Rassias, New trends in general variational inequalities, *Acta Math. Appl.* **170**(1), 981–1046 (2020).

[85] M. A. Noor, K. I. Noor and T. M. Rassias, Some aspects of variational inequalities, *J. Comput. Appl. Math.* **47**, 285–312 (1993).

[86] M. A. Noor, K. I. Noor and T. M. Rassias, Iterative methods for variational inequalities. In D. Andrica and T. M. Rassias (eds.) *Differential and Integral Inequalities*. Springer Optimization and Its Applications, Vol. 151 (2019), pp. 603–618.

[87] M. A. Noor and W. Oettli, On general nonlinear complementarity problems and quasi equilibria, *Le Matematiche* **49**, 313–331 (1994).

[88] M. A. Noor, K. I. Noor and A. Bnouhachem, On a unified implict method for variational inequalities, *J. Comput. Appl. Math.* **249**, 69–73 (2013).

[89] M. A. Noor and E. Al-Said, Change of variable method for generalized complementarity problems, *J. Optim. Theory Appl.* **100**, 389–395 (1999).

[90] M. A. Noor and E. A. Al-Said, Finite difference method for a system of third-order boundary value problems, *J. Optim. Theory Appl.* **112**, 627–637 (2002).

[91] M. A. Noor, Y. J. Wang and N. H. Xiu, Some new projection methods for variational inequalities, *Appl. Math. Comput.* **137**, 423–435 (2003).

[92] S. Pal and T. K. Wong, On exponentially concave functions and a new information geometry, *Ann. Probab.* **46**(2), 1070–1113 (2018).

[93] J. S. Pang, On the convergence of a basic iterative method for the implicit complementarity problems, *J. Optim. Theory Appl.* **37**, 149–162 (1982).
[94] M. Patriksson, *Nonlinear Programming and Variational Inequality Problems: A Unified Approach* (Kluwer Academic Publishers, Dordrecht, 1998).
[95] A. Pervez, A. G. Khan, M. A. Noor and K. I. Noor, Mixed quasi variational inequalities involving four nonlinear operators, *Honam Math. J.* **42**(1), 17–35 (2020).
[96] B. T. Polyak, *Introduction to Optimization* (Optimization Software, New York, 1987).
[97] S. M. Robinson, Normal maps induced by linear transformations, *Math. Oper. Res.* **17**, 691–714 (1992).
[98] R. T. Rockafellar, Monotone operators and the proximal point algorithms, *SIAM J. Control Optim.* **14**, 877–898 (1976).
[99] P. Shi, Equivalence of variational inequalities with Wiener-Hopf equations, *Proc. Am. Math. Soc.* **111**, 339–346 (1991).
[100] M. Sibony, Methodes iteratives pour les equations et inequations aux derivees partielles nonlineaires de type monotone, *Calcolo* **7**, 65–183 (1970).
[101] M. V. Solodov and B. F. Svaiter, A new projection method for variational inequality problems, *SIAM J. Control Optim.* **42**, 309–321 (1997).
[102] M. V. Solodov and P. Tseng, Modified projection type methods for monotone variational inequalities, *SIAM J. Control Optim.* **34**, 1814–1830 (1996).
[103] G. Stampacchia, Formes bilineaires coercivites sur les ensembles convexes, *C. R. Acad.* (Paris) **258**, 4413–4416 (1964).
[104] K. Taji, M. Fukushima and T. Ibaraki, A globally convergent Newton method for solving strongly monotone variational inequalities, *Math. Program.* **58**, 369–383 (1993).
[105] E. Tonti, Variational formulation for every nonlinear problem, *Int. J. Eng. Sci.* **22**, 1343–1371 (1984).
[106] P. Tseng, A modified forward-backward splitting method for maximal monotone mappings, *SIAM J. Control Optim.* **38**, 431–446 (2000).
[107] P. Tseng, On linear convergence of iterative methods for variational inequality problem, *J. Comput. Appl. Math.* **60**, 237–252 (1995).
[108] Y. J. Wang, N. H. Xiu and C. Y. Wang, Unified framework of projection methods for pseudomonotone variational inequalities, *J. Optim. Theory Appl.* **111**, 643–658 (2001).

[109] Y. J. Wang, N. H. Xiu and C. Y. Wang, A new version of extragradient projection method for variational inequalities, *Comput. Math. Appl.* **42**, 969–979 (2001).

[110] Y. S. Xia and J. Wang, On the stability of globally projected dynamical systems, *J. Optim. Theory Appl.* **106**, 129–150 (2000).

[111] N. Xiu, J. Zhang and M. A. Noor, Tangent projection equations and general variational equalities, *J. Math. Anal. Appl.* **258**, 755–762 (2001).

[112] N. H. Xiu and J. Zhang, Some recent advances in projection-type methods for variational inequalities, *J. Comput. Appl. Math.* **152**, 559–585 (2003).

[113] N. M. Xiu and J. Zhang, Local convergence of projection-type algorithms: A unified approach, *J. Optim. Theory Appl.* **115**, 211–230 (2002).

[114] H. Yang and M. G. H. Bell, Traffic restraint, road pricing and network equilibrium, *Transp. Res. B* **31**, 303–314 (1997).

[115] E. A. Youness, E-convex sets, E-convex functions and E-convex programming, *J. Optim. Theory Appl.* **102**, 439–450 (1999).

[116] D. Zhang and A. Nagurney, On the stability of the projected dynamical systems, *J. Optim. Theory Appl.* **85**, 97–124 (1995).

[117] Y. B. Zhao, Extended projection methods for monotone variational inequalities, *J. Optim. Theory Appl.* **100**, 219–231 (1999).

[118] Y. Zhao and D. Sun, Alternative theorems for nonlinear projection equations and applications to generalized complementarity problems, *Nonlinear Anal.* **46**(6), 853–868 (2001).

[119] D. L. Zhu and P. Marcotte, Cocoercivity and its role in the convergence of iterative schemes for solving variational inequalities, *SIAM J. Optim.* **6**, 714–726 (1996).

© 2024 World Scientific Publishing Company
https://doi.org/10.1142/9789811267048_0022

Chapter 22

Biconvex Functions and Bivariational Inequalities

Muhammad Aslam Noor[*,§], Khalida Inayat Noor[*,¶],
Michael Th. Rassias[†,‖], and Waseem Asghar Khan[‡,**]

[*]*Department of Mathematics, COMSATS University Islamabad,
Park Road, Islamabad, Pakistan*
[†]*Institute of Mathematics, University of Zurich,
Zurich, Switzerland*
[‡]*Department of Mathematics, Faculty of Sciences AlZulfi,
Majmaah University, Majmaah, Saudi Arabia*
[§]*noormaslam@gmail.com*
[¶]*khalidan@gmail.com*
[‖]*michail.rassias@math.uzh.ch*
[**]*wa.khan@mu.edu.sa*

In this chapter, we consider and introduce some new concepts of the biconvex functions and monotone operators involving an arbitrary bifunction. Some new relationships among various concepts of biconvex functions have been established. We have shown that the optimality conditions for the biconvex functions can be characterized by a class of bivariational inequalities. Auxiliary principle technique is used to propose proximal point methods for solving bivariational-like inequalities. We also discussed the conversance criteria for the suggested methods under pseudo-monotonicity. Our method of proof is very simple as compared with methods. Several special cases are discussed as applications of our main concepts and results. Our method of proof of convergence is very simple as compared with techniques. This chapter can be viewed as expository nature. It is a challenging problem to explore the applications of the bivariational inequalities in pure and applied sciences.

1. Introduction

Convexity theory is a branch of mathematical sciences with a wide range of applications in industry, physical, social, regional and engineering sciences. The general theory of the convexity started soon after the introduction of differential and integral calculus by Newton and Leibnitz, although some individual optimization problems have been investigated before that. Motivated by the geometrical considerations, Euler deduced his first principle which is now referred to as Euler's differential equation for the determination of maximizing and minimizing arcs. By variational principles, we mean maximum and minimum problems arising in game theory, mechanics, geometrical optics, general relativity theory, economics, transportation, differential geometry, data analysis, machine learning and related areas. It is also known that the variational formulation of field theories allows for a degree of unification absent in their versions in terms of differential equations. Convexity plays an important part in the existence and stability of soliton, which occurs in almost every branch of physics. For more details, see Refs. [1, 2, 5, 13–15, 18, 20] and the references therein. It is worth mentioning that variational inequalities represent the optimality conditions for the differentiable convex functions on the convex sets in normed spaces. which were introduced and considered by Stampacchia [21] in potential theory. In fact, it has been shown that the minimum of a differentiable convex function is characterized by an inequality, which is called the variational inequality. This simple fact has played important role in the development of branches of sciences. Variational inequalities combine both theoretical and algorithmic advances with new and novel domain of applications. Analysis of these problems requires a blend of techniques from convex analysis, functional analysis and numerical analysis. In recent years, considerable interest has been shown in developing various generalizations of variational inequalities and generalized convexity, both for their own sake and their applications.

Noor and Noor [13–15] introduced and studied some new classes of biconvex sets and biconvex functions. These new concepts may be viewed as refinements of the known concepts. They have shown that the biconvex functions enjoy some nice properties, which convex have. For more details, see Refs. [13–15, 18] and the references therein. In this chapter, we mainly review these developments and include the necessary information for the convenience of the interested readers. Several new concepts of monotonicity are introduced and are discussed. We establish the relationship between these classes and derive some new results under some mild conditions. It is shown that the optimality conditions of the differentiable biconvex functions can be characterized by a class of variational inequalities, which is called bivariational inequality. It is well known that the projection-type methods and their invariant forms cannot be used for solving the bivariational inequalities. To overcome this drawback, one usually uses the auxiliary principle technique, which is due to Glowinski et al. [3]. This technique deals with finding a suitable auxiliary problem and proving that the solution of the auxiliary problem is the solution of the original problem by using the fixed-point approach, see Refs. [3, 4, 8–19, 23]. We use this technique to suggest and analyze several methods for solving bivariational inequalities and variational inequalities. It has been shown that a substantial number of numerical methods can be obtained as special cases from this technique. Some iterative methods are suggested for solving bivariational inequalities using the auxiliary principle technique [13–15, 18]. Convergence criteria is also discussed using the pseudomonotonicity which is a weaker condition then monotonicity. It is expected that the ideas and techniques of this chapter may stimulate further research in this field.

2. Preliminary Results

Let K be a nonempty closed set in a real Hilbert space H. We denote by $\langle \cdot, \cdot \rangle$ and $\|\cdot\|$ the inner product and norm, respectively. Let $F: K_\beta \to R$ be a continuous function and let $\beta(.,.): K_\beta \times K_\beta \to R$ be an arbitrary continuous bifunction.

We now recall the known concepts and basic results, which are mainly due to Noor and Noor [13–15].

Definition 1. The set K_β in H is said to be a biconvex set with respect to an arbitrary bifunction $\beta(\cdot - \cdot)$, if

$$u + \lambda\beta(v - u) \in K_\beta, \quad \forall u, v \in K_\beta, \lambda \in [0, 1].$$

The biconvex set K_β is also called a β-connected set. Note that the biconvex set with $\beta(v, u) = v - u$ is a convex set K, but the converse is not true. For example, the set $K_\beta = R - (-\frac{1}{2}, \frac{1}{2})$ is an biconvex set with respect to η, where

$$\beta(v - u) = \begin{cases} v - u, & \text{for } v > 0, u > 0 \text{ or } v < 0, u < 0 \\ u - v, & \text{for } v < 0, u > 0 \text{ or } v < 0, u < 0. \end{cases}$$

It is clear that K_β is not a convex set.

From now onward, K_β is a nonempty closed biconvex set in H with respect to the bifunction $\beta(\cdot - \cdot)$, unless otherwise specified.

We now introduce some new concepts of biconvex functions and their variant forms, which is the main motivation of this chapter.

Definition 2. The function F on the biconvex set K_β is said to be a biconvex with respect to the bifunction $\beta(\cdot - \cdot)$, if

$$F(u + \lambda\beta(v - u)) \leq (1 - \lambda)F(u) + \lambda F(v), \quad \forall u, v \in K_\beta, \lambda \in [0, 1]. \tag{1}$$

The function F is said to be biconcave, if and only if, $-F$ is a biconvex function. Consequently, we have a new concept.

Definition 3. A function F is said to be affine involving an arbitrary bifunction $\beta(\cdot - \cdot)$, if

$$F(u + \lambda\beta(v - u)) = (1 - \lambda)F(u) + \lambda F(v), \quad \forall u, v \in K_\beta, \lambda \in [0, 1]. \tag{2}$$

Note that every convex function is a biconvex, but the converse is not true.

If $\beta(v-u) = v-u$, then the biconvex function becomes a convex function, that is,

$$F(u + \lambda(v-u)) \leq (1-\lambda)F(u) + \lambda F(v), \quad \forall u,v \in K, \ \lambda \in [0,1].$$

For the properties of the convex functions in variational inequalities and equilibrium problems, see Noor [8–12].

Definition 4. The function F on the biconvex set K_β is said to be quasi biconvex with respect to the bifunction $\beta(\cdot - \cdot)$, if

$$F(u + \lambda\beta(v-u)) \leq \max\{F(u), F(v)\}, \quad \forall u,v \in K_\beta, \ \lambda \in [0,1].$$

Definition 5. The function F on the biconvex set K_β is said to be log-biconvex with respect to the bifunction $\beta(\cdot - \cdot)$, if

$$F(u + \lambda\beta(v-u)) \leq (F(u))^{1-\lambda}(F(v))^\lambda, \quad \forall u,v \in K_\beta, \ \lambda \in [0,1],$$

where $F(\cdot) > 0$.

We can rewrite Definition 5 in the following equivalent form:

Definition 6. The function F on the biconvex set K_β is said to be log-biconvex with respect to the bifunction $\beta(\cdot - \cdot)$, if

$$\log F(u + \lambda\beta(v-u)) \leq (1-\lambda)\log F(u) + \lambda \log F(v),$$
$$\forall u,v \in K_\beta, \ \lambda \in [0,1],$$

where $F(\cdot) > 0$.

This equivalent definition can be used to discus the properties of the differentiable log-biconvex functions.

From the above definitions, we have

$$F(u + \lambda\beta(v-u)) \leq (F(u))^{1-\lambda}(F(v))^\lambda$$
$$\leq (1-\lambda)F(u) + \lambda F(v)$$
$$\leq \max\{F(u), F(v)\}.$$

This shows that every log-biconvex function is a biconvex function and every biconvex function is a quasi-biconvex function. However, the converse is not true.

For $\lambda = 1$, Definitions 2 and 5 reduce to the following condition:

Condition A.
$$F(u + \beta(v - u)) \leq F(v), \quad \forall v, u \in K_\beta,$$

which is called Condition A.

We now define the biconvex functions on the interval $K_\beta = I_\beta = [a, a + \beta(b - a)]$.

Definition 7. Let $I = [a, a + \beta(b - a)]$. Then F is a biconvex function, if and only if,

$$\begin{vmatrix} 1 & 1 & 1 \\ a & x & a + \beta(b - a) \\ F(a) & F(x) & F(b) \end{vmatrix} \geq 0; \quad a \leq x \leq a + \beta(b - a).$$

One can easily show that the following are equivalent:

(1) F is a biconvex function.
(2) $F(x) \leq F(a) + \frac{F(b) - F(a)}{\beta(b-a)}(x - a)$.
(3) $\frac{F(x) - F(a)}{x - a} \leq \frac{F(b) - F(a)}{\beta(b-a)}$.
(4) $\frac{F(a)}{(\beta(b-a))(a-x)} + \frac{F(x)}{(x-a-\beta(b-a))(a-x)} + \frac{F(b)}{\beta(b-a)(x-b)} \leq 0$,

where $x = a + \lambda\beta(b - a) \in [a, a + \beta(b - a)]$.

3. Properties of Biconvex Functions

In this section, we consider some basic properties of biconvex functions and their variant forms.

Theorem 1. *Let F be a strictly biconvex function. Then any local minimum of F is a global minimum.*

Proof. Let the biconvex function F have a local minimum at $u \in K_\beta$. Assume the contrary, that is, $F(v) < F(u)$ for some $v \in K_\beta$. Since F is a biconvex function, so

$$F(u + \lambda\beta(v - u)) < \lambda F(v) + (1 - \lambda)F(u), \quad \text{for } 0 < \lambda < 1.$$

Thus,
$$F(u + \lambda\beta(v-u)) - F(u) < \lambda[F(v) - F(u)] < 0,$$
from which it follows that
$$F(u + \lambda\beta(v-u)) < F(u),$$
for arbitrary small $\lambda > 0$, contradicting the local minimum. □

Theorem 2. *If the function F on the convex set K_β is biconvex, then the level set*
$$L_\alpha = \{u \in K : F(u) \leq \alpha, \ \alpha \in R\}$$
is a biconvex set.

Proof. Let $u, v \in L_\alpha$. Then, $F(u) \leq \alpha$ and $F(v) \leq \alpha$.
Now, $\forall \lambda \in (0,1)$, $w = u + \lambda\beta(v-u) \in K_\beta$, since K_β is a biconvex set. Thus, by the biconvexity of F, we have
$$F(u + \lambda\beta(v-u)) \leq (1-\lambda)F(u) + \lambda F(b)$$
$$\leq (1-t)\alpha + t\alpha = \alpha,$$
from which it follows that $u + t\beta(v-u) \in L_\alpha$. Hence, L_α is a biconvex set. □

Theorem 3. *A positive function F is a biconvex, if and only if*
$$\mathrm{epi}(F) = \{(u, \alpha) : u \in K_\beta : F(u) \leq \alpha, \alpha \in R\}$$
is a biconvex set.

Proof. Assume that F is a biconvex function. Let $(u, \alpha), (v, \beta) \in \mathrm{epi}(F)$. Then, it follows that $F(u) \leq \alpha$ and $F(v) \leq \beta$. Thus, $\forall \lambda \in [0,1]$, $u, v \in K_\beta$, we have
$$F(u + \lambda\beta(v-u)) \leq (1-\lambda)F(u) + \lambda F(v)$$
$$\leq (1-t)\alpha + t\beta,$$
which implies that
$$(u + \lambda\beta(v-u), (1-\lambda)\alpha + \lambda\beta) \in \mathrm{epi}(F).$$

Thus, epi(F) is a biconvex set. Conversely, let epi(F) be a biconvex set. Let $u, v \in K_\beta$. Then $(u, F(u)) \in $ epi(F) and $(v, F(v)) \in $ epi(F). Since epi(F) is a biconvex set, we must have

$$(u + \lambda\beta(v - u), (1 - \lambda)F(u) + \lambda F(v)) \in \text{epi}(F),$$

which implies that

$$F(u + \lambda\beta(v - u)) \leq (1 - \lambda)F(u) + \lambda F(v).$$

This shows that F is a biconvex function. $\square$

Theorem 4. *A positive function F is quasi biconvex, if and only if, the level set*

$$L_\alpha = \{u \in K_\beta, \alpha \in R : F(u) \leq \alpha\}$$

is a biconvex set.

Proof. Let $u, v \in L_\alpha$. Then $u, v \in K_\beta$ and $\max(F(u), F(v)) \leq \alpha$. Now for $\lambda \in (0, 1), w = u + \lambda\beta(v - u) \in K_\beta$. We have to prove that $u + \lambda\beta(v - u) \in L_\alpha$. By the quasi biconvexity of F, we have

$$F(u + \lambda\beta(v - u)) \leq \max(F(u), F(v)) \leq \alpha,$$

which implies that $u + \lambda\beta(v - u) \in L_\alpha$, showing that the level set L_α is indeed a biconvex set.

Conversely, assume that L_α is a biconvex set. Then $\forall u, v \in L_\alpha$, $\lambda \in [0, 1]$,

$$u + \lambda\beta(v - u) \in L_\alpha.$$

Let $u, v \in L_\alpha$ for

$$\alpha = \max(F(u), F(v)) \quad \text{and} \quad F(v) \leq F(u).$$

From the definition of the level set L_α, it follows that

$$F(u + \lambda\beta(v - u)) \leq \max(F(u), F(v)) \leq \alpha.$$

Thus F is a quasi-biconvex function. This completes the proof. $\square$

Theorem 5. *Let F be a biconvex function. Let $\mu = \inf_{u \in K_\beta} F(u)$. Then the set $E = \{u \in K_\beta : F(u) = \mu\}$ is a biconvex set of K_β. If F is strictly biconvex, then E is a singleton.*

Proof. Let $u, v \in E$. For $0 < \lambda < 1$, let $w = u + \lambda\beta(v - u)$. Since F is a biconvex function,

$$F(w) = F(u + \lambda\beta(v - u)) \leq (1 - \lambda)F(u) + \lambda F(v)$$
$$= \lambda\mu + (1 - \lambda)\mu = \mu,$$

which implies that $w \in E$ and hence E is a biconvex set. For the second part, assume to the contrary that $F(u) = F(v) = \mu$. Since K_β is a biconvex set, then for $0 < \lambda < 1$, $u + \lambda\beta(v - u) \in K_\beta$. Further, since F is strictly biconvex,

$$F(u + \lambda\beta(v - u)) < (1 - \lambda)F(u) + \lambda F(v)$$
$$= (1 - t)\mu + t\mu = \mu.$$

This contradicts the fact that $\mu = \inf_{u \in K_\beta} F(u)$ and hence the result follows. $\square$

Theorem 6. *If F is a biconvex function such that $F(v) < F(u)$, $\forall u, v \in K_\beta$, then F is a strictly quasi-biconvex function.*

Proof. By the biconvexity of the function F, $\forall u, v \in K_\beta, \lambda \in [0, 1]$, we have

$$F(u + \lambda\beta(v - u)) \leq (1 - \lambda)F(u) + \lambda F(v) < F(u),$$

since $F(v) < F(u)$, which shows that the function F is strictly quasi biconvex. $\square$

We now derive some properties of the differentiable log-biconvex functions.

To derive the main results, we need the following assumptions regarding the bifunction $\beta(\cdot - \cdot)$.

Condition M. The bifunction $\beta(,-,)$ is said to satisfy the assumptions, if

(i) $\beta(\gamma\beta(v - u)) = \gamma\beta(v - u), \quad \forall u, v \in K_\beta, \gamma \in R^n,$

(ii) $\beta(v - u - \gamma\beta(v - u)) = (1 - \gamma)\beta(v - u), \quad \forall u, v \in K_\beta.$

Remark 1. Let $\beta(\cdot - \cdot) : K_\beta \times K_\beta \to H$ satisfy the assumption

$$\beta(v - u) = \beta(v - z) + \beta(z - u), \quad \forall u, v, z \in K_\beta.$$

One can easily show that $\beta(v-u) = 0\ \forall u,v \in K_\beta$. Consequently,

$$\beta(v-u) = 0 \Leftrightarrow v = u, \quad \forall u,v \in K_\beta.$$

Also

$$\beta(v-u) + \beta(u-v) = 0, \quad \forall u,v \in K_\beta.$$

This implies that the bifunction $\beta(.,.)$ is skew symmetric.

Theorem 7. *Let F be a differentiable function on the biconvex set K_β and let the condition M hold. Then the function F is a log-biconvex function, if and only if*

$$\log F(v) - \log F(u) \geq \left\langle \frac{F'(u)}{F(u)}, \beta(v-u) \right\rangle, \quad \forall v,u \in K_\beta. \qquad (3)$$

Proof. Let F be a log-biconvex function. Then,

$$\log F(u + \lambda\beta(v-u)) \leq (1-\lambda)\log F(u) + \lambda \log F(v), \quad \forall u,v \in K_\beta,$$

which can be written as

$$\log F(v) - \log F(u) \geq \left\{ \frac{\log F(u + \lambda\beta(v-u)) - \log F(u)}{\lambda} \right\}.$$

Taking the limit in the above inequality as $\lambda \to 0$, we have

$$\log F(v) - \log F(u) \geq \left\langle \frac{F'(u)}{F(u)}, \beta(v-u) \right\rangle,$$

which is (3), the required result.

Conversely, let (3) hold. Then, $\forall u,v \in K_\beta, \lambda \in [0,1], v_\lambda = u + \lambda\beta(v-u) \in K_\beta$ and using the condition M, we have

$$\log F(v) - \log F(v_\lambda) \geq \left\langle \frac{F'(v_\lambda)}{F(v_\lambda)}, \beta(v-v_\lambda)) \right\rangle$$

$$= (1-\lambda)\left\langle \frac{F'(v_\lambda)}{F(v_\lambda)}, \beta(v-u) \right\rangle. \qquad (4)$$

In a similar way, we have

$$\log F(u) - \log F(v_\lambda) \geq \left\langle \frac{F'(v_\lambda)}{F(v_\lambda)}, \beta(u - v_\lambda) \right\rangle$$

$$= -\lambda \left\langle \frac{F'(v_\lambda)}{F(v_\lambda)}, \beta(v - u) \right\rangle. \quad (5)$$

Multiplying (4) by λ and (5) by $(1 - \lambda)$ and adding the resultant, we have

$$\log F(u + \lambda\beta(v - u)) \leq (1 - \lambda) \log F(u) + \lambda \log F(v),$$

showing that F is a log-biconvex function. $\square$

Remark 2. From (3), we have

$$F(v) \geq F(u) \exp\left\{ \left\langle \frac{F'(u)}{F(u)}, \beta(v - u) \right\rangle \right\}, \quad u, v \in K_\beta.$$

Changing the role of u and v in the above inequality, we also have

$$F(u) \geq F(v) \exp\left\{ \left\langle \frac{F'(v)}{F(v)}, \beta(u - v) \right\rangle \right\}, \quad u, v \in K_\beta.$$

Thus, we can obtain the following inequality:

$$F(u) + F(v) \geq F(v) \exp\left\{ \left\langle \frac{F'(v)}{F(v)}, \beta(u - v) \right\rangle \right\},$$

$$+ F(u) \exp\left\{ \left\langle \frac{F'(u)}{F(u)}, \beta(v - u) \right\rangle \right\} \quad u, v \in K_\beta.$$

Theorem 8. *Let F be a differentiable function on the biconvex set K_β and Condition M hold. Then the function F is a log-biconvex function, if and only if*

$$\left\langle \frac{F'(u)}{F(u)}, \beta(v - u) \right\rangle + \left\langle \frac{F'(v)}{F(v)}, \beta(u - v) \right\rangle \leq 0, \quad \forall v, u \in K_\beta. \quad (6)$$

Proof. Let F be a differentiable function on the biconvex set K_β. Then, from Theorem 7, it follows that

$$\log F(v) - \log F(u) \geq \left\langle \frac{F'(u)}{F(u)}, \beta(v-u) \right\rangle, \quad \forall v, u \in K_\beta. \tag{7}$$

Changing the role of u and v in (7), we have

$$\log F(u) - \log F(v) \geq \left\langle \frac{F'(v)}{F(v)}, \beta(v-u) \right\rangle, \quad \forall v, u \in K_\beta. \tag{8}$$

Adding (7) and (8), we have

$$\left\langle \frac{F'(u)}{F(u)}, \beta(v-u) \right\rangle + \left\langle \frac{F'(v)}{F(v)}, \beta(u-v) \right\rangle \leq 0, \quad \forall v, u \in K_\beta,$$

which is the required (6).

Since K_β is a biconvex set, so, $\forall u, v \in K_\beta, \lambda \in [0,1], v_\lambda = u + \lambda \beta(v-u) \in K_\beta$. Conversely, from (6), we have

$$\left\langle \frac{F'(v_\lambda)}{F(v_\lambda)}, \beta(u - v_\lambda) \right\rangle \leq \left\langle \frac{F'(u)}{F(u)}, \beta(u - v_\lambda) \right\rangle$$

$$= -\lambda \left\langle \frac{F'(u)}{F(u)}, \beta(v-u) \right\rangle, \tag{9}$$

which implies that

$$\left\langle \frac{F'(v_\lambda)}{F(v_\lambda)}, \beta(v-u) \right\rangle \geq \left\langle \frac{F'(u)}{F(u)}, \beta(v-u) \right\rangle. \tag{10}$$

Consider the auxiliary function

$$\xi(\lambda) = \log F(u + \lambda(v-u)),$$

from which we have

$$\xi(1) = \log F(v), \quad \xi(0) = \log F(u).$$

Then, from (10), we have

$$\xi'(\lambda) = \left\langle \frac{F'(v_\lambda)}{F(v_\lambda)}, \beta(v-u) \right\rangle \geq \left\langle \frac{F'(u)}{F(u)}, \beta(v-u) \right\rangle. \tag{11}$$

Integrating (11) between 0 and 1, we have

$$\xi(1) - \xi(0) = \int_0^1 \xi'(t)dt \geq \left\langle \frac{F'(u)}{F(u)}, \beta(v-u) \right\rangle.$$

Thus, it follows that

$$\log F(v) - \log F(u) \geq \left\langle \frac{F'(u)}{F(u)}, \beta(v-u) \right\rangle,$$

which is the required (3). □

Definition 8. An operator $T: K_\beta \to H$ is said to be

(1) β-monotone, iff

$$\langle Tu, \beta(v-u) \rangle + \langle Tv, \beta(u-v) \rangle \leq 0, \quad \forall u, v \in K_\beta,$$

(2) β-pseudomonotone, iff

$$\langle Tu, \beta(v-u) \rangle \geq 0 \implies -\langle Tv, \beta(u-v) \rangle \geq 0, \quad \forall u, v \in K_\beta,$$

(3) relaxed β-pseudomonotone, iff

$$\langle Tu, \beta(v-u) \rangle \geq 0 \implies -\langle Tv, \beta(u-v) \rangle \geq 0, \quad \forall u, v \in K_\beta,$$

(4) strictly β-monotone, iff

$$\langle Tu, \beta(v-u) \rangle + \langle Tv, \beta(u-v) \rangle < 0, \quad \forall\, u, v \in K_\beta,$$

(5) β-pseudomonotone, iff

$$\langle Tu, \beta(v-u) \rangle \geq 0 \implies \langle Tv, \eta(u,v) \rangle \leq 0, \quad \forall u, v \in K_\beta,$$

(6) quasi β-monotone, iff

$$\langle Tu, \beta(v-u) \rangle > 0 \implies \langle Tv, \beta(u-v) \rangle \leq 0, \quad \forall u, v \in K_\beta,$$

(7) strictly β-pseudomonotone, iff

$$\langle Tu, \beta(v-u) \rangle \geq 0 \implies \langle Tv, \beta(u-v) \rangle < 0, \quad \forall u, v \in K_\beta.$$

Definition 9. A differentiable function F on the biconvex set K_η is said to be pseudo β-biconvex function, iff

$$\left\langle \frac{F'(u)}{F(u)}, \beta(v-u) \right\rangle \geq 0 \implies F(v) - F(u) \geq 0, \quad \forall u, v \in K_\beta.$$

Definition 10. A differentiable function F on K_β is said to be quasi-biconvex function, iff

$$F(v) \leq F(u) \Rightarrow \left\langle \frac{F'(u)}{F(u)}, \beta(v-u) \right\rangle \leq 0, \quad \forall u, v \in K_\beta.$$

Definition 11. The function F on the set K_β is said to be pseudo-biconvex, iff

$$\left\langle \frac{F'(u)}{F(u)}, \beta(v-u) \right\rangle \geq 0 \Rightarrow F(v) \geq F(u), \quad \forall u, v \in K_\beta.$$

Definition 12. The differentiable function F on the K_β is said to be quasi-biconvex function, iff

$$F(v) \leq F(u) \Rightarrow \left\langle \frac{F'(u)}{F(u)}, \beta(v-u) \right\rangle \leq 0, \quad \forall u, v \in K_\beta.$$

We remark that the concepts introduced in this chapter represent significant improvement of the previously known ones. All these new concepts may play important and fundamental part in the mathematical programming and optimization.

Theorem 9. *Let F be a differentiable function on the biconvex set K_β in H and let the condition M hold. Then the function F is a biconvex function, if and only if F is a biconvex function.*

Proof. Let F be a biconvex function on the biconvex set K_β. Then,

$$F(u + \lambda\beta(v-u)) \leq (1-\lambda)F(u) + \lambda F(v) \quad \forall u, v \in K_\beta, \lambda \in [0,1],$$

which can be written as

$$F(v) - F(u) \geq \left\{ \frac{F(u + \lambda\beta(v-u)) - F(u)}{\lambda} \right\}.$$

Taking the limit in the above inequality as $\lambda \to 0$, we have

$$F(v) - F(u) \geq \langle F'(u), \beta(v-u) \rangle.$$

This shows that F is a biconvex function.

Conversely, let F be a biconvex function on the biconvex set K_β. Then, $\forall u, v \in K_\beta, \lambda \in [0,1], v_t = u + \lambda\beta(v-u) \in K_\beta$ and using the condition M, we have

$$F(v) - F(u + \lambda\beta(v-u))$$
$$\geq \langle F'(u + \lambda\beta(v-u)), \beta(v - u + \lambda\beta(v-u)) \rangle$$
$$= (1-\lambda)F'(u + \lambda\beta(v-u)), \beta(v-u)\rangle. \quad (12)$$

In a similar way, we have

$$F(u) - F(u + \lambda\beta(v-u))$$
$$\geq \langle F'(u + \lambda\beta(v-u)), \beta(u, u + \lambda\beta(v-u)) \rangle$$
$$= -\lambda F'(u + \lambda\beta(v-u)), \beta(v-u)\rangle. \quad (13)$$

Multiplying (12) by λ and (13) by $(1-\lambda)$ and adding the resultant, we have

$$F(u + \lambda\beta(v-u)) \leq (1-\lambda)F(u) + \lambda F(v),$$

showing that F is a biconvex function. □

Theorem 10. *Let F be a differentiable biconvex function on the biconvex set K_β. If F is a biconvex function, then*

$$\langle F'(u), \beta(v-u))\rangle + \langle F'(v), \beta(u-v)\rangle \leq 0, \quad \forall u, v \in K_\beta. \quad (14)$$

Proof. Let F be a biconvex function on the biconvex set K_β. Then,

$$F(v) - F(u) \geq \langle F'(u), \beta(v-u))\rangle, \quad \forall u, v \in K_\beta. \quad (15)$$

Changing the role of u and v in (15), we have

$$F(u) - F(v) \geq \langle F'(v), \beta(u-v)\rangle, \quad \forall u, v \in K_\beta. \quad (16)$$

Adding (15) and (16), we have

$$\langle F'(u), \beta(v-u))\rangle + \langle F'(v), \beta(u-v)\rangle \leq 0, \quad \forall u, v \in K_\beta, \quad (17)$$

which shows that $F'(.)$ is a β-monotone operator. □

Theorem 11. *If the differential $F'(.)$ is a β-monotone, then*
$$F(v) - F(u) \geq \langle F'(u), \beta(v-u)\rangle.$$

Proof. Let $F'(.)$ be a β-monotone. From (17), we have
$$\langle F'(v), \beta(u-v)\rangle \geq \langle F'(u), \beta(v-u))\rangle. \tag{18}$$

Since K_β is an biconvex set, $\forall u, v \in K_\beta, \lambda \in [0,1]$ $v_t = u + \lambda\beta(v-u) \in K_\beta$. Taking $v = v_\lambda$ in (18) and using Condition M, we have
$$\langle F'(v_\lambda), \beta(u - u - \lambda\beta(v-u)))\rangle \leq \langle F'(u), \eta(u + \lambda\beta(v-u) - u))\rangle$$
$$+ \|\beta(u - u - \lambda\beta(v-u)\|^p\}$$
$$= -\lambda\langle F'(u), \beta(v-u)\rangle,$$

which implies that
$$\langle F'(v_\lambda), \beta(v-u)\rangle \geq \langle F'(u), \beta(v-u)\rangle. \tag{19}$$

Let $\xi(\lambda) = F(u + \lambda\beta(v-u))$. Then, from (19), we have
$$\xi'(\lambda) = \langle F'(u + \lambda\beta(v-u)), \beta(v-u)\rangle$$
$$\geq \langle F'(u), \beta(v-u)\rangle. \tag{20}$$

Integrating (20) between 0 and 1, we have
$$\xi(1) - \xi(0) \geq \langle F'(u), \beta(v-u)\rangle,$$

that is,
$$F(u + \lambda\beta(v-u)) - F(u) \geq \langle F'(u), \beta(v-u)\rangle.$$

By using Condition A, we have
$$F(v) - F(u) \geq \langle F'(u), \beta(v-u)\rangle,$$

the required result. $\square$

We now give a necessary condition for β-pseudo-biconvex function.

Theorem 12. *Let $F'(.)$ be a relaxed β-pseudomonotone operator and Conditions A and M hold. Then F is a β-pseudo-biconvex function.*

Proof. Let F' be a relaxed β-pseudomonotone. Then, $\forall u, v \in K_\beta$,
$$\langle F'(u), \beta(v-u) \rangle \geq 0$$
implies that
$$-\langle F'(v), \beta(u-v) \rangle \geq 0. \tag{21}$$
Since K is an biconvex set, $\forall u, v \in K_\eta$, $\lambda \in [0,1]$, $v_\lambda = u + \lambda\beta(v-u) \in K_\beta$. Taking $v = v_\lambda$ in (21) and using Condition M, we have
$$-\langle F'(u+\lambda\beta(v-u)), \beta(u-v) \rangle \geq 0. \tag{22}$$
Let
$$\xi(\lambda) = F(u+\lambda\beta(v-u)), \quad \forall u, v \in K_\beta, \ \lambda \in [0,1].$$
Then, using (22), we have
$$\xi'(\lambda) = \langle F'(u+\lambda\beta(v-u)), \beta(u-v) \rangle \geq 0.$$
Integrating the above relation between 0 and 1, we have
$$\xi(1) - \xi(0) \geq 0,$$
that is,
$$F(u+\lambda\beta(v-u)) - F(u) \geq 0,$$
which implies, using Condition A,
$$F(v) - F(u) \geq 0,$$
showing that F is a β-pseudo-biconvex function. □

Definition 13. The function F is said to be sharply pseudo-biconvex, if
$$\langle F'(u), \beta(v-u) \rangle \geq 0 \Rightarrow F(v) \geq F(v+\lambda\beta(v-u)),$$
$$\forall u, v \in K_\beta, \ \lambda \in [0,1].$$

Theorem 13. *Let F be a sharply pseudo biconvex function on K_β. Then,*
$$-\langle F'(v), \beta(v-u) \rangle \geq 0, \quad \forall u, v \in K_\beta.$$

Proof. Let F be a sharply pseudo-biconvex function on K_β. Then,

$$F(v) \geq F(v + \lambda\beta(v - u)), \quad \forall u, v \in K_\beta, \ \lambda \in [0, 1],$$

from which we have

$$\frac{F(v + \lambda\beta(v - u)) - F(v)}{\lambda} \leq 0.$$

Taking limit in the above-mentioned inequality, as $\lambda \to 0$, we have

$$-\langle F'(v), \beta(v - u) \rangle \geq 0,$$

the required result. $\square$

Definition 14. A function F is said to be a pseudo-biconvex function with respect to strictly positive bifunction $W(.,.)$, if

$$F(v) < F(u) \Rightarrow F(u + \lambda\beta(v - u)) < F(u) + \lambda(\lambda - 1)W(v, u),$$
$$\forall u, v \in K_\beta, \ \lambda \in [0, 1].$$

Theorem 14. *If the function F is a biconvex function such that $F(v) < F(u)$, then the function F is pseudo-biconvex.*

Proof. Since $F(v) < F(u)$ and F is biconvex function, then $\forall u, v \in K_\eta, \ \lambda \in [0, 1]$, we have

$$\begin{aligned}
F(u + \lambda\beta(v - u)) &\leq F(u) + \lambda(F(v) - F(u)) \\
&< F(u) + \lambda(1 - \lambda)(F(v) - F(u)) \\
&= F(u) + \lambda(\lambda - 1)(F(u) - F(v)) \\
&< F(u) + \lambda(\lambda - 1)W(u, v),
\end{aligned}$$

where $W(u, v) = F(u) - F(v) > 0$. This shows that the function F is a pseudo-biconvex. $\square$

4. Bivariational Inequalities

In this section, we consider the bivariational inequalities and suggest some iterative methods by using the auxiliary principle techniques involving the Bregman distance functions.

For the readers' convenience, we recall some basic properties of the Bregman [1] convex functions. For strongly convex function F, we define the Bregman distance function as

$$B(v,u) = F(v) - F(u) - \langle F'(u), v - u \rangle \geq \alpha \|v - u\|^2, \quad \forall u, v \in K. \tag{23}$$

It is important to emphasize that various types of function F give different Bregman distances. We give the following important examples of some practical important types of function F and their corresponding Bregman distances, see Ref. [22].

Examples

(1) If $f(v) = \|v\|^2$, then $B(v,u) = \|v - u\|^2$, which is the squared Euclidean distance (SE).
(2) If $f(v) = \sum_{i=1}^{n} a_i \log(v_i)$, which is known as Shannon entropy, then its corresponding Bregman distance is given as

$$B(v,u) = \sum i = 1^{n} \left(v_i \log \left(\frac{v_i}{u_i} \right) + u_i - v_i \right).$$

This distance is called Kullback–Leibler (KL) distance and has become a very important tool in several areas of applied mathematics, such as machine learning.
(3) If $f(v) = -\sum_{i=1}^{n} \log(v_i)$, which is called Burg entropy, then its corresponding Bregman distance is given as

$$B(v,u) = \sum_{i=1}^{n} \left(\log \left(\frac{v_i}{u_i} \right) + \frac{v_i}{u_i} - 1 \right).$$

This is called Itakura–Saito (IS) distance, which is very important in information theory, data analysis and machine learning.

Remark 3. It is a challenging problem to explore the applications of Bregman distance functions for other types of nonconvex functions such as biconvex, k-convex functions and harmonic functions.

We now discuss the optimality conditions for the differentiable biconvex functions.

Theorem 15. *Let F be a differentiable biconvex function with modulus $\mu > 0$. $u \in K_\beta$ is the minimum of the function F, if and only if $u \in K_\beta$ satisfies the*

$$\langle F'(u), \beta(v-u)\rangle \geq 0, \quad \forall v \in K_\beta. \tag{24}$$

Proof. Let $u \in K_\beta$ be a minimum of the function F. Then,

$$F(u) \leq F(v), \quad \forall v \in K_\beta. \tag{25}$$

Since K_β is a biconvex set, so, $\forall u, v \in K_\beta$, $\lambda \in [0, 1]$,

$$v_\lambda = u + \lambda \beta(v-u) \in K_\beta.$$

Taking $v = v_\lambda$ in (25), we have

$$0 \leq \lim_{\lambda \to 0} \left\{ \frac{F(u + \lambda \beta((v-u))) - F(u)}{\lambda} \right\} \langle F'(u), \beta(v-u)\rangle, \tag{26}$$

which is inequality (24).

Since F is a differentiable biconvex function, so

$$F(u + \lambda \beta(v-u)) \leq F(u) + \lambda(F(v) - F(u)), \quad \forall u, v \in K_\beta,$$

from which, using (24), we have

$$F(v) - F(u) \geq \lim_{\lambda \to 0} \left\{ \frac{F(u + \lambda \beta(v-u)) - F(u)}{\lambda} \right\}$$
$$= \langle F'(u), \beta(v, u)\rangle \geq 0,$$

from which, we have

$$F(u) \leq F(v), \quad \forall v \in K_\beta, \tag{27}$$

which implies that $u \in K_\beta$ is the minimum of the biconvex functions. □

Remark 4. We would like to mention that, if $u \in K_\beta$ satisfies the inequality

$$\langle F'(u), \beta(v-u)\rangle \geq 0, \quad \forall u, v \in K_\beta, \tag{28}$$

then $u \in K_\beta$ is the minimum of the differentiable preinbex function F.

The inequality of type (28) is called the bivariational inequality and appears to new one.

It is worth mentioning that inequalities of type (28) may not arise as a minimization of the biconvex functions. This motivated us to consider a more general bivariational inequality of which (28) is a special case.

For a given operator T, bifunction $\beta(.-.)$, consider the problem of finding $u \in K_\beta$, such that

$$\langle Tu, \beta(v-u)\rangle \geq 0, \quad \forall v \in K_\beta, \tag{29}$$

which is called bivariational inequality.

It is worth mentioning that for suitable and appropriate choice of the operators, biconvex sets and spaces, one can obtain a wide class of variational inequalities and optimization problems. This shows that the bivariational inequalities are quite flexible and unified ones.

Due to the inherent nonlinearity, the projection method and its variant form cannot be used to suggest the iterative methods for solving these bivariational inequalities. To overcome these drawbacks, one may use the auxiliary principle technique of Glowinski *et al.* [3] as developed by Noor [9–12] and Noor *et al.* [16–19] to suggest and analyze some iterative methods for solving the bivariational-like inequalities (29). This technique does not involve the concept of the projection, which is the main advantage of this technique. We again use the auxiliary principle technique coupled with Bergman functions. These applications are based on the type of convex functions associated with the Bregman distance. We now suggest and analyze some iterative methods for bivariational inequalities (29) using the auxiliary principle technique coupled with Bregman distance functions.

For a given $u \in K_\beta$ satisfying the bivariational inequality (29), we consider the auxiliary problem of finding a $w \in K$ such that

$$\langle \rho Tw, \beta(v-w)\rangle + \langle E'(w) - E'(u), \beta(v-w)\rangle \geq 0, \quad \forall v \in K_\beta, \tag{30}$$

where $\rho > 0$ is a constant and $E'(u)$ is the differential of a strongly biconvex function $E(u)$ at $u \in K_\beta$. Since $E(u)$ is a strongly biconvex function, this implies that its differential E' is strongly β-monotone. Consequently, it follows that problem (29) has an unique solution.

Remark 5. The function $B(w, u) = E(w) - E(u) - \langle E'(u), \beta(w, u) \rangle$ associated with the biconvex function $E(u)$ is called the generalized Bregman function. By the strong biconvexity of the function $E(u)$, the Bregman function $B(.,.)$ is nonnegative and $B(w, u) = 0$, if and only if $u = w, \forall u, w \in K_\beta$. For the applications of the Bregman function in solving variational inequalities and complementarity problems, see Refs. [13–19, 23].

We note that, if $w = u$, then clearly w is a solution of the bivariational inequality (29). This observation enables us to suggest and analyze the following iterative method for solving (29).

Algorithm 1. For a given $u_0 \in H$, compute the approximate solution u_{n+1} by the iterative scheme

$$\langle \rho T u_{n+1}, \beta(v - u_{n+1}) \rangle + \langle E'(u_{n+1})$$
$$- E'(u_n), \beta(v - u_{n+1}) \rangle \geq 0, \quad \forall v \in K_\beta, \tag{31}$$

where $\rho > 0$ is a constant. Algorithm 1 is called the proximal method for solving bivariational inequalities (29). In passing, we remark that the proximal point method was suggested in the context of convex programming problems as a regularization technique.

If $\beta(v - u) = v - u$, then Algorithm 1 collapses to the following:

Algorithm 2. For a given $u_0 \in H$, compute the approximate solution u_{n+1} by the iterative scheme

$$\langle \rho T(u_{n+1}), v - u_{n+1} \rangle + \langle E'(u_{n+1}) - E'(u), v - u_{n+1} \rangle \geq 0, \quad \forall v \in K,$$

for solving the variational inequality.

For suitable and appropriate choice of the operators and the spaces, one can obtain a number of known and new algorithms for solving variational inequalities and related problems.

Theorem 16. *Let the bifunction T be pseudomonotone. Let E be differentiable higher-order strongly biconvex function with module $\nu > 0$ and Condition M hold. If $\rho\mu \leq \nu$, then the approximate solution u_{n+1} obtained from Algorithm 1 converges to a solution $u \in K$ satisfying the bivariational inequality* (29).

Proof. Let $u \in K$ be a solution of bivariational inequality (29). Then,
$$\langle Tu, \beta(v-u)\rangle \geq 0, \quad \forall v \in K_\beta,$$
implies that
$$-\langle Tv, \beta(u-v))\rangle \geq 0, \quad \forall v \in K_\beta, \tag{32}$$
since T is β-pseudomonotone.

Taking $v = u$ in (31) and $v = u_{n+1}$ in (32), we have
$$\langle \rho T(u_{n+1}), \beta(u, u-n+1)\rangle + \langle E'(u_{n+1}) - E'_k(u_n, \beta(u-u_{n+1})\rangle \geq 0. \tag{33}$$
and
$$-\langle Tu_{n+1}, \beta(u-u_{n+1})\rangle \geq 0. \tag{34}$$

We now consider the Bregman distance function
$$B(u,w) = E(u) - E(w) - \langle E'(w), \beta(u-w)\rangle \geq \nu\|\beta(v-u)\|^2, \tag{35}$$
using higher-order strongly biconvexity of E.

Now combining (33), (34) and (35), we have
$$\begin{aligned}
&B(u, u_n) - B(u, u_{n+1})\\
&= E(u_{n+1}) - E(u_n) - \langle E'(u_n), \beta(u-u_n)\rangle\\
&\quad + \langle E'(u_{n+1}), \beta(u-u_{n+1})\rangle\\
&= E(u_{n+1}) - E(u_n) - \langle E'(u_n) - E'(u_{n+1}, \beta(u-u_{n+1})\rangle\\
&\quad - \langle E'(u_n, u_{n+1} - u_n\rangle\\
&\geq \nu\|\beta(u_{n+1} - u_n)\|^2 + \langle E'(u_{n+1}) - E'(u_n), \beta(u-u_{n+1})\rangle\\
&\geq \nu\|\beta(u_{n+1} - u_n)\|^2 - \rho\langle T(u_{n+1}), \beta(u-u_{n+1})\rangle\\
&\quad - \rho\mu\|\beta(u-u_{n+1})\|^2\\
&\geq (\nu - \rho\mu)\|\beta(u_{n+1} - u_n)\|^2.
\end{aligned}$$

If $u_{n+1} = u_n$, then clearly u_n is a solution of problem (29). Otherwise, it follows that $B(u, u_n) - B(u, u_{n+1})$ is non-negative and we must have

$$\lim_{n \to \infty} \|\beta(u_{n+1} - u_n)\| = 0.$$

from which we have

$$\lim_{n \to \infty} \|u_{n+1} - u_n\| = 0.$$

It follows that the sequence $\{u_n\}$ is bounded. Let $\bar{u}$ be a cluster point of the subsequence $\{u_{n_i}\}$, and let $\{u_{n_i}\}$ be a subsequence converging toward $\bar{u}$. Now using the technique of Zhu and Marcotte [23], it can be shown that the entire sequence $\{u_n\}$ converges to the cluster point $\bar{u}$ satisfying the bivariational inequality (29). □

It is well known that to implement the proximal point methods, one has to find the approximate solution implicitly, which is itself a difficult problem. To overcome this drawback, we now consider another method for solving the bivariational inequality (29) using the auxiliary principle technique.

For a given $u \in K_\beta$, find $w \in K_\beta$ such that

$$\langle \rho T(u, \beta(v - w)) \rangle + \langle E'(w) - E', \beta(v - w) \rangle \geq 0, \quad \forall v \in K_\beta, \quad (36)$$

where $E'(u)$ is the differential of a biconvex function $E(u)$ at $u \in K_\beta$. Problem (29) has a unique solution, since E is a strongly biconvex function. Note that problems (31) and (36) are quite different problems.

It is clear that for $w = u$, w is a solution of (29). This fact allows us to suggest and analyze another iterative method for solving the bivariational inequality (29).

Algorithm 3. For a given $u_0 \in H$, compute the approximate solution u_{n+1} by the iterative scheme

$$\langle \rho T u_n, \beta(v - u_{n+1}) \rangle + \langle E'(u_{n+1}) - E'(u_n), \beta(v - u_{n+1}) \rangle \geq 0,$$
$$\forall v \in K_\beta, \quad (37)$$

for solving the bivariational inequality (29).

If $\beta(v,u)) = v - u$, Algorithm 3 collapses to the following:

Algorithm 4. For a given $u_0 \in H$, compute the approximate solution u_{n+1} by the iterative schemes

$$\rho \langle Tu_n, \beta(v - u_{n+1}) \rangle$$
$$+ \langle E'(u_{n+1}) - E'(u_n), \beta(v - u_{n+1}) \rangle \geq 0, \quad \forall v \in K_\beta,$$

for solving the bivariational inequalities which appears to be a new one.

We now again use the auxiliary principle to suggest some more iterative methods for solving bivariational inequalities.

For a given $u \in K_\beta$ satisfying (29), find $w \in K_\beta$ such that

$$\langle \rho T(w, \beta(v - w)) \rangle + \langle w - u + \alpha(u - u), v - w \rangle \geq 0, \quad \forall v \in K_\beta, \tag{38}$$

which is the auxiliary bivariational inequality. We note that, if $w = u$, w is a solution of (29). This fact allows us to suggest and analyze another iterative method for solving the bivariational inequality (29).

Algorithm 5. For a given $u_0 \in H$, compute the approximate solution u_{n+1} by the iterative schemes

$$\rho \langle Tu_{n+1}, \beta(v - u_{n+1}) \rangle$$
$$+ \langle u_{n+1} - u_n \rangle + \alpha(u_n - u_{n-1}), v - u_{n+1} \rangle \geq 0, \quad \forall v \in K_\beta, \tag{39}$$

where α is a constant. Algorithm 5 is called the inertial proximal method for solving the bivariational inequalities (29). For $\alpha = 0$, Algorithm 5 becomes the following:

Algorithm 6. For a given $u_0 \in H$, compute the approximate solution u_{n+1} by the iterative schemes

$$\rho \langle Tu_{n+1}, \beta(v - u_{n+1}) \rangle + \langle u_{n+1} - u_n \rangle, v - u_{n+1} \rangle \geq 0, \quad \forall v \in K_\beta,$$

which is called the proximal method for solving the bivariational inequalities (29).

If $\beta(.-.) = v - u$, then the biconvex set K_β becomes the convex set K. Consequently, Algorithm 5 reduces to the following:

Algorithm 7. For a given $u_0 \in H$, compute the approximate solution u_{n+1} by the iterative schemes

$$\rho\langle Tu_{n+1}, v - u_{n+1}\rangle + \langle u_{n+1} - u_n) + \alpha(u_n - u_{n-1}), v - u_{n+1}\rangle \geq 0,$$
$$\forall v \in K.$$

Algorithm 7 is known as the inertial proximal method for solving variational inequalities.

We now consider the convergence analysis of Algorithm 5.

Theorem 17. *Let $\bar{u} \in K_\beta$ be a solution of (29) and let u_{n+1} be the approximate solution obtained from Algorithm 5. If the $T : H \longrightarrow R$ is pseudo β-monotone, then*

$$\|u_{n+1} - \bar{u}\|^2 \leq \|u_n - \bar{u}\|^2 - \|u_{n+1} - u_n - \alpha_n(u_n - u_{n-1})\|^2$$
$$+ \alpha_n\{\|u_n - \bar{u}\|^2 - \|u_{n-1} - \bar{u}\|^2 + 2\|u_n - u_{n-1}\|^2\}. \quad (40)$$

Proof. Let $\bar{u} \in K_\beta$ be a solution of (29). Then,

$$\langle Tu, \beta(v - u)\rangle \geq 0, \quad \forall v \in K_\beta,$$

implies that

$$-\langle Tv, \beta(\bar{u} - v))\rangle \geq 0, \quad (41)$$

since T is pseudo β-monotone.

Taking $v = u_{n+1}$ in (41), we have

$$\langle Tu_{n+1}, \beta(\bar{u} - u_{n+1})\rangle \geq 0. \quad (42)$$

Now taking $v = \bar{u}$ in (39), we obtain

$$\langle \rho Tu_{n+1}, \beta(\bar{u} - u_{n+1})\rangle$$
$$+\langle u_{n+1} - u_n - \alpha_n(u_n - u_{n-1}), \bar{u} - u_{n+1}\rangle \geq 0. \quad (43)$$

From (42) and (43), we have

$$\langle u_{n+1} - u_n - \alpha_n(u_n - u_{n-1}), \bar{u} - u_{n+1}\rangle$$
$$\geq -\langle \rho Tu_{n+1}, \beta(\bar{u} - u_{n+1})\rangle \geq 0. \quad (44)$$

One can write (44) in the form

$$\langle u_{n+1} - u_n, \bar{u} - u_{n+1} \rangle \geq \alpha_n \langle u_n - u_{n-1}, \bar{u} - u_n + u_n - u_{n+1} \rangle. \tag{45}$$

Using the inequality $2\langle u, v \rangle = \|u + v\|^2 - \|u\|^2 - \|v\|^2, \forall u, v \in H$ and rearranging the terms in (45), one can easily obtain (40), the required result. □

Theorem 18. *Let H be a finite dimensional space. Let u_{n+1} be the approximate solution obtained from Algorithm 5 and $\bar{u} \in K_\beta$ be a solution of (29). If there exists $\alpha \in (0, 1)$ such that $0 \leq \alpha_n \leq \alpha$, $\forall n \in N$ and*

$$\sum_{n=1}^{\infty} \alpha_n \|u_n - u_{n-1}\|^2 \leq \infty,$$

then $\lim_{n \to \infty} u_n = \bar{u}$.

Proof. Let $\bar{u} \in K_\beta$ be a solution of (29). First, we consider the case $\alpha_n = 0$. In this case, we see from (40) that the sequence $\{\|\bar{u} - u_n\|\}$ is nonincreasing and consequently $\{u_n\}$ is bounded. Also from (40), we have

$$\sum_{n=0}^{\infty} \|u_{n+1} - u_n\|^2 \leq \|u_0 - \bar{u}\|^2,$$

which implies that

$$\lim_{n \to \infty} \|u_{n+1} - u_n\| = 0. \tag{46}$$

Let $\hat{u}$ be the cluster point of $\{u_n\}$ and the subsequence $\{u_{n_j}\}$ of the sequence $\{u_n\}$ converge to $\hat{u} \in H$. Replacing u_n by u_{n_j} in (39) and taking the limit $n_j \to \infty$ and using (46), we have

$$\langle T\hat{u}, \beta(v - \hat{u}) \rangle \geq 0, \quad \forall v \in K_\beta,$$

which implies that $\hat{u}$ solves the bihemivariational inequality problem (29) and

$$\|u_{n+1} - u_n\|^2 \leq \|u_n - \bar{u}\|^2.$$

Thus, it follows from the above inequality that the sequence $\{u_n\}$ has exactly one cluster point $\hat{u}$ and $\lim_{n \to \infty} u_n = \hat{u}$.

Now, we consider the case $\alpha_n > 0$. From (39), we have

$$\sum_{n+1}^{\infty} \|u_{n+1} - u_n - \alpha_n(u_n - u_{n-1})\|^2 \leq \|u_0 - \bar{u}\|^2$$

$$+ \sum_{n=1}^{\infty} \{\alpha\|u_n - \bar{u}\|^2 + 2\|u_n - u_{n-1}\|^2\} \leq \infty,$$

which implies that

$$\lim_{n \to \infty} \|u_{n+1} - u_n - \alpha_n(u_n - u_{n-1})\|^2 = 0.$$

Repeating the above arguments as in the case $\alpha_n = 0$, one can easily show that $\lim_{n\to\infty} u_n = \hat{u}$, the required result. □

For a given $u \in K$ satisfying the bivariational inequality (29), consider the auxiliary problem of finding $w \in K_\beta$ such that

$$\langle \rho T u, \beta(v - u) \rangle + \langle w - u, v - w \rangle \geq 0, \quad \forall v \in K_\beta, \qquad (47)$$

where $\rho > 0$ is a constant. Problem (47) is known as the auxiliary bivariational inequality. We note that if $w = u$, then clearly w is a solution of problem (29). This observation enables us to suggest and analyze the following iterative method for solving problem (29).

Algorithm 8. For a given $u_0 \in H$, compute the approximate solution u_{n+1} by the iterative scheme

$$\langle \rho T w_n, \beta(v - w_n) \rangle + \langle u_{n+1} - w_n, v - u_{n+1} \rangle \geq 0, \quad \forall v \in K_\beta, \qquad (48)$$

$$\langle \nu T(u_n), \beta(v - u_n) \rangle + \langle w_n - u_n, v - w_n \rangle \geq 0, \quad \forall v \in K_\beta, \qquad (49)$$

where $\rho > 0$ and $\nu > 0$ are constants. Algorithm 8 is a two-step predictor–corrector method for solving the bivariational inequalities (29).

Remark 6. For suitable and appropriate choice of the operators and the spaces, one can obtain various known and new algorithms for solving bivariational inequality (29) and related optimization problems. Convergence analysis of these new algorithms can be considered and investigated using the above techniques and ideas. It is an interesting problem from both analytical and numerical point of views.

5. Conclusion

In this chapter, we have introduced and studied some new classes of biconvex functions. These concepts are more general and unifying ones. Several new properties of these strongly biconvex functions are discussed and their relations with previously known results are highlighted. It is shown that the optimality conditions of the differentiable biconvex functions can be characterized by a class of bivariational inequalities. This result is used to introduce a more general class of bivariational inequalities (29). Auxiliary principle techniques are used to suggest and analyze some iterative methods for solving the bivariational inequalities. Convergence analysis of the proposed methods is condition using the pseudo-monotonicity which is a weaker condition than monotonicity. Our method of proofs is very simple as compared with other techniques. It is itself an interesting problem to develop some efficient numerical methods for solving bivariational inequalities along with applications in pure and applied sciences. Despite the current activities in these fields, much clearly remains to be done in these fields. It is expected that the ideas and techniques of this chapter may be the starting point for future research activities.

Acknowledgements

We wish to express our deepest gratitude to our teachers, colleagues, students, collaborators and friends, who have direct or indirect contributions to the process of this chapter.

References

[1] L. M. Bregman, The relaxation method of finding the common point of convex sets and its application to the solution of problems in convex programming, *USSR Comput. Math. Math. Phys.* **7**, 200–217 (1967).
[2] G. Cristescu and L. Lupsa, *Non-Connected Convexities and Applications* (Kluwer Academic Publisher, Dordrechet, 2002).
[3] R. Glowinski, J. L. Lions and R. Tremolieres, *Numerical Analysis of Variational Inequalities* (North-Holland, Amsterdam, 1981).
[4] J. L. Lions and G. Stampacchia, Variational inequalities, *Commun. Pure Appl. Math.* **20**, 493–519 (1967).

[5] C. P. Niculescu and L. E. Persson, *Convex Functions and Their Applications* (Springer-Verlag, New York, 2018).
[6] M. A. Noor. On variational inequalities, PhD Thesis, Brunel University, London, UK (1975).
[7] M. A. Noor, General variational inequalities, *Appl. Math. Lett.* **1**, 11–121 (1988).
[8] M. A. Noor, Variational like inequalities, *Optimization* **30**, 323–333 (1994).
[9] M. A. Noor, New approximation schemes for general variational inequalities, *J. Math. Anal. Appl.* **151**, 217–229 (2000).
[10] M. A. Noor, Some developments in general variational inequalities, *Appl. Math. Comput.* **251**, 199–277 (2004).
[11] M. A. Noor, Invex equilibrium problems, *J. Math. Anal. Appl.* **302**, 463–475 (2005).
[12] M. A. Noor, Fundamentals of equilibrium problems, *Math. Inequal. Appl.* **9**(3), 529–566 (2006).
[13] M. A. Noor and K. I. Noor, Higher order strongly exponentially biconvex functions and bivariational inequalities, *J. Math. Anal.* **12**(2) (2021).
[14] M. A. Noor and K. I. Noor, Exponentially biconvex functions and bivariational inequalities. In B. Hazarika, S. Acharjee and H. M. Srivastava (eds.) *Advances in Mathematical Analysis and Multidisciplinary Applications* (CRC Press).
[15] M. A. Noor and K. I. Noor, Higher order strongly biconvex functions and biequilibrium problems, *Adv. Linear Algebra Matrix Theory* **11**(2) (2021).
[16] M. A. Noor, K. I. Noor and E. Al-Said, Auxiliary principle technique for solving bifunction variational inequalities, *J. Optim. Theory Appl.* **149**, 441–445 (2011).
[17] M. A. Noor, K. I. Noor and M. T. Rassias, New trends in general variational inequalities, *Acta Appl. Math.* **170**(1), 981–1046 (2020).
[18] M. A. Noor, K. I. Noor and M. Th. Rassias, Strongly biconvex functions and bivariational inequalities. In P. M. Pardalos and T. M. Rassias (eds.) *Mathematical Analysis, Optimization, Approximation and Applications* (World Scientific Publishing Company, Singapore).
[19] M. A. Noor, K. I. Noor and T. M. Rassias, Some aspects of variational inequalities, *J. Appl. Math. Comput.* **47**, 485–512 (1993).
[20] J. Pecaric, F. Proschan and Y. L. Tong, *Convex Functions, Partial Orderings and Statistical Applications* (Academic Press, New York, 1992).
[21] G. Stampacchia, Formes bilineaires coercivites sur les ensembles convexes, *C. R. Acad.* (Paris) **258** 4413–4416 (1964).

[22] P. Sunthrayuth and P. Cholamjiak, Modified extragradient method with Bregman distance for variational inequalities, *Appl. Anal.* **100** (2021).
[23] D. L. Zhu and P. Marcotte, Co-coercivity and its role in the convergence of iterative schemes for solving variational inequalities, *SIAM J. Optim.* **6**, 714–726 (1966).

© 2024 World Scientific Publishing Company
https://doi.org/10.1142/9789811267048_0023

Chapter 23

Some Properties of a Class of Network Games with Strategic Complements or Substitutes

Mauro Passacantando[*,‡] and Fabio Raciti[†,§]

[*]*Department of Business and Law, University of Milan-Bicocca, Milan, Italy*

[†]*Department of Mathematics and Computer Science, University of Catania, Catania, Italy*

[‡]*mauro.passacantando@unimib.it*
[§]*fabio.raciti@unict.it*

We investigate a class of parametric network games which encompasses both the cases of strategic complements and strategic substitutes. In the case of a bounded strategy space, we derive a representation formula for the unique Nash equilibrium. We also prove a comparison result between the Nash equilibrium and the social optimum and then compute the price of anarchy for some simple test problems.

1. Introduction

This chapter investigates some aspects of a class of network games, within the framework developed in the seminal paper [3], in a socio-economic context. For an excellent review on this topic, the interested reader can refer to Ref. [10]. Here, we recall that the peculiarity of this approach is that each player is identified with the node of a graph and players that can interact directly are connected through links of the graph. The so-called *peers* of a given player can influence her action, according to their proximity in the network of relationships. The influence of peers on a given neighbor player can be of two different types. Roughly speaking, for a given player, if an increase in the action of her peers causes an increase in the player's action, we say that the peers act as strategic complements, and if an increase in the action of her peers causes a decrease in her action, we say that the peers act as strategic substitutes. In order to keep the analysis at a reasonable level of complexity, authors have mainly focused on games with strategic complements or with strategic substitutes, where the type of interaction is the same for all players. The graph structure has thus a prominent role in modeling the interactions among the various players who can represent different kinds of socio-economic agents, depending on the specific application.

As is common in social and economic game-theoretical models, two important concepts are the Nash equilibrium and the social optimum (or welfare) of the game which, in the above-mentioned papers, were connected to graph-algebraic quantities. In particular, in the case of interior solution, a very interesting representation formula has been derived in the seminal paper by Ballester *et al.* [3], which involves the so-called Katz–Bonacich centrality measure [6]. As a matter of fact, a large number of papers devoted to this topic have focused on the case of interior solution and unbounded strategy space, utilizing classical game-theoretical methods, i.e., the best response approach. Only very recently some authors have framed the topic of network games in the theory of variational inequalities, although the variational inequality approach to Nash equilibrium problems was initiated by Gabay and Moulin [9] more than forty years ago. In this respect, we refer the reader to the interesting paper by Parise and Ozdaglar [15], which although comprehensive in many respects, such as uniqueness and sensitivity of equilibrium, does not focus on the Katz–Bonacich representation of the solution or on the comparison with the social optimum. On the other hand, in Ref. [16], the authors started to generalize some classical results to the case where some components of the solution lay on the boundary, while in Ref. [17], the case of a generalized Nash equilibrium has been treated for the first time, within the network games framework. The variational inequality approach has also been applied to a game with global complementarities and global congestion in Ref. [19].

In this work, we extend the results in Ref. [16] where we considered the standard quadratic reference model with strategic complements. Specifically, this chapter is structured as follows. In Section 2, we provide some basic material on graph theory and define the class of network games with strategic complements and substitutes. Moreover, we recall the definition of Nash equilibrium of a game and its relationship with variational inequalities. Section 3 is devoted to the investigation of a class of parametric quadratic utility functions considered in Ref. [1] which encompasses both the classes of strategic complements and substitutes. For both classes, we derive a Katz–Bonacich-like representation formula in the case where the solution has some boundary components. Moreover, in the case of strategic complements, by exploiting the sequential best-response dynamics, we compare the components of the unique Nash equilibrium of the game and the unique social optimum, proving that the Nash equilibrium is component-wise less than or equal to the social optimal solution. Section 4 is devoted to illustrate our findings by means of some numerical experiments, and we also analyze the so-called *price of anarchy*. We touch upon possible future developments in the small concluding section.

2. Basics on Network Games and Variational Inequalities

In network games, players are represented by the nodes of a graph (V, E), where V is the set of nodes and E is the set of arcs formed by ordered pairs of nodes (v, w). In the case where, for all arcs in the network, (v, w) and (w, v) are the same, and there are neither multiple arcs connecting the same pair of nodes, nor loops, the graph is called undirected and simple. In our model, we allow for asymmetric relationships between pairs of players, hence we consider directed graphs.

Two nodes v and w are said to be adjacent if they are connected by an arc, i.e., if (v, w) or (w, v) is an arc. The information about the adjacency of nodes can be stored in the adjacency matrix G whose elements g_{ij} are equal to 1 if (v_i, v_j) is an arc, 0 otherwise. We also consider the more general case where each arc is given a non-negative weight w_{ij}. In this case, G is called the weighted adjacency matrix of the graph. G is thus an asymmetric and zero-diagonal matrix. Given a node v, the nodes connected to v with an arc are called the *neighbors* of v. A *walk* in the graph g is a finite sequence of the form $v_{i_0}, e_{j_1}, v_{i_1}, e_{j_2}, \ldots, e_{j_k}, v_{j_k}$, which consists of alternating nodes and arc of the graph, such that $v_{i_{t-1}}$ and v_{i_t} are end nodes of e_{j_t}. In the case of an unweighted graph, the *length* of a walk is simply the number of its arcs. Let us remark that, in a walk, it is allowed to visit a node or go through an arc more than once. The indirect connections between any two nodes in the graph are described by means of the powers of the adjacency matrix G. Indeed, for an unweighted graph, without loops and multiple arcs, it can be proved that the element $g_{ij}^{[k]}$ of G^k gives the number of walks of length k between v_i and v_j.

In the sequel, the set of players will be denoted by $\{1, 2, \ldots, n\}$ instead of $\{v_1, v_2, \ldots, v_n\}$. We denote with $A_i \subset \mathbb{R}$ the action space of player i, while $A = A_1 \times \cdots \times A_n$. A vector $x = (x_1, \ldots, x_n) \in A$ is called a *profile*. We also use the common notations $x_{-i} = (x_1, \ldots, x_{i-1}, x_{i+1}, \ldots, x_n)$ and $x = (x_i, x_{-i})$ when we wish to distinguish the action of player i from the action of all the other players. Each player i is endowed with a payoff function $u_i : A \to \mathbb{R}$ that she wishes to maximize. The notation $u_i(x, G)$ is often utilized when one wants to emphasize that the utility of player i also depends on the actions taken by her neighbors in the graph.

We now recall the definition of a Nash equilibrium, which is one of the most common solution concepts in game theory.

Definition 1. An action profile $x^* \in A$ is a Nash equilibrium iff for each $i \in \{1, \ldots, n\}$

$$u_i(x_i^*, x_{-i}^*) \geq u_i(x_i, x_{-i}^*), \quad \forall \, x_i \in A_i. \tag{1}$$

Another quantity of interest, in particular in socio-economic application, is the *Welfare* associated with each action profile:

$$W(x) := \sum_{i=1}^{n} u_i(x). \tag{2}$$

In the case where the function W has a unique maximizer x^{so} over A (called social optimum), and the game has a unique Nash equilibrium x^*, it is interesting to compute the following ratio:

$$\gamma = \frac{W(x^*)}{W(x^{so})}, \qquad (3)$$

which, in similar models, is known as the *price of anarchy* (see, e.g., Ref. [20]).

As mentioned in the introduction, it is convenient, for tractability reasons, to consider games where the neighbors of a player influence the player's behavior in the same direction for all players. We make this concept precise with the help of the marginal utility function.

Definition 2. *The network game has the property of strategic complements if*

$$\frac{\partial^2 u_i}{\partial x_j \partial x_i}(x) > 0, \quad \forall\, (i,j) : g_{ij} \neq 0, \quad \forall\, x \in A.$$

Definition 3. *The network game has the property of strategic substitutes if*

$$\frac{\partial^2 u_i}{\partial x_j \partial x_i}(x) < 0, \quad \forall\, (i,j) : g_{ij} \neq 0, \quad \forall\, x \in A.$$

The variational inequality approach to Nash equilibrium problems is recalled in the following theorem. For an account of variational inequalities, the interested reader can refer to Ref. [12, 14].

Theorem 1. *For each $i \in \{1,\ldots,n\}$, let u_i be a continuously differentiable function on A and $u_i(\cdot, x_{-i})$ be concave with respect to its own action x_i, for each $x_{-i} \in A_{-i}$. Moreover, let A be closed and convex. Then, x^* is a Nash equilibrium if and only if it solves the variational inequality $VI(T, A)$: find $x^* \in A$ such that*

$$T(x^*)^\top (x - x^*) \geq 0, \quad \forall\, x \in A, \qquad (4)$$

where the operator

$$[T(x)]^\top := -\left(\frac{\partial u_1}{\partial x_1}(x), \ldots, \frac{\partial u_n}{\partial x_n}(x)\right) \qquad (5)$$

is also called the pseudo-gradient of the game.

We recall here some useful monotonicity properties.

Definition 4. An operator $T : \mathbb{R}^n \to \mathbb{R}^n$ is said to be monotone on A iff
$$[T(x) - T(y)]^\top (x - y) \geq 0, \quad \forall\, x, y \in A.$$
If the equality holds only when $x = y$, T is said to be strictly monotone on A.

T is said to be τ-strongly monotone on A iff there exists $\tau > 0$ such that
$$[T(x) - T(y)]^\top (x - y) \geq \tau \|x - y\|^2, \quad \forall\, x, y \in A.$$

Remark 1. For linear operators on $\mathbb{R}^n$, the two concepts of strict and strong monotonicity coincide and are equivalent to the positive definiteness of the corresponding matrix.

Conditions that ensure the unique solvability of a variational inequality problem are given by the following theorem (see, e.g., Ref. [14]).

Theorem 2. *If $K \subset \mathbb{R}^n$ is a compact convex set and $T : \mathbb{R}^n \to \mathbb{R}^n$ is continuous on K, then the variational inequality problem $VI(T, K)$ admits at least one solution. In the case that K is unbounded, existence of a solution may be established under the following coercivity condition:*
$$\lim_{\|x\| \to +\infty} \frac{[T(x) - T(x_0)]^\top (x - x_0)}{\|x - x_0\|} = +\infty,$$
for $x \in K$ and some $x_0 \in K$.

Furthermore, the solution is unique if T is strictly monotone on K.

3. The Parametric Quadratic Model

Let $A_i = [0, L_i]$ for any $i \in \{1, \ldots, n\}$, hence $A = [0, L_1] \times \cdots \times [0, L_n]$. The payoff of player i is given by
$$u_i(x) = -\frac{\beta}{2} x_i^2 + \alpha_i x_i + \sum_{\substack{j=1 \\ j \neq i}}^n f_{ij}(\alpha) x_i x_j, \quad \alpha, \beta > 0. \tag{6}$$

The last term describes the interaction between player i and her neighbors. The coefficient α_i describes the type of agent and in some economic applications (see, e.g., Ref. [7]) can be interpreted as the household's parental capital. If, for all $j \neq i$ and a fixed value of $\alpha = (\alpha_1, \ldots, \alpha_n)$, $f_{ij}(\alpha) \geq 0$ holds, then the associated game falls in the class of games with strategic complements; if, for all $j \neq i$ and a fixed value of α, $f_{ij}(\alpha) \leq 0$ holds, it falls in the class of games with strategic substitutes. The pseudo-gradient's components of this game are easily computed as follows:

$$T_i(x) = \beta x_i - \alpha_i - \sum_{\substack{j=1 \\ j \neq i}}^{n} f_{ij}(\alpha) x_j, \quad i \in \{1, \ldots, n\},$$

which can be written in compact form as

$$T(x) = [\beta I - \mathcal{F}(\alpha)] x - \alpha, \tag{7}$$

where $\mathcal{F}(\alpha)$ is a zero-diagonal matrix whose off-diagonal entries are equal to $f_{ij}(\alpha)$ and is called the *interaction matrix*.

Throughout this chapter, we posit the symmetry assumption on the interaction matrix:

$$f_{ij}(\alpha) = f_{ji}(\alpha), \quad \forall\, \alpha,\ \forall\, i, j \in \{1, \ldots, n\},\ i \neq j. \tag{S}$$

Remark 2. Under the symmetry assumption (S), the game under consideration also falls in the class of potential games according to the definition introduced by Monderer and Shapley [13]. Indeed, a potential function is given by

$$P(x) = \sum_{i=1}^{n} u_i(x) - \frac{1}{2} \sum_{i=1}^{n} \sum_{\substack{j=1 \\ j \neq i}}^{n} f_{ij}(\alpha) x_i x_j.$$

Applying a result of Monderer and Shapley to our case, we obtain in general that the solutions of the problem $\max_{x \in A} P(x)$ form a subset of the solution set of the Nash game. Therefore, if both problems have a unique solution, it follows that they are equivalent.

We seek Nash equilibrium points by solving the variational inequality:

$$T(x^*)^\top (x - x^*) \geq 0, \quad \forall\, x \in A. \tag{8}$$

The following lemma characterizes the monotonicity of F given in (7).

Lemma 1.

(a) *Fix $\alpha > 0$ and let $f_{ij}(\alpha) \geq 0$ for any i,j. The matrix $\beta I - \mathcal{F}(\alpha)$ is positive definite iff*

$$\beta > \lambda_{\max}(\mathcal{F}(\alpha)) = \rho(\mathcal{F}(\alpha)), \qquad (9)$$

where $\lambda_{\max}(\mathcal{F}(\alpha))$ is the maximum eigenvalue of $\mathcal{F}(\alpha)$ and $\rho(\mathcal{F}(\alpha))$ its spectral radius.

(b) *Let $f_{ij}(\alpha) \leq 0$ for any i,j. The matrix $\beta I - \mathcal{F}(\alpha)$ is positive definite iff*

$$\beta > \lambda_{\max}(\mathcal{F}(\alpha)) \qquad (10)$$

or, equivalently, $\lambda_{\min}(-\mathcal{F}(\alpha)) > -\beta$.

Moreover, the condition $\beta > \rho(\mathcal{F}(\alpha))$ is, in general, stronger than the two equivalent conditions above.

Proof.

(a) Let $f_{ij}(\alpha) \geq 0$ and recall that if M is a non-negative symmetric matrix, a consequence of the Perron–Frobenius theorem is that $\rho(M) = \lambda_{\max}(M)$. Furthermore,

$$\lambda_{\max}(M) = \max_{x \neq 0} \frac{x^\top M x}{x^\top x}.$$

The matrix $\beta I - \mathcal{F}(\alpha)$ is positive definite iff $x^\top [\beta I - \mathcal{F}(\alpha)] x > 0$ for any $x \neq 0$, that is,

$$\beta > \frac{x^\top \mathcal{F}(\alpha) x}{x^\top x}, \quad \forall\, x \neq 0,$$

which is equivalent to $\beta > \lambda_{\max}(\mathcal{F}(\alpha))$, and, as a consequence of the Perron–Frobenius theorem, we finally get that $\beta I - \mathcal{F}(\alpha)$ is positive definite iff $\beta > \lambda_{\max}(\mathcal{F}(\alpha)) = \rho(\mathcal{F}(\alpha))$.

We note that this condition also ensures that the matrix $I - \frac{1}{\beta}\mathcal{F}(\alpha)$ is non-singular and its inverse matrix can be expanded in a power series according to Lemma 2.

(b) Following the same reasoning as in the non-negative case, we get that the matrix $\beta I - \mathcal{F}(\alpha)$ is positive definite if and only if $\beta > \max_{x \neq 0} \frac{x^\top \mathcal{F}(\alpha) x}{x^\top x} = \lambda_{\max}(\mathcal{F}(\alpha)) = -\lambda_{\min}(-\mathcal{F}(\alpha))$. However, the condition $\beta > \rho(\mathcal{F}(\alpha))$ is stronger because

$$\rho(\mathcal{F}(\alpha)) \geq |\lambda_{\max}(\mathcal{F}(\alpha))| \geq \lambda_{\max}(\mathcal{F}(\alpha)).$$

$\square$

In the following lemma, we recall a well-known result about series of matrices.

Lemma 2 (see, e.g., Ref. [2]). Let M be a square matrix and consider the series $\sum_{p=0}^{\infty} M^p$. The series converges provided that $\lim_{p \to \infty} M^p = 0$, which is equivalent to $\rho(M) < 1$. In such case, the matrix $I - M$ is non-singular and we have the power series expansion $(I - M)^{-1} = \sum_{p=0}^{\infty} M^p$.

We now introduce a centrality measure of networks, known as the Katz–Bonacich vector (see, e.g., Ref. [6]), which allows for an interesting interpretation of the Nash equilibrium of network games. Although we confine our analysis to the symmetric case, we give here the definition for the general case of a general matrix G, with entries g_{ij}. Such a matrix can be thought of as the adjacency matrix of a weighted directed graph. The case of an undirected network without self-loops is characterized by $g_{ij} = g_{ji}$, $j \neq i$, $g_{ii} = 0$, and if G is a $0-1$ matrix, the graph in unweighted.

Definition 5. Let w be a non-negative vector. The weighted vector of Katz–Bonacich, of parameter ϕ, in the graph is given by

$$b_w(G, \phi) = [I - \phi G]^{-1} w = \sum_{p=0}^{\infty} \phi^p G^p w. \tag{11}$$

The inverse exists and can be expressed by the series above if the condition $\phi \rho(G) < 1$ is satisfied. We also recall that, if $G \geq 0$, a theorem on non-negative matrices ensures that $[I - \phi G]^{-1}$ is non-negative too. In the simplest case of a $0-1$ adjacency matrix, indeed, the (i, j) entry, $g_{ij}^{[p]}$, of the matrix G^p gives the number of walks of length p between nodes i and j, and if $w = (1, \ldots, 1)$, $b_{w,i}(G, \phi)$ counts the total number of walks in the graph, which start at node i, exponentially damped by ϕ. In the general case, the weights of the links are taken into account, and paths reaching an arbitrary node j are pondered by w_j.

The importance of the Katz–Bonacich vector stems from the fact that, when the strategy space is $\mathbb{R}_+^n$, it is related in a simple manner to the unique Nash equilibrium of the game. Indeed, the relation given in Ref. [3] can be extended in a straightforward fashion to the case of the utility functions (6) as follows.

Theorem 3. *Let $A = \mathbb{R}_+^n$ and consider the utility functions defined in (6), with $f_{ij} \geq 0$ for any i, j. Moreover, let $\beta > \rho(\mathcal{F}(\alpha))$. Then, the unique Nash equilibrium x^* is interior and given by*

$$x^* = \frac{1}{\beta}\left[I - \frac{1}{\beta}\mathcal{F}(\alpha)\right]^{-1}\alpha = \sum_{p=0}^{\infty}\frac{1}{\beta^{p+1}}[\mathcal{F}(\alpha)]^p\alpha = \frac{1}{\beta}b_w\left(\mathcal{F}(\alpha), \frac{1}{\beta}\right). \tag{12}$$

We now recall a proposition due to Ref. [1] which, under some additional assumptions, provides a sufficient condition for $\beta > \rho(\mathcal{F}(\alpha))$ to be true. This condition involves a smallness condition on the variance of α and, roughly speaking, means that a low variability of the types of players, given by α, entails a unique Nash equilibrium.

Proposition 1. *For a given type profile α, consider the game defined in (6) with interaction terms given by $f_{ij}(\alpha) = h_i(\alpha)f(\alpha_i - \alpha_j)$ or $f_{ij}(\alpha) = h_i(\alpha)f(|\alpha_i - \alpha_j|)$, where $f : \mathbb{R} \to \mathbb{R}$ is non-expansive and there is $\delta_0 \in \mathbb{R}$ such that $f(\delta_0) = 0$. Then, this is a game with complementarities that admits a unique equilibrium if the standard deviation of types σ_α satisfies the following inequality:*

$$\beta > n\, h(k)\,(\sqrt{2}\,\sigma_\alpha + |\delta_0|),$$

where $h(k) = \max_{i=1,\ldots,n}|h_i(\alpha)|$.

Proof. The proof can be found in Ref. [1] but we warn the reader that the formula therein differs from ours for the missing coefficient n multiplying δ_0 (probably due to a misprint). □

We now assume that the strategies of each player have an upper bound and derive a Katz–Bonacich-type representation of the solution, in the case where exactly k components take on their maximum value.

Theorem 4. *Let u_i be defined as in (6), $\beta > \rho(\mathcal{F}(\alpha))$, $x_i \in [0, L_i]$ for any $i \in \{1, \ldots, n\}$ and x^* be the unique Nash equilibrium of the game:*

(a) *Assume that $f_{ij}(\alpha) \geq 0$ for any $i, j \in \{1, \ldots, n\}$. We then have that $x_i^* > 0$ for any $i \in \{1, \ldots, n\}$. Moreover, assume that exactly k components of x^* take on their maximum value: $x_{i_1}^* = L_{i_1}, \ldots, x_{i_k}^* = L_{i_k}$, and denote with $\tilde{x}^* = (\tilde{x}_{i_{k+1}}^*, \ldots, \tilde{x}_{i_n}^*)$ the subvector of the non-boundary components of x^*. We then get*

$$\tilde{x}^* = [\beta I_{n-k} - \mathcal{F}_1(\alpha)]^{-1} w = b_w\left(\mathcal{F}_1(\alpha), \frac{1}{\beta}\right), \quad (13)$$

where $\mathcal{F}_1(\alpha)$ is the submatrix obtained from $\mathcal{F}(\alpha)$ choosing the rows $i_{k+1}, \ldots, i_n$ and the columns $i_{k+1}, \ldots, i_n$; $\mathcal{F}_2(\alpha)$ is the submatrix obtained from $\mathcal{F}(\alpha)$ choosing the rows $i_{k+1}, \ldots, i_n$ and the columns $i_1, \ldots, i_k$; $w = \alpha_{n-k} + \mathcal{F}_2(\alpha) L$ and $L = (L_{i_1}, \ldots, L_{i_k})$, $\alpha_{n-k} = (\alpha_{i_{k+1}}, \ldots, \alpha_{i_n})$.

(b) *Assume now that $f_{ij}(\alpha) \leq 0$ for any $i, j \in \{1, \ldots, n\}$, and there are no zero components of the solution x^*, while exactly k components of x^* take on their maximum value. Then, formula (13) also applies to this case. Moreover, $\tilde{x}^*$ can alternatively be expressed as*

$$\tilde{x}^* = \frac{1}{\beta}\left[I_{n-k} + \frac{1}{\beta}\mathcal{F}_1(\alpha)\right] b_w\left(\mathcal{F}_1^2(\alpha), \frac{1}{\beta^2}\right). \quad (14)$$

Proof.

(a) The Nash equilibrium x^* of the game solves the variational inequality

$$\sum_{i=1}^{n} T_i(x^*)^\top (x_i - x_i^*) \geq 0, \quad \forall\, x \in A, \quad (15)$$

where $A = [0, L_1] \times \cdots \times [0, L_n]$. Let us assume that there exists l such that $x_l^* = 0$, and choose in (15) $x = (x_1^*, \ldots, x_{l-1}^*, L_l, x_{l+1}^*, \ldots, x_n^*) \in A$. With this choice, (15) reads

$$0 \leq T_l(x^*)x_l = \left(-\sum_{j \neq l}^{n} f_{lj}(\alpha)x_j^* - \alpha_i\right) L_l < 0$$

which yields a contradiction. Thus, $x_i^* > 0$ for any $i = 1, \ldots, n$. Let $\tilde{A}$ denote the face of A obtained intersecting A with the hyperplanes: $x_{i_1} = L_{i_1}, \ldots, x_{i_k} = L_{i_k}$. Moreover, let $\tilde{x} = (x_{i_{k+1}}, \ldots, x_{i_n})$, $\tilde{x}^* = (\tilde{x}_{i_{k+1}}^*, \ldots, \tilde{x}_{i_n}^*)$ and $\tilde{T} : \mathbb{R}^{n-k} \to \mathbb{R}^{n-k}$ such that $\tilde{T}_{i_l}(\tilde{x})$ is obtained by fixing $x_{i_1} = L_{i_1}, \ldots, x_{i_k} = L_{i_k}$ in $T_{i_l}(a)$. We consider now the restriction of (15) to $\tilde{A}$, which reads

$$\sum_{l=k+1}^{n} \tilde{T}_{i_l}(\tilde{x}^*)(\tilde{x}_{i_l} - \tilde{x}_{i_l}^*) \geq 0, \quad \forall\, \tilde{x} \in \tilde{A}.$$

Since we are assuming that exactly k components of the solution x^* reach their upper bounds, it follows that $\tilde{x}^*$ lies in the interior of $\tilde{A}$, hence

$$\tilde{T}(\tilde{x}^*) = 0,$$

which can be written explicitly as

$$\beta x_{i_l}^* - \sum_{m=k+1}^{n} f_{i_l i_m}(\alpha) x_{i_m}^*$$

$$= \alpha_{i_l} + \sum_{m=1}^{k} f_{i_l i_m}(\alpha) L_{i_m}, \quad l = k+1, \ldots, n,$$

which yields

$$[\beta I_{n-k} - \mathcal{F}_1(\alpha)]\tilde{x}^* = \alpha_{n-k} + \mathcal{F}_2(\alpha)L. \tag{16}$$

Since the matrix $[\beta I_{n-k} - \mathcal{F}_1(\alpha)]$ is non-singular, the thesis is proved.

(b) To prove (14), divide both sides of (16) by β and multiply by the matrix $[I_{n-k} + \frac{1}{\beta}\mathcal{F}_1(\alpha)]$ to get

$$\left[I_{n-k} - \frac{1}{\beta^2}\mathcal{F}_1^2(\alpha)\right] = \frac{1}{\beta}\left[I_{n-k} + \frac{1}{\beta}\mathcal{F}_1(\alpha)\right](\alpha_{n-k} + \mathcal{F}_2(\alpha)L),$$

whence

$$\tilde{x}^* = \frac{1}{\beta}\left[I_{n-k} - \frac{1}{\beta^2}\mathcal{F}_1^2(\alpha)\right]^{-1}\left[I_{n-k} + \frac{1}{\beta}\mathcal{F}_1(\alpha)\right](\alpha_{n-k} + \mathcal{F}_2(\alpha)L)$$

$$= \left[I_{n-k} + \frac{1}{\beta}\mathcal{F}_1(\alpha)\right]\left[I_{n-k} - \frac{1}{\beta^2}\mathcal{F}_1^2(\alpha)\right]^{-1}(\alpha_{n-k} + \mathcal{F}_2(\alpha)L)$$

$$= \frac{1}{\beta}\left[I_{n-k} + \frac{1}{\beta}\mathcal{F}_1(\alpha)\right]b_w\left(\mathcal{F}_1^2(\alpha), \frac{1}{\beta^2}\right).$$

Formula (14) admits the following interpretation, which is better illustrated in case of interior solution, where it reads

$$x^* = \frac{1}{\beta}\left[I + \frac{1}{\beta}\mathcal{F}(\alpha)\right]b_w\left(\mathcal{F}^2(\alpha), \frac{1}{\beta^2}\right).$$

Indeed, it is evident in this case that our solution is obtained by transforming, through the matrix $[I + \frac{1}{\beta}\mathcal{F}(\alpha)]$, the solution of an auxiliary game with strategic complements associated with the interaction matrix $\mathcal{F}^2(\alpha)$. □

The following result shows a relationship between the social optimum and the Nash equilibrium of the game, in the case of strategic complements.

Theorem 5. *Assume that u_i are defined as in (6), $f_{ij}(\alpha) \geq 0$ for any $i,j \in \{1,\ldots,n\}$, $\beta > 2\rho(\mathcal{F}(\alpha))$, and $x_i \in [0, L_i]$ for any $i \in \{1,\ldots,n\}$. Then,*

$$x_i^* \leq x_i^{so} \quad \forall\, i \in \{1,\ldots,n\}, \tag{17}$$

where x^* is the Nash equilibrium and x^{so} is the social optimum of the game.

Proof. Since $\beta > 2\rho(\mathcal{F}(\alpha))$ and the welfare function reads

$$W(x) = -\frac{1}{2} x^\top [\beta I - 2\mathcal{F}(\alpha)] x + \alpha^\top x,$$

Lemma 1 guarantees that there exists a unique Nash equilibrium x^* and a unique social optimum x^{so}. Moreover, x^{so} satisfies the following Karush–Kuhn–Tucker system for some multiplier vectors $\lambda, \mu \in \mathbb{R}_+^n$:

$$\beta x_i^{so} - 2 \sum_{\substack{j=1 \\ j \neq i}}^n f_{ij}(\alpha) x_j^{so} - \alpha_i - \lambda_i + \mu_i = 0 \quad i = 1, \ldots, n,$$

$$x_i^{so} \geq 0, \quad \lambda_i \geq 0, \quad \lambda_i x_i^{so} = 0 \qquad i = 1, \ldots, n,$$

$$x_i^{so} \leq L_i, \quad \mu_i \geq 0, \quad \mu_i (x_i^{so} - L_i) = 0 \qquad i = 1, \ldots, n.$$

It is easy to check that the above system is equivalent to the following system:

$$x_i^{so} = \max \left\{ 0, \ \min \left\{ L_i, \frac{1}{\beta} \left[\alpha_i + 2 \sum_{\substack{j=1 \\ j \neq i}}^n f_{ij}(\alpha) x_j^{so} \right] \right\} \right\}$$

$$= \min \left\{ L_i, \frac{1}{\beta} \left[\alpha_i + 2 \sum_{\substack{j=1 \\ j \neq i}}^n f_{ij}(\alpha) x_j^{so} \right] \right\} \quad i = 1, \ldots, n,$$

where the last equality holds since $\alpha, \beta > 0$ and $f_{ij}(\alpha), x^{so} \geq 0$.

Given any strategy profile $x = (x_i, x_{-i})$, the best response of player i to rivals' strategies x_{-i} is given by

$$B_i(x_{-i}) = \arg \max_{x_i \in [0, L_i]} u_i(\cdot, x_{-i}) = \min \left\{ L_i, \frac{1}{\beta} \left[\alpha_i + \sum_{\substack{j=1 \\ j \neq i}}^n f_{ij}(\alpha) x_j \right] \right\}.$$

We now consider the sequential best response dynamics starting from the social optimum x^{so} that is the sequence $\{x^k\}$ defined as follows:

$$x^0 = x^{so},$$
$$x^1 = \left(B_1(x^0_{-1}),\ x^0_2,\ x^0_3, \ldots, x^0_n\right),$$
$$x^2 = \left(B_1(x^0_{-1}),\ B_2(x^1_{-2}),\ x^0_3, \ldots, x^0_n\right),$$
$$\ldots$$
$$x^n = \left(B_1(x^0_{-1}),\ B_2(x^1_{-2}),\ B_3(x^2_{-3}), \ldots, B_n(x^{n-1}_n)\right),$$
$$x^{n+1} = \left(B_1(x^n_{-1}),\ B_2(x^1_{-2}), \ldots, B_n(x^{n-1}_n)\right),$$
$$x^{n+2} = \left(B_1(x^n_{-1}),\ B_2(x^{n+1}_{-2}),\ B_3(x^2_{-3}), \ldots, B_n(x^{n-1}_n)\right), \ldots.$$

We note that

$$x_1^1 = B_1\left(x^0_{-1}\right) = \min\left\{L_1,\ \frac{1}{\beta}\left[\alpha_1 + \sum_{\substack{j=1 \\ j\neq 1}}^n f_{1j}(\alpha)x_j^0\right]\right\}$$

$$\leq \min\left\{L_1,\ \frac{1}{\beta}\left[\alpha_1 + 2\sum_{\substack{j=1 \\ j\neq 1}}^n f_{1j}(\alpha)x_j^0\right]\right\} = x_1^0,$$

hence $x^1 \leq x^0$. Moreover, we have

$$x_2^2 = B_2\left(x^1_{-2}\right) = \min\left\{L_2,\ \frac{1}{\beta}\left[\alpha_2 + \sum_{\substack{j=1 \\ j\neq 2}}^n f_{2j}(\alpha)x_j^1\right]\right\}$$

$$\leq \min\left\{L_2,\ \frac{1}{\beta}\left[\alpha_2 + \sum_{\substack{j=1 \\ j\neq 2}}^n f_{2j}(\alpha)x_j^0\right]\right\}$$

$$\leq \min\left\{L_2,\ \frac{1}{\beta}\left[\alpha_2 + 2\sum_{\substack{j=1 \\ j\neq 2}}^n f_{2j}(\alpha)x_j^0\right]\right\} = x_2^0 = x_2^1,$$

hence $x^2 \leq x^1$. Similarly, we can prove that $x^n \leq x^{n-1} \leq \cdots \leq x^1 \leq x^0$. Furthermore, we get

$$x_1^{n+1} = B_1(x_{-1}^n) = \min\left\{L_1, \frac{1}{\beta}\left[\alpha_1 + \sum_{\substack{j=1 \\ j\neq 1}}^n f_{1j}(\alpha)x_j^n\right]\right\}$$

$$\leq \min\left\{L_1, \frac{1}{\beta}\left[\alpha_1 + \sum_{\substack{j=1 \\ j\neq 1}}^n f_{1j}(\alpha)x_j^0\right]\right\}$$

$$= B_1\left(x_{-1}^0\right) = x_1^n,$$

hence $x^{n+1} \leq x^n$, and

$$x_2^{n+2} = B_2(x_{-2}^{n+1}) = \min\left\{L_2, \frac{1}{\beta}\left[\alpha_2 + \sum_{\substack{j=1 \\ j\neq 2}}^n f_{2j}(\alpha)x_j^{n+1}\right]\right\}$$

$$\leq \min\left\{L_2, \frac{1}{\beta}\left[\alpha_2 + \sum_{\substack{j=1 \\ j\neq 2}}^n f_{2j}(\alpha)x_j^1\right]\right\}$$

$$= B_2\left(x_{-2}^1\right) = x_2^{n+1},$$

thus $x^{n+2} \leq x^{n+1}$. Following the same argument as before, we can prove that $x^{k+1} \leq x^k$ for any $k \in \mathbb{N}$ and hence, in particular, $x^k \leq x^{so}$ holds for any k. Since the potential function P is strongly concave, the sequence $\{x^k\}$ converges to the unique Nash equilibrium x^* (see, e.g., Ref. [5, Proposition 3.9]), and hence $x^* \leq x^{so}$. □

We remark that inequality (17) does not hold in general in the case of strategic substitutes, as the example in the following section shows.

4. Numerical Experiments

In this section, we show a numerical example for the parametric quadratic game described in Section 3.

Example 1. We consider a game with $n = 5$ players, where $L_i = L = 1$ for any $i \in \{1, \ldots, n\}$, $\alpha = (1, 2, 1, 2, 1)$, $\beta = 2.5$ and the interaction matrix is given by

$$f_{ij}(\alpha) = B|\alpha_i - \alpha_j| \quad \forall\, i, j = 1, \ldots, n.$$

We consider two cases: $B = 0.5$ (strategic complements) and $B = -0.5$ (strategic substitutes). In both cases, the spectral radius of the matrix $\mathcal{F}(\alpha)$ results to be $\rho(\mathcal{F}(\alpha)) \simeq 1.2247$. Since $\beta > 2\rho(\mathcal{F}(\alpha))$, there exists a unique Nash equilibrium and a unique social optimum. Table 1 shows the unconstrained Nash equilibrium (assuming $L = +\infty$, given by formula (12)), the constrained Nash equilibrium (assuming $L = 1$) and the social optimum in the case $B = 0.5$.

Figure 1 shows the price of anarchy of the Nash equilibrium for different values of L and β, in the case $B = 0.5$.

The results suggest that the price of anarchy is a non-increasing function of L; it is constant when either L is small enough (i.e., the Nash equilibrium coincides with the social optimum) or greater than some threshold (i.e., the Nash equilibrium and the social optimum are both interior to the feasible region); the larger the value of β, the larger the asymptotic value of the price of anarchy.

Table 1. Case $B = 0.5$: Unconstrained Nash equilibrium, constrained Nash equilibrium (assuming $L = 1$) and social optimum for Example 1.

Player	Unconstrained NE	Constrained NE	Social Optimum
1	0.9474	0.8000	1.0000
2	1.3684	1.0000	1.0000
3	0.9474	0.8000	1.0000
4	1.3684	1.0000	1.0000
5	0.9474	0.8000	1.0000

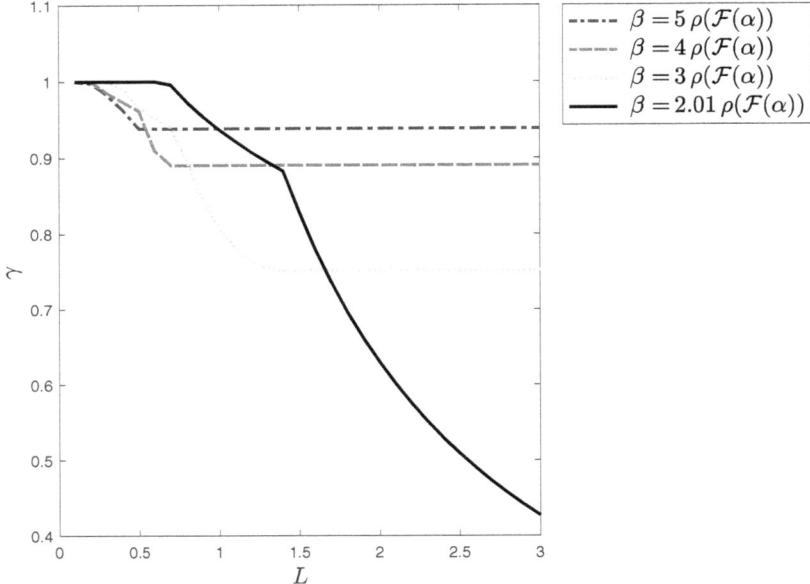

Figure 1. Case $B = 0.5$: Price of anarchy for different values of L and β.

Table 2. Case $B = -0.5$: Unconstrained Nash equilibrium, constrained Nash equilibrium (assuming $L = 1$) and social optimum for Example 1.

Player	Unconstrained NE	Constrained NE	Social Optimum
1	0.1053	0.1053	0.0000
2	0.7368	0.7368	0.8000
3	0.1053	0.1053	0.0000
4	0.7368	0.7368	0.8000
5	0.1053	0.1053	0.0000

The case $B = -0.5$ with strategic substitutes is analyzed in Table 2 and Figure 2.

In particular, Table 2 shows that in the case of strategic substitutes, neither the inequality (17) between the Nash equilibrium and the social optimum nor the opposite inequality applies.

On the other hand, Figure 2 suggests that in the case of strategic substitutes, the price of anarchy is not in general a non-increasing

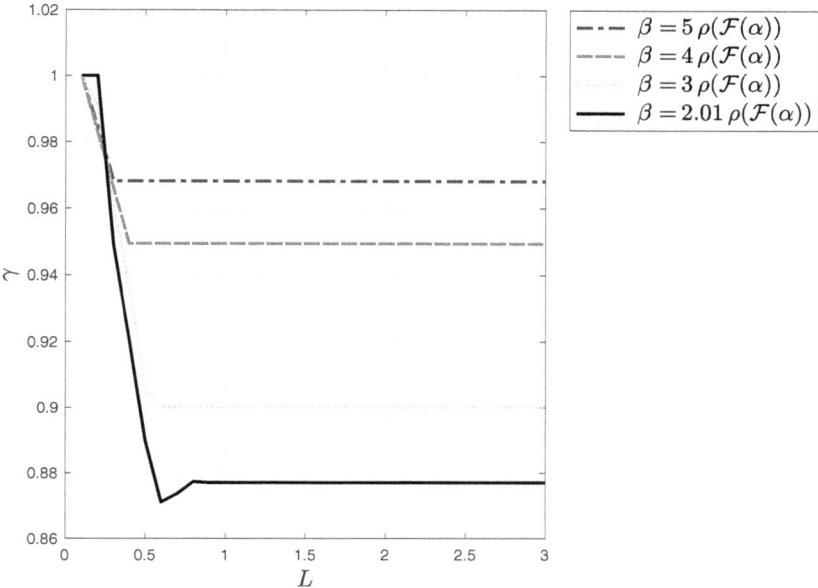

Figure 2. Case $B = -0.5$: Price of anarchy for different values of L and β.

function of L (contrary to the case of strategic complements), while, as in the case of strategic complements, the price of anarchy is constant when either L is small enough or greater than some threshold, and the larger the value of β, the larger its asymptotic value.

5. Conclusions and Future Research Perspectives

In this work, we carried on our research program of applying the variational inequality approach to network game problems. We investigated a class of parametric quadratic utility functions for which we obtained a Katz–Bonacich-like representation formula for the unique solution and studied both theoretically and numerically the price of anarchy. Future research will concern the investigation of games with nonlinear utility function and the inclusion of random data in the model (see, e.g., Refs. [11, 18]). Moreover, the topic of generalized Nash equilibrium problems on networks will be investigated using the theoretical results developed in Ref. [8].

Acknowledgements

This research was partially supported by the research project "Programma ricerca di ateneo UNICT 2020-22 linea 2-OMNIA" of the University of Catania. This support is gratefully acknowledged. The authors are members of the Gruppo Nazionale per l'Analisi Matematica, la Probabilita e le loro Applicazioni (GNAMPA — National Group for Mathematical Analysis, Probability and their Applications) of the Istituto Nazionale di Alta Matematica (INdAM — National Institute of Higher Mathematics).

References

[1] C. Ballester and A. Calvo-Armengol, Interactions with hidden complementarities, *Reg. Sci. Urban Econ.* **40**, 397–406 (2010).
[2] K. Atkinson and W. Han, *Theoretical Numerical Analysis* (Springer, New York, NY, 2007).
[3] C. Ballester, A. Calvo-Armengol and Y. Zenou, Who's who in networks. Wanted: The key player, *Econometrica* **74**, 1403–1417 (2006).
[4] Y. Bramoullé and R. Kranton, Public goods in networks, *J. Econ. Theory* **135**, 478–494 (2007).
[5] D. P. Bertsekas and J. N. Tsitsiklis, *Parallel and Distributed Computation: Numerical Methods* (Athena Scientific, Belmont, Massachusetts 1997).
[6] P. Bonacich, Power and centrality: A family of measures, *Am. J. Soc.* **92**, 1170–1182 (1987).
[7] G. J. Borjas, To Ghetto or not to ghetto: Ethnicity and residential segregation, *J. Urban Econ.* **44**, 228–253 (1998).
[8] F. Faraci and F. Raciti, On generalized Nash equilibrium problems in infinite dimension: The Lagrange multipliers approach, *Optimization* **64**, 321–338 (2015).
[9] D. Gabay and H. Moulin, On the uniqueness and stability of Nash equilibria in noncooperative games. In A. Bensoussan, P. Kleindorfer and C. S. Tapiero (eds.) *Applied Stochastic Control in Econometrics and Management Science* (North-Holland, Amsterdam, 1980), pp. 271–294.
[10] M. O. Jackson and Y. Zenou, Games on networks. In *Handbook of Game Theory with Economic Applications* (Elsevier, Amsterdam, Netherlands, 2015), pp. 95–163.

[11] B. Jadamba and F. Raciti, On the modelling of some environmental games with uncertain data, *J. Optim. Theory Appl.* **167**, 959–968 (2015).
[12] I. Konnov, *Equilibrium Models and Variational Inequalities* (Elsevier, Amsterdam, Netherlands, 2007).
[13] D. Monderer and L. S. Shapley, Potential games, *Games Econ. Behav.* **14**, 124–143 (1996).
[14] A. Nagurney, *Network Economics A Variational Inequality Approach* (Springer, Dordrecht, 1999).
[15] F. Parise and A. Ozdaglar, A variational inequality framework for network games: Existence, uniqueness, convergence and sensitivity analysis, *Games Econ. Behav.* **114**, 47–82 (2019).
[16] M. Passacantando and F. Raciti, A note on network games with strategic complements and the Katz-Bonacich centrality measure. In R. Cerulli *et al.* (eds.) *Optimization and Decision Science.* AIRO Springer Series, Vol. 7 (Springer, Cham, Switzerland, 2021), pp. 51–61.
[17] M. Passacantando and F. Raciti, A note on generalized Nash games played on networks. In T. M. Rassias (ed.) *Nonlinear Analysis, Differential Equations, and Applications.* Springer Optimization and Its Applications, Vol. 173 (Springer, Cham, 2021), pp. 365–380.
[18] M. Passacantando and F. Raciti, Optimal road maintenance investment in traffic networks with random demands, *Optim. Lett.* **15**, 1799–1819 (2021).
[19] M. Passacantando and F. Raciti, A variational inequality approach to a class of network games with local complementarities and global congestion. In L. Amorosi, P. Dell'olmo and I. Lari (eds.). AIRO Springer Series, Vol. 8 (Springer, Cham, Switzerland, 2022), pp. 1–11.
[20] T. Roughgarden and E. Tardos, Bounding the inefficiency of equilibria in nonatomic congestion games, *Games Econ. Behav.* **47**, 389–403 (2004).

© 2024 World Scientific Publishing Company
https://doi.org/10.1142/9789811267048_0024

Chapter 24

Chebyshev Polynomials of the First Kind and Applications in Tomography

Nicholas E. Protonotarios[*,‡], Vangelis Marinakis[†,§], Nikolaos Dikaios[*,¶], and George A. Kastis[*,‖]

[*]*Mathematics Research Center, Academy of Athens, Athens, Greece*

[†]*Department of Civil Engineering, University of the Peloponnese, Patras, Greece*

[‡]*nprotonotarios@academyofathens.gr*
[§]*vmarinakis@uop.gr*
[¶]*ndikaios@academyofathens.gr*
[‖]*gkastis@academyofathens.gr*

The analytical inversion of the celebrated two-dimensional Radon transform of a function involves a certain integral; the computation of this integral requires the derivative of the Hilbert transform of the Radon transform of the initial function. This inversion provides great insight in the field of medical imaging, especially in positron emission tomography. In this chapter, following our previous works based on third-degree splines, we present a novel numerical implementation of the inversion of the Radon transform based on Chebyshev polynomials and the corresponding reconstruction algorithm. The Chebyshev approximation scheme has the advantage of significantly simplifying the mathematical formulas associated with the inversion of the Radon transform in two dimensions.

1. Introduction

The celebrated two-dimensional Radon transform of a function is defined as the set of all its line integrals [1]. Associated with the Radon transform is the corresponding inverse problem, namely to "reconstruct" a two-dimensional function from its line integrals. It is important to note that the Radon transform and its inversion provide the mathematical foundations of computerized tomography [2, 3] (CT), positron emission tomography [4, 5] (PET) and even radial magnetic resonance imaging [6, 7] (MRI), via the Fourier transform.

The analytical inversion of the two-dimensional Radon transform of a function involves a certain integral of the derivative of the Hilbert transform of the function. In 1991, Novikov and Fokas [8] rederived the inversion of the Radon transform [9] by performing *spectral analysis* on the following eigenvalue equation:

$$\left[\frac{1}{2}\left(k+\frac{1}{k}\right)\partial_{x_1} + \frac{1}{2i}\left(k-\frac{1}{k}\right)\partial_{x_2}\right]\mu = f, \quad k \in \mathbb{C}. \quad (1)$$

This analysis encapsulates two significant problems in complex analysis, namely the scalar Riemann–Hilbert (RH) problem and the $\bar{d}$-problem. The inversion of the two-dimensional Radon transform can be performed in a simpler fashion, by utilizing the Fourier transform. However, the advantages of the derivation of Fokas and Novikov were clearly demonstrated in 2002 by Novikov [10]. Novikov showed that the inverse of a certain generalization of the Radon transform can be established by applying a similar analysis to that performed in (1), only this time to a generalization of equation (1), namely

$$\left[\frac{1}{2}\left(k+\frac{1}{k}\right)\partial_{x_1} + \frac{1}{2i}\left(k-\frac{1}{k}\right)\partial_{x_2} - \nu\right]\mu = f, \quad k \in \mathbb{C}. \quad (2)$$

The analysis of the generalizations of the Radon transform and their inversions [11, 12] gave rise to the simplification of the analytical inversion of the two-dimensional Radon transform, especially in its numerical implementation [13].

In this chapter, we present a novel algorithm based on the Chebyshev approximations of the Hilbert transform of certain functions and their derivatives, and the corresponding reconstruction algorithm. The Chebyshev polynomials' numerical implementation has

2. The Two-Dimensional Radon Transform and Its Inversion

A line L on the plane may be specified by the signed distance from the origin ρ, $-\infty < \rho < \infty$, and the angle θ, $0 \leqslant \theta < 2\pi$, see Figure 1. We denote the unit vectors perpendicular and parallel to the line L by $\mathbf{n}$ and $\mathbf{p}$, respectively. These unit vectors are expressed by

$$\mathbf{n} = (-\sin\theta, \cos\theta)^{\mathrm{T}} \qquad (3)$$

and

$$\mathbf{p} = (\cos\theta, \sin\theta)^{\mathrm{T}}. \qquad (4)$$

These vectors are perpendicular to each other, hence

$$\mathbf{n} \cdot \mathbf{p} = 0. \qquad (5)$$

Every point $\mathbf{x} = (x_1, x_2)^{\mathrm{T}}$ on L in Cartesian coordinates can be expressed in terms of the so-called *local coordinates* (ρ, τ) via

$$\mathbf{x} = \rho\mathbf{n} + \tau\mathbf{p},$$

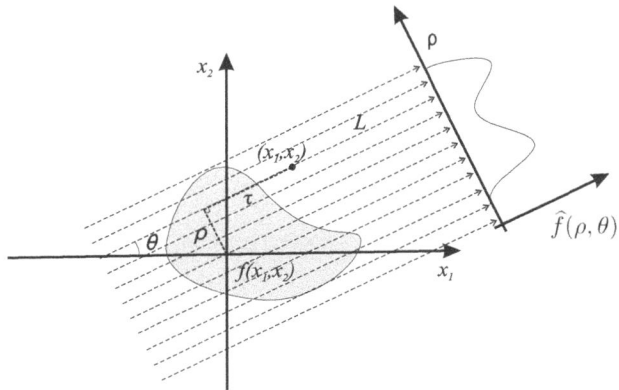

Figure 1. A two-dimensional function $f(x_1, x_2)$ expressed in Cartesian coordinates, and its projections $\widehat{f}(\rho, \theta)$, expressed in local coordinates.

where τ denotes the arc length. We parameterize each point $\mathbf{x} \in L$ as follows:
$$\mathbf{x} := \mathbf{x}(\rho, \tau; \theta) = \begin{bmatrix} x_1(\rho, \tau; \theta) \\ x_2(\rho, \tau; \theta) \end{bmatrix} = \begin{bmatrix} \tau \cos\theta - \rho \sin\theta \\ \tau \sin\theta + \rho \cos\theta \end{bmatrix}. \quad (6)$$

We express the local coordinates (ρ, τ) in terms of Cartesian coordinates (x_1, x_2) and the associated angle θ via equation (6):
$$\begin{bmatrix} \rho \\ \tau \end{bmatrix} := \begin{bmatrix} \rho(x_1, x_2; \theta) \\ \tau(x_1, x_2; \theta) \end{bmatrix} = \begin{bmatrix} x_2 \cos\theta - x_1 \sin\theta \\ x_2 \sin\theta + x_1 \cos\theta \end{bmatrix}. \quad (7)$$

The line integral over all lines L, defined in equation (3), of a two-dimensional Schwartz function $f : \mathbb{R}^2 \to \mathbb{R}$, $f \in S(\mathbb{R}^2)$, is defined as its *two-dimensional Radon transform*, $\mathcal{R}f$. The Radon transform of the function f is stored in the form of the so-called *sinogram*, denoted by $\widehat{f}(\rho, \theta)$, i.e.,
$$\mathcal{R}f = \widehat{f}(\rho, \theta) = \int_L f \, \mathrm{d}s, \quad (8)$$

where $\mathrm{d}s$ and $S(\mathbb{R}^2)$ denote a differential of arc length and the space of Schwartz functions in $\mathbb{R}^2$, respectively. The space $S(\mathbb{R}^2)$ is defined by
$$S(\mathbb{R}^2) = \{f \in C^\infty(\mathbb{R}^2) : \|f\|_{\alpha,\beta} < \infty\} \subset C^\infty(\mathbb{R}^2) \quad (9)$$

and
$$\|f\|_{\alpha,\beta} = \sup_{x \in \mathbb{R}^2} |x^\alpha D^\beta f(x)|, \quad \forall \text{ multi-index } \alpha, \beta,$$
$$|x^\alpha D^\beta f(x)| \to 0, \quad \text{as } |x| \to \infty. \quad (10)$$

The parameterization $\mathbf{x} := \mathbf{x}(\tau)$ of the line L, $\mathbf{x} : \mathbb{R}^2 \to L$, enables us to rewrite equation (8) as follows:
$$\widehat{f}(\rho, \theta) = \int_{-\infty}^{\infty} f(\mathbf{x}(\tau)) \|\mathbf{x}'(\tau)\|_2 \, \mathrm{d}\tau, \quad (11)$$

where $\|\cdot\|_2$ denotes the L^2-norm in $\mathbb{R}^2$. In the above, it is worth noting that
$$\|\mathbf{x}'(\tau)\|_2 = \sqrt{\left(\frac{\mathrm{d}x_1}{\mathrm{d}\tau}\right)^2 + \left(\frac{\mathrm{d}x_2}{\mathrm{d}\tau}\right)^2} = \cos^2\theta + \sin^2\theta = 1. \quad (12)$$

Hence, the parameterization (6) is a *natural parameterization* of the set of all parallel lines L. In this direction, the Radon transform defined in equation (8) may be expressed in local coordinates as

$$\mathcal{R}f = \widehat{f}(\rho,\theta) = \int_{-\infty}^{\infty} f(\tau\cos\theta - \rho\sin\theta, \tau\sin\theta + \rho\cos\theta)\,\mathrm{d}\tau, \quad (13)$$

with $0 \leqslant \theta < 2\pi$ and $-\infty < \rho < \infty$. If we employ a Dirac delta function, namely a line impulse, and take into account equation (7), then the Radon transform defined in equation (13) may be rewritten in the following form:

$$\mathcal{R}f = \widehat{f}(\rho,\theta) = \int_{-\infty}^{\infty}\int_{-\infty}^{\infty} f(x_1, x_2)\delta(\rho + x_1\sin\theta - x_2\cos\theta)\,\mathrm{d}x_1\mathrm{d}x_2, \quad (14)$$

which is proven to be quite useful in the digital signal processing context, especially as it relates with medical imaging applications.

The inverse Radon transform is given by the closed-form expression [14]

$$f(x_1, x_2) = \frac{1}{4\mathrm{i}\pi^2}\left(\frac{\partial}{\partial x_1} - \mathrm{i}\frac{\partial}{\partial x_2}\right)\int_0^{2\pi} e^{\mathrm{i}\theta} H(\rho,\theta)\,\mathrm{d}\theta,$$
$$-\infty < x_{1,2} < \infty, \quad (15)$$

where $H(\rho,\theta)$ denotes the Hilbert transform of the sinogram, $\widehat{f}(\rho,\theta)$,

$$H(\rho,\theta) = \oint_{-\infty}^{\infty} \frac{\widehat{f}(\rho',\theta)}{\rho' - \rho}\,\mathrm{d}\rho', \quad (16)$$

and $\oint$ denotes principal value integral. By utilizing the following relation [4]:

$$\frac{\partial}{\partial x_1} - \mathrm{i}\frac{\partial}{\partial x_2} = -\sin\theta\frac{\partial}{\partial\rho} - \mathrm{i}\cos\theta\frac{\partial}{\partial\rho} = -\mathrm{i}e^{-\mathrm{i}\theta}\frac{\partial}{\partial\rho}, \quad (17)$$

we are able to rewrite equation (15) as

$$f(x_1, x_2) = -\frac{1}{4\pi^2}\int_0^{2\pi} \frac{\partial H(\rho,\theta)}{\partial\rho}\,\mathrm{d}\theta. \quad (18)$$

3. Numerical Implementation via Chebyshev Polynomials

For the numerical implementation of the mathematical setting of the analytical inversion of the Radon transform established in equation (18), we assume the following:

- $f(x_1, x_2)$ has compact support, namely $f(x_1, x_2) = 0$, for all points (x_1, x_2) satisfying $x_1^2 + x_2^2 \geqslant 1$, and
- $\widehat{f}(\rho, \theta)$ is given for every angle $\theta \in [0, 2\pi]$ at the roots $\rho_i^* \in [-1, 1]$ of the corresponding Chebyshev polynomials of the first kind, $T_n(\rho)$, defined in the following.

3.1. Chebyshev polynomials of the first kind

In order to numerically implement the inversion of the Radon transform via equation (18), it is essential to introduce the appropriate mathematical setting of Chebyshev polynomials [15]. Chebyshev polynomials constitute a certain class of polynomials suitable for the approximation of other functions; the Chebyshev polynomial class is broadly utilized in numerical analysis, especially in the numerical solution of partial differential equations [16].

A Chebyshev polynomial of degree n is defined by

$$T_n(x) = \cos(n \arccos x), \quad x \in [-1, 1], \quad n \in \mathbb{N}, \qquad (19)$$

or, equivalently,

$$T_n(\cos \theta) = \cos(n\theta), \quad \theta \in [0, 2\pi], \quad n \in \mathbb{N}, \qquad (20)$$

in the sense that

$$T_n(x) = \cos(n\theta), \quad x = \cos \theta, \quad \theta \in [0, 2\pi]. \qquad (21)$$

Chebyshev polynomials of the first kind, $T_n(x)$, for $n = 0, 1, \ldots, 5$ and $x \in [-1, 1]$ are shown in Figure 2.

We immediately observe that the Chebyshev polynomials $T_n(x)$ have the following properties:

Chebyshev Polynomials of the First Kind and Applications in Tomography

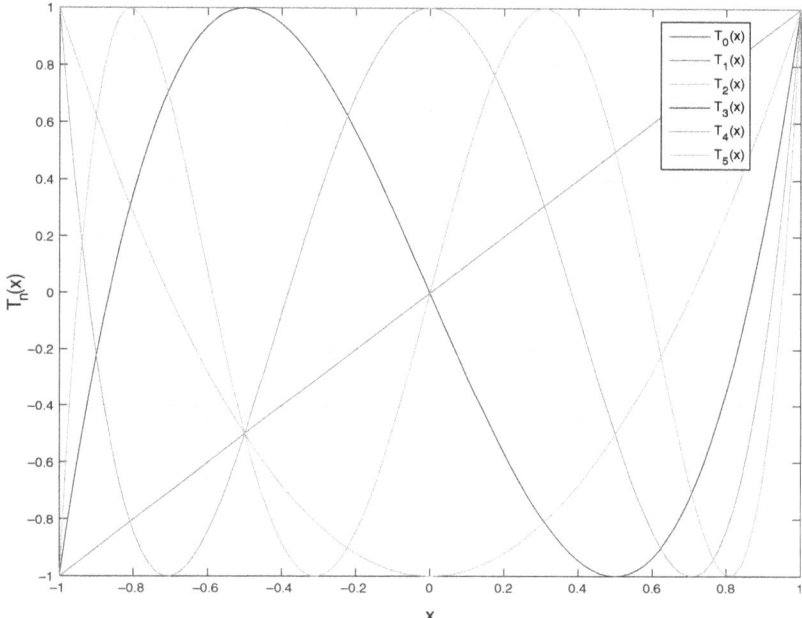

Figure 2. Chebyshev polynomials of the first kind, $T_n(x)$, for $n = 0, 1, \ldots, 5$ and $x \in [-1, 1]$.

(1) distinct zeros at $x = x_m^* \in [-1, 1]$, where

$$x_m^* = \cos\left[\left(\frac{2m-1}{n}\right)\frac{\pi}{2}\right], \quad m = 1, \ldots, n, \quad n \in \mathbb{N}, \qquad (22)$$

(2) extrema located at $x = x_m \in [-1, 1]$

$$x_m = \cos\left(\frac{m\pi}{n}\right), \quad m = 0, \ldots, n, \quad n \in \mathbb{N}. \qquad (23)$$

The value of all extrema is either 1 or -1, i.e., $T_n(x_m) = (-1)^m$,
(3) for all $n \in \mathbb{N}$

$$T_n(1) = 1 \quad \text{and} \quad T_n(-1) = (-1)^n, \qquad (24)$$

(4) for $n = 0$, it follows from definition (19) that

$$T_0(x) = 1, \qquad (25)$$

(5) for $n = 1$, it follows from definition (21) that

$$T_1(x) = x. \tag{26}$$

Furthermore, it is worth mentioning that there exists a recursive definition for the Chebyshev polynomials [17], as the following lemma suggests.

Lemma 1. *The Chebyshev polynomials are given by the following recursive expression:*

$$T_{n+1}(x) = 2xT_n(x) - T_{n-1}(x), \quad n \in \mathbb{N}. \tag{27}$$

Proof. We employ definition (21) for $n + 1$:

$$\begin{aligned} T_{n+1}(x) &= \cos\left[(n+1)\theta\right] = \cos(n\theta + \theta) \\ &= \cos(n\theta)\cos\theta - \sin(n\theta)\sin(\theta). \end{aligned} \tag{28}$$

Similarly, for $n - 1$,

$$\begin{aligned} T_{n-1}(x) &= \cos\left[(n-1)\theta\right] = \cos(n\theta - \theta) \\ &= \cos(n\theta)\cos\theta + \sin(n\theta)\sin(\theta). \end{aligned} \tag{29}$$

Adding equations (28) and (29) yields

$$T_{n+1}(x) + T_{n-1}(x) = 2\cos(n\theta)\cos\theta = 2T_n(x)x, \tag{30}$$

which implies equation (27). □

In the following lemma, we prove an interesting property of the Chebyshev polynomials.

Lemma 2. *The definite integral of a Chebyshev polynomial of the first kind, within its compact support, satisfies*

$$\int_{-1}^{1} T_n(x)\,\mathrm{d}x = \begin{cases} \dfrac{2}{1-n^2}, & n \text{ even}, \\ 0, & n \text{ odd}. \end{cases} \tag{31}$$

Proof. We calculate the antiderivative of $T_n(x)$ as follows:

$$\int T_n(x)dx = \int T_n(\cos\theta)d\cos\theta = -\int \cos(n\theta)\sin\theta d\theta$$

$$= -\frac{1}{2}\int[\sin((n+1)\theta) - \sin((n-1)\theta)]dx$$

$$= \frac{1}{2}\left(\frac{\cos((n+1)\theta)}{n+1} - \frac{\cos((n-1)\theta)}{n-1}\right) + c$$

$$= \frac{1}{2}\left(\frac{T_{n+1}(x)}{n+1} - \frac{T_{n-1}(x)}{n-1}\right) + c, \qquad (32)$$

where c is a constant of integration. From equation (32), it is straightforward to compute the integral on the left-hand-side of equation (31), namely

$$\int_{-1}^{1} T_n(x)dx = \frac{1}{2}\left[\frac{T_{n+1}(x)}{n+1} - \frac{T_{n-1}(x)}{n-1}\right]_{-1}^{1}$$

$$= \frac{1}{2}\left(\frac{1}{n+1} - \frac{1}{n-1}\right) - \frac{1}{2}\left(\frac{(-1)^{n+1}}{n+1} - \frac{(-1)^{n-1}}{n-1}\right)$$

$$= \frac{1+(-1)^n}{1-n^2}, \qquad (33)$$

which yields

$$\int_{-1}^{1} T_n(x)dx = \begin{cases} \frac{2}{1-n^2}, & n \text{ even,} \\ 0, & n \text{ odd.} \end{cases} \qquad (34)$$

□

As it will be evident in the next section, Chebyshev polynomials are proven to be very useful for the computation and the inversion of integral transforms in general, and of the Radon transform in particular.

3.2. Inverse Radon transform via Chebyshev polynomials of the first kind

For the calculation of the Hilbert transform, defined in equation (16), and its corresponding derivative with respect to ρ, in each interval

$[\rho_i, \rho_{i+1}]$, we approximate the sinogram function [18] $\widehat{f}(\rho, \theta)$ by

$$\widehat{f}(\rho, \theta) \simeq \frac{1}{2}c_0(\theta) + \sum_{n=1}^{N} c_n(\theta) T_n(\rho), \qquad (35)$$

where $T_n(\rho)$ are the Chebyshev polynomials [19] of the first kind, defined in equations (19)–(21). We assume that N is sufficiently large.

By utilizing the approximation defined in equation (35), we express the Hilbert transform defined in equation (16) as

$$H(\rho, \theta) = \oint_{-1}^{1} \frac{\widehat{f}(\rho', \theta)}{\rho' - \rho} d\rho' = \int_{-1}^{1} \frac{\widehat{f}(\rho', \theta) - \widehat{f}(\rho, \theta)}{\rho' - \rho} d\rho' \\ + \int_{-1}^{1} \frac{\widehat{f}(\rho, \theta)}{\rho' - \rho} d\rho'. \qquad (36)$$

We proceed as follows: for the first integral of the right-hand side of equation (36), we substitute $\widehat{f}$ by its Chebyshev approximation, see equation (35), while for the second integral, we calculate under the assumption that $\widehat{f}$ has compact support. These yield

$$H(\rho, \theta) = \sum_{n=1}^{N} c_n(\theta) \int_{-1}^{1} \frac{T_n(\rho') - T_n(\rho)}{\rho' - \rho} d\rho' + \widehat{f}(\rho, \theta) \ln\left(\frac{1-\rho}{1+\rho}\right). \qquad (37)$$

We define the interval occurring on the right-hand side of equation (37) as

$$I_n(\rho) = \int_{-1}^{1} \frac{T_n(\rho') - T_n(\rho)}{\rho' - \rho} d\rho'. \qquad (38)$$

In equation (38), for $n = 0$, the integral vanishes, i.e.,

$$I_0(\rho) = 0, \qquad (39)$$

and for $n = 1$, the integral is constant,

$$I_1(\rho) = 2. \qquad (40)$$

It is worth noting that the integral on the right-hand side of equation (38) involves the Hilbert transform of $T_n(\rho')$ with respect to ρ, in the sense that

$$I_n(\rho) = \int_{-1}^{1} \frac{T_n(\rho')\mathrm{d}\rho'}{\rho' - \rho} - T_n(\rho) \int_{-1}^{1} \frac{\mathrm{d}\rho'}{\rho' - \rho}$$
$$= H_n(\rho) - T_n(\rho) \ln\left(\frac{1-\rho}{1+\rho}\right), \tag{41}$$

where $H_n(\rho)$ denotes the Hilbert transform of $T_n(\rho')$.

Proposition 1. *The integral defined in equation (38) is given by the recursive formula*

$$I_{n+1}(\rho) = 2\rho I_n(\rho) - I_{n-1}(\rho) + a_n, \tag{42}$$

where a_n is defined by

$$a_n = 2\int_{-1}^{1} T_n(\rho)\mathrm{d}\rho = \begin{cases} \dfrac{4}{1-n^2}, & n \text{ even}, \\ 0, & n \text{ odd}. \end{cases} \tag{43}$$

Proof. We define $A_n(\rho)$ by

$$A_n(\rho) = I_{n+1}(\rho) - 2\rho I_n(\rho) + I_{n-1}(\rho). \tag{44}$$

In order to prove the proposition, we need to establish that $A_n(\rho) = a_n$. Taking into account the definition (38), we rewrite $A_n(\rho)$ as follows:

$$A_n(\rho) = \int_{-1}^{1} \frac{T_{n+1}(\rho') - T_{n+1}(\rho) - 2\rho T_n(\rho') + 2\rho T_n(\rho) + T_{n-1}(\rho') - T_{n-1}(\rho)}{\rho' - \rho}\mathrm{d}\rho'$$
$$= \int_{-1}^{1} \frac{(T_{n+1}(\rho') + T_{n-1}(\rho')) - (T_{n+1}(\rho) + T_{n-1}(\rho) - 2\rho T_n(\rho)) - 2\rho T_n(\rho')}{\rho' - \rho}\mathrm{d}\rho'$$
$$= \int_{-1}^{1} \frac{2\rho' T_n(\rho') - 2\rho T_n(\rho')}{\rho' - \rho}\mathrm{d}\rho'$$
$$= 2\int_{-1}^{1} T_n(\rho')\mathrm{d}\rho'$$
$$= a_n.$$

In the above, we employed the recursive formula provided in Lemma 1, as well as Lemma 2. □

For the calculation of the inverse of the Radon transform defined in equation (18), we must calculate the partial derivative with respect to ρ of (i) the Hilbert transform of the sinogram, defined in equation (16), and (ii) the sinogram itself. In this direction, we combine equations (37) and (38) to get

$$H(\rho,\theta) = \sum_{n=1}^{N} c_n(\theta) I_n(\rho) + \widehat{f}(\rho,\theta) \ln\left(\frac{1-\rho}{1+\rho}\right). \tag{45}$$

Therefore, the partial derivative of $H(\rho,\theta)$ with respect to ρ may be written as

$$\frac{\partial H(\rho,\theta)}{\partial \rho} = \sum_{n=1}^{N} c_n(\theta) \frac{\partial I_n(\rho)}{\partial \rho} + \frac{\partial \widehat{f}(\rho,\theta)}{\partial \rho} \ln\left(\frac{1-\rho}{1+\rho}\right) + \frac{2}{\rho^2-1} \widehat{f}(\rho,\theta). \tag{46}$$

Equation (46) involves the partial derivatives of $I_n(\rho)$ with respect to ρ for all n. These derivatives are computed via the recursive relation

$$\frac{\partial I_{n+1}(\rho)}{\partial \rho} = 2 I_n(\rho) + 2\rho \frac{\partial I_n(\rho)}{\partial \rho} - \frac{\partial I_{n-1}(\rho)}{\partial \rho}, \tag{47}$$

which follows from the partial differentiation of both sides of equation (42) with respect to ρ. For $n = 0$ and $n = 1$, $I_n(\rho)$ is constant, thus the corresponding derivatives vanish, i.e.,

$$\frac{\partial I_0(\rho)}{\partial \rho} = \frac{\partial I_1(\rho)}{\partial \rho} = 0. \tag{48}$$

Furthermore, for the calculation of the partial derivative of the sinogram, we follow the same pattern as with the sinogram itself; we apply Chebyshev approximation, as in equation (35), in the sense that

$$\frac{\partial \widehat{f}(\rho,\theta)}{\partial \rho} \simeq \frac{1}{2} d_0(\theta) + \sum_{n=1}^{N} d_n(\theta) T_n(\rho), \tag{49}$$

where $d_n(\theta)$ are the corresponding coefficients of the Chebyshev expansion of the partial derivative with respect to ρ of the sinogram and $T_n(\rho)$ are Chebyshev polynomials of the first kind. It is essential to emphasize that, due to the compact support assumption, for every (x_1, x_2), the corresponding local coordinates (ρ, θ), defined in equation (7), are limited to

$$\rho^2 + \tau^2 = x_1^2 + x_2^2 \leqslant 1. \tag{50}$$

Equation (46) enables us to express f, defined in equation (18), in the following form:

$$f(x_1, x_2) = -\frac{1}{4\pi^2} \int_0^{2\pi} \left(\sum_{n=1}^N c_n(\theta) \frac{\partial I_n(\rho)}{\partial \rho} \right.$$
$$\left. + \frac{\partial \widehat{f}(\rho, \theta)}{\partial \rho} \ln\left(\frac{1-\rho}{1+\rho}\right) + \frac{2}{\rho^2 - 1} \widehat{f}(\rho, \theta) \right) d\theta. \tag{51}$$

The computations involved in the analytic inversion of the Radon transform are achieved by employing appropriate polynomial approximation of the sinogram and of its partial derivative with respect to ρ. For the calculation of the inverse Radon transform (18) via equation (51), we proceed as follows:

(i) We assume that $\widehat{f}(\rho, \theta)$ is known for a given number of projection angles θ, i.e.,

$$\theta = \theta_j, \quad j = 1, \ldots, K, \quad \theta_j \in [0, 2\pi], \tag{52}$$

and for a given number ρ values, i.e.,

$$\rho = \rho_i, \quad i = 1, \ldots, L, \quad \rho_i \in [-1, 1]. \tag{53}$$

located at the n roots of the Chebyshev polynomial $T_n(\rho)$, as in equation (22),

(ii) We calculate the coefficients $c_n(\theta)$ and $d_n(\theta)$, occurring in equations (35) and (49), respectively, for all n, using appropriate, commercially available subroutines, such as `chebft` and `chder` from the classic textbook "Numerical Recipes" [20].

(iii) For all (x_1, x_2), for a given θ, we calculate its corresponding local coordinate ρ via equation (7).

(iv) We calculate $\widehat{f}(\rho,\theta)$ via equation (35), using **chebev** from "Numerical Recipes" [20].

(v) We calculate $\frac{\partial \widehat{f}(\rho,\theta)}{\partial \rho}$ via equation (49), using again **chebev** from "Numerical Recipes" [20].

(vi) We calculate I_n and then $\frac{\partial I_n(\rho)}{\partial \rho}$ via the recursive relations (42) and (47), respectively.

(vii) We calculate the sum $\sum_{n=1}^{N} c_n(\theta) \frac{\partial I_n(\rho)}{\partial \rho}$.

(viii) We calculate $\frac{\partial H(\rho,\theta)}{\partial \rho}$ via equations (46) and (47).

(ix) Combine all of the above to integrate $\frac{\partial H(\rho,\theta)}{\partial \rho}$ with respect to θ, via numerical integration, and get $f(x_1, x_2)$ via equation (51).

3.3. Numerical implementation of the inversion of the Radon transform via Chebyshev polynomials of the first kind: Analytic examples

For the evaluation of our algorithm, we employed two simulated phantoms, as in the following two examples. These phantoms were chosen in the basis of their analytic nature: they are two of the most characteristic examples where one can analytically derive their Radon transform, as well as the inverse of their Radon transform, thus performing analytical reconstruction.

Example 1. For the purposes of our simulations, we created a two-dimensional circular phantom of radius $0 < R \leqslant 1$, characterized by a function $f(x_1, x_2)$, namely

$$f(x_1, x_2) = \begin{cases} \frac{1}{2}, & \text{for } x_1^2 + x_2^2 \leqslant R \\ 0, & \text{otherwise} \end{cases}. \tag{54}$$

The above function f represents a circular nonzero region, centered at the origin of the $x_1 x_2$-plane. The Radon transform of the phantom function f, defined in equation (54), corresponds to the following analytically calculated [21] sinogram:

$$\widehat{f}(\rho, \theta) = \begin{cases} \sqrt{R^2 - \rho^2}, & \text{for } |\rho| \leqslant R \\ 0, & \text{otherwise.} \end{cases} \tag{55}$$

The analytic sinogram, defined in equation (55), evaluated for $R = 1$, 180 equally spaced angles θ and 50 ρ at the roots of the corresponding Chebyshev polynomials of the first kind, $T_n(\rho)$ is shown in Figure 3(a), whereas in Figure 3(b), we present the corresponding reconstruction of the sinogram evaluated via equation (18) and steps (i)–(vii) of Section 3.2 for a 50×50 square grid.

Example 2. For the purposes of our simulations, we created a two-dimensional phantom, characterized by a function $f(x_1, x_2)$, namely

$$f(x_1, x_2) = \begin{cases} 1 - (x_1^2 + x_2^2), & \text{for } x_1^2 + x_2^2 \leqslant 1 \\ 0, & \text{otherwise} \end{cases}. \tag{56}$$

The above function f is a relatively smooth function. The Radon transform of the phantom function f, defined in equation (56), corresponds to the following analytically calculated sinogram:

$$\widehat{f}(\rho, \theta) = \begin{cases} \dfrac{4}{3} \left(\sqrt{1 - \rho^2} \right)^3, & \text{for } |\rho| \leqslant 1 \\ 0, & \text{otherwise} \end{cases}. \tag{57}$$

The analytic sinogram, defined in equation (57), evaluated for 180 equally spaced angles θ and 50 ρ at the roots of the corresponding Chebyshev polynomials of the first kind, $T_n(\rho)$ is shown in Figure 4(a), whereas in Figure 4(b), we present the corresponding reconstruction of the sinogram evaluated via equation (18) and steps (i)–(vii) of Section 3.2 for a 50×50 square grid.

For both examples presented above, it is worth mentioning that since all ρ nodes, denoted by $\{\rho_i\}_{i=1}^L$, are located at the roots of the corresponding Chebyshev polynomials of the first kind, as equation (22) suggests, it follows that they all belong to the interval of reference, namely $\rho_i \in [-1, 1]$, $\forall\, i, i = 1, \ldots, L$. Similarly, in the θ-dimension of the sinogram, θ-locations, denoted by $\{\theta_j\}_{j=1}^K$ are equally spaced and span the interval $\theta_j \in [0, 2\pi]$, $\forall\, j, j = 1, \ldots, K$.

Furthermore, we need to emphasize the fact that the Chebyshev-based reconstructions are evaluated at the roots of the Chebyshev polynomials of the first kind, as described above. Although this is

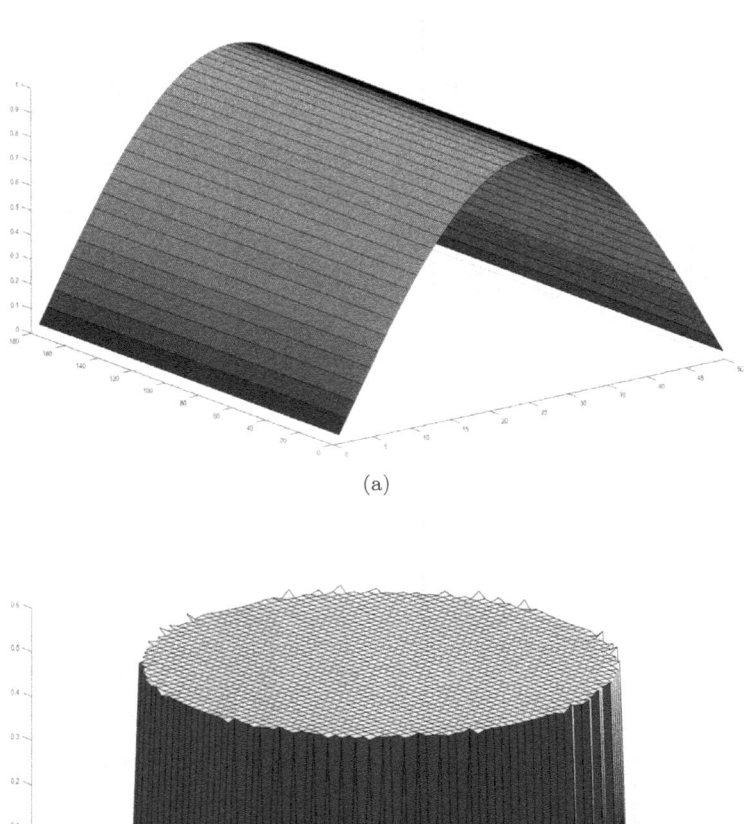

Figure 3. (a) Analytic sinogram defined in equation (55), evaluated for $R = 1$, 180 equally spaced angles θ and at 50 ρ at the roots of the corresponding Chebyshev polynomials of the first kind, $T_n(\rho)$, and (b) reconstruction of sinogram via equation (18) and steps (i)–(vii) of Section 3.2 for a 50×50 square grid.

Chebyshev Polynomials of the First Kind and Applications in Tomography 727

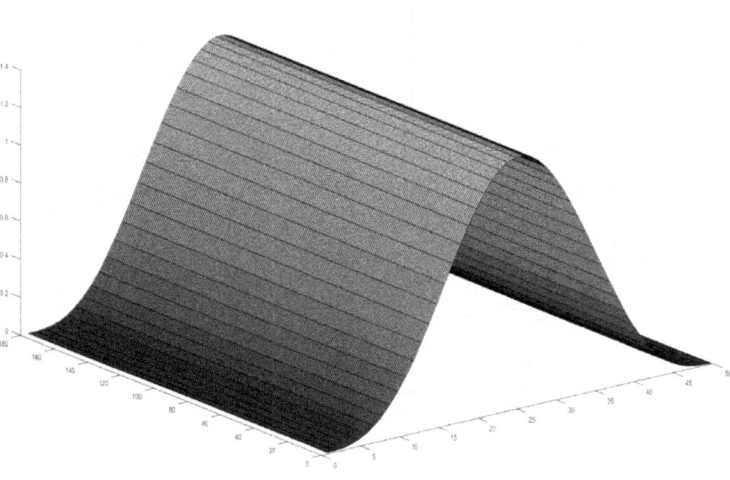

(a)

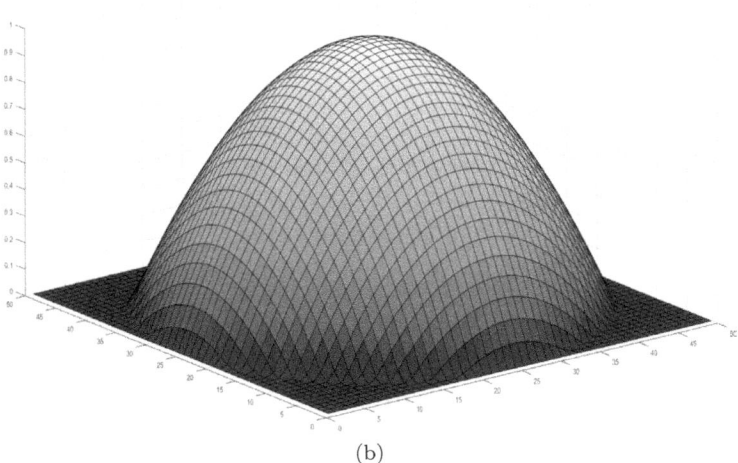

(b)

Figure 4. (a) Analytic sinogram defined in equation (57), evaluated for 180 equally spaced angles θ and at 50 ρ at the roots of the corresponding Chebyshev polynomials of the first kind, $T_n(\rho)$, and (b) reconstruction of sinogram via equation (18) and steps (i)–(vii) of Section 3.2 for a 50×50 square grid.

quite uncommon, as opposed to other analytic reconstruction techniques where equal spacing is usually employed, including the industry default standard of filtered backprojection (FBP) [22] and the more recent spline reconstruction technique (SRT) [23], with our Chebyshev-based approach, we were able to efficiently reconstruct analytically derived sinograms. To further justify our selection of nodes via the Chebyshev roots, we note that Gil, Segura and Temme have established that the monic Chebyshev polynomial of degree $n+1$, denoted by $\widehat{T}_{n+1}(\rho)$ and defined as

$$\widehat{T}_{n+1}(\rho) = \prod_{k=1}^{n+1}(\rho - \rho_k), \tag{58}$$

where ρ_k are the roots of the corresponding Chebyshev polynomial of the first kind, $T_{n+1}(\rho)$, namely

$$\rho_k = \cos\left[\left(\frac{2k-1}{n+1}\right)\frac{\pi}{2}\right], \quad k = 1, \ldots, n+1, \quad n \in \mathbb{N}, \tag{59}$$

is the polynomial of degree $n+1$ with the minimum uniform norm in the interval $[-1, 1]$, i.e.,

$$\left\|\widehat{T}_{n+1}(\rho)\right\| \leqslant \|P_{n+1}(\rho)\|, \tag{60}$$

where $P_{n+1}(\rho)$ denotes monic polynomials of degree $n+1$; for details, see Theorem 3.4 of their book [16]. For a bounded function $h(\rho)$ in $[-1, 1]$, the uniform norm, $\|\cdot\|$, is defined by

$$\|h(\rho)\| := \max_{\rho \in [-1,1]} |h(\rho)|. \tag{61}$$

The above results essentially imply that, whenever the interpolation nodes are chosen at the corresponding roots of the Chebyshev polynomials of the first kind, the approximation error is minimized, in the uniform norm sense. In this context, the selection of the roots of the Chebyshev polynomials of the first kind as the nodes for the corresponding calculations is considered optimal. However, in future studies, we plan to apply sinogram rebinning [24], via interpolation. In this direction, we aim to perform a corresponding sensitivity analysis, and we expect that the analysis will demonstrate insignificant differences.

References

[1] P. Kuchment, *The Radon Transform and Medical Imaging*. CBMS-NSF Regional Conference Series in Applied Mathematics (Society for Industrial and Applied Mathematics, Philadelphia, 2014).

[2] F. Natterer and F. Wuebbeling, *Mathematical Methods in Image Reconstruction*. Monographs on Mathematical Modeling and Computation (Society for Industrial and Applied Mathematics, Philadelphia, 2001).

[3] A. Averbuch, I. Sedelnikov and Y. Shkolnisky, CT reconstruction from parallel and fan-beam projections by a 2D discrete Radon transform, *IEEE Trans. Image Process.* **21**(2), 733–741 (2011). doi: 10.1109/TIP.2011.2164416.

[4] G. A. Kastis, D. Kyriakopoulou, A. Gaitanis, Y. Fernández, B. F. Hutton and A. S. Fokas, Evaluation of the spline reconstruction technique for PET, *Med. Phys.* **41**(4), 042501 (2014). doi: 10.1118/1.4867862.

[5] J. You, The attenuated Radon transform with complex coefficients, *Inverse Probl.* **23**(5), 1963 (2007). doi: 10.1088/0266-5611/23/5/010.

[6] N. Dikaios, N. E. Protonotarios, A. S. Fokas and G. A. Kastis, Quantification of T1, T2 relaxation times from magnetic resonance fingerprinting radially undersampled data using analytical transformations, *Magn. Reson. Imaging.* **80**, 81–89 (2021). doi: 10.1016/j.mri.2021.04.013.

[7] J. Park, J. Lee, J. Lee, S.-K. Lee and J.-Y. Park, Strategies for rapid reconstruction in 3D MRI with radial data acquisition: 3D fast Fourier transform vs two-step 2D filtered back-projection, *Sci. Rep.* **10**(1), 1–11 (2020). doi: 10.1038/s41598-020-70698-4.

[8] A. Fokas and R. Novikov, Discrete analogues of δ-equation and of Radon transform, *C. R. Acad. Sci. Sér. 1, Math.* **313**(2), 75–80 (1991).

[9] J. Radon, Über die bestimmung von funktionen durch ihre integralwerte längs gewisser mannigfaltigkeiten, *Akad. Wiss.* **69**, 262–277 (1917).

[10] R. Novikov, An inversion formula for the attenuated X-ray transformation, *Ark. Mat.* **40**(1), 145–167 (2002). doi: 10.1007/BF02384507.

[11] N. E. Protonotarios, A. S. Fokas, K. Kostarelos and G. A. Kastis, The attenuated spline reconstruction technique for single photon emission computed tomography, *J. R. Soc. Interface.* **15**(148), 20180509 (2018). doi: 10.1098/rsif.2018.0509.

[12] N. E. Protonotarios, G. A. Kastis and A. S. Fokas, A new approach for the inversion of the attenuated Radon transform. In T. M. Rassias and P. M. Pardalos (eds.) *Mathematical Analysis and Applications* (Springer International Publishing, Cham, 2019), pp. 433–457. doi: 10.1007/978-3-030-31339-5_16.

[13] A. S. Fokas, *A Unified Approach to Boundary Value Problems*, Vol. 78 (Society for Industrial and Applied Mathematics, Philadelphia, 2008). doi: 10.1137/1.9780898717068.

[14] A. Fokas, A. Iserles and V. Marinakis, Reconstruction algorithm for single photon emission computed tomography and its numerical implementation, *J. R. Soc. Interface* **3**(6), 45–54 (2006). doi: 10.1098/rsif.2005.0061.

[15] P. Chebyshev, *Théorie des mécanismes connus sous le nom de parallélogrammes* (Imprimerie de l'Académie impériale des Sciences, 1853).

[16] A. Gil, J. Segura and N. M. Temme, *Numerical Methods for Special Functions* (SIAM, Philadelphia, 2007).

[17] R. Piessens, Computing integral transforms and solving integral equations using Chebyshev polynomial approximations, *J. Comput. Appl. Math.* **121**(1–2), 113–124 (2000). doi: 10.1016/S0377-0427(00)00349-6.

[18] A. S. Fokas and V. Marinakis, Reconstruction algorithm for the brain imaging techniques of PET and SPECT, *HERCMA* 25–27 (2003).

[19] P. Butzer and F. Jongmans, PL Chebyshev (1821–1894): A guide to his life and work, *J. Approx. Theory.* **96**(1), 111–138 (1999). doi: 10.1006/jath.1998.3289.

[20] W. H. Press, S. A. Teukolsky, W. T. Vetterling and B. P. Flannery, *Numerical Recipes: The Art of Scientific Computing*, 3rd edn. (Cambridge University Press, New York, 2007).

[21] A. C. Kak and M. Slaney, *Principles of Computerized Tomographic Imaging* (SIAM, Philadelphia, 2001).

[22] C. D. Ramos, Y. E. Erdi, M. Gonen, E. Riedel, H. W. Yeung, H. A. Macapinlac, R. Chisin and S. M. Larson, FDG-PET standardized uptake values in normal anatomical structures using iterative reconstruction segmented attenuation correction and filtered back-projection, *Eur. J. Nucl. Med.* **28**(2), 155–164 (2001). doi: 10.1007/s002590000421.

[23] N. E. Protonotarios, G. A. Kastis, N. Dikaios and A. S. Fokas, Piecewise polynomial inversion of the Radon transform in three space dimensions via plane integration and applications in positron emission tomography. In *Nonlinear Analysis, Differential Equations, and Applications* (Springer, Cham, 2021), pp. 381–396. doi: 10.1007/978-3-030-72563-1_17.

[24] M. Defrise, P. E. Kinahan, D. W. Townsend, C. Michel, M. Sibomana and D. F. Newport, Exact and approximate rebinning algorithms for 3D PET data, *IEEE Trans. Med. Imaging* **16**(2), 145–158 (1997). doi: 10.1109/42.563660.

© 2024 World Scientific Publishing Company
https://doi.org/10.1142/9789811267048_0025

Chapter 25

Bernstein-Type Polynomials Associated with Characteristic Function, Moment Generating Functions of Beta-Type Distribution and Their Approximation Applications

Yilmaz Simsek* and Fusun Yalcin[†]

*Department of Mathematics, Faculty of Science,
Akdeniz University, Antalya, Turkey*

*ysimsek@akdeniz.edu.tr
[†]fusunyalcin@akdeniz.edu.tr

This chapter is a survey of the authors' [49, 50] articles published in recent years, and the sections in these articles have been prepared with the addition of new and applicable results. That is, we survey on the beta-type distribution associated with the Bernstein-type basis functions and the beta function, which were defined by authors [49, 50]. When it comes to the content of this chapter, the keywords that will be covered in this article are the following topics that have vital applications in all applied sciences: beta distribution, Bernstein polynomials, generating function, moment generating function, Stirling numbers, digamma function, beta function, gamma function and expected values for the logarithm of random variables. By blending the topics mentioned above, old and new results are provided. The relationships between these are examined and interpreted. Moreover, the known consequences of Bernstein polynomials are exhibited in the results related to the approximation theory to which they are related.

1. Introduction

Distribution functions, characteristic functions, moment generating functions, etc. are among the most important topics of probability, statistics and combinatorics analysis. In particular, symmetric probability distribution, symmetric random variables and symmetric two independent random variables have also various applications in many applied sciences. As for, symmetric beta distribution has also important applications in related areas because mathematical models of the behavior of random variables can be constructed via the beta distribution. It is well known that the beta distribution and its applications are used in all branches of sciences. (cf. Refs. [1–54] and the references cited therein).

Since expected values for the logarithm of random variable can be used to evaluate of the value for the logarithm of the geometric mean of a distribution, in Refs. [49, 50], using the generating function with their functional equations method, we gave not only many identities and relations of the Bernstein-type basis functions, beta-type distributions, moment generating functions, expected values for the logarithm of random variable, the Stirling numbers, the Apostol–Bernoulli numbers, and the digamma function, but also further remarks and observations about the results.

We start this chapter with the following notations, definitions and relations:

Let $\mathbb{N}$, $\mathbb{Z}$, $\mathbb{Q}$, $\mathbb{R}$ and $\mathbb{C}$ denote the set of natural numbers, the set of integers, the set of rational numbers, the set of real numbers and the set of complex numbers, respectively:

$$\mathbb{N}_0 = \mathbb{N} \cup \{0\}.$$

Let $s \in \mathbb{C}$,

$$\ln(s) = \ln|s| + i \arg(s),$$

where the principal branch of the logarithm is assumed to be taken (cf. Refs. [18, 30, 45, 46] and the references cited therein).

The polygamma function is defined as follows:

$$\psi^{(n)}(s) = \frac{d^{n+1}}{ds^{n+1}} \{\ln(\Gamma(s))\}, \tag{1}$$

where $\Gamma(s)$ denotes the gamma function:

$$\Gamma(s) = \int_0^\infty t^{s-1} e^{-t} dt,$$

where $s = x + iy \in \mathbb{C}$ and $x, y \in \mathbb{R}$ with $x > 0$ (cf. [18, 30, 45, 46]). It is well known that $\Gamma(s)$ has the following functional equation:

$$\Gamma(s+1) = s\Gamma(s)$$

for $\operatorname{Re}(s) > 0$. Substituting $s = n$ with $n \in \mathbb{N}_0$, we have

$$\Gamma(n+1) = n!.$$

Since

$$\Gamma(1) = \int_0^\infty e^{-t} dt = 1,$$

we have

$$\Gamma(1) = 1 = 0!.$$

For $s \neq 0, -1 - 2, \ldots$, and $n = 1$, (1) reduces to

$$\psi(s) = \frac{d}{ds} \{\ln(\Gamma(s))\} = \frac{1}{\Gamma(s)} \frac{d}{ds} \{\Gamma(s)\}. \tag{2}$$

The function $\psi(s)$ can also be given by

$$\psi(s) = -\gamma + \sum_{n=0}^\infty \left(\frac{1}{n+1} - \frac{1}{n+s} \right), \tag{3}$$

where γ denotes the Euler constant, that is,

$$\gamma = \lim_{n \to \infty} (-\log n + H_n),$$

where H_n denotes the Harmonic numbers which are defined by

$$H_n = \sum_{j=1}^{n} \frac{1}{j},$$

such that $H_0 = 0$ (cf. Refs. [18, 30, 45, 46, 53]). It is time to give the following generalized Harmonic numbers of order r:

$$-\frac{\ln(1-x)}{x(1-x)^r} = \sum_{n=1}^{\infty} H_n^{(r)} x^{n-1} \qquad (4)$$

(cf. Refs. [7, 43, 46]).

For $r = 1$, the numbers $H_n^{(1)}$ reduce to the Harmonic numbers H_n.

The λ-array polynomials $S_k^n(x;\lambda)$ are defined by the following generating function:

$$\frac{(\lambda e^t - 1)^k e^{tx}}{k!} = \sum_{n=0}^{\infty} S_k^n(x;\lambda) \frac{t^n}{n!}, \qquad (5)$$

where $k \in \mathbb{N}_0$ and $\lambda \in \mathbb{C}$ (cf. Ref. [39]).

Substituting $x = 0$ and $\lambda = 1$ into (5), we have generating functions for the Stirling numbers of the second kind:

$$\frac{(e^t - 1)^k}{k!} = \sum_{n=0}^{\infty} S_2(n,k) \frac{t^n}{n!} \qquad (6)$$

(cf. Refs. [39, 42, 45, 46]).

The Stirling numbers of the second kind are given by the following finite sum:

$$x^n = \sum_{k=0}^{n} x(x-1)(x-2)\ldots(x-k+1) S_2(n,k), \qquad (7)$$

where $k, n \in \mathbb{N}_0$, and $S_2(n, k) = 0$ if $k > n$ (cf. Refs. [39, 42, 45, 46] and the references cited therein).

The Stirling numbers of the first kind $S_1(n, k)$ are also defined by the following generating functions:

$$F_{S_1}(t, k) = \frac{(\log(1+t))^k}{k!} = \sum_{n=0}^{\infty} S_1(n, k) \frac{t^n}{n!}. \tag{8}$$

The Stirling numbers of the first kind are given by the following finite sum:

$$x(x-1)(x-2)\ldots(x-n+1) = \sum_{k=0}^{n} S_1(n, k) x^k, \tag{9}$$

where $k, n \in \mathbb{N}_0$, and $S_1(n, k) = 0$ if $k > n$ (cf. Refs. [31, 39, 42, 46] and the references cited therein).

Let $t \in \mathbb{C}$, $m \in \mathbb{N}$ and $x \in [a, b]$ with $a \neq b$. The Bernstein-type basis functions $Y_k^n(x; a, b, m)$ are defined by means of the following generating functions:

$$\frac{t^k (x-a)^k e^{-xt}}{(b-a)^m k!} = \sum_{n=0}^{\infty} Y_k^n(x; a, b, m) \frac{e^{-bt} t^n}{n!}, \tag{10}$$

where $k = 0, 1, \ldots, n$ (cf. Ref. [40]).

Using (10) yields

$$Y_k^n(x; a, b, m) = \binom{n}{k} \frac{(x-a)^k (b-x)^{n-k}}{(b-a)^m}, \tag{11}$$

where $k = 0, 1, \ldots, n$ and $x \in [a, b]$ and also $0 \leq \left|\frac{x-a}{b-a}\right| \leq 1$, $0 \leq \left|\frac{b-x}{b-a}\right| \leq 1$ and

$$\binom{n}{k} = \frac{n!}{k!(n-k)!}$$

(cf. Ref. [40]; see also Refs. [22, 32–41]).

Putting $a = 0$ and $b = 1$ in (11), we have

$$B_k^n(x) = \binom{n}{k} x^k (1-x)^{n-k}, \tag{12}$$

where $B_k^n(x)$ denotes the Bernstein basis functions. This basis function can be studied by many authors. For instance, Farouki [9], Goldman [11], Kim [20], Bayad *et al.* [4], Li and Goldman [12], Acikgoz and Araci [2], Simsek [32] and Simsek [34–41]. Generating functions for the Bernstein basis functions can also be derived from (10), that is,

$$\frac{t^k x^k e^{-xt}}{k!} = \sum_{n=0}^{\infty} \mathbb{Y}_k^n(x;0,1,n) \frac{e^{-t}t^n}{n!}.$$

Hence,

$$B_k^n(x) := \mathbb{Y}_k^n(x;0,1,n).$$

The Bernstein basis functions have also been studied intensively by Refs. [5, 24], Simsek [34–43] and also Acikgoz and Araci [2], Goldman *et al.* [11, 12], Kim *et al.* [4, 8, 9, 20–23], Simsek [32] and the references cited therein.

2. Beta Distributions and Their Properties

Many interesting analogs of the beta distribution were investigated by Johnson *et al.* [18, Equation (25.1), p. 210]. A family of beta distributions, which is composed of all distributions with probability density functions, is defined by

$$P_Y(x;a,b;p,q) = \frac{1}{B(p,q)} \frac{(x-a)^{p-1}(b-x)^{q-1}}{(b-a)^{p+q-1}}, \tag{13}$$

where $a \leq x \leq b$, $p \geq 0$, $q \geq 0$ and

$$B(p,q) = \int_0^1 u^{p-1}(1-u)^{q-1}\,du$$

and

$$B(p,q) = \frac{\Gamma(p)\Gamma(q)}{\Gamma(p+q)}, \tag{14}$$

where $B(p,q)$ denotes the beta function, which is a symmetric function and related to the gamma function $\Gamma(p)$ with

$$\Gamma(k+1) = k!$$

for $k \in \mathbb{N}_0$ (cf. Ref. [18]; see also Refs. [29, 30, 37, 45, 46]).

By implementing the following Stirling's approximation (also known as Stirling's formula), which is not only an approximation for factorials but also a good approximation, leading to accurate results even for small values of n,

$$n! \sim \sqrt{2\pi n} \left(\frac{n}{e}\right)^n,$$

we get not only

$$\Gamma(n+1) \sim \sqrt{2\pi n} \left(\frac{n}{e}\right)^n$$

for $n \to \infty$ but also the following well-known asymptotic formula or approximation of the beta function

$$B(p,q) \sim \sqrt{2\pi} \frac{p^{p-\frac{1}{2}} q^{q-\frac{1}{2}}}{(p+q)^{q+p-\frac{1}{2}}} \tag{15}$$

for large p and large q. It is also known that if large p and q is fixed, then we have

$$B(p,q) \sim \Gamma(q) p^{-q} \tag{16}$$

(cf. Refs. [1, 30, 46, 54]).

Substituting $q = 1$ into (13), we have the power-function distribution.

For $y = \frac{x-a}{b-a}$, the distribution $P_Y(x; a, b; p, q)$ reduces to the following well-known beta probability density function $P_B(y; p, q)$:

$$y(1-y) P_B(y; p, q) = \frac{y^p (1-y)^q}{B(p,q)}, \tag{17}$$

which is also known as the standard form of the beta distribution with parameters p and q (cf. Ref. [18]).

It is known that the function $P_B(y; p, q)$ has reflection symmetry property. The beta function $B(p, q)$ is a symmetric function, that is,

$$B(p,q) = B(q,p),$$

we have

$$P_B(y; p, q) = P_B(1-y; q, p).$$

For $q = 1$, the function $P_B(y; p, 1)$ gives us the standard power-function:

$$yP_B(y; p, 1) = py^p$$

(cf. Ref. [18, p. 210]).

The Beta distribution, which is a member of continuous probability distributions that depend on two positive parameters p and q and is also defined on the intervals $[0, 1]$ or $(0, 1)$, has very important applications in probability and statistics theory. This distribution has been studied in the various fields of work of many researchers in recent years. Since, this distribution is one of the most important distributions applied in a wide variety of disciplines to model revailing the behavior of random variables limited to finite length intervals. In fact, it is well known that this distribution is also a suitable model for the random behavior of percentages and ratios. In addition, the other basis of this distribution in Bayesian inference, Bernoulli, binomial, negative binomial and conjugate priority probability distribution for geometric distributions. It is also well known that the generalization of more than one variable or multivariate beta distribution is also called Dirichlet distribution.

3. Beta-Type Distributions and Their Properties

In this section, we define beta-type distributions in terms of the Bernstein-type basis functions. We investigate some properties of these distributions. We also give symmetry property of these distributions. Moreover, using binomial theorem, we give series representations for these distributions. We give some series and integral representations for these distributions. Combining (11) with (13), we define the following family of the beta type distributions:

$$F(x; a, b; n, m, k) = \frac{(b-a)^{m-n-1} \binom{n}{k}^{-1} \mathbb{Y}_k^n(x; a, b, m)}{B(n-k+1, k+1)}, \qquad (18)$$

where $a \neq b$, $b \geq a$; $n, m \in \mathbb{N}_0$, $k \in \{0, 1, \ldots, n\}$ and $x \in [a, b]$ (cf. Ref. [49]).

In Refs. [49, 50], we studied the beta-type probability density function with their moment generating function and expected value.

It is time to give some properties of the beta-type distributions which is represented by (18).

Putting $m = n$ into equation (18), which was given in Ref. [49], we have

$$F(x; a, b; n, n, k) := F(x; a, b; n, k).$$

By implementing (11), and using the symmetry property our the Bernstein basis functions, $\mathbb{Y}_k^n(x; a, b, m)$, we have

$$\mathbb{Y}_k^n(a + b - x; a, b, m) - \mathbb{Y}_{n-k}^n(x; a, b, m) = 0.$$

In addition to the following symmetry property of the beta function,

$$B(n - k + 1, k + 1) = B(k + 1, n - k + 1),$$

we show that the distribution $F(x; a, b; n, m, k)$ has a symmetry property. That is,

$$F(a + b - x; a, b; n, m, k) = F(x; a, b; n, m, n - k)$$

(cf. Ref. [50]).

For all $x \in [a, b]$, we have

$$F(x; a, b; n, m, k) \geq 0$$

and

$$\int_a^b F(x; a, b; n, m, k) \, dx = 1. \qquad (19)$$

For the proof of equation (19), we need to use the following well-known integral formula:

$$\int_a^b (x - a)^{\alpha - 1} (b - x)^{\beta - 1} \, dx = (b - a)^{\alpha + \beta - 1} B(\alpha, \beta) \qquad (20)$$

(cf. Refs. [30, 41, 45]). Substituting (18) into (19), we have

$$\int_a^b F(x; a, b; n, m, k) \, dx = \frac{\Gamma(n + 2)}{(b - a)^{n - m + 1} \binom{n}{k} \Gamma(n - k + 1) \Gamma(k + 1)}$$

$$\int_a^b \mathbb{Y}_k^n(x; a, b, m) dx.$$

Combining the above equation with (11) yields

$$\int_a^b F(x;a,b;n,m,k)\,dx = \frac{(n+1)\binom{n}{k}}{(b-a)^{n+1}} \int_a^b (x-a)^k (b-x)^{n-k}\,dx. \tag{21}$$

Joining (21) with (20) yields

$$\int_a^b F(x;a,b;n,m,k)\,dx = (n+1)\binom{n}{k} B(k+1, n-k+1).$$

After some elementary calculations in the above equation, we arrive at the desired result.

Substituting $k = u$, $n - k = v$ into (18), we have

$$F_1(x;a,b;u+v,u) = \frac{(x-a)^u (b-x)^v}{(b-a)^{u+v+1} B(v+1, u+1)}, \tag{22}$$

where $a \neq b$, $b \geq a$, $x \in [a,b]$, $u, v \in (0, \infty)$.

Remark 1. Putting $a = 0$, $b = 1$ in (18), it is easy to see that

$$F(x;0,1;n,m,k) = \frac{\mathbb{Y}_k^n(x;0,1,m)}{\binom{n}{k} B(n-k+1, k+1)}.$$

Joining the above equation with (12), for

$$0 \leq x \leq 1$$

and

$$k = 0, 1, 2, \ldots, n,$$

we have

$$(n+1)\binom{n}{k} B(n-k+1, k+1) F(x; 0, 1; n, n, k) = B_k^n(x),$$

which is also denoted as the Newton distribution (cf. Refs. [18, 24, 32, 33, 37, 40]).

When $u, v \in \mathbb{N}_0$, and $a \neq b$, with the aid of (14), (22) we give the following function:

$$F_1(x; a, b; u+v, u) = \frac{u!v!(x-a)^u(b-x)^v}{(b-a)^{u+v+1}(v+u+1)!}$$

and

$$F_1(x; a, b; u+v, u) = \frac{(x-a)^u(b-x)^v}{(b-a)^{u+v+1}(v+u+1)\binom{u+v}{v}},$$

and since $\left|\frac{a}{b}\right| < 1$, we also have

$$F_1(x; a, b; u+v, u)$$
$$= \frac{(x-a)^u(b-x)^v}{b^{u+v+1}(v+u+1)\binom{u+v}{u}} \sum_{j=0}^{\infty} \binom{u+v+j}{j}\left(\frac{a}{b}\right)^j.$$

3.1. Series representations

Interesting formulas are given here with the help of Laplace transform and series representations. These formulas may have very qualified applications in applied sciences. Applying binomial theorem to the sum of generating functions, we give some series representations for the function $F(x; a, b; n, m, k)$.

Joining (18) with (24) and (11) yields

$$\sum_{n=k}^{\infty} \frac{F(x; a, b; n, m, k)}{n+1} = \frac{(x-a)^k}{(b-a)^{k+1}} \sum_{n=0}^{\infty} \left(\frac{b-x}{b-a}\right)^n.$$

If $\left|\frac{b-x}{b-a}\right| < 1$, then the above equation reduces to

$$\sum_{n=k}^{\infty} \frac{F(x; a, b; n, m, k)}{n+1} = \frac{(x-a)^k}{(b-a)^{k+1}} \left(\frac{1}{1-\frac{b-x}{b-a}}\right).$$

Therefore,

$$\sum_{n=k}^{\infty} \frac{F(x; a, b; n, m, k)}{(n+1)(x-a)} = \left(\frac{x-a}{b-a}\right)^k.$$

When $a = 0$ and $b = 1$, the previous equation reduces to
$$\sum_{n=k}^{\infty} \frac{F(x;0,1;n,m,k)}{(n+1)} = x^{k+1} - x^k.$$

Using (11) and (18) yields
$$\sum_{n=0}^{\infty} \frac{F(x;a,b;n,m,k)}{\binom{n}{k}} = \frac{\left(\frac{x-a}{b-a}\right)^k}{(b-a)} \sum_{n=0}^{\infty} (n+1) \left(\frac{b-x}{b-a}\right)^n.$$

From the above equation, assuming that $\left|\frac{b-x}{b-a}\right| < 1$, we have
$$\sum_{n=0}^{\infty} \frac{F(x;a,b;n,m,k)}{\binom{n}{k}} = \frac{\left(\frac{x-a}{b-a}\right)^k}{(b-a)} \left(\sum_{n=0}^{\infty} n \left(\frac{b-x}{b-a}\right)^n + \frac{b-a}{x-a}\right).$$

Combining the above equation with the following well-known formulas
$$\sum_{n=1}^{\infty} n w^n = \frac{w}{(1-w)^2},$$

where $|w| < 1$, after some elementary calculations, we arrive at the following result, which can be proved by the authors [49, Theorem 4] with the same method of Simsek [36–39] involving the Laplace transform to the generating functions for the Bernstein basis functions and sum of generating function for this basis:

Theorem 1.
$$\sum_{n=0}^{\infty} \frac{F(x;a,b;n,m,k)}{\binom{n}{k}} = \frac{(b-a)(x-a)^{k-2}}{(b-x)^k}, \tag{23}$$

where $\left|\frac{b-x}{b-a}\right| < 1$, $x \neq a$ and $x \neq b$.

By using not only
$$\sum_{n=k}^{\infty} \binom{n}{k} x^{n-k} = \frac{1}{(1-x)^{k+1}} \tag{24}$$

(cf. Ref. [13, Equation (1.9)]), but also with similar processing steps and the method given by Simsek [36–39], the following results can be given to the readers:

Theorem 2 (cf. Ref. [49, Theorem 4]).
$$\sum_{n=k}^{\infty} \frac{F(x; a, b; n, m, k)}{n+1} = \frac{1}{x-a}, \qquad (25)$$
where $\left|\frac{b-x}{b-a}\right| < 1$, $x \neq a$ and $x \neq b$.

Substituting $m = n$, $a = 0$ and $b = 1$ into (23) and (25), we have the following known result, respectively:
$$\sum_{n=k}^{\infty} \frac{F(x; 0, 1; n, n, k)}{\binom{n}{k}} = \frac{1}{(1-x)^2}$$
and
$$\sum_{n=k}^{\infty} \frac{F(x; 0, 1; n, n, k)}{n+1} = \frac{1}{x}$$
(cf. Refs. [36–39]).

4. Moment Generating Functions and Characteristic Function

In probability theory and statistics, it is well know that the moment generating function of a real-valued random variable is an expected value of e^{xt}. This function is denoted by $M_x(t)$. Mean, variance, skewness, kurtosis, etc. when studied with statistical moments (continuous or discrete for a given probability distribution) such as moment generating functions play a vital role. Moment generating functions are also useful to find explicit values the moments of a random variable and expected value or mean and also variance, etc.

If x is a discrete random variable, then
$$M_x(t) = \mathbb{E}(e^{xt}) = \sum_{j=0}^{\infty} e^{jt} f(x),$$
where $f(x)$ denotes the probability density functions of x.

If x is a continuous random variable x, then

$$M_x(t) = \mathbb{E}(e^{xt}) = \int_{-\infty}^{+\infty} e^{jt} f(x) dx,$$

where $f(x)$ denotes the probability distribution of x. Thus, there are many applications of the moment generating functions and their properties (cf. Refs. [10, 25, 32, 33, 40, 44, 50] and the references cited therein).

4.1. *Moment generating function for beta-type distributions*

In Refs. [49, 50], we introduced moment generating function for beta-type distributions. We gave moments and expected value for the logarithm of random variable of the beta-type distributions involving generalized harmonic number, the Stirling numbers and the polygamma function. In Refs. [49, 50], we also gave numerical values of the expected values and the logarithm of the geometric mean of a distribution with random variables, and also the arithmetic mean of $E(\ln(X))$. It is well known that $E(\ln(X))$ has also been many applications in mathematical models, information quantities (entropy), Bayesian inference and probability integral transformation.

We [49, 50] gave the following moment generating function for the distribution $F(x; a, b; n, m, k)$:

$$M_x(t; a, b; n, m, k) = \int_a^b F(x; a, b; n, m, k) e^{xt} dx$$

or

$$M_x(t; a, b; n, m, k)$$
$$= \sum_{j=0}^{\infty} \left(\sum_{c=0}^{k} (-1)^{k-c} \binom{k}{c} \frac{B(c+j+1, n-k+1) - a^{k-c} b^{c+j+n-k+1}}{(b-a)^{n+1} B(n-k+1, k+1)} \right) \frac{t^j}{j!}.$$

By using the function $M_x(t; a, b; n, m, k)$, we have

$$\mu_l(a, b; n, m, k) = \mathbb{E}(X^l) = \frac{\partial^l}{\partial t^l} \{M_x(t; a, b; n, m, k)\}|_{t=0}$$

(cf. Refs. [49, 50]).

The expected value or the mean of a beta-type distribution random variable X with 5 parameters a, b, n, m and k is given by

$$\mu_1(a,b;n,m,k) = \mathbb{E}(X) = \frac{\partial}{\partial t}\{M_x(t;a,b;n,m,k)\}|_{t=0}.$$

The expected value or the variance of a beta-type distribution random variable X with 5 parameters a, b, n, m and k is given by

$$\sigma^2(a,b;n,m,k) = \mu_2(a,b;n,m,k) - \mu_1^2(a,b;n,m,k).$$

The harmonic mean H_X of a beta-type distribution $F(x;a,b;n,m,k)$ with shape parameters p and q is given by

$$H_X(a,b;n,m,k) = \frac{1}{\mathbb{E}\left(\frac{1}{X};a,b;n,m,k\right)}$$

$$= \frac{1}{\int_a^b \frac{F(x;a,b;n,m,k)}{x}dx}.$$

The expected value or the mean of a beta-type distribution for $e^{-t\ln(x-a)}$ is given by the following theorem:

Theorem 3 (cf. Refs. [49, 50]). *Let $n, m \in \mathbb{N}_0$, $k \in \{0, 1, \ldots, n\}$ and $x \in [a,b]$. Then, we have*

$$\mathbb{E}\left(e^{-t\ln(x-a)}\right) = \sum_{v=0}^{\infty}\sum_{j=0}^{v}\sum_{l=0}^{v}(-1)^{v-l+j}\binom{v}{l}$$

$$\times \frac{(b-a)^{m+l}S_1(v,j)B(k+l+1,n-k+1)}{v!B(n-k+1,k+1)}t^j,$$

(26)

where $x - a > 0$ and $b \neq a$.

By using (26), we have the following result:

Theorem 4 (cf. Refs. [49, 50]). *Let $n, m \in \mathbb{N}_0$, $k \in \{0,1,\ldots,n\}$. Then, we have*

$$\sum_{d=0}^{\infty}\sum_{k=0}^{d}(-1)^k\binom{d}{k}\frac{(a+1)^{d-k}\mu_k}{d!}t(t+1)\ldots(t+d-1)$$

$$= \sum_{v=0}^{\infty} \sum_{j=0}^{v} \sum_{l=0}^{v} (-1)^{v-l+j} \binom{v}{l}$$

$$\times \frac{(b-a)^{m+l} S_1(v,j) B(k+l+1, n-k+1)}{v! B(n-k+1, k+1)} t^j, \quad (27)$$

where $b \neq a$ and

$$\mu_d = \mathbb{E}(x^d; a, b; n, m, k).$$

We now give the following mean absolute difference for the Beta type distribution:

$$MA_B(a, b; n, m, k) := \int_a^b \int_a^b F(x; a, b; n, m, k) F(y; a, b; n, m, k)$$
$$\times |x-y|\, \mathrm{d}x\mathrm{d}y.$$

The following formula gives us mean absolute difference for the beta distribution:

$$MA_B(p, q) = \int_0^1 \int_0^1 P_B(x; p, q) P_B(y; p, q) |x-y|\, \mathrm{d}x\mathrm{d}y$$

(cf. Ref. [52]), which yields

$$MA_B(p, q) = \frac{4 B(p+q, p+q)}{(p+q) B(p,p) B(q,q)}.$$

We also give the following formula:

$$MA_B(p, q) = \frac{(2\Gamma(p+q))^2 \Gamma(2p) \Gamma(2q)}{(p+q) \Gamma(2p+2q) (\Gamma(p)\Gamma(q))^2}.$$

When $p = c+1$ and $q = d+1$ with $c, d \in \mathbb{N}$, we have the following interesting formula:

$$MA_B(c, d) = \frac{4(2c+1)!(2d+1)!}{(c+d+2)(2c+2d+3)!} \left(\frac{(c+d)!}{c!d!}\right)^2$$

and

$$MA_B(c, d) = \frac{4(2c+1)(2d+1)}{(c+d+2)(2c+2d+3)(2c+2d+2)(2c+2d+1)}$$
$$\times \frac{\binom{2c}{c}\binom{2d}{d}}{\binom{2c+2d}{c+d}}.$$

4.2. Formulas expected value of beta-type distributions

In Refs. [49, 50], we gave many formulas including the polygamma functions, the digamma function and the expected value for the logarithm of random variable of the beta-type distributions.

The following theorem gives us the logarithm of the geometric mean of the beta-type distribution with random variable X is the arithmetic mean of $\ln(x - a)$.

Theorem 5. *Let $x - a > 0$, $b \neq a$ and $j \in \mathbb{N}_0$. Then, we have*

$$\mathbb{E}\left((\ln(x-a))^j\right) = \int_a^b \ln(x-a))^j F(x; a, b; n, m, k) \, dx. \tag{28}$$

In Refs. [49, 50], we gave explicit values of (28), by the following theorem:

Theorem 6 (cf. Refs. [49, 50]). *Let $j \in \mathbb{N}_0$. Then, we have*

$$\mathbb{E}\left((\ln(x-a))^j\right) = \frac{1}{B(v+1, u+1)} \sum_{m=0}^{j} \binom{j}{m} \ln^m (b-a) \frac{\partial^{j-m}}{\partial u^{j-m}}$$
$$\times \{B(u+1, v+1)\}, \tag{29}$$

where $x - a > 0$ and $b \neq a$.

Combining the above equation with the following known result (cf. Ref. [52])

$$\mathbb{E}(\ln(1-x)) = \psi(u) - \psi(u+v), \tag{30}$$

we get

$$\mathbb{E}(\ln(1-x)) = \sum_{n=1}^{\infty} H_n(\mu_{n+1} - \mu_n), \tag{31}$$

$$\mathbb{E}\left(\ln(x-x^2)\right) = \psi(u) + \psi(v) - 2\psi(u+v)$$

and

$$\mathbb{E}\left(\ln\left(\frac{1-x}{x}\right)\right) = \psi(u) - \psi(v).$$

Combining the above equation with the following equation:

$$\mathbb{E}\left(\ln(1-x)\right) = \sum_{n=1}^{\infty} \frac{\mu_n}{n}. \tag{32}$$

Combining (31) with (32) yields the following result:

Corollary 1.

$$\sum_{n=1}^{\infty} \frac{\mu_n}{n} = \sum_{n=1}^{\infty} (\mu_{n+1} - \mu_n) H_n.$$

We also have

$$\mathbb{E}\left(\ln\left(\frac{1-x}{x}\right)\right) = \varkappa(u) - \varkappa(v),$$

where $u, v \notin \mathbb{Z}^- = \{0, -1, -2, \ldots\}$ and

$$\varkappa(u) = \sum_{n=0}^{\infty} \left(\frac{1}{n+1} - \frac{1}{n+u}\right)$$

and

$$\varkappa(v) = \sum_{n=0}^{\infty} \left(\frac{1}{n+1} - \frac{1}{n+v}\right).$$

For $j = 1$, we have

$$\mathbb{E}\left(\ln(x-a)\right) = \ln(b-a) + \psi(u+1) - \psi(u+v+2).$$

For $a = 0$ and $b = 1$, the above equation reduces to the the expected value of $\ln x$ for the beta distribution. That is, with the aid of (17) and (29), we get

$$\mathbb{E}\left(\ln(x)\right) = \int_0^1 \ln x P_B(x; p, q) \mathrm{d}x$$

or

$$\mathbb{E}\left(\ln(x)\right) = \psi(p) - \psi(p+q)$$

(cf. Refs. [18, 52]).

By applying the Newton–Mercator series for $\ln(x)$, we have
$$\mathbb{E}\left(\ln(x)\right) = \sum_{j=1}^{\infty} \frac{(-1)^{j-1}}{j} \sum_{k=0}^{j} (-1)^{j-k} \binom{j}{k} \mathbb{E}\left(x^k\right).$$
Since
$$\mathbb{E}\left(x^k\right) = \mu_k,$$
we get the following result:

Corollary 2.
$$\psi(p) - \psi(p+q) = \sum_{j=1}^{\infty} \frac{(-1)^{j-1}}{j} \sum_{k=0}^{j} (-1)^{j-k} \binom{j}{k} \mu_k.$$

The harmonic mean H_X of a beta distribution with shape parameters p and q is given by
$$H_X = \frac{1}{\mathbb{E}\left(\frac{1}{X}\right)}$$
$$= \frac{1}{\int_0^1 \frac{P_B(x;p,q)}{x} dx}.$$
After some elementary calculations, we have
$$H_X = \frac{p-1}{p+q-1},$$
where $p > 1$ and $q > 0$ (cf. Refs. [18, 52]).

Remark 2. By implementing discrete random variable and the Bernstein basis functions, binomial distribution and moment generating functions have been given [12, 24, 25, 32, 33, 40]. Using Equation (10), Simsek [32] gave the following moment generating functions:
$$B_k^n(x; a, b) = \binom{n}{k} \left(\frac{x-a}{b-a}\right)^k \left(\frac{b-x}{b-a}\right)^{n-k},$$
where $k = 0, 1, \ldots, n$ and $x \in [a, b]$ and also $0 \leq \left|\frac{x-a}{b-a}\right| \leq 1$, $0 \leq \left|\frac{b-x}{b-a}\right| \leq 1$, that is, $M_x(t, x : n; a, b)$ is given by
$$M_x(t, x : n; a, b) = \left(e^t \frac{x-a}{b-a} + \frac{b-x}{b-a}\right)^n.$$

Remark 3. Substituting $a = 0$ and $b = 1$ into the above function, we have the following well-known formula for the moment generating function of the Binomial distribution:

$$M_x(t, x : n; 0, 1) = \left(e^t x + 1 - x\right)^n$$

(cf. Refs. [11, 24, 25, 32, 33, 40]). By using (18), the authors [49] gave the following interesting formula for the moments $\mu_l(a, b; n, k)$:

$$\mu_l(a, b; n, k) = \sum_{c=0}^{k} (-1)^{k-c} \binom{k}{c} a^{k-c} b^{l+c+n-k+1} (b-a)^{-n-1}$$

$$\times \frac{B(c+l+1, n-k+1)}{B(n-k+1, k+1)}.$$

4.3. *Formula for moment generating function for the distribution $F(x; a, b; n, k)$*

The authors [49] gave the following moment generating function for the distribution $F(x; a, b; n, k)$

$$M_x(t; a, b; n, k)$$

$$= \sum_{j=0}^{\infty} \sum_{d=0}^{k} \frac{\binom{k}{d}(-a)^{k-d} b^{d+j+n+1-k} B(d+j+1, n-k+1)}{j!(b-a)^{n+1} B(n-k+1, k+1)} t^j$$

(33)

and

$$M_x(vt; a, b; n, k) = \sum_{l=0}^{v} \binom{v}{l} l! \sum_{j=0}^{\infty} S_2(j, l) \frac{t^j}{j!} \int_a^b x^j F(x; a, b; n, k) \, dx.$$

Since

$$\mu_l(a, b; n, m, k) = \int_a^b x^l F(x; a, b; n, k) \, dx \qquad (34)$$

(cf. Ref. [49, p. 9]), we have

$$M_x(vt; a, b; n, k) = \sum_{j=0}^{\infty} \mu_j(a, b; n, k) \sum_{l=0}^{v} v(v-1) \dots (v-l+1) S_2(j, l) \frac{t^j}{j!}.$$

(35)

5. Characteristic Function for Beta-Type Distributions

In probability theory and statistics, the characteristic function of any real-valued random variable describes its probability distribution. It is well known that if a random variable accepts a probability density function, the characteristic function is the Fourier transform of the probability density function. It is also well known that, unlike the moment-generating function, the characteristic function always exists when considered as a function of a real-valued argument. There are relationships between the behavior of a distribution's characteristic function and properties of the distribution, such as the existence of moments and the existence of the density function.

Let X be a random variable of the probability distribution $f(x)$. The characteristic function and moment generating functions of the random variable X are given, respectively, by

$$K_x(t) = \mathbb{E}(e^{itx})$$

and

$$M_x(t) = \mathbb{E}(e^{tx}),$$

where $\mathbb{E}(X)$ denotes the expected value or mean of the random variable X (cf. Refs. [25, 32, 33, 44] and the references cited therein). In Ref. [25], for distribution function $f(x)$, Lukacs gave various proprieties of the characteristic function $K_x(t)$:

$$K_x(t) = \int_{-\infty}^{\infty} f(x) \exp(itx) \mathrm{d}x. \qquad (36)$$

By using (36), we have $K(0) = 1$. It is easy to see that

$$|K(t)| \leq 1$$

and

$$K_x(-t) = \overline{K_x(t)},$$

where $\overline{K_x(t)}$ is a the complex conjugate of $K_x(t)$. $K_x(t)$ is uniformly continuous on the field of $\mathbb{R}$ (cf. Refs. [18, 25, 32, 33, 44]).

Since $\mathbb{E}(e^{itx})$ is a linear function with respect to its variable, we have

$$K_x(t) + M_x(t) = \mathbb{E}(e^{itx} + e^{tx}).$$

Combining the following Euler formula with the above equation, we get

$$K_x(t) + M_x(t) = \mathbb{E}(\cos(tx) + e^{tx}) + i\mathbb{E}(\sin(tx)).$$

Thus,

$$K_x(t) = \mathbb{E}(\cos(tx)) + i\mathbb{E}(\sin(tx)).$$

In Ref. [32], Simsek gave many novel identities and results involving the characteristic function, moment generating functions for the Bernstein basis functions and other special numbers and polynomials. Here, we can use some of them.

We [50] gave characteristic function of beta-type distribution $F(x; a, b; n, m, k)$. Here, we also investigate some properties of this function.

By applying equation (36) to equation (18), we get

$$K_x(t; a, b, n, k) = \int_a^b F(x; a, b; n, m, k) e^{itx} dx, \qquad (37)$$

where $a, b \in \mathbb{R}$ with $a \neq b$, $b > a$; $t \in \mathbb{C}$; $n, m \in \mathbb{N}_0$, $k \in \{0, 1, \ldots, n\}$. After some calculations, we obtain

$$K_x(t; a, b, n, k) = \sum_{v=0}^{k} \sum_{d=0}^{n-k} \frac{(-1)^{n-d-v} \binom{k}{v} \binom{n-k}{d} b^d (-a)^{k-v}}{(b-a)^{n+1} B(n-k+1, k+1)}$$

$$\times \int_a^b x^{n+v-k-d} e^{itx} dx$$

(cf. Ref. [50]). Joining the above equation with the following known integral formula

$$\int x^n e^{cx} dx = e^{cx} \sum_{j=0}^{n} \frac{(-1)^{n-j} n!}{j! c^{n-j+1}} x^j$$

(cf. Refs. [14, 26, 51]), we yields

$$K_x(t; a, b, n, k)$$
$$= \sum_{v=0}^{k}\sum_{d=0}^{n-k}(-1)^{n-d-v}\frac{\binom{k}{v}\binom{n-k}{d}b^d(-a)^{k-v}}{(b-a)^{n+1}B(n-k+1,k+1)}$$
$$\times \sum_{j=0}^{n+v-k-d}\frac{(-1)^{n+v-k-d+j}(n+v-k-d)!(b^j e^{itb} - a^j e^{ita})}{j!(it)^{n+v-k-d+1-j}},$$
(38)

where $a, b \in \mathbb{R}$ with $a \neq b$, $b > a$, $t \in \mathbb{C}$, $n, m \in \mathbb{N}_0$, $k \in \{0, 1, \ldots, n\}$ (cf. Ref. [50]).

Substituting $a = 0$ and $b = 1$ into (38), we have

$$K_x(t; 0, 1, n, k) = (-1)^k \frac{(n+1)!}{k!} \sum_{d=0}^{n-k} \frac{(n-d)!(e^{it}-1)}{(it)^{n-d+1}d!(n-k-d)!} \quad (39)$$

(cf. Ref. [50]).

Combining (39) with (6) yields the following result:

Corollary 3.

$$K_x(t; 0, 1, n, k) = (-1)^k \frac{(n+1)!}{k!} \sum_{a=0}^{\infty}\sum_{d=0}^{n-k} \frac{(n-d)! S_2(a,1) i^n}{(it)^{n-d+1} d!(n-k-d)!} \frac{t^n}{n!}.$$

Substituting $a = 0$ and $b = 1$ into (38), after some elementary calculations, we have the following characteristic functions of the Beta probability density function:

$$\mathbb{E}(e^{it}) = \int_0^1 e^{itx} P_B(x; p, q) dx = 1 + \sum_{k=0}^{\infty}\prod_{r=0}^{k-1} \frac{p+r}{p+q} \frac{(it)^k}{k!}$$

(cf. Refs. [6, 7, 10, 52]).

Theorem 7 (cf. Ref. [50]). *Let $a, b \in \mathbb{R}$ with $a \neq b$, $b > a, t \in \mathbb{C}, n \in \mathbb{N}_0, k \in \{0, 1, \ldots, n\}$. Then we have*

$$K_x(t; a, b, n, k) = \sum_{v=0}^{k} \sum_{d=0}^{n-k} (-1)^{n-d-v} \binom{k}{v} \binom{n-k}{d}$$

$$\times \frac{b^d(-a)^{k-v}}{(b-a)^{n+1} B(n-k+1, k+1)}$$

$$\times \sum_{j=0}^{n+v-k-d} \frac{(-1)^j j! \binom{n+v-k-d}{j} (e^{itb} b^{n+v-k-d-j} - e^{ita} a^{n+v-k-d-j})}{(it)^{j+1}}.$$

Putting $a = 0$ and $b = 1$ in Theorem 7, we have

$$K_x(t; 0, 1, n, k) = (n+1) \binom{n}{k} \sum_{d=0}^{n-k} \frac{(-1)^k (n-d)! \binom{n-k}{d} (e^{it} - 1)}{(it)^{n-d+1}}$$

(cf. Ref. [50]).

Here we also note that the probability density function of any continuous random variable x is obtained uniquely by its characteristic function. That is,

$$2\pi f(x) = \int_{-\infty}^{+\infty} e^{-itx} K_x(t) dt$$

(cf. Refs. [10, 25, 44]).

Consequently, in line with the previous formula and (37), we reach the following interesting result:

Theorem 8.

$$F(x; a, b; n, m, k) = \frac{1}{2\pi} \int_a^b K_x(t; a, b, n, m, k) e^{-itx} dt.$$

6. Approximation Properties of the Bernstein Polynomials

Approximation theory is one of the important branches of mathematics that is frequently used in fields, such as functional analysis, probability theory and functions theory. There are many methods of approach. It is well known in the literature that *"approximation"* is a method often used in cases where an exact form or an exact numerical number is unknown or difficult to obtain. Therefore, some known forms may exist and represent the true form without significant deviation. In addition, regarding the approximation of functions or the approximation of functions with polynomials, the asymptotic value of a function, that is, the value of one or more parameters of a function, becomes arbitrarily large. Generally, $\sim$ is used to mean $\approx$ or asymptotically equal.

As for the approximation properties of the Bernstein polynomials by the aid of probabilistic tools, we need the following definitions and relations:

Throughout of this chapter, we assume that f is a real function defined on the interval $[0,1] \subset \mathbb{R}$, and $P_{n,x}$ is a binomial random variable with parameters n and x, and $\mathbb{E}[X]$ denotes the expected value of X. The Bernstein polynomials is defined by

$$B_{k,n}[f(x)] = \sum_{j=0}^{n} f\left(\frac{j}{n}\right) B_j^n(x) := \mathbb{E}\left[f\left(\frac{P_{n,x}}{n}\right)\right]$$

to the limit function f, when the f being approximated is a Lipschitz continuity. The rate $\mathcal{O}\left(n^{\frac{1}{3}}\right)$ obtained with Bernstein's classic probabilistic proof, where all that is used is Chebyshev's inequality, was improved to $\mathcal{O}\left(\left(\frac{\ln n}{n}\right)^{\frac{1}{2}}\right)$ (cf. Refs. [8, 16, 24, 30] and the references cited therein). In Ref. [47], Totik gave the following elegant formula: Since $f \in C[0,1]$, and for step-weight function

$$\beta(x) = \sqrt{x(1-x)}$$

and

$$w_\beta(f, \alpha) = \sup_{0 \leq t \leq \alpha} \sup_x |f(x - t\beta(x)) - 2f(x) + f(x + t\beta(x))|$$

good the approximation via the Bernstein polynomials is given by

$$\|B_{k,n}[f(x)] - f(x)\| \sim w_\beta(f, \alpha),$$

where $\sim$ means that the ratio of the two sides lies in between independently of $f \in C[0,1]$ and n. It is well known that Bernstein polynomials give one way to prove the Weierstrass approximation theorem that every real-valued continuous function on a real interval $[a,b]$ can be uniformly approximated by polynomial functions over interval $(-\infty, +\infty) := \mathbb{R}$. That is, $|B_{k,n}[f(x)] - f(x)| \to 0$ as $n \to \infty$ (cf. Refs. [8, 16, 24, 30, 38]).

Recently, the approximation properties of the beta distribution have been studied (cf. Ref. [27]).

There are many methods and techniques in the literature regarding the approximation of some special functions with Bernstein polynomials. Our aim here is not to study this subject in depth but only to provide the reader with brief information about this subject. In our opinion, it can only be emphasized here that studying the approximation of some special functions with Bernstein-type polynomials and giving results on them is within the scope of a separate project, and for this purpose, it would be appropriate to study more additional information of the approximation theory.

We note that in the light of equations (15) and (16), after some basic operations and analysis methods are performed, the approximation formulas and asymptotic formula for the beta probability density function and the beta-type distributions can be studied in detail. Applications of this approach formulae may be applied for different fields. These results could also naturally be left to the researchers.

References

[1] M. Abramowitz and I. A. Stegun, *Handbook of Mathematical Functions with Formulas, Graphs, and Mathematical Tables*. National Bureau of Standards Applied Mathematics Series, Vol. 55 (Tenth Printing, Dover Publication INC., 1972).

[2] M. Acikgoz and S. Araci, On generating function of the Bernstein polynomials, *Numer. Anal. Appl. Math. Am. Inst. Phys. Conf. Proc.* **CP1281**, 1141–1143 (2010).

[3] F. Bagarello, F. Gargano, S. Spagnolo and S. Triolo, Coordinate representation for non-Hermitian position and momentum operators, *Proc. R. Soc. A Math. Phys. Eng. Sci.* **20170434**, 1–13 (2017).

[4] A. Bayad, T. Kim, S. H. Lee and D. V. Dolgy, A note on the generalized Bernstein polynomials, *Honam Math. J.* **33**, 1–11 (2011).
[5] S. N. Bernstein, Démonstration du théorème de Weierstrass fondée sur la calcul des probabilités, *Comm. Soc. Math. Charkow Sér.* **13**, 1–2 (1912).
[6] C. A. Charalambides, *Enumerative Combinatorics* (Chapman and Hall/CRC, Boca Raton, London, New York, 2018).
[7] L. Comtet, *Advanced Combinatorics: The Art of Finite and Infinite Expansions* (Springer Science and Business Media, Dordrecht, Holland, 1974).
[8] R. T. Farouki, *Pythagorean-Hodograph Curves: Algebra and Geometry Inseparable* (Springer, Berlin, 2008).
[9] R. T. Farouki, The Bernstein polynomial basis: A centennial retrospective, *Comput. Aided Geom. Des.* **29**, 379–419 (2012).
[10] W. Feller, *An Introduction to Probability Theory and Its Applications*, Vol. 1, 3rd edn. (Wiley, New York, 1968).
[11] R. N. Goldman, Identities for the univariate and bivariate Bernstein basis functions. In *Graphics Gems V* (Academic Press, Elsevier, 1995), pp. 149–162.
[12] R. Goldman, P. Simeonov and Y. Simsek, Generating functions for the q-Bernstein bases, *SIAM J. Discrete Math.*, **28**(3), 1009–1025 (2014).
[13] H. W. Gould, *Fundamentals of Series: Table II: Examples of Series Which Appear in Calculus.* In J. Quaintance (ed.) (Philadelphia, 2010), https://math.wvu.edu/~hgould/Vol.2.PDF (8 June 2024).
[14] I. S. Gradshteyn and I. M. Ryzhik, *Table of Integrals, Series, and Products* (Academic Press, New York, 2014).
[15] A. K. Gupta and S. Nadarajah, *Handbook of Beta Distribution and Its Applications*, 1st edn. (CRC Press, Boca Raton, 2004).
[16] H. Gzyl and J. L. Palacios, On the approximation properties of Bernstein polynomials via probabilistic tools, *Bol. Asoc. Mat. Venez.* **10**(1), 5–13 (2003).
[17] L. C. Jang, W. J. Kim and Y. Simsek, A study on the-adic integral representation on associated with Bernstein and Bernoulli polynomials, *Adv. Differ. Equ.* **2010**, 1–6 (2010).
[18] N. L. Johnson, S. Kotz and N. Balakrishnan, *Continuous Univariate Distributions; Volume 2* (John Wiley & Sons, Hoboken, New Jersey, 1995).
[19] A. M. Khidr and B. S. El-Desouky, A symmetric sum involving the Stirling numbers of the first kind. *Eur. J. Comb.* **5**, 51–54 (1984).
[20] M. S. Kim, D. Kim and T. Kim, On the q-Euler numbers related to modified q-Bernstein polynomials, *Abstr. Appl. Anal.* **820109**, 1–15 (2010).

[21] W. J. Kim, D. S. Kim, H. Y. Kim and T. Kim, Some identities of degenerate Euler polynomials associated with degenerate Bernstein polynomials, *J. Inequal. Appl.* **2019**, 160, 1–11 (2019). https://doi.org/10.1186/s13660-019-2110-y.
[22] I. Kucukoglu and Y. Simsek, A note on generating functions for the unification of the Bernstein-type basis functions, *Filomat* **30**(4), 985–992 (2016).
[23] I. Kucukoglu, B. Simsek and Y. Simsek, Multidimensional Bernstein polynomials and Bezier curves: Analysis of machine learning algorithm for facial expression recognition based on curvature, *Appl. Math. Comput.* **344**, 150–162 (2019).
[24] G. G. Lorentz, *Bernstein Polynomials* (Chelsea Publishing Company, New York, 1986).
[25] E. Lukacs, *Characteristic Function*, 2nd edn. (Charles Griffin & Company Limited, London, 1970).
[26] V. H. Moll, *Numbers and Functions. From a Classical-Experimental Mathematician's Point of View*, Vol. 65 (American Mathematical Society, Providence, Rhode Island, 2012).
[27] V. M. Nawa and S. Nadarajah, New closed form estimators for the beta distribution, *Mathematics* **11**, 2799, 1–20 (2023).
[28] D. C. D. Oguamanam, H. R. Martin and J. P. Huissoon, On the application of the beta distribution to gear damage analysis, *Appl. Acoust.* **45**, 247–261 (1993).
[29] G. Rahman, S. Mubeen, A. Rehman and M. Naz, On k-gamma and k-beta distributions and moment generating functions, *J. Probab. Stat.* **982013**, 1–6 (2014).
[30] E. D. Rainville, *Special Functions* (Macmillan Company, New York, 1960).
[31] J. Riordan, *Introduction to Combinatorial Analysis* (Princeton University Press, Princeton, 1958).
[32] B. Simsek, Formulas derived from moment generating functions and Bernstein polynomials, *Appl. Anal. Discret. Math.* **13**, 839–848 (2019).
[33] B. Simsek and B. Simsek, The computation of expected values and moments of special polynomials via characteristic and generating functions, *AIP Conf. Proc.* **1863**, 300012 (2017).
[34] Y. Simsek, Interpolation function of generalized q-Bernstein type polynomials and their application. Lecture Notes in Computer Science, Vol. 6920 (Springer-Verlag, Berlin, 2011), pp. 647–662.
[35] Y. Simsek, Construction a new generating function of Bernstein-type polynomials, *Appl. Math. Comput.* **218**, 1072–1076 (2011).
[36] Y. Simsek, On q-deformed Stirling numbers, *Int. J. Math. Comput.* **15**, 1–11 (2012).

[37] Y. Simsek, q-beta polynomials and their applications, *Appl. Math. Inf. Sci.* **7**, 2539–2547 (2013).
[38] Y. Simsek, Functional equations from generating functions: A novel approach to deriving identities for the Bernstein basis functions, *Fixed Point Theory Appl.* **2013**, 1–13 (2013).
[39] Y. Simsek, Generating functions for generalized Stirling type numbers, array type polynomials, Eulerian type polynomials and their applications, *Fixed Point Theory Appl.* **87**, 1–28 (2013).
[40] Y. Simsek, Generating functions for the Bernstein type polynomials: A new approach to deriving identities and applications for the polynomials, *Hacet. J. Math. Stat.* **43**, 1–14 (2014).
[41] Y. Simsek, Analysis of the Bernstein basis functions: An approach to combinatorial sums involving binomial coefficients and Catalan numbers, *Math. Meth. Appl. Sci.* **38**, 3007–3021 (2015).
[42] Y. Simsek, Explicit formulas for p-adic integrals: Approach to p-adic distributions and some families of special numbers and polynomials, *Montes Taurus J. Pure Appl. Math.* **1**, 1–76 (2019).
[43] Y. Simsek, Some classes of finite sums related to the generalized harmonic functions and special numbers and polynomials, *Montes Taurus J. Pure Appl. Math.* **4**(3), 61–79 (2022).
[44] T. T. Soong, *Fundamentals of Probability and Statistics for Engineers* (John Wiley & Sons Ltd., Chichester, England, 2004).
[45] H. M. Srivastava and J. Choi, *Series Associated with the Zeta and Related Functions* (Kluwer Academic Publishers, Dordrecht, 2001).
[46] H. M. Srivastava and J. Choi, *Zeta and q-Zeta Functions and Associated Series and Integrals* (Elsevier Science Publishers, Amsterdam, 2012).
[47] V. Totik, Approximation by Bernstein polynomials, *Am. J. Math.* **116**(4), 995–1018 (1994).
[48] C. Trapani, S. Triolo and F. Tschinkei, Distribution frames and bases, *J. Fourier Anal. Appl.* **25**(4), 2109–2140 (2019).
[49] F. Yalcin and Y. Simsek, A new class of symmetric beta type distributions constructed by means of symmetric Bernstein type basis functions, *Symmetry*, **12**(5), 779 (2020).
[50] F. Yalcin and Y. Simsek, Formulas for characteristic function and moment generating functions of beta type distribution. *Rev. Real Acad. Cienc. Exactas FÃsicas Nat. Ser. A. Mat.* **116**(2), 86 (2022).
[51] https://en.wikipedia.org/wiki/List_of_integrals_of_exponential_functions. 25.11.2021.
[52] https://en.wikipedia.org/wiki/Beta_distribution. 01.12.2021.
[53] https://en.wikipedia.org/wiki/Digamma_function. 01.12.2021.
[54] https://en.wikipedia.org/wiki/Beta_function. 20.09.2023.

© 2024 World Scientific Publishing Company
https://doi.org/10.1142/9789811267048_0026

Chapter 26

On Removing Diverse Data for Training Machine Learning Models

Kim Thuyen Ton[*], Daniel Aloise[†], and Claudio Contardo[‡]

Polytechnique Montréal and GERAD, Montréal, Canada
Concordia University and GERAD, Montréal, Canada

[*] kim-thuyen.ton@polymtl.ca
[†] daniel.aloise@polymtl.ca
[‡] claudio.contardo@concordia.ca

Providing the right data to a machine learning model is an important step to insure the performance of the model. Noncompliant training data instances may lead to wrong predictions yielding models that cannot be used in production. Instance or prototype selection methods are often used to curate training sets thus leading to more reliable and efficient models. In this work, we investigate if diversity is helpful as a criterion for choosing which instances to remove from a given training set. We test our hypothesis against a random selection method and Mahalanobis outlier selection, using benchmark datasets with different data characteristics. Our computational experiments demonstrate that selection by diversity achieves better classification performance than random selection and can hence be considered as an alternative data selection criterion for effective model training.

1. Introduction

Machine learning (ML) has often been perceived as requiring the largest possible amount of data to gain accuracy in predicting a behavior. Typically, there are three stages for an ML model: preprocessing, training and decision/prediction [1]. During preprocessing, the provided training set might be transformed before being fed to the ML model. In the sequel, during the training stage, the model processes the training set to generalize rules and formulas for prediction with a minimum amount of classification errors. Finally, at the prediction stage, new unlabeled data instances are given to the ML model which must predict a class or a value for them.

Nowadays, several preprocessing methods exist to ensure that only the right data are given to an ML model — a concern that accompanies the ML field since its origin [2, 3]. For example, a model that overfits or underfits the training data will result in poor predicting capabilities as they lose their ability to generalize over unseen data [4]. In addition, being able to reduce the amount of data needed to correctly train an ML model is crucial to speed up its training process and save memory resources.

Preprocessing techniques can be mainly categorized into feature selection, instance selection and outlier detection methods [2, 3]. All of them seek to decrease the amount of data fed to a classification model. Feature selection methods reduce the training dataset by decreasing the number of used features, therefore the dimension of the data. Those methods weight the features in order of relevance and remove the least important ones [5]. Both outlier detection methods and instance selection methods work by reducing the amount of data instances. A method can either focus on removing noisy instances, superfluous data instances or both. Noisy or outlier data instances deteriorate the performance of classifiers when added to the training set, while superfluous instances do not impact the performance when removed [6]. Outlier detection methods, as the name indicates, focus on removing outliers from the dataset [7].

The goal of an instance selection method is to speedup the model training by reducing the size of the training set without impacting the model's performance [6, 8, 9]. An instance selection method can either start with an empty training set and add data instances, or start with all the data then remove instances. The selection criterion is usually based on a performance metric or a selection formula. With a metric performance, the methods reduce the training set as long as the classification performance stays above a predefined threshold [9]. With a selection formula, the stopping condition is typically defined by the user, e.g., number of needed instances and logical tests. Multiple criteria can be combined together to achieve a more complex method [10].

In this chapter, we investigate if *diversity* can be used as an effective criterion for removing instances within selection methods. Our concept of diversity is related to variety among the data instances, which is quantified by the observed dissimilarities among them [11]. Our research hypothesis is that the removed data instances are diverse, representing instances less likely to belong together to the same class. Thus, given that one decides to reduce the training size of an ML model, these data instances are rather selected to be suppressed.

We test our approach on classifying eight different benchmark datasets, comparing it with two baseline methods. The first one selects data instances for suppression completely at random, whereas the second consists of the classical Mahalanobis outlier detection method [6, 12].

This chapter is organized as follows. In Section 2, we present the *maximum diversity problem* which is optimized to decide the data instances to be removed from the available training set. In Section 3, we explain our instance selection method based on maximum diversity. Section 4 describes the experimental methodology used to test our research hypothesis. In Section 5, we present and discuss the performed computational results. Finally, in Section 6, we present our concluding remarks.

2. The Maximum Diversity Problem

Given a set of n data instances $U = \{u_1, \ldots, u_n\}$ for which a symmetric dissimilarity matrix $D = \{d_{ij} : 1 \leq i, j \leq n\}$ is defined such that $d_{ii} = 0$ and $d_{ij} \geq 0$ for every $1 \leq i < j \leq n$, the *maximum diversity problem* (MDP) consists of selecting a subset $P \subset U$ of size $p < n$ such as the sum of dissimilarities between the elements of P is maximum. The problem is formalized as

$$\max \sum_{i=1}^{n-1} \sum_{j=i+1}^{n} d_{ij} x_i x_j$$

$$\text{subject } \sum_{i=1}^{n} x_i = p$$

$$x_i \in \{0, 1\} \quad \forall i = 1, \ldots, n. \tag{1}$$

The MDP arises in many real-life applications. For example, in facility location, one may be interested in locating competing stores in a city as far as possible, or to place trash/pollutant storage as to not concentrate exposure in one area of the town [13, 14]. The MDP is also applied in biology for deciding about ecosystems' re-population or for genetic engineering to produce more resilient plants [15–19], or for product design where companies want to have products that are different from their competitor [20]. The problem was shown to be NP-hard by Kuo et al. [11].

Several exact and heuristic methods have been proposed in the literature to solve the MDP [14, 21, 22]. The state-of-the-art exact method for the MDP is due to Martí et al. [23] who proposed a branch-and-bound able to optimally solve medium-size instances with $n = 100$ in 1 hour of CPU time. Regarding heuristics, Martí et al. [24] have very recently performed an exhaustive comparison of state-of-the-art heuristics on the MDPLIB 2.0 — Maximum Diversity Problem Library available at https://www.uv.es/rmarti/paper/mdp.html. Among the compared methods, the OBMA method of Zhou et al. [25] emerges as the best heuristic.

3. Instance Selection by Maximum Diversity

Using the MDP as underlying optimization model, we propose a new instance selection method which removes p data instances from the training set. The so-called *Max Diversity Instance Selection Method* (MaxDivSelec) is described in Algorithm 1. MaxDivSelec proceeds by removing a total of p data instances from the classes of the training dataset. For that, it solves an MDP in each class. The algorithm starts by initializing the training set with all labeled data instances (line 1). After that, the algorithm iterates (lines 2–8) over each class label $c = 1, \ldots, k$ of the training dataset. In line 3, the data instances of class c are isolated in $X_c \subseteq X$, and then the covariance matrix X_c is computed in line 4. Then, in line 5, a matrix D_c of distances is computed for each pair of instances in class c. In our case, $D = (d_{ij})$ are computed as Mahalanobis distances, i.e.,

$$d_{ij} = \sqrt{(x_i - x_j)^T \Sigma^{-1} (x_i - x_j)}. \qquad (2)$$

We note that Σ is approximated by singular value decomposition (SVD) factorization if it is singular [26]. In the sequel, an MDP solver — in our case, OBMA — is called to solve an MDP problem for D_c, selecting the $\bar{p}_c$ points of maximum diversity in class c. More details on how $\bar{p}_c$ is computed are given in Section 4.3. The algorithm then removes the selected data instances from the training set in line 7. Finally, the reduced training set T is returned in line 9.

Algorithm 1 MaxDivSelec

Input: X: labelled dataset of dimension $n \times s$,
$\bar{p}$: array of dimension $1 \times k$ with the number of instances to be removed per class

1: $T \leftarrow X$
2: **for** $c = 1, \ldots, k$ **do**
3:    Let $X_c \subseteq X$ be the matrix of dimension $n_c \times s$ composed by the data instances of class c in X
4:    Compute the covariance matrix Σ_c of X_c
5:    Compute a distance matrix D_c of dimension $n_c \times n_c$
6:    $R \leftarrow \text{SolveMDP}(D_c, \bar{p}_c)$
7:    $T \leftarrow T \setminus R$
8: **end for**
9: **return** T

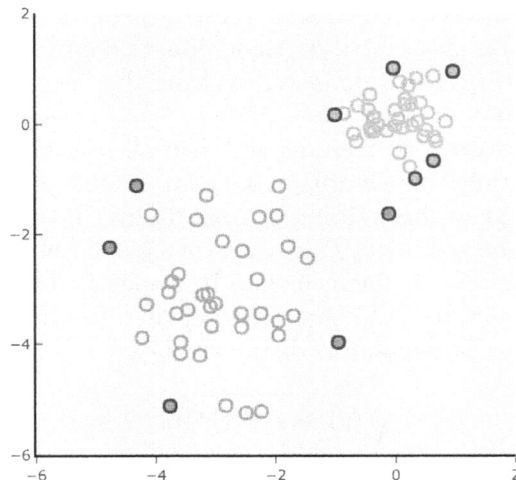

Figure 1. Illustration of the data instances selected by MaxDivSelec for $\bar{p}_1 = 5$ and $\bar{p}_2 = 5$, which corresponds to 12.5% of the whole dataset.

Let $X_c \subseteq X$ be the matrix of dimension $n_c \times s$ composed by the data instances of class c in X Compute the covariance matrix Σ_c of X_c Compute a distance matrix D_c of dimension $n_c \times n_c$ $R \leftarrow$ SolveMDP($D_c, \bar{p}_c$) $T \leftarrow T \setminus R$.

Figure 1 illustrates the use of MaxDivSelec on a 2D synthetic dataset consisting of two Gaussians with 40 data instances each. The first Gaussian is generated with $\mu = 0$ and $\sigma = 0.5$, while the second has $\mu = -3$ and $\sigma = 1$. In the example, five data instances are removed from each Gaussian.

4. Experimental Methodology

4.1. K-nearest neighbors

In order to prove our hypothesis about effectiveness of MaxDivSelec as a data instance selection method for classification models, we had to choose one representative ML model from which our conclusions could be better generalized.

The K-nearest neighbor (KNN) model is a simple yet effective supervised classifier [27]. It predicts the class of an unseen instance by finding its K closest data instances from the training set. The unlabeled instance is then assigned to the majority class among them. The *KNN* classification model was a natural choice for our experiments for three reasons:

(i) It relies on a distance metric — as well as the MDP.
(ii) It is quite tolerant to outliers and noisy data.
(iii) Its classification performance, memory usage and computing times are tightly linked to the number of data instances used for training.[a]

4.2. *Baseline methods*

We compared MaxDivSelec against two other selection methods. The first method, called `Random`, corresponds to our null hypothesis. It simply chooses p data instances to remove at random, with equal probability.

The second method, denoted here as `Mahalanobis`, is well known in the literature [29]. It removes outliers by computing the Mahalanobis distance (2) from each data instance to the centroid of the class it belongs. Larger values of the Mahalanobis distance indicate a greater outlier likelihood. The method aims to improve model classification by removing from the training dataset the instances which are too different to be statistically part of a class.

Algorithm 2 presents the pseudo-code of method `Mahalanobis` which returns a set T of data instances for model training. The method computes for each available labeled data instance its Mahalanobis distance to the centroid of the class to which it belongs. In the sequel, the algorithm removes, from each class c, $\bar{p}_c$ data instances whose Mahalanobis distances are the largest computed.

Figure 2 illustrates the application of the `Mahalanobis` method over the same synthetic example of the two the last section. Here, again, five data instances are removed from each Gaussian. We remark that the selections performed by `MaxDivSelec` and `Mahalanobis` differ of two data instances only.

---

[a] *KNN* makes use of the so-called *lazy training* or *instance-based learning*. It simply queries over the data to make a prediction [28].

Algorithm 2 Mahalanobis method

Input: X: labelled dataset of dimension $n \times s$,
$\bar{p}$: array of dimension $1 \times k$ with the number of instances to be removed per class

1: $T \leftarrow X$
2: **for** $c = 1, \ldots, k$ **do**
3:     Let $X_c \subseteq X$ be the matrix of dimension $n_c \times s$ composed by the data instances of class c in X
4:     Compute the covariance matrix Σ_c of X_c
5:     **for** each data instance x_ℓ of class c **do**
6:         Compute the Mahalanobis distance $d_\ell = \sqrt{(x_\ell - \mu_c)^{\mathrm{T}} \Sigma_c^{-1} (x_\ell - \mu_c)}$ between x_ℓ and the centroid μ_c of class c
7:     **end for**
8:     $R \leftarrow$ the $\bar{p}_c$ instances of class c with largest d
9:     $T \leftarrow T \setminus R$
10: **end for**
11: **return** T

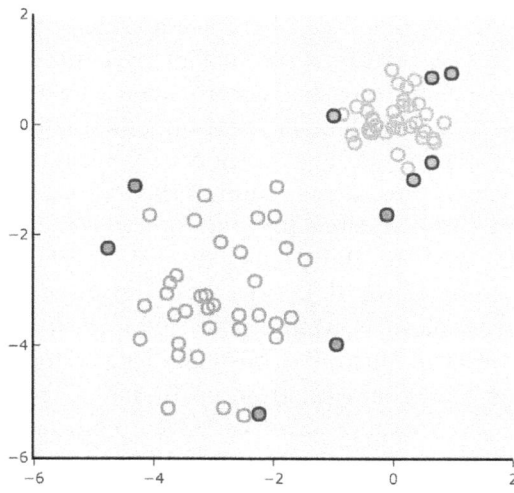

Figure 2. Illustration of the data instances selected by Mahalanobis for $\bar{p}_1 = 5$ and $\bar{p}_2 = 5$, which corresponds to 12.5% of the whole dataset.

4.3. *The α parameter*

The α parameter controls the percentage of data instances to be removed from the training set. For example, for $\alpha = 50\%$ and $n = 100$, 50 data instances are suppressed from the training dataset

($p = 50\% \times 100$). This amount is split proportionally across the provided classes. Thus, the method removes $\bar{p}_c = \alpha \times n_c$ data instances from each class $c = 1, \ldots, k$, rounding it to the closest integer. At the end, some adjustments must be performed so that $\sum_{c=1}^{k} \bar{p}_c = p$. Considering $p' = p - \sum_{c=1}^{k} \bar{p}_c$ as the number of adjustments, we either remove or add a data instance to the final training set depending on whether p' is positive or negative. The adjustments performed in each class are limited to one and are performed from the class with the largest amount of data instances to the least populated class. To illustrate, for a training set with five classes such that $n_1 = 10$, $n_2 = 10$, $n_3 = 10$, $n_4 = 30$, $n_5 = 40$, for $\alpha = 50\%$ (that is, $p = 50$), we obtain $\bar{p}_1 = 5$, $\bar{p}_2 = 5$, $\bar{p}_3 = 5$, $\bar{p}_4 = 15$, $\bar{p}_5 = 20$. Since $p' = 0$, there is no need to adjust the values. For the same training set, by taking $\alpha = 12.5\%$ (that is, $p = 13$), we have $\bar{p}_1 = 1$, $\bar{p}_2 = 1$, $\bar{p}_3 = 1$, $\bar{p}_4 = 4$, $\bar{p}_5 = 5$. Since, in this case, $p' = 13 - 12 = 1$, $\bar{p}_5$ is adjusted to 6, making a total of $p = 13$ data instances.

5. Computational Experiments

5.1. *Datasets*

We compare the presented methods over different real-world benchmark datasets. The used datasets are shown in Table 1. The different number of classes and attributes across them are aimed to test how well the compared methods handle complex classification problems. All datasets were numerically normalized in each attribute dimension before use.

Table 1. Table with datasets' characteristics.

Dataset	n	#classes	#attributes
Iris [30]	150	3	4
Seeds [31]	210	3	7
Dermatology	358	6	34
Ionosphere [32]	351	2	34
Breast cancer Wisc. [33]	683	2	9
Mammographic [34]	830	2	5
Contraceptive [35]	1473	3	9
Abalone [36]	4177	29	8

5.2. Evaluation

We used different classification performance metrics depending on whether the classification problem was (i) binary or multiclass and (ii) balanced or unbalanced. A binary classification problem is one in which prediction is done for two classes only, while a multiclass problem involves more than two classes. A balanced problem supposes that the number of data instances of each class is approximately the same, while an unbalanced classification task has the majority of the data instances belonging to a subset of the provided classes. To accommodate those different categories of problems, three performances metrics were used, namely accuracy, RMSE and the F1-score.

The accuracy score is a classification performance metric often used for supervised classification problems [37]. It compares the prediction to the ground-truth class thus computing the ratio of right predictions. In a binary classification problem, there exist four possible cases for a given prediction: a True Positive (TP), a False Positive (FP), a True Negative (TN) and a False Negative (FN) [38]. The two *true* cases happen when the model predicts the correct class (positive or negative). Conversely, the *false* cases happen when the model predicts the opposite class. For example, an FP occurs when the model predicts a positive class for an actual negative data instance. The accuracy score is given by

$$\text{Accuracy} = \frac{\text{TP} + \text{TN}}{\text{TP} + \text{TN} + \text{FP} + \text{FN}}. \quad (3)$$

An accuracy value of 1 means that the model predicted 100% of the classes correctly, while a score of 0 means that none of the predictions was correct. The accuracy metric can also be generalized for k classes as

$$\text{Accuracy} = \frac{\sum_{c=1}^{k} T_c}{\sum_{c=1}^{k} T_c + \sum_{c=1}^{k} F_c}, \quad (4)$$

where T_c is the number of TP for class c and F_i the number of FP for that same class.

The Root Mean Squared Error (RMSE) score is a performance metric commonly used for multiclass models [39] for which the classes are ordered somehow. Thus, for an expected value of 0, predicting 1 is less "wrong" than predicting 10, for instance. The RMSE is equal to the squared root of the mean of squared errors between the predictions and the ground-truth values. The formula with y' and y as the predicted and ground-truth values, respectively, and n as the number of predictions is given by

$$\text{RMSE} = \sqrt{\frac{1}{n}\sum_{i=1}^{n}(y_i - y'_i)^2}. \qquad (5)$$

The score is represented by how much the model is off on average from the expected values. A score of 0 means that no error was made and the classifier is perfect.

The last and more complex performance metric is the F1-score also called the F-measure. It is used for unbalanced binary classification problems. It is a suitable score for when the model has to predict well one class in particular among others. The *recall* refers to the model's capacity to detect the positive class of interest among the total amount of positive samples, while the *precision* is the model's capacity to well classify TP data instances over the total amount of instances predicted as members of the positive class [38].

The recall and the precision of a model are computed as

$$\text{recall} = \frac{\text{TP}}{\text{TP} + \text{FN}} \qquad \text{precision} = \frac{\text{TP}}{\text{TP} + \text{FP}}. \qquad (6)$$

The F1-score is finally calculated as the geometric mean of both measures:

$$\text{F1-score} = 2 \cdot \frac{\text{recall} \cdot \text{precision}}{\text{recall} + \text{precision}}. \qquad (7)$$

The Iris, Seeds, Dermatology and Contraceptive datasets are assessed according to the accuracy score since they are balanced. The Abalone dataset is an unbalanced dataset with more than two ordered classes. Consequently, KNN's classification performance is evaluated according to the RMSE score for that dataset. Finally, Ionosphere, Breast cancer Wisconsin and Mammographic datasets are evaluated according to F1-score since they consist of binary labeled data, with one majority class.

5.3. *Cross-validation*

The three methods `MaxDivSelec`, `Random` and `Mahalanobis` are tested with the KNN classifier for $K \in \{3, 5, 10\}$ and $\alpha \in \{0.125, 0.25, 0.5\}$, which yields a total of 9 combinations of parameters to be tested. The KNN classifier uses the Euclidean distance. To generalize our results, a 5-fold cross-validation is used to produce multiple test sets. A 5-fold cross-validation separates the data into 5 sets where each set is used as the test set while the rest is used as the training set [40]. That means that 80% of the dataset is used for training, while the other 20% is used for the testing. For this experiment, ten 5-fold cross-validation processes are made to produce a total of 50 pairs of training/test sets. The instance selection methods are employed on the training set of each fold. Since `Random` is a stochastic method, it is executed 20 times with a different seed (0–19) for each fold.

Table 2 reports the benchmark performance scores of the KNN classifier for each dataset. They correspond to the classifier's mean performance when using the whole set of labeled data instances for training, i.e., without instance selection. The datasets are grouped in the table according to the used performance metric.

Table 2. Mean benchmark performance of KNN [Accuracy, F1-score, RMSE] for each tested dataset.

Dataset	$k = 3$	$k = 5$	$k = 10$
Iris	0.960	0.961	0.964
Seed	0.920	0.929	0.922
Dermatology	0.954	0.956	0.958
Contraceptive	0.458	0.482	0.501
Ionosphere	0.716	0.722	0.707
Breast cancer Wisconsin	0.945	0.952	0.953
Mammographic	0.770	0.788	0.791
Abalone	2.855	2.801	2.691

6. Results and Discussion

Our computational results for methods `Random`, `Mahalanobis` and `MaxDivSec` are displayed as box plots to focus on the classification performance distributions of the methods over the tested folds. Besides, we present line charts of the mean performance obtained by each method. We present here a subset of the results, but all box plots can be checked at https://ktton.github.io/master-research/. Results are grouped by the number of used neighbors K and by the resulting training size after instance selection. By grouping the results, we can better evaluate how the parameters K and α affect classification performance.

The methods are compared regarding their general behavior and also on their worst-case result. The worst-case result is the lowest performance result achieved by the method for a given data fold. Regarding accuracy and F1-score that corresponds to the lowest obtained score for a tested fold, whereas for the RMSE, that corresponds to the highest obtained score. The results are further analyzed by means of a Wilcoxon statistical test with a confidence level of 5%. That test tells us if the results achieved by our method are statistically different from those obtained by the baseline methods `Random` and `Mahalanobis`.

6.1. *Classification performance results*

First, we checked the general performance of the instance selection methods for each dataset. For the smallest datasets (regarding n), the three methods presented similar performance. To illustrate that, Figures 3(a) and 4(a) show the results for the dataset Seeds. Moreover, the obtained means are not far from the benchmark KNN performance, which means that using instance selection methods for small datasets does not incur significant losses of classification performance.

Regarding the largest datasets Breast cancer, Mammographic masses, Ionosphere, Contraceptive and Abalone, we observe a major difference between the classification metrics obtained by the different methods. Performing instance selection with `Random` appears to incur more varied classification performance than by using `Mahalanobis`

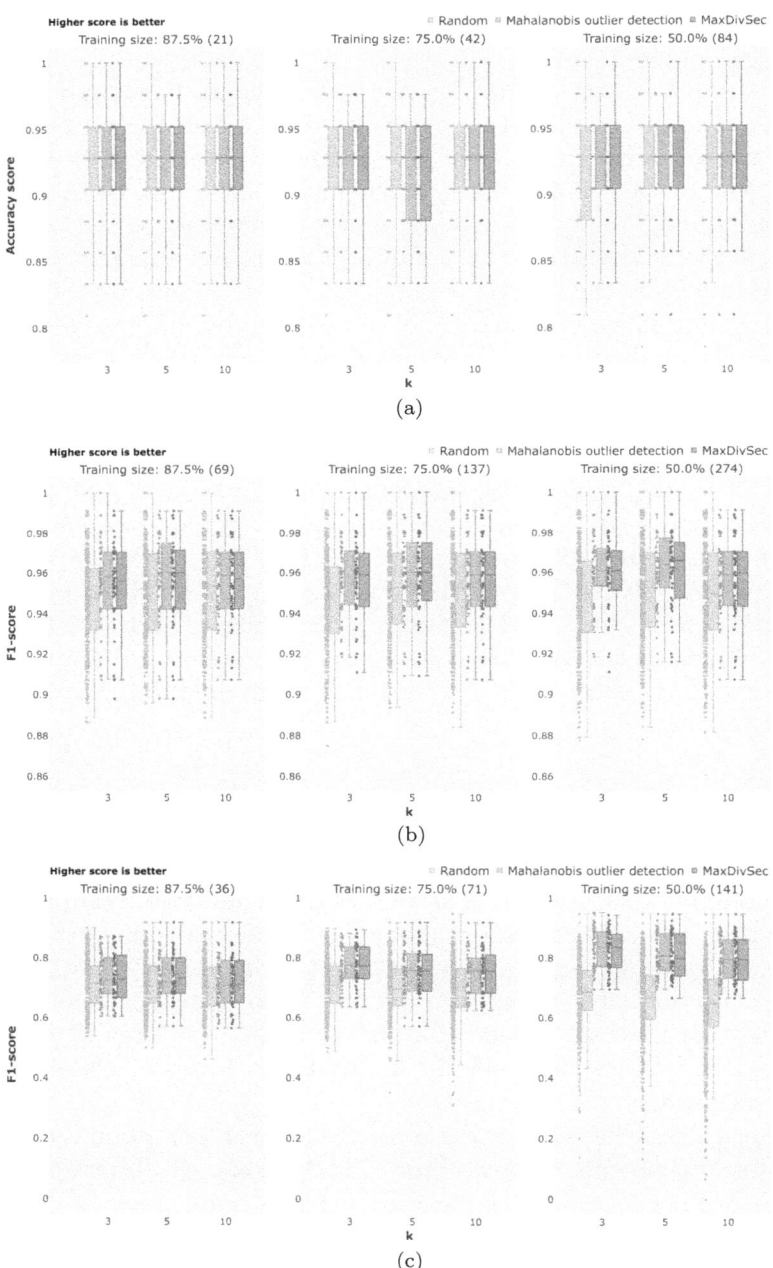

Figure 3. Box plot results grouped by training size: (a) Seeds dataset, (b) Cancer dataset, (c) Ionosphere dataset and (d) Abalone dataset.

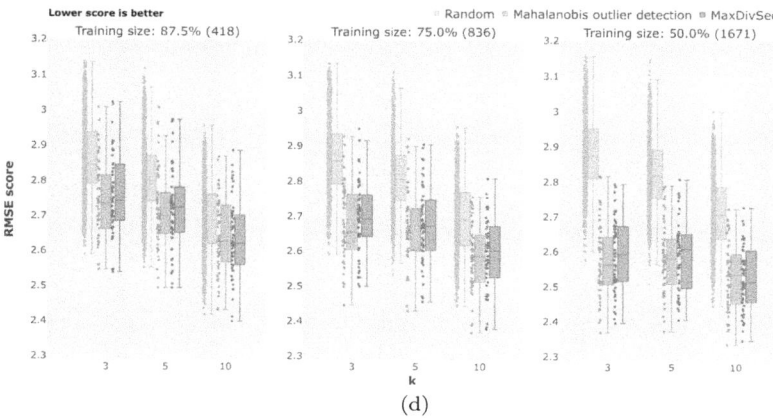

(d)

Figure 3. (*Continued*)

and MaxDivSec, as shown in the box plots of Figures 3(b)–3(d). Figures 4(b)–4(d) show the mean performance of each method for the same three datasets.

The Mahalanobis and MaxDivSec methods outperform the Random method particularly for Ionosphere (Figure 4(c)) and Abalone (Figure 4(d)). Besides, we note across the plots that the Mahalanobis and MaxDivSelec methods obtain better mean classification scores than those obtained by Random. These score differences become larger as more instances are removed from the training sample. In fact, for these instances, the Mahalanobis and MaxDivSelec methods perform better than the benchmark performance obtained by the KNN classifier using the whole data for training. We can hence conclude that for these datasets, restricting the training data to relevant instances is important for increasing the generalization capability of the model.

Regarding the worst-case performance, MaxDivSelec always obtains better or equal worst-case classification results than Random. Having a better worst-case scenario means that our instance selection method is more robust regarding the posterior classification performance of the classifier when predicting the labels of unseen data. When compared to Mahalanobis, our method appears to have similar worst-case performance.

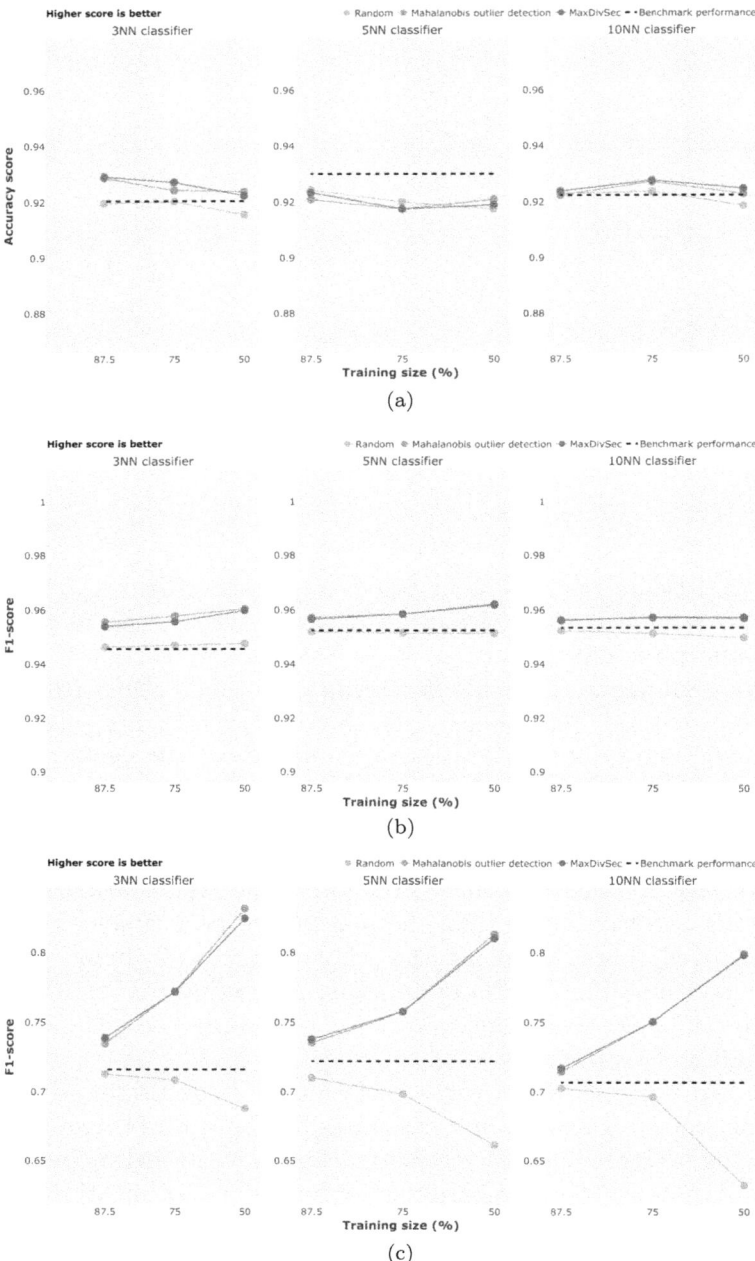

Figure 4. Mean classification results grouped by K: (a) Seeds dataset, (b) Cancer dataset, (c) Ionosphere dataset and (d) Abalone dataset.

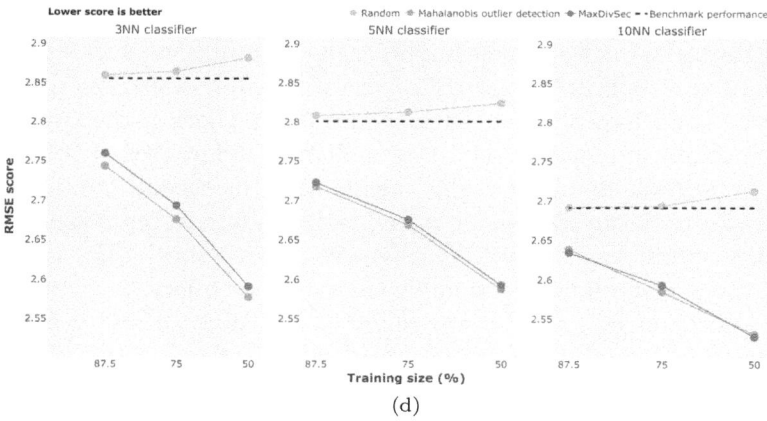

Figure 4. (*Continued*)

Finally, we also analyze the classification performance for varied values of α and the number of neighbors K used by the KNN classifier. Our conclusions are as follows:

- For the smallest datasets and breast cancer Wisconsin, the classification performance is lightly affected by K and α.
- For both the Abalone and Contraceptive datasets, the classification performance improves as K increases for methods `MaxDivSelec`, `Mahalanobis` and `Random`.
- For the Ionosphere dataset, the classification performance decreases as K increases for methods `MaxDivSelec`, `Mahalanobis` and `Random`, especially with $\alpha = 50\%$.
- For the Abalone, Contraceptive, Mammographic and Ionosphere datasets, the classification performance of `MaxDivSelec` and `Mahalanobis` increases as α gets larger, i.e., as more data instances are removed from the training set. They are actually better than the benchmark performance presented in Table 2 except for the Mammographic dataset.

6.2. Wilcoxon tests

This section presents Wilcoxon signed-ranks tests [41] in order to compare the obtained results in terms of statistical significance.

For each dataset, we compared the different methods Random, Mahalanobis and MaxDivSelec on each combination of $K = 3, 5$ and 10, and $\alpha = 0.25, 0.50$ and 0.75, totalizing nine Wilcoxon tests per dataset. Two hypothesis are tested. First, we check if the two result distributions are similar (i.e., the median of differences $= 0$). If that first hypothesis is rejected, this means that the second hypothesis is true, i.e., that the methods obtain statistically different results (the median of differences < 0). We used a confidence level of 5% meaning that the p-value must be smaller than 0.05 to reject an hypothesis. The Wilcoxon test results are reported in Tables 3 and 4 for each dataset.

Table 3. Wilcoxon test results with a confidence level of 5% for comparing MaxDivSec and Mahalanobis with Random.

Dataset	MaxDivSec Different	MaxDivSec Better	Mahalanobis Different	Mahalanobis Better
Iris	5/9	4/5	7/9	5/7
Seed	4/9	4/4	2/9	2/2
Dermatology	5/9	5/5	6/9	6/6
Ionosphere	9/9	9/9	9/9	9/9
Cancer	8/9	8/8	8/9	8/8
Mammographic	8/9	4/8	6/9	4/6
Contraceptive	9/9	9/9	8/9	8/8
Abalone	9/9	9/9	9/9	9/9

Table 4. Wilcoxon test results with a confidence level of 5% for comparing MaxDivSec with Mahalanobis.

Dataset	Different	Better
Iris	1/9	1/1
Seed	0/9	0/0
Dermatology	0/9	0/0
Ionosphere	0/9	0/0
Cancer	1/9	0/1
Mammographic	7/9	3/7
Contraceptive	8/9	8/8
Abalone	3/9	0/3

Table 3 shows the results of the comparisons of the methods MaxDivSec and Mahalanobis with the method Random. We observe that our instance selection method MaxDivSelec is statistically different from the Random method for most of K and α combinations for all the datasets. We can also verify the same behavior with the Mahalanobis method except for the Seed dataset. The Mahalanobis method is only different for two combinations. Moreover, when our method is different from the Random method, it is most of the time better.

In Table 4, we show the Wilcoxon results obtained when comparing our method MaxDivSec with the Mahalanobis method. We note that MaxDivSec is seldom different or better than Mahalanobis except for two datasets: Contraceptive and Mammographic. However, such difference does not mean that the first is necessarily better than the later. In most of the cases, MaxDivSelec is not statistically different from the Mahalanobis method.

7. Concluding Remarks

This chapter proposed to investigate the use of diversity for selecting data instances for model training. With that purpose, we proposed MaxDivSec, an algorithm that proceeds by removing from the training set of a machine learning model the subset of data instances for which its associated diversity is maximum. We compared MaxDivSec, regarding the classification performance of a target classifier, with two other baseline instance selection methods: one that randomly selects data instances for suppression and another based on the removal of data outliers. Our results demonstrated that diversity is actually a good criterion for data instance selection as the obtained results by MaxDivSec led to superior classification performance in the vast majority of the tested scenarios when compared to the random approach. However, the proposed method was not shown to be significantly different from the method based on the suppression of outliers. Finally, although we demonstrated by our experiments that maximum diversity is effective on selecting data instances for model training, its computation still requires the solution of an NP-hard problem either exactly or heuristically.

Acknowledgments

The authors would like to thank professors Eduardo Pardo and Abraham Duarte for providing us the OBMA code and executable. This research was partially funded by Natural Sciences and Engineering Research Council of Canada (NSERC) under grants DG-2017–05617 and DG-2020–06311 for its financial support.

References

[1] C. M. Bishop, Chapter: Introduction. In *Pattern Recognition and Machine Learning*. Information Science and Statistics (Springer-Verlag, Berlin, 2006), pp. 1–66.

[2] D. Pyle, *Data Preparation for Data Mining*, 1st edn. (Morgan Kaufmann Publishers Inc., San Francisco, 1999).

[3] S. Zhang, C. Zhang and Q. Yang, Data preparation for data mining, *Appl. Artif. Intell.* **17**(5–6), 375–381 (2003). https://doi.org/10.1080/713827180.

[4] G. I. Webb, Chapter O: Overfitting. In C. Sammut and G. I. Webb (eds.) *Encyclopedia of Machine Learning* (Springer US, Boston, 2010), pp. 744–744.

[5] L. C. Molina, L. Belanche and A. Nebot. Feature selection algorithms: A survey and experimental evaluation. In *2002 IEEE International Conference on Data Mining, 2002. Proceedings* (2002), pp. 306–313. doi: 10.1109/ICDM.2002.1183917.

[6] S. Garcia, J. Derrac, J. Cano and F. Herrera, Prototype selection for nearest neighbor classification: Taxonomy and empirical study, *IEEE Trans. Pattern Anal. Machine Intell.* **34**(3), 417–435 (2012). doi: 10.1109/TPAMI.2011.142.

[7] T. Dasu and T. Johnson, *Exploratory Data Mining and Data Cleaning*, 1st edn. (John Wiley & Sons Inc., USA, 2003).

[8] J. Nalepa and M. Kawulok, Selecting training sets for support vector machines: A review, *Artif. Intell. Rev.* **52**(2), 857–900 (2019). https://doi.org/10.1007/s10462-017-9611-1.

[9] J. A. Olvera-López, J. A. Carrasco-Ochoa, J. F. Martínez-Trinidad and J. Kittler, A review of instance selection methods, *Artif. Intell. Rev.* **34**(2), 133–143 (2010). doi: 10.1007/s10462-010-9165-y.

[10] M. Blachnik, Ensembles of instance selection methods: A comparative study, *Int. J. Appl. Math. Comput. Sci.* **29**(1), 151–168 (2019). http://dx.doi.org/10.2478/amcs-2019-0012.

[11] C.-C. Kuo, F. Glover and K. S. Dhir, Analyzing and modeling the maximum diversity problem by zero-one programming*, *Decis. Sci.* **24**(6), 1171–1185 (1993). https://onlinelibrary.wiley.com/doi/abs/10.1111/j.1540-5915.1993.tb00509.x.
[12] A. Zimek and P. Filzmoser, There and back again: Outlier detection between statistical reasoning and data mining algorithms, *WIREs Data Mining Knowl. Discov.* **8**(6), e1280 (2018). https://onlinelibrary.wiley.com/doi/abs/ 10.1002/widm.1280.
[13] M. J. Kuby, Programming models for facility dispersion: The p-dispersion and maxisum dispersion problems, *Geogr. Anal.* **19** (4), 315–329 (1987). https://onlinelibrary.wiley.com/doi/abs/10.1111/j.1538-4632.1987.tb00133.x.
[14] F. Glover, C.-C. Kuo and K. S. Dhir, Heuristic algorithms for the maximum diversity problem, *J. Inform. Optim. Sci.* **19**(1), 109–132 (1998). https://doi.org/10.1080/02522667.1998.10699366.
[15] F. Glover, K. Ching-Chung and K. S. Dhir, A discrete optimization model for preserving biological diversity, *Appl. Math. Modell.* **19**(11), 696–701 (1995). https://doi.org/10.1016/0307-904X(95)00083-V. https://www.sciencedirect.com/science/article/pii/0307904X9500083V.
[16] A. Duarte and R. Martí, Tabu search and grasp for the maximum diversity problem, *Eur. J. Oper. Res.* **178**(1), 71–84 (2007). https://doi.org/10.1016/j.ejor.2006.01.021. https://www.sciencedirect.com/science/article/pii/S0377221706000634.
[17] K. Ralls, P. Sunnucks, R. C. Lacy and R. Frankham, Genetic rescue: A critique of the evidence supports maximizing genetic diversity rather than minimizing the introduction of putatively harmful genetic variation, *Biol. Conserv.* **251**, 108784 (2020). https://doi.org/10.1016/j.biocon.2020.108784. https://www.sciencedirect.com/science/article/pii/S0006320720308429.
[18] A. A. Hoffmann, A. D. Miller and A. R. Weeks, Genetic mixing for population management: From genetic rescue to provenancing, *Evol. Appl.* **14**(3), 634–652 (2021). https://onlinelibrary.wiley.com/doi/abs/10.1111/eva.13154.
[19] T. Leinster and M. W. Meckes, Maximizing diversity in biology and beyond, *Entropy* **18**(3) (2016). doi: 10.3390/e18030088. https://www.mdpi.com/1099-4300/18/3/88.
[20] T. von Ghyczy, Product diversity and proliferation as a new mode of competing in the motor car, *Int. J. Veh. Des.* **6**(4/5), 423–425 (1985).

[21] M. Lozano, D. Molina and C. García-Martínez, Iterated greedy for the maximum diversity problem, *Eur. J. Oper. Res.* **214**(1), 31–38 (2011). https://doi.org/10.1016/j.ejor.2011.04.018. https://www.sciencedirect.com/science/article/pii/S0377221711003626.
[22] R. Martí, M. Gallego and A. Duarte, Optsicom project (2010). http://grafo.etsii.urjc.es/optsicom/mdp/.
[23] R. Martí, M. Gallego and A. Duarte, A branch and bound algorithm for the maximum diversity problem, *Eur. J. Oper. Res.* **200**(1), 36–44 (2010).
[24] R. Martí, A. Martínez-Gavaraa, S. Pérez-Pelób and J. Sánchez-Orob, Discrete diversity and dispersion maximization: A review and an empirical analysis from an OR perspective (2021). Submitted March 2021.
[25] Y. Zhou, J.-K. Hao and B. Duval, Opposition-based memetic search for the maximum diversity problem, *IEEE Trans. Evol. Comput.* **21**(5), 731–745 (2017).
[26] G. Strang, Chapter 7: The singular value decomposition (SVD). In *Introduction to Linear Algebra*, 5th edn. (Wellesley-Cambridge Press, Wellesley, 2009), pp. 364–400.
[27] T. Cover and P. Hart, Nearest neighbor pattern classification, *IEEE Trans. Inform. Theory* **13**(1), 21–27 (1967). doi: 10.1109/TIT.1967.1053964.
[28] E. Keogh, Chapter I: Instance-based learning. In *Encyclopedia of Machine Learning* (Springer US, Boston, 2010), pp. 549–550. https://doi.org/10.1007/978-0-387-30164-8_409.
[29] J. Fernández Pierna, F. Wahl, O. de Noord and D. Massart, Methods for outlier detection in prediction, *Chemom. Intell. Lab. Sys.* **63**(1), 27–39 (2002). https://doi.org/10.1016/S0169-7439(02)00034-5. http://www.sciencedirect.com/science/article/pii/S0169743902000345.
[30] R. Fisher, UCI machine learning repository: Iris data set (1936). https://archive.ics.uci.edu/ml/datasets/iris.
[31] M. Charytanowicz, J. Niewczas, P. Kulczycki, P. A. Kowalski, S. Łukasik and S. Żak, UCI machine learning repository: Seeds data set (2012). https://archive.ics.uci.edu/ml/datasets/seeds.
[32] V. G. Sigillito, S. P. Wing, L. V. Hutton and K. B. Baker, UCI machine learning repository: Ionosphere data set (1989). https://archive.ics.uci.edu/ml/datasets/ionosphere.
[33] W. H. Wolberg and O. L. Mangasarian, Multisurface method of pattern separation for medical diagnosis applied to breast cytology, *Proc. Natl. Acad. Sci. USA* **87**(23), 9193–9196 (1990).

[34] M. Elter, R. Schulz-Wendtland and T. Wittenberg, UCI machine learning repository: Mammographic mass data set (2007). http://archive.ics.uci.edu/ml/datasets/Mammographic+Mass.
[35] T.-S. Lim, W.-Y. Loh and Y.-S. Shih. UCI machine learning repository: Contraceptive method data set (1999). http://archive.ics.uci.edu/ml/datasets/Contraceptive+Method+Choice.
[36] W. J. Nash, T. L. Sellers, S. R. Talbot, A. J. Cawthorn and W. B. Ford, UCI machine learning repository: Abalone data set (1994). https://archive.ics.uci.edu/ml/datasets/Abalone.
[37] C. Sammut and G. I. Webb (eds.) Chapter A: Accuracy. In C. Sammut and G. I. Webb (eds.) *Encyclopedia of Machine Learning* (Springer US, Boston, 2010), pp. 9–10.
[38] K. M. Ting, Chapter P: Precision and recall. In C. Sammut and G. I. Webb (eds.) *Encyclopedia of Machine Learning* (Springer US, Boston, 2010), pp. 781–781.
[39] C. Sammut and G. I. Webb (eds.) Chapter M: Mean squared error. In C. Sammut and G. I. Webb (eds.) *Encyclopedia of Machine Learning* (Springer US, Boston, 2010), pp. 653–653.
[40] C. Sammut and G. I. Webb (eds.) Chapter C: Cross-validation. In C. Sammut and G. I. Webb (eds.) *Encyclopedia of Machine Learning* (Springer US, Boston, 2010), pp. 249–249.
[41] F. Wilcoxon, Individual comparisons by ranking methods, *Biom. Bull.* 1(6), 80–83 (1945). http://www.jstor.org/stable/3001968.

© 2024 World Scientific Publishing Company
https://doi.org/10.1142/9789811267048_0027

Chapter 27

Wardowski Maps Modeled by Low-Oscillation Functions

Mihai Turinici

*A. Myller Mathematical Seminar,
A. I. Cuza University, Iași, Romania*

mturi@uaic.ro

All Wardowski-type contractions modeled by low-oscillation functions are, in fact, Matkowski ones. Technical connections with some particular statements in the area obtained by Dung and Hang (*Vietnam J. Math.* 43:743–755, 2015) and Fulga and Proca (*Adv. Theory Nonlinear Anal. Appl.* 1:57–63, 2017) are also discussed.

1. Introduction

Let X be a nonempty set. Call the subset Y of X, *almost singleton* (in short, *asingleton*), provided $[y_1, y_2 \in Y$ implies $y_1 = y_2]$, and *singleton* if, in addition, Y is nonempty; note that in this case $Y = \{y\}$, for some $y \in X$. Further, take a *metric* $d : X \times X \to R_+ := [0, \infty[$ over X; the couple (X, d) will be called a *metric space*. Given the sequence $(x_n; n \geq 0)$ in X and the point $x \in X$, let us say that (x_n) d-converges to x [written as $x_n \xrightarrow{d} x$] iff $d(x_n, x) \to 0$ as $n \to \infty$, i.e.,

$$\forall \varepsilon > 0, \exists i = i(\varepsilon): i \leq n \implies d(x_n, x) \leq \varepsilon,$$

also referred to as x is the d-limit of $(x_n; n \geq 0)$. Denote by $\lim_n(x_n)$ the set of all such elements; if it is nonempty, then $(x_n; n \geq 0)$ is called d-convergent. Clearly, as $(d =$ metric$)$, we have that (X, d) is separated: $\lim_n(x_n)$ is an asingleton, for each sequence (x_n) of X. Further, given the sequence $(x_n; \geq 0)$ in X, call it d-Cauchy provided $d(x_m, x_n) \to 0$ as $m, n \to \infty$, $m < n$, i.e.,

$$\forall \varepsilon > 0, \exists j = j(\varepsilon): j \leq m < n \implies d(x_m, x_n) \leq \varepsilon,$$

the class of all these being indicated as $Cauchy(d)$. By definition, the couple $((\xrightarrow{d}), Cauchy(d))$ is called the *conv-Cauchy structure* attached to d. Note that (as $d =$ metric), (X, d) is *regular*: any d-convergent sequence is d-Cauchy; if the reciprocal holds too, then (X, d) is called *complete*.

Having these precise, let $T \in \mathcal{F}(X)$ be a selfmap of X. [Here, for each couple A, B of nonempty sets, $\mathcal{F}(A, B)$ denotes the class of all functions from A to B; when $A = B$, we write $\mathcal{F}(A)$ in place of $\mathcal{F}(A, A)$.] Denote Fix$(T) = \{x \in X; x = Tx\}$; each point of this set is referred to as *fixed* under T. In the following, sufficient conditions are given for the existence and/or uniqueness of elements in Fix(T). The determination of such points is carried out in the following context, comparable with the one in Rus [23, Chapter 2, Section 2.2]:

(**pic-1**) We say that T is a *Picard operator* (modulo d) if, for each $x \in X$, the iterative sequence $(T^n x; n \geq 0)$ is d-Cauchy.

(**pic-2**) We say that T is a *strong Picard operator* (modulo d) if, for each $x \in X$, $(T^n x; n \geq 0)$ is d-convergent with $\lim_n(T^n x) \in$ Fix(T).

(**pic-3**) We say that T is *fix-asingleton* if Fix(T) is asingleton; and *fix-singleton*, provided Fix(T) is singleton.

To solve our posed problem along the precise directions, a lot of *functional*-type contractive requirements is needed. Given $\varphi \in \mathcal{F}(R_+)$, let us say that T is φ-contractive provided

(M-con) $d(Tx,Ty) \leq \varphi(d(x,y))$, $x,y \in X$, $d(x,y) > 0$,
$d(Tx,Ty) > 0$;

when φ is generic here, we then say that T is a *Matkowski contraction*. The functions to be considered here are as follows. Denote

$\mathcal{F}(\text{re})(R_+)$ = the class of all *regressive* $\varphi \in \mathcal{F}(R_+)$
(in the sense: $\varphi(0) = 0$ and $\varphi(t) < t$, for all $t > 0$);
$\mathcal{F}(\text{in})(R_+)$ = the class of all *increasing* $\varphi \in \mathcal{F}(R_+)$;

then put $\mathcal{F}(\text{re, in})(R_+) = \mathcal{F}(\text{re})(R_+) \cap \mathcal{F}(\text{in})(R_+)$. Finally, define the property, for each function φ in $\mathcal{F}(\text{re})(R_+)$

(M-ad) φ is *Matkowski admissible*: $\varphi^n(t) \to 0$ as $n \to \infty$,
$\forall t \in R_+$.

[Here, for each $n \geq 0$, φ^n stands for the nth iterate of φ.] The following answer to the posed problem [referred to as Matkowski fixed-point theorem; in short (M-fpt)] is our starting point.

Theorem 1. *Suppose that T is φ-contractive, for some regressive function $\varphi \in \mathcal{F}(R_+)$. Further, assume that (X,d) is complete. Then,*

(11-a) *T is strong Picard (modulo d) if*
 φ is increasing and Matkowski admissible,
(11-b) *in addition, T is fix-asingleton (hence, fix-singleton).*

Note that a similar conclusion is to be reached when $\varphi \in \mathcal{F}(\text{re})(R_+)$ fulfills

(BW-ad) φ is *Boyd–Wong admissible*: $\limsup_{t \to s+} \varphi(t) < s$,
$\forall s > 0$,

and Theorem 1 is just the one in Boyd and Wong [4]. In particular, when $\varphi \in \mathcal{F}(\text{re, in})(R_+)$, the admissible condition above writes

(s-M-ad) φ is *super Matkowski admissible*: $\varphi(s+0) < s$, for each $s > 0$,

and these results coincide; we do not give details.

In particular, when φ is *linear* (i.e., $\varphi(t) = \alpha t$, $t \in R_+$, for some $\alpha \in [0,1[$), the second conclusion above is necessarily retainable, and the Matkowski fixed point theorem is just Banach's contraction mapping principle [2].

This result, obtained in 1975 by Matkowski [14], found many interesting applications in operator equations theory, so, it was the subject of various extensions. A natural way of doing this is to consider some *rational*-type contractions like below. Denote, for $x, y \in X$,

$$Q_1(x,y) = d(x,Tx),\ Q_2(x,y) = d(x,y),\ Q_3(x,y) = d(x,Ty),$$
$$Q_4(x,y) = d(Tx,y),\ Q_5(x,y) = d(Tx,Ty),\ Q_6(x,y) = d(y,Ty),$$
$$\mathcal{Q}(x,y) = (Q_1(x,y), Q_2(x,y), Q_3(x,y), Q_4(x,y), Q_5(x,y),$$
$$Q_6(x,y)),$$
$$K_1(x,y) = d(x, T^2 x),\ K_2(x,y) = d(Tx, T^2 x),$$
$$K_3(x,y) = d(T^2 x, y),\ K_4(x,y) = d(T^2 x, Ty),$$
$$\mathcal{K}(x,y) = (K_1(x,y), K_2(x,y), K_3(x,y), K_4(x,y)).$$

Given $v \in \mathcal{F}(R_+^6 \times R_+^4, R_+)$, let us consider the composed function

$(V \in \mathcal{F}(X \times X, R_+))$: $V(x,y) = v(\mathcal{Q}(x,y), \mathcal{K}(x,y))$, $x, y \in X$.

Then, the announced general class of contractions writes

(M-r-con) $d(Tx, Ty) \leq \varphi(V(x,y))$,
$x, y \in X$, $d(Tx, Ty) > 0$, $V(x,y) > 0$.

These will be referred to as T is *Matkowski* $(\varphi; V)$-*contractive*, and, if $(\varphi; V)$ is generic, we say that T is a *Matkowski rational contraction*. Some basic contributions in this area were obtained in Leader [13] and Reich [20]; see also the survey paper by Rhoades [21]. Further, a way of extending this class of contractions is the implicit one, based on *functional* contractive conditions

(f-con) $W(x,y) := w(\mathcal{Q}(x,y), \mathcal{K}(x,y)) \leq 0$, $\forall x, y \in X$,

where $w : R_+^6 \times R_+^4 \to R$ is an appropriate function. These will be referred to as T is W-*contractive*, and, if W is generic, we say that T is a *functional contraction*. Some recent statements of this type may be found in Akkouchi [1], or Berinde and Vetro [3]; see also Nashine et al. [16]. We stress that in almost all papers based on implicit techniques — including the ones we just quoted — it is claimed that the starting point in the area is represented by the 1999 contribution due to Popa [18]. Unfortunately, this claim is not true: fixed-point results based on implicit techniques were obtained more than two decades ago in Turinici [25, 26]. But, some partial aspects have been discussed in the 1969 Meir–Keeler fixed-point principle [15].

Concerning this class, a natural problem to be posed is that of an implicit contraction (f-con) being reducible to a Matkowski rational one. As we will see, a basic class of such objects for which these reduction techniques apply is written as

(W-r-con) $F(d(Tx,Ty)) \leq F(V(x,y)) - C$,
$x, y \in X$, $d(Tx,Ty) > 0$, $V(x,y) > 0$;

where F is an increasing function over $\mathcal{F}(R_+^0, R)$, $C > 0$ is a constant, and V is as before. Since the first object of this type was introduced in 2012 by Wardowski [33], the above described property will be referred to as T is *Wardowski $(F;C;V)$-contractive*, and, if $(F;C;V)$ is generic, we say that T is a *Wardowski rational contraction*. According to the observations in Turinici [30, Paper-1–5] or Găvruţa and Manolescu [10], these contractions are, essentially, Meir–Keeler ones [15]; so that — from a methodological perspective — their theory is more or less complete. Despite this, the described line of research developed in a tremendous way: in less than a decade, hundreds of contributions have been dedicated to it, according to the recent survey papers by Wardowski [34], Vujaković et al. [32] and Karapinar et al. [11]. Concerning this aspect, one might believe that — as a compensation — most of these are ultimately original from a technical viewpoint. Unfortunately, this impression is not (entirely) realistic. Precisely, in a pretty large number of concrete circumstances, the continuity — or, at least, the low-oscillation condition — for the ambient function F is a must in solving the considered fixed-point question. But — as proved in what follows — the problem we deal with is then reducible to the rational Matkowski results we just sketched. This means that — from a technical perspective as well — the metrical type studies devoted to this topic seem to have been attained the maximum of their potential, and forces us to conclude that genuine future developments of Wardowski-type contractions theory are to be sought within the applications area. Further aspects are discussed elsewhere.

2. Matkowski Associated Maps

In the following, an associated map construction is given for any couple (F, C), where $F \in \mathcal{F}(R_+^0, R)$ is increasing and $C > 0$ is a constant.

Let $\mathcal{F}(\text{in})(R_+^0, R)$ stand for the class of all increasing functions in $\mathcal{F}(R_+^0, R)$. For any such function F and each $b > 0$, denote

(left-osc) $\text{osc}(-)(F; b) = F(b) - F(b - 0)$
(the *left oscillation* of F at b)
(right-osc) $\text{osc}(+)(F; b) = F(b + 0) - F(b)$
(the *right oscillation* of F at b)
(bila-osc) $\text{osc}(\pm)(F; b) = F(b + 0) - F(b - 0)$
(the *bilateral oscillation* of F at b).

Given some (real) constant $C > 0$, and some $b > 0$, let us introduce the properties

(left-bo) $(F, C; b)$ is *left bounded oscillating*: $\text{osc}(-)(F; b) < C$,
(right-bo) $(F, C; b)$ is *right bounded oscillating*: $\text{osc}(+)(F; b) < C$,
(bila-bo) $(F, C; b)$ is *bilateral bounded oscillating*: $\text{osc}(\pm)(F; b) < C$.

When these hold for all $b > 0$, the resulting property is referred to, respectively, as

$(F; C)$ is *left (right, bilateral) bounded oscillating*.

Let us say that $F \in \mathcal{F}(\text{in})(R_+^0, R)$ is a *semi-Wardowski function*, if

(s-W) $F(0 + 0) := \lim_{t \to 0+} F(t) = -\infty$.

Note that, as a direct consequence (see, e.g., Turinici [27])

(for each sequence (t_n) in R_+^0): $F(t_n) \to -\infty$ iff $t_n \to 0+$.

Denote for simplicity

$\mathcal{F}(\text{in-s-W})(R_+^0, R)$ = the class of all $F \in \mathcal{F}(R_+^0, R)$
with the increasing and semi-Wardowski properties.

Then, let F be an element of this class, and fix a strictly positive number $C > 0$. Define the multivalued map $\mathcal{L} := \mathcal{L}(F, C)$ in $\mathcal{F}(R_+^0, \exp(R_+^0))$, according to

$$\mathcal{L}(t) = \{s \in R_+^0; F(s) \leq F(t) - C\}, t \in R_+^0.$$

Clearly, $\mathcal{L}$ is well defined by the imposed properties. Further,

(hered) $\mathcal{L}(.)$ is *hereditary*: $t > 0$ and $s \in \mathcal{L}(t)$ imply $]0, s] \subseteq \mathcal{L}(t)$,

(s-incr) $\mathcal{L}(.)$ is *set increasing*: $(0<)t_1 \le t_2$ implies $L(t_1) \subseteq L(t_2)$,
(s-uni) $\mathcal{L}$ is *subunitary*: $\mathcal{L}(t) \subseteq]0,t[$, for all $t > 0$.

Having these precise, define a function $\varphi \in \mathcal{F}(R_+^0)$ as

$$\varphi(t) = \sup \mathcal{L}(t), \ t > 0.$$

As $\mathcal{L}(.) =$ subunitary, the definition is consistent; in addition,

(pro-1) φ is weakly regressive: $\varphi(t) \le t, \forall t > 0$,
(pro-2) φ is increasing (as $\mathcal{L}(.)$ is set increasing).

Finally, note that by the definition of supremum,

$$(\forall t > 0): s \in \mathcal{L}(t) \implies s \le \varphi(t); \text{ hence, } \mathcal{L}(t) \subseteq]0, \varphi(t)].$$

The reciprocal inclusion is not in general true, for, given $t > 0$, $\varphi(t)$ may be not an element of $\mathcal{L}(t)$. However, we do have (by the hereditary property)

$$(\forall t > 0):]0, \varphi(t)[\subseteq \mathcal{L}(t); \text{ hence, }]0, \varphi(t)[\subseteq \mathcal{L}(t) \subseteq]0, \varphi(t)].$$

In fact, let $s > 0$ be such that $s < \varphi(t)$. By the definition of supremum, there exists $q \in \mathcal{L}(t)$ with $s < q$, and then, by the hereditary property, $s \in]0, q] \subseteq \mathcal{L}(t)$.

Completing this convention with $\varphi(0) = 0$, we may take the underlying function as an element of $\mathcal{F}(\text{in})(R_+)$. By definition, $\varphi = \varphi(F,C)$ is referred to as the *attached to* (F,C) *function* in $\mathcal{F}(R_+)$. Further properties of this object are obtainable from the following auxiliary fact:

Proposition 1. *Let the increasing semi-Wardowski function* $F \in \mathcal{F}(R_+^0, R)$ *and the constant* $C > 0$ *be as before. Then, the associated to* (F,C) *function* $\varphi \in \mathcal{F}(R_+)$ *enjoys the following (conditional) properties*:

(21-1) φ *is regressive* ($\varphi(t) < t, \forall t > 0$) *provided (in addition)* $(F;C)$ *is left bounded oscillating*,
(21-2) φ *is Matkowski admissible* ($\lim_n \varphi^n(t) = 0$, *for all* $t > 0$) *provided (in addition to the preceding)* $(F;C)$ *is right bounded oscillating*,
(21-3) φ *is super Matkowski admissible* ($\varphi(t+0) < t, \forall t > 0$) *provided (in addition to both these)* $(F;C)$ *is bilateral bounded oscillating*.

Proof. **Step 1.** Suppose, by contradiction, that $\varphi(t) = t$, for some $t > 0$. By definition, there exists a strictly ascending sequence (t_n) in R_+^0, with

(p1) (t_n) is in $\mathcal{L}(t)$; that is, $F(t_n) \leq F(t) - C$, for all n,
(p2) $t_n \to t$ (hence, $t_n \to t-$) as $n \to \infty$.

Combining these yields (by a limit process)

$$F(t-0) \leq F(t) - C; \text{ so that, } C \leq \text{osc}(-)(F;t),$$

in contradiction with the left bounded oscillating property.

Step 2. Fix $t_0 > 0$ and put $(t_n = \varphi^n(t_0); n \geq 0)$. Clearly, this is a strictly descending sequence; whence, $t := \lim_n t_n$ exists, with

$$t_n > t, \forall n; \text{ hence, } t_n \to t+.$$

In this case, by a preceding observation

$$(\forall n): t < t_{n+1} = \varphi(t_n) \text{ implies } F(t) \leq F(t_n) - C.$$

This yields (by a limit process)

$$F(t) \leq F(t+0) - C; \text{ so that, } C \leq \text{osc}(+)(F;t),$$

in contradiction with the right bounded oscillating property. Hence, $\lim_n t_n = 0$, and this, by the arbitrariness of t_0, proves our claim.

Step 3. By the preceding stage, we have for the moment

$$\varphi(t+0) \leq t, \quad \text{for each } t > 0.$$

Suppose by contradiction that

(equ) $(\varphi(t) <) t = \varphi(t+0)$, for some $t > 0$.

Take a strictly descending sequence (t_n) in R_+^0, with

$t_n \to t+$ as $n \to \infty$; hence, $(s_n := \varphi(t_n); n \geq 0)$ is descending and
$s_n \to t (= \varphi(t+0))$ as $n \to \infty$.

Two alternatives occur.
  (Alter-1) Suppose that

$$A(t) := \{n \in N; s_n = t\} \text{ is nonempty.}$$

Without loss — passing to a subsequence if necessary — we may assume that

$$0 \in A(t); \text{ hence } s_n = t, \quad \forall n.$$

Let (σ_n) be a strictly ascending sequence with

$$\sigma_n \to t-; \text{ hence, } (\sigma_n < t, \forall n); \text{ wherefrom, } \sigma_n < s_i = \varphi(t_i), \forall n, \forall i.$$

This, by definition, tells us that

$$F(\sigma_n) \leq F(t_i) - C, \quad \forall n, \forall i.$$

Passing to limit as $n \to \infty$, $i \to \infty$ gives

$$F(t-0) \leq F(t+0) - C; \text{ that is, } C \leq \mathrm{osc}(\pm)(F;C),$$

in contradiction with the bilateral bounded oscillating property.

(Alter-2) Suppose that

$$A(t) := \{n \in N; s_n = t\} \text{ is empty: } (\varphi(t_n) =)s_n > t, \text{ for each } n.$$

By definition, this yields

$$F(t) \leq F(t_n) - C, \quad \text{for all } n.$$

Passing to limit as $n \to \infty$, we derive

$$(F(t-0) \leq)F(t) \leq F(t+0) - C, \text{ that is, } C \leq \mathrm{osc}(\pm)(F;C),$$

in contradiction with the bilateral bounded oscillating property. Summing up, our working assumption (equ) cannot be accepted; and the conclusion follows. □

Note that, an early version of this result was obtained in Turinici [27]. In fact, it can be directly adapted to the class of low-oscillating functions, so as to get the underlying statement; we do not give further details.

A basic completion of the above developments is to be given along the following lines. Let the function $F \in \mathcal{F}(\text{in-s-W})(R_+^0, R)$ and the constant $C > 0$ be endowed with the properties

$(F;C)$ is left bounded oscillating and right bounded oscillating.

By the preceding result, the associated to (F,C) function $\varphi \in \mathcal{F}(R_+)$ fulfills

(P-1) φ is increasing and regressive ($\varphi(t) < t$, for all $t > 0$),
(P-2) φ is Matkowski admissible ($\lim_n \varphi^n(t) = 0$, for all $t > 0$).

Let $\chi = I - \varphi$, where $(I(t) = t; t \in R_+)$ (the identity function of $\mathcal{F}(R_+)$) stands for the *complement* function attached to φ. Clearly,

$$\chi(0) = 0, \text{ and } \chi \text{ is regressive (because } \varphi(R_+^0) \subseteq R_+^0),$$

but, in general, χ is not increasing. Remember that the *coerciveness* property of this last function means

(coerc-1) $(t - \varphi(t) =)\chi(t) \to \infty$ as $t \to \infty$.

By a direct argument, this is nothing else than

(coerc-2) $\forall \alpha > 0, \exists \beta > \alpha, (\forall t): \chi(t) \leq \alpha \Longrightarrow t \leq \beta$,

and this, from the perspective of φ, writes

$$H(\alpha) := \{\beta > \alpha; t \leq \alpha + \varphi(t) \Longrightarrow t \leq \beta\} \neq \emptyset, \forall \alpha > 0,$$

referred to as the following: φ is *complement-coercive*.

Now, a natural question is that of determining the conditions upon the generating couple (F, C) so that the associated to (F, C) function $\varphi = \varphi(F, C)$ be complement coercive. An appropriate answer to this is the following one:

Proposition 2. *Let the increasing semi-Wardowski function $F \in \mathcal{F}(R_+^0, R)$ and the constant $C > 0$ be such that*

(22-I) $(F; C)$ *is left bounded oscillating and right bounded oscillating.*

In addition, assume that

(22-II) F *is quasi-asymptotic:*
$F(s + \alpha) - F(s) \to 0$ *as $s \to \infty$, for each $\alpha > 0$.*

Then, the associated to (F, C) function $\varphi \in \mathcal{F}(R_+)$ is complement coercive.

Proof. Suppose by absurd that this property is not true, that is, for some $\alpha > 0$:

for each $\beta > \alpha$, there exists $t > \beta$ with $t \leq \alpha + \varphi(t)$.

As a consequence of this, there exists a sequence (t_n) in R_+^0, with (t_n) is strictly ascending, $t_n \to \infty$ and $[t_n \leq \alpha + \varphi(t_n), \forall n]$.

Let $\gamma > \alpha$ be arbitrary fixed (and sufficiently close α). By the preceding relations, there exists some index $n(\gamma)$ such that

$$s_n := t_n - \gamma > 0, \quad \text{for all } n \geq n(\gamma).$$

Without loss (passing to a subsequence if necessary), one may assume $n(\gamma) = 0$ so that $0 < s_n := t_n - \gamma < t_n - \alpha \leq \varphi(t_n), \forall n$.

By a previous observation involving the properties of φ, one gets

$(\forall n): F(t_n - \gamma) \leq F(t_n) - C$, or, equivalently (by our notation) $(0 <)C \leq F(s_n + \gamma) - F(s_n)$.

This, along with $\lim_n s_n = \infty$, yields (by the imposed condition)

$$(0 <)C \leq \liminf_n [F(s_n+\gamma) - F(s_n)] = \lim_n [F(s_n+\gamma) - F(s_n)] = 0,$$

a contradiction, proving the desired conclusion. □

Note finally that in all these constructions the increasing property of F is essential. So, it is natural asking whether an extension of these facts is possible beyond the increasing framework; we conjecture that the answer is positive.

3. Main Results

Let (X, d) be a metric space, and $T : X \to X$ be a selfmap of X. As precise, we are interested to get sufficient conditions for the existence and/or uniqueness of elements in Fix(T).

The proposed problem is developed in the (iterative) setting of

(it-seq) the fixed points of T are ultimately chosen among the limit points (if any) $T^\omega x_0 := \lim_n (T^n x_0)$, where $x_0 \in X$ is arbitrary fixed.

(Here, ω is the first transfinite ordinal.) To do this, a lot of technical facts about iterative sequences are needed.

Let x_0 be some point in X. By an *orbital* (or *iterative*) sequence attached to x_0 (and T), we mean any sequence $X_0 := (x_n; n \geq 0)$ (or, simply, $X_0 = (x_n)$) defined as $(x_n = T^n x_0; n \geq 0)$. When x_0 is generic here, the resulting object will be referred to as an *orbital sequence* (in short: *o-sequence*) in X.

Fix in the following such an object $X_0 = (x_n)$ and denote

$[X_0] = \{x_n; n \geq 0\}$ (the *trajectory* attached to $X_0 = (x_n)$),
$[[X_0]] := \mathrm{cl}([X_0])$ (the *complete trajectory* attached to $X_0 = (x_n)$).

Further, let us construct the function

(def-B) $B(x,y) = \min\{d(x,y), d(Tx, Ty)\}$, $(x, y) \in X \times X$

and introduce the relation (over X)

(rela-B) $(B > 0) = \{(x, y) \in X \times X; B(x, y) > 0\}$.

The position of iterative sequence $X_0 = (x_n)$ with respect to this relation is described by the following alternatives:

(Alt-1) The orbital sequence $X_0 = (x_n)$ is *telescopic*, in the sense

there exists $h \geq 0$, such that $d(x_h, x_{h+1}) = 0$, i.e., $x_h = x_{h+1}$.

By the iterative definition of our sequence, one derives

$x_h = x_n$, for all $n \geq h$; whence, $z := x_h$ is an element of $\mathrm{Fix}(T)$.

Consequently, this case is completely clarified from the fixed point perspective.

(Alt-2) The orbital sequence $X_0 = (x_n)$ is *nontelescopic*, in the sense

$r_n := d(x_n, x_{n+1}) > 0$, for all n.

Note that, in this case,

(x_n) is $(B > 0)$-ascending: $B(x_n, x_{n+1}) > 0$, for all n,
referred to as (x_n) is *B-ascending*.

We then say that the iterative sequence $X_0 = (x_n)$ has the *B-property* or, equivalently, that $X_0 = (x_n)$ is (B-o).

Having these precise, the determination of fixed points is carried out in the following context, comparable with the one in Rus [23, Chapter 2, Section 2.2]:

(pic-1) We say that the (B-o) sequence $X_0 = (x_n)$ is *semi-Picard* (modulo $(d;T)$) when (x_n) is d-asymptotic, in the sense that $(r_n := d(x_n, x_{n+1}); n \geq 0)$ is zero-convergent $[r_n \to 0$ as $n \to \infty]$.

(pic-2) We say that the (B-o) sequence $X_0 = (x_n)$ is *descending semi-Picard* (modulo $(d;T)$), when (x_n) is descending d-asymptotic: $(r_n := d(x_n, x_{n+1}); n \geq 0)$ is strictly descending and zero-convergent.

(pic-3) We say that the (B-o) sequence $X_0 = (x_n)$ is *Picard* (modulo $(d;T)$) if (x_n) is d-Cauchy.

(pic-4) We say that the (B-o) sequence $X_0 = (x_n)$ is *strong Picard* (modulo $(d;T)$) if it (x_n) is d-convergent and $x_\omega := \lim_n(x_n)$ exists as an element of Fix(T)

(pic-5) We say that T is *fix-asingleton* if Fix(T) is an asingleton; and *fix-singleton*, provided Fix(T) is a singleton.

The regularity conditions to be used for getting such properties are introduced as follows. Let us say that the (B-o) sequence $Y_0 = (y_n)$ is *full*, provided

$$n \mapsto y_n \quad \text{is injective } (i \neq j \text{ implies } y_i \neq y_j).$$

In this case, $Y_0 = (y_n)$ is referred to as a *(B-o-f) sequence*.

(reg-1) We say that X is *(B-o-f,d)-complete*, provided, for any (B-o-f) sequence $Y_0 = (y_n; n \geq 0)$, one has (y_n) is d-Cauchy implies (y_n) is d-convergent.

(reg-2) We say that T is *(B-o-f,d)-continuous*, if, for any (B-o-f) sequence $Y_0 = (y_n; n \geq 0)$, one has $y_n \xrightarrow{d} z$ implies $Ty_n \xrightarrow{d} Tz$.

As a completion of these, we now describe the class of contractive-type requirements to be used. Let $V \in \mathcal{F}(X \times X, R_+)$ be a map. Usually, this may be taken according to the composition rule

$$V(x,y) = v(\mathcal{Q}(x,y), \mathcal{K}(x,y)), \quad x, y \in X$$

with respect to some function $v \in \mathcal{F}(R_+^6 \times R_+^4, R_+)$, but this is not effective for us in the sequel. The regularity condition imposed upon the underlying map is

(posi) V is *B-positive*: $B(x,y) > 0$ implies $V(x,y) > 0$.

Further, let $\varphi \in \mathcal{F}(R_+)$ be a function. The contractions to be considered are

(M-r-con) $\quad d(Tx, Ty) \leq \varphi(V(x,y)), \ x, y \in X, \ B(x,y) > 0.$

These are referred to as follows: T is *Matkowski* $(B > 0; \varphi; V)$-*contractive*; and, if $(\varphi; V)$ is generic, we say that T is a *Matkowski rational contraction*.

It is our aim in the following to formulate the general and specific conditions upon these data under which the fixed-point problem we just exposed has a solution. To do this, we need some preliminary facts.

(I) The setting of our problem is

(s-M-ad) $\quad \varphi \in \mathcal{F}(\text{re, in})(R_+)$ is super Matkowski admissible.

A useful property of the functions in this class may be stated as follows. Let us say that the couple (t_n), (s_n) of sequences in R_+^0 is *lower diagonal compatible*, provided

$t := \lim_n t_n, \ s := \lim_n s_n$ exist, with $t > 0$ and $t \geq s$.

Proposition 3. *Let the function φ be taken as before. Then,*

> *φ is asymptotic upper diagonal: there are no couples of lower diagonal compatible sequences (t_n) and (s_n) in R_+^0 with $[t_n \leq \varphi(s_n), \forall n]$.*

Proof. Suppose by contradiction that

> there is a couple of sequences (t_n) and (s_n) in R_+^0 with such a property.

By the second half of working condition and φ = regressive,

$t_n < s_n, \ \forall n$; whence (passing to limit) $t \leq s$,

and this, combined with the lower diagonal compatible property, gives $t = s > 0$. Define the associated sequence $(S_n = \max\{s_n, t(1+2^{-n})\}; n \geq 0)$; clearly,

$(s_n \leq S_n, \forall n)$ and $S_n \to t+$ as $n \to \infty$.

Again by the second half of working condition, we derive ($t_n \leq \varphi(S_n); n \geq 0$), and this, by a limit process gives (as $\varphi =$ super Matkowski admissible) $t \leq \varphi(t+0) < t$, contradiction. Hence, the underlying working assumption is not acceptable, and φ is asymptotic upper diagonal, as asserted. □

(II) To solve the semi-Picard question, a new convention is added. Let $(a, b) \mapsto a * b$ be some binary relation over R_+; we call it *semi regular*, provided

(sreg-1) $(a \leq a * b)$, $(c * c = c)$, $\forall a, b, c \in R_+$,
(sreg-2) $(a, b) \mapsto a * b$ is continuous in its arguments.

Given such an object, call it

(reg) *regular*, when ($b < a * b$ implies $b < a$),
or, equivalently, ($a \leq b$ implies $a * b \leq b$),
(s-reg) *strongly regular*, when ($b \neq a * b$ implies $b < a$),
or, equivalently, ($a \leq b$ implies $a * b = b$).

A basic example of such an operation is

$a \top b = \max\{a, b\} = a + \max\{0, b - a\}$, $a, b \in R_+$;
clearly, ($\top$) is strongly regular.

Some other examples are provided by the following statement. Let $I(.)$ stand for the unitary function of $\mathcal{F}(R_+)$ defined as ($I(t) = t$; $t \geq 0$). Given the increasing $f \in \mathcal{F}(R_+)$, define the binary operations

$a(f)b = a + \max\{0, f(b) - f(a)\}$, $a, b \in R_+$,
$a[f]b = a + |f(b) - f(a)|$, $a, b \in R_+$.

Clearly, f and $f - \alpha$ (where $0 \leq \alpha \leq f(0)$) generate the same binary operations, so, without loss, one may assume that $f(0) = 0$.

Proposition 4. *Let the function* $f \in \mathcal{F}(R_+)$ *be continuous increasing with* $f(0) = 0$. *Then,*

(32-1) *the binary operation* (f) *is regular when* $I - f$ *is increasing,*
(32-2) *the binary operation* $[f]$ *is regular whenever* $I - f$ *is increasing,*
(32-3) *the binary operation* (f) *is strongly regular iff* $f = I$; *moreover, in this case,* $(a \top b = a(f)b, \forall a, b \in R_+)$,

(32-4) *the binary operation $[f]$ is strongly regular iff $f = I$; moreover, in this case, $(a\top b \leq a[f]b, \forall a, b \in R_+)$.*

Proof. (i) Clearly, (f) is semiregular. Further, let $a, b \in R_+$ be such that

$a \leq b$; hence, $f(b) \geq f(a)$; wherefrom $a(f)b = a + f(b) - f(a)$.

The posed relation $a(f)b \leq b$ becomes

$a - f(a) \leq b - f(b)$; evident, via $I - f =$ increasing.

(ii) As before, $[f]$ is semiregular. Further, let $a, b \in R_+$ be such that

$a \leq b$; hence, $f(b) \geq f(a)$; wherefrom $a(f)b = a + f(b) - f(a)$.

The posed relation $a[f]b \leq b$ becomes

$a - f(a) \leq b - f(b)$; evident, via $I - f =$ increasing.

(iii) Let $a, b \in R_+$ be such that

$a \leq b$; hence, $f(b) \geq f(a)$; wherefrom $a(f)b = a + f(b) - f(a)$.

The posed relation $a(f)b = b$ becomes

$a - f(a) = b - f(b)$ (for all such (a, b)).

But then (combining with $f(0) = 0$), $f = I$, and the assertion follows.

(iv) Let $a, b \in R_+$ be such that

$a \leq b$; hence, $f(b) \geq f(a)$; wherefrom $a[f]b = a + f(b) - f(a)$.

The posed relation $a[f]b = b$ becomes

$a - f(a) = b - f(b)$ (for all such (a, b)).

But then (combining with $f(0) = 0$), $f = I$, and the assertion follows. □

(III) Passing to the Picard question, the following concept is useful. Let $Y_0 = (y_n)$ be a d-asymptotic (B-o-f) sequence. By a *Boyd–Wong Y_0-system,* we mean any triple $[\varepsilon; ((m(k)), (n(k)))]$, where $\varepsilon \in R_+^0$ and $[(m(k)), (n(k))]$ is a couple of rank sequences with the properties

(BW-1) $k \leq m(k) < m(k) + 3 < n(k), \quad \forall k,$

(BW-2) $\lim_k d(y_{m(k)}, y_{m(k)+u}) = \lim_k d(y_{n(k)}, y_{n(k)+v}) = 0$,
$\forall u, v \in N$,
(BW-3) $d(y_{m(k)+s}, y_{n(k)+t}) \to \varepsilon+$ as $k \to \infty$, $\forall s, t \in N[0, 3]$,
(BW-4) $d(y_{m(k)+p}, y_{n(k)+q}) \to \varepsilon$ as $k \to \infty$, for all $p, q \in N$.

Then, define the following property (for the B-positive function V as before):

V is *right asymptotic*: for each d-asymptotic (B-o-f) sequence $Y_0 = (y_n)$ and each attached Boyd–Wong Y_0-system $[\varepsilon; (m(k)), (n(k))]$, the sequence $(s_k := V(y_{m(k)}, y_{n(k)}); k \geq 0)$ is in R_+^0 and d-convergent, with $\varepsilon \geq \lim_k s_k$.

(IV) Further, passing to the strong Picard question, the following convention is needed. Given the (B-o-f) sequence $Y_0 = (y_n)$ and the point $z \in X$, let us call $(Y_0; z)$ *limit separated* when

$$y_n \xrightarrow{d} z \ (B(y_n, z) > 0, \text{ for all } n), \text{ and } d(z, Tz) > 0.$$

Then, let us say that (the B-positive map) V is *point asymptotic*, provided

for each limit separated couple $(Y_0; z)$ as before, we have that the sequence $(p_n := V(y_n, z); n \geq 0)$ is in R_+^0 and d-convergent, with $d(z, Tz) \geq \lim_n p_n$.

(V) Finally, let us say that (the B-positive map) V is Fix(T)-*nonexpansive*, if

$$z_1, z_2 \in \text{Fix}(T) \text{ and } z_1 \neq z_2 \text{ imply } V(z_1, z_2) \leq d(z_1, z_2).$$

The following is our first main result in this exposition.

Theorem 2. *Let the selfmap $T \in \mathcal{F}(X)$ be such that*

(31-i) *T is Matkowski $(B > 0; \varphi; V)$-contractive (see above), where $\varphi \in \mathcal{F}(R_+)$ and $V \in \mathcal{F}(X \times X, R_+)$ are taken so as*
(31-ii) *φ is regressive, increasing, and super Matkowski admissible,*
(31-iii) *V is B-positive: $B(x, y) > 0$ implies $V(x, y) > 0$.*

Further, assume that X is (B-o-f,d)-complete. Then, for the iterative sequence $X_0 = (x_n)$ with
(31-iv) $X_0 = (x_n)$ is (B-o) (hence, $r_n := d(x_n \cdot x_{n+1}) > 0$, $\forall n$), the following conclusions hold:

(31-a) $X_0 = (x_n)$ is descending semi-Picard (modulo $(d;T)$) (hence, in particular, full), provided there exists a regular binary operation $(*)$, with the property

V is weakly $(*)$-iterative ($V^n := V(x_n, x_{n+1}) \leq r_n * r_{n+1}$, $\forall n$).

(31-b) $X_0 = (x_n)$ is Picard (modulo $(d;T)$) if, in addition, V is right asymptotic,
(31-c) $X_0 = (x_n)$ is strong Picard (modulo $(d;T)$) if (in addition to these requirements) V is point asymptotic,
(31-d) T is fix-asingleton if V is Fix(T)-nonexpansive.

Proof. The argument consists in a number of parts.
Part 1. From the choice of $X_0 = (x_n)$ and weak iterative condition,

$$0 < d(Tx_n, Tx_{n+1}) = r_{n+1}, \ 0 < V^n := V(x_n, x_{n+1}) \leq r_n * r_{n+1}, \ \forall n.$$

The contractive condition is applicable to (x_n, x_{n+1}), for all n, and gives

$(\forall n)$: $r_{n+1} \leq \varphi(r_n * r_{n+1})$; whence (as φ = regressive) $r_{n+1} < r_n * r_{n+1}$.

Combining with the regularity of $(*)$ gives

$(r_{n+1} < r_n, \forall n)$: so, (r_n) is strictly descending.

As a first consequence of this, $X_0 = (x_n)$ is full. In fact, suppose by contradiction that there exist $i, j \in N$ such that $i < j$, $x_i = x_j$. Then, $x_{i+1} = x_{j+1}$; whence: $r_i = r_j$; impossible, via $(r_n) =$ strictly descending. Hence, our working hypothesis cannot hold; and the assertion follows. As a second consequence of this, $r := \lim_n (r_n)$ exists in R_+. Suppose by contradiction that $r > 0$. For the moment, we have

$r_n \to r+$ and $r_n * r_{n+1} \to r+$, as $n \to \infty$,

where the second relation is deductible by the first one, via

$(r_n * r_{n+1} \geq r_n, \forall n)$ and $r_n * r_{n+1} \to r * r = r$, as $n \to \infty$.

On the other hand, by the contractive condition,

$(t_n \leq \varphi(s_n), \forall n);$ where $(t_n := r_{n+1}; n \geq 0)$, $(s_n := r_n * r_{n+1}; n \geq 0)$.

This, along with $\lim_n t_n = \lim_n s_n = r > 0$ yields a contradiction with respect to φ = asymptotic upper diagonal, and our assertion follows.

Part 2. For the moment, $X_0 = (x_n)$ is full, d-asymptotic. Suppose by contradiction that $X_0 = (x_n)$ is not d-Cauchy. By a result in Turinici [31, Section 14], there exists at least one Boyd–Wong X_0-system $[\varepsilon; (m(k)), (n(k))]$, where $\varepsilon \in R_+^0$ and $[(m(k); k \geq 0), (n(k); k \geq 0)]$ is a couple of rank sequences with the precise properties. As V is B-positive, right asymptotic, we have that

$(s_k := V(x_{m(k)}, x_{n(k)}); k \geq 0)$ is a d-convergent sequence in R_+^0, with (in addition) $\varepsilon \geq \lim_k s_k$.

On the other hand, by the same properties of our system

$(t_k := d(x_{m(k)+1}, x_{n(k)+1}); k \geq 0)$ is a sequence in R_+^0 with (in addition) $t_k \to \varepsilon+$ as $k \to \infty$.

This, along with $(t_k \leq \varphi(s_k); k \geq 0)$, yields a contradiction with φ = asymptotic upper diagonal, and we are done.

Part 3. Summing up, $X_0 = (x_n)$ is a d-Cauchy (B-o-f) sequence. From the completeness assumption,

$x_n \xrightarrow{d} z$ as $n \to \infty$, for some $z \in X$.

By the full property of $X_0 = (x_n)$,

(bd) $H := \{n \in N; (x_{n+1} =)Tx_n = Tz\}$ is a singleton (hence, bounded).

As a consequence, there exists an index $i = i(z) \in N$ such that

$Tx_n \neq Tz$ (hence, $B(x_n, z) > 0$), for all $n \geq i$.

We show that, necessarily, $z = Tz$. Suppose not: $d(z, Tz) > 0$. By the obtained relations, $Y_0 = (y_n)$, where $(y_n = x_{n+i}; n \geq 0)$ is a (B-o-f) sequence with

$$y_n \xrightarrow{d} z \quad \text{and} \quad (B(y_n, z) > 0, \text{ for all } n),$$

and this, along with $d(z, Tz) > 0$, tells us that $(Y_0; z)$ is a limit separated couple. Combining with $V = B$-positive, point asymptotic, one derives that

the sequence $(p_n := V(y_n, z); n \geq 0)$ is in R_+^0 and d-convergent, with $d(z, Tz) \geq \lim_n p_n$.

On the other hand, by the choice of $Y_0 = (y_n)$ and a metrical property of d,

the sequence $(q_n := d(Ty_n, Tz); n \geq 0)$ is in R_+^0 and d-convergent, with $\lim_n q_n = d(z, Tz)$.

This, along with $(q_n \leq \varphi(p_n); n \geq 0)$, is in contradiction with $\varphi = $ asymptotic upper diagonal, and our assertion follows.

Part 4. Let $z_1, z_2 \in \text{Fix}(T)$ be such that $z_1 \neq z_2$; hence, $B(z_1, z_2) > 0$. As a first consequence of this, we have

$0 < V(z_1, z_2) \leq d(z_1, z_2)$ (as V is B-positive and Fix(T)-nonexpansive).

Moreover, as a second consequence of this, the contractive condition applies to (z_1, z_2) and gives (in combination with the preceding fact)

$$d(z_1, z_2) \leq \varphi(V(z_1, z_2)) < V(z_1, z_2) \leq d(z_1, z_2).$$

The contradiction at which we arrived shows that our working assumption is not acceptable, and the assertion follows. □

Having these established, let $V \in \mathcal{F}(X \times X, R_+)$ be a map. Usually, one may take it according to the composition rule

$$V(x, y) = v(\mathcal{Q}(x, y), \mathcal{K}(x, y)), \quad x, y \in X$$

with respect to some function $v \in \mathcal{F}(R_+^6 \times R_+^4, R_+)$, but this will be not effective for us in the sequel. The regularity condition imposed upon the underlying map is

(posi) V is B-positive: $B(x, y) > 0$ implies $V(x, y) > 0$.

Further, letting $F \in \mathcal{F}(R_+^0, R)$ and $C > 0$, define the class of contractions

(W-r-con) $F(d(Tx,Ty)) \leq F(V(x,y)) - C$, $x, y \in X$, $B(x,y) > 0$.

This is referred to as: T is *Wardowski* $(B > 0; F, C; V)$-*contractive*; and, if $(F, C; V)$ is generic, we say that T is a *Wardowski rational contraction*.

Now, by simply combining the preceding statement with the reduction-type developments, one gets our second main result of this exposition.

Theorem 3. *Let the selfmap* $T \in \mathcal{F}(X)$ *be such that*

(32-i) T *is Wardowski* $(B > 0; F, C; V)$-*contractive (see above) where the function* $F \in \mathcal{F}(R_+^0, R)$, *the constant* $C > 0$, *and the function* V *satisfy*

(32-ii) F *is increasing semi-Wardowski and* (F, C) *is bilateral bounded oscillating*

(32-iii) V *is* B-*positive:* $B(x, y) > 0$ *implies* $V(x, y) > 0$.

Further, assume that X *is* $(B$-o-$f,d)$-*complete. Then, for the iterative sequence* $X_0 = (x_n)$ *with*

(32-iv) $X_0 = (x_n)$ *is* $(B$-$o)$ *(hence,* $r_n := d(x_n, x_{n+1}) > 0$, $\forall n)$, *the following conclusions hold:*

(32-a) $X_0 = (x_n)$ *is descending semi-Picard (modulo* $(d; T))$ *(hence, in particular, full) provided there exists a regular binary operation* $(*)$, *with the following:*

V *is weakly* $(*)$-*iterative* $(V^n := V(x_n, x_{n+1}) \leq r_n * r_{n+1}, \forall n)$.

(32-b) $X_0 = (x_n)$ *is Picard (modulo* $(d; T))$ *if, in addition,* V *is right asymptotic*

(32-c) $X_0 = (x_n)$ *is strong Picard (modulo* $(d; T))$ *if (in addition to these requirements)* V *is point asymptotic.*

(32-d) T *is fix-asingleton if* V *is* $\text{Fix}(T)$-*nonexpansive.*

Proof. (Sketch) By the imposed upon (F, C) hypotheses, the associated to (F, C) function $\varphi \in \mathcal{F}(R_+)$ satisfies

(P-1) φ is regressive, increasing, super Matkowski admissible and the initial contractive condition yields

(P-2) $d(Tx, Ty) \leq \varphi(V(x, y))$, $x, y \in X$, $B(x, y) > 0$, that is, T is Matkowski $(B > 0; \varphi; V)$-contractive.

This, along with the first main result, gives us all needed conclusions. □

Note that a reduction statement of this type is working as well for a large class of contractions (including the preceding ones) introduced by Krasnoselskii and Stetsenko [12], re-discovered by Rhoades [22] and refined by Dutta and Choudhury [8]. We discuss these facts elsewhere.

4. Particular Aspects

In the following, some particular cases of our main result above are discussed.

Let (X, d) be a metric space, and $T : X \to X$ be a mapping. Define the *elementary* mappings in $\mathcal{F}(X \times X, R_+)$: for each $x, y \in X$,

$A_0(x, y) = d(x, y)$, $A_1(x, y) = \max\{d(x, Tx), d(y, Ty)\}$,
$A_2(x, y) = (1/2)[d(x, Ty) + d(Tx, y)]$,
$A_3(x, y) = (1/2)[d(x, T^2x) + d(T^2x, Ty)]$,
$A_4(x, y) = \max\{d(Tx, T^2x), d(T^2x, y), d(T^2x, Ty)\}$,
$A_5(x, y) = \max\{d(Tx, T^2x), d(T^2x, y)\}$,
$A_6(x, y) = \max\{d(x, Tx) + d(T^2x, Ty), d(y, Ty) + d(Tx, y)\}$,
$E(x, y) = d(x, y) + |d(x, Tx) - d(y, Ty)|$.

Remember that the binary operations on R_+

$$a \top b = \max\{a, b\},\ a * b = a + |a - b|,\ a, b \in R_+$$

enjoy the properties:

both ($\top$) and ($*$) are strongly regular (hence, regular), with $a \top b \leq a * b$, for all $a, b \in R_+$.

In addition, we have by definition

$$(\forall x \in X): E(x, Tx) = d(x, Tx) * d(Tx, T^2 x), \quad x \in X;$$
whence, $E(x, Tx) \geq A_0(x, Tx)$.

The idea of introducing the function $E(.,.)$ is due to Popescu [19].

Now, as a maximum-type combination of these, we construct our mappings to be used as examples in our main result: for $x, y \in X$,

$V_0(x, y) = A_0(x, y)$,
$V_1(x, y) := \max\{A_0, A_1, A_2\}(x, y)$,
$V_2(x, y) := \max\{E, A_1, A_2\}(x, y)$,
$V_3(x, y) := \max\{A_0, A_1, A_2, A_3, A_4\}(x, y)$,
$V_4(x, y) := \max\{E, A_1, A_2, A_3, A_4\}(x, y)$,
$V_5(x, y) := \max\{A_0, A_2, A_3, A_5, A_6\}(x, y)$,
$V_6(x, y) := \max\{E, A_2, A_3, A_5, A_6\}(x, y)$.

The verification of working conditions in the underlying result upon $V \in \{V_0, \ldots, V_6\}$ is carried out in what follows:

(Check-0) [V is B-positive].
Let $x, y \in X$ be such that

$B(x, y) > 0$; hence, $A_0(x, y) = d(x, y) > 0$, $d(Tx, Ty) > 0$.

According to the definitions above,

$V_0(x, y) = A_0(x, y) > 0$, $V_1(x, y) \geq A_0(x, y) > 0$,
$V_2(x, y) \geq E(x, y) \geq A_0(x, y) > 0$, $V_3(x, y) \geq A_0(x, y) > 0$,
$V_4(x, y) \geq E(x, y) \geq A_0(x, y) > 0$, $V_5(x, y) \geq A_0(x, y) > 0$,
$V_6(x, y) \geq E(x, y) \geq A_0(x, y) > 0$,

and from this

each $V \in \{V_0, \ldots, V_6\}$ is B-positive.

(Check-1) [V is weakly $(*)$-iterative].
Let $X_0 = (x_n)$ be an iterative sequence with

$X_0 = (x_n)$ is (B-o) (hence, $r_n := d(x_n, x_{n+1}) > 0$, $\forall n$).

By definition, we have

$(\forall n): A_0(x_n, x_{n+1}) = r_n \leq \max\{r_n, r_{n+1}\}$,
$A_1(x_n, x_{n+1}) = \max\{r_n, r_{n+1}\}$,
$A_2(x_n, x_{n+1}) = (1/2)d(x_n, x_{n+2}) \leq \max\{r_n, r_{n+1}\}$,

$$A_3(x_n, x_{n+1}) = (1/2)d(x_n, x_{n+2}) \leq \max\{r_n, r_{n+1}\},$$
$$A_4(x_n, x_{n+1}) = r_{n+1} \leq \max\{r_n, r_{n+1}\},$$
$$A_5(x_n, x_{n+1}) = r_{n+1} \leq \max\{r_n, r_{n+1}\},$$
$$A_6(x_n, x_{n+1}) = \max\{r_n, r_{n+1}\},$$
$$E(x_n, x_{n+1}) = r_n * r_{n+1} \geq \max\{r_n, r_{n+1}\},$$

and, from this

$$(\forall n): V_i(x_n, x_{n+1}) \leq r_n * r_{n+1}, \ i \in \{0, 1, 2, 3, 4, 5, 6\},$$

or, in other words,

each $V \in \{V_0, \ldots, V_6\}$ is weakly $(*)$-iterative.

(Check-2) [V is right asymptotic].
Let the d-asymptotic (B-o-f) sequence $Y_0 = (y_n)$ be given; as well as a Boyd–Wong Y_0-system $[\varepsilon; (m(k)), (n(k))]$ attached to it. By the definition of our maps,

$$\lim_k A_0(y_{m(k)}, y_{n(k)}) = \varepsilon, \quad \lim_k A_1(y_{m(k)}, y_{n(k)}) = 0,$$
$$\lim_k A_2(y_{m(k)}, y_{n(k)}) = \varepsilon, \quad \lim_k A_3(y_{m(k)}, y_{n(k)}) = \varepsilon/2,$$
$$\lim_k A_4(y_{m(k)}, y_{n(k)}) = \varepsilon, \quad \lim_k A_5(y_{m(k)}, y_{n(k)}) = \varepsilon,$$
$$\lim_k A_6(y_{m(k)}, y_{n(k)}) = \varepsilon, \quad \lim_k E(y_{m(k)}, y_{n(k)}) = \varepsilon,$$

and this gives

$$\lim_k V_0(y_{m(k)}, y_{n(k)}) = \varepsilon, \quad \lim_k V_1(y_{m(k)}, y_{n(k)}) = \varepsilon,$$
$$\lim_k V_2(y_{m(k)}, y_{n(k)}) = \varepsilon, \quad \lim_k V_3(y_{m(k)}, y_{n(k)}) = \varepsilon,$$
$$\lim_k V_4(y_{m(k)}, y_{n(k)}) = \varepsilon, \quad \lim_k V_5(y_{m(k)}, y_{n(k)}) = \varepsilon,$$
$$\lim_k V_6(y_{m(k)}, y_{n(k)}) = \varepsilon,$$

which tells us that

each $V \in \{V_0, \ldots, V_6\}$ is right asymptotic.

(Check-3) [V is point asymptotic].
Let the (B-o-f) sequence $Y_0 = (y_n)$ and the point $z \in X$ be such that $(Y_0; z)$ is limit separated, i.e.,

$$y_n \xrightarrow{d} z, \ (B(y_n, z) > 0, \text{ for all } n), \text{ and } b := d(z, Tz) > 0.$$

We have, by definition,

$$\lim_n A_0(y_n, z) = 0, \quad \lim_n A_1(y_n, z) = b,$$
$$\lim_n A_2(y_n, z) = b/2, \quad \lim_n A_3(y_n, z) = b/2,$$

$$\lim_n A_4(y_n, z) = b, \quad \lim_n A_5(y_n, z) = 0,$$
$$\lim_n A_6(y_n, z) = b, \quad \lim_n E(y_n, z) = b,$$

and this gives

$$\lim_n V_0(y_n, z) = 0, \quad \lim_n V_1(y_n, z) = b,$$
$$\lim_n V_2(y_n, z) = b, \quad \lim_n V_3(y_n, z) = b,$$
$$\lim_n V_4(y_n, z) = b, \quad \lim_n V_5(y_n, z) = b,$$
$$\lim_n V_6(y_n, z) = b,$$

which tells us that

each $V \in \{V_0, \ldots, V_6\}$ is point asymptotic.

(Check-4) [V is Fix(T)-nonexpansive].
Let z_1 and z_2 be two fixed points of T; and assume that $a := d(z_1, z_2) > 0$. By definition,

$$A_0(z_1, z_2) = a, \ A_1(z_1, z_2) = 0, \ A_2(z_1, z_2) = a,$$
$$A_3(z_1, z_2) = a/2, \ A_4(z_1, z_2) = a, \ A_5(z_1, z_2) = a,$$
$$A_6(z_1, z_2) = a, \ E(z_1, z_2) = a,$$

and this gives

$$V_0(z_1, z_2) = a, \ V_1(z_1, z_2) = a, \ V_2(z_1, z_2) = a,$$
$$V_3(z_1, z_2) = a, \ V_4(z_1, z_2) = a, \ V_5(z_1, z_2) = a, \ V_6(z_1, z_2) = a,$$

proving that

each $V \in \{V_0 \ldots, V_6\}$ is Fix(T)-nonexpansive.

By the second main result, we then derive the following fixed point statement with a methodological value.

Theorem 4. *Let the selfmap $T \in \mathcal{F}(X)$ and the index $m \in \{0, 1, 2, 3, 4, 5, 6\}$ be taken in accordance with*

(41-i) *T is Wardowski $(B > 0; F, C; V_m)$-contractive, where the function $F \in \mathcal{F}(R_+^0, R)$ and the constant $C > 0$ satisfy*

(41-ii) *F is increasing semi-Wardowski and (F, C) is bounded oscillating and the functional family $\{V_0, \ldots, V_6\}$ is the above described one. Further, assume that X is $(B\text{-}o\text{-}f, d)$-complete. Then, for the iterative sequence $X_0 = (x_n)$, with*

(41-iii) *$X_0 = (x_n)$ is $(B\text{-}o)$ (hence, $r_n := d(x_n, x_{n+1}) > 0, \forall n$),* the following conclusions hold:

- **(41-a)** $X_0 = (x_n)$ is descending semi-Picard (modulo $(d; T)$) (hence, in particular, full).
- **(41-b)** $X_0 = (x_n)$ is Picard (modulo $(d; T)$).
- **(41-c)** $X_0 = (x_n)$ is strong Picard (modulo $(d; T)$).
- **(41-d)** T is fix-asingleton.

Some remarks are in order.

- **(Rem-0)** An alternate method of getting the stated conclusions is that of using the implicit methods in Turinici [29]. We do not give details.
- **(Rem-1)** The particular case of $V = V_0$ yields a result comparable with the one in Wardowski [33]. Clearly, this last result contains the particular case in question. However, a large number of Wardowski-type contractions where (F, C) fulfills the bounded oscillation requirement above are in fact reducible to the case $V = V_0$, so, these are Matkowski-type contractions.
- **(Rem-2)** The particular case of $V = V_1$ yields a result comparable with the one in Wardowski and Dung [35].
- **(Rem-3)** The particular case of $V = V_3$ yields a result extending the continuous portion of the one in Dung and Hang [7].
- **(Rem-4)** The particular case of $V = V_5$ yields a result extending the continuous portion of the one in Piri and Kumam [17].
- **(Rem-5)** The particular case of $V \in \{V_2, V_4\}$ yields a couple of results extending the one in Fulga and Proca [9].

In fact, there are many other statements of this type fulfilling the conditions of our main result, such as the ones in Secelean and Zhou [24]. Further aspects are delineated elsewhere.

5. Wardowski–Dung Maps

Let (X, d) be a metric space, and $T \in \mathcal{F}(X)$ be a selfmap of X. By the developments above, it follows that for a pretty large number of mappings $V \in \mathcal{F}(X \times X, R_+)$ the attached Wardowski-type contractions admit Picard points. However, there exist mappings

$V \in \mathcal{F}(X \times X, R_+)$ for which the precise techniques do not work. A basic example is to be constructed as follows. Define

$(V_\infty \in \mathcal{F}(X \times X, R_+))$: $V_\infty(x,y) = \text{diam}\{x, Tx, T^2x, y, Ty\}$,
$x, y \in X$,

where *diam* means the *diameter* function. Note that, under the notation

(def-B) $B(x,y) = \min\{d(x,y), d(Tx, Ty)\}$, $x, y \in X$,

the following property holds:

$V_\infty(x,y) \geq B(x,y)$, $x, y \in X$; hence, V_∞ is B-positive.

Moreover, this function is representable in the standard composition way we just described. Precisely, denote for $x, y \in X$

$Q_1(x,y) = d(x, Tx)$, $Q_2(x,y) = d(x,y)$, $Q_3(x,y) = d(x, Ty)$,
$Q_4(x,y) = d(Tx, y)$, $Q_5(x,y) = d(Tx, Ty)$, $Q_6(x,y) = d(y, Ty)$,
$\mathcal{Q}(x,y) = (Q_1(x,y), Q_2(x,y), Q_3(x,y), Q_4(x,y), Q_5(x,y),$
$\qquad Q_6(x,y))$,
$K_1(x,y) = d(x, T^2x)$, $K_2(x,y) = d(Tx, T^2x)$,
$K_3(x,y) = d(T^2x, y)$, $K_4(x,y) = d(T^2x, Ty)$,
$\mathcal{K}(x,y) = (K_1(x,y), K_2(x,y), K_3(x,y), K_4(x,y))$.

Then, let $v_\infty : R_+^6 \times R_+^4 \to R_+$ be the function

$v_\infty(t_1, \ldots, t_6; s_1, \ldots, s_4) = \max\{t_1, \ldots, t_6; s_1, \ldots, s_4\}$,
$\qquad (t_1, \ldots, t_6) \in R_+^6$, $(s_1, \ldots, s_4) \in R_+^4$;

it is not hard to see that the representation formula is holding

$V_\infty(x,y) = v_\infty(\mathcal{Q}(x,y), \mathcal{K}(x,y))$, $x, y \in X$.

Now, conditions of Theorem 3 are not applicable here because

V_∞ is not (in general) weakly $(*)$-iterative,
    as results from
Q_3 is not (in general) weakly $(*)$-iterative.

So, we may ask of under which extra conditions are the conclusions of Theorem 3 available. It is our aim to show that a positive answer to this is possible by means of some global techniques, developed — for the associated Matkowski-type contraction — in the 2014 fixed-point statement obtained by Turinici [28], extending the well known 1974 result due to Ćirić [5].

Take some function $\varphi \in \mathcal{F}(R_+)$, and define the class of contractions

(M-D) $\quad d(Tx, Ty) \leq \varphi(V_\infty(x, y))$, $x, y \in X$, $B(x, y) > 0$.

This is referred to as: T is *Matkowski–Dung $(B > 0; \varphi)$-contractive*; and, if φ is generic, we say that T is a *Matkowski–Dung contraction*. A motivation of this convention is that the mapping V_∞ is related with the developments in Dung [6].

We are now in position to state our basic result of this section, referred to as *Matkowski–Dung fixed point principle* (in short: (MD-fpp)).

Theorem 5. *Suppose that T is Matkowski–Dung $(B > 0; \varphi)$-contractive, for some function $\varphi \in \mathcal{F}(R_+)$, with the following:*

(51-i) φ *is regressive, increasing, Matkowski admissible.*
(51-ii) φ *is complement-coercive:*

$H(\alpha) := \{\beta > \alpha; t \leq \alpha + \varphi(t) \implies t \leq \beta\} \neq \emptyset$, $\forall \alpha > 0$.

Further, assume that X is $(B\text{-}o,d)$-complete, and let $X_0 = (x_n)$ be such that

(51-iii) $\quad X_0 = (x_n)$ *is $(B\text{-}o)$ (hence, $r_n := d(x_n, x_{n+1}) > 0$, $\forall n$).*

Then

(51-a) X_0 *is Picard (modulo $(d; T)$).*
(51-b) $X_0 = (x_n)$ *is strong Picard (modulo $(d; T)$), when in addition T is $(B\text{-}o,d)$-continuous.*
(51-c) $X_0 = (x_n)$ *is strong Picard (modulo $(d; T)$), when in addition φ is super Matkowski admissible.*
(51-d) T *is fix-asingleton.*

The argument is, essentially, the one in Turinici [28]; however, for completeness reasons, we provide it, with some modifications.

Proof. Denote, for simplicity,

$(<; N) = \{(i, j) \in N \times N; i < j\}$, $(\leq; N) = \{(i, j) \in N \times N; i \leq j\}$;

these are nothing else than the graphs of the relations $(<)$ and $(\leq)$ over N, respectively. Then, let us introduce the following notations for the (finite and infinite) orbital segments of the (B-o) sequence $X_0 = (x_n)$:

$$x[p, q] = \{x_p, \ldots, x_q\}, (p, q) \in (\leq; N);$$
$$x[p, \infty[= \{x_p, x_{p+1}, \ldots\}, p \in N.$$

From the (B-o) property of $X_0 = (x_n)$, the following property holds:

(rela-1) $\operatorname{diam}(x[i, j]) \geq r_i > 0,$ for each $(i, j) \in (<; N)$.

This in turn gives (by the very definition of $V_\infty(.,.)$)

(rela-2) $V_\infty(x_i, x_j) \geq \operatorname{diam}(x[i, i+2]) \geq r_i > 0, \forall (i, j) \in (<; N)$.

There are several steps to be passed.

Step 1. We start with the following useful evaluation:

Proposition 5. *Under the introduced notations,*

$$d(x_i, x_j) \leq \varphi(\operatorname{diam}(x[i-1, j])) < \operatorname{diam}(x[i-1, j]),$$
whenever $1 \leq i \leq j$ *(that is,* $0 \leq i - 1 < j$*)*.

Proof. (Proposition 5) Let the couple (i, j) be as before. The second inequality above is clear, via (rela-1) and φ = regressive, so, it remains to establish the first inequality. The case of $d(x_i, x_j) = 0$ is immediate, via $\varphi(R_+) \subseteq R_+$, so, we may assume that $d(x_i, x_j) > 0$; whence, $1 \leq i < j$. From (rela-1),

$$\operatorname{diam}(x[i-1, j]) \geq r_{i-1} > 0; \text{ because } i - 1 < i < j.$$

Further, by the same choice of (i, j),

$$x_{i-1} \neq x_{j-1} \text{ (in view of } x_i \neq x_j); \text{ whence, } B(x_{i-1}, x_{j-1}) > 0.$$

The contractive condition applies to (x_{i-1}, x_{j-1}) and gives (by the above)

$$d(x_i, x_j) \leq \varphi(V_\infty(x_{i-1}, x_{j-1})).$$

In addition, by (rela-2) (and $i + 1 \leq j$),

$$0 < V_\infty(x_{i-1}, x_{j-1}) = \operatorname{diam}(x[i-1, i+1] \cup x[j-1, j]) =$$
$$\operatorname{diam}\{x_{i-1}, x_i, x_{i+1}, x_{j-1}, x_j\} \leq \operatorname{diam}(x[i-1, j]),$$

by a discussion of the alternatives $i < j - 1$ (hence, $i + 1 \leq j - 1$) and $i = j - 1$ (hence, $i + 1 = j$). Combining these ends the argument. □

Step 2. The following consequence of this fact is to be noted.

Proposition 6. *Pick some $\beta \in H(r_0)$ (hence, $\beta > r_0 > 0$; see above). Then,*

$$\mathrm{diam}(x[0,n]) \leq \beta, \text{ for all } n \geq 1, \text{ so, necessarily,}$$
$$\mathrm{diam}(x[0,\infty[) \leq \beta.$$

Proof. (Proposition 6) The case $n = 1$ is clear, via $\beta > r_0$, so, we may assume that $n \geq 2$. For each (i,j) with $1 \leq i \leq j \leq n$, we have by Proposition 5,

$$d(x_i, x_j) < \mathrm{diam}(x[i-1,j]) \leq \mathrm{diam}(x[0,n])$$

so that, necessarily,

$$0 < r_0 \leq \mathrm{diam}(x[0,n]) = d(x_0, x_k), \quad \text{for some } k \in \{1, \ldots, n\},$$

where the first inequality follows from (rela-1) [and the choice of k is a consequence of it]. By Proposition 5, we have an evaluation like

$$d(x_1, x_k) \leq \varphi(\mathrm{diam}(x[0,k])) \leq \varphi(\mathrm{diam}(x[0,n])).$$

Putting these together yields, by the triangle inequality,

$$\mathrm{diam}(x[0,n]) = d(x_0, x_k) \leq d(x_0, x_1) + d(x_1, x_k) \leq$$
$$d(x_0, x_1) + \varphi(\mathrm{diam}(x[0,n])) = r_0 + \varphi(\mathrm{diam}(x[0,n])),$$
$$\text{wherefrom (as } \beta \in H(r_0)\text{), } \mathrm{diam}(x[0,n]) \leq \beta,$$

and the claim follows. □

Step 3. The following d-Cauchy property of $X_0 = (x_n)$ is now available.

Proposition 7. *With the same notations as before, one has*

$$\mathrm{diam}(x[n,\infty[) \leq \varphi^n(\beta), \text{ for all } n \geq 0;$$

hence, the (B-o) sequence $X_0 = (x_n)$ is d-Cauchy.

Proof. (Proposition 7) The case $n = 0$ is established in Proposition 6, so, we may assume that $n \geq 1$. By Proposition 5, (rela-i-j) $d(x_i, x_j) \leq \varphi(\text{diam}(x[i-1, j])) \leq \varphi(\text{diam}(x[n-1, \infty[))$, for each (i, j) with $n \leq i \leq j$.

On the other hand, by definition,

$\text{diam}(x[n, \infty[) = $ the supremum of all $d(x_i, x_j)$ with $n \leq i \leq j$.

So, passing to supremum over all such (i, j) in (rela-i-j) yields

$\text{diam}(x[n, \infty[) \leq \varphi(\text{diam}(x[n-1, \infty[))$.

After n steps, one thus gets

$\text{diam}(x[n, \infty[) \leq \varphi^n(\text{diam}(x[0, \infty[)) \leq \varphi^n(\beta)$,

and conclusion follows. □

Step 4. As X is (B-o,d)-complete, $x_n \xrightarrow{d} z$, for some (uniquely determined) $z \in X$. There are two cases to be discussed.

(Case-1) Suppose that T is (B-o,d)-continuous. Then, $y_n := Tx_n \xrightarrow{d} z$. On the other hand, $(y_n = x_{n+1}; n \geq 0)$ is a subsequence of $(x_n; n \geq 0)$ so that $y_n \xrightarrow{d} z$. As d is separated, this yields $z = Tz$.

(Case-2) Suppose that φ is super Matkowski admissible. To get the desired fact, we use a *reductio ad absurdum* argument. Namely, assume that $z \neq Tz$, i.e., $b := d(z, Tz) > 0$. Two alternatives occur.

(Alter-1) Suppose that

$H := \{n \in N; Tx_n = Tz\}$ is unbounded (in N).

By a direct procedure, there may be obtained a strictly ascending sequence of ranks $(i(n); n \geq 0)$, such that

$(\forall n): Tx_{i(n)} = Tz$; hence, $a_n := x_{i(n)+1} = Tz$.

But, $(a_n; n \geq 0)$ is a subsequence of $(x_n; n \geq 0)$ so that $\lim_n a_n = z$. Passing to limit in the relation above gives $z = Tz$.

(Alter-2) Suppose that

the subset H is bounded (in N).

This tells us that

$\exists i = i(z) \in N$, such that $n \geq i$ implies $Tx_n \neq Tz$ (hence, $B(x_n, z) > 0$).

Denote for simplicity $(u_n = x_{n+i}; n \geq 0)$; clearly, by the preceding relation,

(non-id) $(\forall n)$: $Tu_n \neq Tz$ (hence, $B(u_n, z) > 0$).

The contractive property is thus applicable to (u_n, z), for all n, and gives

(contr-pr) $d(u_{n+1}, Tz) \leq \varphi(V_\infty(u_n, z))$, for each n.

By the very definition above, the sequence $(\sigma_n := V_\infty(u_n, z); n \geq 0)$ fulfills

$(\sigma_n \geq b, \forall n)$ and $\sigma_n \to b$, as $n \to \infty$.

Further, by the d-continuity of the map $(x, y) \mapsto d(x, y)$, the sequence $(\tau_n := d(u_{n+1}, Tz); n \geq 0)$ is in R_+^0 and fulfills $\tau_n \to b$ as $n \to \infty$. This, along with $(\tau_n \leq \varphi(\sigma_n); n \geq 0)$, yields a contradiction with the asymptotic diagonal property of φ (obtainable via φ = super Matkowski admissible). Summing up, our working assumption $b > 0$ is not acceptable, and then, $z = Tz$.

Step 5. Take the points $z_1, z_2 \in \text{Fix}(T)$, according to

$d(z_1, z_2) > 0$; hence, $B(z_1, z_2) > 0$.

The contractive condition applies to (z_1, z_2) and yields

$d(z_1, z_2) = d(Tz_1, Tz_2) \leq \varphi(V_\infty(z_1, z_2))$.

On the other hand, by the very definition of our map, $V_\infty(z_1, z_2) = d(z_1, z_2) > 0$. Combining these, one derives

$d(z_1, z_2) \leq \varphi(d(z_1, z_2)) < d(z_1, z_2)$; a contradiction.

Hence, $z_1 = z_2$, and the fix-asingleton property follows. □

Now, a basic particular case of this result is obtainable by the reduction procedures we already introduced. Letting $F : R_+^0 \to R$ be an increasing function and $C > 0$ be a constant, define the class of contractions

(W-D) $F(d(Tx, Ty)) \leq F(V_\infty(x, y)) - C$, $x, y \in X$, $B(x, y) > 0$.

This is referred to as: T is *Wardowski–Dung* $(B > 0;$ $F, C)$-*contractive*; and, if (F, C) is generic, we say that T is a

Wardowski–Dung contraction. As before, a motivation of this convention comes from the mapping V_∞ appearing here is strongly related with Dung's contributions in this area [6].

As a direct application of our main result, we then have the following:

Theorem 6. *Suppose that T is Wardowski–Dung $(B > 0; F, C)$-contractive, for some function $F : R_+^0 \to R$ and some constant $C > 0$, with*

(52-i) *F is increasing semi-Wardowski and (F, C) is left- and right-bounded oscillating*
(52-ii) *F is quasi-asymptotic:*
$F(s + \alpha) - F(s) \to 0$ *as $s \to \infty$, for each $\alpha > 0$.*

In addition, let X be $(B$-$o,d)$-complete, and let $X_0 = (x_n)$ be such that

(52-iii) *$X_0 = (x_n)$ is $(B$-$o)$: (hence, $r_n := d(x_n, x_{n+1}) > 0$, $\forall n$).*

Then,

(52-a) *$X_0 = (x_n)$ is Picard (modulo $(d; T)$).*
(52-b) *$X_0 = (x_n)$ is strong Picard (modulo $(d; T)$), when in addition the selfmap T is $(B$-$o,d)$-continuous.*
(52-c) *$X_0 = (x_n)$ is strong Picard (modulo $(d; T)$), when in addition $(F; C)$ is bilateral bounded oscillating.*
(52-d) *T is fix-singleton.*

Proof. (Sketch) Let $\varphi \in \mathcal{F}(R_+)$ stand for the associated to (F, C) function. By an auxiliary fact above,

(P-1) *φ is regressive, increasing, Matkowski admissible when $(F; C)$ is left and right bounded oscillating,*
(P-2) *in addition, φ is super Matkowski admissible when $(F; C)$ is bilateral bounded oscillating.*

Moreover, in terms of this function, the initial contractive condition yields

(P-2) *$d(Tx, Ty) \le \varphi(V_\infty(x, y))$, $x, y \in X$, $B(x, y) > 0$, that is, T is Matkowski–Dung $(B > 0; \varphi)$-contractive.*

This, along with the previous main result, gives us all needed conclusions. □

The obtained statement yields a positive answer to an open problem formulated in Vujaković *et al.* [32]. It would be interesting to determine whether a proof of the underlying statement is obtainable, by avoiding the Matkowski-type reduction; we conjecture that this is possible. Further aspects will be delineated elsewhere.

References

[1] M. Akkouchi, Common fixed points for weakly compatible maps satisfying implicit relations without continuity, *Dem. Math.* **44**, 151–158 (2011).

[2] S. Banach, Sur les opérations dans les ensembles abstraits et leur application aux équations intégrales, *Fund. Math.* **3**, 133–181 (1922).

[3] V. Berinde and F. Vetro, Common fixed points of mappings satisfying implicit contractive conditions, *Fixed Point Theory Appl.* **2012**, 105 (2012).

[4] D. W. Boyd and J. S. W. Wong, On nonlinear contractions, *Proc. Am. Math. Soc.* **20**, 458–464 (1969).

[5] L. B. Ćirić, A generalization of Banach's contraction principle, *Proc. Am. Math. Soc.* **45**, 267–273 (1974).

[6] N. V. Dung, A simple generalization of Ciric fixed point theorem (2013). *Arxiv* 1309-5589-v1. 22 September 2013.

[7] N. V. Dung and V. I. Hang, A fixed point theorem for generalized F-contraction on complete metric space, *Vietnam J. Math.* **43**, 743–755 (2015).

[8] P. N. Dutta and B. S. Choudhury, A generalisation of contraction principle in metric spaces, *Fixed Point Theory Appl.* **2008** (2008). Article ID 406368.

[9] A. Fulga and A. Proca, A new generalization of Wardowski fixed point theorem in complete metric spaces, *Adv. Theory Nonlinear Anal. Appl.* **1**, 57–63 (2017).

[10] P. Găvruţa and L. Manolescu, New classes of Picard operators (2020). *Arxiv* 2009-13157-v1. 28 September 2020.

[11] E. Karapinar, A. Fulga and R. P. Agarwal, A survey: F-contractions with related fixed point results, *J. Fixed Point Theory Appl.* 22:69 (2020).

[12] M. A. Krasnoselskii and V. Y. Stetsenko, Towards a theory of equations with concave operators (Russian), *Sib. Math. Zh.* **10**, 565–572 (1969).
[13] S. Leader, Fixed points for general contractions in metric spaces, *Math. Jpn.* **24**, 17–24 (1979).
[14] J. Matkowski, Integrable solutions of functional equations, *Diss. Math.* (The Polish Academy of Sciences, Warsaw) **127** (1975).
[15] A. Meir and E. Keeler, A theorem on contraction mappings, *J. Math. Anal. Appl.* **28**, 326–329 (1969).
[16] H. K. Nashine, Z. Kadelburg and P. Kumam, Implicit-relation-type cyclic contractive mappings and applications to integral equations, *Abstr. Appl. Anal.* **2012** (2012). Article ID 386253.
[17] H. Piri and P. Kumam, Wardowski type fixed point theorems in complete metric spaces, *Fixed Point Theory Appl.* **2016**, 45 (2016).
[18] V. Popa, Some fixed point theorems for compatible mappings satisfying an implicit relation, *Demonstr. Math.* **32**, 157–163 (1999).
[19] O. Popescu, A new type of contractive mappings in complete metric spaces. Preprint.
[20] S. Reich, Fixed points of contractive functions, *Boll. Un. Mat. Ital.* **5**, 26–42 (1972).
[21] B. E. Rhoades, A comparison of various definitions of contractive mappings, *Trans. Am. Math. Soc.* **226**, 257–290 (1977).
[22] B. E. Rhoades, Some theorems on weakly contractive maps, *Nonlinear Anal.* **47**, 2683–2693 (2001).
[23] I. A. Rus, *Generalized Contractions and Applications* (Cluj University Press, Cluj-Napoca, 2001).
[24] N. A. Secelean and M. Zhou, Generalized F-contractions on product of metric spaces, *Mathematics* **7**, 1040 (2019).
[25] M. Turinici, Fixed points of implicit contraction mappings, *An. Şt. Univ. "Al. I. Cuza" Iaşi Mat.* **22**, 177–180 (1976).
[26] M. Turinici, Fixed points of implicit contractions via Cantor's intersection theorem, *Bul. Inst. Polit. Iaşi (Sect I: Mat. Mec. Teor. Fiz.)* **26**(30), 65–68 (1980).
[27] M. Turinici, Wardowski implicit contractions in metric spaces (2013). Arxiv 1211-3164-v2. 15 September 2013.
[28] M. Turinici, Contractive maps in Mustafa-Sims metric spaces, *Int. J. Nonlinear Anal. Appl.* **5**, 36–53 (2014).
[29] M. Turinici, Implicit contractive maps in ordered metric spaces. In T. M. Rassias and L. Toth (eds.) *Topics in Mathematical Analysis and Applications* (Springer International Publishing, Switzerland, 2014), pp. 715–746.

[30] M. Turinici, *Modern Directions in Metrical Fixed Point Theory* (Pim Editorial House, Iaşi, 2016).

[31] M. Turinici, *Reports in Metrical Fixed Point Theory* (Pim Editorial House, Iaşi, 2020).

[32] J. Vujaković, S. Mitrović, M. Pavlović, and S. Radenović, On recent results concerning F-contraction in generalized metric spaces, *Mathematics* **8**, 767 (2020).

[33] D. Wardowski, Fixed points of a new type of contractive mappings in complete metric spaces, *Fixed Point Theory Appl.* **2012**, 94 (2012).

[34] D. Wardowski, Solving existence problems via F-contractions, *Proc. Am. Math. Soc.* **146**, 1585–1598 (2018).

[35] D. Wardowski and N. V. Dung, Fixed points of F-weak contractions on complete metric spaces, *Demonstr. Math.* **47**, 146–155 (2014).

© 2024 World Scientific Publishing Company
https://doi.org/10.1142/9789811267048_0028

Chapter 28

Contractive Maps on Relational MC-Quasimetric Spaces

Mihai Turinici

A. I. Cuza University, Iaşi, Romania

mturi@uaic.ro

Some fixed-point results are given for a class of implicit contractions acting on relational metrical-compatible quasimetric spaces. The connections with a related statement in Turinici (*Modern Directions Metrical Fixed Point Theory*, 2023) are also being discussed.

1. Introduction

Let X be a nonempty set. Call the subset Y of X, *almost singleton* (in short *asingleton*,) provided $[y_1, y_2 \in Y$ implies $y_1 = y_2]$, and *singleton* if, in addition, Y is nonempty; note that in this case $Y = \{y\}$, for some $y \in X$. Further, let $d : X \times X \to R_+ := [0, \infty[$ be a *metric* over X; the couple (X, d) will be termed a *metric space*. Finally, let $T \in \mathcal{F}(X)$ be a selfmap of X. [Here, for each couple A, B of nonempty sets, $\mathcal{F}(A, B)$ stands for the class of all functions from

A to B; when $A = B$, we write $\mathcal{F}(A)$ in place of $\mathcal{F}(A, A)$.] Denote $\mathrm{Fix}(T) = \{x \in X; x = Tx\}$; each point of this set is referred to as *fixed* under T. In the metrical fixed-point theory, such points are to be determined according to the following context, comparable with the one described in Rus [46, Chapter 2, Section 2.2]:

(**pic-0**) We say that T is *fix-asingleton*, if $\mathrm{Fix}(T)$ is an asingleton; and *fix-singleton*, if $\mathrm{Fix}(T)$ is a singleton.

(**pic-1**) We say that $x \in X$ is a *semi-Picard point* (modulo (d, T)) when $(T^n x; n \geq 0)$ is *d-asymptotic* ($d(T^n x, T^{n+1} x) \to 0$ as $n \to \infty$). If this property holds for all $x \in X$, we say that T is a *semi-Picard operator* (modulo d).

(**pic-2**) We say that $x \in X$ is a *Picard point* (modulo (d, T)) when $(T^n x; n \geq 0)$ is *d-Cauchy* ($d(T^m x, T^n x) \to 0$ as $m, n \to \infty$, $m < n$). If this property holds for all $x \in X$, we say that T is a *Picard operator* (modulo d).

(**pic-3**) We say that $x \in X$ is a *strongly Picard point* (modulo (d, T)) when $(T^n x; n \geq 0)$ is *d-convergent* with $\lim_n (T^n x) \in \mathrm{Fix}(T)$. If this property holds for all $x \in X$, we say that T is a *strongly Picard operator* (modulo d).

In this perspective, a basic answer to our question is the *Banach contraction principle*. Given $\alpha \geq 0$, let us say that T is *Banach* $(d; \alpha)$-*contractive*, provided

(B-contr) $\quad d(Tx, Ty) \leq \alpha d(x, y)$, for all $x, y \in X$.

Theorem 1. *Suppose that T is Banach $(d; \alpha)$-contractive, for some $\alpha \in [0, 1[$. In addition, let X be d-complete. Then,*

(11-a) $\quad T$ *is fix-singleton:* $\mathrm{Fix}(T) = \{z\}$, *for some* $z \in X$,
(11-b) $\quad T$ *is a strongly Picard operator (modulo d), precisely,* $T^n x \xrightarrow{d} z$ *as* $n \to \infty$, *for each* $x \in X$.

This result — obtained in 1922 by Banach [7] — found a multitude of applications in operator equations theory, so, it was the subject of many extensions. The most general ones have the (set) *implicit* form

(imp-set) $\quad (d(x, Tx), d(x, y), d(x, Ty), d(Tx, y), d(Tx, Ty), d(y, Ty)) \in \mathcal{M}$, for all $x, y \in X$, $x \mathcal{R} y$, where $\mathcal{M} \subseteq R_+^6$ is a (nonempty) subset, and $\mathcal{R} \subseteq X \times X$ is a *relation* over X. In particular, when $\mathcal{M}$ is the zero-section of a certain function $F: R_+^6 \to R$, that is,

$$\mathcal{M} = \{(t_1, t_2, t_3, t_4, t_5, t_6) \in R_+^6; F(t_1, t_2, t_3, t_4, t_5, t_6) \le 0\},$$

the implicit contractive condition above has the functional form:

(imp-fct) $F(d(x, Tx), d(x, y), d(x, Ty), d(Tx, y), d(Tx, Ty),$
$d(y, Ty)) \le 0$, for all $x, y \in X$, $x\mathcal{R}y$.

Sometimes, an *explicit* version of this contractive condition is available, by solving this inequality with respect to its fifth variable.

Concerning the existing work in this area, several main directions are of interest:

(I) Suppose that $\mathcal{R} = X \times X$ (the *trivial relation* of X). Some outstanding explicit results have been established in Boyd and Wong [13], Reich [41] and Matkowski [34]; see also the survey paper by Rhoades [42]. And, for the implicit functional version above, certain technical aspects have been considered by Argoubi et al. [6], Meir and Keeler [35], Khojasteh et al. [32], Leader [33] and Turinici [51].

(II) Suppose that $\mathcal{R}$ is a *(partial) order* (i.e., reflexive, transitive, antisymmetric relation) over X. An appropriate extension of Matkowski's fixed-point theorem was obtained in the 1986 paper by Turinici [52]; two decades later, this result has been re-discovered — at the level of Banach contractive maps — by Ran and Reurings [40]; see also Nieto and Rodriguez-Lopez [38]. Then, an extension — to the same framework — of Leader's contribution was performed in Agarwal et al. [2]. Further extensions of these facts may be found in Chandok et al. [15], Harjani et al. [21], Karapinar et al. [27, 28] and Turinici [57].

(III) Suppose that $\mathcal{R}$ is a *relation* over X. A first result in the area was obtained in 2008 by Jachymski [23] in metric spaces endowed with a graph. Further, in 2012, Samet and Turinici [47] established a lot of Leader-type results over relational metric spaces. Note that, as shown in Roldán and Shahzad [45], these approaches are, practically, identical. Some notable contributions in this direction are those in Batra and Vashistha [9], Beg et al. [10], or Chakraborty et al. [16];

see also Bojor [14], Karapinar and Fulga [25], Samet *et al.* [48] and Turinici [58, Section 21].

(IV) Suppose that $d(.,.)$ is a *quasimetric* on X and $\mathcal{R}$ is a *relation* over X. The case of $\mathcal{R}$ = quasi-order was discussed in Turinici [58, Section 41]. The remaining situations (relative to $\mathcal{R}$) were considered in the papers by Agarwal *et al.* [3], Alsulami *et al.* [4] and Karapinar *et al.* [24].

Having these precise, it is our aim in the following to get further statements belonging to this last class — involving contractive maps acting on *relational mc-quasimetric spaces*. The obtained facts may be viewed as a completion of the related statement due to Ahmed and Fulga [5], having as contractive support the recent contributions in Gornicki [20] and Proinov [39]. Some coincidence point extensions of these statements are obtainable under the lines in Berzig [12] and Roldan *et al.* [43, 44]; we will discuss them elsewhere.

2. Dependent Choice Principles

Throughout this exposition, the axiomatic system in use is Zermelo–Fraenkel's (abbreviated ZF), as described by Cohen [19, Chapter 2]. The notations and basic facts to be considered in this system are more or less standard. Some important ones are discussed below.

(A) Let X be a nonempty set. By a *relation* over X, we mean any (nonempty) part $\mathcal{R} \subseteq X \times X$; then, $(X, \mathcal{R})$ will be referred to as a *relational structure*. For simplicity, we sometimes write $(x, y) \in \mathcal{R}$ as $x\mathcal{R}y$. Note that $\mathcal{R}$ may be regarded as a mapping between X and $\exp[X]$ (= the class of all subsets in X). In fact, denote

$$X(x, \mathcal{R}) = \{y \in X; x\mathcal{R}y\} \text{ (the } section \text{ of } \mathcal{R} \text{ through } x), x \in X;$$

then, the desired mapping representation is $[\mathcal{R}(x) = X(x, \mathcal{R}); x \in X]$.

A basic example of relational structure is to be constructed as follows. Let $N = \{0, 1, \ldots\}$ be the set of *natural* numbers, endowed with the usual addition and (partial) order ($\leq$); note that

$(N, \leq)$ is *well ordered*: any (nonempty) subset of N has a first element.

Further, denote for $p, q \in N$, $p \leq q$,

$$N[p,q] = \{n \in N; p \leq n \leq q\}, \quad N]p,q[= \{n \in N; p < n < q\},$$
$$N[p,q[= \{n \in N; p \leq n < q\}, \quad N]p,q] = \{n \in N; p < n \leq q\},$$

as well as, for $r \in N$,

$$N[r,\infty[= \{n \in N; r \leq n\}, \quad N]r,\infty[= \{n \in N; r < n\}.$$

For each $r \in N$, the section $N[0, r[= N(r, >)$ is referred to as the *initial interval* (in N) induced by r. Any set P with $P \sim N$ (in the sense that there exists a *bijection* from P to N) will be referred to as *effectively denumerable*. In addition, given some natural number $n \geq 1$, any set Q with $Q \sim N(n, >)$ will be said to be *n-finite*; when n is generic here, we say that Q is *finite*. Finally, the (nonempty) set Y is called (at most) *denumerable* iff it is either effectively denumerable or finite.

Let X be a nonempty set. By a *sequence* in X, we mean any mapping $x : N \to X$, where $N = \{0, 1, \ldots\}$ is the set of *natural numbers*. For simplicity reasons, it will be useful to denote it as $(x(n); n \geq 0)$, or $(x_n; n \geq 0)$; moreover, when no confusion can arise, we further simplify this notation as $(x(n))$ or (x_n), respectively. Also, any sequence $(y_n := x_{i(n)}; n \geq 0)$ with

$(i(n); n \geq 0)$ is strictly ascending (hence, $i(n) \to \infty$ as $n \to \infty$)

will be referred to as a *subsequence* of $(x_n; n \geq 0)$. Note that, under such a convention, the relation "subsequence of" is transitive, i.e.,

$(z_n) =$ subsequence of (y_n) and $(y_n) =$ subsequence of (x_n)
imply $(z_n) =$ subsequence of (x_n).

(B) Remember that, an outstanding part of (ZF) is the *Axiom of Choice* (abbreviated as (AC)); which, in a convenient manner, may be written as follows:

(AC) For each couple (J, X) of nonempty sets and each function $F : J \to \exp(X)$, there exists a *(selective) function* $f : J \to X$ (in the sense that $f(\nu) \in F(\nu)$, for each $\nu \in J$).

(Here, $\exp(X)$ stands for the class of all nonempty elements in $\exp[X]$.) Sometimes, when the index set J is denumerable, the existence of such a selective function may be determined by using a weaker form of (AC), called *Dependent Choice* principle (in short: (DC)). Call the relation $\mathcal{R}$ over X *proper* when

$$(X(x,\mathcal{R}) =)\mathcal{R}(x) \text{ is nonempty, for each } x \in X.$$

Then, $\mathcal{R}$ is to be viewed as a mapping between X and $\exp(X)$, and the couple $(X, \mathcal{R})$ will be referred to as a *proper relational structure*. Further, given $a \in X$, let us say that the sequence $(x_n; n \geq 0)$ in X is $(a; \mathcal{R})$-*iterative*, provided

$$x_0 = a, \text{ and } x_n \mathcal{R} x_{n+1} \text{ (i.e., } x_{n+1} \in \mathcal{R}(x_n)\text{), for all } n.$$

Proposition 1. *Let the relational structure* $(X, \mathcal{R})$ *be proper. Then, for each* $a \in X$ *there is at least one* $(a, \mathcal{R})$-*iterative sequence in* X.

This principle — proposed, independently, by Bernays [11] and Tarski [50] — is deductible from (AC) but not conversely; cf. Wolk [59]. Moreover, by the developments in Moskhovakis [37, Chapter 8] and Schechter [49, Chapter 6], the *reduced system* (ZF-AC+DC) is comprehensive enough so as to cover the *usual* mathematics; see also Moore [36, Appendix 2].

A basic consequence of (DC) is the so-called *Denumerable Axiom of Choice* [in short (AC(N))].

Proposition 2. *Let* $F : N \to \exp(X)$ *be a function. Then, for each* $a \in F(0)$ *there exists a function* $f : N \to X$ *with* $f(0) = a$ *and* $[f(n) \in F(n), \forall n \in N]$.

Proof. Denote $Q = N \times X$, and let us introduce the (proper) relation $\mathcal{R}$ over it:

$$\mathcal{R}(n, x) = \{n+1\} \times F(n+1), \quad n \in N, x \in X.$$

Then, an application of (DC) to the proper relational structure $(Q, \mathcal{R})$ yields the desired conclusion; we do not give details. $\square$

As a consequence of the above facts,

(DC) $\implies$ (AC(N)) in the *strongly reduced system* (ZF-AC),
   or, equivalently,
(AC(N)) is deductible in the *reduced system* (ZF-AC+DC).

The reciprocal of this inclusion is not true; see Moskhovakis [37, Chapter 8, Section 8.25] for details.

3. Classes of Pseudometric Spaces

In the following, some preliminary facts about convergent and Cauchy sequences in certain pseudometric spaces are being discussed.

Let X be a nonempty set. Call the subset Y of X *almost-singleton* (in short *a-singleton*) provided $[y_1, y_2 \in Y$ implies $y_1 = y_2]$, and *singleton* if, in addition, Y is nonempty; note that, in this case, $Y = \{y\}$, for some $y \in X$.

By a *pseudometric* over X, we mean any map $a : X \times X \to R_+$. For an easy reference, we list the conditions to be used:

(ref) $a(.,.)$ is *reflexive*: $a(x,x) = 0$, for each $x \in X$,
(suf) $a(.,.)$ is *sufficient*: $x, y \in X$, $a(x,y) = 0$ imply $x = y$,
(sym) $a(.,.)$ is *symmetric*: $a(x,y) = a(y,x)$, for all $x, y \in X$,
(tri) $a(.,.)$ is *triangular*: $a(x,z) \leq a(x,y) + a(y,z)$, $\forall x, y, z \in X$.

In fact, there are many other properties to be considered; see, for instance, the 2001 PhD Thesis by Hitzler [22, Chapter 1]; but, for the following developments, this will suffice.

(A) Let $a(.,.)$ be a reflexive pseudometric (in short r-pseudometric) on X. A convergence and Cauchy structure over X may be introduced according to the lines below.

We say that the sequence (x_n) in X *a-converges* to $x \in X$ (and write $x_n \xrightarrow{a} x$) iff $a(x_n, x) \to 0$ as $n \to \infty$, that is,

$$\forall \delta > 0, \exists p = p(\delta), \forall n: (p \leq n \implies a(x_n, x) < \delta).$$

Then x is called an *a-limit* of (x_n); the set of all these is denoted as $a\text{-}\lim_n(x_n)$ or $\lim_n(x_n)$ when $a(.,.)$ is understood. When $\lim_n(x_n)$ is nonempty, we say that (x_n) is *a-convergent*; the class of all these is denoted as $Conv(a)$.

Clearly, the introduced convergence ($\xrightarrow{a}$) has the properties

(conv-1) (($\xrightarrow{a}$) is *hereditary*)
  if $x_n \xrightarrow{a} x$, then $y_n \xrightarrow{a} x$, for each subsequence (y_n) of (x_n),
(conv-2) (($\xrightarrow{a}$) is *reflexive*) for each $u \in X$,
  the constant sequence $(x_n = u; n \geq 0)$ fulfills $x_n \xrightarrow{a} u$,

so, it fulfills the general requirements in Kasahara [30]. Note that, by the imposed upon $a(.,.)$ conditions, $\lim_n(x_n)$ may have more than one element. Concerning this aspect, let us introduce the property

(sep) ($\xrightarrow{a}$) is *separated*:
$\lim_n(x_n)$ is an asingleton, for each sequence $(x_n; n \geq 0)$ in X;

this will be also referred to as $a(.,.)$ is *separated*. In this case, given the a-convergent sequence (x_n), we must have $\lim_n(x_n) = \{x\}$ (for some $x \in X$); this is written as $\lim_n(x_n) = x$.

Remark 1.
Under this larger framework, we have

(suf) $a(.,.)$ is *sufficient*: $a(x,y) = 0$ implies $x = y$.

In fact, let $u, v \in X$ be such that $a(u,v) = 0$. The constant sequence $(x_n = u; n \geq 0)$ fulfills $x_n \xrightarrow{a} u$, $x_n \xrightarrow{a} v$ so that (by the separated property) $u = v$.

Further, let us say that (x_n) is *a-Cauchy*, provided $a(x_m, x_n) \to 0$ as $m, n \to \infty$, $m < n$, that is,

$$\forall \delta > 0, \exists q = q(\delta), \forall (m,n): (q \leq m < n \implies a(x_m, x_n) < \delta).$$

The class of all these is indicated as $Cauchy(a)$; some basic properties of it are described in the following:

(Cauchy-1) ($Cauchy(a)$ is *hereditary*)
  (x_n) is a-Cauchy implies (y_n) is a-Cauchy,
  for each subsequence (y_n) of (x_n)
(Cauchy-2) ($Cauchy(a)$ is *reflexive*) for each $u \in X$,
  the constant sequence $(x_n = u; n \geq 0)$ is a-Cauchy,

so, this concept fulfills the general requirements in Turinici [54].
Finally, let us say that (x_n) is *a-asymptotic*, provided

$$a(x_n, x_{n+1}) \to 0 \text{ as } n \to \infty.$$

The class of all these is indicated as $Asy(a)$. Clearly,

(Asy-1) ($Asy(a)$ is *reflexive*) for each $u \in X$,
  the constant sequence $(x_n = u; n \geq 0)$ is a-asymptotic.

Unfortunately, a property like

(Asy-2) ($Asy(a)$ is *hereditary*)
  (x_n) is a-asymptotic implies (y_n) is a-asymptotic,
  for each subsequence (y_n) of (x_n)

is not in general true.

Finally, we say that $(Conv(a), Cauchy(a))$ is a *Conv-Cauchy structure* induced by $a(.,.)$. Note that, by the ambient reflexive setting, properties like

$(\forall(x_n))$: (x_n) fulfills (P) implies (x_n) fulfills (Q),
where $P, Q \in \{Conv(a), Cauchy(a)\}$,

are not in general true.

Likewise, $(Conv(a), Cauchy(a), Asy(a))$ is a *Conv-Cauchy-Asy structure* induced by $a(.,.)$. Note that, by the ambient reflexive setting, many properties like

$(\forall(x_n))$: (x_n) fulfills (P) implies (x_n) fulfills (Q),
where $P, Q \in \{Conv(a), Cauchy(a), Asy(a)\}$,

are not in general true. However, we do have

(C-asy) $(\forall(x_n))$: (x_n) is a-Cauchy implies (x_n) is a-asymptotic.

Since the verification is immediate, we do not give details.

(B) Let (X, b) be a pseudometric space, with

$b(.,.)$ is reflexive, sufficient and triangular.

Note that $b(.,.)$ has all properties of a metric, excepting symmetry; we call it a *quasimetric* on X; and the couple (X, b) is referred to as a *quasimetric space*. As a consequence of this, the conjugate (to b) pseudometric

$c(x, y) = b(y, x)$, $x, y \in X$

is also a quasimetric on X. Finally, the symmetric cover of (b and c)

$d(x, y) = \max\{b(x, y), c(x, y)\}$, $x, y \in X$
(in short: $d = \max\{b, c\}$)

is a symmetric quasimetric — hence, a metric — over X, and the couple (X, d) is referred to as a *metric space*.

The concept of quasimetric seems to have a pretty long tradition in metrical spaces theory. For example, in his 2001 PhD Thesis, Hitzler [22, Chapter 1, Section 1.2] introduced this notion as a useful tool for the topological study of logic semantics. Later, in a 2004 paper, Turinici [53] used the same concept [referred to as *reflexive triangular sufficient pseudometric*] with the aim of establishing a Caristi–Kirk fixed-point theorem over such structures. Further aspects may be found in Cobzaş [18].

Let the quasimetric b (and its associated maps $\{c,d\}$) be given as before. In addition, let $\mathcal{G} \subseteq X \times X$ be a (nonempty) *relation* over X; the triple $(X, b, \mathcal{G})$ will be referred to as a *relational quasimetric space*.

Remember that, given the sequence (x_n) in X, its b-Cauchy property writes

(b-C) $\forall \varepsilon \in R_+^0, \exists i = i(\varepsilon) \in N: i \leq m < n \implies b(x_m, x_n) < \varepsilon$.

In the following, a basic b-Cauchy criterion will be stated over the relational quasimetric space $(X, b, \mathcal{G})$. Some preliminaries are needed.

Let us say that $(x_n; n \geq 0)$ is b-*asymptotic*, provided

$$b(x_n, x_{n+1}) \to 0 \quad \text{as } n \to \infty.$$

Clearly, each b-Cauchy sequence is b-asymptotic too; the reciprocal of this is not in general true. This tells us that the b-Cauchy criteria we are looking for are to be sought in the class of b-asymptotic sequences. The general scheme of handling these is as follows. Suppose that the following inclusion is valid

(inc-1) (for each b-asymptotic sequence (x_n)):
(x_n) is not b-Cauchy implies (x_n) has the property (π).

Then, the underlined b-Cauchy criterion writes

(inc-2) (for each b-asymptotic sequence (x_n)):
(x_n) is not endowed with the property (π) implies (x_n) is b-Cauchy.

To get a property of this type, we need two classes of (regularity type) conditions.

(I) The former of these is a global condition, involving the initial structure $(X, b, \mathcal{G})$. Let us introduce the concept

(mc-qm) $b(.,.)$ is *metric compatible quasimetric* (in short: *mc-quasimetric*): the couple (b, d) is $(Conv\text{-}Cauchy\text{-}Asy)$-equivalent.

In order words, $b(.,.)$ is a mc-quasimetric, provided

(mc-1) (b, d) is $Conv$-equivalent: $x_n \xrightarrow{b} x \iff x_n \xrightarrow{d} x$,
(mc-2) (b, d) is $Cauchy$-equivalent:
(x_n) is b-Cauchy $\iff (x_n)$ is d-Cauchy,

(mc-3) (b,d) is *Asy*-equivalent:
(x_n) is *b*-asymptotic $\iff$ (x_n) is *d*-asymptotic.

In this case, the structure (X, b) will be referred to as an *mc-quasimetric space*.

Actually, the second property is deductible from the third one, as results from the statement below.

Proposition 3. *Under the described setting, we have the following:*

$((b,d)$ *is Asy-equivalent*$)$ *implies* $((b,d)$ *is Cauchy-equivalent*$)$.

Proof. Suppose that $((b,d)$ is *Asy*-equivalent). It will suffice proving that

$(\forall(x_n))$: (x_n) is *b*-Cauchy $\implies$ (x_n) is *d*-Cauchy.

So, let (x_n) be *b*-Cauchy. As $d =$ metric, the *d*-Cauchy property of (x_n) writes

(d-C-equi) $\forall \varepsilon > 0$, $\exists p = p(\varepsilon) \in N$, $(\forall n)$: $p < n$ implies $d(x_p, x_n) < \varepsilon$.

Suppose, by absurd, that (x_n) is not *d*-Cauchy. From the preceding characterization, there exist $\varepsilon > 0$ and a strictly ascending rank sequence $(i(n); n \geq 0)$, such that the subsequence $(y_n = x_{i(n)}; n \geq 0)$ of (x_n) fulfills

$d(y_n, y_{n+1}) \geq \varepsilon$, for each n; so, (y_n) is not *d*-asymptotic.

On the other hand, clearly,

(y_n) is *b*-asymptotic (because (x_n) is *b*-Cauchy),

and this, by the imposed premise, gives us that (y_n) is *d*-asymptotic; impossible, by the above deduced fact. This proves our assertion. □

A basic example of such an object is the following. Let us say that $b(.,.)$ is *linearly mc-quasimetric*, if

(l-mc-qm) $d(x,y) \leq \lambda b(x,y)$, $x, y \in X$, for some $\lambda \geq 1$.

Note that, an equivalent way of expressing this is

(l-mc-qm-c) $c(x,y) \leq \lambda b(x,y)$, $x, y \in X$, for some $\lambda \geq 1$.

Proposition 4. *Suppose that $b(.,.)$ is a linearly mc-quasimetric. Then, necessarily, $b(.,.)$ is an mc-quasimetric.*

Proof. By the imposed condition (and $b \leq d$),
$$b(x,y) \leq d(x,y) \leq \lambda b(x,y), \quad x,y \in X, \text{ for some } \lambda \geq 1.$$
And this, by the very definition of the (Conv-Cauchy-Asy) structures attached to b and d, gives the desired facts. □

A natural question to be posed is that of the reciprocal inclusion

$b(.,.)$ is an mc-quasimetric implies $b(.,.)$ is a linearly mc-quasimetric

being true. The answer is negative. In fact, given $\varphi \in \mathcal{F}(R_+)$, define the concept

(b,d) is φ-compatible: $d(x,y) \leq \varphi(b(x,y))$, $x,y \in X$.

Proposition 5. *Suppose that (b,d) is φ-compatible, where $\varphi \in \mathcal{F}(R_+)$ fulfills*

φ *is zero-continuous:* $\lim_{t \to 0+} \varphi(t) = 0 = \varphi(0)$.

Then, necessarily, $b(.,.)$ is an mc-quasimetric.

Proof. There are three steps to be passed.

Step 1. Let the sequence (x_n) in X and the point $x \in X$ be such that $x_n \xrightarrow{b} x$. From the compatible property,
$$d(x_n, x) \leq \varphi(b(x_n, x)), \quad \text{for all } n.$$
Passing to limit yields $\lim_n d(x_n, x) = 0$; whence (b,d) is *Conv*-equivalent.

Step 2. Let the sequence (x_n) in X be such that $b(x_n, x_{n+1}) \to 0$ as $n \to \infty$. From the compatible property,
$$d(x_n, x_{n+1}) \leq \varphi(b(x_n, x_{n+1})), \quad \text{for all } n.$$
Passing to limit as $n \to \infty$ yields $\lim_n d(x_n, x_{n+1}) = 0$, proving that (b,d) is *Asy*-equivalent.

Step 3. As already precise, the property we just obtained tells us that (b,d) is *Cauchy*-equivalent, and, from this, we are done. □

Remark 2.
In particular, when

φ is *linear*: $\varphi(t) = \lambda t$, $t \in R_+$, for some $\lambda \geq 1$,

it follows that any quasimetric $b(.,.)$ for which (b,d) is φ-compatible is a linearly mc-quasimetric. So, it is legitimate to ask whether for any mc-quasimetric $b(.,.)$ it is the case that (b,d) is φ-compatible, where $\varphi \in \mathcal{F}(R_+)$ is zero-continuous. The answer is negative; an appropriate example is given a bit further.

Returning to the general setting above, the following basic properties of such quasimetrics are valid:

Proposition 6. *Let (X,b) be an mc-quasimetric space. Then, we have the following:*

(34-1) *The mapping $(x,y) \mapsto b(x,y)$ is d-Lipschitz, in the sense*
$$|b(x,y) - b(u,v)| \leq d(x,u) + d(y,v), \quad \forall (x,y), (u,v) \in X \times X.$$

(34-2) *As a consequence, $b(.,.)$ is first variable continuous, in the sense*
$$x_n \xrightarrow{b} x \text{ implies } b(x_n, y) \to b(x,y).$$

(34-3) *In addition, $b(.,.)$ is second variable continuous, in the sense*
$$y_n \xrightarrow{b} y \text{ implies } b(x, y_n) \to b(x,y).$$

Proof. (i) Let $(x,y), (u,v) \in X \times X$ be given. By the triangular inequality (and $b \leq d$), one gets
$$b(x,y) \leq b(x,u) + b(u,v) + b(v,y) \leq b(u,v) + d(x,u) + d(y,v),$$
$$b(u,v) \leq b(u,x) + b(x,y) + b(y,v) \leq b(x,y) + d(x,u) + d(y,v),$$
and, from this, we are done.

(ii) Let the sequence (x_n) in X and the points $x, y \in X$ be such that
$$x_n \xrightarrow{b} x; \text{ hence, } x_n \xrightarrow{d} x, \text{ as } b(.,.) \text{ is mc-quasimetric.}$$
By the preceding stage,
$$|b(x_n, y) - b(x,y)| \leq d(x_n, x), \quad \forall n.$$

Passing to limit as $n \to \infty$ yields the desired fact.

(iii) Let the sequence (y_n) in X and the points $x, y \in X$ be such that

$$y_n \xrightarrow{b} y; \text{ hence, } y_n \xrightarrow{d} y, \text{ as } b(.,.) \text{ is mc-quasimetric.}$$

By the preceding stage,

$$|b(x, y_n) - b(x, y)| \le d(y_n, y), \quad \forall n.$$

Passing to limit as $n \to \infty$ we are done. □

II: The latter of these consists in a group of local conditions involving our starting sequence (x_n), in the relational mc-pseudometric space $(X, b, \mathcal{G})$.

Denote, for each $p, q \in N$, $p \le q$,

$$x[p, q] = \{x_i; i \in N[p, q]\}, \; x]p, q[= \{x_i; i \in N]p, q[\},$$
$$x]p, q] = \{x_i; i \in N]p, q]\}, \; x[p, q[= \{x_i; i \in N[p, q[\}.$$

Then, let us introduce the convention

$$(\le; N) = \{(m, n) \in N \times N; m \le n\}, \; (<; N) = \{(m, n) \in N \times N; m < n\};$$

these are just the graph over N of the relations $(\le)$ and $(<)$, respectively. Given $h \in N(1, \le)$, let us say that (x_n) is $(\mathcal{G}, h)$-*uniformly admissible* (in short $(\mathcal{G}, h)$-*uadmissible*), provided

for each $(m, n) \in (<; N)$, we have $\mathcal{G}(x_m) \cap x]n, n + h] \ne \emptyset$;
that is: there exists $p \in N$ such that $x_m \mathcal{G} x_p$ and $n < p \le n+h$.

The class of all such elements h will be denoted as $\mathrm{ua}(x_n)$ (the *uadmissible region* of (x_n)). Clearly,

$\mathrm{ua}(x_n)$ is *hereditary*: $h \in \mathrm{ua}(x_n)$, $q \ge h$ imply $q \in \mathrm{ua}(x_n)$;

when $\mathrm{ua}(x_n)$ is nonempty, we say that (x_n) is $\mathcal{G}$-*uadmissible*. A characterization of this property may be given as follows. Denote by $\widehat{\mathcal{G}}$ the (x_n)-*trace* of $\mathcal{G}$ over N

$(m, n \in N)$: $m\widehat{\mathcal{G}}n$ iff $x_m \mathcal{G} x_n$.

Then, let $\widetilde{\mathcal{G}}$ stand for the *nondiagonal* portion of $\widehat{\mathcal{G}}$:

$(m, n \in N)$: $m\widetilde{\mathcal{G}}n$ iff $m < n$ and $(m,n) \in \widehat{\mathcal{G}}$,
or, equivalently, $m\widetilde{\mathcal{G}}n$ iff $m < n$ and $x_m \mathcal{G} x_n$.

Then, by this very definition,

(x_n) is $(\mathcal{G}, h)$-uadmissible iff $\widetilde{\mathcal{G}}$ is h-uadmissible:
$\forall (m,n) \in (<; N)$, we have $\widetilde{\mathcal{G}}(m) \cap N]n, n+h] \neq \emptyset$, that is,
$\exists p \in N$ such that $m\widetilde{\mathcal{G}}p$ and $n < p \le n+h$.

Remember that our sequence (x_n) is taken so as

(x_n) is b-asymptotic: $b(x_n, x_{n+1}) \to 0$, as $n \to \infty$.

Note that, as $b(.,.)$ is mc-quasimetric, this is equivalent with

(x_n) is d-asymptotic: $d(x_n, x_{n+1}) \to 0$, as $n \to \infty$.

Given $\varepsilon > 0$, let us say that $i \in N(1, \le)$ is ε-regular, provided

$i \le n$ implies $d(x_n, x_{n+1}) < \varepsilon$.

The class $\mathcal{Z}(\varepsilon)$ of all these numbers is nonempty so that

$(\forall \varepsilon > 0)$: $Z(\varepsilon) := \min \mathcal{Z}(\varepsilon)$ is well defined, as an element of
$N(1, \le)$, with, in addition: $d(x_n, x_{n+1}) < \varepsilon$, for all $n \ge Z(\varepsilon)$.

Further, note that by this very definition,

$(\forall j \ge 1)$: $\Delta_n(j) := \mathrm{diam}(x[n, n+j]) < \varepsilon$, $\forall n \ge Z(\varepsilon/j)$,
whence, $\Delta_n(j) \to 0$ as $n \to \infty$.

Here, for each (nonempty) subset Y in X, $\mathrm{diam}(Y) = \sup\{d(x,y); x, y \in Y\}$ stands for its *diameter* (modulo d). In fact, for each (nonempty) subset M of N, we have the simpler representation of $Y := x(M)$, as

$\mathrm{diam}(x(M)) = \sup\{d(x_i, x_j); i, j \in M, i < j\}$.

The following auxiliary fact represents our basic step towards an appropriate answer to the posed question.

Theorem 2. *Let the sequence (x_n) in X and the number $h \in N(1, \le)$ fulfill*

(31-i) (x_n) *is b-asymptotic (or, equivalently: d-asymptotic),*
(31-ii) (x_n) *is $(\mathcal{G}, h)$-uadmissible.*

Further, let $\widetilde{\mathcal{G}}$ stand for the (x_n)-trace of $\mathcal{G}$ over $(<; N)$ (see above). Then, (x_n) is b-Cauchy, if and only if it fulfills the relational condition:

(b-C-rela) $\forall \varepsilon \in R_+^0$, $\exists j = j(\varepsilon) \in N$, such that
$(j \leq m < n,\ m\widetilde{\mathcal{G}}n)$ implies $b(x_m, x_n) \leq \varepsilon$.

Proof. By definition, the b-Cauchy property of (x_n) writes

(b-C) $\forall \varepsilon \in R_+^0$, $\exists i = i(\varepsilon) \in N$: $i \leq m < n \implies b(x_m, x_n) \leq \varepsilon$.

We claim that, by the imposed conditions, this is equivalent with its relational form (b-C-rela). In fact, the inclusion (b-C) $\implies$ (b-C-rela) is clear, so, we have to establish the reverse inclusion: (b-C-rela) $\implies$ (b-C). To this end, let $\varepsilon > 0$ be arbitrary fixed. By (b-C-rela), there exists some index $j_1(\varepsilon) \in N$, such that

(b-C-rela-1) $(j_1(\varepsilon) \leq m < n,\ m\widetilde{\mathcal{G}}n)$ imply $b(x_m, x_n) \leq \varepsilon/2$.

On the other hand, by the d-asymptotic property of (x_n), one may construct the index $j_2(\varepsilon) = Z(\varepsilon/2h)$; which, by definition, fulfills

(d-asy) $q \geq j_2(\varepsilon)$ implies $\Delta_q(h) := \operatorname{diam}(x[q, q+h]) < \varepsilon/2$.

Denote $j(\varepsilon) = \max\{j_1(\varepsilon), j_2(\varepsilon)\}$, and let the couple of ranks (m, n) be such that $j(\varepsilon) \leq m < n$. If $n \leq m + h$, then, by $j_2(\varepsilon) \leq m < n$ and (d-asy),

$$b(x_m, x_n) \leq d(x_m, x_n) \leq \Delta_m(h) = \operatorname{diam}(x[m, m+h]) < \varepsilon/2 < \varepsilon,$$

and we are done. Suppose now that $n > m+h$ (that is, $n-h > m$). By the h-uadmissible property of $\widetilde{\mathcal{G}}$, we must have $\widetilde{\mathcal{G}}(m) \cap N]n-h, n] \neq \emptyset$. Letting p be some point in this intersection, we have (from the above choice)

$(m\widetilde{\mathcal{G}}p$ and) $n - h < p \leq n$; whence, $m < n - h < p \leq n < p + h$.

Then, by $j(\varepsilon) \leq m < p$ and (b-C-rela-1),

$$b(x_m, x_n) \leq b(x_m, x_p) + b(x_p, x_n) \leq b(x_m, x_p) + d(x_p, x_n) \leq$$
$$\varepsilon/2 + \Delta_p(h) < \varepsilon/2 + \varepsilon/2 = \varepsilon,$$

and the assertion follows. □

We are now in position to state the desired answer. Let us say that the subset Θ of R^0_+ is $(>)$-*cofinal* in R^0_+, when

for each $\varepsilon \in R^0_+$, there exists $\theta \in \Theta$ with $\varepsilon > \theta$.

Further, given the sequence $(r_n; n \geq 0)$ in R and the point $r \in R$, let us write

$r_n \to r+$ (also expressed as: $\lim_n r_n = r+$),
if $r_n \to r$ and $(r_n > r$, for all $n \geq 0)$.

The following statement (referred to as Boyd–Wong relational non-Cauchy Criterion; in short (BW-r-n-CC)) is now available, over the relational mc-quasimetric space $(X, b, \mathcal{G})$.

Theorem 3. *Let the sequence* $(x_n; n \geq 0)$ *in* X *and the natural number* $h \geq 1$ *be such that*

(32-i) (x_n) *is b-asymptotic (or, equivalently: d-asymptotic),*
(32-ii) (x_n) *is* $(\mathcal{G}, h)$-*uadmissible,*
(32-iii) (x_n) *is not b-Cauchy.*

Further, let $\widetilde{\mathcal{G}}$ *stand for the* (x_n)-*trace of* $\mathcal{G}$ *over* $(<; N)$, *and take a subset* Θ *of* R^0_+ *with the* $(>)$-*cofinal property. There exist then a number* $\gamma \in \Theta$, *a rank* $J := J(\gamma, h) \geq 1$ *and a couple* $(m(k); k \geq 0)$, $(n(k); k \geq 0)$ *of rank sequences, with*

(32-a) $(\forall k \geq 0)$: $J + k \leq m(k) < m(k) + 3h < n(k)$
    (hence, $m(k) \to \infty$, $n(k) \to \infty$, *as* $k \to \infty$*),*
(32-b) $x_{m(k)} \mathcal{G} x_{n(k)}$ *and* $b(x_{m(k)}, x_{n(k)}) > \gamma$, *for all* $k \geq 0$,
(32-c) $(\forall k \geq 0)$: $m(k) < p(k) =$ *the predecessor of* $n(k)$ *in* $\widetilde{\mathcal{G}}(m(k))$, *and* $b(x_{m(k)}, x_{p(k)}) \leq \gamma + \Delta_{m(k)}(3h) + \Delta_{n(k)-2h}(5h)$,
(32-d) $U_k := b(x_{m(k)}, x_{n(k)}) \to \gamma+$ *as* $k \to \infty$,
(32-e) *for each* $s, t \in N[0, 3h]$, *we have*
    $V_k(s, t) := b(x_{m(k)+s}, x_{n(k)+t}) \to \gamma+$, *as* $k \to \infty$,
(32-f) *for each* $p, q \in N$, *we have*
    $V_k(p, q) := b(x_{m(k)+p}, x_{n(k)+q}) \to \gamma$, *as* $k \to \infty$.

Proof. There are several steps to be passed.

Step 1. By a previous auxiliary fact, the b-Cauchy property of (x_n) may be written, under a relational form, as

(b-C-rela) $\forall \varepsilon \in R_+^0$, $\exists j(\varepsilon) \in N$, such that
$(j(\varepsilon) \leq m < n, m\widetilde{\mathcal{G}}n)$ imply $b(x_m, x_n) \leq \varepsilon$.

On the other hand, as Θ is a $(>)$-cofinal part in R_+^0, the relational b-Cauchy property (b-C-rela) may be also written as

(b-C-rela-Th) $\forall \theta \in \Theta$, $\exists j(\theta) \in N$, such that
$(j(\theta) \leq m < n, m\widetilde{\mathcal{G}}n)$ imply $b(x_m, x_n) \leq \theta$.

Step 2. Suppose now that

(x_n) is not b-Cauchy; hence, the negation of (b-C-rela-Th) holds.

By definition, there exists $\beta \in \Theta$ such that

(rela-1) $E(j) := \{(m, n) \in (<; N); j \leq m < n, m\widetilde{\mathcal{G}}n,$
$b(x_m, x_n) > \beta\}$ is a nonempty relation over N, for all $j \geq 0$.

As $(x_n; n \geq 0)$ is d-asymptotic, we may construct the map $Z : R_+^0 \to N(1, \leq)$ with the precise properties. Let the number $\gamma \in \Theta$ be given according to

$\beta > 3\gamma$ (possible, since Θ is $(>)$-cofinal in R_+^0).

Further, let us take

(rela-2) $J(\gamma, h) = Z(\gamma/3h)$; hence, in particular
$\Delta_q(3h) := \mathrm{diam}(x[q, q+3h]) < \gamma$, $\forall q \geq J(\gamma, h)$.

Denote for simplicity $J := J(\gamma, h)$, and put

$(A(k) = E(J+k); k \geq 0)$; hence, by definition,
$A(k) := \{(m, n) \in (<; N); J+k \leq m < n, m\widehat{\mathcal{G}}n, b(x_m, x_n) > \beta\}$, $k \geq 0$.

Clearly, by the preceding nonemptiness property (rela-1),

$A(k)$ is a nonempty relation over N, for each $k \geq 0$.

From the d-Lipschitz property of $(x, y) \mapsto b(x, y)$, we have for all $k \geq 0$,

$b(x_{m+s}, x_{n+t}) \geq b(x_m, x_n) - d(x_m, x_{m+s}) - d(x_n, x_{n+t}) >$
$\beta - \gamma - \gamma = \beta - 2\gamma > \gamma$, $\forall s, t \in N[0, 3h]$, $\forall (m, n) \in A(k)$,

which tells us that

$B(k) := \{(m,n) \in (<;N),$
$\quad J+k \leq m < n, m\widetilde{\mathcal{G}}n, b(x_{m+s}, x_{n+t}) > \gamma, \forall s,t \in N[0,3h]\}$
is a nonempty relation over N, for all $k \geq 0$.

Having this precise, denote for each $k \geq 0$

$$m(k) = \min \text{Dom}(B(k)), \ n(k) = \min B(k)(m(k)).$$

By this very definition, we get

(pro-1) $(\forall k \geq 0): J+k \leq m(k) < n(k), m(k)\widetilde{\mathcal{G}}n(k)$
(so: $x_{m(k)}\mathcal{G}x_{n(k)}$),
(pro-2) $(\forall k \geq 0): b(x_{m(k)+s}, x_{n(k)+t}) > \gamma, \forall s,t \in N[0,3h].$

Step 3. We claim that the couple (γ, J) and the couple of rank sequences $(m(k); k \geq 0)$ and $(n(k); k \geq 0)$ fulfill all conclusions in our statement.

(i) For the moment, the first and second parts of conclusion (32-a) are fulfilled. Concerning the third part of the same, suppose by contradiction that

$$(m(k) <)n(k) \leq m(k) + 3h, \text{ for some } k \geq 0.$$

Then by (rela-2) (and $m(k) \geq J+k \geq J$),

$$d(x_{m(k)}, x_{n(k)}) \leq \Delta_{m(k)}(3h) < \gamma,$$

in contradiction with (pro-2); whence, the third part of (32-a) holds too.

(ii) Further, by (pro-1) an (pro-2), it is clear that conclusion (32-b) follows.

(iii) Let $k \geq 0$ be arbitrary fixed. By definition,

$$(n(k) \in \widetilde{\mathcal{G}}(m(k)) \text{ and } d(x_{m(k)+s}, x_{n(k)+t}) > \gamma, \forall s,t \in N[0,3h].$$

On the other hand, by the preceding step,

$$m(k) < n(k) - 2h < n(k) - h < n(k),$$

and this yields (as $\widetilde{\mathcal{G}}$ is h-uadmissible)

$$\widetilde{\mathcal{G}}(m(k)) \cap N]n(k)-2h, n(k)-h] \neq \emptyset;$$
$$\text{whence, } \widetilde{\mathcal{G}}(m(k)) \cap N]n(k)-2h, n(k)[\neq \emptyset.$$

In this case, $p(k) = \max \widetilde{\mathcal{G}}(m(k)) \cap N]n(k) - 2h, n(k)[$ exists as the predecessor of $n(k) \in \widetilde{\mathcal{G}}(m(k))$; precisely,

(pred-1) $p(k), n(k) \in \widetilde{\mathcal{G}}(m(k))$ and $\widetilde{\mathcal{G}}(m(k)) \cap N]p(k), n(k)[= \emptyset$,
(pred-2) in addition, $m(k) < n(k) - 2h < p(k) < n(k)$;
hence, $n(k) - 2h < p(k) < p(k) + 3h < n(k) + 3h$.

This, along with $[J+k \leq m(k) < p(k)]$, tells us by the very definition of $n(k)$ [as minimal element in $B(k)(m(k))$] that

(pro-3) $b(x_{m(k)+u}, x_{p(k)+v}) \leq \gamma$, for some $u, v \in N[0, 3h]$.

Combining with the triangular inequality gives (again from the d-Lipschitz property of $(x,y) \mapsto b(x,y)$)

(pro-4) $b(x_{m(k)}, x_{p(k)}) \leq b(x_{m(k)+u}, x_{p(k)+v})$
$+ d(x_{m(k)}, x_{m(k)+u}) + d(x_{p(k)}, x_{p(k)+v}) \leq \gamma + \Delta_{m(k)}(3h)$
$+ \Delta_{n(k)-2h}(5h), \quad \forall k \geq 0$,

and (32-c) is proved.

(iv) By the definition of $(B(k); k \geq 0)$, we have $(U_k > \gamma, \forall k)$. Further, by the very definition of these ranks,

$m(k) < n(k) - 2h < p(k) < n(k), \quad \forall k \geq 0.$

Combining with the triangular inequality and (pro-4) gives

$\gamma < b(x_{m(k)}, x_{n(k)}) \leq b(x_{m(k)}, x_{p(k)}) + b(x_{p(k)}, x_{n(k)})$
$\leq b(x_{m(k)}, x_{p(k)}) + d(x_{p(k)}, x_{n(k)})$
$\leq \gamma + \Delta_{m(k)}(3h) + \Delta_{n(k)-2h}(5h) + \Delta_{n(k)-2h}(2h), \quad \forall k \geq 0.$

Passing to limit as $k \to \infty$ yields (32-d).

(v) Let $s, t \in N[0, 3h]$ be arbitrary fixed. By the definition of $(B(k); k \geq 0)$, we have $(V_k(s,t) > \gamma, \forall k)$. Moreover, from the d-Lipschitz property of $(x,y) \mapsto b(x,y)$,

$|b(x_{m(k)}, x_{n(k)}) - b(x_{m(k)+s}, x_{n(k)+t})|$
$\leq d(x_{m(k)}, x_{m(k)+s}) + d(x_{n(k)}, x_{n(k)+t})$
$\leq \Delta_{m(k)}(3h) + \Delta_{n(k)}(3h)$, for all $k \geq 0$.

Passing to limit in the relation between the extremal members of this relation gives (by the above) the conclusion (32-e).

(vi) Let $p, q \in N$ be arbitrary fixed. By the same d-Lipschitz property of the mapping $(x,y) \mapsto b(x,y)$ we just evoked, one may write

$$|b(x_{m(k)}, x_{n(k)}) - b(x_{m(k)+p}, x_{n(k)+q})|$$
$$\leq d(x_{m(k)}, x_{m(k)+p}) + d(x_{n(k)}, x_{n(k)+q})$$
$$\leq \Delta_{m(k)}(p) + \Delta_{n(k)}(q), \text{ for all } k \geq 0.$$

Passing to limit in the relation between the extremal members of this relation gives (32-f) and completes the argument. □

By definition, the quadruple $[\gamma; J; (m(k); k \geq 0); (n(k); k \geq 0)]$ given by this result is referred to as a *Boyd–Wong* (Θ, h)-*system* attached to (x_n). Consequently, the following characterization of the b-Cauchy condition is available.

Theorem 4. *Let the sequence* $(x_n; n \geq 0)$ *in* X *be b-asymptotic and $\mathcal{G}$-uadmissible. Then, the following conditions/properties are equivalent:*

(33-a) $(x_n; n \geq 0)$ *is not d-Cauchy,*
(33-b) *for each* $(>)$-*cofinal subset* Θ *of* R_+^0 *and each* $h \in \mathrm{ua}(x_n)$, *there is at least one Boyd–Wong* $[\Theta, h]$-*system attached to* (x_n).

Proof. By the preceding result, the first condition includes the second one. Conversely, if the second condition holds, then fix some $(>)$-cofinal subset Θ of R_+^0 as well as some $h \in \mathrm{ua}(x_n)$. From the imposed hypothesis, there must be some Boyd–Wong $[\Theta, h]$-system $[\gamma; J; (m(k); k \geq 0); (n(k); k \geq 0)]$ attached to (x_n); hence (from the written conclusions),

$$(\gamma \in \Theta \text{ and }) \; d(x_{m(k)}, x_{n(k)}) \to \gamma+ \quad \text{as } k \to \infty.$$

This necessarily gives us that $(x_n; n \geq 0)$ is not b-Cauchy; otherwise — by the very definition of b-Cauchy sequence — it follows that

$$b(x_{m(k)}, x_{n(k)}) \to 0 \quad \text{as } k \to \infty,$$

in contradiction with the limiting property above. □

In particular, when $\Theta = R_+^0$, $b(.,.)$ is linearly mc-quasimetric and

(x_n) is *totally ascending* $(x_i \mathcal{G} x_j$, whenever $i < j)$,

Theorem 3 includes directly the statement in Karapinar et al. [26]. On the other hand, when $\Theta = R_+^0$, $b(.,.)$ is a metric and $\mathcal{G} = X \times X$ (the trivial relation over X), the same statement covers the 1969 one in Boyd and Wong [13], so, it is natural that Theorem 3 be referred to in the proposed way. Further aspects may be found in Abtahi [1].

4. Statement of the Problem

Let X be a nonempty set. Take an *mc-quasimetric* $b : X \times X \to R_+$ on X, and let $\mathcal{G} \subseteq X \times X$ be a relation over X; the triple $(X, b, \mathcal{G})$ will be said to be a *relational mc-quasimetric space*. Given the relations $\mathcal{A}, \mathcal{B}$ on X, denote for $Z \in \exp(X)$

$$(\mathcal{A}; \mathcal{B}; Z) = \mathcal{A} \cap \mathcal{B} \cap (Z \times Z), \quad (\mathcal{A}; Z) = (\mathcal{A}; \mathcal{A}; Z);$$

in this case, $(\mathcal{A}; \mathcal{B}; X) := \mathcal{A} \cap \mathcal{B}$ is denoted as $(\mathcal{A}; \mathcal{B})$, and $(\mathcal{A}; X)$ is just $\mathcal{A}$.

Let $T \in \mathcal{F}(X)$ be a selfmap of X. In the following, sufficient conditions are given for the existence of elements in $\text{Fix}(T)$.

The metrical way of solving this problem is *sequential* in nature; precisely, the points of $\text{Fix}(T)$ are chosen among the limit points (if any)

$$T^\omega x_0 := \lim_n (T^n x_0), \quad \text{where } x_0 \in X \text{ is appropriately given.}$$

(Here, ω is the first transfinite ordinal.) To do this, a lot of technical facts about *iterative processes* must be introduced and discussed.

(4-I) Let x_0 be some point in X. By an *orbital* (or *iterative*) sequence attached to x_0 (and T), we mean any sequence $X_0 := (x_n; n \geq 0)$ (or, simply, $X_0 = (x_n)$) defined as $(x_n = T^n x_0; n \geq 0)$. When x_0 is generic here, the resulting object is referred to as an *orbital sequence* (in short *o-sequence*) on X; the class of all these is denoted as $o(T)$. Then, call the o-sequence $X_0 = (x_n)$, $\mathcal{G}$-*ascending* (in short *ascending*) if

(x_n) is $\mathcal{G}$-ascending (in short: *ascending*): $x_n \mathcal{G} x_{n+1}$, for all n.

As before, when X_0 is generic here, we talk about *ascending orbital sequences* (in short *(a-o)-sequences*) on X; the class of all these is denoted as $\text{ao}(T)$.

Let $X_0 = (x_n)$ be an (a-o)-sequence in X. Given the natural number $h \geq 1$, let us say that $X_0 = (x_n)$ is $(\mathcal{G}, h)$-*uadmissible*, provided

$$\forall (m, n) \in (<; N), \exists p \in N, \text{ such that } n < p \leq n + h \text{ and } x_m \mathcal{G} x_p.$$

The set of all such elements h is denoted as $\text{ua}(X_0)$; when it is nonempty, we say that $X_0 = (x_n)$ is $\mathcal{G}$-*uadmissible* (in short *uadmissible*). Finally, let $\text{aoua}(T)$ stand for the intersection of the classes above; that is,

$X_0 = (x_n)$ is in aoua(T) provided it is ascending, orbital and uadmissible.

(4-II) For the rest of our exposition, the following condition is essential:

(aoua-reg) T is *(aoua)-regular*: aoua(T) is nonempty.

Concerning this synthetic requirement, a basic problem to be noted is that of getting elements of aoua(T) by starting from elements of ao(T). Fix in the following an ascending orbital sequence $X_0 = (x_n)$, and define

$$\text{spec}(x_0) = \{i \in N(1, \leq); x_0 \mathcal{G} T^i x_0\} \text{ [the } spectrum \text{ of } x_0 \text{ (modulo } (\mathcal{G}, T))].$$

Clearly, $1 \in \text{spec}(x_0)$, but the alternative $\text{spec}(x_0) = \{1\}$ cannot be avoided. Then, let us consider the condition

(t-comp) $X_0 = (x_n)$ is *translation compatible*:
$x_i \mathcal{G} x_j$ implies $x_{i+s} \mathcal{G} x_{j+s}$, for all $s \in N$.

Clearly, this holds when

(incr) T is $\mathcal{G}$-*increasing* ($x\mathcal{G}y \implies Tx\mathcal{G}Ty$);

however, we do not take such a particular setting into account. Given the natural number $h \geq 1$, call the subset M of $N(1, \leq)$, h-*admissible* provided

(M-adm) for each $n \in N$ there exists $r \in M$ with $n < r \leq n+h$.

When such numbers $h \geq 1$ exist, we say that M is *admissible*.

Proposition 7. *Let the ascending orbital sequence $X_0 = (x_n)$ fulfill*

(41-I) $X_0 = (x_n)$ *is translation compatible*,
(41-II) $\text{spec}(x_0)$ *is admissible (see above).*

Then, necessarily, $X_0 = (x_n)$ is $\mathcal{G}$-uadmissible.

Proof. Fix $(m, n) \in (<; N)$. By hypothesis, there exists $h \in N(1, \leq)$, with

$\text{spec}(x_0) = \{i \in N(1, \leq); x_0 \mathcal{G} x_i\}$ is h-admissible; wherefrom there exists $r \in \text{spec}(x_0)$ (hence, $x_0 \mathcal{G} x_r$), with $n - m < r \leq n - m + h$.

In this case, under the notation $p = m + r$, we have

$n < p \leq n + h$ and $x_m \mathcal{G} x_{m+r}$ (that is: $x_m \mathcal{G} x_p$),

if we remember that $X_0 = (x_n)$ is translation compatible. Hence, the orbital sequence $X_0 = (x_n)$ is $\mathcal{G}$-uadmissible, as claimed. $\square$

In the following, some other example of such objects is given. The following auxiliary fact is our starting point.

Proposition 8. *Let the ascending sequence $(z_n; n \geq 0)$ in X and the number $h \in N(2, \leq)$ be such that (under the notation $Z := \{z_n; n \geq 0\}$)*

$(\mathcal{G}; Z)$ *is h-transitive:* $(\mathcal{G}; Z)^h \subseteq (\mathcal{G}; Z)$.

Then, the regular property holds, for each $r \geq 0$:

(Reg;r) $(z_i, z_{i+1+r(h-1)}) \in \mathcal{G}$, *for all $i \geq 0$.*

Proof. We use the induction with respect to r. First, by the $\mathcal{G}$-ascending property of our sequence,

$(z_i, z_{i+1}) \in (\mathcal{G}; Z) \subseteq \mathcal{G}$, $\forall i \geq 0$; whence, the relation (Reg;0) holds.

This, by definition (and the h-transitive hypothesis), yields

$(z_i, z_{i+h}) \in (\mathcal{G}; Z)^h \subseteq (\mathcal{G}; Z) \subseteq \mathcal{G}$, $\forall i \geq 0$;

whence, (Reg;1) holds too. Suppose that (Reg;r) holds for all $r \in \{0, \ldots, s\}$, where $s \geq 1$; we claim that it (Reg;s + 1) holds as well. In fact, let $i \geq 0$ be arbitrary fixed. Again by the $\mathcal{G}$-ascending property of our sequence,

$(z_{i+1+s(h-1)}, z_{i+1+(s+1)(h-1)}) \in (\mathcal{G}; Z)^{h-1}$

so that by the inductive hypothesis (and properties of relational product)

$(z_i, z_{i+1+(s+1)(h-1)}) \in (\mathcal{G}; Z) \circ (\mathcal{G}; Z)^{h-1} = (\mathcal{G}; Z)^h \subseteq (\mathcal{G}; Z) \subseteq \mathcal{G}$;

hence the claim. The proof is thereby complete. $\square$

Remark 3.
In case of $h = 2$, the conclusion in our statement above becomes

(Reg) $(z_i, z_{i+1+r}) \in \mathcal{G}$, $\forall i \geq 0$, $\forall r \geq 0$; or, equivalently:
$(z_i, z_j) \in \mathcal{G}$, whenever $i < j$; referred to as: (z_n) is *totally ascending*.

Technically speaking, this is a trivialization of the ascending concept; however, it may be sometimes useful in practice.

Given the number $h \in N(2, \leq)$, call the ascending orbital sequence $X_0 = (x_n)$ in X $(\mathcal{G}, h)$-*transitive* when

the restriction $(\mathcal{G}; [X_0])$, where $[X_0] := \{x_n; n \geq 0\}$, is h-transitive.

When $h \in N(2, \leq)$ is generic in this convention, we then say that the $\mathcal{G}$-ascending orbital sequence $X_0 = (x_n)$ is $\mathcal{G}$-*transitive*.

Proposition 9. *Let the ascending orbital sequence $X_0 = (x_n)$ be $\mathcal{G}$-transitive. Then, necessarily, $X_0 = (x_n)$ is uadmissible.*

Proof. By definition, there exists $h \in N(2, \leq)$ such that $X_0 = (x_n)$ is $(\mathcal{G}, h)$-transitive. Let $(m, n) \in (<; N)$ be arbitrary fixed. By the preceding result,

$(x_m, x_{m+1+r(h-1)}) \in \mathcal{G}$, for all $r \in N$.

As $m + 1 \leq n$, there exists at least one $r \in N(1, \leq)$ with $n < m + 1 + r(h - 1) \leq n + h$. It will suffice putting $p = m + 1 + r(h - 1)$ to end the reasoning. □

Remark 4.
When $h = 2$, the conclusion in this statement becomes (see above)

(Reg-var) $X_0 = (x_n)$ *totally ascending*: $(x_m, x_n) \in \mathcal{G}$, $\forall (m, n) \in (<; N)$.

This tells us that the uadmissible problem involving our (a-o)-sequence $X_0 = (x_n)$ is solved in a trivial way.

(4-III) Having these precise, we may now pass to the essential part of our developments. For each $x, y \in X$, define the functions

(n-ant) $Q_1(x,y) = b(x,Tx)$, $Q_2(x,y) = b(x,y)$,
$Q_3(x,y) = b(x,Ty)$, $Q_4(x,y) = b(Tx,y)$,
$Q_5(x,y) = b(Tx,Ty)$, $Q_6(x,y) = b(y,Ty)$,
$\mathcal{Q}(x,y) = (Q_1(x,y), Q_2(x,y), Q_3(x,y), Q_4(x,y), Q_5(x,y),$
$Q_6(x,y))$.

Further, let us construct the family of functions [for $x, y \in X$]

$B_1(x,y) = \min\{Q_1, Q_2, Q_5, Q_6\}(x,y)$,
$B_2(x,y) = \min\{Q_1, Q_2, Q_3, Q_4, Q_5, Q_6\}(x,y)$,

as well as the associated relations (over X)

$(B_i > 0) = \{(x,y) \in X \times X; B_i(x,y) > 0\}$, $i \in \{1,2\}$.

Remember that the following essential condition was imposed:

(aoua-reg) T is *(aoua)-regular*: aoua(T) is nonempty.

According to it, there exist ascending, orbital, uadmissible sequences (in short (aoua)-sequences) in X; let $X_0 = (x_n)$ be one of these. Note that, without loss, one may arrange for the extra property being fulfilled:

(B-asc) $X_0 = (x_n)$ is $(B_1 > 0)$-ascending: $B_1(x_n, x_{n+1}) > 0$, for all n.

In fact, given the initial (aoua)-sequence $X_0 = (x_n)$, we have two alternatives:

(Alt-1) The underlying sequence $X_0 = (x_n)$ is *telescopic*, in the sense

there exists $h \geq 0$, such that $b(x_h, x_{h+1}) = 0$ [that is, $x_h = x_{h+1}$].

By the iterative definition of our sequence, one derives

$x_h = x_n$, for all $n \geq h$; whence, $z := x_h$ is an element of Fix(T).

Consequently, this case is completely clarified from the fixed-point perspective.

(Alt-2) The underlying sequence $X_0 = (x_n)$ is *non-telescopic*, in the sense

$(b(x_n, x_{n+1}) > 0, \forall n)$; wherefrom, $(B_1(x_n, x_{n+1}) > 0, \forall n)$.

In other words, the (aoua)-sequence $X_0 = (x_n)$ is, in addition, $(B_1 > 0)$-ascending; this is referred to as $X_0 = (x_n)$ is an *(Baoua)-sequence*. Denote also

$$[X_0] = \{x_n; n \geq 0\}, \ [[X_0]] = \mathrm{cl}([X_0]).$$

Here, cl means the *b*-closure operator; note that, by the mc-quasimetric property of *b*, this is identical with the *d*-closure operator.

Having these precise, the basic directions under which the investigations are conducted are described in the following list:

(rpic-1) We say that the (Baoua)-sequence $X_0 = (x_n)$ is *semi Picard* (modulo $(b, \mathcal{G}; T)$) when (x_n) is *b*-asymptotic ($b(x_n, x_{n+1}) \to 0$ as $n \to \infty$).

(rpic-2) We say that the (Baoua)-sequence $X_0 = (x_n)$ is *Picard* (modulo $(b, \mathcal{G}; T)$) when (x_n) is *b*-Cauchy ($b(x_m, x_n) \to 0$ as $m, n \to \infty$, $m < n$).

(rpic-3) We say that the (Baoua)-sequence $X_0 = (x_n)$ is *strongly Picard* (modulo $(b, \mathcal{G}; T)$) when $x_\omega := \lim_n(x_n)$ exists, with $x_\omega \in \mathrm{Fix}(T)$.

(rpic-4) We say that the (Baoua)-sequence $X_0 = (x_n)$ is *semi Bellman Picard* (modulo $(b, \mathcal{G}; T)$) when $x_\omega := \lim_n(x_n)$ exists, with $x_\omega \in \mathrm{Fix}(T)$ and (in addition) $(x_n; n \geq 0)\mathcal{G}\mathcal{G}x_\omega$.

Here, for each sequence $(z_n; n \geq 0)$ in X and each point $z \in X$, we denoted

$(z_n; n \geq 0)\mathcal{G}z$ iff $z_n\mathcal{G}z$, for all $n \geq 0$
$(z_n; n \geq 0)\mathcal{G}\mathcal{G}z$ iff there exists a subsequence $(w_n = z_{i(n)}; n \geq 0)$ of $(z_n; n \geq 0)$ such that $(w_n; n \geq 0)\mathcal{G}z$.

In particular, when $\mathcal{G}$ is a quasi-order on X, these conventions are comparable with the ones in Turinici [55], which, in case of $\mathcal{G} = X \times X$ (the trivial relation over X) reduce to the ones in Rus [46, Chapter 2, Section 2.2], because, in this setting, $X(T, \mathcal{G}) = X$.

The sufficient (regularity) conditions for such properties are obtainable within the class of all (Baoua)-sequences $X_0 = (x_n)$, with the additional property

$X_0 = (x_n)$ is *full*: $n \mapsto x_n$ is injective ($i \neq j$ implies $x_i \neq x_j$),

referred to as *(Baoua-f) sequences*.

(reg-1) Call X, *(Baoua-f,b)-complete* provided (for each (Baoua-f)-sequence) *b*-Cauchy $\Longrightarrow$ *b*-convergent.

(**reg-2**) Let us say that T is *(Baoua-f,b)-continuous*, if $[(z_n) =$ (Baoua-f)-sequence and $z_n \xrightarrow{b} z]$ imply $T z_n \xrightarrow{b} T z$.

(**reg-3**) Call $\mathcal{G}$, *(Baoua-f,b)-almost-selfclosed* when $[(z_n) =$ (Baoua-f)-sequence and $z_n \xrightarrow{b} z]$ imply $(z_n; n \geq 0)\mathcal{G}\mathcal{G}z$.

(**4-IV**) As a completion of the developments above, we list the metrical contractive requirements upon our data. So, let $X_0 = (x_n)$ be a (Baoua)-sequence in X. Denote $[X_0] = \{x_n; n \geq 0\}$, $[[X_0]] = \text{cl}([X_0])$; here, cl is the *b*-closure (or, equivalently, *d*-closure) operator.

Let Υ be a nonempty subset of R_+^6 and $\mathcal{R} \in \{(B_i > 0); i \in \{1,2\}\}$ be some relation. Define the concept

(GR-Ups) T is $(\mathcal{G}; \mathcal{R}; \Upsilon)$-*contractive*: $\mathcal{Q}(x,y) \in \Upsilon$, $\forall (x,y) \in (\mathcal{G}; \mathcal{R})$.

The natural choice is $\mathcal{R} = (B_1 > 0)$. Note that $B_1 \geq B_2$, and this yields

T is $(\mathcal{G}; (B_1 > 0); \Upsilon)$-contractive implies T is $(\mathcal{G}; (B_2 > 0); \Upsilon)$-contractive.

Having these precise, we may now list the specific conditions to be fulfilled by these data.

(**I**) The first condition is of *descending* type. We say that the vector $t := (t_1, \ldots, t_6)$ in R_+^6 is *normal*, provided

$t_1 = t_2 > 0$, $t_5 = t_6 > 0$, $(t_3 \leq t_1 + t_6)$, and $(t_4 = 0)$.

Call Υ, *descending* on $[X_0]$ when

(desc) each normal vector $t := (t_1, \ldots, t_6)$ in R_+^6 with $t \in \mathcal{Q}(\mathcal{G}; (B_1 > 0); [X_0]) \cap \Upsilon$, satisfies $t_1 > t_6$.

(**II**) The second condition to be considered is related to *semi-right* properties. We say that the sequence $(t^n := (t_1^n, \ldots, t_6^n); n \geq 0)$ in R_+^6 is *normal*, provided

$(\forall n)$: t^n is normal, that is,
$t_1^n = t_2^n > 0$, $t_5^n = t_6^n > 0$, $(t_3^n \leq t_1^n + t_6^n)$, and $(t_4^n = 0)$.

Take some point $e = (e_1, \ldots, e_6)$ in R_+^6. We say that the normal sequence $(t^n := (t_1^n, \ldots, t_6^n); n \geq 0)$ in R_+^6 is *semi-right* at e, if

(s-r-e) $t_i^n \to e_i+$, $\forall i \in \{1,2,5,6\}$, $(\limsup_n t_3^n \leq e_3)$, and $(t_4^n = e_4, \forall n)$.

[Clearly, these impose certain restrictions upon e; but this is not essential to us.] Given $\rho > 0$, let us say that Υ is *nsright* at ρ on $[X_0]$, if

(nsright) $\forall$ normal sequence $(t^n; n \geq 0)$ in $\mathcal{Q}(\mathcal{G}; (B_1 > 0); [X_0]) \cap \Upsilon$,
the semi-right property at $(\rho, \rho, 2\rho, 0, \rho, \rho)$ is not true.

The class of all these $\rho > 0$ is denoted as $\operatorname{nsright}(\Upsilon; [X_0])$. In this case, define

(n-s-r) Υ is *nsright* on $[X_0]$, if $\operatorname{nsright}(\Upsilon; [X_0])$ is identical with R_+^0.

(III) The third condition to be considered is related to *total-right* properties. Take some point $e = (e_1, \ldots, e_6)$ in R_+^6. Call the sequence $(t^n := (t_1^n, \ldots, t_6^n); n \geq 0)$ in R_+^6, *total-right* at e, if

(t-r-e) $t_i^n \to e_i+$, for each $i \in \{1, \ldots, 6\}$.

Given $\lambda > 0$, let us say that Υ is *ntright* at λ on $[X_0]$, if

(ntright) $\forall$ sequence $(t^n; n \geq 0)$ in $\mathcal{Q}(\mathcal{G}; B_2 > 0; [X_0]) \cap \Upsilon$,
the total-right property at $(0, \lambda, \lambda, \lambda, \lambda, 0)$ is not true.

The class of all these $\lambda > 0$ is denoted as $\operatorname{ntright}(\Upsilon; [X_0])$. In this case, we say that

(a-n-r) Υ is *almost ntright* on $[X_0]$, if $\Theta := \operatorname{ntright}(\Upsilon; [X_0])$ is $(>)$-cofinal in R_+^0 (for each $\varepsilon \in R_+^0$ there exists $\theta \in \Theta$ with $\varepsilon > \theta$)
(n-r) Υ is *ntright* on $[X_0]$, if $\operatorname{ntright}(\Upsilon; [X_0])$ is identical with R_+^0.

(IV) The fourth condition involves *point* properties. Take some point $e = (e_1, \ldots, e_6)$ in R_+^6. We say that the sequence $(t^n := (t_1^n, \ldots, t_6^n); n \geq 0)$ in R_+^6 is *point* at e, if

(p-e) $(t_i^n \to e_i$, for each $i \in \{1, \ldots, 6\})$ and $[t_6^n = e_6, \forall n]$.

Given $\mu > 0$, let us say that Υ is *npoint* at μ on $[[X_0]]$, if

(npoint) $\forall$ sequence $(t^n; n \geq 0)$ in $\mathcal{Q}(\mathcal{G}; B_2 > 0; [[X_0]]) \cap \Upsilon$,
the point property at $(0, 0, \mu, 0, \mu, \mu)$ is not true.

The class of all these $\mu > 0$ will be denoted as npoint$(\Upsilon; [[X_0]])$. Then, define

(n-p) Υ is *npoint* on $[[X_0]]$, if npoint$(\Upsilon; [[X_0]])$ is identical with R_+^0.

Some concrete examples of such functions are given a bit further.

5. Main Result

Let X be a nonempty set. Take an mc-quasimetric $b : X \times X \to R_+$ on X, and let $\mathcal{G} \subseteq X \times X$ be a relation over X; the triple $(X, b, \mathcal{G})$ is said to be a *relational mc-quasimetric space*.

Further, let $T \in \mathcal{F}(X)$ be a selfmap of X with

(aoua-reg) T is *(aoua)-regular*: aoua(T) is nonempty.

According to it, there exist ascending orbital uadmissible sequences (in short (aoua)-sequences) in X; let $X_0 = (x_n)$ be one of these. Note that, without loss, one may arrange for the extra property being fulfilled:

(B-asc) $X_0 = (x_n)$ is $(B_1 > 0)$-ascending: $B_1(x_n, x_{n+1}) > 0$, for all n;

this is referred to as $X_0 = (x_n)$ is a *(Baoua)-sequence*. Denote also

$$[X_0] = \{x_n; n \geq 0\}, \ [[X_0]] = \text{cl}([X_0]);$$

here, cl is the b-closure (or, equivalently, d-closure) operator.

Define the sequences

$$(r_n = d(x_n, x_{n+1}); n \geq 0), \ (p_n = d(x_n, x_{n+2}); n \geq 0),$$
$$(E_n := (r_n, r_n, p_n, 0, r_{n+1}, r_{n+1}); n \geq 0).$$

Clearly, $(r_n := r_n(X_0))$ and $(p_n := p_n(X_0))$ and sequences in R_+^0 and R_+, respectively; in addition,

$$(p_n \leq r_n + r_{n+1}, \forall n); \text{ hence, } \limsup_n p_n \leq 2 \limsup_n r_n;$$

these are referred to referred to as the *telescopic sequences* attached to X_0. This, by a previous convention, tells us that $(E_n = E_n(X_0))$ is a normal sequence in R_+^6; it is called the *normal telescopic sequence* attached to $X_0 = (x_n)$.

The basic directions and regularity conditions relative to X_0 under which the problem of determining fixed points of T is to be solved were already listed; and the contractive-type framework (involving X_0, $[X_0]$ and $[[X_0]]$) was settled.

Our main result in this exposition is as follows:

Theorem 5. *Assume that the selfmap T is $(\mathcal{G};(B_1 > 0);\Upsilon)$-contractive, where the subset $\Upsilon \in \exp(R_+^6)$ and the relation $(B_1 > 0)$ over X were introduced under the precise conventions. In addition, suppose that X is (Baoua-f,b)-complete. Then,*

- **(51-a)** $X_0 = (x_n)$ *is a (Baoua-f) sequence whenever Υ is descending on $[X_0]$,*
- **(51-b)** $X_0 = (x_n)$ *is (Baoua-f) and semi Picard (modulo $(b,\mathcal{G};T)$), whenever (in addition) Υ is nsright on $[X_0]$,*
- **(51-c)** $X_0 = (x_n)$ *is (Baoua-f) and Picard (modulo $(b,\mathcal{G};T)$), whenever (in addition to the setting of (51-a)+(51-b)) Υ is almost ntright on $[X_0]$,*
- **(51-d)** $X_0 = (x_n)$ *is (Baoua-f) and strongly Picard (modulo $(b,\mathcal{G};T)$), whenever (in addition to the setting of (51-a)+(51-b)) T is (Baoua-f,b)-continuous,*
- **(51-e)** $X_0 = (x_n)$ *is (Baoua-f) and semi-Bellman Picard (modulo $(b,\mathcal{G};T)$), whenever (in addition to the setting of (51-a)+(51-b)) $\mathcal{G}$ is (Baoua-f,b)-almost-selfclosed and Υ is npoint on $[[X_0]]$.*

Proof. The argument is divided into a number of steps:

Step 1. Let $(r_n := r_n(X_0))$, $(p_n := p_n(X_0))$ and $(E_n = E_n(X_0))$ be the telescopic sequences attached to the (Baoua)-sequence $X_0 = (x_n)$:

$(r_n = d(x_n, x_{n+1}); n \geq 0)$, $(p_n = d(x_n, x_{n+2}); n \geq 0)$,
$(E_n := (r_n, r_n, p_n, 0, r_{n+1}, r_{n+1}); n \geq 0)$.

Clearly, the following property holds:

$(\forall n): (p_n \leq r_n + r_{n+1}, \forall n)$; hence, $\limsup_n p_n \leq 2\limsup_n r_n$.

This, on the one hand, tells us that

$(E_n; n \geq 0)$ is a normal sequence in R_+^6.

On the other hand, by the very definition of the involved components,

$(\forall n)$: $E_n = \mathcal{Q}(x_n, x_{n+1}) \in \mathcal{Q}(\mathcal{G}; (B_1 > 0); [X_0]) = \mathcal{Q}(\mathcal{G}; (B_1 > 0); [X_0]) \cap \Upsilon$,

where the last equality is a consequence of contractive condition. By the descending property of Υ at $[X_0]$, one derives

(s-desc) $(r_n > r_{n+1}, \forall n)$; meaning that: (r_n) is strictly descending.

We claim that, necessarily, $X_0 = (x_n)$ is a full sequence. In fact, suppose by contradiction that $x_i = x_j$, for some $i, j \in N$ with $i < j$. This, by the iterative construction of our sequence, gives $x_{i+1} = x_{j+1}$; whence, $r_i = r_j$, in contradiction with $r_i > r_j$ (deduced by the strict descending property of (r_n)).

Step 2. From the conclusions of our preceding step, $\rho := \lim_n r_n$ exists in R_+, with $(r_n > \rho, \forall n)$. We claim that $\rho = 0$. Suppose not: $\rho > 0$; clearly, by the accepted hypothesis, $\rho \in \text{nsright}(\Upsilon; [X_0])$. By a previous conclusion,

the vectorial sequence $(E_n := (r_n, r_n, p_n, 0, r_{n+1}, r_{n+1}); n \geq 0)$ fulfills
(see above) $(E_n \in (\mathcal{G}; (B_1 > 0); [X_0]) \cap \Upsilon, \forall n)$.

On the other hand, by the imposed conditions,

$(E_n := (r_n, r_n, p_n, 0, r_{n+1}, r_{n+1}); n \geq 0)$ is normal and
$r_n \to \rho+$, $r_{n+1} \to \rho+$, $\limsup_n p_n \leq \limsup_n (r_n + r_{n+1}) \leq 2\rho$,

proving that (E_n) is semi-right at ρ. This contradicts the choice of ρ as element of $\text{nsright}(\Upsilon; [X_0])$ and proves our assertion.

Step 3. Summing up, $X_0 = (x_n)$ is (Baoua-f) and b-asymptotic. On the other hand, as Υ is almost ntright, $\Theta := \text{ntright}(\Upsilon; [X_0])$ appears as $(>)$-cofinal in R_+^0. Finally, by the very choice of our sequence, there exists an index $h \geq 1$, such that $X_0 = (x_n)$ is $(\mathcal{G}, h)$-uadmissible. We show that, under these conditions,

$X_0 = (x_n)$ is b-Cauchy; hence, Picard (modulo $(b, \mathcal{G}; T)$).

Suppose that this is not true. As a consequence,

(p-1) $X_0 = (x_n)$ is orbital, $(\mathcal{G}; (B_1 > 0))$-ascending, b-asymptotic,
(p-2) $X_0 = (x_n)$ is $(\mathcal{G}, h)$-uadmissible, and not b-Cauchy.

By a previous auxiliary statement, there exist a number $\gamma \in \Theta$, a rank $J := J(\gamma, h) \geq 1$ and a couple of rank-sequences $(m(k); k \geq 0)$, $(n(k); k \geq 0)$, with

(prop-1) $(\forall k)$: $J + k \leq m(k) < m(k) + 3h < n(k)$, and $x_{m(k)} \mathcal{G} x_{n(k)}$,
(prop-2) $\forall s, t \in N[0, 3h]$: $V_k(s,t) := b(x_{m(k)+s}, x_{n(k)+t}) \to \gamma+$ as $k \to \infty$.

By the full and b-asymptotic properties of $X_0 = (x_n)$,

$(t_1^k := r_{m(k)}; k \geq 0)$, $(t_6^k := r_{n(k)}; k \geq 0)$

are sequences in R_+^0 with $t_1^k, t_6^k \to 0$ as $k \to \infty$.

Moreover, taking (prop-2) into account yields

$(t_2^k := V_k(0,0); k \geq 0)$, $(t_3^k := V_k(0,1); k \geq 0)$,
$(t_4^k := V_k(1,0); k \geq 0)$, $(t_5^k := V_k(1,1); k \geq 0)$

are sequences in R_+^0 with $t_2^k, t_3^k, t_4^k, t_5^k \to \gamma+$ as $k \to \infty$; hence, putting these together,

(ga-tright) the vectorial sequence $(t^k = (t_1^k, \ldots, t_6^k); k \geq 0)$ is total-right at $(0, \gamma, \gamma, \gamma, \gamma, 0)$.

Concerning the localization of the sequence $(t^k; k \geq 0)$, note that by the same properties, we must have

$(\forall k)$: $x_{m(k)} \mathcal{G} x_{n(k)}$, and $B_2(x_{m(k)}, x_{n(k)}) = \min\{t_1^k, \ldots, t_6^k\} > 0$; so that,
$t^k = \mathcal{Q}(x_{m(k)}, x_{n(k)}) \in \mathcal{Q}(\mathcal{G}; (B_2 > 0); [X_0]) = \mathcal{Q}(\mathcal{G}; (B_2 > 0); [X_0]) \cap \Upsilon$,

where the last equality is a consequence of contractive property. This, via (ga-tright), contradicts the choice of γ as element of $\Theta := \text{ntright}(\Upsilon; [X_0])$. Hence, our working assumption is not acceptable, and the assertion follows.

Step 4. From these developments, $X_0 = (x_n)$ is (Baoua-f) and b-Cauchy. As X is (Baoua-f,b)-complete, we must have

X_0 is b-convergent: $x_n \xrightarrow{b} z_0$ as $n \to \infty$, for some $z_0 \in X$.

Two possible cases — treated in the following steps — are to be discussed.

Step 5. Suppose that T is (Baoua-f,b)-continuous at X_0. Then, $y_n := Tx_n \xrightarrow{d} Tz_0$ as $n \to \infty$. On the other hand, $(y_n = x_{n+1}; n \geq 0)$ is a subsequence of $(x_n; n \geq 0)$; whence $y_n \xrightarrow{d} z_0$ as $n \to \infty$. Combining with b = separated yields $z_0 = Tz_0$, that is, $z_0 \in \text{Fix}(T)$.

Step 6. Suppose that $\mathcal{G}$ is (Baoau-f,b)-almost-selfclosed and Υ is npoint on $[[X_0]]$. By the full property of $X_0 = (x_n)$,

$$H_1 := \{n \in N; Tx_n = Tz_0\}, \ H_2 := \{n \in N; x_n = Tz_0\} \text{ are asingletons;}$$

and this tells us that there exists $s = s(z_0) \in N$, such that

$$m \geq s \text{ implies } Tx_m \neq Tz_0 \text{ (hence, } x_m \neq z_0 \text{) and } x_m \neq Tz_0;$$

note that, by the very choice of our iterative sequence $X_n = (x_n)$, we must have

(posi-1) $m \geq s$ implies $b(x_m, Tx_m) > 0$, $b(x_m, z_0) > 0$, $b(x_m, Tz_0) > 0$, $b(Tx_m, z_0) > 0$, $b(Tx_m, Tz_0) > 0$.

We now show that, a hypothesis like

(posi-2) $z_0 \neq Tz_0$; that is: $\mu := b(z_0, Tz_0) > 0$

yields a contradiction. For the moment, by the very conventions above, $\mu \in R_+^0 = \text{npoint}(\Upsilon; [[X_0]])$. Then, as a direct consequence of the preceding relations, we have

(posi-3) $B_2(x_n, z) > 0$, for all $n \geq s$.

As (x_n) is a (Baoua-f)-sequence with $x_n \xrightarrow{b} z_0$ and $\mathcal{G}$ is (Baoua-f,b)-almost-selfclosed, there must be some subsequence $(u_n := x_{i(n)}; n \geq 0)$ of (x_n), with

$(u_n, z) \in \mathcal{G}$, for all n.

Moreover, as $\lim_n i(n) = \infty$, one may assume, without loss, that

$i(n) \geq s, \forall n$; whence, (by the above) $(u_n, z) \in (\mathcal{G}; B_2 > 0; [[X_0]]), \forall n$.

From the above positivity relations and the convergence property of (u_n), we have

(posi-2) $(a_1^n := Q_1(u_n, z_0); n \geq 0)$, $(a_2^n := Q_2(u_n, z_0); n \geq 0)$, $(a_3^n := Q_3(u_n, z_0); n \geq 0)$, $(a_4^n := Q_4(u_n, z_0); n \geq 0)$, $(a_5^n := Q_5(u_n, z_0); n \geq 0)$, $(a_6^n := Q_6(u_n, z_0); n \geq 0)$,

are sequences in R_+^0 with the convergence properties
$$(a_1^n, a_2^n, a_4^n \to 0), (a_3^n, a_5^n \to \mu), (a_6^n = \mu, \forall n);$$
hence, putting these together,

(mu-point) the vectorial sequence $(a^n = (a_1^n, \ldots, a_6^n); n \geq 0)$ is point at $(0, 0, \mu, 0, \mu, \mu)$.

Concerning the localization of this sequence, note that (by the relations above)
$$(\forall n): a^n = \mathcal{Q}(u_n, z_0) \in \mathcal{Q}(\mathcal{G}; (B_2 > 0); [[X_0]]) = \mathcal{Q}(\mathcal{G}; (B_2 > 0); [[X_0]]) \cap \Upsilon;$$
where the last inclusion is a consequence of contractive property. This, via (mu-point), contradicts the relation $\mu \in \text{npoint}(\Upsilon; [[X_0]])$. Hence, our working assumption $\mu > 0$ is not acceptable, and then, $\mu = 0$, that is, $z_0 \in \text{Fix}(T)$. □

In particular, when $b = $ metric and $\mathcal{G}$ is a quasi-order on X, this result includes (in a partial way) the related statement in Turinici [56], obtained via similar procedures. Some multivalued extensions of these facts are available under the lines in Karapinar et al. [26, 29]; we will discuss them elsewhere. Finally, some structural aspects of the discussed facts may be found in Chen et al. [17].

6. Ahmed–Fulga Approach

In the following, a particular version of our main result is given, so as to compare it with a related statement in Ahmed and Fulga [5].

Let X be a nonempty set. Take a *mc-quasimetric* $b : X \times X \to R_+$ on X, and let $\mathcal{G} \subseteq X \times X$ be a relation over X; the triple $(X, b, \mathcal{G})$ is said to be a *relational mc-quasimetric space*. Further, let $T \in \mathcal{F}(X)$ be a selfmap of X with

(aoua-reg) T is *(aoua)-regular*: aoua(T) is nonempty.

According to it, there exist ascending orbital uadmissible sequences (in short (aoua)-sequences) in X; let $X_0 = (x_n)$ be one of these. Note that, without loss, one may arrange for the extra property being fulfilled:

(B-asc) $X_0 = (x_n)$ is $(B_1 > 0)$-ascending: $B_1(x_n, x_{n+1}) > 0$, for all n;

this is referred to as $X_0 = (x_n)$ is an *(Baoua)-sequence*. Denote also

$$[X_0] = \{x_n; n \geq 0\}, \; [[X_0]] = \mathrm{cl}([X_0]),$$

where cl is the *b*-closure (or, equivalently, *d*-closure) operator.

The basic directions and regularity conditions relative to X_0 under which the problem of determining fixed points of T is to be solved were already listed, and the contractive-type framework (involving X_0, $[X_0]$ and $[[X_0]]$) was settled. As a by-product of these, we established our main result in this exposition, Theorem 5. It is our aim in the following to derive a certain particular variant of it, in view of a better comparison between our main result and a related statement in the area due to Ahmed and Fulga [5].

For each $x, y \in X$, define the functions

(n-ant) $Q_1(x,y) = b(x, Tx)$, $Q_2(x,y) = b(x,y)$,
$Q_3(x,y) = b(x, Ty)$, $Q_4(x,y) = b(Tx, y)$,
$Q_5(x,y) = b(Tx, Ty)$, $Q_6(x,y) = b(y, Ty)$,
$\mathcal{Q}(x,y) = (Q_1(x,y), Q_2(x,y), Q_3(x,y), Q_4(x,y), Q_5(x,y), Q_6(x,y))$.

To formulate the contractive condition, define, for $x, y \in X$,

$A_1(x,y) = 2Q_4(x,y) + Q_5(x,y) + Q_6(x,y)$,
$A_2(a,y) = Q_1(x,y) + Q_5(x,y) + Q_6(x,y)$.

Note that all these may be viewed as composition functions, in the sense

$$A_i(x,y) = a_i(\mathcal{Q}(x,y)), \; x, y \in X, \; i \in \{1, 2\},$$

where the functions $a_1, a_2 \in \mathcal{F}(R_+^6, R)$ are introduced as

$$a_1(t) = 2t_4 + t_5 + t_6, \; a_2(t) = t_1 + t_5 + t_6, \; t = (t_1, \ldots, t_6) \in R_+^6.$$

Letting $\psi, \varphi, \chi \in \mathcal{F}(R_+, R)$ and $\mathcal{R} \in \{(B_1 > 0), (B_2 > 0)\}$, define the concept (comparable with the one in Ahmed and Fulga [5])

(GR-AF) T is $(\mathcal{G}; \mathcal{R}; \psi, \varphi, \chi)$-*contractive*:
$\psi(Q_5(x,y)) \leq \varphi(A_1(x,y)) + \chi(A_2(x,y)), \; \forall (x,y) \in (\mathcal{G}; \mathcal{R})$.

The natural choice is $\mathcal{R} = (B_1 > 0)$. Note that $B_1 \geq B_2$, and this yields

T is $(\mathcal{G};(B_1>0);\psi,\varphi,\chi)$-contractive implies
T is $(\mathcal{G};(B_2>0);\psi,\varphi,\chi)$-contractive.

This contraction is a (GR-Ups) one, where $\Upsilon \in \exp(R_+^6)$ is introduced as

$\Upsilon=$ the class of all $t=(t_1,\ldots,t_6) \in R_+^6$ with $\psi(t_5) \leq \varphi(a_1(t))+\chi(a_2(t))$.

So, applying the main result to such contractions amounts to establish of to what extend are the general specific conditions upon Υ and $(\mathcal{G};(B_1>0))$ in our main result fulfilled by our concrete data. For an easy reference, we list the conditions to be considered here. Some preliminaries are needed.

For any $\varphi \in \mathcal{F}(R_+,R)$ and any $s \in R_+^0$, put

$\Lambda^+\varphi(s) = \inf_{0<\varepsilon<s} \Phi(s+)(\varepsilon)$, where $\Phi(s+)(\varepsilon) = \sup\varphi(]s,s+\varepsilon[)$,
$\Lambda^\pm\varphi(s) = \inf_{0<\varepsilon<s} \Phi(s\pm)(\varepsilon)$, where $\Phi(s\pm)(\varepsilon) = \sup\varphi(]s-\varepsilon, s+\varepsilon[)$,
$\Lambda_+\varphi(s) = -\Lambda^+(-\varphi)(s)$, $\Lambda_\pm\varphi(s) = -\Lambda^\pm(-\varphi)(s)$.

By definition, these limit quantities fulfill

$-\infty \leq \Lambda_+\varphi(s) \leq \Lambda^+\varphi(s) \leq \Lambda^\pm\varphi(s) \leq \infty$,
$-\infty \leq \Lambda_+\varphi(s) \leq \Lambda_\pm\varphi(s) \leq \Lambda^\pm\varphi(s) \leq \infty$, $\forall s \in R_+^0$,

but the case of extremal values being attained cannot be avoided.

The following properties of introduced concepts will be useful. (Since the verification is immediate, we do not give details.)

Proposition 10. *Given $\varphi \in \mathcal{F}(R_+,R)$ and $s \in R_+^0$, we have in the reduced system (ZF-AC+DC),*

(61-1) $\limsup_n(\varphi(t_n)) \leq \Lambda^+\varphi(s)$,
for each sequence (t_n) in R_+^0 with $t_n \to s+$
(61-2) $\limsup_n(\varphi(t_n)) \leq \Lambda^\pm\varphi(s)$,
for each sequence (t_n) in R_+^0 with $t_n \to s$
(61-3) $\liminf_n(\varphi(t_n)) \geq \Lambda_+\varphi(s)$,
for each sequence (t_n) in R_+^0 with $t_n \to s+$
(61-4) $\liminf_n(\varphi(t_n)) \geq \Lambda_\pm\varphi(s)$,
for each sequence (t_n) in R_+^0 with $t_n \to s$.

For an easy reference, we may now give the conditions to be imposed upon the triple (ψ,φ,χ) over $\mathcal{F}(R_+,R)$.

(C-12) The first two conditions are global ones, and write

(Cond-1) $\psi(s) > \varphi(2s) + \chi(3s)$, for each $s > 0$,
(Cond-2) χ is increasing over R_+.

Note that, in such a case,

$(\forall s \in R_+^0): \Lambda_+\chi(s) = \Lambda^+\chi(s) = \chi(s+0)$,
$\Lambda^\pm\chi(s) = \max\{\chi(s+0), \chi(s), \chi(s-0)\}$,
$\Lambda_\pm\chi(s) = \min\{\chi(s+0), \chi(s), \chi(s-0)\}$;

hence, all these quantities are finite.

(C-345) The next three conditions are limit global ones. Let us introduce the subsets of R_+^0 (related to the underlying triple):

nsright$(\psi, \varphi, \chi) := \{s \in R_+^0; \Lambda_+\psi(s) > \Lambda^+\varphi(2s) + \Lambda^+\chi(3s)\}$,
ntright$(\psi, \varphi, \chi) := \{s \in R_+^0; \Lambda_+\psi(s) > \Lambda^+\varphi(3s) + \Lambda^+\chi(s)\}$,
npoint$(\psi, \varphi, \chi) := \{s \in R_+^0; \Lambda_\pm\psi(s) > \Lambda^\pm\varphi(2s)\} + \Lambda^\pm\chi(2s)\}$.

In this case, the announced conditions write

(Cond-3) nsright(ψ, φ, χ) is identical with R_+^0,
(Cond-4) ntright(ψ, φ, χ) is cofinal in R_+^0,
(Cond-5) ntright(ψ, φ, χ) is identical with R_+^0.

The following is our essential step for solving the posed problem.

Proposition 11. *Let the accepted notations and conditions be in use. Then,*
  (62-1) Υ *is descending on* $[X_0]$ *under (Cond-1)+(Cond-2),*
  (62-2) Υ *is nsright on* $[X_0]$ *under (Cond-3),*
  (62-3) Υ *is almost ntright on* $[X_0]$ *under (Cond-4),*
  (62-4) Υ *is npoint on* $[[X_0]]$ *under (Cond-5).*

Proof. There are several step to be passed.

Step 1. [Υ is descending on $[X_0]$ under (Cond-1)+(Cond-2).]

Remember that the vector $t := (t_1, \ldots, t_6)$ in R_+^6 is *normal*, provided

$t_1 = t_2 > 0$, $t_5 = t_6 > 0$, $(t_3 \leq t_1 + t_6)$ and $(t_4 = 0)$.

Let us consider the concept

(Ups-desc) Υ is *descending* on $[X_0]$: each normal vector
$t := (t_1, \ldots, t_6)$ in R_+^6 with $t \in \mathcal{Q}(\mathcal{G}; (B_1 > 0); [X_0]) \cap \Upsilon$
satisfies $t_1 > t_6$.

We have to establish that, under (Cond-1)+(Cond-2), this property holds. Assume not: there exists a vector $t := (t_1, \ldots, t_6)$ in R_+^6, with

(p-1) t is *normal*, in the sense:
$t_1 = t_2 > 0$, $t_5 = t_6 > 0$, $(t_3 \leq t_1 + t_6)$, and $(t_4 = 0)$,
(p-2) $t \in \mathcal{Q}(\mathcal{G}; (B_1 > 0); [X_0]) \cap \Upsilon$,
(p-3) t satisfies $t_1 \leq t_6$.

From the second property above

$t \in \Upsilon$, that is, $\psi(t_5) \leq \varphi(a_1(t)) + \chi(a_2(t))$,

and this, along with the structure of (a_1, a_2), gives

$$\psi(t_5) \leq \varphi(2t_4 + t_5 + t_6)) + \chi(t_1 + t_5 + t_6).$$

Combining with the normal properties of t, one gets

$$\psi(t_6) \leq \varphi(2t_6) + \chi(t_1 + 2t_6); \text{ wherefrom } \psi(t_6) \leq \varphi(2t_6) + \chi(3t_6),$$

if we use the third property above and (Cond-2). This, however, is impossible via (Cond-1), and our assertion follows.

Step 2. [Υ is nsright on $[X_0]$ under (Cond-3).]
Remember that the sequence $(t^n := (t_1^n, \ldots, t_6^n); n \geq 0)$ in R_+^6 is *normal*, provided

($\forall n$): t^n is normal, that is,
$t_1^n = t_2^n > 0$, $t_5^n = t_6^n > 0$, $(t_3^n \leq t_1^n + t_6^n)$, and $(t_4^n = 0)$.

Taking some point $e = (e_1, \ldots, e_6)$ in R_+^6, we say that the normal sequence $(t^n := (t_1^n, \ldots, t_6^n); n \geq 0)$ is *semi-right* at e, if

(s-r-e) $t_i^n \to e_i+$, $\forall i \in \{1, 2, 5, 6\}$, $(\limsup_n t_3^n \leq e_3)$ and $(t_4^n = e_4, \forall n)$.

Given $\rho > 0$, let us say that Υ is *nsright* at ρ on $[X_0]$, if

(nsright) $\forall$ normal sequence $(t^n; n \geq 0)$ in $\mathcal{Q}(\mathcal{G}; (B_1 > 0); [X_0]) \cap \Upsilon$,
the semi-right property at $(\rho, \rho, 2\rho, 0, \rho, \rho)$ is not true.

The class of all these $\rho > 0$ is denoted as $\mathrm{nsright}(\Upsilon; [X_0])$. In this case, we say that Υ is *nsright* on $[X_0]$, if $\mathrm{nsright}(\Upsilon; [X_0]) = R_+^0$.

Under these preliminaries, we claim that

$$\mathrm{nsright}(\psi, \varphi, \chi) \subseteq \mathrm{nsright}(\Upsilon; [X_0]),$$

and this, via (Cond-3), gives the desired assertion. Suppose that this is not true:

there exists $\rho \in \mathrm{nsright}(\psi, \varphi, \chi)$ not belonging to $\mathrm{nsright}(\Upsilon; [X_0])$.

This, by definition, tells us that there exists a sequence $(t^n := (t_1^n, \ldots, t_6^n); n \geq 0)$ in R_+^6, with

(q-1) $(\forall n)$: t^n is normal: $t_1^n = t_2^n > 0$, $t_5^n = t_6^n > 0$, $t_3^n \leq t_1^n + t_6^n$, $t_4^n = 0$,
(q-2) $t^n \in \mathcal{Q}(\mathcal{G}; (B_1 > 0); [X_0]) \cap \Upsilon$, $\forall n$,
(q-3) $(t^n; n \geq 0)$ has the semi-right property at $(\rho, \rho, 2\rho, 0, \rho, \rho)$: $t_i^n \to \rho+$, $\forall i \in \{1, 2, 5, 6\}$, $(\limsup_n t_3^n \leq 2\rho)$, and $(t_4^n = 0, \forall n)$.

As a consequence of these, we have that, for the underlying sequence $(t^n; n \geq 0)$, the contractive condition applies and gives

(contr) $\psi(t_5^n)) \leq \varphi(a_1(t^n)) + \chi(a_2(t^n))$, $\forall n$,

where, according to definition,

$$(\forall n): a_1(t^n) = 2t_4^n + t_5^n + t_6^n,\ a_2(t^n) = t_1^n + t_5^n + t_6^n.$$

But, from the semi-right property of (t^n) and the limit properties above,

$\lim_n(t_5^n) = \rho+$,
$\limsup_n(a_1(t^n)) = \lim_n(2t_4^n + t_5^n + t_6^n) = 2\rho+$,
$\lim_n(a_2(t^n)) = \lim_n(t_1^n + t_5^n + t_6^n) = 3\rho+$.

Passing to inferior/superior limit as $n \to \infty$, in (contr), we derive

$\Lambda_+\psi(\rho) \leq \liminf_n(\psi(t_5^n)) \leq \limsup_n(\psi(t_5^n)) \leq$
$\limsup_n(\varphi(a_1(t^n))) + \limsup_n \chi(a_2(t^n))) \leq \Lambda^+\varphi(2\rho) + \Lambda^+\chi(3\rho),$

in contradiction with the choice of ρ. Hence, $R_+^0 = \text{nsright}(\Upsilon;[X_0])$, and all is clear.

Step 3. [Υ is ntright on $[X_0]$ under (Cond-4).]
Let $e = (e_1,\ldots,e_6)$ in R_+^6. We say that the sequence $(t^n := (t_1^n,\ldots,t_6^n); n \geq 0)$, is *total-right* at e, if

(t-r-e) $t_i^n \to e_i+, \quad \forall i.$

Given $\lambda > 0$, let us say that Υ is *ntright* at λ on $[X_0]$, if

(ntright) $\forall$ sequence $(t^n; n \geq 0)$ in $\mathcal{Q}(\mathcal{G};(B_2 > 0);[X_0]) \cap \Upsilon$, the total-right property at $(0,\lambda,\lambda,\lambda,\lambda,0)$ is not true.

The class of all these $\lambda > 0$ is denoted as $\text{ntright}(\Upsilon;[X_0])$. In this case, we say that

(a-ntr) Υ is *almost ntright* on $[X_0]$, if $\text{ntright}(\Upsilon;[X_0])$ is $(>)$-cofinal in R_+^0.
(ntr) Υ is *ntright* on $[X_0]$, if $\text{ntright}(\Upsilon;[X_0]) = R_+^0$.

Under these conventions, we claim that

$$\text{ntright}(\psi,\varphi,\chi) \subseteq \text{ntright}(\Upsilon;[X_0]),$$

and this, via (Cond-4), assures us that Υ is almost ntright on $[X_0]$. Suppose that this is not true:

there exists $\lambda \in \text{ntright}(\psi,\varphi,\chi)$ not belonging to ntright $(\Upsilon;[X_0])$.

This, by definition, tells us that there exists a sequence $(t^n := (t_1^n,\ldots,t_6^n); n \geq 0)$ in R_+^6, with

(r-1) $t^n \in \mathcal{Q}(\mathcal{G};(B_2 > 0);[X_0]) \cap \Upsilon$, $\forall n$,
(r-2) $(t^n; n \geq 0)$ has the total-right property at $(0,\lambda,\lambda,\lambda,\lambda,0)$:
$t_i^n \to \lambda+, \forall i \in \{2,3,4,5\}, t_i^n \to 0+, \forall i \in \{1,6\}$.

As a consequence of these, we have that, for the underlying sequence $(t^n; n \geq 0)$, the contractive condition applies and gives

(contr) $\psi(t_5^n)) \leq \varphi(a_1(t^n)) + \chi(a_2(t^n)), \quad \forall n$,

where, according to definition,

$(\forall n)$: $a_1(t^n) = 2t_4^n + t_5^n + t_6^n$, $a_2(t^n) = t_1^n + t_5^n + t_6^n$.

But, from the total-right property of our sequence,

$\lim_n(t_5^n) = \lambda+,$
$\limsup_n(a_1(t^n)) = \limsup_n(2t_4^n + t_5^n + t_6^n) = 3\lambda+,$
$\lim_n(a_2(t^n)) = \lim_n(t_1^n + t_5^n + t_6^n) = \lambda+.$

Passing to inferior/superior limit as $n \to \infty$, in (contr), we derive

$$\Lambda_+\psi(\lambda) \leq \liminf_n(\psi(t_5^n)) \leq \limsup_n(\psi(t_5^n))$$
$$\leq \limsup_n(\varphi(a_1(t^n))) + \limsup_n \chi(a_2(t^n)))$$
$$\leq \Lambda^+\varphi(3\lambda) + \Lambda^+\chi(\lambda),$$

in contradiction with the choice of λ. Hence, $\mathrm{ntright}(\Upsilon; [X_0])$ is cofinal in R_+^0, and all is clear.

Step 4. [Υ is npoint on $[[X_0]]$ under (Cond-5).]

Take some point $e = (e_1, \ldots, e_6)$ in R_+^6. We say that the sequence $(t^n := (t_1^n, \ldots, t_6^n); n \geq 0)$, is *point* at e, if

(p-e) $(t_i^n \to e_i, \forall i)$ and $[t_6^n = e_6, \forall n]$.

Given $\mu > 0$, let us say that Υ is *npoint* at μ on $[[X_0]]$, if

(npoint) $\forall$ sequence $(t^n; n \geq 0)$ in $\mathcal{Q}(\mathcal{G}; (B_2 > 0); [[X_0]]) \cap \Upsilon$, the point property at $(0, 0, \mu, 0, \mu, \mu)$ is not true.

The class of all these $\mu > 0$ is denoted as $\mathrm{npoint}(\Upsilon; [[X_0]])$. Then, define

(n-p) Υ is *npoint* on $[[X_0]]$, if $\mathrm{npoint}(\Upsilon; [[X_0]]) = R_+^0$.

Under these preliminaries, we claim that

$$\mathrm{npoint}(\psi, \varphi, \chi) \subseteq \mathrm{npoint}(\Upsilon; [[X_0]]),$$

and this, via (Cond-5), assures us that Υ is npoint on $[[X_0]]$. Suppose not:

there exists $\mu \in \mathrm{npoint}(\psi, \varphi, \chi)$ not belonging to $\mathrm{npoint}(\Upsilon; [[X_0]])$.

This, by definition, tells us that there exists a sequence $(t^n := (t_1^n, \ldots, t_6^n); n \geq 0)$ in R_+^6, with

(s-1) $t^n \in \mathcal{Q}(\mathcal{G}; (B_2 > 0); [[X_0]]) \cap \Upsilon, \forall n$,
(s-2) $(t^n; n \geq 0)$ has the point property at $(0, 0, \mu, 0, \mu, \mu)$: $t_i^n \to 0, \forall i \in \{1, 2, 4\}, t_i^n \to \mu, \forall i \in \{3, 5\}, (t_6^n = \mu, \forall n)$.

As a consequence of these, we have that, for the underlying sequence $(t^n; n \geq 0)$, the contractive condition applies and gives

(contr) $\quad \psi(t_5^n)) \leq \varphi(a_1(t^n)) + \chi(a_2(t^n)), \quad \forall n,$

where, according to definition,

$(\forall n): a_1(t^n) = 2t_4^n + t_5^n + t_6^n, \; a_2(t^n) = t_1^n + t_5^n + t_6^n.$

But, from the point property of our sequence,

$\lim_n(t_5^n) = \mu,$
$\limsup_n(a_1(t^n)) = \limsup_n(2t_4^n + t_5^n + t_6^n) = 2\mu,$
$\lim_n(a_2(t^n)) = \lim_n(t_1^n + t_5^n + t_6^n) = 2\mu.$

Passing to inferior/superior limit as $n \to \infty$, in (contr), we derive

$\Lambda_\pm\psi(\mu) \leq \liminf_n \psi(t_5^n) \leq \limsup_n(\psi(t_5^n))$
$\leq \limsup_n \varphi(a_1(t^n)) + \limsup_n \chi(a_2(t^n)) \leq \Lambda^\pm\varphi(2\mu) + \Lambda^\pm\chi(2\mu),$

in contradiction with the choice of μ so that $R_+^0 = \text{npoint}(\Upsilon; [X_0])$. $\square$

Finally, by simply combining this with our main result, one gets the following particular fixed point principle over such structures.

Theorem 6. *Assume that the selfmap T is $(\mathcal{G}; (B_1 > 0); \psi, \varphi, \chi)$-contractive, for some triple (ψ, φ, χ) over $\mathcal{F}(R_+, R)$, fulfilling (Cond-1)–(Cond-5). In addition, suppose that X is (Baoua-f,b)-complete. Then, the following are valid:*

(61-a) $X_0 = (x_n)$ *is a (Baoua-f) sequence,*
(61-b) $X_0 = (x_n)$ *is (Baoua-f) and semi Picard (modulo $(b, \mathcal{G}; T)$),*
(61-c) $X_0 = (x_n)$ *is (Baoua-f) and Picard (modulo $(b, \mathcal{G}; T)$),*
(61-d) $X_0 = (x_n)$ *is (Baoua-f) and strongly Picard (modulo $(b, \mathcal{G}; T)$), whenever (in addition to the introduced setting) T is (Boaua-f,b)-continuous,*
(61-e) $X_0 = (x_n)$ *is (Baoua-f) and semi Bellman Picard (modulo $(b, \mathcal{G}; T)$), whenever (in addition to the introduced setting) $\mathcal{G}$ is (Baoua-f,b)-almost-selfclosed.*

In particular, when $X_0 = (x_n)$ is taken so as

$X_0 = (x_n)$ is b-asymptotic and completely $\mathcal{G}$-ascending $(x_i \mathcal{G} x_j,$ for $i < j)$,

Theorem 6 extends in a direct way the related statement in Ahmed and Fulga [5]. Some extensions of these facts to multivalued maps may be carried out under the lines in Karapinar *et al.* [26] and Khojasteh and Rakočević [31]; we will discuss them elsewhere.

7. Nonlinear Aspects

Let (X, b) be a pseudometric space, where

$$b: X \times X \to R_+ \text{ is reflexive, sufficient and triangular.}$$

Note that $b(.,.)$ has all properties of a metric, excepting symmetry; we call it a *quasimetric* on X, and the couple (X, b) is referred to as a *quasimetric space*. As a consequence of this, the conjugate (to b) pseudometric

$$c(x, y) = b(y, x), \ x, y \in X$$

is also a quasimetric on X. Finally, the symmetric cover of (b and c)

$$d(x, y) = \max\{b(x, y), c(x, y)\}, \ x, y \in X \text{ (in short: } d = \max\{b, c\})$$

is a symmetric quasimetric — hence, a metric — over X, and the couple (X, d) is referred to as a *metric space*.

(A) Given the quasimetric space (X, b), we introduced the concept

(mc-qm) $b(.,.)$ is *metric compatible quasimetric* (in short *mc-quasimetric*): the (Conv-Cauchy-Asy) structures attached to b and d are equivalent.

In order words, $b(.,.)$ is an mc-quasimetric, provided

$$(\forall(x_n), \forall x): [x_n \xrightarrow{b} x \iff x_n \xrightarrow{d} x], \ [(x_n) \text{ is } b\text{-Cauchy} \iff (x_n) \text{ is } d\text{-Cauchy}], \ [(x_n) \text{ is } b\text{-asymptotic} \iff (x_n) \text{ is } d\text{-asymptotic}].$$

In this case, the structure (X, b) is referred to as an *mc-quasimetric space*.

In particular, let us say that b is *linearly mc-quasimetric* if

(l-mc-qm) $d(x, y) \le \lambda b(x, y), \ x, y \in X$, for some $\lambda \ge 1$.

By the imposed condition (and $b \le d$)

$$b(x,y) \le d(x,y) \le \lambda b(x,y), \ x,y \in X, \text{ for some } \lambda \ge 1.$$

And this, by the very definition of the (Conv-Cauchy-Asy) structures attached to b and d, yields a conclusion like

$b(.,.)$ is a linearly mc-quasimetric implies $b(.,.)$ is a mc-quasimetric.

This class of quasimetrics was exclusively used until now in many papers; see, for instance, Ahmed and Fulga [5], Karapinar et al. [26] and the references therein. So, it is natural to ask whether the underlined objects exhaust the whole class of all mc-quasimetrics. As precise, the answer to this is negative; a functional class of such objects were given in a previous place. It is our aim in the following to extend this conclusion to a class of "nonlinear" mc-quasimetrics that is not reducible to the above discussed ones.

(B) Let the locally Riemann integrable function $a : R_+ \to R_+^0$ be given and $A : R_+ \to R_+$ stand for its *primitive*:

$$A(t) = \int_0^t a(s) \mathrm{d}s, \ t \in R_+.$$

Suppose in the following that

(norm) (a, A) is *normal*: $a(.)$ is decreasing and $A(\infty) = \infty$.

In particular, we have the (integral) representation

(int-rep) $\int_p^q a(\xi) \mathrm{d}\xi = (q-p) \int_0^1 a(p + \tau(q-p)) \mathrm{d}\tau$, when $0 \le p < q < \infty$.

Some basic facts involving this couple are being collected in the statement below.

Theorem 7. *The following are valid:*

(71-a) $A(.)$ *is a continuous order isomorphism of* R_+*; hence, so is* $A^{-1}(.)$,

(71-b) $a(s) \le (A(s) - A(t))/(s-t) \le a(t), \forall t, s \in R_+, t < s$,

(71-c) $A(.)$ *is almost concave:*
$t \mapsto [A(t+s) - A(t)]$ *is decreasing on* $R_+, \forall s \in R_+$,

(71-d) $A(.)$ *is concave:* $A(t + \lambda(s-t)) \ge A(t) + \lambda(A(s) - A(t))$,
for all $t, s \in R_+$ *with* $t < s$ *and all* $\lambda \in [0, 1]$,

(71-e) $A(.)$ *is sub-additive (hence* $A^{-1}(.)$ *is super-additive).*

The proof is immediate, by the normality conditions above. Note that the properties (71-c) and (71-d) are equivalent to each other, under (71-a). This follows at once from the (nondifferential) mean value theorem in Bantaş and Turinici [8].

Now, let X be some nonempty set and $g : X \times X \to R_+$ be a metric over it. Further, let the map $\Gamma : X \to R_+$ be taken as

(a-nexp) Γ is g-*nonexpansive*: $|\Gamma(x) - \Gamma(y)| \leq g(x, y)$, $\forall x, y \in X$.

Define a pseudometric $b := [g; A, \Gamma]$ over X as

(b-expli) $b(x, y) = A(\Gamma(x) + g(x, y)) - A(\Gamma(x))$, $x, y \in X$.

This may be viewed as an "explicit" formula; the "implicit" version of it is

(b-impli) $g(x, y) = A^{-1}(A(\Gamma(x)) + b(x, y)) - \Gamma(x)$, $x, y \in X$.

We establish some properties of this map, useful in the sequel.

First, the "metrical" nature of $(x, y) \mapsto b(x, y)$ is of interest.

Theorem 8. *The mapping $(x, y) \mapsto b(x, y)$ is a quasimetric over X.*

Proof. The reflexivity and sufficiency are clear, by Theorem 7 (first part), so, it remains to establish the triangular property. Let $x, y, z \in X$ be arbitrary fixed. The triangular property of $g(.,.)$ yields [via Theorem 7 (first part)]

$$b(x, z) \leq A(\Gamma(x) + g(x, y) + g(y, z)) - A(\Gamma(x) + g(x, y)) + b(x, y).$$

On the other hand, the g-nonexpansiveness of Γ gives

$\Gamma(x) + g(x, y) \geq \Gamma(y)$, so (by Theorem 7 (third part))
$A(\Gamma(x) + g(x, y) + g(y, z)) - A(\Gamma(x) + g(x, y)) \leq b(y, z).$

Combining with the previous relation yields our desired conclusion. $\square$

By definition, $b(.)$ is the *Zhong quasimetric* attached to the triple (g, A, Γ).

A natural problem to be posed is that of establish whether $b(.,.)$ is effectively a quasimetric. A positive answer to this is available, in the following sense:

$b(.,.)$ is asymmetric (hence, not a metric) for many choices of $(g; A, \Gamma)$.

This is shown from the following:

Example 1.
Suppose that

(con-1) $a(.)$ is strictly decreasing and continuous,
(con-2) Γ is strictly g-nonexpansive:
$|\Gamma(x) - \Gamma(y)| < g(x,y)$, $\forall x, y \in X$, $x \neq y$,
(con-3) $\Gamma(X)$ is not a singleton.

From the last condition, there exist $u, v \in X$ with

$(u \neq v$ and) $\Gamma(v) < \Gamma(u)$; hence,
$\Gamma(v) < \Gamma(u) < \Gamma(v) + g(u,v) < \Gamma(u) + g(u,v)$,

if one takes the strict nonexpansive property of Γ into account. We now claim that $b(u,v) = b(v,u)$ is impossible. Suppose, by absurd, that $b(u,v) = b(v,u)$. From the very definition of our quasimetric, one gets

$A(\Gamma(u) + g(u,v)) - A(\Gamma(u)) = A(\Gamma(v) + g(u,v)) - A(\Gamma(v))$,
that is,
$A(\Gamma(u) + g(u,v)) - A(\Gamma(v) + g(u,v)) = A(\Gamma(u)) - A(\Gamma(v))$.

This, by the integral representation (int-rep), yields (after simplifications)

$\int_0^1 a(\Gamma(v) + g(u,v) + \tau(\Gamma(u) - \Gamma(v)))d\tau = \int_0^1 a(\Gamma(v) + \tau(\Gamma(u) - \Gamma(v)))d\tau$,
or, equivalently,
$\int_0^1 [a(\Gamma(v) + g(u,v) + \tau(\Gamma(u) - \Gamma(v))) - a(\Gamma(v)) + \tau(\Gamma(u) - \Gamma(v)))] d\tau = 0$.

Combining with the strict decreasing continuous property of $a(.)$ and the Lagrange mean value theorem gives a contradiction and proves our initial assertion.

The following properties of the couple (g, b) are immediate (via Theorem 7):

Proposition 12. *Under the prescribed conventions,*

(71-1) $(\forall x, y \in X)$: $a(\Gamma(x) + g(x,y))g(x,y) \leq b(x,y) \leq a(\Gamma(x))g(x,y)$,

(71-2) $(\forall x, y \in X)$: $b(x,y) \leq a(0)g(x,y)$; hence, $d(x,y) \leq a(0)g(x,y)$.

(C) We may now pass to the announced answer to our problem. To reach the desired conclusions, some preliminaries are needed.

As already precise, $A^{-1}(.)$ is a homeomorphism of R_+ to itself. In particular, given the (nonempty) compact M of R_+,

$A^{-1}: M \to R_+$ is continuous on M and $A^{-1}(M) =$ compact.

By a well-known result, A^{-1} is uniformly continuous on M; hence, the function $\Phi = \Phi(A^{-1}, M)$ in $\mathcal{F}(R_+)$ introduced as

$$\Phi(\eta) = \sup\{|A^{-1}(t) - A^{-1}(s)|; t, s \in M, |t-s| \leq \eta\}, \eta \in R_+$$

is well defined, with the properties

(p-1) Φ is increasing and zero-continuous: $\Phi(0+) = 0 = \Phi(0)$
(p-2) $|A^{-1}(t) - A^{-1}(s)| \leq \Phi(|t-s|)$, for all $t, s \in M$;

it is referred to as the *uniform continuity indicator function* attached to (A^{-1}, M).

Proposition 13. *The Cauchy properties attached to the triple (b, g, d) are identical, in the sense that*

$(\forall(x_n))$: (x_n) is b-Cauchy $\iff$ (x_n) is g-Cauchy $\iff$ (x_n) is d-Cauchy.

Proof. Clearly, it suffices proving that

$(\forall(x_n))$: (x_n) is b-Cauchy $\implies$ (x_n) is g-Cauchy $\implies$ (x_n) is d-Cauchy.

The second half of this inequality chain is immediate, via Proposition 12 (second part), so, it remains to establish the first half of the same. Assume that (x_n) is a b-Cauchy sequence in X. As $b =$ triangular,

$b(x_i, x_j) \leq \mu$, for all (i,j) with $i \leq j$, and some $\mu > 0$.

This, along with the implicit definition of b, yields

$$g(x_0, x_i) = A^{-1}(A(\Gamma(x_0)) + b(x_0, x_i)) - \Gamma(x_0)$$
$$\leq A^{-1}(A(\Gamma(x_0)) + \mu) - \Gamma(x_0), \forall i \geq 0,$$

wherefrom (by the choice of Γ)

$$\Gamma(x_i) \leq \Gamma(x_0) + g(x_0, x_i) \leq A^{-1}(A(\Gamma(x_0)) + \mu)$$
(hence $A(\Gamma(x_i)) \leq A(\Gamma(x_0)) + \mu$), for all $i \geq 0$.

Putting these facts together yields (again via implicit definition of b)
$$\Gamma(x_i) + g(x_i, x_j) = A^{-1}(A(\Gamma(x_i)) + b(x_i, x_j))$$
$$\leq \nu := A^{-1}(A(\Gamma(x_0)) + 2\mu), \text{ for all } (i,j) \text{ with } i \leq j.$$

And this, via Proposition 12 (first part), gives (for the same pairs (i,j))
$$b(x_i, x_j) \geq a(\Gamma(x_i) + g(x_i, x_j))g(x_i, x_j) \geq a(\nu)g(x_i, x_j).$$

But then, the g-Cauchy property of (x_n) is clear, and the proof is complete. □

For the moment, the only condition imposed upon Γ was its g-nonexpansive property. In the following, a stronger condition is needed, namely

(cond) Γ is (g-nonexpansive and) bounded: $\gamma := \Gamma(X) < \infty$.

A useful example of this type is that of

($\Gamma(x) = 1/(1 + \Delta(x))$, $x \in X$), where $\Delta : X \to R_+$ is g-nonexpansive;

since the verification is immediate, we do not give details.

Proposition 14. *Suppose that Γ is taken as in (cond). Then, the convergence properties attached to the triple (b, g, d) are identical:*

$$(\forall (x_n), \forall x): x_n \xrightarrow{b} x \iff x_n \xrightarrow{g} x \iff x_n \xrightarrow{d} x.$$

Proof. Clearly, it suffices proving that

$$(\forall (x_n), \forall x): x_n \xrightarrow{b} x \implies x_n \xrightarrow{g} x \implies x_n \xrightarrow{d} x.$$

The second half of this inequality chain is immediate, via Proposition 12 (second part), so, it remains to establish the first half of the same. Assume that (x_n) and x are such that $x_n \xrightarrow{b} x$. We claim that

$b(x_i, x) \leq \mu$, for all i, and some $\mu > 0$.

In fact, by the convergence property, it follows that, given $\alpha > 0$, there exists some index $p = p(\alpha)$, such that

(con-1) $b(x_q, x) \leq \alpha$, for all $q \geq p$.

On the other hand, evidently,

(con-2) $\beta := \sup\{b(x_i, x_p); i \leq p\} < \infty$,

and this gives (by the triangular property)

$$b(x_i, x) \leq b(x_i, x_p) + b(x_p, x) \leq \alpha + \beta, \quad \forall i \leq p,$$

wherefrom, the desired relation follows with $\mu = \alpha + \beta$.

Combining with the boundedness property of Γ yields

$$0 \leq A(\Gamma(x_i)) \leq A(\Gamma(x_i)) + b(x_i, x) \leq A(\gamma) + \mu, \quad \forall i.$$

Put $M = [0, A(\gamma) + \mu)]$, and let $\Phi = \Phi(A^{-1}, M)$ be the uniform continuity indicator function attached to (A^{-1}, M). By the implicit definition of $b(.)$, one gets

$$g(x_i, x) = A^{-1}(A(\Gamma(x_i)) + b(x_i, x)) - \Gamma(x_i)$$
$$= A^{-1}(A(\Gamma(x_i)) + b(x_i, x)) - A^{-1}(A(\Gamma(x_i))) \leq \Phi(b(x_i, x)), \quad \forall i,$$

and this, combined with the initial choice of (x_n), yields $x_n \xrightarrow{g} x$, as desired. □

Finally, we show that, under the same assumptions as above, one may give a positive answer to the asymptotic question.

Proposition 15. *Suppose that Γ is taken as in (cond). Then, the asymptotic properties attached to the triple (b, g, d) are identical:*

$(\forall(x_n))$: (x_n) *is b-asymptotic* $\iff$ (x_n) *is g-asymptotic*
$\iff$ (x_n) *is d-asymptotic.*

Proof. Clearly, it suffices proving that

$(\forall(x_n))$: (x_n) is b-asymptotic $\implies$ (x_n) is g-asymptotic
$\implies$ (x_n) is d-asymptotic.

The second half of this inequality chain is immediate, via Proposition 12 (second part), so, it remains to establish the first half of the same. Assume that (x_n) is b-asymptotic. Clearly,

$b(x_i, x_{i+1}) \leq \mu$, for all i, and some $\mu > 0$.

Combining with the boundedness property of Γ yields

$$0 \leq A(\Gamma(x_i)) \leq A(\Gamma(x_i)) + b(x_i, x_{i+1}) \leq A(\gamma) + \mu, \quad \forall i.$$

Put $M = [0, A(\gamma) + \mu)]$, and let $\Phi = \Phi(A^{-1}, M)$ be the uniform continuity indicator function attached to (A^{-1}, M). By the implicit definition of $b(.)$, one gets

$$\begin{aligned} g(x_i, x_{i+1}) &= A^{-1}(A(\Gamma(x_i)) + b(x_i, x_{i+1})) - \Gamma(x_i) \\ &= A^{-1}(A(\Gamma(x_i)) + b(x_i, x_{i+1})) - A^{-1}(A(\Gamma(x_i))) \\ &\leq \Phi(b(x_i, x_{i+1})), \quad \forall i, \end{aligned}$$

and this, via initial choice of (x_n), yields $(x_n) = g$-asymptotic, as desired. □

Putting these together, one gets the following synthetic result. Let the couple (a, A) be taken as before and $g(.,.)$ be a metric over X. Let in addition $\Gamma : X \to R_+$ be a g-nonexpansive bounded mapping.

Theorem 9. *Under the described setting, the pseudometric $b(.,.)$ attached to (g, A, Γ) is an mc-quasimetric.*

(D) Now, by simply combining this with our preceding developments, one gets the following "nonlinear" variant of our main result above. Let the data (a, A) and (g, Γ) be taken as before and $b = b(g, A, \Gamma)$ stand for the attached mc-quasimetric on X. Then, let $\mathcal{G} \subseteq X \times X$ be a relation over X. In addition to these, let $T \in \mathcal{F}(X)$ be a selfmap of X with

(aoua-reg) T is *(aoua)-regular*: aoua(T) is nonempty.

According to it, there exist ascending orbital uadmissible sequences (in short (aoua)-sequences) in X; let $X_0 = (x_n)$ be one of these. Note that, without loss, one may arrange for the extra property being fulfilled:

(B-asc) $X_0 = (x_n)$ is $(B_1 > 0)$-ascending: $B_1(x_n, x_{n+1}) > 0$, for all n;

this is referred to as $X_0 = (x_n)$ is an *(Baoua)-sequence*. Denote also

$[X_0] = \{x_n; n \geq 0\}$, $[[X_0]] = \text{cl}([X_0])$.

The following variant of our main result over such structures is now available.

Theorem 10. *Assume that the selfmap T is $(\mathcal{G}; (B_1 > 0); \Upsilon)$-contractive, where the subset $\Upsilon \in \exp(R_+^6)$ and the relation $(B_1 > 0)$ over X were introduced under the precise conventions. In addition,* suppose that X is *(Baoua-f,b)-complete*. Then, the following are valid:

(74-a) $X_0 = (x_n)$ is a *(Baoua-f)* sequence whenever Υ is descending on $[X_0]$,
(74-b) $X_0 = (x_n)$ is *(Baoua-f)* and semi Picard (modulo $(b, \mathcal{G}; T)$), whenever (in addition) Υ is nsright on $[X_0]$,
(74-c) $X_0 = (x_n)$ is *(Baoua-f)* and Picard (modulo $(b, \mathcal{G}; T)$), whenever (in addition to the setting of (74-a)+(74-b)) Υ is almost ntright on $[X_0]$,
(74-d) $X_0 = (x_n)$ is *(Baoua-f)* and strongly Picard (modulo $(b, \mathcal{G}; T)$), whenever (in addition to the setting of (74-a)+(74-b)) T is *(Baoua-f,b)-continuous*,
(74-e) $X_0 = (x_n)$ is *(Baoua-f)* and semi Bellman Picard (modulo $(b, \mathcal{G}; T)$), whenever (in addition to the setting of (74-a)+(74-b)) $\mathcal{G}$ is *(Baoua-f,b)-almost-selfclosed* and Υ is npoint on $[[X_0]]$.

A corresponding version of this result with respect to the Ahmed–Fulga statement we just exposed is now immediate; so, further details are not needed.

References

[1] M. Abtahi, Fixed point theorems for Meir-Keeler type contractions in metric spaces, *Fixed Point Theory* **17**, 225–236 (2016).
[2] R. P. Agarwal, M. A. El-Gebeily and D. O'Regan, Generalized contractions in partially ordered metric spaces, *Appl. Anal.* **87**, 109–116 (2008).
[3] R. P. Agarwal, E. Karapinar and A. Roldán, Fixed point theorems in quasi-metric spaces and applications to multidimensional fixed point theorems on G-metric spaces, *J. Nonlinear Convex Anal.* **2014**, 36 (2014).
[4] H. H. Alsulami, E. Karapinar, F. Khojasteh and A. Roldán, A proposal to the study of contractions in quasi metric spaces, *Discrete Dyn. Nat. Soc.* **2014** (2014). Article ID: 269286.
[5] A. El-Sayed Ahmed and A. Fulga, The Gornicki-Proinov type contraction on quasi-metric spaces, *AIMS Math.* **6**, 8815–8834 (2021).
[6] H. Argoubi, B. Samet and C. Vetro, Nonlinear contractions involving simulation functions in a metric space with a partial order, *J. Nonlinear Sci. Appl.* **8**, 1082–1094 (2015).

[7] S. Banach, Sur les opérations dans les ensembles abstraits et leur application aux équations intégrales. *Fund. Math.* **3**, 133–181 (1922).

[8] G. Bantaş and M. Turinici, Mean value theorems via division methods, *An. Şt. Univ. "Al. I. Cuza" Iaşi Mat.* **40**, 135–150 (1994).

[9] R. Batra and S. Vashistha, Fixed points of an F-contraction on metric spaces with a graph, *Int. J. Comput. Math.* **91**, 2483–2490 (2014).

[10] I. Beg, A. R. Butt and S. Radojević, The contraction mapping principle for set valued mappings on a metric space with a graph, *Comput. Math. Appl.* **60**, 1214–1219 (2010).

[11] P. Bernays, A system of axiomatic set theory: Part III. Infinity and enumerability analysis, *J. Symb. Log.* **7**, 65–89 (1942).

[12] M. Berzig, Coincidence and common fixed point results on metric spaces endowed with an arbitrary binary relation and applications, *J. Fixed Point Theory Appl.* **12**, 221–238 (2013).

[13] D. W. Boyd and J. S. W. Wong, On nonlinear contractions, *Proc. Am. Math. Soc.* **20**, 458–464 (1969).

[14] F. Bojor, Fixed points of Kannan mappings in metric spaces endowed with a graph, *An. Şt. Univ. Ovidius Constanţa Mat.* **20**, 31–40 (2012).

[15] S. Chandok, B. S. Choudhury and N. Metiya, Fixed point results in ordered metric spaces for rational type expressions with auxiliary functions, *J. Egypt. Math. Soc.* **23**, 95–101 (2015).

[16] P. Chakraborty, B. S. Choudhury and M. De la Sen, Relation theoretic fixed point theorems for generalized weakly contractive mappings, *Symmetry* **12**, 29 (2020).

[17] C. M. Chen, E. Karapinar, V. Rakočević, Existence of periodic fixed point theorems in the setting of generalized quasi-metric spaces, *J. Appl. Math.* **2014** (2014). Article ID: 353765.

[18] S. Cobzaş, Fixed points and completeness in metric and in generalized metric spaces (2019). Arxiv 1508-05173-v5. 7 October 2019.

[19] P. J. Cohen, *Set Theory and the Continuum Hypothesis* (Benjamin, New York, 1966).

[20] J. Gornicki, Remarks on asymptotic regularity and fixed points, *J. Fixed Point Theory Appl.* **21** (2019). Art No: 29.

[21] J. Harjani, B. Lopez and K. Sadarangani, A fixed point theorem for mappings satisfying a contractive condition of rational type on a partially ordered metric space, *Abstr. Appl. Anal.* **2010** (2010). Article ID: 190701.

[22] P. Hitzler, Generalized metrics and topology in logic programming semantics, PhD Thesis, National University of Ireland, University College Cork (2001).

[23] J. Jachymski, The contraction principle for mappings on a metric space with a graph, *Proc. Am. Math. Soc.* **136**, 1359–1373 (2008).

[24] E. Karapinar, M. De la Sen and A. Fulga, A Note on the Gornicki-Proinov type contraction, *J. Funct. Spaces* **2021** (2021). Article ID: 6686644.
[25] E. Karapinar and A. Fulga, On hybrid contractions via simulation function in the context of quasi-metric spaces, *J. Nonlinear Convex Anal.* **21**, 2115–2124 (2020).
[26] E. Karapinar, A. Fulga and S. S. Yeşilkaya, Fixed points of Proinov type multivalued mappings on quasimetric spaces, *J. Funct. Spaces* **2022** (2022). Article ID: 7197541.
[27] E. Karapinar, P. Kumam and P. Salimi, On $\alpha - \psi$-Meir-Keeler contractive mappings, *Fixed Point Theory Appl.* **2013**, 94 (2013).
[28] E. Karapinar, A. F. Roldan and B. Samet, Matkowski theorems in the context of quasi-metric spaces and consequences on G-metric spaces, *An. Şt. Univ. Ovidius Constanţa Mat.* **24**, 309–333 (2016).
[29] E. Karapinar, S. Romaguera and P. Tirado, Contractive multivalued maps in terms of Q-functions on complete quasimetric spaces, *Fixed Point Theory Appl.* **2014**, 53 (2014).
[30] S. Kasahara, On some generalizations of the Banach contraction theorem, *Publ. Res. Inst. Math. Sci. Kyoto Univ.* **12**, 427–437 (1976).
[31] F. Khojasteh and V. Rakočević, Some new common fixed point results for generalized contractive multi-valued non-self-mappings, *Appl. Math. Lett.* **25**, 287–293 (2012).
[32] F. Khojasteh, S. Shukla and S. Radenović, Formulization of many contractions via the simulation functions (2013). *Arxiv* 1109-3021-v2. 13 August 2013.
[33] S. Leader, Fixed points for general contractions in metric spaces, *Math. Jpn.* **24**, 17–24 (1979).
[34] J. Matkowski, Integrable solutions of functional equations, *Diss. Math.* (Polish Scientific Publishers, Warsaw) **127** (1975).
[35] A. Meir and E. Keeler, A theorem on contraction mappings, *J. Math. Anal. Appl.* **28**, 326–329 (1969).
[36] G. H. Moore, *Zermelo's Axiom of Choice: Its Origin, Development and Influence* (Springer, New York, 1982).
[37] Y. Moskhovakis, *Notes on Set Theory* (Springer, New York, 2006).
[38] J. J. Nieto and R. Rodriguez-Lopez, Contractive mapping theorems in partially ordered sets and applications to ordinary differential equations, *Order* **22**, 223–239 (2005).
[39] P. D. Proinov, Fixed point theorems for generalized contractive mappings in metric spaces, *J. Fixed Point Theory Appl.* **22** (2020). Art No: 21.

[40] A. C. M. Ran and M. C. Reurings, A fixed point theorem in partially ordered sets and some applications to matrix equations, *Proc. Am. Math. Soc.* **132**, 1435–1443 (2004).
[41] S. Reich, Fixed points of contractive functions, *Boll. Un. Mat. Ital.* **5**, 26–42 (1972).
[42] B. E. Rhoades, A comparison of various definitions of contractive mappings, *Trans. Am. Math. Soc.* **226**, 257–290 (1977).
[43] A. Roldán, E. Karapinar and M. De La Sen, Coincidence point theorems in quasi-metric spaces without assuming the mixed monotone property and consequences in G-metric spaces, *Fixed Point Theory Appl.* **2014**, 184 (2014).
[44] A.-F. Roldán, E. Karapinar, C. Roldán and J. Martinez-Moreno, Coincidence point theorems on metric spaces via simulation method, *Fixed Point Theory Appl.* **2015**, 98 (2015).
[45] A. F. Roldán and N. Shahzad, From graphical metric spaces to fixed point theory in binary related distance spaces, *Filomat* **31**, 3209–3231 (2017).
[46] I. A. Rus, *Generalized Contractions and Applications* (Cluj University Press, Cluj-Napoca, 2001).
[47] B. Samet and M. Turinici, Fixed point theorems on a metric space endowed with an arbitrary binary relation and applications, *Commun. Math. Anal.* **13**, 82–97 (2012).
[48] B. Samet, C. Vetro and P. Vetro, Fixed point theorems for $(\alpha - \psi)$-contractive type mappings, *Nonlinear Anal. Theory Methods Appl.* **75**, 2154–2165 (2012).
[49] E. Schechter, *Handbook of Analysis and Its Foundation* (Academic Press, New York, 1997).
[50] A. Tarski, Axiomatic and algebraic aspects of two theorems on sums of cardinals, *Fund. Math.* **35**, 79–104 (1948).
[51] M. Turinici, Fixed points of implicit contraction mappings, *An. Şt. Univ. "Al. I. Cuza" Iaşi Mat.* **22**, 177–180 (1976).
[52] M. Turinici, Abstract comparison principles and multivariable Gronwall-Bellman inequalities, *J. Math. Anal. Appl.* **117**, 100–127 (1986).
[53] M. Turinici, Pseudometric versions of the Caristi-Kirk fixed point theorem, *Fixed Point Theory* (Cluj-Napoca) **5**, 147–161 (2004).
[54] M. Turinici, Function pseudometric VP and applications, *Bul. Inst. Polit. Iaşi (Sect. Mat. Mec. Teor. Fiz.)* **53**(57), 393–411 (2007).
[55] M. Turinici, Ran-Reurings theorems in ordered metric spaces, *J. Indian Math. Soc.* **78**, 207–214 (2011).

[56] M. Turinici, Implicit contractive maps in ordered metric spaces. In T. M. Rassias and L. Tóth (eds.) *Topics in Mathematical Analysis and Applications* (Springer International Publishing, Switzerland, 2014), pp. 715–746.

[57] M. Turinici, Contraction maps in pseudometric structures. In T. M. Rassias and P. M. Pardalos (eds.) *Essays in Mathematics and Its Applications* (Springer International Publishing, Switzerland, 2016), pp. 513–562.

[58] M. Turinici, *Modern Directions Metrical Fixed Point Theory*, Revised edn. (Pim Editorial House, Iași, 2023).

[59] E. S. Wolk, On the principle of dependent choices and some forms of Zorn's lemma, *Canad. Math. Bull.* **26**, 365–367 (1983).

© 2024 World Scientific Publishing Company
https://doi.org/10.1142/9789811267048_0029

Chapter 29

Hyers–Ulam–Rassias Stability of the Nonlinear Fractional Differential Equations with ρ-Fractional Derivative

Chun Wang

Department of Mathematics, Changzhi University, Changzhi City, China

wangchun12001@163.com

In this chapter, the Hyers–Ulam–Rassias stability of the nonlinear fractional differential equations with the ρ-fractional derivative on the continuous function space is investigated by using the weighted space method. Some sufficient conditions are obtained in order that the nonlinear fractional differential equations are stable on the continuous function space. The results improve and extend some recent results.

1. Introduction

In recent years, many researchers have focused on the study of the Hyers–Ulam–Rassias stability of the differential equations and systems. Cădariu and Găvruţa proposed the weighted space method for the stability study of some nonlinear equations [6, 7]. By using the Aboodh transform method, Murali investigated the Hyers–Ulam stability, Hyers–Ulam–Rassias stability, Mittag–Leffler–Hyers–Ulam stability and Mittag–Leffler–Hyers–Ulam–Rassias stability of the second-order linear differential equations [13]. Mohanapriya *et al.* presented the results on the Mittag–Leffler–Hyers–Ulam and Mittag–Leffler–Hyers–Ulam–Rassias stability of linear differential equations of first, second and nth order by the Fourier transform method [12]. By using the Laplace transform method (via the Wright function), Liu *et al.* [11] investigated the Hyers–Ulam stability of linear Caputo–Fabrizio fractional differential equations with Mittag–Leffler kernel. Existence, uniqueness and generalized Hyers–Ulam–Rassias stability results for nonlinear problems are established. Alam and Shah [4] investigated the existence, uniqueness and stability of coupled implicit fractional integro-differential equations with Riemann–Liouville derivatives and they also presented different types of stabilities by using the classical technique of functional analysis. Zhou *et al.* [15] studied a nonlinear ψ-Hilfer fractional integro-differential equation on finite interval $[a, b]$. Sufficient conditions for the existence of solution and stability for the initial value problem are obtained, and the Ulam–Hyers–Rassias stability, Ulam–Hyers stability and semi-Ulam–Hyers–Rassias stability of the system are discussed. Ahmadova [3] established stability results in Ulam–Hyers sense for the nonlinear fractional stochastic neutral differential equations system with the aid of weighted maximum norm and Itô's isometry in finite-dimensional stochastic settings. Vu and Hoa [14] investigated the Hyers–Ulam stability and the Hyers–Ulam–Rassias stability of fuzzy fractional Volterra integral equations involving the kernel ψ-function. They proposed conditions for the existence, uniqueness and stability of FFVIE by employing the method of successive approximation. Andersona and Onitsuka [5] clarified the Hyers–Ulam stability of certain first-order linear constant coefficient dynamic equations on time scales. Zada *et al.* [16] proved the Hyers–Ulam stability and the Hyers–Ulam–Rassias stability of a class of higher-order nonlinear delay differential equations with multiple bounded variable delays on a compact interval. Abdo *et al.* [2] investigated the qualitative theory for nonlinear fractional functional differential equations of arbitrary order with infinite delay involving generalized Hilfer fractional derivative. Some new and recent results of existence and

Ulam–Hyers–Mittag–Leffler stability of solution for the proposed problem are also discussed.

A new simple well-behaved fractional calculus is investigated by Abdeljawad [1]. The author proposed and discussed the fractional version of chain rule, exponential functions, Gronwall's inequality, integration by parts, Laplace transforms and linear differential systems. A new fractional integral is presented by Katugampola [8], which generalized the Riemann–Liouville and Hadamard fractional integrals into a single form. The author gave some conditions for such a fractional integration operator to be bounded in an extended Lebesgue measurable space and proved the semigroup property for the integral operator. Katugampola [9] also presented a new fractional derivative which generalized the familiar Riemann–Liouville and Hadamard fractional derivatives to a single form. These fractional integrals and derivatives may be called the generalized fractional integrals and the generalized fractional derivatives, or ρ-fractional integrals and ρ-fractional derivatives, respectively.

However, as far as we know, no work has been done on the Hyers–Ulam–Rassias stability of the nonlinear fractional differential equations and systems with the ρ-fractional derivatives.

This chapter deals with the Hyers–Ulam–Rassias stability of the following nonlinear fractional Cauchy type problem

$$\left(_aD^{\alpha,\rho}\right)(t) = f(t, y(t)), \quad \alpha > 1, \quad t \in [a, b] \tag{1}$$

with initial conditions

$$\left(_aD^{\alpha-k,\rho}y\right)|_{t=a} = b_k, \quad b_k \in \mathbb{R} \ (k = 1, 2, \ldots, n-1), \quad b_n = 0, \tag{2}$$

where $f(t, y(t)) \in C_\gamma[a,b], 0 \le \gamma < 1$, and

$$\left(_aD^{\alpha-k,\rho}y\right)|_{t=a} = \lim_{t \to a^+}\left(_aD^{\alpha-k,\rho}y\right)(t), 1 \le k \le n-1,$$

$$b_n = \lim_{t \to a^+}\left(_aI^{n-\alpha,\rho}y\right)(t), n = -[-\alpha],$$

$_aD^{\alpha,\rho}$ and $_aI^{\alpha,\rho}$ are ρ-fractional derivative and integral of order α, respectively.

2. Preliminary

In this section, some definitions and lemmas are presented which are used further for our main results in this chapter.

Definition 1. The ρ-left fractional integral of order α of a function v is defined by

$$\left(_aI^{\alpha,\rho}v\right)(t) = \frac{1}{\Gamma(\alpha)} \int_a^t \left(\frac{t^\rho - s^\rho}{\rho}\right)^{\alpha-1} v(s) \frac{\mathrm{d}s}{s^{1-\rho}},$$

where $\rho > 0$.

Definition 2. The ρ-left fractional derivative of order $\alpha > 0$ of a function v is defined by

$$\left(_aD^{\alpha,\rho}v\right)(t) = \gamma^n \left(_aI^{n-\alpha,\rho}v\right)(t)$$
$$= \frac{\gamma^n}{\Gamma(n-\alpha)} \int_a^t \left(\frac{t^\rho - s^\rho}{\rho}\right)^{n-\alpha-1} v(s) \frac{\mathrm{d}s}{s^{1-\rho}},$$

where $\rho > 0, n = [\Re(\alpha)] + 1$ and $\gamma = t^{1-\rho}\frac{\mathrm{d}}{\mathrm{d}t}$.

It should be mentioned that when $\rho = 1$, the integral and the derivative become the Riemann–Liouville fractional integral and derivative, respectively. And in case one takes the limit as $\rho \to 0$, the ρ-fractional integral and derivative become the Hadamard fractional integral and derivative, respectively.

Let $[a, b]$ be a finite interval and $AC[a, b]$ be the set of absolute continuous functions on $[a, b]$. Then, we define function space

$$AC_\rho^n[a, b] = \left\{y : [a, b] \to \mathbb{C} \text{ and } \left(t^{1-\rho}\frac{\mathrm{d}}{\mathrm{d}t}\right)^{n-1} y(t) \in AC[a, b]\right\}.$$

We also define the function space

$$C_{\gamma,\rho}[a, b] = \{f : (a, b] \to \mathbb{R}, \ (t^\rho - a^\rho)^\gamma f(t) \in C[a, b]\}$$

endowed with the norm $\|f\|_{C_{\gamma,\rho}} = \|(t^\rho - a^\rho)^\gamma f(t)\|_{C[a,b]}$.

Lemma 1. If $\Re(\alpha) > 0, n = -[-\Re(\alpha)], y(t) \in L[a,b]$ and ${}_aI^{\alpha,\rho}y \in AC^n_\rho[a,b]$, then

$$\frac{1}{\Gamma(\alpha+1)} \int_a^t \left(\frac{t^\rho - s^\rho}{\rho}\right)^\alpha \frac{d}{ds}\left[\left(s^{1-\rho}\frac{d}{ds}\right)^{n-1} {}_aI^{n-\alpha,\rho}y\right](s)ds$$

$$= \frac{1}{\Gamma(\alpha-n+1)} \int_a^t \left(\frac{t^\rho - s^\rho}{\rho}\right)^{\alpha-n} {}_aI^{n-\alpha,\rho}y(s)\frac{ds}{s^{1-\rho}}$$

$$- \sum_{k=1}^n \frac{{}_aI^{k-\alpha,\rho}y(a^+)}{\Gamma(\alpha+2-k)} \left(\frac{t^\rho - a^\rho}{\rho}\right)^{\alpha-k+1}.$$

Proof. It is easy to get

$$\frac{1}{\Gamma(\alpha+1)} \int_a^t \left(\frac{t^\rho - s^\rho}{\rho}\right)^\alpha \frac{d}{ds}\left[\left(s^{1-\rho}\frac{d}{ds}\right)^{n-1} {}_aI^{n-\alpha,\rho}y\right](s)ds$$

$$= \frac{1}{\Gamma(\alpha+1)} \int_a^t \left(\frac{t^\rho - s^\rho}{\rho}\right)^\alpha d\left[\left(s^{1-\rho}\frac{d}{ds}\right)^{n-1} {}_aI^{n-\alpha,\rho}y\right](s).$$

Applying integration by parts, we get

$$\frac{1}{\Gamma(\alpha+1)} \int_a^t \left(\frac{t^\rho - s^\rho}{\rho}\right)^\alpha d\left[\left(s^{1-\rho}\frac{d}{ds}\right)^{n-1} {}_aI^{n-\alpha,\rho}y\right](s)$$

$$= \frac{1}{\Gamma(\alpha+1)} \Bigg\{ -\left(\frac{t^\rho - a^\rho}{\rho}\right)^\alpha \left[\left(s^{1-\rho}\frac{d}{ds}\right)^{n-1} {}_aI^{n-\alpha,\rho}y\right](a^+)$$

$$+ \alpha \int_a^t \left(\frac{t^\rho - s^\rho}{\rho}\right)^{\alpha-1} d\left[\left(s^{1-\rho}\frac{d}{ds}\right)^{n-2} {}_aI^{n-\alpha,\rho}y\right](s)\Bigg\}.$$

Applying integration by parts for the second term, we get

$$\int_a^t \left(\frac{t^\rho - s^\rho}{\rho}\right)^{\alpha-1} d\left[\left(s^{1-\rho}\frac{d}{ds}\right)^{n-2} {}_aI^{n-\alpha,\rho}y\right](s)$$

$$= -\left(\frac{t^\rho - a^\rho}{\rho}\right)^{\alpha-1} \left[\left(s^{1-\rho}\frac{d}{ds}\right)^{n-2} {}_aI^{n-\alpha,\rho}y\right](a^+)$$

$$+ (\alpha-1) \int_a^t \left(\frac{t^\rho - s^\rho}{\rho}\right)^{\alpha-2} d\left[\left(s^{1-\rho}\frac{d}{ds}\right)^{n-3} {}_aI^{n-\alpha,\rho}y\right](s).$$

Applying the same produce n times, we get

$$\frac{1}{\Gamma(\alpha+1)} \int_a^t \left(\frac{t^\rho - s^\rho}{\rho}\right)^\alpha \frac{\mathrm{d}}{\mathrm{d}s}\left[\left(s^{1-\rho}\frac{\mathrm{d}}{\mathrm{d}s}\right)^{n-1} {_aI^{n-\alpha,\rho}y}\right](s)\mathrm{d}s$$

$$= \frac{1}{\Gamma(\alpha+1)}\left\{ -\left(\frac{t^\rho - a^\rho}{\rho}\right)^\alpha \left[\left(s^{1-\rho}\frac{\mathrm{d}}{\mathrm{d}s}\right)^{n-1} {_aI^{n-\alpha,\rho}y}\right](a^+) \right.$$

$$-\alpha\left(\frac{t^\rho - a^\rho}{\rho}\right)^{\alpha-1} \left[\left(s^{1-\rho}\frac{\mathrm{d}}{\mathrm{d}s}\right)^{n-2} {_aI^{n-\alpha,\rho}y}\right](a^+)$$

$$-\alpha(\alpha-1)\left(\frac{t^\rho - a^\rho}{\rho}\right)^{\alpha-2} \left[\left(s^{1-\rho}\frac{\mathrm{d}}{\mathrm{d}s}\right)^{n-3} {_aI^{n-\alpha,\rho}y}\right](a^+)$$

$$- \cdots$$

$$- \alpha(\alpha-1)(\alpha-2)\ldots(\alpha-(n-2))$$
$$\cdot \left(\frac{t^\rho - a^\rho}{\rho}\right)^{\alpha-(n-1)} \left[\left(s^{1-\rho}\frac{\mathrm{d}}{\mathrm{d}s}\right)^{n-n} {_aI^{n-\alpha,\rho}y}\right](a^+)$$

$$+ \alpha(\alpha-1)(\alpha-2)\ldots(\alpha-(n-1))$$
$$\left. \cdot \int_a^t \left(\frac{t^\rho - s^\rho}{\rho}\right)^{\alpha-n} \frac{\mathrm{d}}{\mathrm{d}s}\left[\left(s^{1-\rho}\frac{\mathrm{d}}{\mathrm{d}s}\right)^{n-(n+1)} {_aI^{n-\alpha,\rho}y}\right](s)\mathrm{d}s \right\}$$

$$= \frac{1}{\Gamma(\alpha-n+1)} \int_a^t \left(\frac{t^\rho - s^\rho}{\rho}\right)^{\alpha-n} ({_aI^{n-\alpha,\rho}y})(s)\frac{\mathrm{d}s}{s^{1-\rho}}$$

$$- \sum_{k=1}^n \frac{{_aI^{k-\alpha,\rho}y}(a^+)}{\Gamma(\alpha+2-k)} \left(\frac{t^\rho - a^\rho}{\rho}\right)^{\alpha-k+1}.$$

The proof is complete. □

Lemma 2. *If* $\Re(\alpha) > 0, n = -[-\Re(\alpha)], y \in L[a,b]$ *and* ${_aI^{\alpha,\rho}y} \in AC_\rho^n[a,b]$, *then*

$$({_aI^{\alpha,\rho}}{_aD^{\alpha,\rho}y})(t) = y(t) - \sum_{k=1}^n \frac{{_aI^{k-\alpha,\rho}y}(a^+)}{\Gamma(\alpha-k+1)} \left(\frac{t^\rho - a^\rho}{\rho}\right)^{\alpha-k}.$$

Proof. It is easy to get

$$\left({}_aI^{\alpha,\rho}{}_aD^{\alpha,\rho}y\right)(t)$$

$$= \frac{1}{\Gamma(\alpha)} \int_a^t \left(\frac{t^\rho - s^\rho}{\rho}\right)^{\alpha-1} \left[\left(s^{1-\rho}\frac{d}{ds}\right)^n {}_aI^{n-\alpha,\rho}y\right](s)\frac{ds}{s^{1-\rho}}$$

$$= \left(t^{1-\rho}\frac{d}{dt}\right) \left\{\frac{1}{\Gamma(\alpha+1)} \int_a^t \left(\frac{t^\rho - s^\rho}{\rho}\right)^\alpha \right.$$

$$\times \left.\left[\left(s^{1-\rho}\frac{d}{ds}\right)^n {}_aI^{n-\alpha,\rho}y\right](s)\frac{ds}{s^{1-\rho}}\right\}$$

$$= \left(t^{1-\rho}\frac{d}{dt}\right) \left\{\frac{1}{\Gamma(\alpha+1)} \int_a^t \left(\frac{t^\rho - s^\rho}{\rho}\right)^\alpha \frac{d}{ds}\right.$$

$$\times \left.\left[\left(s^{1-\rho}\frac{d}{ds}\right)^{n-1} {}_aI^{n-\alpha,\rho}y\right](s)ds\right\}.$$

By using Lemma 1, we have

$$\left({}_aI^{\alpha,\rho}{}_aD^{\alpha,\rho}y\right)(t)$$

$$= \left(t^{1-\rho}\frac{d}{dt}\right) \left\{\frac{1}{\Gamma(\alpha-n+1)} \int_a^t \left(\frac{t^\rho - s^\rho}{\rho}\right)^{\alpha-n} ({}_aI^{n-\alpha,\rho}y)(s)\frac{ds}{s^{1-\rho}}\right.$$

$$\left. - \sum_{k=1}^n \frac{{}_aI^{k-\alpha,\rho}y(a^+)}{\Gamma(\alpha+2-k)} \left(\frac{t^\rho - a^\rho}{\rho}\right)^{\alpha-k+1}\right\}$$

$$= \left(t^{1-\rho}\frac{d}{dt}\right) \left\{{}_aI^{\alpha-n+1,\rho}({}_aI^{n-\alpha,\rho}y)(t)\right.$$

$$\left. - \sum_{k=1}^n \frac{{}_aI^{k-\alpha,\rho}y(a^+)}{\Gamma(\alpha+2-k)} \left(\frac{t^\rho - a^\rho}{\rho}\right)^{\alpha-k+1}\right\}$$

$$= \left(t^{1-\rho}\frac{d}{dt}\right) \left\{{}_aI^{1,\rho}y(t) - \sum_{k=1}^n \frac{{}_aI^{k-\alpha,\rho}y(a^+)}{\Gamma(\alpha+2-k)} \left(\frac{t^\rho - a^\rho}{\rho}\right)^{\alpha-k+1}\right\}$$

$$= y(t) - \sum_{k=1}^n \frac{{}_aI^{k-\alpha,\rho}y(a^+)}{\Gamma(\alpha-k+1)} \left(\frac{t^\rho - a^\rho}{\rho}\right)^{\alpha-k}.$$

The proof is complete. □

Corollary 1. *When $0 < \alpha < 1$, then*

$$\left({}_aI^{\alpha,\rho}{}_aD^{\alpha,\rho}y\right)(t) = y(t) - \frac{{}_aI^{1-\alpha,\rho}y(a^+)}{\Gamma(\alpha)}\left(\frac{t^\rho - a^\rho}{\rho}\right)^{\alpha-1}.$$

Lemma 3. *If $\gamma \in \mathbb{R}, 0 \leq \gamma < 1, \alpha > 0$, then the generalized integrable operator ${}_aI^{\alpha,\rho}$ is bounded on the space $C_{\gamma,\rho}[a,b]$, and*

$$\|{}_aI^{\alpha,\rho}y\|_{C_{\gamma,\rho}} \leq \left(\frac{b^\rho - a^\rho}{\rho}\right)^\alpha \frac{\Gamma(1-\gamma)}{\Gamma(1+\alpha-\gamma)}\|y\|_{C_{\gamma,\rho}}.$$

Proof. First, we note that $\|y\|_{C_{\gamma,\rho}} = \|(t^\rho - a^\rho)^\gamma y(t)\|_C$. It is easy to get

$$\|{}_aI^{\alpha,\rho}y\|_{C_{\gamma,\rho}}$$
$$= \|(t^\rho - a^\rho)^\gamma {}_aI^{\alpha,\rho}y\|_C$$
$$= \left\|(t^\rho - a^\rho)^\gamma \frac{1}{\Gamma(\alpha)}\int_a^t \left(\frac{t^\rho - s^\rho}{\rho}\right)^{\alpha-1} y(s)\frac{\mathrm{d}s}{s^{1-\rho}}\right\|_C$$
$$= \left\|\frac{1}{\Gamma(\alpha)}\int_a^t (t^\rho - a^\rho)^\gamma \left(\frac{t^\rho - s^\rho}{\rho}\right)^{\alpha-1} y(s)\frac{\mathrm{d}s}{s^{1-\rho}}\right\|_C$$
$$= \max_{t\in[a,b]}\left|\frac{1}{\Gamma(\alpha)}\int_a^t (t^\rho - a^\rho)^\gamma \left(\frac{t^\rho - s^\rho}{\rho}\right)^{\alpha-1} y(s)\frac{\mathrm{d}s}{s^{1-\rho}}\right|$$
$$= \max_{t\in[a,b]}\left|\frac{1}{\Gamma(\alpha)}\int_a^t \frac{(t^\rho - a^\rho)^\gamma}{(s^\rho - a^\rho)^\gamma}\cdot\frac{(t^\rho - s^\rho)^{\alpha-1}}{\rho^{\alpha-1}}(s^\rho - a^\rho)^\gamma y(s)\frac{\mathrm{d}s}{s^{1-\rho}}\right|$$
$$\leq \max_{t\in[a,b]}\frac{1}{\Gamma(\alpha)}\int_a^t \frac{(t^\rho - a^\rho)^\gamma}{(s^\rho - a^\rho)^\gamma}\cdot\frac{(t^\rho - s^\rho)^{\alpha-1}}{\rho^{\alpha-1}}|(s^\rho - a^\rho)^\gamma y(s)|\frac{\mathrm{d}s}{s^{1-\rho}}$$
$$\leq \max_{t\in[a,b]}\frac{1}{\Gamma(\alpha)}\|y\|_{C_{\gamma,\rho}}\int_a^t \frac{(t^\rho - a^\rho)^\gamma}{(s^\rho - a^\rho)^\gamma}\cdot\frac{(t^\rho - s^\rho)^{\alpha-1}}{\rho^{\alpha-1}}\frac{\mathrm{d}s}{s^{1-\rho}}$$
$$= \max_{t\in[a,b]}\frac{1}{\Gamma(\alpha)}\|y\|_{C_{\gamma,\rho}}\int_a^t \frac{(t^\rho - a^\rho)^{\gamma+\alpha-1}}{(s^\rho - a^\rho)^\gamma}\cdot\frac{1}{\rho^{\alpha-1}}\cdot\left(\frac{t^\rho - s^\rho}{t^\rho - a^\rho}\right)^{\alpha-1}\frac{\mathrm{d}s}{s^{1-\rho}}.$$

Applying the substitution of variables $\eta = \frac{t^\rho - s^\rho}{t^\rho - a^\rho}$ and the definition of the beta function, we have

$$\|_aI^{\alpha,\rho}y\|_{C_{\gamma,\rho}}$$

$$= \max_{t \in [a,b]} \frac{1}{\Gamma(\alpha)} \|y\|_{C_{\gamma,\rho}} \left(\frac{t^\rho - a^\rho}{\rho}\right)^\alpha \int_0^1 (1-\eta)^{-\gamma} \eta^{\alpha-1} d\eta$$

$$= \max_{t \in [a,b]} \frac{1}{\Gamma(\alpha)} \|y\|_{C_{\gamma,\rho}} \left(\frac{t^\rho - a^\rho}{\rho}\right)^\alpha B(\alpha, 1-\gamma)$$

$$\leq \left(\frac{b^\rho - a^\rho}{\rho}\right)^\alpha \frac{\Gamma(1-\gamma)}{\Gamma(1+\alpha-\gamma)} \|y\|_{C_{\gamma,\rho}}.$$

The proof is complete. □

Corollary 2. *If $\alpha > 0$, then the generalized integrable operator $_aI^{\alpha,\rho}$ is bounded on the space $C[a,b]$ with respect to the norm $\|u\| = \sup_{t \in [a,b]} |u(t)|$, and*

$$\|_aI^{\alpha,\rho}u\| \leq A\|u\|,$$

where $A = \left(\frac{b^\rho - a^\rho}{\rho}\right)^\alpha \cdot \frac{1}{\Gamma(1+\alpha)}$. Particularly, when $\rho = 1$, then $A = \frac{(b-a)^\alpha}{\Gamma(1+\alpha)}$.

Lemma 4. *The Cauchy-type problems (1)–(2) are equivalent to the following Volterra nonlinear integral equation of the second kind:*

$$y(t) = \sum_{j=1}^{n-1} \frac{b_j}{\Gamma(\alpha - j + 1)} \left(\frac{t^\rho - a^\rho}{\rho}\right)^{\alpha - j}$$

$$+ \frac{1}{\Gamma(\alpha)} \int_a^t \left(\frac{t^\rho - s^\rho}{\rho}\right)^{\alpha - 1} f(s, y(s)) \frac{ds}{s^{1-\rho}}.$$

The proof of the result is similar to the proof of Theorem 3.10 in the book written by Kilbas et al. [10].

Lemma 5. Let $y \in C_{\gamma,\rho}[a,b]$. Assume that $\varphi : [a,b] \to (0,\infty)$ is a bounded function. Then the set

$$\mathcal{Y} = \left\{ u \in C_{\gamma,\rho}[a,b] : \sup_{t \in [a,b]} \frac{|(t^\rho - a^\rho)^\gamma (u(t) - y(t))|}{\varphi(t)} < +\infty \right\}$$

is a complete metric space with the weighted metric

$$d(u_1, u_2) = \sup_{t \in [a,b]} \frac{(t^\rho - a^\rho)^\gamma |u_1(t) - u_2(t)|}{\varphi(t)}.$$

Proof. It is clear that $d(\cdot,\cdot)$ is a metric on $\mathcal{Y}$. Obviously, $d(u_1, u_2) \geq 0$ and $d(u_1, u_2) = d(u_2, u_1)$. We can get

$$\begin{aligned}
d(u_1, u_2) &= \sup_{t \in [a,b]} \frac{(t^\rho - a^\rho)^\gamma |u_1(t) - u_2(t)|}{\varphi(t)} \\
&\leq \sup_{t \in [a,b]} \frac{(t^\rho - a^\rho)^\gamma [|u_1(t) - u_3(t)| + |u_3(t) - u_2(t)|]}{\varphi(t)} \\
&\leq \sup_{t \in [a,b]} \frac{(t^\rho - a^\rho)^\gamma |u_1(t) - u_3(t)|}{\varphi(t)} \\
&\quad + \sup_{t \in [a,b]} \frac{(t^\rho - a^\rho)^\gamma |u_2(t) - u_3(t)|}{\varphi(t)} \\
&= d(u_1, u_3) + d(u_2, u_3).
\end{aligned}$$

Next, we show that $(\mathcal{Y}, d)$ is complete. Let $\{u_n\}$ be a Cauchy sequence in $(\mathcal{Y}, d)$. We get that for any $\varepsilon > 0$, there exists an integer $N(\varepsilon) > 0$ such that $d(u_n, u_m) \leq \varepsilon$ for all $n, m \geq N(\varepsilon)$. Using the definition of $d(\cdot,\cdot)$, we get

$$\forall \varepsilon > 0, \quad \exists N(\varepsilon) > 0, \quad \forall n, m \geq N(\varepsilon),$$
$$\forall t \in [a,b] : \frac{(t^\rho - a^\rho)^\gamma |u_n(t) - u_m(t)|}{\varphi(t)} \leq \varepsilon.$$

For a fixed $t \in [a,b]$, since φ is a bounded function, the relation $(t^\rho - a^\rho)^\gamma |u_n(t) - u_m(t)| \leq \varphi(t)\varepsilon$ shows that $\{u_n(t)\}$ is a Cauchy sequence in $\mathbb{R}$. Hence, $\{u_n(t)\}$ converges for $\forall t \in (a, b]$. Thus, one can define a function $u : (a, b] \to \mathbb{R}$ by

$$u(t) = \lim_{n \to \infty} u_n(t).$$

Let $m \to \infty$, we obtain

$$\forall \varepsilon > 0, \quad \exists N(\varepsilon) > 0, \quad \forall n \geqslant N(\varepsilon),$$
$$\forall t \in [a, b] : |(t^\rho - a^\rho)^\gamma u_n(t) - (t^\rho - a^\rho)^\gamma u(t)| \leqslant \varphi(t)\varepsilon.$$

Since φ is bounded, we show that $(t^\rho - a^\rho)^\gamma u_n(t)$ converges uniformly to $(t^\rho - a^\rho)^\gamma u(t)$. Hence, $(t^\rho - a^\rho)^\gamma u \in C[a, b]$, we get $u(t) \in C_{\gamma,\rho}[a, b]$. We also have

$$\frac{(t^\rho - a^\rho)^\gamma |u(t) - y(t)|}{\varphi(t)}$$
$$\leqslant \frac{(t^\rho - a^\rho)^\gamma |u(t) - u_n(t)|}{\varphi(t)} + \frac{(t^\rho - a^\rho)^\gamma |u_n(t) - y(t)|}{\varphi(t)}$$
$$\leqslant \varepsilon + \frac{(t^\rho - a^\rho)^\gamma |u_n(t) - y(t)|}{\varphi(t)} < +\infty,$$

we get $u \in \mathcal{Y}$.

Thus, we show that the Cauchy sequence $\{u_n\}$ converges to u in $(\mathcal{Y}, d)$. With this the proof is now complete. □

The following theorem is a fundamental result in the fixed-point theory.

Theorem 1 (Banach's contraction principle). *Let (X, d) be a complete metric space, and consider a mapping $\Lambda : X \to X$, which is strictly contractive, that is,*

$$d(\Lambda x, \Lambda y) \leqslant L d(x, y), \quad \forall x, y \in X,$$

for some constant $0 \leqslant L < 1$. Then

(i) *the mapping Λ has a unique fixed point $x^* = \Lambda x^*$,*
(ii) *the fixed point x^* is globally attractive, namely for any starting point $x \in X$, the following relations holds:*

$$\lim_{n \to \infty} \Lambda^n x = x^*,$$

(iii) *we have the following estimation inequalities:*

$$d(\Lambda^n x, x^*) \leq L^n d(x, x^*), \quad \forall n \geq 0, \ \forall x \in X;$$

$$d(\Lambda^n x, x^*) \leq \frac{1}{1-L} d(\Lambda^n x, \Lambda^{n+1} x), \quad \forall n \geq 0, \ \forall x \in X;$$

$$d(x, x^*) \leq \frac{1}{1-L} d(x, \Lambda x), \quad \forall x \in X.$$

3. Main Results

In this section, we study the Hyers–Ulam–Rassias stability of (1)–(2).

Theorem 2. *Let $\alpha > 1$, $f(t, y(t)) \in C_\gamma[a, b]$ and $f(t, y)$ satisfy a Lipschitz-type condition with respect to y:*

$$|f(t, y_1) - f(t, y_2)| \leq L(t^\rho - a^\rho)^\gamma |y_1 - y_2| \tag{3}$$

for any $t \in [a, b]$. If a function $y \in C[a, b]$ satisfies (2) and

$$|({}_aD^{\alpha,\rho}y)(t) - f(t, y(t))| \leq \varphi(t) \tag{4}$$

for any $t \in [a, b]$, where $\varphi : [a, b] \to (0, \infty)$ is a bounded function with

$$\left| \frac{1}{\Gamma(\alpha)} \int_a^t \left(\frac{t^\rho - s^\rho}{\rho} \right)^{\alpha-1} (t^\rho - a^\rho)^\gamma \varphi(s) \frac{ds}{s^{1-\rho}} \right| \leq K\varphi(t) \tag{5}$$

for each $t \in [a, b]$. If $0 < KL < 1$, then there exists a unique $y_0 \in C[a, b]$ such that

$$y_0(t) = \sum_{j=1}^{n-1} \frac{b_j}{\Gamma(\alpha - j + 1)} \left(\frac{t^\rho - a^\rho}{\rho} \right)^{\alpha-j}$$

$$+ \frac{1}{\Gamma(\alpha)} \int_a^t \left(\frac{t^\rho - s^\rho}{\rho} \right)^{\alpha-1} f(s, y_0(s)) \frac{ds}{s^{1-\rho}}$$

(consequently, y_0 is a solution to (1)) and

$$|y(t) - y_0(t)| \leq \frac{K}{1 - KL} \frac{\varphi(t)}{(t^\rho - a^\rho)^\gamma}$$

for all $t \in [a, b]$.

Proof. We consider the space

$$\mathcal{Y} = \left\{ u \in C_{\gamma,\rho}[a,b] : \sup_{t \in [a,b]} \frac{|(t^\rho - a^\rho)^\gamma (u(t) - y(t))|}{\varphi(t)} < +\infty \right\}$$

with the weighted metric

$$d(u_1, u_2) = \sup_{t \in [a,b]} \frac{(t^\rho - a^\rho)^\gamma |u_1(t) - u_2(t)|}{\varphi(t)}.$$

Now we define the nonlinear operator

$$(Tu)(t) = \sum_{j=1}^{n-1} \frac{b_j}{\Gamma(\alpha - j + 1)} \left(\frac{t^\rho - a^\rho}{\rho}\right)^{\alpha - j}$$

$$+ \frac{1}{\Gamma(\alpha)} \int_a^t \left(\frac{t^\rho - s^\rho}{\rho}\right)^{\alpha - 1} f(s, u(s)) \frac{\mathrm{d}s}{s^{1-\rho}}, \quad u \in \mathcal{Y}.$$

By Lemma 3, the integral in the right-hand side of the above equation is a continuous function on $[a,b]$, and the first term in the right-hand side of the above equation is also a continuous function on $[a,b]$. So $(Tu)(t) \in C[a,b]$.

For $u_1, u_2 \in \mathcal{Y}$, we have

$d(Tu_1, Tu_2)$

$$= \sup_{t \in [a,b]} \frac{(t^\rho - a^\rho)^\gamma \left| \frac{1}{\Gamma(\alpha)} \int_a^t \left(\frac{t^\rho - s^\rho}{\rho}\right)^{\alpha-1} [f(s,u_1(s)) - f(s,u_2(s))] \frac{\mathrm{d}s}{s^{1-\rho}} \right|}{\varphi(t)}$$

$$\leq \sup_{t \in [a,b]} \frac{(t^\rho - a^\rho)^\gamma \frac{1}{\Gamma(\alpha)} \int_a^t \left(\frac{t^\rho - s^\rho}{\rho}\right)^{\alpha-1} L(s^\rho - a^\rho)^\gamma |u_1(s) - u_2(s)| \frac{\mathrm{d}s}{s^{1-\rho}}}{\varphi(t)}$$

$$= \sup_{t \in [a,b]} \frac{(t^\rho - a^\rho)^\gamma \frac{1}{\Gamma(\alpha)} \int_a^t \left(\frac{t^\rho - s^\rho}{\rho}\right)^{\alpha-1} L \frac{(s^\rho - a^\rho)^\gamma |u_1(s) - u_2(s)|}{\varphi(s)} \varphi(s) \frac{\mathrm{d}s}{s^{1-\rho}}}{\varphi(t)}.$$

Applying the weighted metric definition, we get

$$\sup_{t\in[a,b]} \frac{(t^\rho-a^\rho)^\gamma \frac{1}{\Gamma(\alpha)}\int_a^t \left(\frac{t^\rho-s^\rho}{\rho}\right)^{\alpha-1} L\frac{(s^\rho-a^\rho)^\gamma |u_1(s)-u_2(s)|}{\varphi(s)}\varphi(s)\frac{ds}{s^{1-\rho}}}{\varphi(t)}$$

$$\leq \sup_{t\in[a,b]} \frac{d(u_1,u_2)(t^\rho-a^\rho)^\gamma \frac{L}{\Gamma(\alpha)}\int_a^t \left(\frac{t^\rho-s^\rho}{\rho}\right)^{\alpha-1}\varphi(s)\frac{ds}{s^{1-\rho}}}{\varphi(t)}$$

$$\leq KLd(u_1,u_2). \tag{6}$$

On the other hand, from (4) we get

$$-\varphi(t) \leq ({}_aD^{\alpha,\rho}y)(t) - f(t,y(t)) \leq \varphi(t). \tag{7}$$

Applying the operator ${}_aI^{\alpha,\rho}$ to both sides of (7), one can obtain

$$-{}_aI^{\alpha,\rho}\varphi(t) \leq {}_aI^{\alpha,\rho}({}_aD^{\alpha,\rho}y)(t) - {}_aI^{\alpha,\rho}f(t,y(t)) \leq {}_aI^{\alpha,\rho}\varphi(t).$$

By Lemma 2, we have

$$-{}_aI^{\alpha,\rho}\varphi(t) \leq y(t) - \sum_{k=1}^{n}\frac{{}_aD^{\alpha-k,\rho}y(a^+)}{\Gamma(\alpha-k+1)}\left(\frac{t^\rho-a^\rho}{\rho}\right)^{\alpha-k}$$
$$- {}_aI^{\alpha,\rho}f(t,y(t))$$
$$\leq {}_aI^{\alpha,\rho}\varphi(t), \tag{8}$$

where ${}_aD^{\alpha-k,\rho}y(a^+) = b_k$, $b_k \in \mathbb{R}$ ($k=1,\ldots,n$). By using (2), we reduce (8) to the following:

$$-{}_aI^{\alpha,\rho}\varphi(t) \leq y(t) - \sum_{k=1}^{n-1}\frac{b_k}{\Gamma(\alpha-k+1)}\left(\frac{t^\rho-a^\rho}{\rho}\right)^{\alpha-k} - {}_aI^{\alpha,\rho}f(t,y(t))$$
$$\leq {}_aI^{\alpha,\rho}\varphi(t). \tag{9}$$

From (9), we have

$$|y(t) - (Ty)(t)| \leq \frac{1}{\Gamma(\alpha)}\int_a^t \left(\frac{t^\rho-s^\rho}{\rho}\right)^{\alpha-1}\varphi(s)\frac{ds}{s^{1-\rho}} \tag{10}$$

for each $t \in [a,b]$. From (10), we get

$$(t^\rho-a^\rho)^\gamma |y(t)-(Ty)(t)|$$
$$\leq \frac{1}{\Gamma(\alpha)}\int_a^t \left(\frac{t^\rho-s^\rho}{\rho}\right)^{\alpha-1}(t^\rho-a^\rho)^\gamma \varphi(s)\frac{ds}{s^{1-\rho}}. \tag{11}$$

By (5), one can get the following relation

$$(t^\rho - a^\rho)^\gamma |y(t) - (Ty)(t)| \leq K\varphi(t) \tag{12}$$

for each $t \in [a, b]$. From (12), we obtain

$$d(y, Ty) \leq K. \tag{13}$$

By using (6) and (13), we have

$$d(Tu, y) \leq d(Tu, Ty) + d(Ty, y)$$
$$\leq KLd(u, y) + K < +\infty.$$

It follows that if $u \in \mathcal{Y}$, then $Tu \in \mathcal{Y}$, hence the operator $T : \mathcal{Y} \to \mathcal{Y}$ is well defined. Thus, we obtain that T is a strictly contractive operator on $\mathcal{Y}$ with constant $0 < KL < 1$.

Using Theorem 1 on the weighted space $\mathcal{Y}$, we get the existence of a function $y_0 \in \mathcal{Y}$ (y_0 is also in $C_\gamma[a, b]$) such that

(i) y_0 is the unique fixed point of T, that is,

$$y_0(t) = \sum_{j=1}^{n-1} \frac{b_j}{\Gamma(\alpha - j + 1)} \left(\frac{t^\rho - a^\rho}{\rho}\right)^{\alpha - j}$$
$$+ \frac{1}{\Gamma(\alpha)} \int_a^t \left(\frac{t^\rho - s^\rho}{\rho}\right)^{\alpha - 1} f(s, y_0(s)) \frac{ds}{s^{1-\rho}};$$

(ii) $\lim_{n \to \infty} d(T^n y, y_0) = 0$, that is,

$$y_0(t) = \lim_{n \to \infty} (T^n y)(t), \quad \forall t \in [a, b];$$

(iii) $d(y, y_0) \leq \frac{1}{1-KL} d(y, Ty)$, that is,

$$d(y, y_0) \leq \frac{K}{1 - KL},$$

which implies the estimation

$$|y(t) - y_0(t)| \leq \frac{K}{1 - KL} \frac{\varphi(t)}{(t^\rho - a^\rho)^\gamma}.$$

The proof is complete. □

Theorem 3. *Let* $\alpha > 1, n = [\alpha] + 1, 1 < p < \alpha, \frac{1}{p} + \frac{1}{q} = 1, \rho > 0$, *and* $[a,b]$ *be a closed interval on* $\mathbb{R}$. *Set* $M = \frac{1}{\Gamma(\alpha)} \left[\frac{1}{p(\alpha-1)+1} \left(\frac{b^\rho - a^\rho}{\rho} \right)^{p(\alpha-1)+1} \right]^{\frac{1}{p}}$. *Assume that* $f(t, y(t)) \in C_\gamma[a,b]$, *and* $f(t,y)$ *satisfies a Lipschitz-type condition with respect to* y:

$$|f(t, y_1) - f(t, y_2)| \leq L(t^\rho - a^\rho)^\gamma |y_1 - y_2| \qquad (14)$$

for any $t \in [a,b]$, *where* $0 \leq \gamma < 1$. *If a function* $y \in C[a,b]$ *satisfies* (2) *and*

$$|(_aD^{\alpha,\rho}y)(t) - f(t, y(t))| \leq \varphi(t) \qquad (15)$$

for any $t \in [a,b]$, *where* $\varphi : [a,b] \to (0, \infty)$ *is a bounded function with*

$$\left(\int_a^t [(t^\rho - a^\rho)^\gamma \varphi(s)]^q \frac{ds}{s^{1-\rho}} \right)^{\frac{1}{q}} \leq K\varphi(t) \qquad (16)$$

for each $t \in [a,b]$, *and* $0 < KLM < 1$, *then there exists a unique* $y_0 \in C[a,b]$ *such that*

$$y_0(t) = \sum_{j=1}^{n-1} \frac{b_j}{\Gamma(\alpha - j + 1)} \left(\frac{t^\rho - a^\rho}{\rho} \right)^{\alpha - j}$$

$$+ \frac{1}{\Gamma(\alpha)} \int_a^t \left(\frac{t^\rho - s^\rho}{\rho} \right)^{\alpha - 1} f(s, y_0(s)) \frac{ds}{s^{1-\rho}}$$

(*consequently,* y_0 *is a solution to* (1)) *and*

$$|y(t) - y_0(t)| \leq \frac{MK}{1 - KLM} \frac{\varphi(t)}{(t^\rho - a^\rho)^\gamma} \qquad (17)$$

for all $t \in [a,b]$.

Proof. We consider the set

$$\mathcal{Y} = \left\{ u \in C_{\gamma,\rho}[a,b] : \sup_{t \in [a,b]} \frac{(t^\rho - a^\rho)^\gamma |u(t) - y(t)|}{\varphi(t)} < +\infty \right\}.$$

By Lemma 5, we know that $\mathcal{Y}$ is a complete metric space with the weighted metric

$$d(u_1, u_2) = \sup_{t \in [a,b]} \frac{(t^\rho - a^\rho)^\gamma |u_1(t) - u_2(t)|}{\varphi(t)}.$$

We define the nonlinear operator

$$(Tu)(t) = \sum_{j=1}^{n-1} \frac{b_j}{\Gamma(\alpha - j + 1)} \left(\frac{t^\rho - a^\rho}{\rho}\right)^{\alpha - j}$$

$$+ \frac{1}{\Gamma(\alpha)} \int_a^t \left(\frac{t^\rho - s^\rho}{\rho}\right)^{\alpha - 1} f(s, u(s)) \frac{ds}{s^{1-\rho}}.$$

By Lemma 3, we have that $(Tu)(t) \in C[a,b]$. For $u_1, u_2 \in \mathcal{Y}$, we obtain

$d(Tu_1, Tu_2)$

$$= \sup_{t \in [a,b]} \frac{(t^\rho - a^\rho)^\gamma \left| \frac{1}{\Gamma(\alpha)} \int_a^t \left(\frac{t^\rho - s^\rho}{\rho}\right)^{\alpha - 1} [f(s, u_1(s)) - f(s, u_2(s))] \frac{ds}{s^{1-\rho}} \right|}{\varphi(t)}$$

$$\leq \sup_{t \in [a,b]} \frac{(t^\rho - a^\rho)^\gamma \frac{1}{\Gamma(\alpha)} \int_a^t \left(\frac{t^\rho - s^\rho}{\rho}\right)^{\alpha - 1} L(s^\rho - a^\rho)^\gamma |u_1(s) - u_2(s)| \frac{ds}{s^{1-\rho}}}{\varphi(t)}$$

$$= \sup_{t \in [a,b]} \frac{(t^\rho - a^\rho)^\gamma \frac{1}{\Gamma(\alpha)} \int_a^t \left(\frac{t^\rho - s^\rho}{\rho}\right)^{\alpha - 1} L \frac{(s^\rho - a^\rho)^\gamma |u_1(s) - u_2(s)|}{\varphi(s)} \varphi(s) \frac{ds}{s^{1-\rho}}}{\varphi(t)}$$

$$\leq \sup_{t \in [a,b]} \frac{(t^\rho - a^\rho)^\gamma \frac{L}{\Gamma(\alpha)} d(u_1, u_2) \int_a^t \left(\frac{t^\rho - s^\rho}{\rho}\right)^{\alpha - 1} \varphi(s) \frac{ds}{s^{1-\rho}}}{\varphi(t)}. \quad (18)$$

By Hölder inequality, we get

$$\int_a^t \left(\frac{t^\rho - s^\rho}{\rho}\right)^{\alpha - 1} (t^\rho - a^\rho)^\gamma \varphi(s) \frac{ds}{s^{1-\rho}}$$

$$\leq \left[\frac{1}{p(\alpha - 1) + 1} \left(\frac{b^\rho - a^\rho}{\rho}\right)^{\alpha - 1 + \frac{1}{p}}\right]^{\frac{1}{p}}$$

$$\cdot \left[\int_a^t [(t^\rho - a^\rho)^\gamma \varphi(s)]^q \frac{ds}{s^{1-\rho}} \right]^{\frac{1}{q}}$$

$$\leq \left[\frac{1}{p(\alpha-1)+1} \left(\frac{b^\rho - a^\rho}{\rho} \right)^{\alpha-1+\frac{1}{p}} \right]^{\frac{1}{p}} \cdot K\varphi(t). \tag{19}$$

By (18) and (19), we obtain

$$d(Tu_1, Tu_2) \leq KL \frac{1}{\Gamma(\alpha)} \left[\frac{1}{p(\alpha-1)+1} \left(\frac{b^\rho - a^\rho}{\rho} \right)^{\alpha-1+\frac{1}{p}} \right] d(u_1, u_2)$$

$$= KLM d(u_1, u_2). \tag{20}$$

On the other hand, from (15) we get

$$-\varphi(t) \leq (_aD^{\alpha,\rho}y)(t) - f(t, y(t)) \leq \varphi(t). \tag{21}$$

Applying the operator $_aI^{\alpha,\rho}$ to both sides of (21), one can obtain

$$-_aI^{\alpha,\rho}\varphi(t) \leq {_aI^{\alpha,\rho}}(_aD^{\alpha,\rho}y)(t) - {_aI^{\alpha,\rho}}f(t, y(t)) \leq {_aI^{\alpha,\rho}}\varphi(t).$$

By Lemma 2, we have

$$-_aI^{\alpha,\rho}\varphi(t) \leq y(t) - \sum_{k=1}^n \frac{_aD^{\alpha-k,\rho}y(a^+)}{\Gamma(\alpha-k+1)} \left(\frac{t^\rho - a^\rho}{\rho} \right)^{\alpha-k}$$

$$-_aI^{\alpha,\rho}f(t, y(t)) \leq {_aI^{\alpha,\rho}}\varphi(t). \tag{22}$$

We have

$$-_aI^{\alpha,\rho}\varphi(t) \leq y(t) - \sum_{k=1}^{n-1} \frac{b_k}{\Gamma(\alpha-k+1)} \left(\frac{t^\rho - a^\rho}{\rho} \right)^{\alpha-k}$$

$$- \frac{1}{\Gamma(\alpha)} \int_a^t \left(\frac{t^\rho - s^\rho}{\rho} \right)^{\alpha-1} f(s, y(s)) \frac{ds}{s^{1-\rho}} \leq {_aI^{\alpha,\rho}}\varphi(t). \tag{23}$$

By (23), we get

$$-_aI^{\alpha,\rho}\varphi(t) \leq y(t) - (Ty)(t) \leq {_aI^{\alpha,\rho}}\varphi(t).$$

From the above inequality, one gets

$$|y(t) - (Ty)(t)| \leq {}_aI^{\alpha,\rho}\varphi(t) \leq \frac{1}{\Gamma(\alpha)} \int_a^t \left(\frac{t^\rho - s^\rho}{\rho}\right)^{\alpha-1} \varphi(s) \frac{\mathrm{d}s}{s^{1-\rho}}.$$

By Hölder inequality, we get

$$(t^\rho - a^\rho)^\gamma |y(t) - (Ty)(t)|$$
$$\leq \frac{1}{\Gamma(\alpha)} \int_a^t \left(\frac{t^\rho - s^\rho}{\rho}\right)^{\alpha-1} (t^\rho - a^\rho)^\gamma \varphi(s) \frac{\mathrm{d}s}{s^{1-\rho}}$$
$$\leq \frac{1}{\Gamma(\alpha)} \left[\frac{1}{p(\alpha-1)+1} \left(\frac{b^\rho - a^\rho}{\rho}\right)^{\alpha-1+\frac{1}{p}}\right]$$
$$\cdot \left[\int_a^t [(t^\rho - a^\rho)^\gamma \varphi(s)]^q \frac{\mathrm{d}s}{s^{1-\rho}}\right]^{\frac{1}{q}}$$
$$\leq \frac{1}{\Gamma(\alpha)} \left[\frac{1}{p(\alpha-1)+1} \left(\frac{b^\rho - a^\rho}{\rho}\right)^{\alpha-1+\frac{1}{p}}\right] \cdot K\varphi(t)$$
$$\leq MK\varphi(t).$$

From the above equation, we obtain

$$d(y, Ty) \leq MK. \tag{24}$$

By using (20) and (24), we have

$$d(Tu, y) \leq d(Tu, Ty) + d(Ty, y)$$
$$\leq KLMd(u, y) + MK < +\infty.$$

It follows that if $u \in \mathcal{Y}$, then $Tu \in \mathcal{Y}$. Hence, the operator $T : \mathcal{Y} \to \mathcal{Y}$ is well defined.

From (20), we obtain that T is a strictly contractive operator on $\mathcal{Y}$ with a constant $0 < KLM < 1$. Applying Theorem 1 on the weighted space $\mathcal{Y}$, we get the existence of a function $y_0 \in \mathcal{Y}$ (y_0 is also in $C[a, b]$) such that

(i) y_0 is the unique fixed point of T, that is,

$$y_0(x) = \sum_{j=1}^{n-1} \frac{b_j}{\Gamma(\alpha-j+1)} \left(\frac{t^\rho - a^\rho}{\rho}\right)^{\alpha-j}$$

$$+ \frac{1}{\Gamma(\alpha)} \int_a^t \left(\frac{t^\rho - s^\rho}{\rho}\right)^{\alpha-1} f(s, y_0(s)) \frac{ds}{s^{1-\rho}};$$

(ii) $\lim_{n \to \infty} d(T^n y, y_0) = 0$, that is,

$$y_0(t) = \lim_{n \to \infty} (T^n y)(t), \quad \forall t \in [a, b];$$

(iii) $d(y, y_0) \leq \frac{1}{1-KLM} d(y, Ty)$, that is,

$$d(y, y_0) \leq \frac{MK}{1 - KLM},$$

which implies

$$|y(t) - y_0(t)| \leq \frac{MK}{1 - KLM} \frac{\varphi(t)}{(t^\rho - a^\rho)^\gamma}.$$

The proof is complete. □

Theorem 4. *Let $\alpha > 1, n = [\alpha]+1, 1 < q < \alpha, \frac{1}{p} + \frac{1}{q} = 1, \rho > 0$, and $[a,b]$ be a closed interval on $\mathbb{R}$, the function $l(x)$ satisfies*

$$\int_a^b l^q(s) \frac{ds}{s^{1-\rho}} < +\infty.$$

Set

$$M' = \frac{1}{\Gamma(\alpha)} \left[\frac{1}{p(\alpha-1)+1} \left(\frac{b^\rho - a^\rho}{\rho}\right)^{p(\alpha-1)+1}\right]^{\frac{1}{p}}$$

$$\cdot (b^\rho - a^\rho)^\gamma \left(\int_a^b l^q(s) \frac{ds}{s^{1-\rho}}\right)$$

and $0 < M' < 1$. Suppose that $f(t, y(t)) \in C_\gamma[a, b]$. Assume $f(t, y)$ satisfies a Lipschitz-type condition with respect to y:

$$|f(t, y_1) - f(t, y_2)| \leq l(t)(t^\rho - a^\rho)^\gamma |y_1 - y_2| \tag{25}$$

for any $t \in [a, b]$, and $0 \leq \gamma < 1$. If a function $y \in C[a, b]$ satisfies (2) and

$$|({}_aD^{\alpha,\rho}y)(t) - f(t, y(t))| \leq \varepsilon \tag{26}$$

for any $t \in [a, b]$ and for some $\varepsilon > 0$, then there exists a unique $y_0 \in C[a, b]$ such that

$$y_0(t) = \sum_{j=1}^{n-1} \frac{b_j}{\Gamma(\alpha - j + 1)} \left(\frac{t^\rho - a^\rho}{\rho}\right)^{\alpha - j}$$

$$+ \frac{1}{\Gamma(\alpha)} \int_a^t \left(\frac{t^\rho - s^\rho}{\rho}\right)^{\alpha - 1} f(s, y_0(s)) \frac{ds}{s^{1-\rho}}$$

(consequently, y_0 is a solution to (1)) and

$$|y(t) - y_0(t)| \leq \frac{1}{1 - M'} \frac{1}{\Gamma(\alpha + 1)} \left(\frac{b^\rho - a^\rho}{\rho}\right)^\alpha \varepsilon$$

for all $t \in [a, b]$.

Proof. For some $\varepsilon > 0$, and $y \in C_{\gamma, \rho}[a, b]$, we consider the set

$$\mathcal{Y} = \left\{ u \in C_{\gamma, \rho}[a, b] : \sup_{t \in [a, b]} \frac{(t^\rho - a^\rho)^\gamma |u(t) - y(t)|}{\varepsilon} < +\infty \right\}.$$

It is easy to prove that $\mathcal{Y}$ is a complete metric space with the weighted metric

$$d(u_1, u_2) = \sup_{t \in [a, b]} \frac{(t^\rho - a^\rho)^\gamma |u_1(t) - u_2(t)|}{\varepsilon}.$$

We define the nonlinear operator T on $\mathcal{Y}$ by

$$(Tu)(t) = \sum_{j=1}^{n-1} \frac{b_j}{\Gamma(\alpha - j + 1)} \left(\frac{t^\rho - a^\rho}{\rho}\right)^{\alpha - j}$$

$$+ \frac{1}{\Gamma(\alpha)} \int_a^t \left(\frac{t^\rho - s^\rho}{\rho}\right)^{\alpha - 1} f(s, u(s)) \frac{ds}{s^{1-\rho}}.$$

By Lemma 3, $(Tu)(t)$ is a continuous function on $[a, b]$, so $(Tu)(t) \in C[a, b]$.

For $u_1, u_2 \in \mathcal{Y}$, we have

$d(Tu_1, Tu_2)$
$$= \sup_{t \in [a,b]} \frac{(t^\rho - a^\rho)^\gamma |Tu_1(t) - Tu_2(t)|}{\varepsilon}$$
$$\leq \sup_{t \in [a,b]} \frac{(t^\rho - a^\rho)^\gamma \frac{1}{\Gamma(\alpha)} \int_a^t \left(\frac{t^\rho - s^\rho}{\rho}\right)^{\alpha-1} l(s)(s^\rho - a^\rho)^\gamma |u_1(s) - u_2(s)| \frac{ds}{s^{1-\rho}}}{\varepsilon}$$
$$\leq \sup_{t \in [a,b]} d(u_1, u_2)(t^\rho - a^\rho)^\gamma \frac{1}{\Gamma(\alpha)} \int_a^t \left(\frac{t^\rho - s^\rho}{\rho}\right)^{\alpha-1} l(s) \frac{ds}{s^{1-\rho}}.$$

By Hölder inequality, we get

$$\sup_{t \in [a,b]} d(u_1, u_2)(t^\rho - a^\rho)^\gamma \frac{1}{\Gamma(\alpha)} \int_a^t \left(\frac{t^\rho - s^\rho}{\rho}\right)^{\alpha-1} l(s) \frac{ds}{s^{1-\rho}}$$
$$\leq \sup_{t \in [a,b]} d(u_1, u_2) \frac{1}{\Gamma(\alpha)} \left[\int_a^t \left(\frac{t^\rho - s^\rho}{\rho}\right)^{p(\alpha-1)} \frac{ds}{s^{1-\rho}}\right]^{\frac{1}{p}}$$
$$\cdot \left[\int_a^t [(t^\rho - a^\rho)^\gamma l(s)]^q \frac{ds}{s^{1-\rho}}\right]^{\frac{1}{q}}$$
$$\leq \sup_{t \in [a,b]} d(u_1, u_2) \frac{1}{\Gamma(\alpha)} \left[\int_a^t \left(\frac{t^\rho - s^\rho}{\rho}\right)^{p(\alpha-1)} \frac{ds}{s^{1-\rho}}\right]^{\frac{1}{p}}$$
$$\cdot \left[\int_a^b (b^\rho - a^\rho)^{q\gamma} l^q(s) \frac{ds}{s^{1-\rho}}\right]^{\frac{1}{q}}$$
$$\leq M' d(u_1, u_2). \tag{27}$$

From (27), we obtain that T is a strictly contractive operator on $\mathcal{C}$ with a constant $0 < M' < 1$.

On the other hand, from (26) we get

$$-\varepsilon \leq (_aD^{\alpha,\rho}y)(t) - f(t, y(t)) \leq \varepsilon. \tag{28}$$

Applying the operator $_aI^{\alpha,\rho}$ to both sides of (28), one can obtain

$$-_aI^{\alpha,\rho}\varepsilon \leq {_aI^{\alpha,\rho}}(_aD^{\alpha,\rho}y)(t) - {_aI^{\alpha,\rho}}f(t,y(t)) \leq {_aI^{\alpha,\rho}}\varepsilon.$$

By Lemma 2, we have

$$-{}_aI^{\alpha,\rho}\varepsilon \leq y(t) - \sum_{k=1}^{n} \frac{{}_aD^{\alpha-k,\rho}y(a^+)}{\Gamma(\alpha-k+1)}\left(\frac{t^\rho - a^\rho}{\rho}\right)^{\alpha-k}$$
$$- {}_aI^{\alpha,\rho}f(t,y(t)) \leq {}_aI^{\alpha,\rho}\varepsilon. \tag{29}$$

By the definition of the operator T and the initial conditions, we have

$$-{}_aI^{\alpha,\rho}\varepsilon \leq y(t) - (Ty)(t) \leq {}_aI^{\alpha,\rho}\varepsilon. \tag{30}$$

From (30), we have

$$|y(t) - (Ty)(t)| \leq {}_aI^{\alpha,\rho}\varepsilon \leq \frac{\varepsilon}{\Gamma(\alpha+1)}\left(\frac{b^\rho - a^\rho}{\rho}\right)^{\alpha}. \tag{31}$$

By (31), we obtain

$$\frac{(t^\rho - a^\rho)^\gamma |y(t) - (Ty)(t)|}{\varepsilon} \leq \frac{1}{\Gamma(\alpha+1)}\frac{(b^\rho - a^\rho)^{\alpha+\gamma}}{\rho^\alpha}. \tag{32}$$

By the definition of the weighted metric on $\mathcal{Y}$, we have

$$d(y, Ty) \leq \frac{1}{\Gamma(\alpha+1)}\frac{(b^\rho - a^\rho)^{\alpha+\gamma}}{\rho^\alpha}. \tag{33}$$

Since

$$d(Tu, y) \leq d(Tu, Ty) + d(Ty, y)$$
$$\leq M'd(u,y) + \frac{1}{\Gamma(\alpha+1)}\frac{(b^\rho - a^\rho)^{\alpha+\gamma}}{\rho^\alpha} < +\infty,$$

we obtain $Tu \in \mathcal{Y}$, the operator T is well defined.

Using Theorem 1 on the weighted space $\mathcal{Y}$, we get the existence of a function $y_0 \in \mathcal{Y}$ such that

(i) y_0 is the unique fixed point of T, that is,

$$y_0(t) = \sum_{j=1}^{n-1} \frac{b_j}{\Gamma(\alpha-j+1)}\left(\frac{t^\rho - a^\rho}{\rho}\right)^{\alpha-j}$$
$$+ \frac{1}{\Gamma(\alpha)}\int_a^t \left(\frac{t^\rho - s^\rho}{\rho}\right)^{\alpha-1} f(s, y_0(s))\frac{ds}{s^{1-\rho}};$$

(ii) $\lim_{n\to\infty} d(T^n y, y_0) = 0$, that is,
$$y_0(t) = \lim_{n\to\infty} (T^n y)(t), \quad \forall t \in [a,b];$$

(iii) $d(y, y_0) \leq \frac{1}{1-M'} d(y, Ty)$, that is,
$$d(y, y_0) \leq \frac{1}{1-M'} \frac{1}{\Gamma(\alpha+1)} \frac{(b^\rho - a^\rho)^{\alpha+\gamma}}{\rho^\alpha}.$$

It follows that
$$|y(t) - y_0(t)| \leq \frac{1}{1-M'} \frac{1}{\Gamma(\alpha+1)} \left(\frac{b^\rho - a^\rho}{\rho}\right)^\alpha \varepsilon,$$

which completes the proof. □

References

[1] T. Abdeljawad, On conformable fractional calculus, *J. Comput. Appl. Math.* **279**, 57–66 (2015).
[2] M. S. Abdo, S. K. Panchal and H. A. Wahash, Ulam-Hyers-Mittag-Leffler stability for a ψ-Hilfer problem with fractional order and infinite delay, *Results Appl. Math.* **7**, 100115 (2020).
[3] A. Ahmadova and N. Mahmudov, Ulam-Hyers stability of Caputo type fractional stochastic neutral differential equations, *Stat. Probab. Lett.* **168**, 108949 (2021).
[4] M. Alam and D. Shah, Hyers-Ulam stability of coupled implicit fractional integro-differential equations with Riemann-Liouville derivatives, *Chaos Solitons Fractals* **150**, 111122 (2021).
[5] D. R. Andersona and M. Onitsuka, Hyers-Ulam stability for a discrete time scale with two step sizes, *Appl. Math. Comput.* **344–345**, 128–140 (2019).
[6] L. Cădariu, L. Găvruţa and P. Găvruţa, Weighted space method for the stability of some nonlinear equations, *Appl. Anal. Discrete Math.* **6**, 126–139 (2012).
[7] P. Găvruţa and L. Găvruţa, A new method for the generalized Hyers-Ulam-Rassias stability, *Int. J. Nonlinear Anal. Appl.* **1**, 11–18 (2010).
[8] U. N. Katugampola, New approach to a generalized fractional integral, *Appl. Math. Comput.* **218**, 860–865 (2011).
[9] U. N. Katugampola, A new approach to generalized fractional derivatives, *Bull. Math. Anal. Appl.* **6**, 1–15 (2014).

[10] A. A. Kilbas, H. M. Srivastava and J. J. Trujillo, *Theory and Applications of Fractional Differential Equations* (Elsevier, Amsterdam, 2006).

[11] K. Liu, J. Wang, Y. Zhou, et al. Hyers-Ulam stability and existence of solutions for fractional differential equations with Mittag-Leffler kernel, *Chaos Solitons Fractals* **132**, 109534 (2020).

[12] A. Mohanapriya, C. Park, A. Ganesh, et al. Mittag-Leffler-Hyers-Ulam stability of differential equation using Fourier transform, *Adv. Differ. Equ.* **2020**, 389 (2020).

[13] R. Murali, A. P. Selvan, C. Park, et al. Aboodh transform and the stability of second order linear differential equations, *Adv. Diff. Equ.* **2021**, 296 (2021).

[14] H. Vu and N. V. Hoa, Hyers-Ulam stability of fuzzy fractional Volterra integral equations with the kernel ψ-function via successive approximation method, *Fuzzy Sets Syst.* **419**, 67–98 (2021).

[15] J. Zhou, S. Zhang and Y. He, Existence and stability of solution for a nonlinear fractional differential equation, *J. Math. Anal. Appl.* **498**, 124921 (2021).

[16] A. Zada, W. Ali and C. Park, Ulam's type stability of higher order nonlinear delay differential equations via integral inequality of Grönwall-Bellman-Bihari's type, *Appl. Math. Comput.* **350**, 60–65 (2019).

© 2024 World Scientific Publishing Company
https://doi.org/10.1142/9789811267048_0030

Chapter 30

Out-of-Plane Equilibrium Points in the Restricted Three-Body Problem with Radiation Pressure, Poynting–Robertson Drag and Angular Velocity Variation

Aguda Ekele Vincent[*,§], Angela E. Perdiou[†,¶], and Jagadish Singh[‡,∥]

[*]*Department of Mathematics, School of Basic Sciences, Nigeria Maritime University, Okerenkoko, Delta State, Nigeria*
[†]*Department of Civil Engineering, University of Patras, Patras, Greece*
[‡]*Department of Mathematics, Faculty of Physical Sciences, Ahmadu Bello University, Zaria, Nigeria*
[§]*vincentekele@yahoo.com, vincent.aguda@nmu.edu.ng*
[¶]*aperdiou@upatras.gr*
[∥]*jgds2004@yahoo.com*

This chapter deals with the motion of an infinitesimal body near the out-of-plane equilibrium points in the restricted problem of three bodies under the radial component of Poynting–Robertson (P–R) drag and radiation pressure of the primaries as well as their angular velocity. In particular, the out-of-plane equilibria are first determined analytically and it is found that their existence and positions depend on the perturbing forces involved in the equations of motion. Due to the symmetry of the problem, these points appear in pairs and, depending on the parameter values, their number may be zero, two ($L_{6,7}$) or four ($L_{6,7}$ and $L_{8,9}$). Finally, the effects of the parameters are shown on the positions of the out-of-plane equilibrium points for the binary systems Kruger-60 and Achird. An investigation of the stability of the out-of-plane equilibrium points shows that they are unstable for both binary systems.

1. Introduction

The study of the equilibrium state of an infinitesimal test particle and their stability with regard to the dynamical system of the restricted three-body problem (R3BP), without any doubt, remains one of the most interesting and intriguing aspects in the study of celestial mechanics and dynamical astronomy. The R3BP consists of two finite bodies, known as primaries which rotate in circular orbits around their common center of mass and a test particle which moves in the plane of motion of the primaries under their gravitational attraction and does not affect their motion.

In the classical R3BP, it is well known that there are five equilibrium points: three of them lie on the x-axis and are called collinear (or Eulerian) while the other two are away from the x-axis and are called triangular (or Lagrangian) equilibrium points [1]. The same situation also holds in the more general case of the three-body problem [2] or may occur when certain perturbations are taken into account in this classical model (see, e.g., Refs. [3–8]). However, in the perturbed few body problem, apart from these five equilibria, additional equilibrium points have also been found and analyzed for different dynamical systems in the literature (e.g., the R3BP with negative values for radiation factors [9–14], the R3BP with oblate or prolate primaries [15–17], the R3BP with three-body interactions [18, 19] or with quantum corrections [20], the restricted four-body problem [21–23] and the restricted problem of five bodies [24–28]). When the Poynting–Robertson (P–R) of primaries (one or both) is taken into account, the existence of out-of-plane equilibrium points, locating out of the orbital plane of primaries, is declared [29–33]. Chermnykh's problem, in which the angular velocity variation of the primaries is also considered in the model, has also been discussed in the context of three-dimensional case with regard to equilibrium points out of the plane of motion of the primaries and the dynamics of orbits around these new equilibrium points [34]. We remark that Nan et al. [35] show analytically that out-of-plane equilibrium points do not physically exist in the R3BP when one primary is a rotational ellipsoid. They note that the same conclusion can be drawn if both primaries are rotational ellipsoids.

In this chapter, we aim to make an extension to the work of Kalantonis et al. [34] by also taking into account P–R drag of the primaries as well as the relativistic ones of the photogravitational R3BP (see Ref. [30]) and continue to study numerically the existence, location and stability of the out-of-plane equilibrium points. As an application in this study, we consider the Kruger-60 and Achird binary systems. However, the labeling for the equilibrium points as L_i, $i = 6, 7, 8, 9$, is taken from Ref. [30].

This chapter is organized as follows: In Section 2, the dynamical equations that involve the parameters of the infinitesimal particle in the binary systems under consideration are obtained. Section 3 determines analytically both the existence and location of the out-of-plane equilibrium points and establishes their linear stability analysis. In Section 4, we study numerically the locations and stability of the out-of-plane equilibria for values of the parameters of the problem corresponding to the binary systems Kruger-60 and Achird. This chapter ends in Section 5 in which the obtained results and conclusions are discussed.

2. Equations of Motion

We consider a barycentric coordinate system $Oxyz$ rotating relative to an inertial reference system with angular velocity ω about a common z-axis. Let m_1 and m_2 be the masses of the bigger and smaller body (known as primaries), respectively, with force of radiation pressure parameters $q_i \leqslant 1$, $i = 1, 2$, correspondingly. The two bodies perform circular orbits around the center of mass with angular velocity $\omega \geqslant 0$ of the system and are always on the Ox-axis. We let (x, y, z) be the coordinates of an infinitesimal mass m_3 and we assume that the distance between the two primaries as well as the sum of the stars masses are equal to unity. Then the positions of the bigger and smaller primaries are $(-\mu, 0, 0)$ and $(1 - \mu, 0, 0)$, respectively. Accordingly, the equations of motion of the infinitesimal mass (test particle) moving in the radiation, subject to the radial component of P–R drag and gravitational field of the binary system, are given as [30, 34]

$$\ddot{x} - 2\omega\dot{y} = \omega^2 x - \frac{Q_1(x+\mu)}{r_1^3} - \frac{Q_2(x+\mu-1)}{r_2^3}$$
$$- \frac{W_1}{r_1^2}\left\{\frac{(x+\mu)}{r_1^2}[(x+\mu)\dot{x}+y\dot{y}+z\dot{z}] + \dot{x} - y\right\}$$
$$- \frac{W_2}{r_2^2}\left\{\frac{(x+\mu-1)}{r_2^2}[(x+\mu-1)\dot{x}+y\dot{y}+z\dot{z}] + \dot{x} - y\right\},$$

$$\ddot{y} + 2\omega\dot{x} = \left(\omega^2 - \frac{Q_1}{r_1^3} - \frac{Q_2}{r_2^3}\right)y - \frac{W_1}{r_1^2}\left\{\frac{y}{r_1^2}[(x+\mu)\dot{x}+y\dot{y}+z\dot{z}]\right.$$
$$\left. + \dot{y} + (x+\mu)\right\}$$
$$- \frac{W_2}{r_2^2}\left\{\frac{y}{r_2^2}[(x+\mu-1)\dot{x}+y\dot{y}+z\dot{z}] + \dot{y} + (x+\mu-1)\right\},$$

$$\ddot{z} = \left(-\frac{Q_1}{r_1^3} - \frac{Q_2}{r_2^3}\right)z - \frac{W_1}{r_1^2}\left\{\frac{z}{r_1^2}[(x+\mu)\dot{x}+y\dot{y}+z\dot{z}] + \dot{z}\right\}$$
$$- \frac{W_2}{r_2^2}\left\{\frac{z}{r_2^2}[(x+\mu-1)\dot{x}+y\dot{y}+z\dot{z}] + \dot{z}\right\},$$
(1)

where

$$r_1^2 = (x+\mu)^2 + y^2 + z^2, \quad r_2^2 = (x+\mu-1)^2 + y^2 + z^2 \quad (2)$$

and

$$Q_1 = q_1(1-\mu), \quad Q_2 = q_2\mu, \quad W_1 = \frac{(1-\mu)(1-q_1)}{c_d},$$
$$W_2 = \frac{\mu(1-q_2)}{c_d}. \quad (3)$$

Here, r_i, $i = 1, 2$, are the distances of the third body from the two primaries m_1 and m_2, respectively, W_1, W_2 ($W_i \ll 1$, $i = 1, 2$) stand for P–R drag due to the radiation of the first and second primary bodies, respectively, and c_d is the nondimensional velocity of light. Then, $\mu = m_2/(m_1+m_2)$ is the mass parameter with $0 < \mu \leqslant 1/2$ and $m_1 > m_2$ while we denote the radiation pressure parameters by

$q_i = 1 - b_i$, $i = 1, 2$, where b_1 and b_2 are the ratios of the radiation force F_{ri} to the gravitational force F_{gi} which results from the gravitation due to the primary bodies m_1 and m_2, respectively.

We note here that in the previous studies on three-body problem a necessary condition in order to exist critical points out-of-plane is that the radiation parameter must be negative (see Refs. [9, 10, 30, 34]). So, following the above studies, we can safely assume possible values for the reduction factor to be taken in $q_i \in (-\infty, 1]$, $i = 1, 2$, while for $q_i = 0$, the radiation pressure forces balance the gravitational attractions, a case which is not be considered in our study.

3. Positions and Stability of the Out-of-Plane Equilibrium Points

The positions of the out-of-plane equilibrium points can be determined from the equations of motion (1) by imposing the conditions $\dot{x} = \dot{y} = \dot{z} = \ddot{x} = \ddot{y} = \ddot{z} = 0$ and solving the resulting system for x, y and z. The second equation of (1) is not satisfied for $y = 0$, so we must solve the three equations for $x, y, z \neq 0$; thus, (x_0, y_0, z_0) is the position of the out-of-plane equilibrium point which results from the solution of the nonlinear equations:

$$\omega^2 x_0 - \frac{Q_1(x_0 + \mu)}{r_{10}^3} - \frac{Q_2(x_0 + \mu - 1)}{r_{20}^3} + \left(\frac{W_1}{r_{10}^2} + \frac{W_2}{r_{20}^2}\right) y_0 = 0, \quad (4)$$

$$\left(\omega^2 - \frac{Q_1}{r_{10}^3} - \frac{Q_2}{r_{20}^3}\right) y_0 - \frac{W_1}{r_{10}^2}(x_0 + \mu) - \frac{W_2}{r_{20}^2}(x_0 + \mu - 1) = 0, \quad (5)$$

$$\left(\frac{Q_1}{r_{10}^3} + \frac{Q_2}{r_{20}^3}\right) z_0 = 0, \quad (6)$$

with

$$r_{10} = \sqrt{(x_0 + \mu)^2 + y_0^2 + z_0^2}, \quad r_{20} = \sqrt{(x_0 + \mu - 1)^2 + y_0^2 + z_0^2}, \quad (7)$$

where equation (5) leads to the condition $y \neq 0$ due to the presence of the P–R drag, indicating that the out-of-plane equilibrium points are in the space $Oxyz$.

From equation (6), we have that

$$\frac{r_{20}}{r_{10}} = \left(-\frac{Q_2}{Q_1}\right)^{\frac{1}{3}}, \qquad (8)$$

whereby, the solution of equations (4)–(6) shall depend on the parameters μ, ω, c_d, q_i, W_i, $i = 1, 2$, of the problem. From equation (8), it can be easily seen that, for the existence of any real solution, the following condition is necessary to hold:

$$q_1 \, q_2 < 0, \qquad (9)$$

which means that the radiation pressure force of just one of the primaries has overwhelmed its gravitational attraction. Note that for $\omega = 1$ equations (4)–(6) correspond to the one studied in full detail in Ref. [30] and if $W_1 = W_2 = 0$ (there is no P–R effect acting on the particle) and $\omega > 0$, we obtain the equations studied extensively by Kalantonis et al. [34], while for $\omega = 0$, the particle moves under the influence of radiation pressure and P–R drag. From now on, we shall consider the case $q_1, q_2 \neq 0$, which corresponds to radiation pressure forces with P–R drag, while the angular velocity ω varies continuously in the interval $\omega \in (0, \infty]$.

So, the out-of-plane equilibria of the primaries lay on the $Oxyz$-space, and in this case, with the help of equations (6) and (7), as in Ref. [30], we have

$$x_0 = \frac{1}{2} - \mu + \frac{1}{2}\left[1 - \left(-\frac{Q_2}{Q_1}\right)^{\frac{2}{3}}\right] r_{10}^2. \qquad (10)$$

Substituting equation (10) in equation (5) together with the help of (6), we have

$$y_0 = \frac{1}{2\omega^2}\left\{W_1 - W_2 - \left[W_1\left(-\frac{Q_2}{Q_1}\right)^{\frac{2}{3}} - W_2\left(-\frac{Q_1}{Q_2}\right)^{\frac{2}{3}}\right]\right\}$$
$$+ \frac{1}{2\omega^2}\left[W_1 - W_2\left(-\frac{Q_1}{Q_2}\right)^{\frac{2}{3}}\right]\frac{1}{r_{10}^2}. \qquad (11)$$

So, by combining equations (11), (10) and (6) with (4), we obtain that the distance from any existing equilibrium point to the bigger primary m_1 must satisfy the sixth-degree polynomial equation in r_{10}:

$$p(r_{10}) = b_6 r_{10}^6 + b_4 r_{10}^4 + b_2 r_{10}^2 + b_1 r_{10} + b_0 = 0, \qquad (12)$$

where we have abbreviated the quantities:

$$b_6 = \frac{\omega^2}{2}\left[1-\left(-\frac{Q_2}{Q_1}\right)^{\frac{2}{3}}\right], \quad b_4 = \omega^2\left(\frac{1}{2}-\mu\right),$$

$$b_2 = \frac{1}{2\omega^2}\left[W_1 + W_2\left(-\frac{Q_1}{Q_2}\right)^{\frac{2}{3}}\right]$$

$$\times \left\{W_1 - W_2 - \left[W_1\left(-\frac{Q_2}{Q_1}\right)^{\frac{2}{3}} - W_2\left(-\frac{Q_1}{Q_2}\right)^{\frac{2}{3}}\right]\right\},$$

$$b_1 = -Q_1, \quad b_0 = \frac{1}{2\omega^2}\left[W_1^2 - \left(-\frac{Q_1}{Q_2}\right)^{\frac{4}{3}}W_2^2\right].$$

It is obvious from equation (12) that the distance of any existing equilibrium point with respect to the first primary is dependent on all the involved parameters. Thus, the type of an equilibrium point can be determined through the number of positive real roots of the polynomial equation (12) for certain values of the parameters, which is the sought value of r_{10} by using any numerical method.

From equation (12), any positive root together with the corresponding value of r_{20} obtained from (6) must fulfill the triangular inequalities [30]:

$$r_{10} + r_{20} > 1 \quad \text{or} \quad |r_{10} - r_{20}| < 1. \tag{13}$$

Specifically, if there are no positive real roots of (12), then there are no equilibrium points; if there is exactly one positive real root of (12), then there exists at most one pair of equilibrium points; if two positive real roots of (12) exist, then there are at most two pairs of equilibrium points while they always satisfy the condition (13). Consequently, the positions of the out-of-plane equilibria can be calculated from equations (10), (11) and the z_0-coordinates, resulting from the first equation of (7), expressed as follows:

$$z_0 = \pm\sqrt{r_{10}^2 - (x_0+\mu)^2 - y_0^2}. \tag{14}$$

This proves the existence of out-of-plane equilibria in pairs with members of each pair located symmetrically with respect to the axis Ox.

Now in the photogravitational restricted three-body problem, we found that there are combinations of the parameters q_1, q_2 and μ which lead to one or two pairs of equilibria (see, e.g., Refs. [30, 34]). More precisely, the one or two pairs of equilibria on the Oxz plane can exist only if for the parameters hold $q_2 > 0$ and $q_1 < 0$ or $q_2 < 0$ and $q_1 > 0$, respectively. It becomes obvious that the parameters q_1 and q_2 do not display the same strength of influence on the total number of the equilibria. Hence, we claim that, the effect of the parameter q_2 will be very strong in the dynamical motion of the two bodies, compared with the parameter q_1 which represent the radiation factor of the bigger primary. For this reason, and from now on, we consider sets of (q_1, q_2, μ, ω) which satisfy the conditions (9) and (13).

We remark that equations (10), (11) and (14) are the coordinates of the out-of-plane equilibrium points under radiation pressure of the stars with P–R drag components and angular velocity variation of the primaries. The coordinates differ from those in Ref. [30] because of angular velocity parameter ω. Evidently, when $\omega = 1$, the coordinates are fully analogous with those obtained by Ragos et al. [30].

To determine the linear stability of an out-of-plane equilibrium point (x_0, y_0, z_0), we transfer the origin to this point by introducing the new variables (ξ, η, ζ) and linearize the equations of motion to first-order terms arriving at the variational equations in the form [30, 33]:

$$\ddot{\xi} - 2\omega\dot{\eta} = A_2\xi + A_4\eta + A_6\zeta + A_1\dot{\xi} + A_3\dot{\eta} + A_5\dot{\zeta},$$
$$\ddot{\eta} + 2\omega\dot{\xi} = B_2\xi + B_4\eta + B_6\zeta + B_1\dot{\xi} + B_3\dot{\eta} + B_5\dot{\zeta}, \quad (15)$$
$$\ddot{\zeta} = C_2\xi + C_4\eta + C_6\zeta + C_1\dot{\xi} + C_3\dot{\eta} + C_5\dot{\zeta}.$$

Explicitly, the partial derivatives of system (15) at the equilibrium point yield

$$A_1 = \Omega_{x\dot{x}}^{(0)} = -\left[\frac{W_1(x_0+\mu)^2}{r_{10}^4} + \frac{W_2(x_0+\mu-1)^2}{r_{20}^4} + \frac{W_1}{r_{10}^2} + \frac{W_2}{r_{20}^2}\right],$$

$$A_2 = \Omega_{xx}^{(0)} = \omega^2 - \frac{Q_1}{r_{10}^3} - \frac{Q_2}{r_{20}^3} + 3\left[\frac{Q_1(x_0+\mu)^2}{r_{10}^5} + \frac{Q_2(x_0+\mu-1)^2}{r_{20}^5}\right]$$
$$- 2\left[\frac{W_1(x_0+\mu)}{r_{10}^4} + \frac{W_2(x_0+\mu-1)}{r_{20}^4}\right]y_0,$$

$$A_3 = \Omega^{(0)}_{x\dot{y}} = -\left[\frac{W_1(x_0+\mu)}{r_{10}^4} + \frac{W_2(x_0+\mu-1)}{r_{20}^4}\right]y_0,$$

$$A_4 = \Omega^{(0)}_{xy} = 3\left[\frac{Q_1(x_0+\mu)}{r_{10}^5} + \frac{Q_2(x_0+\mu-1)}{r_{20}^5}\right]y_0 + \frac{W_1}{r_{10}^2} + \frac{W_2}{r_{20}^2}$$
$$- 2\left[\frac{W_1}{r_{10}^4} + \frac{W_2}{r_{20}^4}\right]y_0^2,$$

$$A_5 = \Omega^{(0)}_{x\dot{z}} = -\left[\frac{W_1(x_0+\mu)}{r_{10}^4} + \frac{W_2(x_0+\mu-1)}{r_{20}^4}\right]z_0,$$

$$A_6 = \Omega^{(0)}_{xz} = 3\left[\frac{Q_1(x_0+\mu)}{r_{10}^5} + \frac{Q_2(x_0+\mu-1)}{r_{20}^5}\right]z_0$$
$$- 2\left[\frac{W_1}{r_{10}^4} + \frac{W_2}{r_{20}^4}\right]y_0 z_0,$$

$$B_1 = \Omega^{(0)}_{y\dot{x}} = -\left[\frac{W_1(x_0+\mu)}{r_{10}^4} + \frac{W_2(x_0+\mu-1)}{r_{20}^4}\right]y_0,$$

$$B_2 = \Omega^{(0)}_{yx} = 3\left[\frac{Q_1(x_0+\mu)}{r_{10}^5} + \frac{Q_2(x_0+\mu-1)}{r_{20}^5}\right]y_0 - \frac{W_1}{r_{10}^2} - \frac{W_2}{r_{20}^2}$$
$$+ 2\left[\frac{W_1(x_0+\mu)^2}{r_{10}^4} + \frac{W_2(x_0+\mu-1)^2}{r_{20}^4}\right],$$

$$B_3 = \Omega^{(0)}_{y\dot{y}} = -\frac{W_1}{r_{10}^2} - \frac{W_2}{r_{20}^2} - \left[\frac{W_1}{r_{10}^4} + \frac{W_2}{r_{20}^4}\right]y_0^2,$$

$$B_4 = \Omega^{(0)}_{yy} = \omega^2 - \frac{Q_1}{r_{10}^3} - \frac{Q_2}{r_{20}^3} + 3\left[\frac{Q_1}{r_{10}^5} + \frac{Q_2}{r_{20}^5}\right]y_0^2$$
$$+ 2\left[\frac{W_1(x_0+\mu)}{r_{10}^4} + \frac{W_2(x_0+\mu-1)}{r_{20}^4}\right]y_0,$$

$$B_5 = \Omega^{(0)}_{y\dot{z}} = -\left[\frac{W_1}{r_{10}^4} + \frac{W_2}{r_{20}^4}\right]y_0 z_0,$$

$$B_6 = \Omega^{(0)}_{yz} = 3\left[\frac{Q_1}{r_{10}^5} + \frac{Q_2}{r_{20}^5}\right]y_0 z_0 + 2\left[\frac{W_1(x_0+\mu)}{r_{10}^4} + \frac{W_2(x_0+\mu-1)}{r_{20}^4}\right]z_0,$$

$$C_1 = \Omega_{z\dot{x}}^{(0)} = -\left[\frac{W_1(x_0+\mu)}{r_{10}^4} + \frac{W_2(x_0+\mu-1)}{r_{20}^4}\right]z_0,$$

$$C_2 = \Omega_{zx}^{(0)} = 3\left[\frac{Q_1(x_0+\mu)}{r_{10}^5} + \frac{Q_2(x_0+\mu-1)}{r_{20}^5}\right]z_0,$$

$$C_3 = \Omega_{z\dot{y}}^{(0)} = -\left[\frac{W_1}{r_{10}^4} + \frac{W_2}{r_{20}^4}\right]y_0 z_0, \quad C_4 = \Omega_{zy}^{(0)} = 3\left[\frac{Q_1}{r_{10}^5} + \frac{Q_2}{r_{20}^5}\right]y_0 z_0,$$

$$C_5 = \Omega_{z\dot{z}}^{(0)} = -\frac{W_1}{r_{10}^2} - \frac{W_2}{r_{20}^2} - \left[\frac{W_1}{r_{10}^4} + \frac{W_2}{r_{20}^4}\right]z_0^2,$$

$$C_6 = \Omega_{zz}^{(0)} = -\frac{Q_1}{r_{10}^3} - \frac{Q_2}{r_{20}^3} + 3\left[\frac{Q_1}{r_{10}^5} + \frac{Q_2}{r_{20}^5}\right]z_0^2,$$

with

$$r_{10}^2 = (x_0+\mu)^2 + y_0^2 + z_0^2, \quad r_{20}^2 = (x_0+\mu-1)^2 + y_0^2 + z_0^2,$$

where we have denoted the right-hand side of (1) by Ω_x, Ω_y and Ω_z, respectively, while A_i, B_i and C_i, $i = 1, 2, \ldots, 6$, are arbitrary constants.

The characteristic equation corresponding to system (15) is given by

$$\lambda^6 + f_5\lambda^5 + f_4\lambda^4 + f_3\lambda^3 + f_2\lambda^2 + f_1\lambda + f_0 = 0, \tag{16}$$

where

$$f_5 = -(A_1 + B_3 + C_5), \quad f_0 = A_6B_4C_2 - A_4B_6C_2 - A_6B_2C_4$$
$$+ A_2B_6C_4 + A_4B_2C_6 - A_2B_4C_6,$$

$$f_4 = -A_2 - A_3B_1 + A_1B_3 - B_4 - A_5C_1 - B_5C_3 + A_1C_5$$
$$+ B_3C_5 - C_6 + 2\omega A_3 - 2B_1\omega + 4\omega^2,$$

$$f_3 = -A_4B_1 - A_3B_2 + A_2B_3 + A_1B_4 - A_6C_1 + A_5B_3C_1 - A_3B_5C_1$$
$$- A_5C_2 - A_5B_1C_3 + A_1B_5C_3 - B_6C_3 - B_5C_4 + A_2C_5$$
$$+ A_3B_1C_5 - A_1B_3C_5 + B_4C_5 + A_1C_6 + B_3C_6 + 2\omega A_4 - 2\omega B_2$$
$$- 2\omega B_5C_1 + 2\omega A_5C_3 - 2\omega A_3C_5 + 2\omega B_1C_5 - 4\omega^2 C_5,$$

$$f_2 = -A_4B_2 + A_2B_4 + A_6B_3C_1 + A_5B_4C_1 - A_4B_5C_1 - A_3B_6C_1$$
$$- A_6C_2 + A_5B_3C_2 - A_3B_5C_2 - A_6B_1C_3 - A_5B_2C_3 + A_2B_5C_3$$
$$+ A_1B_6C_3 - A_5B_1C_4 + A_1B_5C_4 - B_6C_4 + A_4B_1C_5 + A_3B_2C_5$$
$$- A_2B_3C_5 - A_1B_4C_5 + A_2C_6 + A_3B_1C_6 - A_1B_3C_6 + B_4C_6$$
$$- 2\omega B_6C_1 - 2\omega B_5C_2 + 2\omega A_6C_3 + 2\omega A_5C_4 - 2\omega A_4C_5$$
$$+ 2\omega B_2C_5 - 2\omega A_3C_6 + 2\omega B_1C_6 - 4\omega^2 C_6,$$
$$f_1 = A_6B_4C_1 - A_4B_6C_1 + A_6B_3C_2 + A_5B_4C_2 - A_4B_5C_2 - A_3B_6C_2$$
$$- A_6B_2C_3 + A_2B_6C_3 - A_6B_1C_4 - A_5B_2C_4 + A_2B_5C_4$$
$$+ A_1B_6C_4 + A_4B_2C_5 - A_2B_4C_5 + A_4B_1C_6 + A_3B_2C_6$$
$$- A_2B_3C_6 - A_1B_4C_6 - 2\omega B_6C_2 + 2\omega A_6C_4 - 2\omega A_4C_6$$
$$+ 2\omega B_2C_6,$$

which is a polynomial of sixth degree in λ and f_i, $i = 0, 1, \ldots, 5$, are arbitrary constants. The eigenvalues of (16) determine the stability or instability of the respective equilibria. An equilibrium point is linearly stable only when all roots of the characteristic equation for λ are pure imaginary quantities or complex values with negative real parts. Otherwise, the equilibrium is unstable. We remark that when $W_1 = W_2 = 0$, these equations reduce to those in Ref. [34].

4. Numerical Application

In this section, we study numerically the locations and stability of the out-of-plane equilibrium points by taking into account the mass parameter, radiation pressure, P–R drag and angular velocity variation of the primaries. For the purpose of computations in this chapter, the astrophysical data of the two binary systems Kruger-60 (AB) and Achird (see Table 1) are borrowed from NASA ADS and Singh and Simeon [36]. First, for comparison reasons and checking purposes, we present the positions of the out-of-plane equilibria of Ref. [30] when the effect of angular velocity variation ω is neglected in this study. For this special case, the numerical solution depicted in Figure 1 agrees very well with the result presented in Ref. [30] (see Figures 4(a), 4(b) and 5). Note that the different

Table 1. Numerical data for the binary Kruger-60 (AB) and Achird systems.

Binary	Mass ($M_\otimes$) m_1	m_2	Luminosity (L_{SUN}) L_1	L_2	Binary separation a	Dimensionless speed of light c_d	Mass ratio μ
Kruger-60	0.271	0.176	0.01	0.0034	9.5	46393.84	0.3937
Achird	0.95	0.62	1.29	0.06	71	67675.52	0.3949

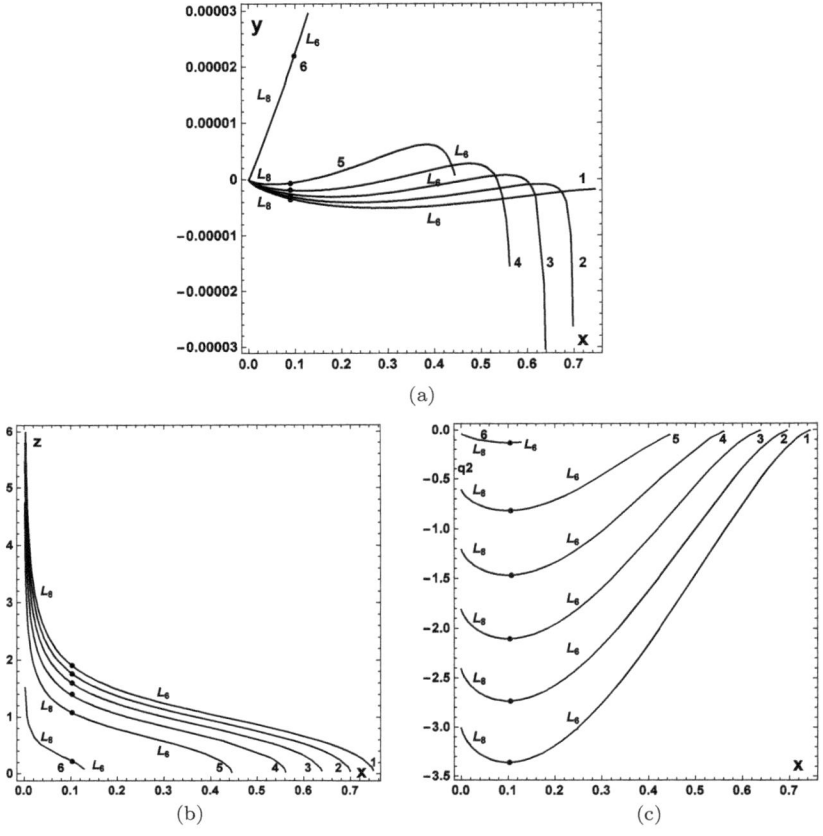

Figure 1. Positions of the two positive out-of-plane equilibria L_6, L_8 for binary system Kruger-60 ($\mu = 0.25$) as a function of radiation factor q_2 in the interval $-3.5 \leqslant q_2 < 0$ and various values of the radiation factor q_1: (1) $q_1 = 1$, (2) $q_1 = 0.8$, (3) $q_1 = 0.6$, (4) $q_1 = 0.4$, (5) $q_1 = 0.2$, (6) $q_1 = 0.01$. (a) Variation of y versus x-coordinate, (b) variation of z versus x-coordinate and (c) variation of x-coordinate versus q_2. The values of $\mu = 0.25$ and $C_d = 48002.33$ are fixed in all cases [30].

scales for the horizontal and the vertical axes are due to the machine precision.

In what follows, we shall investigate how the angular velocity ω and the radiation parameters q_1 and q_2 influence the positions of the out-of-plane equilibria, when they vary in the intervals $\omega \in (0, 2]$, $q_i \in (-\infty, 1]$, $i = 1, 2$, while they always satisfy the conditions (9) and (13). Table 2 shows the effect of radiation pressure and angular velocity variation on the location of out-of-plane equilibria for the binary system Kruger-60. The sample results given here are obtained with $\omega \in (0, 2]$ and $q_2 \in [-1, 1]$. Only two values of $q_1 = 0.6$ and $q_1 = -0.05$ are listed which can be seen as representative for other cases. We observe that the equilibrium points $L_{6(7)}$ exist for certain ranges of q_1 and q_2 for the binary system. More so, certain values of q_1 and q_2 exist for which it is possible to have four equilibria (L_i, $i = 6, 7, 8, 9$). Evidently, the coordinates x, y and z of the two or four critical points, namely $L_{6(7)}$ and $L_{6(7)}$, $L_{8(9)}$ of the binary system, depend on the parameters ω, q_1 and q_2 as they increase or decrease for various values of the angular velocity ω and the radiation pressure. From this table, it can be seen that as the radiation pressure of smaller primary q_2 and angular velocity parameter ω increase for fixed $q_1 = 0.6$, the x-coordinates of L_6 and L_8 increase and decrease, respectively, the z-coordinates of L_6 and L_8 decrease and increase, respectively, while at the same time the y-coordinates of L_6 and L_8 increase and decrease, correspondingly. However, in the case where $q_1 = -0.05$, an increase in radiation pressure of the smaller primary q_2 and angular velocity parameter ω results in a decrease in the x-coordinates of L_6, increase in the z-coordinates of L_6 and increase or decrease in the y-coordinates of L_6.

The positions of the out-of-plane equilibrium points for Kruger-60 are shown graphically in Figures 2 and 3. Figure 2 introduces the positions of the two positive out-of-plane equilibria L_6 and L_8 in (x, q_2), (y, q_2) and (z, q_2) planes, correspondingly, as a function of q_2 in the interval $[-0.9846, 0)$ when the angular velocity parameter takes the values $\omega = 0.5$ (solid line), 0.75 (dot line), 1 (dash line) and 1.3 (dot-dash line) for arbitrary fixed value of $q_1 = 0.6$. We can see that as the radiation pressure q_2 and angular velocity ω increase, the positions of the out-of-plane equilibria L_6 and L_8 move toward the center of mass (i.e., approach the origin). It is also observed that as ω increases, the ranges of q_2 for fixed $q_1 > 0$ get bigger. In Figure 3, we plot the positions of the out-of-plane equilibrium point L_6 (the

Table 2. Positions $(x_0, y_0, \pm z_0)$ of the two (or four, depending on q_1, q_2) out-of-plane equilibrium points $L_{6(7)}$ and $L_{8(9)}$ for Kruger-60 ($\mu = 0.3937$) and for varying the values of radiation parameters q_1, q_2 and angular velocity ω.

		$q_1 = 0.6$	
q_2	ω	$L_{6(7)}$	$L_{8(9)}$
−0.1	0.75	$(0.541384, 3.50517 \times 10^{-6}, \pm 0.501464)$	—
−0.3		$(0.444849, 8.49224 \times 10^{-7}, \pm 0.761684)$	—
−0.5		$(0.356264, -1.04242 \times 10^{-6}, \pm 0.96210)$	—
−0.93		$(0.098735, -2.89611 \times 10^{-6}, \pm 1.80508)$	$(0.00192672, -2.86660 \times 10^{-7}, \pm 6.93835)$
−0.94		$(0.083834, -2.73538 \times 10^{-6}, \pm 1.91733)$	$(0.00935199, -8.02879 \times 10^{-7}, \pm 4.08476)$
−0.945		$(0.074396, -2.60786 \times 10^{-6}, \pm 2.00216)$	$(0.01549260, -1.10322 \times 10^{-6}, \pm 3.44471)$
−0.1	1.0	$(0.444445, -2.59586 \times 10^{-6}, \pm 0.415353)$	—
−0.3		$(0.368058, -1.59410 \times 10^{-6}, \pm 0.641843)$	—
−0.5		$(0.298118, -2.00635 \times 10^{-6}, \pm 0.814442)$	—
−0.93		$(0.101230, -2.40790 \times 10^{-6}, \pm 1.449540)$	$(0.00107071, -1.60313 \times 10^{-7}, \pm 6.96688)$
−0.94		$(0.091901, -2.33130 \times 10^{-6}, \pm 1.505500)$	$(0.00491917, -4.37283 \times 10^{-7}, \pm 4.17842)$
−0.945		$(0.086664, -2.28203 \times 10^{-6}, \pm 1.539940)$	$(0.00770270, -5.84341 \times 10^{-7}, \pm 3.59230)$
−0.1	1.3	$(0.373791, -4.74022 \times 10^{-6}, \pm 0.321141)$	—
−0.3		$(0.312192, -2.40975 \times 10^{-6}, \pm 0.531209)$	—
−0.5		$(0.255960, -2.20469 \times 10^{-6}, \pm 0.684758)$	—
−0.93		$(0.102762, -2.02993 \times 10^{-6}, \pm 1.179260)$	$(0.00062961, -9.45790 \times 10^{-8}, \pm 6.98154)$
−0.94		$(0.096475, -1.98754 \times 10^{-6}, \pm 1.211290)$	$(0.00282305, -2.54989 \times 10^{-7}, \pm 4.22198)$
−0.945		$(0.093106, -1.96280 \times 10^{-6}, \pm 1.229410)$	$(0.00433350, -3.37509 \times 10^{-7}, \pm 3.65429)$

$q_1 = -0.05$

q_2	ω	$L_{6(7)}$
1.0	0.5	$(-0.640039, -4.09892 \times 10^{-5}, \pm 0.518830)$
0.8		$(-0.566594, -3.11972 \times 10^{-5}, \pm 0.572625)$
0.6		$(-0.479576, -2.11694 \times 10^{-5}, \pm 0.626485)$
0.4		$(-0.369420, -1.11122 \times 10^{-5}, \pm 0.689388)$
0.1		$(-0.054973, 9.56409 \times 10^{-7}, \pm 1.256910)$
0.09		$(-0.030683, 8.77271 \times 10^{-7}, \pm 1.538790)$
1.0	0.75	$(-0.446387, -5.26154 \times 10^{-6}, \pm 0.491428)$
0.8		$(-0.393446, -2.36089 \times 10^{-6}, \pm 0.515486)$
0.6		$(-0.330782, 3.22275 \times 10^{-7}, \pm 0.542538)$
0.4		$(-0.251604, 2.45090 \times 10^{-6}, \pm 0.581211)$
0.1		$(-0.031076, 1.09172 \times 10^{-6}, \pm 1.145410)$
0.09		$(-0.016138, 6.75561 \times 10^{-7}, \pm 1.446270)$
1.0	1.0	$(-0.342799, 3.51880 \times 10^{-6}, \pm 0.442609)$
0.8		$(-0.300899, 4.38786 \times 10^{-6}, \pm 0.455965)$
0.6		$(-0.251361, 4.96705 \times 10^{-6}, \pm 0.473119)$
0.4		$(-0.188918, 4.94463 \times 10^{-6}, \pm 0.503349)$
0.1		$(-0.019862, 8.39027 \times 10^{-7}, \pm 1.088980)$
0.09		$(-0.009827, 4.62243 \times 10^{-7}, \pm 1.404180)$
1.0	2.0	$(-0.173898, 6.08865 \times 10^{-6}, \pm 0.274826)$
0.8		$(-0.150343, 5.62360 \times 10^{-6}, \pm 0.277920)$
0.6		$(-0.122677, 4.94225 \times 10^{-6}, \pm 0.287788)$
0.4		$(-0.088270, 3.86913 \times 10^{-6}, \pm 0.318330)$
0.1		$(-0.005920, 2.95345 \times 10^{-7}, \pm 1.014280)$
0.09		$(-0.000270, 1.41638 \times 10^{-7}, \pm 1.355060)$

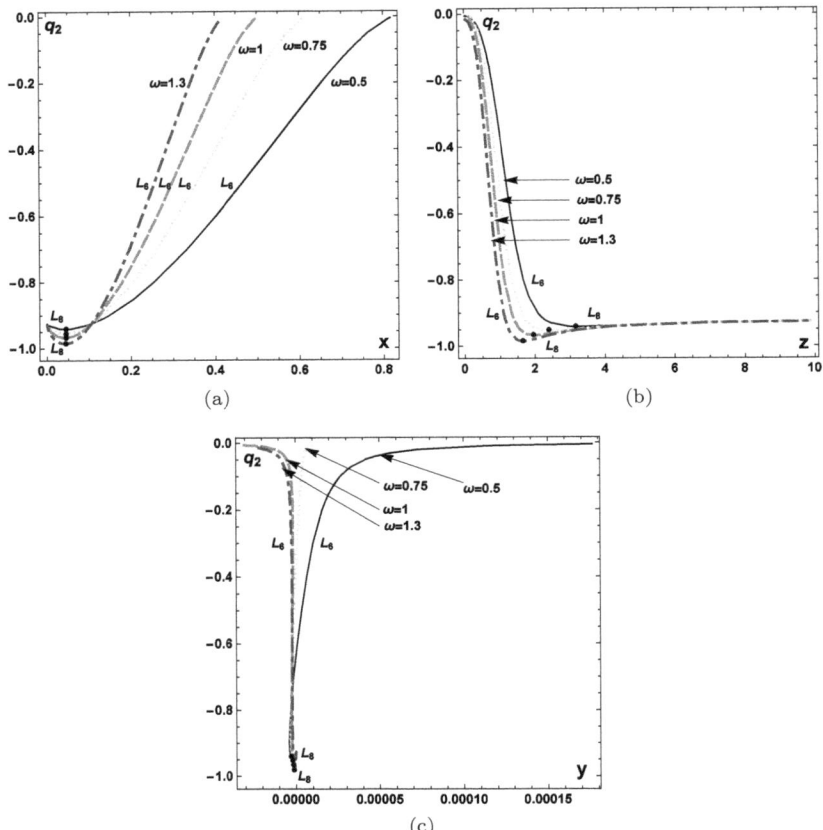

Figure 2. Frames (a) and (b): Positions of the two positive out-of-plane equilibria L_6, L_8 for Kruger-60 ($\mu = 0.3937$) in the (x, z) plane as a function of q_2 in the interval $[-0.9846, 0)$ for $\omega = 0.5$ (black color), 0.75 (green color), 1 (red color), 1.3 (blue color) for fixed $q_1 = 0.6$. (c) The variation of y-coordinate versus q_2 for the same values of ω when $q_1 = 0.6$. Symbol "•" signifies where L_6 coincides with L_8.

situation is same at the symmetric point L_7) in (x, q_2), (y, q_2) and (z, q_2) planes, correspondingly, as a function of q_2 in the interval $[1, 0.078]$ when the angular velocity parameter takes the values $\omega = 0.5$ (solid line), 0.75 (dotted line), 1 (dash line) and 2 (dot-dash line) for arbitrary fixed value of $q_1 = -0.05$. We can see that the out-of-plane equilibria move toward the origin as the radiation pressure q_2 and angular velocity ω increase; this means that the third body will be closer to the primaries. Recall here that radiation pressure

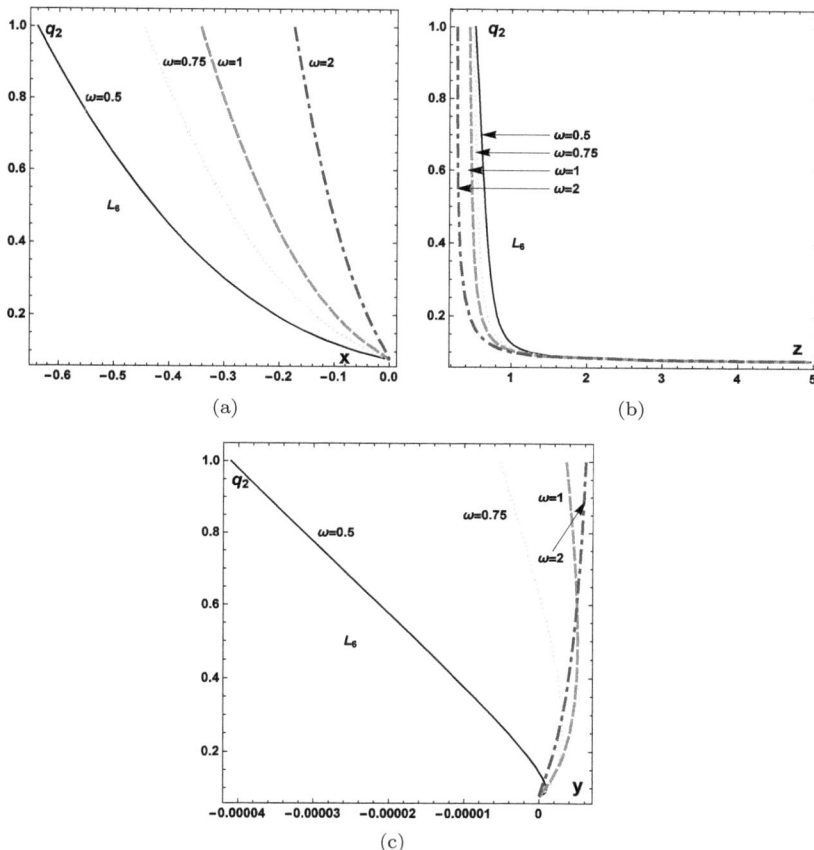

Figure 3. Frames (a) and (b): Positions of the positive out-of-plane equilibrium L_6 (the situation is same at the symmetric point L_7) for Kruger-60 ($\mu = 0.3937$) in the (x, z) plane as a function of q_2 in the interval $[1, 0.078]$ for $\omega = 0.5$ (black color), 0.75 (green color), 1 (red color), 2 (blue color) for fixed $q_1 = -0.05$. (c) The variation of y-coordinate versus q_2 for same numerical values of ω when $q_1 = -0.05$.

increases when $q_{1,2}$ decreases. It is also observed from Table 2 as well as Figures 2 and 3 that, when the angular velocity parameter $\omega < 1$, it has a greater effect on the third particle's dynamic behavior as the positions of the equilibria are grown than the case $\omega \geq 1$ where the positions of the equilibria are shrink.

The roots of the characteristic equation (16) for Kruger-60 binary system are presented in Table 3 for a wide range of values of the parameters q_1, q_2 and ω. It is obvious that all the equilibrium points

Table 3. The eigenvalues $\lambda_{1,2}$, $\lambda_{3,4}$, $\lambda_{5,6}$ of equation (16) and the corresponding positions $(x_0, y_0, \pm z_0)$ of the out-of-plane equilibria $L_{6,7}$ and $L_{8,9}$ for Kruger-60 ($\mu = 0.3937$).

| q_2 | ω | $q_1 = 0.6$ | |
		$L_{6(7)}$	$L_{8(9)}$
-0.1	0.5	0.428532, ± 0.464588i $-0.428568, \pm 0.464565$i $-0.00002192, \pm 0.660036$i $(0.72257, 2.66239 \times 10^{-5}, \pm 0.59046)$	— — — —
-0.5	0.75	0.415434, ± 0.532366i $-0.415454, \pm 0.532357$i $-0.00001267, \pm 0.950458$i $(0.35626, -1.04242 \times 10^{-5}, \pm 0.9621)$	— — — —
-0.93	1.0	$4.55549 \times 10^{-6}, \pm 0.840093$i $-0.0000104, \pm 0.195514$i $-0.0000125135, \pm 1.12072$i $(0.10123, -2.4079 \times 10^{-6}, \pm 1.44954)$	± 0.00368304 $5.90976 \times 10^{-7} \pm 0.997355$i $-1.03463 \times 10^{-6} \pm 1.00264$i $(0.00107, -1.60313 \times 10^{-7}, \pm 6.96688)$
-0.945	1.3	$1.68418 \times 10^{-5}, \pm 1.07625$i $-0.000014130, \pm 0.257944$i $-0.000027291, \pm 1.468040$i $(0.093106, -1.9628 \times 10^{-6}, \pm 1.22941)$	± 0.01738 0.0000197352 ± 1.28689i -0.0000213422 ± 1.3131i $(0.004334, -3.37509 \times 10^{-7}, \pm 3.65429)$
-0.97	1.3	$1.57557 \times 10^{-5}, \pm 1.13614$i $-0.0000255322, \pm 1.43695$i $-1.08989 \times 10^{-5}, \pm 0.156104$i $(0.072210, -1.77366 \times 10^{-6}, \pm 1.36169)$	± 0.04370 0.0000169235 ± 1.25829i $-0.0000209461 \pm 1.34112$i $(0.017045, -8.06321 \times 10^{-7}, \pm 2.29222)$

		$q_1 = -0.05$
1.0	0.5	0.379692, ±0.483129i
		−0.379735, ±0.483096i
		−0.00003983, ±0.567071i
		($-0.64004, -4.0989 \times 10^{-5}, \pm 0.51883$)
0.6	1.0	0.514463, ±0.859245i
		−0.51449, ±0.859238i
		−0.00009228, ±1.02605i
		($-0.25136, 4.96705 \times 10^{-6}, \pm 0.473119$)
0.2	2.0	0.208696, ±1.98563i
		−0.20875, ±1.98614i
		−0.00004709, ±0.446813i
		($-0.04012, 1.92075 \times 10^{-6}, \pm 0.451873$)

are unstable due to the existence of at least a positive real root and complex root with positive part for each particular set of values. It may be observed that the equilibrium point $L_{8(9)}$ is more unstable than $L_{6(7)}$ since there is always a positive real number in the three pair of roots. However, in the absence of P–R drag, it is possible to have linearly stable motion around $L_{6(7)}$ for certain values of the radiation factors and angular velocity variation contrary to the present result (see also Ref. [34]).

Similarly, for the investigation of the influence of radiation pressure and angular velocity variation on the location of out-of-plane equilibria of Achird, we set the two values $q_1 = 0.6$ and $q_1 = -0.05$, while the angular velocity ω and the radiation parameter of the second primary q_2 vary in the intervals $\omega \in (0, 2]$ and $q_2 \in (-\infty, 1]$, respectively, while they always satisfy conditions (9) and (13). The coordinates of the determined out-of-plane equilibria are shown in Table 4 for fixed values $q_1 = 0.6$ and $q_1 = -0.05$ and different values of ω and q_2 (same ω, q_2 as the previous case). In this case, similar results are observed in the coordinates of the equilibria with regard to the Kruger-60 described previously. However, coordinates of equilibria are now decreased with regard to the corresponding values of Kruger-60.

The locations of out-of-plane equilibrium points L_i, $i = 6, 7, 8, 9$, for Achird are shown graphically in Figures 4 and 5. Figure 4 introduces the positions of the two positive out-of-plane equilibrium points L_6 and L_8 in (x, q_2), (y, q_2) and (z, q_2) planes, correspondingly, as a function of q_2 in the interval $[-0.9786, 0)$ when the angular velocity parameter takes the values $\omega = 0.5$ (solid line), 0.75 (dotted line), 1 (dash line) and 1.3 (dash dotted line) for arbitrary fixed value of $q_1 = 0.6$ while Figure 5 shows the positions of the out-of-plane equilibrium points L_6 (the situation is same at the symmetric point L_7) in (x, q_2), (y, q_2) and (z, q_2) planes, correspondingly, as a function of q_2 in the interval $[1, 0.078]$ when the angular velocity parameter takes the values $\omega = 0.5$ (solid line), 0.75 (dotted line), 1 (dash line) and 2 (dash dotted line) for arbitrary fixed value of $q_1 = -0.05$. Evidently, the variational trend of the corresponding positions is similar to the variational trend of the positions of equilibria of Kruger-60 even though the orbital properties are different.

However, from Figures 2 and 3 to Figures 4 and 5, we can observe two main results as the mass parameter increases. The first is a change of range of radiation factor q_2 which corresponds to all

Table 4. Positions $(x_0, y_0, \pm z_0)$ of the two (or four, depending on q_1, q_2) out-of-plane equilibrium points $L_{6(7)}$ and $L_{8(9)}$ for Achird ($\mu = 0.3949$) and for varying the values of radiation parameters q_1, q_2 and angular velocity ω.

		$q_1 = 0.6$	
q_2	ω	$L_{6(7)}$	$L_{8(9)}$
−0.1	0.75	$(0.539943, 2.38024 \times 10^{-6}, \pm 0.502399)$	—
−0.3		$(0.443079, 5.58829 \times 10^{-7}, \pm 0.763425)$	—
−0.5		$(0.354091, -7.37168 \times 10^{-6}, \pm 0.96499)$	—
−0.93		$(0.090891, -1.92928 \times 10^{-6}, \pm 1.85971)$	$(0.0048500, -3.60018 \times 10^{-7}, \pm 5.08984)$
−0.94		$(0.073337, -1.77525 \times 10^{-6}, \pm 2.01084)$	$(0.0154718, -7.55106 \times 10^{-7}, \pm 3.44385)$
−0.945		$(0.060056, -1.62315 \times 10^{-6}, \pm 2.15941)$	$(0.0256254, -1.02482 \times 10^{-6}, \pm 2.90099)$
−0.1	1.0	$(0.443134, -1.78768 \times 10^{-6}, \pm 0.416267)$	—
−0.3		$(0.366477, -1.10290 \times 10^{-6}, \pm 0.643491)$	—
−0.5		$(0.296210, -1.38490 \times 10^{-6}, \pm 0.817113)$	—
−0.93		$(0.095748, -1.61937 \times 10^{-6}, \pm 1.480170)$	$(0.00264035, -1.99366 \times 10^{-7}, \pm 5.14614)$
−0.94		$(0.085572, -1.55542 \times 10^{-6}, \pm 1.545950)$	$(0.00767964, -3.99553 \times 10^{-7}, \pm 3.59335)$
−0.945		$(0.079679, -1.51270 \times 10^{-6}, \pm 1.588470)$	$(0.01122920, -5.08932 \times 10^{-7}, \pm 3.15977)$
−0.1	1.3	$(0.372561, -3.25046 \times 10^{-6}, \pm 0.322157)$	—
−0.3		$(0.310740, -1.65545 \times 10^{-6}, \pm 0.532854)$	—
−0.5		$(0.254240, -1.51447 \times 10^{-6}, \pm 0.687314)$	—
−0.93		$(0.098638, -1.37172 \times 10^{-6}, \pm 1.198600)$	$(0.00153727, -1.17069 \times 10^{-7}, \pm 5.17401)$
−0.94		$(0.091983, -1.33852 \times 10^{-6}, \pm 1.234210)$	$(0.00431887, -2.30704 \times 10^{-7}, \pm 3.65594)$
−0.945		$(0.088379, -1.31889 \times 10^{-6}, \pm 1.25470)$	$(0.00613802, -2.89795 \times 10^{-7}, \pm 3.24633)$

(Continued)

Table 4. (*Continued*)

$q_1 = -0.05$

q_2	ω	$L_{6(7)}$	
1.0	0.5	$(-0.641879, -2.82077 \times 10^{-5},$	$\pm 0.517499)$
0.8		$(-0.568360, -2.14836 \times 10^{-5},$	$\pm 0.571393)$
0.6		$(-0.481263, -1.45963 \times 10^{-5},$	$\pm 0.625251)$
0.4		$(-0.371015, -7.69276 \times 10^{-5},$	$\pm 0.687956)$
0.1		$(-0.056452, 6.45949 \times 10^{-7},$	$\pm 1.244210)$
0.09		$(-0.032115, 6.05716 \times 10^{-7},$	$\pm 1.513280)$
1.0	0.75	$(-0.447883, -3.63297 \times 10^{-6},$	$\pm 0.490514)$
0.8		$(-0.394886, -1.63959 \times 10^{-6},$	$\pm 0.514518)$
0.6		$(-0.332158, 2.05318 \times 10^{-7},$	$\pm 0.541436)$
0.4		$(-0.252900, 1.67144 \times 10^{-6},$	$\pm 0.579773)$
0.1		$(-0.032068, 7.61457 \times 10^{-7},$	$\pm 1.131370)$
0.09		$(-0.017003, 4.79027 \times 10^{-7},$	$\pm 1.418500)$
1.0	1.0	$(-0.344104, 2.41181 \times 10^{-6},$	$\pm 0.441757)$
0.8		$(-0.302157, 3.01008 \times 10^{-6},$	$\pm 0.455005)$
0.6		$(-0.252561, 3.41014 \times 10^{-6},$	$\pm 0.471962)$
0.4		$(-0.190042, 3.39835 \times 10^{-6},$	$\pm 0.501771)$
0.1		$(-0.020569, 5.90075 \times 10^{-7},$	$\pm 1.073870)$
0.09		$(-0.010396, 3.30760 \times 10^{-7},$	$\pm 1.374950)$
1.0	2.0	$(-0.174851, 4.19115 \times 10^{-6},$	$\pm 0.273508)$
0.8		$(-0.151254, 3.87276 \times 10^{-6},$	$\pm 0.276350)$
0.6		$(-0.123536, 3.40596 \times 10^{-6},$	$\pm 0.285842)$
0.4		$(-0.089050, 2.67025 \times 10^{-6},$	$\pm 0.315727)$
0.1		$(-0.006173, 2.09930 \times 10^{-7},$	$\pm 0.997020)$
0.09		$(-0.002875, 1.02420 \times 10^{-7},$	$\pm 1.323590)$

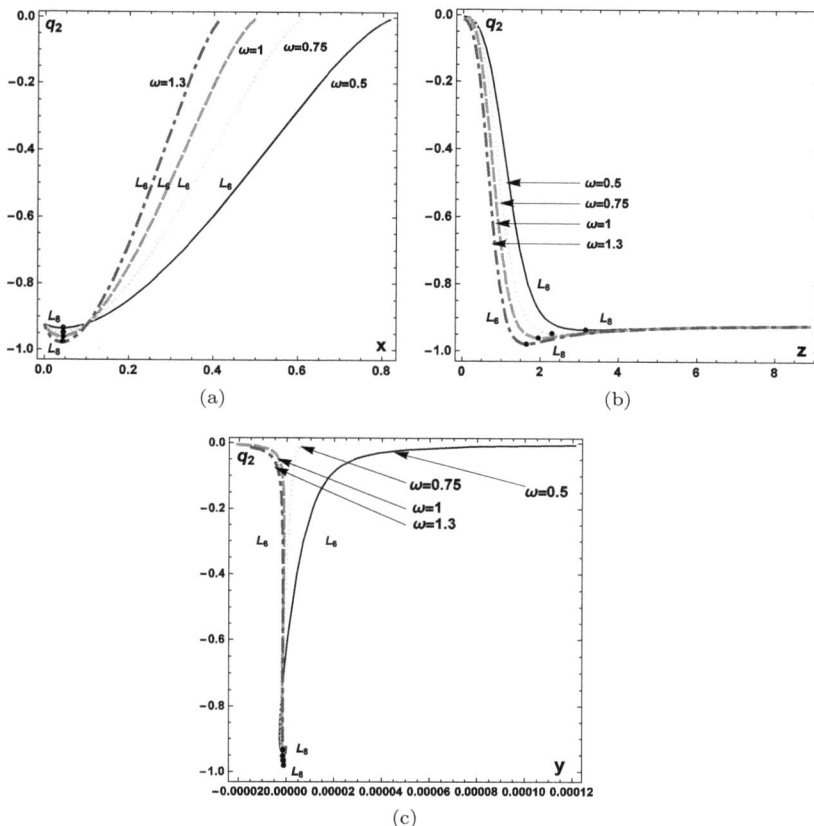

Figure 4. Frames (a) and (b): Positions of the two positive out-of-plane equilibrium points L_6 and L_8 for Achird ($\mu = 0.3949$) in the (x, z) plane as a function of q_2 in $[-0.9786, 0)$ for $\omega = 0.5$ (solid line), 0.75 (dot line), 1 (dash line), 1.3 (dash dotted line) for fixed $q_1 = 0.6$. (c) The variation of y-coordinate versus q_2 for the same values of ω when $q_1 = 0.6$. Symbol "•" signifies where L_6 coincides with L_8.

equilibrium points. The second is a change in the coordinates of all the equilibria. These are directly observable from the tables (Tables 2 and 4) that as the mass parameter increases for fixed $q_1 = 0.6$ and various values of radiation pressure q_2 and angular velocity ω, the x-coordinates of L_6 and L_8 decrease and increase, respectively, the z-coordinates of L_6 and L_8 increase and decrease, respectively, while at the same time, the y-coordinates of L_6 and L_8 both decrease, correspondingly. However, in the case where $q_1 = -0.05$, it is seen that an increase in radiation pressure of the smaller primary q_2 and angular velocity parameter ω results in an increase in the x-coordinates of L_6,

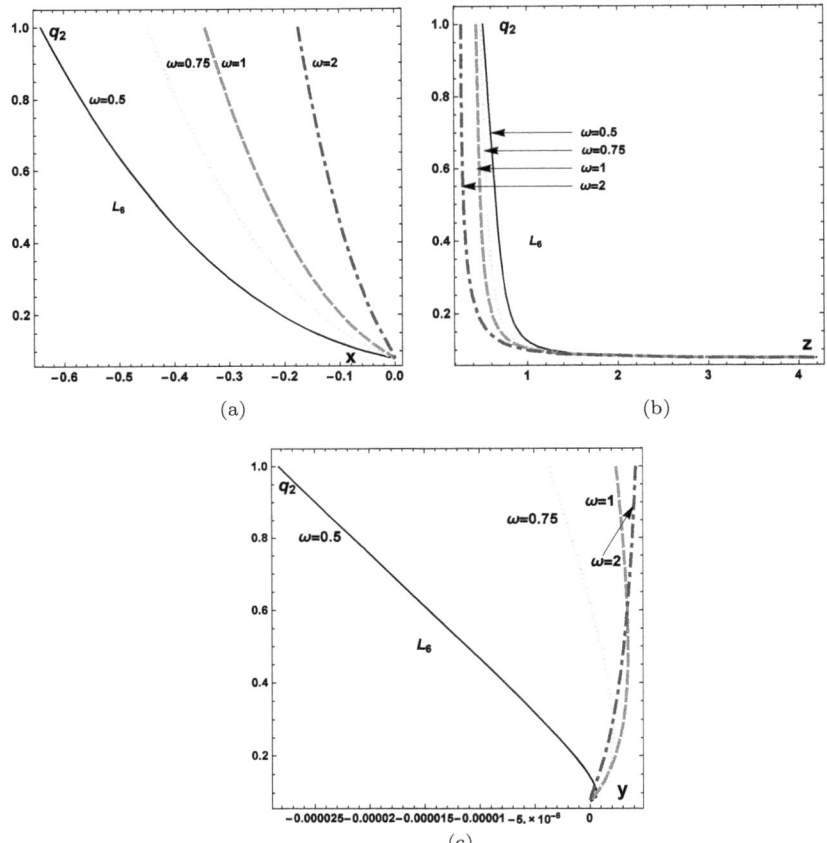

Figure 5. Frames (a) and (b): Positions of the positive out-of-plane equilibrium point L_6 (the situation is same at the symmetric point L_7) for Achird ($\mu = 0.3949$) in the (x, z) plane as a function of q_2 in the interval $[1, 0.078]$ for $\omega = 0.5$ (solid line), 0.75 (dot line), 1 (dash line), 2 (dash dotted line) for fixed $q_1 = -0.05$. (c) The variation of y-coordinate versus q_2 for same values of ω when $q_1 = -0.05$.

decrease in the z-coordinates of L_6 and decrease in the y-coordinates of L_6. Note the different values of the coordinates of equilibria of the two figures (Figures 2 and 4, Figures 3 and 5). However, these effects on the x-coordinate are directly observable from the tables (see Tables 2 and 4).

We compute the characteristic roots which are shown in Table 5 for a wide range of the various parameters. An analysis of Table 5

Table 5. The eigenvalues $\lambda_{1,2}$, $\lambda_{3,4}$, $\lambda_{5,6}$ of equation (16) and the corresponding positions $(x_0, y_0, \pm z_0)$ of the out-of-plane equilibria $L_{6,7}$ and $L_{8,9}$ for Achird ($\mu = 0.3949$).

q_2	ω	$L_{6(7)}$	$L_{8(9)}$
-0.1	0.5	$0.428074, \pm 0.464543i$ $-0.428099, \pm 0.464527i$ $-0.000014994, \pm 0.659492i$ $(0.720896, 1.81876 \times 10^{-5}, \pm 0.591643)$	—
-0.5	0.75	$0.413087, \pm 0.531722i$ $-0.413101, \pm 0.531716i$ $-0.0000086465, \pm 0.949128i$ $(0.354091, -7.37168 \times 10^{-7}, \pm 0.964985)$	—
-0.93	1.0	$2.93699 \times 10^{-6}, \pm 0.851751i$ $-0.0000673355, \pm 0.177345i$ $-0.00000833719, \pm 1.11493i$ $(0.095748, -1.61937 \times 10^{-6}, \pm 1.48017)$	± 0.0076011 $6.00377 \times 10^{-7} \pm 0.994412i$ $-1.15719 \times 10^{-6} \pm 1.00559i$ $(0.0026404, -1.9937 \times 10^{-7}, \pm 5.14614)$
-0.945	1.3	$1.13633 \times 10^{-5}, \pm 1.09152i$ $-0.0000185107, \pm 1.46062i$ $-0.0000913702, \pm 0.234889i$ $(0.0883789, -1.31889 \times 10^{-6}, \pm 1.2547)$	± 0.02247 $0.000013049 \pm 1.28254i$ $-0.0000144423 \pm 1.31742i$ $(0.006138, -2.89795 \times 10^{-7}, \pm 3.24633)$
-0.97	1.3	$1.07642 \times 10^{-5}, \pm 1.1557i$ $-0.000171279, \pm 1.42448i$ $-6.79669 \times 10^{-6}, \pm 0.123327i$ $(0.0645485, -1.15535 \times 10^{-6}, \pm 1.42058)$	± 0.04777 $0.0000113915 \pm 1.24813i$ $-0.0000146633 \pm 1.35072i$ $(0.0220064, -6.44671 \times 10^{-7}, \pm 2.09616)$

(Continued)

Table 5. (*Continued*)

		$q_1 = -0.05$	
1.0	0.5	0.380013, ±0.483038i −0.380043, ±0.483016i −0.0000274039, ±0.567638i (−0.641879, −2.8208 × 10^{-5}, ±0.51750)	—
0.6	1.0	0.516453, ±0.859049i −0.516472, ±0.859045i −0.000063231, ±1.02837i (−0.25256, 3.41014 × 10^{-6}, ±0.471962)	—
0.2	2.0	0.20838, ±1.98527i −0.20842, ±1.98562i −0.0000326029, ±0.450411i (−0.04072, 1.33236 × 10^{-6}, ±0.447319)	—

reveals no case in which all the roots are pure imaginary which means that the equilibria are unstable.

5. Discussion and Conclusion

We examined the positions and stability of out-of-plane equilibrium points in Chermnykh's restricted three-body problem under the influence of radiation pressure and Poynting–Robertson drag from the primaries and found that there are combinations of the parameters q_1, q_2 and μ which lead to one or two pairs of such equilibria. Our analytical results about the out-of-plane equilibrium points L_i, $i = 6, 7, 8, 9$, showed that their positions are affected by the mass parameter, radiation factors and the angular velocity parameter. Our stability analysis revealed that, in the absence of Poynting–Robertson drag, it is possible to have linearly stable motion around $L_{6(7)}$ for certain values of the radiation factors and angular velocity variation.

As application, the effects of radiation pressure and angular velocity parameter on the locations of the out-of-plane equilibrium points were considered for the binary systems Kruger-60 and Achird, respectively. In the case of Kruger-60, we observed that as the radiation pressure and angular velocity effects increase, the coordinates of the out-of-plane equilibrium points increase or decrease, while the positions of these points shift toward the origin as the radiation pressure and angular velocity increase. In the case of Achird, the equilibria exhibit similar variational trend with equilibria of Kruger-60. Despite the differences in the orbital properties of the binary systems, the positions of the out-of-plane equilibria move in the same direction when perturbed. We also studied the stability of the out-of-plane equilibrium points for both binary systems and found that the respective characteristic polynomial possesses at least one positive real root or a complex root with positive part, therefore, the motion of the infinitesimal mass body around the out-of-plane equilibrium points L_i, $i = 6, 7, 8, 9$, of the binary systems Kruger-60 and Achird is unstable.

A natural extension of this chapter would be to study periodic orbits around these points. Periodic solutions together with their relevant equilibria are of significant importance and constitute the backbone of the dynamical behavior for any dynamical system. Also, homoclinic and heteroclinic connections between these orbits or points could be determined in order to establish appropriate connections among them which are useful in space mission design.

Acknowledgment

This research was supported by the Tertiary Education Trust Fund (tetfund), Nigeria. The first author, Aguda E. Vincent, has received this research grant.

References

[1] V. Szebehely, *Theory of Orbits: The Restricted Problem of Three Bodies* (Academic Press, New York, 1967).
[2] E. A. Perdios and V. S. Kalantonis, Symmetric doubly asymptotic orbits at collinear equilibrium points in the general three-body problem, *Astron. Astrophys.* **394**, 323–328 (2002).
[3] O. Ragos, E. A. Perdios, V. S. Kalantonis and M. N. Vrahatis, On the equilibrium points of the relativistic restricted three-body problem, *Nonlinear Anal. Theory Methods Appl.* **47**, 3413–3418 (2001).
[4] V. S. Kalantonis, E. A. Perdios and O. Ragos, Asymptotic and periodic orbits around L_3 in the photogravitational restricted three-body problem, *Astrophys. Space Sci.* **301**, 157–165 (2006).
[5] E. I. Abouelmagd, H. M. Asiri and M. A. Sharaf, The effect of oblateness in the perturbed restricted three-body problem, *Meccanica* **48**, 2479–2490 (2013).
[6] S. M. Elshaboury, E. I. Abouelmagd, V. S. Kalantonis and E. A. Perdios, The planar restricted three-body problem when both primaries are triaxial rigid bodies: Equilibrium points and periodic orbits *Astrophys. Space Sci.* **361**, 315 (2016).
[7] N. Bello, U. Aishetu and A. A. Hussain, Locations of $L_{4,5}$ of a dust grain type II comet tail in Solar-Jupiter system in the photogravitational relativistic R3BP, *IOP Sci. Notes* **2**, 045001 (2021).
[8] O. Leke and J. Singh, Out-of-plane equilibrium points of extra-solar planets in the central binaries PSR B1620-26 and Kepler-16 with clusters of material points and variable masses, *New Astron.* **99**, 101958 (2023).
[9] V. V. Radzievskii, The spatial photogravitational restricted three-body problem, *Astron. Zhurnal* **30**, 265–273 (1953).

[10] J. F. L. Simmons, A. J. C. McDonald and J. C. Brown, The restricted 3-body problem with radiation pressure, *Celest. Mech.* **35**, 145–187 (1985).

[11] Z. Xuentang and Y. Lizhong, Photogravitationally restricted three-body problem and coplanar libration point, *Chin. Phys. Lett.* **10**, 61–64 (1993).

[12] O. Ragos and C. Zagouras, The zero velocity surfaces in the photogravitational restricted three-body problem, *Earth Moon Planets* **41**, 257–278 (1988).

[13] O. Ragos and C. Zagouras, Periodic solutions about the out of plane equilibrium points in the photogravitational restricted three-body problem, *Celest. Mech. Dynam. Astr.* **44**, 135–154 (1988).

[14] A. E. Vincent, Motion around the out-of-plane equilibrium points in the photogravitational Copenhagen elliptic restricted three-body problem with oblateness, *Int. J. Space Sci. Eng.* **5**(3), 269–286 (2019).

[15] M. S. Suraj, S. Alhowaity, R. Aggarwal, M. C. Asique, A. Mittal and M. Jain, On the topology of basins of convergence linked to libration points in the modified R3BP with oblateness, *New Astron.* **94**, 101776 (2022).

[16] H. I. Alrebdi, F. L. Dubeibe and E. E. Zotos, Equilibrium dynamics of the restricted 3-body problem with prolate primaries, *Results Phys.* **48**, 106406 (2023).

[17] E. M. Moneer, Y. M. Allawi, M. M. Alanazi and E. E. Zotos, Revealing the properties of the out-of-plane points of equilibrium of the restricted 3-body problem with non-spherical radiating bodies, *New Astron.* **103**, 102061 (2023).

[18] C. N. Douskos, Effect of three-body interaction on the number and location of equilibrium points of the restricted three-body problem, *Astrophys. Space Sci.* **356**, 251–268 (2015).

[19] M. S. Suraj, R. Aggarwal, M. C. Asique, A. Mittal, M. Jain and V. K. Paliwal, Effect of three-body interaction on the topology of basins of convergence linked to the libration points in the R3BP, *Planet. Space Sci.* **205**, 105281 (2021).

[20] A. A. Alshaery and E. I. Abouelmagd, Analysis of the spatial quantized three-body problem, *Results Phys.* **17**, 103067 (2020).

[21] J. P. Papadouris and K. E. Papadakis, Equilibrium points in the photogravitational restricted four body problem, *Astrophys. Space Sci.* **344**, 21–38 (2013).

[22] J. Singh and A. E. Vincent, Out-of-plane equilibrium points in the photogravitational restricted four-body problem, *Astrophys. Space Sci.* **359**, 38 (2015).

[23] S. Muhammad, F. Z. Duraihem, W. Chen and E. E. Zotos, On the equilibria of the planar equilateral restricted four-body problem with radiation pressure, *Adv. Appl. Math. Mech.* **13**, 966–981 (2021).

[24] M. S. Suraj, E. I. Abouelmagd, R. Aggarwal and A. Mittal, The analysis of restricted five–body problem within frame of variable mass, *New Astron.* **70**, 12–21 (2019).

[25] A. E. Vincent, Out of plane equilibria in the restricted five body problem with radiation pressure, *Int. J. Space Sci. Eng.* **5**, 105–122 (2019).

[26] E. I. Abouelmagd, A. A. Ansari, M. S. Ullah and J. L. G Guirao, A planar five-body problem in a framework of heterogeneous and mass variation effects, *Astron. J.* **160**, 216 (2020).

[27] S. Muhammad, F. Z. Duraihem, W. Chen and E. E. Zotos, On the nature of equilibrium points in the axisymmetric five-body problem, *J. Comput. Nonlinear Dyn.* **16**, 091002 (2021).

[28] P. Sachan, M. S. Suraj, R. Aggarwal, A. Mittal and M. C. Asique, A study of the axisymmetric restricted five-body problem within the frame of variable mass: The concave case, *Astron. Rep.* **67**, 404–423 (2023).

[29] Y. A. Chernikov, The photogravitational restricted three-body problem, *Sov. Astron.* **14**, 176–181 (1970).

[30] O. Ragos, F. A. Zafiropoulos and M. N. Vrahatis, A numerical study of the influence of Poynting Robertson effect on the equilibrium points of the photogravitational restricted three-body problem II. Out of plane case, *Astron. Astrophys.* **300**, 579–590 (1995).

[31] M. K. Das, P. Narang, S. Mahajan and M. Yussa, On out of plane equilibrium points in photogravitational restricted three-body problem, *J. Astrophys. Astron.* **30**, 177–185 (2009).

[32] A. Chakraborty and A. Narayan, Influence of Poynting-Robertson drag and oblateness on existence and stability of out-of-plane equilibrium points in spatial elliptic restricted three-body problem, *J. Inform. Math. Sci.* **10**, 55–72 (2018).

[33] J. Singh and T. O. Amuda, Out-of-plane equilibrium points in the photogravitational CRTBP with oblateness and P-R drag, *Int. J. Astron. Astrophys.* **36**, 291–305 (2015).

[34] V. S. Kalantonis, E. A. Vincent, J. M. Gyegwe and E. A. Perdios, Periodic solutions around the out-of-plane equilibrium points in the restricted three-body problem with radiation and angular velocity variation. In T. M. Rassias and P. M. Pardalos (eds.) *Nonlinear Analysis and Global Optimization*. Springer Optimization and Its Applications, Vol. 167 (Springer Nature Switzerland AG, 2021), pp. 251–275.

[35] W. Nan, X. Wang and L. Y. Zhou, Comment on "Out-of-plane equilibrium points in the restricted three-body problem with oblateness (research note)". *Astron. Astrophys.* **614**, A67 (2018).

[36] J. Singh and A. M. Simeon, Motion around the triangular equilibrium points in the circular restricted three-body problem under triaxial luminous primaries with Poynting-Robertson drag, *Int. Front. Sci. Lett.* **12**, 1–21 (2017).

Index

A

(aoua)-regular, 843, 850
a-Cauchy, 828
a-asymptotic, 828
a-convergent, 827
Aboodh transform, 588, 595–600, 603–608, 610–611
Achird, 922–923, 925–927
adaptive mesh, 139
additive mapping, 79
admissible representation, 287, 843
almost concave, 865
almost ntright, 849, 858, 861
alternating direction method, 175–176, 178–179, 196
analytic function, 225, 231, 236–237, 249, 251, 253, 262, 264
angular velocity, 903–905, 908, 910, 913, 915–916, 919, 922–923, 925, 929
antisymmetric, 15, 17, 22–25, 30, 33, 38, 42, 44, 67
Apostol–Bernoulli numbers, 492, 497
Apostol–Euler numbers, 495, 497
Apostol–Genocchi numbers, 496–497
Appolonius law, 78
approximation of the beta function, 737

approximation properties of the Bernstein polynomials, 754–755
array polynomials, 734
ascending orbital sequences, 16, 842
ascending valued, 19, 52, 60
asingleton, 785, 821
asymptotic upper diagonal, 798
auxiliary principle technique, 617–618, 640, 643, 674, 681, 685
auxiliary problem, 659
auxiliary variational inequality, 620
axiom of choice, 7, 22, 825

B

(Baoua-f) sequences, 847, 856, 871
(Baoua-f,b)-almost-selfclosed, 848
(Baoua-f,b)-complete, 847
(Baoua-f,b)-continuous, 848
B-ascending, 796
B-positive, 797
B-property, 796
Banach algebra, 226, 228, 230
Banach Lie algebra, 571–573, 580
Banach space, 99, 229
Banach's contraction principle, 822, 887
Bernoulli numbers, 493–495, 497–498, 504, 508

Bernstein basis functions, 500
Bernstein-type, 731–732, 735, 738, 756
best-response dynamics, 691
beta distribution, 731–732, 736–738, 746, 748–749, 756
beta function, 489–490, 500
beta probability density function, 737, 753, 756
beta-type distribution, 731–732, 738–739, 744–745, 747, 751–752, 756
biconvex function, 657, 659–663, 665–666, 671, 677–678, 685
bilateral bounded oscillating, 790
bilateral oscillation, 790
binary Kruger-60 (AB) and Achird systems, 914
binary operation, 7, 79
binary systems Kruger-60 (AB) and Achird, 904, 913–915, 929
binomial coefficient, 491, 524
bivariational inequalities, 657, 659, 674, 677–678, 680–681, 685
Blachman–Stam inequality, 475
Bochner integrable, 101
bounded, 869
box product, 8
Boyd–Wong (Θ, h)-system, 841
Boyd–Wong relational non-Cauchy criterion, 837
Brøndsted relation, 9
buried obstacles, 371, 389

C

$(>)$-cofinal, 837
$(\mathcal{G}; \mathcal{R}; \Upsilon)$-contractive, 848
(q, t)-Catalan numbers, 505
max-complete, 39–40
sup-completeness, 42
sup-complete proset, 45
φ-closed, 25
Cantor relation, 9
Catalan numbers, 485–489, 500–506, 509–510, 512–513, 526

Catalan–Daehee numbers, 505, 510
Catalan–Qi function, 486, 506, 508
Catalan–Qi numbers, 505–506, 508
Catalan-type number and polynomial of higher order, 511
Cauchy integral formula, 524
Cauchy numbers, 498
Cauchy problem, 468
Cauchy–Schwarz inequality, 257
Cauchy–Schwarz–Bunyakovsky inequality, 541
Cauchy-type numbers, 518
Changhee numbers, 523, 531
characteristic functions, 731–732, 743, 751–754
Chebyshev polynomials, 712, 716, 718, 720
Chu–Vandermonde identity, 491
closure operation, 2, 24–25, 29, 31–33, 36, 43–45, 50, 70
cofinal, 39
compactness-type condition, 279
compartment model, 299–300
complement, 7
complement-coercive, 794
complementarity problem, 618, 626
complete trajectory, 796
composition relation, 7
concave, 865
context space, 9
continuous function, 17, 397, 399, 403, 407, 412, 416, 422
continuous order isomorphism, 865
conv-Cauchy structure, 786
convergence analysis, 644
convex, 100
convex functions, 392
convexity theory, 658
corelations, 7
Coulomb's friction, 202
countable operation, 81
cover (partition), 9
Cramer–Rao inequality, 473
critical point, 278

D

Index 937

d-Cauchy, 83
d-Lipschitz, 833
d-convergent, 83
Daehee numbers, 522
damage, 201–204, 206, 208, 211, 213
Debreu integrable, 101
decreasing, 16, 19, 865
decreasingly normal, 27
decreasingly regular, 28
dense sets, 16
denumerable axiom of choice, 826
dependent choice principle, 826
descending, 848, 858–859
descending semi-Picard, 797
descending valued, 20, 53, 60
diameter, 811
direct scattering problem, 371–373, 389
Dirichlet boundary condition, 372, 383, 389
Discontinuous Galerkin, 139
distribution functions, 81
domain, 6
dual, 10
dynamical system, 620, 635–636, 639, 649, 904, 930
dynamical system technique, 635

E

elastic waves, 371–372
elasticity, 541
energy, 479
entropy inequality, 472
entropy-type information, 472
equations of motion, 904–905, 907, 910
equilibrium point, 904–905, 908–909, 913
equilibrium problem, 617, 642
equivalence (partial order) relation, 8
equivariant deformation theorem, 276
equivariant linking pair, 283
equivariant minimax principle, 276
Euler numbers, 495, 497, 509

Euler's gamma function, 490, 500–501, 508
evolutionary algorithmic heuristics, 111
exponential entropy, 476
extended general convex function, 623–624
extended general equilibrium problem, 642
extended general variational inequalities, 617–618, 620, 625, 643, 649
extension, 6
extensive, 22–23, 25–26, 31, 34, 65
extensive maps, 23
extensive operation, 24
external functions, 469

F

ρ-fractional derivative, 879
ρ-fractional integrals, 879
factorization method, 371–372, 389
falling factorial, 490–491, 503
far-field operator, 384
far-field pattern, 384, 386–387
fat sets, 16
Fibonacci-type polynomials in two variables, 525
finite element method, 139
first variable continuous, 833
fix-asingleton, 786, 797, 822
fix-singleton, 786, 822
fixed point, 887, 891, 896, 899
fixed point method, 97, 571, 573
fixed-point formulation, 629–630
fixed-point problem, 619, 629
fluid mechanics, 139
formal context, 9
fountain theorem, 287
Fourier transform, 471
fractional integrals, 422
Fubini numbers, 497–498, 529
full, 797, 847
function, 1, 6, 22, 26, 51
functional contraction, 788

Fuss–Catalan numbers, 505–507

G

g-nonexpansive, 866
g-normal, 36
Galois adjoint, 3
Galois connections, 2, 12, 26, 33
game theory, 112, 428
Gauss's hypergeometric function, 507
general complementarity problem, 626
general convex set, 622
general variational inequality, 629
general equilibrium problems, 642
general variational inequalities, 621
generalized complete metric, 83
generalized extended general variational inequality, 647
generalized hypergeometric function, 507
generalized integrable operator, 884
generalized metric, 98
generalized Nash, 112
generalized uniformities, 3
generalized Vandermonde's convolution, 491
generating function, 485–487, 491–496, 498–500, 505, 507, 509–511, 513, 517, 525–526
generating functions for the Bernstein basis functions, 742, 752
Genocchi numbers, 496–497
goset, 1, 9, 16, 18, 22, 26, 53
gross inequality, 468

H

h-transitive, 844
Hölder's inequality, 551
Harmonic numbers, 734, 744
Hausdorff distance, 100
heat equation, 477
hereditary, 790
Herglotz wave function, 384
heuristic algorithms, 113
Hilbert space, 225, 228, 261

Hunsaker–Lindgren relation, 9
Hyers–Ulam stability, 571, 573, 580, 588–592, 594, 597, 603, 605
Hyers–Ulam–Rassias stability, 877–879
hyperrelation, 7

I

idempotent, 24–26
identifying entire solution sets of generalized Nash games, 111
identity relation, 6, 10
images, 6
inclusion-increasing, 18
increasing, 16–20, 26, 29, 35, 38–40, 42, 44–45, 52–53, 56, 60, 63–64, 67, 69–71
increasing regular functions, 29
increasingly g-normal function, 27, 32–33, 35–38 47–48, 51–52
increasingly Φ-regular, 60
increasingly Φ_F-regular, 62–63, 65, 70–71
increasingly φ-normal, 34
increasingly φ-regular function, 28, 30–32, 34, 36, 43–44, 49–50
increasingly φ-semiregular, 43
increasingly g_F-normal, 52, 53–58, 63, 66–67, 70
increasingly left g-seminormal, 27, 32, 36–38, 47, 51
increasingly left φ-regular, 43
increasingly left φ-semiregular, 28, 30–31, 43–44, 49–50
increasingly left Φ-semiregular, 59
increasingly left Φ_F-semiregular, 61, 65
increasingly normal functions, 2, 26–27, 29, 39–41, 45, 52, 70–71
increasingly regular, 28, 44–45, 60, 71
increasingly right g-seminormal, 26, 32, 37, 46–47, 51
increasingly right φ-semiregular, 28, 31, 43, 48, 50

Index

increasingly right G_F-seminormal, 54–55, 57–58, 66
increasingly right Φ-semiregular, 59
increasingly right Φ_F-semiregular, 61–62, 65
inertial proximal method, 647
inf-complete, 15, 24
infinitesimal mass, 929
infinitesimal particle, 905
infinitesimal test particle, 904
inhomogeneous medium, 371–373, 383, 389
injective, 16, 19, 34–35
instance selection method, 762
integrable functions, 392
integral inequalities, 391, 395, 397, 407, 422, 424
intensive, 22, 32–34
interior operation, 33
inverse elastic problem, 389
inverse problem, 712
inverse scattering problem, 371
inversion algorithm, 371
inversion compatible, 8
invex set, 391, 393–396
involution, 37, 47–48
involution operation, 23, 34, 39
isometric representation, 277

J

Jordan–von Neumann identity, 78

K

Katz–Bonacich centrality, 690
KKT systems, 112
Kruger-60, 916, 918, 920, 922
Kruger-60 and Achird binary systems, 905
Kruger-60 binary system, 919
Kupradze radiation condition, 373, 379

L

ρ-left fractional derivative, 880
ρ-left fractional integral, 880

Lebesgue measure, 468
left bounded oscillating, 790
left oscillation, 790
left semi-idempotent, 22, 25, 30
left semi-involutive, 22
left semi-modification operation, 24, 31
Lie bracket derivations, 571–573, 575, 578–580, 582–584
limit separated, 801
linear differential equations, 588–590, 594, 597, 599, 601, 603, 610–611
linear operators, 226, 228
linearly mc-quasimetric, 831, 864
Lipschitz condition, 83, 99
Lipschitz constant, 83, 99
Lipschitz-type condition, 888, 892, 896
little Schroder numbers, 505
Liu chaotic system, 356
Liu system, 343–345, 351–352, 368
logarithmic-quadratic proximal method, 176, 197
lower diagonal compatible, 798
Luxemburg–Jung theorem, 83

M

machine learning, 762
Markov chains, 467
mass parameter, 906, 913, 922, 925, 929
Matkowski admissible, 787
Matkowski contraction, 787
Matkowski rational contraction, 788, 798
Matkowski–Dung contraction, 812
maximal element, 81
maximum diversity problem, 764
mc-quasimetric space, 831, 864
medical imaging, 715
metric, 785
metric compatible quasimetric, 830, 864
metric space, 821
min-completeness, 15, 22, 23

Mittag–Leffler function, 344, 350, 366, 368–369
Mittag–Leffler form function, 351
Mittag–Leffler–Hyers–Ulam stability, 588, 590–591, 593–594, 599, 603, 607, 609, 612
model of Malthus, 301
moment generating functions, 731–732, 738, 743–744, 749–752
momentum, 470
Motzkin numbers, 505

N

n-array operation, 80
g-normal, 33, 35
Narayana numbers, 505
Nash equilibrium, 431–432, 434, 689
natural, 824
natural order, 19
Navier–Stokes equations, 139
negative-order Bernoulli numbers, 494–495
neighborhoods, 6
network games, 689
non-telescopic, 846
nonexpansive, 801
nonlinear fractional differential equations, 879
nonlinear operator, 889, 893, 897
nonpartial relation, 6
nontelescopic, 796
normal, 848, 858–859, 865
normal distribution, 466
normal equations, 480
normal functions, 3
npoint, 849, 858, 862
nsright, 849, 858, 860
ntright, 849, 861
numerical radius, 226, 237, 240, 270–271

O

o-sequence, 796
operator, 99

optimal constant, 473
optimization-based methods, 112
orbital (or *iterative*) sequence, 842
order-increasingness, 18
orthonormal basis, 228, 230, 249, 252, 255, 260
out-of-plane equilibria, 908–909, 914–915, 918, 920, 927
out-of-plane equilibrium points, 903–905, 907, 910, 913, 915–916, 919, 922–923, 925–926, 929

P

p-adic fermionic integral, 503, 509
p-Schatten norm, 226, 229
P–R drag, 905–907, 910, 913, 922
P–R effect, 908
parallelogram identity, 78
partial orders, 19, 34–36
partially ordered, 81
Pervin relation, 8
Picard operator, 786, 797, 822, 847
Picard point, 822
Planck constant, 470
Pochhammer symbol, 489–490
Poincaré–Steklov type operator, 376
Poincare's inequality, 479
point, 849, 862
point asymptotic, 801
polygamma function, 733, 744, 747
poset, 1, 9, 51
positive-order Bernoulli numbers, 493, 495
positron emission tomography, 712
potential games, 695
power set, 2, 10
Poynting–Robertson (P–R) drag, 903–904, 929
preclosure operation, 24
preinvex, 393, 397
preorder, 19, 30, 32–34, 39, 43–44, 48, 50
preorder (tolerance) relation, 8
preorders, 2, 31, 36, 40, 45

Index

prey–mesopredator–predator systems, 298–300
prey–predator models, 299
prey–predator systems, 297–298
price of anarchy, 689
primitive, 865
projected differential equations, 112
projected dynamical system, 636
projection methods, 619, 628
projection operation, 23
projection, Wiener–Hopf, decomposition, 640
proset, 9, 45, 67
proximal interior, 19
proximity spaces, 3
pseudo-gradient of the game, 693
pseudometric, 8, 827

Q

q-Catalan numbers, 505
quadratic functional equation, 98
quadratic mapping, 79
quadrature formula, 395
quantum calculus, 321–322
quasi-asymptotic, 794
quasi-increasing, 64–66, 68, 70
quasi-variational inequalities, 112
quasimetric space, 829, 864

R

radiation factor, 904, 914
radiation parameters, 916, 923
radiation pressure, 903, 905–908, 910, 913, 915, 922, 925, 929
Radon transform, 711, 713, 715
Raina's function, 321, 323, 326–327, 340
Raney numbers, 505
range, 6
rectifiable path, 237, 249, 251, 253
recurrence relation, 487, 493–494, 496–497, 513
reduced system, 826

reflexive, 8, 15–17, 24–25, 29–30, 32–33, 37–38, 41, 43–44, 52–53, 66, 827
regular, 45, 799
regular functions, 3
relation, 1, 6, 10, 19, 69, 824
relational mc-quasimetric space, 850
relational quasimetric space, 830
relational space, 9
relational structure, 824
relator, 2
residual subsets, 16
residuated mappings, 26
restricted, 910
restricted three-body problem (R3BP), 904, 905, 929
restriction, 6, 53, 60
right g-seminormal, 38
right asymptotic, 801
right bounded oscillating, 790
right oscillation, 790
right semi-idempotent, 22, 24
right semi-involutive, 22

S

g-seminormal, 2
φ-semiregular, 2
Schroder–Hipparchus numbers, 505
Schwarz inequality, 226
second variable continuous, 833
selection, 37–38
selection function, 7, 22
self-closure and interior relations, 8
semi Bellman Picard, 847
semi Picard, 847, 797
semi regular, 799
semi-closure operation, 24
semi-idempotent, 23
semi-involutive, 23
semi-Picard operator, 822
semi-Picard point, 822
semi-right, 848, 859
semi-Wardowski function, 790
semiclosure operation, 30

separated, 828
sequence, 825
set increasing, 791
set-valued function, 7, 18, 97
simple relator space, 9
singleton, 7, 785, 821
Sobolev inequality, 467
social dilemmas, 427–428, 432, 434–437, 452–454, 461–462
social optimum, 689
special means, 391, 395, 422, 424
spectrum, 843
square quadratic proximal method, 177
stability, 97–98
standard deviation, 470
Stirling numbers, 731–732, 734–735, 744
Stirling numbers of the first kind, 491
Stirling numbers of the second kind, 492, 498, 504, 527, 529
Stirling's approximation for factorials, 504, 516
Stokes problem, 139
strictly contractive mapping, 83
strictly contractive operator, 99
strictly increasing function, 16, 22–23, 397, 399, 403, 407, 412, 416, 422
strong Picard operator, 786, 797, 822, 847
strongly Picard point, 822
strongly reduced system, 826
strongly regular, 799
sub-additive, 865
subgoset, 10
subsequence, 825
subunitary, 791
sufficient, 827
sup-complete proset, 15, 17, 25, 41
super Matkowski admissible, 798
super relations, 7, 12, 51
super relators, 10
super-additive, 865
super-Catalan numbers, 505

support function, 100
symmetric, 8, 827
symmetric biadditive mapping, 79

T

$\mathcal{G}$-transitive, 845
t-norm, 79
tangent numbers, 498, 527
telescopic, 796, 846
three-body problem, 910
three-compartment model, 308, 310
topological G-space, 277
topological structures, 3
topological tools, 13
topologically open, 16
toset, 9
total, 8, 15–16
total-right, 849, 861
totally ascending, 841
trace, 228–229, 249
trajectory, 796
transitive, 8, 25, 29–32, 38, 41, 44, 52–53, 60, 65–66
translation compatible, 843
triangular, 827
two-dimensional linear elasticity, 372

U

$\mathcal{G}$-uadmissible, 834
uadmissible, 842
Ulam–Hyers–Rassias, 97–98
unary operation, 7
uncertainly principle, 470
uniform continuity indicator function, 868
unilateral contact, 201–202, 205
union-preserving, 2, 67–70
union-reversing, 12
universal relation, 10

V

variational formulation, 377
variational inequalities, 175, 617–619, 658, 661

Index 943

variational inequality, 618, 622, 693
variational method, 371–372, 389
Verhulst model, 302–303, 305, 311, 313
Volkenborn (p-adic bosonic) integral, 502, 508

W

w-Catalan numbers, 509
w-Catalan–Daehee number, 510–511
Wardowski rational contraction, 789
Wardowski–Dung contraction, 817
weakly ($*$)-iterative, 805
weighted metric, 889, 897
weighted space method, 877
Weil surrounding, 8
well-ordered sets, 9, 824
well-posedness, 371–372, 389
Wiener–Hopf equations, 619, 633
woset, 9

Z

Zhong quasimetric, 866

Milton Keynes UK
Ingram Content Group UK Ltd.
UKHW021614091024
449435UK00029B/17